持续推动高炉热风炉科技发展和耐材技术进步

安耐克锥柱旋切顶燃式热风炉
发明专利号：ZL 2013 1 0283577.3

郑州安耐克实业有限公司成立于2003年9月，美国上市，股票代码：ANNC，总部位于河南省新密市。

安耐克是以输出高风温、长寿命、低能耗、低排放、低投资高炉热风炉技术、配套全系列耐材制造及EPC总承包工程为核心业务的高新技术企业。拥有富良、富华、富钢、富宝4个生产公司，以及安耐可（北京）工程技术公司、安耐克热风炉工程技术公司、安耐克窑炉工程公司；建立了顶燃式热风炉技术研发中心及CFD数模冷态、热态仿真综合实验室等科研平台，以及河南省院士工作站、河南省博士后创新实践基地、河南省工程技术中心等多个研发机构；在顶燃式热风炉领域获得省部级科技成果6项、专利及专有技术72项，主持起草热风炉和耐火材料国家及行业标准10项；具备热风炉技术研发、试验与设计能力，热风炉用全系列耐材制造与检测能力，冶金工程施工总承包能力等，着力推动顶燃式热风炉技术向高风温、长寿命、低能耗、低排放、智慧化发展。

安耐克锥柱旋切顶燃式热风炉

技术特点：高风温、长寿命、低能耗、低排放、低投资。

知识产权：集成安耐克专利及专有技术60余项，其中发明专利13项，国际专利3项。

解决难题：高温区炉壳开裂、燃烧器煤气喷口错位、燃烧室剥落开裂、格子砖锅底状下沉、热风出口变形坍塌、热风管道高温等。

应用业绩：已应用600~3000m³高炉200余座热风炉，部分业绩见下表。

客户名称	高炉容积，m³	客户名称	高炉容积，m³
宝武鄂钢	2800	晋钢集团	2×1580
中天南通	2×2400	南阳汉冶	1530
广西翅冀	2060	江阴兴澄	1500
山西高义	1680	武安新兴铸管	1280
河钢塞尔维亚	1595	福建大东海	2×1260
河北东海	2×1580	越南和发	4×1080

地址：河南省新密市产业集聚区　电话：0371-69938988, 69999038
传真：0371-69938888　邮箱：annecsale@annec.com.cn
网址：www.annec.com.cn

巩义市第五耐火材料有限公司

推行安全长寿高炉炉缸炉底内衬及结构新技术

远离炉缸烧穿

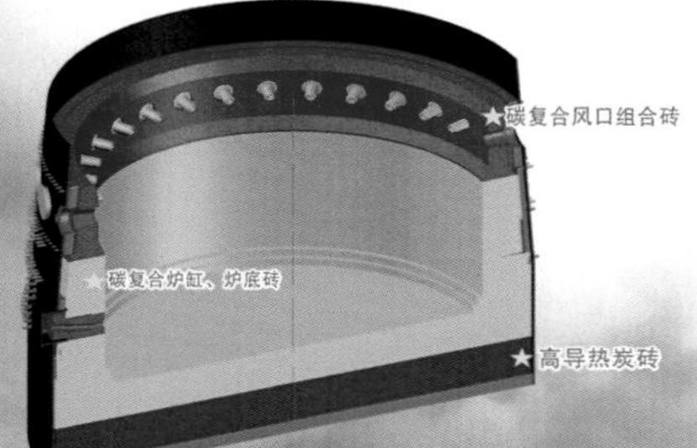

安全长寿炉底炉缸设计方案

安全长寿炉底炉缸结构是以本公司与北京科技大学、河南科技大学共同研制的碳复合砖为基础材料，重点解决炭砖的抗铁水溶蚀性、抗氧化性和抗锌侵蚀性三大问题，以实现高炉的安全生产。碳复合砖的导热性能与高炉微孔炭砖相当，而抗铁水溶蚀性和抗氧化性分别为高炉微孔炭砖的20倍和10倍以上，可以代替炭砖作为高炉炉底炉缸砖衬使用，也可以作为陶瓷杯砖衬使用。

采用安全长寿炉底炉缸结构，可有效防止局部漏水后，因炭砖抗铁水溶蚀性差、抗氧化性差而导致的短时间内炉缸“老鼠洞”形式的烧穿，即提高了高炉炉缸操作的安全性，同时也延长了高炉的寿命，该技术已于2013年9月7日通过河南省科技厅组织的成果鉴定，结论为：国内领先，达到国际先进水平。

安全长寿炉底炉缸结构典型实绩

山西通才2号高炉（410m³）采用安全长寿炉底炉缸结构，于2007年投产。2013年11月，为检查验证安全长寿炉底炉缸结构的使用效果，五耐与通才公司合作，利用2号高炉炉身上部更换冷却壁的机会，将炉底炉缸料扒净，对炉底炉缸破损情况进行了观察（见下图）。结果表明，2号高炉运行7年多时间，仅铁口区碳复合砖有轻微侵蚀，其余炉缸均呈原砖面基本没有侵蚀，同时既没有出现炉缸“象脚状”蚀损，也没有出现风口上翘、炉壳开裂等现象，单位炉容产铁已超过1.4万吨，目前该高炉仍在正常运行。

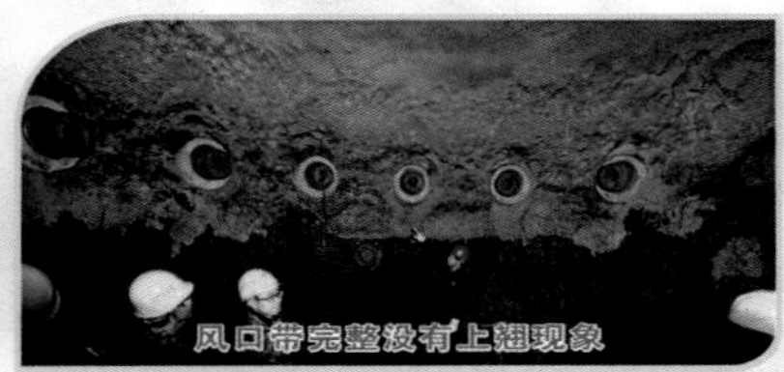

安全长寿炉底炉缸结构目前已推广应 用于通才410m³高炉和1860m³高炉，新金山550m³高炉，黎城太行450m³高炉、山西高义1080m³高炉、扬州恒润1080m³、山西晋南1860m³、河北武安烘熔钢铁1260m³、陕西龙钢850m³和4#1991m³、广西柳钢1580m³、本钢北营以及俄罗斯马钢1580m³等100多座高炉。

科学技术成果鉴定证书

豫科鉴委字［2013］第1486号

成果名称：安全长寿高炉炉缸炉底内衬及结构新技术

完成单位：巩义市五耐科技开发有限公司
北京科技大学
河南科技大学

鉴定形式：会议鉴定

组织鉴定单位：河南省科学技术厅(盖章)

鉴定日期：2013年9月7日

鉴定批准日期：2013年9月7日

河南省科学技术厅

（二〇一二年制）

《安全长寿高炉炉缸炉底内衬及结构新技术》鉴定委员会名单

序号	鉴定会职务	姓名	工作单位	所学专业	现从事专业	职称职务	签名
1	主任委员	刘云彩	（原）首钢总公司	钢铁	炼铁	教授级高工/副总工程师	
2	副主任委员	吴启常	（原）中冶京诚工程技术有限公司	钢铁冶金	炼铁	教授级高工/设计大师	
3	副主任委员	张鑫	河南省耐火材料行业协会	冶金	冶金	教授级高工	
4	委员	汤清华	（原）鞍钢股份公司	钢铁冶金	炼铁	教授级高工/副总工程师	
5	委员	张福明	北京首钢国际工程技术有限公司	钢铁冶金	钢铁冶金	教授级高工	
6	委员	陈良玉	东北大学	冶金机械	冶金设备	教授/博导	
7	委员	全强	中冶京诚工程技术有限公司	炼铁	炼铁	教授级高工	
8	委员	靳亲国	河南省耐火材料行业协会	耐火材料	耐火材料	教授级高工	
9	委员	马金芳	首钢迁钢公司	钢铁冶金	炼铁	教授级高工	
10	委员	柳萌	中冶南方工程技术有限公司	钢铁冶金	炼铁	教授级高工/总设计师	
11	委员	舒文虎	武钢钢铁集团公司	钢铁冶金	炼铁	教授级高工/总工程师	
12							

鉴定意见

受河南省科技厅委托，郑州市科技局于2013年9月7日组织有关专家召开了巩义市五耐科技开发有限公司、北京科技大学和河南科技大学共同完成的“安全长寿高炉炉缸炉底内衬及结构新技术”项目科技成果鉴定会。鉴定委员会听取了课题组的工作汇报，审阅了技术资料，经质疑和讨论，形成鉴定意见如下：

1. 项目鉴定技术资料齐全，内容翔实，数据可靠，符合鉴定要求。

2. 项目将碳、硅等组分合理地引入到刚玉基材料中，采用多元结合和微孔技术实现陶瓷材料与炭素材料的有效复合，成功开发出碳复合砖，具有导热性高、抗铁水熔蚀性好、膨胀率低、强度高等优点，为高炉炉缸炉底的安全长寿提供了新材料。

3.该项目所研发的碳复合砖及胶泥等配套的辅助材料，组成高炉炉缸炉底新型结构，为高炉的安全长寿提供了有力的保障。

4.“安全长寿高炉炉缸炉底内衬及结构新技术”已在山西建邦集团有限公司、襄汾县新金山特钢有限公司、河北纵横钢铁集团有限公司、俄罗斯马钢等高炉上长期成功应用，效果良好，经济和社会效益显著。碳复合砖已批量生产，工艺技术成熟，产品质量稳定，具有良好的推广应用前景。

5.该项目技术先进，属国内首创，达到国际先进水平。

建议：

（1）进一步完善“安全长寿高炉炉缸炉底内衬及结构新技术”技术标准，加大技术推广力度。

（2）项目名称改为“安全长寿高炉炉缸炉底内衬新材料及结构新技术”。

鉴定委员会主任： 副主任：

2013年9月7日

公司简介　**巩义市第五耐火材料有限公司**（原巩义市第五耐火材料总厂）创建于1968年，是炼铁系统用耐火材料为主的现代化大型专业生产企业，年产各种耐火材料25万余吨。从1986年起先后与武汉科技大学、武钢研究院、河南科技大学、北京科技大学、东北大学等院校和科研单位建立了广泛的合作关系，具有较强的自主研发能力。公司先后在热风炉用低蠕变系列、高炉用陶瓷杯系列、铁水预处理用铝碳化硅炭砖等方面多次填补国内空白，通过省部级鉴定十余项，拥有国内和国际专利20余项。在高炉、热风炉和鱼雷式铁水罐的长寿方面有丰富的实践经验和较强的技术实力。

地址：河南省巩义市南郊　**电话：0371-56595699**　**传真：0371-56595558**

邮编：451250　**E-mail：29651@126.com**　**网址：http//www.hnwnjt.com**

耐指纹涂层

优势

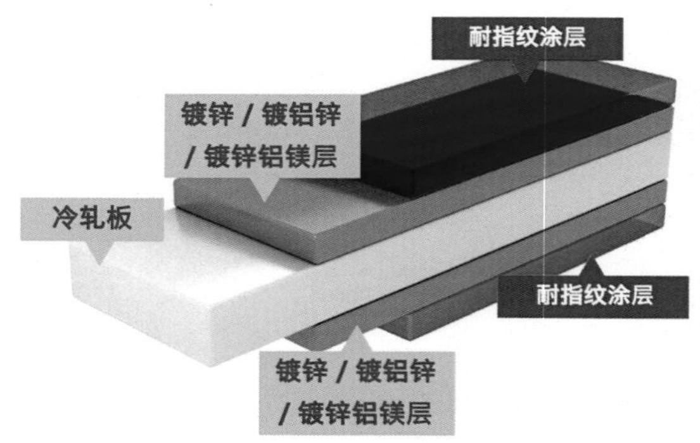

耐指纹

长期抗腐蚀

自润滑性能

出色的辊压成型能力

耐碱性 *

良好的油漆附着力 *

* 适用于部分产品

产品

- BONDERITE O-TO 是一系列有机无铬涂层。
- 可用于制造耐指纹钢板，具有很好的抗腐蚀性能。
- 使用简单，膜层为透明覆盖物，与基材接合力好。
- 环保标准高，满足欧盟 REACH 和 ROHS 检测认证。

产品名称	产品类型	基材种类				应用行业
		HDG	GL	EG	ZAM	
BONDERITE O-TO 5708	无铬耐指纹涂层	√	x	x	x	家电
BONDERITE O-TO 5827		x	x	√	x	家电
BONDERITE O-TO 5816		x	x	x	√	家电（低铝）
BONDERITE O-TO 2700B		x	x	x	√	建筑（高铝）
BONDERITE O-TO 630		x	√	x	x	家电
BONDERITE O-TO 230F		x	√	x	x	家电
BONDERITE O-TO 233		x	√	x	x	建筑

可持续战略

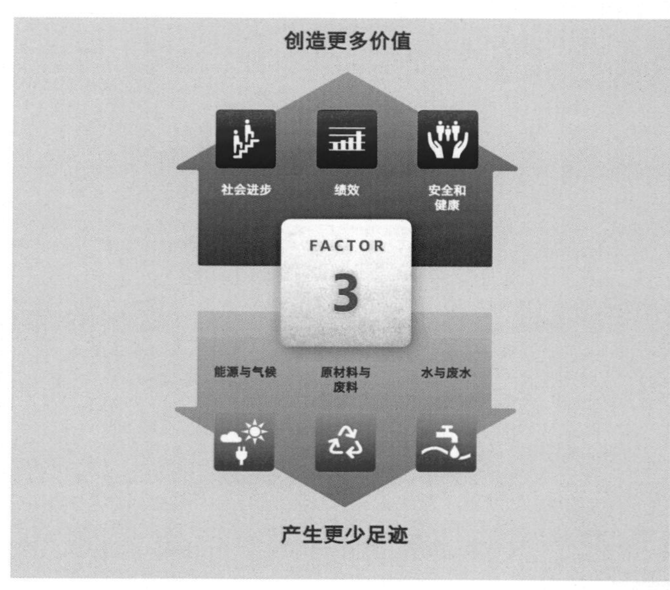

亚太地区 | 中国

汉高中国管理中心
中国上海市江湾城路 99 号尚浦中心 7 号楼
邮编：201203
电话：+86 21 2891 8000

汉高亚太研发总部
中国上海浦东张江高科技园区张衡路 928 号
邮编：201203
电话：+86-21-2891 8000

中国金属学会　编

第十三届
中国钢铁年会论文集
（摘要）

Proceedings of the 13th CSM Steel Congress

北　京
冶　金　工　业　出　版　社
2021

内 容 简 介

本论文集共收录 1046 篇特邀报告和论文摘要、582 篇论文全文，共 821 万字。全书内容包括矿业工程、炼焦化学、炼铁、炼钢、连铸、轧制与热处理、表面与涂镀、金属材料深加工、汽车用钢、管线钢、低合金钢、特殊钢、电工钢、不锈钢、非晶合金、粉末冶金、冶金能源、环保与资源利用、冶金设备与工程技术、冶金自动化与智能化、冶金物流运输、冶金流程工程学、高温合金、低碳发展等方面，全面反映了近两年来我国及世界钢铁行业科研、生产、管理等方面的最新成果，是一本内容全面、新颖，具有较高学术水平的专业论文集。

本论文集以纸质图书和电子版方式出版。纸质图书为收录论文的摘要集（含特邀报告），电子版为收录论文的全文。本书可供钢铁行业的科研人员、管理人员、工程技术人员、高校师生等学习参考。

图书在版编目(CIP)数据

第十三届中国钢铁年会论文集：摘要 / 中国金属学会编. —北京：冶金工业出版社, 2021.10

ISBN 978-7-5024-8948-9

Ⅰ.①第… Ⅱ.①中… Ⅲ.①钢铁工业—学术会议—文集 Ⅳ.① TF4-53

中国版本图书馆 CIP 数据核字 (2021) 第 206464 号

出 版 人　苏长永

地　　址　北京市东城区嵩祝院北巷 39 号　邮编　100009　电话　(010)64027926

网　　址　www.cnmip.com.cn　电子信箱　yjcbs@cnmip.com.cn

责任编辑　李培禄　美术编辑　彭子赫　版式设计　孙跃红

责任校对　王永欣　责任印制　李玉山

ISBN 978-7-5024-8948-9

冶金工业出版社出版发行；各地新华书店经销；北京虎彩文化传播有限公司印刷

2021 年 10 月第 1 版, 2021 年 10 月第 1 次印刷

880mm×1230mm　1/16；53.5 印张；8 彩页；1684 千字；780 页

350.00 元（全套）

冶金工业出版社　投稿电话　(010)64027932　投稿信箱　tougao@cnmip.com.cn

冶金工业出版社营销中心　电话　(010)64044283　传真　(010)64027893

冶金工业出版社天猫旗舰店　yjgycbs.tmall.com

(本书如有印装质量问题，本社营销中心负责退换)

第十三届中国钢铁年会
组 委 会

年会主席　干　勇

执行主席　赵　沛

委　　员（以姓氏笔画为序）

于　勇　王新江　左　良　曲　阳
杨仁树　沈　彬　张少明　陈德荣
赵民革　赵　继　戴志浩

年会秘书长　王新江

年会秘书处　中国金属学会学术工作部

《第十三届中国钢铁年会论文集》编委会

前　言

第十三届中国钢铁年会于2021年10月25～26日在北京召开。2021年是中国共产党成立100周年，我国正向着全面建成社会主义现代化强国的第二个百年奋斗目标迈进，作为国民经济支撑产业和制造业强国建设的基础产业，我国钢铁工业近两年来，虽然受到新冠肺炎疫情的严重影响，依然取得了辉煌的成就和举世瞩目的国际影响，钢铁科技发展也收获了累累硕果。我国碳达峰与碳中和目标的提出，关乎经济社会发展方式的转变，也给钢铁工业带来了巨大的挑战和机遇，在解决绿色低碳对行业发展制约的同时，将会大力推动钢铁行业的高质量发展。本届年会以“**建设绿色、低碳、智能、可持续发展的钢铁工业**”为主题，围绕钢铁生产全流程的基础理论、工艺技术、产品设计、制造和应用技术，研讨钢铁领域科技创新的方向和路径，促进钢铁行业向绿色化、智能化、品牌化发展，加快钢铁强国建设，把中华民族伟大复兴的历史伟业推向前进。

本届年会的征文工作得到全国冶金及材料领域的专家、学者和广大科技人员的积极支持，共收到1012篇投稿论文和189篇特邀报告摘要。经专家评审并根据作者意愿，年会摘要集收录1046篇，全文集收录582篇。内容包括：矿业工程、炼焦化学、炼铁、炼钢、连铸、轧制与热处理、表面与涂镀、金属材料深加工、汽车用钢、管线钢、低合金钢、特殊钢、电工钢、不锈钢、非晶合金、粉末冶金、冶金能源、环保与资源利用、冶金设备与工程技术、冶金自动化与智能化、冶金物流运输、冶金流程工程学、高温合金、低碳发展等方面，在此表示感谢！年会全文以电子版方式出版，特邀报告和录用论文摘要以纸质版方式出版。

由于论文集出版、编辑时间较紧，难免有疏漏与错误之处，恳请读者批评指正。

中国金属学会

2021年10月

目　　录

大会特邀报告

分会场特邀报告

1 矿业工程

2 炼铁与原燃料

2.1 炼焦化学

2.2　烧结、球团

2.3 炼　铁

3 炼钢与连铸

3.1 炼 钢

3.2 连 铸

4　轧制与热处理

4.1　板　　材

4.2　长材和钢管

5 表面与涂镀

6 金属材料深加工

7　先进钢铁材料及其应用

7.1　汽车用钢

7.2　管线钢

7.3 低合金钢

7.4　特殊钢

7.5　高温合金

7.6　电工钢

7.7 不锈钢

7.8 非晶合金

8 粉末冶金

9 能源、环保与资源利用

9.1 冶金能源

9.2 环保与资源利用

10　冶金设备与工程技术

11 冶金自动化与智能化

12 冶金物流运输

13　其　他

大会特邀报告

A Future Sustainable Steel Industry

Edwin Basson

(The World Steel Association)

Abstract: A future sutainable society will continue to require a steel industry that meets future requirements for environmental sustainability. According to recent global evaluation (IEA, 2020), the steel industry is fortunate to be able to draw on multiple approaches towards environmental sustainability. Efficiency, different raw material inputs and new future technoplogies will all need to play a role over a significant length of time. In addition, it is clear that no single approach exist, and significant differences may develop between regions based on existing energy systems and available raw materials.

While the steel industry will adjust to improve its environmental sustainability, requirements from steel using sectors also change. The steel industry are adjusting along multiple axis directions in order to meet these challenges.

The presentation will highlight the trends towards environmental and product sustainability in the global steel industry.

首钢钢铁业高质量发展的探索与展望

张功焰

（首钢集团有限公司，北京 100041）

摘 要：2003 年，首钢实施了史无前例的搬迁调整，成为我国第一个由中心城市搬迁至沿海发展的大型钢铁企业。这既是落实京津冀协同发展战略的一次生动实践，也是开启首钢高质量发展新征程的重要起点。

“十三五”以来，首钢坚持贯彻新发展理念、构建新发展格局。一是以绿色制造助推可持续发展。首钢实施全流程污染物控制，率先实现超低排放，首钢股份公司、京唐公司均被评为环境绩效 A 级企业。首钢京唐公司建立了多目标优化的低碳清洁冶炼炼铁系统，3 座 5500m^3超大型高炉球团矿比例达到 55%，年降低 CO_2 排放 226 万吨。首钢京唐公司建立了炼钢转炉大量消纳二氧化碳和转炉煤气制备乙醇两条示范线。年消纳二氧化碳 5 万吨、生产乙醇 4.5 万吨。二是以智能制造提升全要素生产率。通过云-边-端技术架构，搭建了赋能技术、智能装备、智能车间、智能工厂和协同生态的五层业务体系架构。三是以高品质制造赢得用户信赖。全力打造“产品、质量、技术、服务、成本”五大优势，持续做优做强钢铁主业。

“十四五”期间，首钢将聚焦绿色制造、智能制造和高效制造，突破一批关键核心技术，打造一批高端引领产品，加快推动技术领先成为首钢的核心竞争力，努力成为钢铁高质量发展的引领者，技术创新的新高地，城市复兴的新地标。

日本制铁为实现零碳钢的挑战

小野山修平

（日本制铁株式会社）

摘　要：日本制铁将“挑战零碳钢”作为下期经营计划的主要支柱之一，推进技术开发。报告介绍了我司至今所推行的节能技术，今后对于零碳钢的构想，并将说明以COURSE50项目和Super COURSE50为代表的高炉氢还原技术的概要及课题。

北京科技大学服务钢铁行业发展的探索与实践——学科发展、人才培养、科技创新和人文交流

杨仁树

（北京科技大学，北京　100083）

摘　要：北京科技大学因钢而生、因钢而兴，是共和国建立的第一所钢铁工业高等学府，享有“钢铁摇篮”的美誉。近年来，北京科技大学将学科建设、人才培养、科技创新和文化传承等办学治校重点单元与服务钢铁行业发展有机结合，推进本科生全程导师制、本硕贯通培养模式、教师“双走”战略、科研政策“减增放”等一系列举措，探索出一条多元协同发展的服务行业进步“北科模式”。未来，北京科技大学将继续在创新钢铁技术、培养钢铁人才、传承钢铁文化等方面持续发力，承担起“钢铁强国、科教兴邦”的时代使命。

提升材料品质　促进低碳发展

田志凌

（中国钢研科技集团有限公司，北京　100081）

摘　要：发展和应用高强韧、耐腐蚀、抗疲劳等高性能长寿命新材料技术是促进钢铁行业低碳发展的重要途径。近年来，钢铁研究总院面向量大面广或苛刻环境使用的基础设施和高端装备用钢开展基础研究和新技术、新产品研发。深入研究钒氮微合金钢理论和技术推广应用，形成了钒氮复合微合金化理论体系、VCN析出控制的晶粒细化技术和系列低成本高性能钒氮微合金钢品种技术，实现高效率生产；其中，高强度抗震钢筋由400MPa向500MPa、600MPa方向发展，400MPa级及以上高强度钢筋比例超过90%，据国外研究该成果2019年促进中国总体CO2排放减少1亿吨。针对不同海洋区域以及沿海、近海和远海腐蚀环境，发展了系列耐蚀钢筋技术，包括高性能/经济型双相不锈钢钢筋技术、不锈钢-低合金钢复合钢筋技术、中高铬系合金耐蚀钢筋技术、Cu-P-Cr-Ni系低合金耐蚀钢筋技术。针对南海海洋大气腐蚀环境，开发了Ni系高耐候钢，揭示了其在西沙、文昌海洋大气下的腐蚀规律与机理。针对深井/超深井页岩油气开采对长寿命压裂泵的需求，提出了“未溶第二相诱发锯齿状原奥晶界和细化马氏体亚结构+高温回火奥氏体逆相变和纳米相析出”的组织调控新方法，开发出新型中锰压裂泵阀箱用钢，强韧性、疲劳性能、耐蚀性能较现役钢种均大幅度提高，该方法对于新一代高强韧钢将具有普遍意义。我国齿轮钢质量稳定性与国外先进水平有一定差距，严重制约了齿轮基础件的性能，针对带状组织、淬透带宽、齿坯硬度波动性等质量问题，基于淬透性大数据，并考虑元素交互作用，修正了淬透性预测模型，预测偏差2HRC以内，同时预测成分精确控制范围，建立了基于转炉和电炉两种不同生产工艺的高精确冶炼技术和高均匀凝固技术，形成了≤4HRC超窄淬透性带宽控制技术，经200多炉次 的工业验证圆满完成任务目标。

质量能力分级。中国制造业正在经历从大到强、从量到质的转变。传统的质量评价体系核心是标准门槛，“合格”

之上的质量差异被忽视。质量能力分级旨在将质量全要素量化评价与产品大数据结合，变标准门槛为分级评价，可使优秀企业在合格率 100%基础上能够满足用户的差异化需求。2016 年钢铁研究总院作为第三方专业机构，开展钢铁行业质量能力分级试点工作，建立了质量分级模型，其核心是三组数学模型：价值函数模型将传统的二值评价变为连续评价，工序评价模型将传统的样品考核扩展到工序能力考核，质量遗传模型引入了传统评价中缺失的对原材料和产线组合的考核。模型全部采用客观数据，已连续 3 年发布钢铁产品质量分级评价排名，在行业内形成了良好反响。2019 年，钢铁研究总院承担了工信部《钢铁产品质量分级评价与示范应用》专项，对钢筋、轴承钢、船舶海工用钢等领域开展质量分级示范。目前，钢铁质量能力分级在中石化、中海油、中船等多个大型用户企业成功推广应用。发布了钢铁行业 T/CISA 团体标准，并被工信部推荐在建材、有色、石化等多个原材料领域推广应用。

分会场特邀报告

大会特邀报告
★ 分会场特邀报告
矿业工程
炼铁与原燃料
炼钢与连铸
轧制与热处理
表面与涂镀
金属材料深加工
先进钢铁材料及其应用
粉末冶金
能源、环保与资源利用
冶金设备与工程技术
冶金自动化与智能化
冶金物流运输
其他

深部地热与金属矿产资源共采关键技术

蔡美峰

（北京科技大学，北京 100083）

摘 要：随着浅部金属矿产资源的逐步减少和枯竭，深部开采将是金属矿开采的普遍趋势。但深部开采面临着深部高地应力、高温、深部岩性恶化以及深井提升等问题，使得深部开采的成本大幅提升。而深部矿产资源开采向深部地热开发的延伸，就大幅降低了采矿降温成本，增加了开采效益，为解决深部采矿的经济性问题开辟了有效途径。基于循环水热交换技术的深部矿井高温巷道降温系统的研究，便给深部采矿降温和岩体地热开采相结合提供了实现的可能。未来，深部无人采矿关键工程科技的战略研究和深部采矿向深部地热开发的延伸与结合，必将推动我国在深部采矿领域取得具有国际领先的重要突破，使我国成为未来世界的采矿强国。

金属矿膏体充填绿色开采技术

吴爱祥

（北京科技大学，北京 100083）

摘 要：金属矿开采产生了大量尾砂和采空区，导致了严重的安全、环境问题。膏体充填技术将全尾砂料浆进行深度浓密制备成不分层、不离析、不脱水的膏体充填至井下采空区，实现“一废治两害”，具有安全、环保、经济、高效的显著优势。膏体流变学是膏体充填的基础理论，是具有高度行业特殊性与复杂性的流变学分支，研究膏体充填各工艺环节中膏体料浆或充填体在应力、应变、温度等条件下与时间因素有关流动与变形的规律。同时，目前膏体充填技术在国内外已经得到了广泛的应用，并取得了很好的经济与社会效益。

我国铁矿石资源开发现状与未来

李新创

（冶金工业规划研究院，北京 100029）

摘 要：中国是世界第一钢铁生产大国，同时也是钢材的第一消费大国。但是中国国内铁矿石富矿少，多为贫矿，矿石品位不高，多元素复合矿多，目前主要依靠进口铁矿石来满足国内的生产需求，进口矿石占比超过 80%。近 20 年来，铁矿石价格呈 N 形大幅波动，严重影响钢铁行业稳定安全运行。新冠肺炎疫情以来，铁矿价格突破历史新高，缓解了国内铁矿石的供应问题，但是未来一段时期我国对进口矿的高度依赖局面不会改变。而双循环格局及“双碳”目标的提出，将进一步促进国内高品位矿的开发，带动铁矿产业铁矿产品供应结构的变化。未来，必须在加快国内铁矿石开发基础上，加快推进境外铁矿石开发，形成国内国外双循环铁矿保障格局，并逐步改变全球铁矿供需过于集中的格局。

我国金属矿智能化开采现状与发展

王李管

（中南大学，湖南长沙 410083）

摘 要：矿山智能化是实现矿山企业生产过程的安全、高效、低成本的有效途径，也是矿业发展的必然趋势。目前我国智能矿山建设已经初具规模，但是由于受到矿床赋存条件、技术和装备水平的限制，矿山智能化建设难度较大，建设水平有待提升。从未来采矿趋势发展来看，生产技术数字化、生产装备与系统智能化、安全与生产管理信息化将是金属矿智能化开采持续推进的主要方向，金属矿的开采将通过装备的智能化、系统的智能化和自动化、整个企业的智能化这三个层级逐步完成智能化的建设。

科技赋能、智享未来—宝钢资源建设“安全、绿色、智慧、高效”智慧矿山的实践与探索

张 华

（宝钢资源（国际）有限公司，上海 200126）

摘 要：从智慧矿山建设必要性、指导思路、主要成效以及后续工作措施等方面，详细介绍宝钢资源在智慧矿山建设方面的探索、实践以及后续工作的思考。宝钢资源围绕“安全、绿色、智慧、高效”，从“矿山资源数字化、生产作业智能化、经营管理智慧化、基础设施标准化”四个发展方向，四位一体打造可持续的竞争力，推进智慧矿山建设。目前，宝钢资源在“矿山资源数字化以及与生产计划、地质管理协同”、“采矿设备大型化、遥控化、智能化”、“有轨、无轨运输设备遥控化、无人化”、“选矿系统自动化、智能化”、“公辅设备无人化、集控化”以及集控中心的建设等方面积极实践、有所收获。

钒页岩资源利用技术进展

张一敏

（武汉科技大学，湖北武汉 430081）

摘 要：面向国家产业结构调整和长江经济带绿色发展的重大需求，报告聚焦我国战略优势矿产钒页岩开发利用现状，围绕生产过程中金属提取难度高、固废排量大、产品附加值低等共性关键问题，总结回顾我国钒页岩技术工业化发展历程，介绍钒页岩最新技术进展，以及国家双碳目标下的技术发展趋势。通过源头减量、过程控制、末端利

用的全产业链整套技术研发，打造出具有国际领先水平的钒页岩资源高效清洁利用核心方案，推动产业化集成示范建设，支撑国家战略性多金属页岩绿色发展。

关键词：钒页岩；高效清洁利用；全产业链；低碳发展

Technology Progress of Vanadium-bearing Shale Utilization

Zhang Yimin

(Wuhan University of Science and Technology, Wuhan 430081, China)

Abstract: Under the background of national industry restructuring and green developing of Yangtze River economic belt, the report focuses on the situation of the development and utilization about vanadium-bearing shale which is one of the predominant strategic mineral resources in China. Concentrating on the key common problems involved by metal extraction, solid waste emission, and product added value in its production process, the report reviews the industrialization journey of vanadium-bearing shale utilization and further provides a complete introduction to the latest technologies and future development under the national target of peaking carbon dioxide emissions and carbon neutral. The formation of whole industry chain integrated with source reduction, process control, and end use, is building up an international leading core solution of high-efficient and eco-friendly utilization of vanadium-bearing shale and promoting its industrial demonstration, which will support the green development of strategic polymetallic shale in China.

Key words: vanadium-bearing shale; high-efficient and eco-friendly utilization; whole industry chain; low carbon development

高效脱杂磁选机的设计思路

孙仲元

（中南大学，湖南长沙 410083）

摘 要：当前强磁分选存在机械夹杂脉石导致精矿品位偏低的问题。本文提出了几种高效脱杂磁选机的设计思路以提高磁精矿品位。这几种磁选机是：（1）永磁式旋转磁极带式磁选机；（2）永磁振动桶式磁选机；（3）周期鼓动流膜平环磁选机；（4）短磁路全铠装磁平环磁选机；（5）脉冲给矿超导磁选机。期望汇聚各位专家的智慧，完善现有设计思路，争取早日实现工业应用。

关键词：脱杂；鼓动；振动；磁翻；短磁路；全铠装

The Design Idea of High Gangue Removal Efficiency Magnetic Separator

Sun Zhongyuan

(Central South University, Changsha 410083, China)

Abstract: At present, the problem of gangue mechanical entrainment in high-gradient magnetic separation lead to low concentrate grade.For sloving this problem, several design ideas of high gangue removal efficiency magnetic separator will be introduced. These magnetic separators are:

(1) Permanent magnet rotating magnetic pole belt magnetic separator.
(2) Permanent magnet vibrating drum magnetic separator.
(3) Periodic reciprocating flowing membrane flat ring magnetic separator.
(4) Short magnetic circuit full armor flat ring magnetic separator.
(5) Pulse feeder superconducting magnetic separator.

Key words: gangue removal; reciprocating; vibration; magnetic turn; short magnetic circuit; full armor

氢还原高磷鲕状赤铁矿还原技术

郭培民

（钢铁研究总院，北京　100081）

摘　要：对高磷鲕状赤铁矿的氢还原进行了热力学研究，结果表明：氢气还原过程磷不会发生还原反应，因此磷不会进入还原铁中或以气态挥发分离，仍然以磷酸钙形式存在脉石相中。对氢还原过程主要工艺技术参数进行研究，得到对于还原+磁选路线，氢还原温度需要大于 1000℃以上，入炉温度要比 Midrex 工艺高；高磷铁矿的入炉氢气量为正常富矿富氢还原的 1 倍水平；为了降低氢气入炉流量，应适度提高入炉氢气温度；对于纯氢反应，加热过程氢气会与耐热不锈钢表面的氢气反应生成 CH_4，产生微孔，氢气进一步扩散到耐热钢内部，发生侵蚀反应，为了解决高富氢还原氢侵蚀问题，一种可供考虑的方式，在氢气中添加少量的 CH_4，即可抑制氢侵蚀问题；对于少量的零碳或超低碳钢要求，则要考虑其它的加热方式。

关键词：高磷鲕状赤铁矿；氢还原；低磷金属铁；磁选

Reduction Technology of High Phosphorus Oolitic Hematite by Hydrogen Reduction

Guo Peimin

(Central Iron & Steel Research Institute, Beijing 100081, China)

Abstract: The thermodynamic study on the hydrogen reduction of high phosphorus oolitic hematite is carried out. The results show that phosphorus will not undergo reduction reaction in the process of hydrogen reduction, so phosphorus will not enter the reduced iron or volatilize and separate in gaseous state, but still exists in gangue phase in the form of calcium phosphate. The main technological parameters of hydrogen reduction process are studied. For the reduction + magnetic separation route, the hydrogen reduction temperature needs to be more than 1000 ℃, and the furnace inlet temperature is higher than that of Midrex process; The amount of hydrogen into the furnace of high phosphorus iron ore is twice the level of hydrogen rich reduction of normal rich ore; In order to reduce the flow of hydrogen into the furnace, the temperature of hydrogen into the furnace should be moderately increased; For pure hydrogen reaction, during heating, hydrogen will react with hydrogen on the surface of heat-resistant stainless steel to produce CH_4 and micropores, and hydrogen will further diffuse into the heat-resistant steel to produce corrosion reaction. In order to solve the problem of reducing hydrogen corrosion with high hydrogen enrichment, a way to be considered is to add a small amount of CH_4 to hydrogen to inhibit hydrogen corrosion; For a small amount of zero carbon or ultra-low carbon steel, other heating methods shall be considered.

Key words: high phosphorus oolitic hematite; hydrogen reduction; low phosphorus metal iron; magnetic separation

复杂难选铁矿选矿技术进展

陈 雯

（长沙矿冶研究院有限责任公司，湖南长沙 410012）

摘 要：通过回顾我国铁矿选矿技术发展历程，总结我国铁矿选矿发展历程中对行业技术进步有显著推动作用的关键技术。重点针对低贫矿、菱铁矿褐铁矿、微细粒铁矿等难选矿种的选矿技术现状和当前面临的主要问题展开讨论。提出在国家双碳战略背景下铁矿选矿发展方向。

关键词：低贫矿；微细粒铁矿；菱铁矿；褐铁矿

Progress in Beneficiation Technology for Complex Refractory Iron Ore

Chen Wen

(Changsha Research Institute of Mining and Metallurgy Co., Ltd., Changsha 410012, China)

Abstract: After reviewing the development history of iron ore beneficiation technologies in China, this paper presents a summary of the key technologies that ever played an important role in driving the technical progress of this industry. Then, the present situation of beneficiation technologies and the main problems existed in processing of refractory iron ore, including lean ore, siderite and limonite, as well as fine-grained iron ore are discussed. Finally, the development direction for iron ore beneficiation technologies is put forward under the background of China's "dual carbon" goals.

Key words: lean ore; fine-grained iron ore; siderite; limonite

贫细铁矿石节能减碳碎磨新技术的现状及动向

孙炳泉，沈进杰

（中钢集团马鞍山矿山研究总院股份有限公司，安徽 马鞍山 243000）

摘 要：评述了近 20 年来高压辊磨、塔磨（立式搅拌磨）等破碎磨矿新技术在贫细铁矿石选矿领域的典型应用研究成果，认为这两项关键技术的推广应用将对于贫细铁矿石进一步实现高效节能减碳利用具有重要作用，并结合最新研究动态提出了今后碎磨技术的重点发展方向。

关键词：贫细铁矿石选矿；高压辊磨；塔磨（立式搅拌磨）；节能减碳

Present Situation and Trend of New Crushing and Grinding Technology for Energy Saving and Carbon Reduction of Lean Fine Iron Ore

Sun Bingquan, Shen Jinjie

(Sinosteel Maanshan General Institute of Mining Research Co., Ltd., Maanshan 243000, China)

Abstract: The typical application research results of new crushing and grinding technologies such as high pressure roller mill and tower mill (vertical stirring mill) in the field of beneficiation of lean fine iron ore in recent 20 years are reviewed. It is considered that the popularization and application of these two key technologies will play an important role in further realizing high efficiency, energy saving and carbon reduction utilization of lean fine iron ore. Combined with the latest research trends, the key development direction of crushing and grinding technology in the future is put forward.

Key words: beneficiation of lean fine iron ore; high pressure roller mill; tower mill (vertical stirring mill); energy saving and carbon reduction

先进选矿装备技术的创新和发展

张国旺[1,2]，刘　瑜[1,2]，石　立[1,2]，龙　渊[1,2]，肖　骁[1,2]，谢睿宁[1]

（1. 长沙矿冶研究院有限责任公司，湖南长沙　410012；
2. 湖南金磨科技有限责任公司，湖南长沙　410012）

摘　要：介绍了近年来铁矿石磨矿分级、磁选和压滤等先进选矿装备的创新和典型工业应用。多碎少磨、预先抛尾、选择性细磨、强磁选、细磨及超细磨提质等装备技术的应用，对于难选贫细杂铁矿石的开发应用作出了贡献。面对“碳中和、碳达峰”目标的实现，高效节能环保的先进智能选矿装备技术将更会更加重视并得到广泛应用。

关键词：高压辊磨；超细磨；搅拌磨；强磁选；低碳经济

Innovation and Development of Advanced Mineral Processing Equipment and Technology

Zhang Guowang[1,2], Liu Yu[1,2], Shi Li[1,2], Long Yuan[1,2], Xiao Xiao[1,2], Xie Ruining[1]

(1. Changsha Research Institute of Mining & Metallurgy Co., Ltd., Changsha 410012, China;
2. Hunan Jinmo Technology Co., Ltd., Changsha 410012, China)

Abstract: The innovation and typical industrial application of advanced mineral processing equipment for grinding and classification, magnetic separation and pressure filtration of iron ore in recent years were introduced. The application of equipment for the technique of more crushing and less grinding process, pre-discarding of tails, selective fine-grinding, high intensity magnetic separation, fine grinding and ultra-fine grinding for quality improvement, has contributed to the exploitation and utilization of refractory lean and fine-grained iron ore. In response to the China's “dual carbon” target, i.e. peaking carbon emissions by 2030 and achieving carbon neutrality by 2060, the advanced intelligent mineral processing

equipment and technologies with characteristics of high efficiency, energy-conservation and eco-friendliness will catch more attention and find wide application.
Key words: high pressure roller mill; ultrafine grinding; stirring mill; high-intensity magnetic separation; low carbon economy

基于全流程系统优化理念的焦化行业发展模式探讨

石岩峰

(中国炼焦行业协会，北京 100120)

摘 要：为正确认识新常态、促进新发展,系统分析了我国焦化行业现状及所面临的主要发展瓶颈,阐述了国家对焦化行业环保的总体要求及其对整个行业发展所带来的影响,指出节能、环保、供给侧结构性改革是新时代焦化行业的重点发展方向。在此基础上,提出全流程系统优化理念下的焦化行业发展模式,分析了焦化行业全流程系统优化发展的意义,介绍全流程系统优化的工作方法。同时,基于实际调研及企业反馈,提出当前我国焦化行业所存在的共性技术问题及初步优化方向。
关键词：焦化行业； 化解过剩产能；共性技术问题；全流程系统优化；发展模式

Discussion on the Development Model of Coking Industry based on the Concept of Whole Process System Optimization

Shi Yanfeng

(China Coking Industry Association (CCIA), Beijing 100120, China)

Abstract: To correctly comprehend the new normal and promote new development in the new era, this paper thoroughly analyzes the current situation and main development bottlenecks of China's coking industry, discusses the overall national requirements for environmental protection of coking industry and its impact on the development of the whole industry, and points out that energy conservation, environmental protection and supply side structural reform are the key development direction of coking industry in the new era. On this basis, the development model of coking industry under the concept of whole process system optimization is put forward, the significance of whole process system optimization development of coking industry is analyzed, and the working methods of whole process system optimization are introduced. Meanwhile, based on the actual investigation and enterprise feedback, this paper also puts forward the technical issues in common existing in China's coking industry and the preliminary optimizing orientation.
Key words: coking industry; tackling overcapacity; technical issues in common; whole process system optimization; development model

焦化碳排放分析及碳减排技术

王明登，李 超，许 为

（中冶焦耐（大连）工程技术有限公司，辽宁大连 116085）

摘　要：“碳达峰、碳中和”将对焦化行业产生重大而深刻的影响。本报告分析了我国焦化工业发展现状及双碳战略对行业近期和远期的影响；定量分析焦化生产过程的碳排放量，提出焦化生产碳排放量影响因素；分析了焦化工业碳减排实现路径，并结合我国能源结构变革趋势，提出焦化生产碳减排技术；结合钢铁工业的碳减排技术路径，提出支撑绿色低碳冶金的焦化技术研发方向；最后建议重点关注富氢高炉所需高强度高反应性焦炭生产、低成本工业化制富氢还原气、钢化联产固碳、低碳炼焦等革命性焦化技术。

关键词：焦化；碳排放分析；碳减排技术

Analysis on Carbon Emission and Carbon Emission Reduction Technology in Coking Industry

Wang Mingdeng, Li Chao, Xu Wei

(ACRE Coking & Refractory Engineering Consulting Corporation (Dalian), MCC, Dalian 116085, China)

Abstract: The strategy of “Carbon Emission Peaking” and “Carbon Neutrality” will have a significant and profound impact on the coking industry. This paper analyzes the current status of China’s coking industry and the impact of this strategy on the coking industry in short-term and long-term. Quantitative analysis of the carbon emissions of coking process is made, the influencing factors of carbon emissions of coking production are put forward; Trailblazing is analyzed for carbon emissions reduction in the coking industry and its carbon emissions reduction technology is proposed based on the transformation trend of energy structure in China; The orientation of green coking technology is put forward according to the carbon emissions reduction in steel industry. Finally it is suggested to focus to the revolutionary coking technology such as high strength & high CRI coke production for rich hydrogen BF, low-cost industrial production of rich hydrogen reducing gas, CO_2-fixation through steelmaking & by-products plant as well as low-carbon cokemaking, etc..

Key words: coking; carbon emission analysis; carbon emissions reduction technology

适应高炉低碳炼铁的焦炭质量与优化配煤研究

孟庆波

（中钢集团鞍山热能研究院有限公司，辽宁鞍山　114044）

摘　要：中国钢铁行业碳排放量占全国碳排放总量的 15%。中国钢铁行业以高炉炼铁为代表的长流程占比高达近 90%，其中高炉冶炼占钢铁工业碳排放量的 70%以上，钢铁工业减碳重点在高炉工序。高炉降低碳排放主要措施，一是降低高炉燃料比，二是使用低碳燃料--炉顶煤气 CO 分离回用、H2 或富氢气体入炉用作还原气。焦比下降使焦炭在高炉中停留时间增加，焦炭受到的劣化作用增大；还原性气体的引入使高炉温度场、气体分布、热交换等均发生明显变化，对焦炭质量提出新的要求，焦炭在高炉下部高温区域抵抗高温和碱金属等破坏、渣铁侵蚀、鼓风机械破坏的能力变得更加重要。模拟高炉冶炼条件的研究显示，传统焦炭质量指标相近的焦炭，由于配煤结构的不同或/和工艺的不同抵抗上述破坏的能力不同。研究高炉低碳炼铁对焦炭质量的本质要求，构建焦炭质量评价新体系是保障高炉低碳冶炼的首要任务。加强炼焦煤评价研究，构建炼焦煤应用性分类，根据炼焦机理及配煤研究新成果，开展优化配煤研究，实现炼焦煤资源合理利用与满足高炉低碳炼铁相协调。

关键词：高炉低碳炼铁；焦炭质量；炼焦煤应用性分类；优化配煤

Study on Coke Quality and Optimized Coal Blending for BF Low-carbon Ironmaking

Meng Qingbo

(Anshan Thermal Energy Research Institute Co., Ltd., SINOSTEEL, Anshan 114044, China)

Abstract: According to statistics, the carbon emissions from China's steel industry account for 15% of China's total carbon emissions. In China's iron and steel industry, the long process represented by blast furnace ironmaking accounts for nearly 90%, of which blast furnace smelting is higher than 70% of the carbon emission of the steel industry. Blast furnace production plays a key role in carbon reduction in steel industry. The BF carbon reduction has two measures, one is to decrease the BF fuel rate, and another is to utilize low-carbon fuels, i.e. the CO recovered from the furnace top, and H2 or rich hydrogen returned to the furnace as a reducing gas. The lower coke rate results in longer retention time of the coke inside the blast furnace and more degradation effect against the coke; furthermore, the reducing gases into the furnace would pose a new challenge to the coke quality due to an obvious change on the blast furnace temperature field, gas distribution and heat exchanging, etc. so that it requires higher performances of the coke in heat-resistance, alkali metal corrosion, iron slag erosion, and air blasting damage. The BF simulation analysis shows that the coke of similar properties varies with different performance to withstand the abovementioned damages due to different coal blending and/or cokemaking processes. To study the fundamental requirement for coke quality by low-carbon ironmaking and to establish a new assessment system of the coke quality is a paramount task to ensure the BF low-carbon ironmaking. This paper also points out that to enhance coking coal assessment study, to establish application assortment of coking coals, and to carry out optimal coal blending research according to cokemaking mechanism and new outcome of coal blending can realize rational utilization of coking coal resources coordinated with BF low-carbon ironworking.

Key words: BF low-carbon ironmaking; coke quality; application assortment of coking coals; optimized coal blending

大型焦炉热平衡测试与节能分析

甘秀石[1]，郝 博[1]，李卫东[1]，边子峰[2]，王 磊[2]

（1. 鞍钢集团钢铁研究院，辽宁鞍山 114009；2. 鞍钢股份炼焦总厂，辽宁鞍山 114021）

摘 要：总结分析了常规大型焦炉热平衡测试和计算结果，定量掌握焦炉能量消耗和分配情况，结合焦炉自身情况找出炼焦工序节能降耗的途径。

关键词：焦炉；热平衡；节能降耗

Heat Balance Test and Energy Conservation Analysis on Large-capacity Coke Oven

Gan Xiushi[1], Hao Bo[1], Li Weidong[1], Bian Zifeng[2], Wang Lei[2]

(1. Iron & Steel Institutes of Ansteel Group Corporation, Anshan 114009, China;
2. General Coking Plant of Angang Steel Co., Ltd., Anshan 114021, China)

Abstract: This paper summaries and analyses the results of heat balance determination and calculation on large-capacity coke oven. Energy consumption and distribution in cokemaking can be monitored in a quantitative manner. According to the current situation of the battery in operation, the approaches of minimizing the heat loss and achieving energy saving in cokemaking process can be obtained.

Key words: coke oven; heat balance; energy conservation

脱硫废液制酸工艺在新泰正大焦化的应用

张洪波，刘　宏，吕文才，王　涛

（山东新泰正大焦化有限公司，山东泰安　271200）

摘　要： 介绍了新泰正大焦化采用“浆液进料、富氧焚烧、两转两吸”制酸工艺，处理焦炉煤气氨法湿式催化氧化脱硫副产低纯硫磺及脱硫废液，制取硫酸的工艺流程、技术特点、投产后生产运行及改进情况。

关键词： 脱硫废液；制酸工艺；生产运行；改进建议

Application of Sulfuric Acid Making Process by Using Waste Desulfurization Liquid

Zhang Hongbo, Liu Hong, Lv Wencai, Wang Tao

(Shandong Xintai Zhengda Coking Co., Ltd., Tai'an 271200, China)

Abstract: This paper introduced sulfuric acid making process of “sulfur slurry feeding, combustion with rich oxygen air and two-conversion + two-absorption” in Xintai Zhengda Coking Plant, i.e. the process flow of sulfuric acid making through treatment for the low purity sulfur and waste desulfurization liquid by wet catalytic oxidation process. It also includes its technical features, operation conditions after commissioning and its improvement.

Key words: waste desulfurization liquid; acid-making process; operation; improvement suggestion

碳中和背景下的低碳炼铁与氢冶金

张建良

（北京科技大学冶金与生态工程学院，北京　100083）

摘　要： 中国钢铁工业自建国以来经历了快速发展，但在碳中和新时代的背景下，迫切需要寻求变革。钢铁行业的年二氧化碳排放量占全球碳排放总量的6.7%，而在中国，这一比例达到了15%。作为中国碳排放量最高的制造业，钢铁行业将在外部政策监管和内部技术突破的协调下找到自己的出路。使用氢代替碳作为还原剂在热力学和动力学方面都有很大的优势。本文从理论角度介绍低碳冶金发展的当前潜力，并总结当前研究领域的最新发展。然后列举、评价和比较了目前全世界典型的低碳炼铁工艺，对炼铁行业未来发展提出展望。新的时代背景下我们要重视炼铁工业在低碳方面的多元化发展。

关键词： 碳中和；低碳炼铁；氢冶金；非高炉炼铁

Low-carbon Ironmaking and Hydrogen Metallurgy under the Background of Carbon Neutrality

Zhang Jianliang

(School of Metallurgical and Ecological Engineering, University of Science & Technology Beijing, Beijing 100083, China)

Abstract: China's steel industry has experienced rapid development since the founding of the People's Republic of China, but in the context of the new era of carbon neutrality, it is urgent to seek change. The annual carbon dioxide emissions of the steel industry account for 6.7% of the total global carbon emissions, while in China, this proportion has reached 15%. As the manufacturing industry with the highest carbon emissions in China, the steel industry will find its own way out under the coordination of external policy supervision and internal technological breakthroughs. The use of hydrogen instead of carbon as a reducing agent has great advantages in thermodynamics and kinetics. This article introduces the current potential of the development of low-carbon metallurgy from a theoretical perspective, and summarizes the latest developments in the current research field. Then it enumerates, evaluates and compares the typical low-carbon ironmaking processes in the world at present, and puts forward the prospects for the future development of the ironmaking industry. Under the background of the new era, we must attach importance to the diversified development of the ironmaking industry in low-carbon aspects.

Key words: carbon neutrality; low-carbon ironmaking; hydrogen metallurgy; non-blast furnace ironmaking

Through the Looking Glass: Comparison of Blast Furnace Operational Practices in China and Europe

Geerdes Maarten[1], Sha Yongzhi[2]

(1. Geerdes Advies, The Netherlands; 2. China Iron and Steel Research Institute, Beijing, China)

Abstract: A comparison is made of blast furnace operation practices in China and Europe and North America from benchmark data and site visits. Productivity in China is generally higher than in Europe, but in Europe furnaces are operated at lower coke rates. The data are analyzed from the perspective that the most critical step in blast furnace operation is the melting of the ferrous burden. Melting is more difficult at higher primary slag basicity (as in China) and can be optimized by using burden distribution tools as is done in Europe.

Key words: blast furnace; coke rate; coal injection; slag basicity; burden distribution; productivity; melting of ferrous burden

关于两段式煤粉喷吹工艺的开发与实践

沈峰满[1]，吴钢生[2]，郑海燕[1]，白文广[2]，姜　鑫[1]，席　军[2]，高强建[1]

（1. 东北大学，辽宁沈阳　110819；2. 内蒙古包钢钢联股份有限公司，内蒙古包头　014010）

摘　要： 周知，大喷煤有利于进一步降低炼铁成本、同时也有利于环保。然而随着喷煤量的增加，炉内将积存大量未燃煤粉，加之焦炭用量减少以及焦炭在高炉内存在强度劣化问题，使得高炉料柱透气透液性下降，料柱压差升高，高炉生产操作困难，甚至还可能发生悬料等事故，限制了煤粉量。因此，为了寻求提高喷煤量的新途径，经一系列的理论研究、技术开发、装备设计，成功地研发了两段式煤粉喷吹工艺，并将其应用于大、中、小高炉，取得了显著的经济与社会效益。

本文将重点介绍，以有效抑制焦炭强度劣化、降低料柱压差为核心的两段式喷吹煤粉工艺理论，进而介绍依据该理论开发的两段式喷吹煤粉工艺、相关技术、设备和操作系统，以及二段煤枪位置、喷枪数量、二段适宜喷吹量、载气压力等煤粉喷吹工艺参数的确定方法。并以大中小各类高炉应用实例为例，介绍该工艺能够有效抑制焦炭在高炉内强度劣化、降低高炉压差以及对能源消耗、高炉操作稳定性、CO_2排放方面的积极作用等应用效果。

关键词：高炉喷煤；两段式喷煤；焦炭劣化；高炉压差

Development and Practice of Bi-PCI Process

Shen Fengman[1], Wu Gangsheng[2], Zheng Haiyan[1], Bai Wenguang[2], Jiang Xin[1], Xi Jun[2], Gao Qiangjian[1]

(1. Northeastern University, Shenyang 110819, China;
2. Inner Mongolia Baotou Steel Union Co., Ltd., Baotou 014010, China)

Abstract: It is well known that a large amount of pulverized coal injection is conducive to further reduce ironmaking cost and environmental protection. However, a large amount of unburned pulverized coal will be accumulated in the furnace due to the increase of coal injection amount. In addition, the reduction of coke consumption and the strength deterioration of coke in the blast furnace will reduce the air and liquid permeability of the blast furnace charge column, increase the pressure difference of the charge column, make the production and operation of the blast furnace difficult, and even lead to suspension accidents, which limits the amount of pulverized coal. In order to find a new way to improve the coal injection amount, bi-segment pulverized coal injection process (Bi-PCI) has been successfully developed and applied to all-size blast furnaces by a series of theoretical analyses, technical development and equipment design and remarkable economic and social benefits have been achieved.

This paper will focus on the theory of Bi-PCI process with the core aims of effectively inhibiting the deterioration of coke strength and reducing the pressure difference of charge burden, and then introduce Bi-PCI process, its related technology, equipment and operating system, as well as the determination method of pulverized coal injection process parameters such as position of bi-segment coal lance, the number of the lance, the appropriate injection volume of bi-segment coal lance, and carrier gas pressure. Taking the application of all-size blast furnaces as the examples, the strength deterioration of coke in blast furnace can be effectively inhibited and pressure difference can be reduced which show that Bi-PCI technology plays a positive role in energy consumption, blast furnace operation stability and CO_2 emission and the application effect of Bi-PCI process is remarkable.

Key words: coal injection in blast furnace; bi-segment pulverized coal injection (Bi-PCI); deterioration of coke; pressure difference

面向低碳绿色的现代高炉炼铁技术发展现状与未来

张福明[1,2]

（1. 首钢集团有限公司总工程师室，北京　100041；
2. 北京市冶金三维仿真工程技术研究中心，北京　100043）

摘 要：现代高炉炼铁经过近 200 年的发展演进，实现了高炉大型化、高炉喷煤等重大关键技术突破。回顾分析了 20 年来我国高炉炼铁技术的发展成就和运行实绩。面向全球碳减排和碳中和的发展态势，炼铁工业和传统工艺必须最大限度减少对碳素能源的依赖，降低碳素消耗和 CO_2 排放，这将成为至本世纪中叶的主要发展命题。研究认为，结合我国炼铁工业的资源和能源供给条件、技术装备和操作运行条件，到本世纪中叶我国高炉炼铁工艺流程仍将占有一定比率。面向未来，提出了焦化+烧结+球团+高炉长流程实现低碳冶金和碳-氢耦合冶金的技术路线和技术发展途径。针对高炉工艺革新和实现低碳、超低碳炼铁技术的发展提出了建议，指出了至 2050 年我国高炉炼铁技术的主要发展理念和技术途径。

关键词：低碳；绿色化；炼铁；高炉；氢冶金；碳减排

Developing Situation and Future on Modern Blast Furnace Ironmaking Technology Facing Low Carbon and Greenization

Zhang Fuming

(Chief Engineer Office of Shougang Group Co., Ltd., Beijing100041, China)

Abstract: After recent 200 years of development and evolution, modern blast furnace ironmaking has achieved major key technological breakthroughs such as blast furnace volume enlargement and pulverized coal injection. The development achievements and operation performances of China'sblast furnace ironmaking technology in the past 20 years are reviewed and analyzed. Facing the global carbon emission reduction and carbon neutral development trend, the ironmaking industry and traditional processes must minimize the dependence on carbon energy and reduce carbon consumption and CO_2 emissions, which will become the main development proposition by the middle of this century. The investigation believes that combined with the resources and energy supply conditions, technical equipment and operation conditions, the ironmaking process of blast furnace in China will still occupy a certain proportion in the middle of this century. For the future, the technical route and development approach of the long process of coking, sintering, pelletizing and blast furnace to realize low carbon metallurgy and carbon-hydrogen coupled metallurgy are proposed. It makes suggestions for the innovation of blast furnace technology and the development of low-carbon and ultra-low carbon ironmaking technology, and points out the main development concept and technological approach of blast furnace ironmaking technology in China by 2050.

Key words: low caborn; greenization; ironmaking; blast furnace; hydrogen metallurgy; carbon emission reduction

超高料层均热烧结研究与应用进展

姜 涛，李光辉，范晓慧，刘会波，徐良平

（中南大学，湖南长沙 410083）

摘 要：高料层是实现优质低耗烧结的有效途径。但当料层超过 700 mm 以后，出现烧结矿质量严重不均、生产率下降的问题。针对这一问题，国内外开发了生石灰强化制粒、热风烧结、喷吹焦炉煤气等技术。这些技术一定程度上提高了料层、降低了烧结能耗，但由于对超高料层的实际热状态、热量分布及其对烧结过程的影响缺乏定量、清晰的了解，还没有最大限度挖掘超高料层烧结优质、节能的潜力。本文作者团队从定量研究料层热量分布入手，提出均热高料层烧结技术思想，构建实现均热烧结的料层结构模式，开发均热超高料层烧结技术，在宝钢实现示范生产并在国内大型钢企推广应用，获得大幅降低烧结能耗、提高难处理资源利用率的效果。

Investigation and Application of Super-high Bed Sintering of Iron Ores

Jiang Tao, Li Guanghui, Fan Xiaohui, Liu Huibo, Xu Liangping

Abstract: High-bed sintering is an effective routeonsuperior quality and low consumption production. The unhomogeneous quality and lower productivity arise when the bed depth exceeds 700 mm. To solve above problem, technologies including quicklime intensifying granulation, hot air sintering and coke oven gas injection sintering have been developed.These technologies reduced the energy consumption of high-bed sintering in a certain extent. However, the heat condition of super-high-bed sinter material during sintering is not clear yet, especially lacking quantitative research. The potential capacity of super-high-bed sintering on quality-improving and energy-saving production is locked.In this paper, the heat condition of sinter bed was investigated quantitatively, the author put forward heat-homogenizing sintering and constructed feasible bed structure for heat-homogenizing sintering, finally developed the super-high-bed sintering technology and demonstrated in Baosteel, also applied to other steel plants in China andobtained a significant effect onquality-improving and energy-saving.

钒钛磁铁矿非高炉冶炼技术现状及发展趋势

朱庆山

（中国科学院过程工程研究所，北京　10089）

摘　要：钒钛磁铁矿是我国的优势资源，其高效清洁利用对我国具有重要意义。由于铁钒钛等组分紧密共伴生，钒钛磁铁矿难选、难冶、难利用。当前我国通过高炉-转炉流程冶炼钒钛磁铁精矿，实现了铁和钒的规模化利用，但其中的钛未获利用。直接还原-熔分非高炉流程可在提铁的同时富集钛，被认为是实现铁钒钛综合利用的可行流程。本文将对国内外钒钛磁铁精矿竖炉、回转窑、流化床、转底炉等直接还原技术现状进行系统介绍，在此基础上，分析磁铁矿直接还原需突破的关键基础及技术问题，包括：还原气体平衡转化率低；还原动力学慢；熔分钛渣品位低、难直接利用等。介绍了作者团队在提高钒钛磁铁矿直接还原效率等方面的探索，包括：通过氧化将 $FeTiO_3$、$FeTi_2O_4$ 等物相转化 Fe_2O_3 和 TiO_2，将铁钛化合物的还原转化为 Fe_2O_3 的还原，以期绕过直接还原过程的热力学障碍；还原过程调控，以防止 TiO_2 与 FeO 重新化合、强化还原过程；熔渣冷却过程调控，以实现定向析出黑钛石及提高黑钛石中钛的含量等。最后对在双碳背景下，钒钛磁铁矿非高炉冶炼发展前景进行展望。

Present Situation and Development Trend of Non Blast Furnace Smelting of Vanadium Titanium Magnetite

Zhu Qingshan

(Institute of Process Engineering, Chinese Academy of Sciences, Beijing 100089, China)

Abstract: Vanadium titanomagnetite is the dominant resource in China, and its efficient and clean utilization is of great significance to China.Vanadium-titanium magnetite is difficult to separate, smelt and utilize due to the close coexisting of iron, vanadium-titanium and other components.At present, the large-scale utilization of iron and vanadium has been realized by smelting vanadium-titanium magnetite concentrate through blast furnace - converter process in China, but titanium has not been utilized.The direct reduction-melting non blast furnace process can enrich titanium at the same time of iron extraction and is considered to be a feasible process for comprehensive utilization of iron, vanadium and titanium.In this paper, the status quo of direct reduction technology for vanadium and titanium magnetite concentrate at home and abroad will be systematically introduced, such as shaft furnace, rotary kiln, fluidized bed, rotary hearth furnace and so on. On this basis, the key basic and technical problems to be overcome in direct reduction of magnetite are analyzed, including: low reduction gas balance conversion rate;Reduction kinetics is slow;Melting titanium slag has low grade and is difficult to use directly.In order to avoid the thermodynamic obstacles in the direct reduction process, the author's team's exploration on improving the direct reduction efficiency of vanadium titanium magnetite is introduced, including: converting $FeTiO_3$ and $FeTi_2O_4$ into Fe_2O_3 and TiO_2 by oxidation;The reduction process is regulated to prevent the recombination of TiO_2 and FeO and strengthen the reduction process;The cooling process of slag is regulated to realize directional precipitation of black titanite and increase titanium content in black titanite.Finally, the development prospect of vanadium-titanomagnetite non-bf smelting under double carbon background is prospected.

高炉大波动的恢复与高炉智能控制的开发

陈令坤

（宝钢股份中央研究院炼铁所，湖北武汉 430081）

摘 要：目前高炉智能控制系统的开发方兴未艾，各厂家、人员出于不同目标，开发了各具特色的智能控制系统，这些系统一般包含不少数学模型，涉及的信息也较多，对高炉稳定运行发挥了一定的作用，但在高炉长时间失常过程中，高炉恢复多靠人工调整，很少听到靠智能控制系统的指导把高炉从长时间异常状态恢复到正常生产，当前运行的智能控制系统能否应对高炉长时间失常的炉况，是一个需要思考也研究的问题，本文将结合 2021 年初武钢有限 5、8 号高炉波动恢复过程，对高炉智能控制系统的架构，内容，运转模式等提出看法，涉及开发模式，内容构成，用户需求差异等内容。相关系统将在大修的武钢有限 6 高炉运行。

Recovery of Big Fluctuation in Blast Furnace and Development of Blast Furnace Intelligent Control

Chen Lingkun

(Ironmaking Institute of Baosteel Central Research Institute, Wuhan 430081, China)

Abstract: At present, the development of intelligent control systems for blast furnaces is on the wave. Various manufacturers and personnel have developed intelligent control systems with their own characteristics for different goals. These systems generally contain a lot of mathematical models and involve a lot of information. It has played a certain role, but in the process of long-term abnormality of the blast furnace, the blast furnace is mostly restored by manual adjustment, and the guidance of the intelligent control system is rarely heard to restore the blast furnace from a long-term abnormal state to normal production. Whether the current operating intelligent control system can cope with the long-term abnormal furnace conditions of the blast furnace is a problem that needs to be considered and studied. This article will combine the fluctuation recovery process of WISCO's No. 5 and 8 blast furnaces in early 2021, and propose the structure, content, and operation mode of the blast furnace intelligent control system, involving development models, content composition, differences in user needs, etc. The relevant system will be operated in No. 6 blast furnace of Wuhan Iron and Steel Co., Ltd. which has been overhauled.

铁前超低排放与低碳工艺技术

贾国立

（首钢迁钢，河北迁安 064400）

摘 要：北京首钢股份有限公司位于钢铁工业集中地唐山迁安。近年来，首钢股份铁前积极响应国家产业政策，推动全流程超低排放和低碳工艺技术升级改造。秉持“源头限制，过程管控，末端治理”全方位立体式提升的方针，先后实施了高炉均压煤气全回收、料场全封闭、气力输灰、球团 SNCR 脱硝技术等超低排放工艺技术，以及水冲渣余热回收、烧结低温点火、烧结烟气双循环利用、全流程固废综合利用、铁钢界面技术、环冷机高效密封等低碳工艺，并持续压减烧结矿产能和入炉配比，探索高球比低碳冶炼路线，全力打造绿色发展示范企业，连续两年被评为河北省唯一环保 A 类企业，2019 年被评为全国首家，钢铁行业唯一一家通过全工序超低排放评估验收的企业。成为世界上首家全流程实现超低排放企业。

Iron Making Technology of Ultra Low Emission and Low Carbon Emission

Jia Guoli

Abstract: Beijing Shougang Co., Ltd. is located in Qian'an, Tangshan, where the steel industry is concentrated. In recent years, Shougang Co., Ltd. has actively responded to the national industrial policy and promoted the upgrading and transformation of the whole process with ultra-low emissions and low-carbon process technology. Adhering to the principle of “source limitation, process control and end treatment”, the company has successively implemented ultra-low emission technology such as full recovery of equalized blast furnace gas, full closure of the stock yard, pneumatic ash transport and pellet SNCR denitration technology. And slag waste heat recovery, Sintering low temperature ignition, double recycling of sintering gas, the whole process of solid waste comprehensive utilization of flue gas, iron steel interface technologies, such as efficient sealing ring cooler low carbon technology, and can continue to reduce sintering mineral and charging ratio, explore golf than low-carbon smelting route, to create green development demonstration enterprise, for two consecutive years was rated as the only environmental protection class A enterprises in hebei province, In 2019, it was rated as the first and only enterprise in the iron and steel industry that passed the whole process ultra-low emission assessment and acceptance. To become the first enterprise in the world to achieve ultra-low emissions throughout the whole process.

高炉模型的开发与应用

余艾冰

(ARC Research Hub for Computational Particle Technology, Department of Chemical Engineering, Monash University, VIC 3800, Australia; Centre for Simulation and Modelling of Particulate Systems, Southeast University-Monash University Joint Research Institute, Suzhou 215123, China)

摘　要：为了高炉炼铁的可持续发展，其优化设计及控制显得尤为重要，特别是在越来越大的经济及环境压力之下。为此，理解高炉在不同条件下其内部的多相流动、传热、传质规律及综合性能指标显得非常重要。数学模型，通常与物理模型相结合，在这个方面发挥了重要作用。本次汇报将介绍SIMPAS在过去多年来在高炉模拟与仿真方面的研究成果，主要专注于两个方面：模型开发及模型应用。模型开发工作将主要从模型建立、模型特征及模型验证三个方面进行介绍，特别是最近在模拟高炉层状结构软熔带、颗粒粉化行为及综合模型建立等方面的工作。其次，高炉模型的有效性将通过各个层面的模型应用得以体现，包括优化高炉布料、高炉喷煤及高炉几何形状，以及探索高炉炼铁新工艺，如氧气高炉、氢气喷吹等。最后，高炉炼铁的未来发展也会简要讨论。

Blast Furnace Ironmaking: Modelling and Application

Yu Aibing

(ARC Research Hub for Computational Particle Technology, Department of Chemical Engineering, Monash University, VIC 3800, Australia; Centre for Simulation and Modelling of Particulate Systems, Southeast University-Monash University Joint Research Institute, Suzhou 215123, China)

Abstract: The design and control of blast furnace (BF) ironmaking must be optimized in order to be competitive and sustainable, particularly under the more and more demanding and tough economic and environmental conditions. To achieve this, it is necessary to understand the complex multiphase flow, heat and mass transfer, and global performance of a BF under different conditions. Mathematical modelling, often coupled with physical modelling, plays an important role in this area. This talk will present an overview of modelling and simulation of industrial BFs in our laboratory, focused on two aspects: model development and model application. The model development will be discussed in terms of model formulation, new features and model validation. Our recent efforts in modelling layered cohesive zone and particle size reduction and developing an integrated BF model will be highlighted. Then, the usefulness of the BF models will be demonstrated through various model applications in optimizing burden distribution, pulverised coal injection and BF profile, as well as exploring new ironmaking technologies such as oxygen BF and hydrogen injection. Finally, areas for future development will be briefly discussed.

高效长寿高炉冷却结构与冷却强度的浅议

汤清华[1]，史志苗[2]

（1. 鞍钢股份有限公司，辽宁鞍山　114000；2. 江阴兴澄特种钢铁有限公司，江苏江阴　214400）

摘　要：在高炉改造、设计、建设、生产操作与维护中，经常遇到高炉冷却工艺、结构和强度上的一些问题，如冷却壁冷却比表面积的大小、冷却壁体结构、冷却水质、水速的选择等。笔者在本文中提出自己的看法，并与同仁分享、探讨和请教，希望起到抛砖引玉的效果。

关键词：高炉；长寿；冷却结构；冷却强度

Discussion on the Cooling Structure and Cooling Intensity of High Efficiency and Long Campaign Blast Furnace

Tang Qinghua[1], Shi Zhimiao[2]

(1. Angang Steel Company Limited, Anshan 114000, China;
2. Jiangyin Xingcheng Special Steel Works Co., Ltd., Jiangyin 214400 China)

Abstract: In the process of transformation, design, construction, production operation and maintenance of blast furnace, some problems about cooling structure and cooling intensity are often encountered, such as the specific surface area of the cooling stave, the selection of cooling stave structure and cooling water quality, the speed of cooling water, etc. The author puts forward his views in this article, and shares, discusses and asks for advice with colleagues, hoping to play a role in attracting new ideas.

Key words: blast furnace; long campaign; cooling structure; cooling intensity

焦炭不同层次结构性质及其在高炉中的作用

汪　琦

（辽宁科技大学，辽宁鞍山　114051）

摘　要： 高炉内焦炭溶损及劣化主要发生在 900~1300℃的软融带区域。目前，炼铁工作者对焦炭物理化学性质的认识还停留在以石墨为标准态的气化反应热力学、气-固反应动力学，采用焦炭反应性 CRI 及反应后强度 CSR 指标判断焦炭热性能对高炉冶炼过程的影响。普遍认为高炉使用低 CRI 及高 CSR 焦炭冶炼能实现高生产率及低焦比，但是直到现在很难应用已有的 CRI 及 CSR 数据推算高炉生产指标。本文从高炉炼铁原理和实践分析了焦炭溶损反应在高炉还原、热交换和透气性方面的作用，阐明 CRI 及 CSR 试验模拟性问题。建议将焦炭气孔、光学组织、微晶和碳片层堆砌等的不同层次结构、气-液-固溶损反应热力学和动力学等新认识和研究方法融入到炼铁学，为揭示焦炭热性能、构建新的焦炭热性能评价方法提供理论基础。

Coke Properties for Structures wtih Different Levels and Its Role in BF

Wang Qi

(University of Science and Technology Liaoning, Anshan 114051, China)

Abstract: The cokesolution-loss and degradation in blast furnace mainly occur in the soft melting zone of 900 ~ 1300℃. At present, the understanding of coke physical and chemical propertiesstill maintains gasification reaction thermodynamics and gas-solid reaction kineticswith graphite as the standard state.The influence of coke thermal properties on blast furnace's judged by coke reactivity index CRI and coke strength after reaction index CSR.It is generally believed that high productivity and low coke ratio can be achieved by using low CRI and high CSR coke in blast furnace, but up till now, it is difficult to predict the production index of blast furnace by using the existing CRI and CSRdata. Based on the principle and practice of blast furnace ironmaking, this paper analyzes the role of coke solution-loss reaction in reductionprocess, heat exchange and permeability, and expounds the simulation problems of CRI and CSR tests.It is suggested that new understanding and research methods of different hierarchical structuresincluding coke pores, optical structure, microcrystalline and carbon sheet layer stacking,and thermodynamics and kinetics of gas-liquid-solid solution-loss reactionshould be integrated into side ology, so as to provide a theoretical basis for revealing coke thermal properties and constructing a new evaluation method of coke thermal properties.

高炉用焦炭热性能新评价及炼焦配煤新技术

邬虎林[1]，汪　琦[2]，孟庆波[3]

（1. 包头钢铁（集团）有限责任公司；2. 辽宁科技大学；3. 鞍山热能研究院有限公司）

摘　要：基于包钢不同容积高炉冶炼实践及炼铁原理分析，采用矿焦耦合实验研究方法，揭示了焦炭 CRI 及 CSR 指标对高炉内焦炭溶损劣化的模拟性不足。开发了反映高炉内焦炭溶损特性的焦炭综合热性能评定新方法，用 CSR25 和 CSR25-T 评价焦炭热性能；开发了煤的炭化关联性测定新方法，用于评价炼焦煤的结焦性；建立了用煤炭化关联性指数作为煤炼焦性能指标的焦炭质量预测模型，该模型已成功用于生产实践。为炼焦煤资源的合理利用和降本增效提供了技术支撑。

New Evaluation Method of Coke Thermal Performance and New Technology of Coal Blending for Coking

Wu Hulin[1], Wang Qi[2], Meng Qingbo[3]

(1. Baotou Iron and Steel Co., Ltd.; 2. University of Science and Technology Liaoning; 3. Sinosteel Anshan Research Institute of Thermo-energy Co., Ltd.)

Abstract: Based on the smelting practice and ironmaking principle analysis of blast furnaces with different volumes in Baotou Iron and Steel Co., the insufficient simulation of coke CRI and CSR indexes on coke solution-loss degradation in blast furnace is revealed by using the iron orecoke coupling experimental method. A new method for evaluating the comprehensive thermal properties of coke reflecting the solution-loss characteristics of coke in blast furnace is developed and the thermal properties of coke are evaluated by CSR25 and CSR25-T. A new method for determining the carbonization correlation of coal is developed to evaluate the coking property of coking coal. A coke quality prediction model using carbonization correlation index as coal coking performance index is established and successfully used in production practice. It provides technical support for the rational utilization of coking coal resources, cost-saving and profit-increasing.

高炉的高效与未来的发展

邹忠平

（中冶赛迪）

摘　要：炉缸的活跃是高炉稳定顺行的关键，合适的炉腹煤气指数、较高的鼓风动能、煤气与物料水当量的匹配以保持料柱的疏松，是维持高炉高产低耗的有力措施。良好的设计，有效的施工质量的保障，加上操作上注重：在高炉炉缸活跃的前提下，保持合理的边缘和中心的两道气流，保持炉墙合理的热负荷水平，防止气隙以保证炉缸传热体系的持续有效，将各部位的温度和热负荷控制在允许的临界值以内，是实现高炉长寿的有效保障。高炉仍将是在未来相当长一段时间内的炼铁主流装置，高炉的低碳将对钢铁行业的低碳影响巨大。高炉的减碳路径主要是还原介质的喷吹、高还原性物料的利用、高炉的高效低耗运行，高炉煤气分离出CO再回喷高炉，将是高炉实现减碳的最有效途径。

Highly-efficient Operation and Future Development of Blast Furnace

Zou Zhongping

Abstract: An active hearth plays a crucial role in ensuring a smooth blast furnace operation. One of the main powerful measures for achieving desired blast furnace performances featuring a high output yet a low consumption is to create a loose stock by implementing an appropriate bosh gas index, a pretty high blast kinetic energy and a well-matched gas and material's water equivalent. With an optimized design and effective construction quality control, a blast furnace can be ensured a long service life, in addition to its operation with special attention – operating a blast furnace with an active hearth requires to keep optimized peripheral and central gas flow and maintain a fairly-adequate thermal load level of the furnace walls. These methods can prevent the gas gap from damaging the hearth's effective heat transfer system, and enable the temperatures and thermal loads of various parts to be controlled within the allowable critical values. Considering that the blast furnace remains a predominant facility for ironmaking production for a quite long time in future, the blast furnace's low-carbon production will exert a great impact on the steel industry's low-carbon developments. A carbon reduction route for blast furnace ironmaking focuses on injecting the reducing medium, charging the materials of high reductibility and operating in a high-efficiency, low-consumption way. The most effective solution to reduce carbon is proposed to separate carbon monoxide from the blast furnace gas and blow it back into the blast furnace.

高炉高效长寿技术的创新与应用

段国建

（中冶京诚工程技术有限公司，北京　100176）

摘　要：本文介绍了天津钢铁 3200m^3 高炉大修改造工程采用的高炉高效及长寿技术。包括优化炉型；采用新型炉体冷却结构：炉体采用全铸铁冷却壁，炉腹、炉腰关键部位镶嵌铜冷却条；炉缸应用导热设计理念，采用大块炭砖结合导热质陶瓷杯形成的完全导热炉缸体系；炉缸关键部位采用 SGL 超微孔炭砖，陶瓷垫外环及陶瓷杯采用碳复合砖；采用防炉底板上翘及密封技术，对炉底结构进行加强优化。生产实践证明，高炉生产指标先进、炉底炉缸炭砖温度、水温差、冷却壁温度、铜条热负荷等稳定，高炉采用的各项新技术可保证高炉一代炉役高效长寿。

关键词：高炉；炉型；冷却结构；炉缸结构；炉底板结构

Innovation and Application of High Efficiency and Long Life Technology of Blast Furnace

Duan Guojian

(MCC Jingcheng Engineering & Technology Co., Ltd., Beijing 100176, China)

Abstract: This paper introduces the long campaign and efficientproduction technology of 3200m3 blast furnace overhaul in TISCO. Including the optimization of the internal dimension, the cast iron cooling wall is used for the cooling equipment, the new combined cooling structure is used for the bosh and belly, the heat conduction design concept is used for the hearth, the large carbon brick and the heat conduction ceramic cup are used to form the complete heat conduction type hearth structure, the SGL ultra microporous carbon brick is used for the key parts, and the carbon composite brick is used for the ceramic cup;The furnace bottom structure is strengthened and optimized by using anti-warping and sealing technology.The production practice has proved that the production index of the blast furnace is advanced, the temperature of carbon brick at the bottom and hearth, the water temperature difference, the temperature of cooling stave and the heat load of copper strip are stable. The new technologies adopted by the blast furnace can ensure the high efficiency and long campaign of the next generation of blast furnace.

Key words: blast furnace; internal dimension; cooling structure; hearth structure; bottom plate structure

沙钢铁前多级别产线全流程协同匹配与稳定生产关键技术创新

杜　屏，赵华涛，卢　瑜，张少波，翟　明，韩　旭，周　磊

（江苏省（沙钢）钢铁研究院 炼铁与环境研究室，江苏张家港　215625）

摘　要：沙钢基于实时大数据挖掘、数学模型及实验室科研资源自主开发了铁前一体化管控平台。管控平台、生产管理和技术研发三位一体，提升管控效率，挖掘降本空间。基于该平台开展铁前多产线生产全局匹配和一体化降本工作；搭建铁前全流程异常预警和原因追溯体系，开发铁前关键设备异常在线监测和诊断、仓位监控和原料保供系统，保障全流程生产稳定；在澳煤紧张、非主流矿比例升高的背景下开发配煤、配矿优化模型和精度监控系统提高焦炭和烧结矿质量，优化高炉上、下部制度匹配，降低燃料比，同时开发低碳护炉技术降低护炉成本。通过铁前管控平台的应用，沙钢铁前原料质量及高炉稳定性大幅提高，5800 高炉在炉役后期连续稳定运行 700 天，铁前一体化降本连续突破 100 元/t，炼铁厂焦比平均降低 4.7kg/t。

关键词：多产线；全流程；平台；异常预警；原因追溯；稳定生产；降本

The Key Technology Innovation of Overall Process Coordination for Stable Operation in Multi-production Lines of Ironmaking Process in Shasteel

Du Ping, Zhao Huatao, Lu Yu, Zhang Shaobo, Zhai Ming, Han Xu, Zhou Lei

(Ironmaking & Environment Research Group, Institute of Research of Iron & Steel, Shasteel, Zhangjiagang 215625, China)

Abstract: The management platform of overall ironmaking process in multi-production lines was developed based on online real time data mining technology, mathematical model and laboratory research resources in Shasteel. Forming a pattern of trinity, deep integration and mutual promotion, improving the management efficiency and tapping the cost reduction potential. Relying on the platform, the models for overall coordination of multi-production lines and cost reduction activities in ironmaking process were developed. In addition, the abnormal alarming and cause tracing system in overall process of ironmaking, abnormal detecting and online diagnosis system for key machines working condition, was well as the bin level monitoring and raw material feeding system were developed to guarantee the stable operation of overall ironmaking process. Moreover, the coal and ore blending optimization and accuracy improvement were performed to enhance the raw material quality under the background of the Australia coal shortage and non-main stream ore increase, the blast furnace operation philosophy was optimized to lower down the fuel ratio, and the low-Ti hearth protection technology was developed to reduce the hearth protection cost. Via the application of the platform, the raw material quality and the blast furnace stability was greatly improved, as the result, the 5800m3 blast furnace was running in stable condition for 700 days successively in later period of first campaign, the iron-production cost-down was breaking 100RMB/t，and the coke ratio of ironmaking plant was decreased by 4.7kg/t.

Key words: multi-production line; overall process; platform; abnormal alarming; cause tracing; stable operation; cost down

炼铁工序几项节能降碳关键技术的探讨

张玉柱

（华北理工大学）

摘　要：（1）高炉渣在线成纤余热利用理论与技术。（2）气基竖炉氢还原冶金理论与技术。（3）多因素耦合最优炉料结构低配碳模型的建立。

Discussions on Several Key Technologies of Energy Saving and Carbon Reduction in Ironmaking Process

Zhang Yuzhu

Abstract: (1) Theory and technology of waste heat utilization on online fiber forming of blast furnace slag. (2) Theory and technology of hydrogen reduction metallurgy in gas-based shaft furnace. (3) Establishment of low carbon distribution model of multi-factor coupling optimal burden structure.

首钢球团技术进步及低碳发展实践与展望

青格勒[1]，田筠清[1]，吴小江[2]，刘长江[3]，马　丽[1]，张　彦[2]

（1. 首钢集团有限公司技术研究院；2. 首钢京唐钢铁联合有限责任公司；3. 首钢股份有限公司）

摘　要：三十多年来，首钢球团经历了链篦机-回转窑工艺生产金属化球团，回转窑截窑改造生产氧化球团矿，建立大型带式焙烧机球团生产线，多功能球团矿及熔剂性球团矿开发及高炉高比例应用等过程，球团生产技术水平不断提高，推动球团产量及高炉炼铁中使用比例的增加，为长流程钢铁企业低碳绿色发展提供了可借鉴的途径。本文介绍了首钢球团技术装备及生产运行情况，重点阐述了大型带式焙烧机的投产运行，镁钛低硅新型球团矿的开发及应用，低硅碱性球团矿的制备和大型高炉高比例使用，富矿资源用于球团矿生产，球团工艺的智能化等技术的研究与应用。基于低碳绿色发展需求，探讨了球团工艺的优劣势，展望了球团工艺技术的发展方向。

关键词： 球团；带式焙烧机；低碳绿色；多功能球团矿；资源

Shougang Pelletizing Technology Progress and Low Carbon Development Practice and Prospect

Qing Gele[1], Tian Yunqing[1], Wu Xiaojiang[2], Liu Changjiang[3], Ma Li[1], Zhang Yan[2]

(1. Shougang Research Institute of Technology, Beijing 100043, China; 2. Shougang Jingtang United Iron & Steel Co., Ltd., Tangshan 063200, China; 3. Beijing Shougang Co., Ltd., Beijing 100041, China)

Abstract: Over threedecades, Shougang pelletizing has experienced the production of metallized pellets by grate rotary kiln, transformation of rotary kiln intercepted to produce oxidized pellets, establishment of large scale straight grate indurating machine, development of multifunctional pellets and low-silicafluxed pellets andhigh ratio application in the blast furnace. The continuous improvement of pelletizing technologypromotes the high proportion application of pellets in the blast furnace ironmaking, which provides a certain reference for the low-carbon green development for steel industries with long process.This paper introduces the technical equipment and production operation of Shougangpellet plant, focusing on the production and operation of large scalestraight grate indurating machine, the development and application of new magnesium-titanium bearing pellet with low silica, the preparation of low silicafluxed pellets and the high proportion application in the blast furnace, pellet production using rich iron ore resources, and the intellectualization of pelletizing process. Based on the demand for low-carbon and green development, the advantages and disadvantages of pelletizing process are discussed, and the developing direction of pelletizing technology is prospected.

Key words: iron ore pellet; straight grate indurating machine; low-carbon and green development; multifunctional pellets; resources

高炉渣镁铝比控制理论及其应用

沈峰满，姜 鑫，郑海燕，高强健

（东北大学，辽宁沈阳 110819）

摘 要： 炼铁工序能耗占钢铁冶炼全流程的 70%以上，节能降耗是绿色炼铁的核心方针。进入 21 世纪，我国进口矿石量增多，2019 年约 11 亿吨，约 65%为来源于澳洲高 Al_2O_3 矿石。高炉渣 Al_2O_3 含量随之增大，达到 15%～17%，高炉渣的冶金性能受到影响，如炉渣粘度变大，流动性和脱硫能力降低等。为了突破高 Al_2O_3 渣系、低能源消耗的低碳冶炼技术瓶颈，首先围绕 Al_2O_3、MgO 在烧结-高炉全流程的利与弊展开系统的基础理论研究，提出了高炉渣镁铝比（MgO/Al_2O_3，质量百分比）的操作参数；然后结合相图等热力学基础理论和试验研究，提出高炉渣适宜镁铝比的分段管控理论，即根据炉渣 Al_2O_3 含量的不同将炉渣分为三个区间，针对每一个区间设定适宜的镁铝比；最后在实验室研究的基础上，进行了现场应用的生产案例分析，验证了高炉渣镁铝比参数的正确性和可应用性。本报告旨在从根本上填补炼铁领域关于高炉渣适宜镁铝比的理论空白，促进炼铁的技术进步，达到了科学高效地使用炼铁资源与能源的冶炼目的。

关键词：炼铁；高炉；炉渣；镁铝比；应用

Theory and Application on MgO/Al_2O_3 Ratio in Blast Furnace Slag

Shen Fengman, Jiang Xin, Zheng Haiyan, Gao Qiangjian

(Northeastern University, Shenyang 110819, China)

Abstract: The energy consumption of ironmaking process accounts for more than 70% of total in the whole iron and steel production. Energy saving and consumption reduction are the key policies of green ironmaking production. In 21 century, more and more imported iron ores with high Al_2O_3 content were used in China, approximately 1.1 billion tons in 2019, which increased the Al_2O_3 content in blast furnace (BF) slag, up to 15-17%, and resulted in poor metallurgical properties of slag, e.g., increasing the viscosity and decreasing fluidity and desulphurization ability of BF slag. In order to achieve low carbon ironmaking for high-Al_2O_3 slag, first, the effects of Al_2O_3 and MgO on the whole ironmaking process were investigated, and a new parameter of MgO/Al_2O_3 ratio were proposed. Then, a sectional control theory for MgO/Al_2O_3 ratio according to Al_2O_3 content in slag was built based on phase diagram analyses and experiments. Finally, pilot-scale tests were carried out, and the better operation indices were achieved. Thus, the correctness and applicability of sectional control theory on MgO/Al_2O_3 were verified. The aim of present work is to fill the the knowledge gap on the MgO/Al_2O_3 ratio in BF slag, and to promote the technology progress in ironmaking field, and eventually reasonable and effective using ironmaking resources and energy.

Key words: ironmaking; blast furnace; slag; MgO/Al_2O_3 ratio; application

中冶赛迪智慧炼铁思考与实践

谢　皓[1]，王　刚[2]，孙小东[1]

（1. 中冶赛迪重庆信息技术有限公司，重庆　401122；
2. 工作单位，中冶赛迪工程技术股份有限公司，重庆　401122）

摘　要：结合炼铁行业的发展及现有智能化技术的开发与实际应用情况，对智慧炼铁的技术体系架构、关键技术等进行介绍。智慧炼铁的总体技术架构，包括智慧决策、智慧操控、智慧作业三个层次，三个方面相互联动和支撑，构建起了智能炼铁的整体框架。针对智慧炼铁的实践，介绍了中冶赛迪多个铁区集控项目的情况及效果，以期为行业的智能化升级提供参考。

关键词：炼铁；智能化；集控

The Practice of Intelligent Ironmaking of CISDI

Xie Hao[1], Wang Gang[2], Sun Xiaodong[1]

(1. CISDI Chongqing Information Technology Co., Ltd., Chongqing 401122, China;
2. CISDI Engineering Co., Ltd., Chongqing 401122, China)

Abstract: Combined with the development of iron making industry and the development and practical application of existing intelligent technology, the technical system and key technologies of intelligent iron making are introduced. The overall technical system of intelligent iron making includes three levels: intelligent decision-making, intelligent control and intelligent operation. The interaction and support of the three aspects constitute the overall system of intelligent iron making. In view of the practice of intelligent ironmaking, the situation and effects of several centralized control projects of CISDI are introduced, in order to provide reference for the intelligent upgrading of the industry.

Key words: ironmaking; intelligent; centralized control

山东墨龙 HIsmelt 熔融还原技术创新与应用

张冠琪，张晓峰，王金霞，魏召强

（山东墨龙石油机械股份有限公司，山东寿光 262700）

摘 要：非高炉炼铁技术竞争力主要体现在原料的适用性、燃料的广泛性和环保的优越性，随着钢铁行业冶炼资源环境的变化、环保要求日趋严格和低碳冶金发展趋势，非高炉炼铁技术得到行业广泛关注。HIsmelt 熔融还原技术经过 40 余年的理论研究和试验探索，目前已经在山东墨龙实现连续工业化运行 157 天，在熔融还原领域的生产稳定性、成本经济性和环保先进性方面具有显著优势。报告介绍了 HIsmelt 技术研发历程、技术优势和发展现状，结合山东墨龙生产实践分析了 HIsmelt 技术在钢铁生产过程中的优势，对于我国钢铁产业结构调整、国内铁矿资源的开发利用和钢铁产品质量升级具有重要意义。

关键词：HIsmelt；熔融还原；技术创新；生产实践

Innovation and Application of HIsmelt Reduction Technology of Shandong Molong

Zhang Guanqi, Zhang Xiaofeng, Wang Jinxia, Wei Zhaoqiang

(Shandong Molong Petroleum Machinery Co., Ltd., Shouguang 262700, China)

Abstract: The competitiveness of non-blast furnace iron-making technology is mainly reflected in the applicability of raw materials, the universality of fuel and the superiority of environmental protection. With the change of smelting resources and environment in iron and steel industry, increasingly strict environmental requirements and the development trend of low-carbon metallurgy, non-blast furnace iron-making technology has been widely concerned in the industry. After more than 40 years of theoretical research and experimental exploration, HIsmelt melt reduction technology has been in Continuous industrial operation for 157 days in Shandong Molong. It has significant advantages in production stability, cost economy and advanced environmental protection in the field of melting reduction. The report introduces the research and development history, technical advantages and development status of HIsmelt technology, and analyzes the advantages of HIsmelt technology in the steel production process based on the production practice of Shandong Molong, it has a great significance to the structural adjustment of China's iron and steel industry, the utilization of domestic iron ore resources and the upgrading of iron and steel product quality.

Key words: HIsmelt; smelting reduction; technical innovation; production practice

焦炉煤气制还原气“氢冶金”工艺探索和项目介绍

范晋峰，白明光，季爱兵

（中晋太行矿业有限公司，山西晋中　030600）

摘　要：报告研究工作的目的、方法、结果和结论，而重点是结果和结论。本报告重点介绍焦炉制备还原气用于氢冶金竖炉生产；焦炉煤气中杂质含量的去除、净化工艺；在转化炉内以 CO_2 重整为主的反应存在的问题和解决的措施；特别是如何有效的防止积炭的措施和方法；并简单介绍一下中晋太行矿业有限公司“氢冶金”的试生产情况。

关键词：焦炉煤气；氢冶金；积炭

Process Exploration of “Hydrogen Metallurgy” and Project Introduction for Producing Reducing Gas from Coke Oven Gas

Fan Jinfeng, Bai Mingguang, Ji Aibing

(Zhongjin Taihang Mining Co., Ltd., Jinzhong 030600, China)

Abstract: The report introduces the purpose, methods, results and conclusions of the research work, and the focus is on the results and conclusions. This report should focus on the production of reducing gas from coke ovens for hydrogen metallurgical shaft furnace production; the removal and purification process of impurities in coke oven gas; the problems and solutions in the CO_2 reforming-based reaction in the reformer. In particular; describe the measures and methods to effectively prevent carbon deposition; and briefly introduce the trial production of "Hydrogen Metallurgy" of Zhongjin Taihang Mining Co., Ltd.

Key words: coke oven gas; hydrogen metallurgy; coke deposit

炉外精炼工艺与中高硫含量特殊钢 MnS 夹杂物控制

王新华

（河钢集团邯郸钢铁公司技术中心）

摘　要：钢中硫对绝大多数钢种属于有害元素，但某些特殊性能要求的钢材品种却需要含较高硫，例如日本制铁公司生产的重载铁路钢轨，硫含量控制在 0.009%～0.015%较高含量水平，目的是利用钢中 MnS 夹杂形成更多“氢陷阱”，以防止钢轨服役过程发生氢致断裂。又如日本制铁公司生产的动车车轮等特殊钢，将硫控制在 0.008%～0.02%较高含量水平，以铝脱氧生成的微小 Al_2O_3 作为钢中 MnS 夹杂依附析出的核心，生成“MnS 包裹 Al_2O_3”类夹杂物，既可减轻硬质 Al_2O_3 夹杂物对钢材抗疲劳性能的不利影响，又能减小 MnS 夹杂尺寸并使其在钢中更均匀分布，从而改善钢材切削加工性能。

国内钢厂在中高硫含量特殊钢生产方面目前存在一些问题，例如：（1）生产高速轨、重载轨等钢种，提高其硫含量后，钢材中 MnS 类夹杂午评级不合比率显著增加，因此只能将钢中硫控制在较低含量水平（0.003%～0.007%）；(2)生产汽车齿轮、曲轴、连杆等中高硫含量特殊钢，钢中氧化物类夹杂主要为 $CaO-MgO-Al_2O_3$ 系（而非单一 Al_2O_3），难以形成"MnS 包裹 Al_2O_3"类夹杂物。且由于钢液、炉渣的氧势低，对钢液增硫后会生成 CaS 类夹杂，显著降低钢液可浇性，造成连铸水口粘接、堵塞等问题。

国内中高硫特殊钢生产方面存在的上述问题与炉外精炼过度依赖 LF 精炼工艺方法有关，这是因为：（1）生产高速轨、重载轨钢种，LF 精炼炉渣碱度通常控制在 1.7～2.2，(Fe_tO）含量控制在 0.6%以下，由此造成钢液中[Al]s 含量偏高，大多在 0.0020%～0.0035%范围（日本制铁产重载轨[Al]s 含量在 0.0010%左右），钢液[O]含量低，导致钢中以氧化物夹杂作为依附核心而析出的 MnS 夹杂量少，MnS 因此较粗大，评级易超标。（2）生产汽车齿轮、曲轴、连杆等中高硫特殊钢种，因为采用传统的超低氧特殊钢炉外精炼工艺，造成 LF 精炼深度脱硫，再喂硫线增硫，且易生成 CaS 夹杂导致钢液可浇性降低等问题，且由于钢中氧化物夹杂不是单一 Al_2O_3，也无法生成"MnS 包裹 Al_2O_3 系"结构的夹杂物。

对中高硫特殊钢可尝试采用新的炉外精炼工艺，其要点为：（1）对于高速轨、重载轨钢品种，采用"铁水脱硫-转炉炼钢-RH 精炼"工艺，在转炉出钢过程加入合金进行合金化，采用 RH 真空精炼进行脱氢、合金成分微调、去除夹杂物等，如需要可在 RH 精炼中采用 Al-OB 工艺对钢水加热升温；亦可采用"铁水脱硫-转炉炼钢-LF 精炼-RH 精炼"工艺，但 LF 精炼的主要任务是对钢水加热升温和合金化，且不进行造"白渣"操作。（2）对于铝脱氧中高硫含量特殊钢品种，采用"铁水脱硫-转炉（或电炉）炼钢-喂硫线增硫（如需要）-RH 精炼"冶金工艺，在 RH 精炼开始时向加铝对钢液脱氧，并通过 RH 精炼实现钢液超低氧含量控制，如需要可采用 Al-OB 工艺对钢水加热升温；亦可采用"铁水脱硫-转炉（或电炉）炼钢-LF 精炼-喂硫线增硫-RH 精炼"冶金工艺，但 LF 精炼主要任务是对加热升温和合金化，渣量应尽可能少（满足埋弧需要即可），且保持炉渣的氧化性，在 RH 精炼开始时向钢液加铝充分脱氧，并利用 RH 精炼实现钢液超低氧含量控制。

The Secondary Refining Process for Control of MnS Inclusions in Medium-High Sulfur Special Steels

Wang Xinhua

(R&D Center, HBIS Han Steel Co., Ltd.)

21 世纪炼钢技术的发展与展望

刘　浏

（江苏冶金研究院）

摘　要：21 世纪，人类社会进入了信息时代。历史经验证明，钢铁工业作为社会发展的最主要基础材料，随着历史发展和社会的变革，必然引起重大的技术革命：农业社会开创了铁器时代；工业革命孕育了钢时代；而信息社会必然会推动钢铁工业由高品质制造用钢的生产向绿色、低碳和智能化方向发展，建立 21 世纪新一代钢铁工业。本文结合时代的发展与需求，进一步阐述了高效低成本洁净生产新流程、绿色低碳炼钢工艺以及智能炼钢技术的发展前景与急待解决的技术问题。

二氧化碳绿色洁净炼钢技术及应用

朱 荣[1,2]，董 凯[1,2]，姜娟娟[1]

（1. 北京科技大学冶金与生态工程学院，北京 100083；
2. 北京科技大学二氧化碳科学研究中心，北京 100083）

摘 要：二氧化碳绿色洁净炼钢技术从抑制烟尘、高效脱磷、稳定脱氮、强化控氧和底吹长寿等方面入手，以 CO_2 利用、固废减量、钢质洁净、降本增效为目标，开创了工业制造行业大规模高效有价消纳 CO_2 的先河，激发了企业主动消纳利用 CO_2 的积极性，为温室气体减排开辟了新途径。本技术已在多家钢铁企业成功应用，近三年惠及钢产量 3879.2 万吨，实现工业 CO_2 利用 31.5 万吨。

关键词：二氧化碳；二氧化碳利用；炼钢；降尘；洁净控制

New Science and Technology Progress: Carbon Dioxide Green and Clean Steelmaking Technology and Its Application

Zhu Rong[1,2], Dong Kai[1,2], Jiang Juanjuan[1]

(1. School of Metallurgical and Ecological Engineering, University of Science and Technology Beijing, Beijing 100083, China; 2. Research Center of Carbon Dioxide Science, University of Science and Technology Beijing, Beijing 100083, China)

Abstract: Carbon dioxide green and clean steel-making technology starts from the aspects of dust suppression, efficient dephosphorization, stable denitrification, enhanced oxygen control and bottom blowing longevity, aiming at CO_2 utilization, solid waste reduction, steel cleanliness, cost reduction and efficiency increase. It has created a precedent for large-scale, efficient and valuable consumption of CO_2 in industrial manufacturing industry, stimulated the enthusiasm of enterprises to actively consume and utilize CO_2 and opened up a new way for greenhouse gas emission reduction. This technology has been successfully applied in many iron and steel enterprises, benefiting the steel output of 38.792 million tons and realizing the utilization of 315000 tons of industrial CO_2 in recent three years.

Key words: carbon dioxide; CO_2 utilization; steelmaking; reduce dust; clean control

钢质缺陷全流程遗传性研究

张丙龙，刘延强，韩 乐，周东瑾，刘 浩，杜金磊，乔焕山，李向奎

（首钢京唐钢铁联合有限责任公司制造部，河北唐山 063200）

摘 要：通过对炼钢-热轧-冷轧全流程跟踪，在热轧工序主要有裂纹、卷渣和翘皮缺陷，占炼钢原因缺陷比例为 90%以上，其中翘皮和卷渣缺陷主要由保护渣导致，裂纹缺陷主要由于板坯纵裂纹和横裂纹导致；在冷轧工序主要为线状缺陷和翘皮类缺陷，占炼钢原因缺陷比例 85%以上，由于保护渣卷入、大尺寸氧化铝夹杂物等炼钢原因导致，通

过对线状缺陷大量取样分析，缺陷形成原因主要包含卷渣、簇状 Al_2O_3、气泡和气泡+Al_2O_3 及翘皮缺陷。裂纹及翘皮类在热轧和冷轧卷上具有较好的遗传性，通过热轧、酸洗及镀锌后仍不能消除。

关键词：冷轧钢板；表面质量；卷渣；气泡；Al_2O_3；遗传性

Study on the Heredity of Steel Defects in the Whole Process

Zhang Binglong, Liu Yanqiang, Han Le, Zhou Dongjin,
Liu Hao, Du Jinlei, Qiao Huanshan, Li Xiangkui

(Manufacture Division ShougangJingtang United Iron and Steel Co., Ltd., Tangshan 063200, China)

Abstract: Through tracking the whole process of steelmaking, hot rolling and cold rolling, the main defects in hot rolling process are cracks, flux entrapment and blisters, which account for more than 90% of steelmaking defects. The blisters and flux entrapment defects are mainly caused by mold flux, and the cracks are mainly caused by longitudinal cracks and transverse cracks of slab; In the cold rolling process, the main defects are linear defects and blister defects, which account for more than 85% of the steel-making defects. Due to the involvement of mold flux, large-size alumina inclusions and other steel-making reasons, through a large number of sampling analysis of linear defects, the main causes of defects include slag entrapment, cluster like Al_2O_3, bubbles and bubbles + Al_2O_3 and warping defects.The occurrence rate of the defects of the mold powder entrapment with the fluctuation of the mould liquid level≥5mm on the hot-rolled coil is 100%. The part of mold powder entrapmentcan be eliminated after pickling, which will not affect the surface quality in the subsequent rolling process. The mold powder entrapmentwith spills and the spills have good heredity on the cold-rolled coil , which not be eliminatedthrough pickling , rolling and galvanizing. The cracks and blisters on hot rolled and cold rolled coils have good heredity, which can not be eliminated after hot rolling, pickling and galvanizing.

Key words: cold-rolled plat; surface quality; flux entrapment; blisters; Al_2O_3; heredity

FeV50 合金强化还原及分离理论与应用研究

余 彬，孙朝晖，景 涵，潘 成，杜光超

（钒钛资源综合利用国家重点实验室，四川攀枝花 617000）

摘 要：针对传统直筒炉一步法及倾翻炉多期法钒铁 FeV50 合金制备工艺存在的还原剂添加量与合金成分、冶炼收率的矛盾以及渣金高效分离等传统高温热还原制备贵重金属/合金的共性技术难题。分析了电铝热法 FeV50 合金制备过程不同冶炼阶段热渣中钒的理论赋存状态及渣金 VO-Al 热力学平衡，得到了不同配铝系数条件下铝热反应还原极限，基于此提出了多期梯度配铝 FeV50 合金强化还原工艺思路，考察了原料配比、加料制度和多期配铝系数对渣中钒损及合金成分的影响。铝热还原渣金热力学平衡分析结果表明 FeV50 渣中 TV 含量随合金 Al 含量的升高而降低，当合金 Al 含量为 2.0%时，渣中理论 TV 含量为 0.27%；若要使渣中理论 TV 含量降低到 0.10%以下，合金 Al 含量需达 18.8%以上。在保证单位合金配铝系数≤1.05 的条件下，采用多期梯度配铝工艺后，热渣平均 TV 含量从 1.85%降低到 0.69%，合金 Al 含量从 1.6%降低到 0.4%，从理论上解决了钒的还原极限问题。

针对冷渣 TV 含量显著高于冶炼终点热渣 TV 含量的问题，分析了冷渣中钒的实际赋存状态及分布规律，通过无限大流体重力沉降分析，得到了适用于钒铁合金工业化生产的斯托克斯沉降理论体系及其影响因素，并分别考察了熔渣特性、颗粒尺寸、过热度、沉降时间等工艺参数对渣中钒损的影响。冶炼渣物相分析结果表明：渣中钒的赋存状态主要包括未被还原的钒以及一定铁钒比的初级合金，其中钒氧化物在冷却过程中逐渐与渣中 CaO、MgO 结合形成稳定性更强的复合尖晶石结构，初级合金主要以类球状微细颗粒存在，主要成分为 V 和 Fe，V 质量分数一般≥50.0%。合金熔体重力沉降分析结果表明：合金沉降速度随合金粒度的增加而增大，随熔渣黏度的增加而减小。1850℃条件下，当渣层厚度为 50 mm，熔渣质量分数为 65.2% Al_2O_3、15.5% CaO、14.6% MgO、1.9% Fe_2O_3、0.9% SiO_2 时，粒径为 100μm 的合金沉降时间及熔渣上浮时间分别为 24.9 min 和 1.2 min。通过对熔渣特性、出渣制度、保温制度等工艺参数进行优化后，浇铸冷渣平均 TV 含量能够进一步降低至 0.58%。

Study on Theory and Application of Strengthening Reduction and Separation of FeV50 Alloy

Yu Bin, Sun Zhaohui, Jing Han, Pan Cheng, Du Guangchao

(State Key Laboratory of V and Ti Resources Comprehensive Utilization, Panzhihua 617000, China)

大型转炉高效率、长寿命冶炼技术

杨利彬[1]，邓　勇[2]，蒋晓放[3]，田　勇[4]

（1. 钢铁研究总院，北京　100081；2. 马钢，安徽马鞍山　243000；
3. 宝钢集团中央研究院，上海　201999；4. 鞍钢股份有限公司，辽宁鞍山　114000）

摘　要：大型转炉装备水平高，生产钢种多位高品质、高附加值钢种。国内大型转炉生产中存在主要矛盾相互影响，导致冶炼过程效率低、消耗大、能耗高，并且生产不稳定。钢铁研究总院与马钢、宝钢、鞍钢等组建研发团队，对大型转炉进行了多年的技术研究和应用，形成了大型转炉高效率、长寿命顶底复合吹炼技术，技术应用后取得了良好的技术效果和效益[1-3]。

对大型转炉研究得出冶炼过程成渣及冶金反应规律。经分析总结出冶炼过程不同区间的特征、主任务及工艺要点。冶炼过程炉渣氧位高于元素氧化所需氧位，元素氧化存在选择性竞争。在吹炼前期，P-Si 选择性氧化是脱磷的限制性环节。在冶炼后期，P-C 选择性氧化是脱磷的限制性环节。

氧枪大流量供氧技术：在冶炼过程成渣及脱磷、喷溅机理及控制的基础上进行氧枪喷头改进。结果表明，供氧强度提高到 3.5~3.8Nm^3/(t.min)，能兼顾冶炼前、中、后期成渣、脱磷需求，平稳冶炼，同时保证冶炼效率。

转炉高强度底吹技术：随着底吹供气强度增加，死区面积迅速减小，尤其是提高到 0.2Nm^3/t.min 以上，熔池搅拌效率提高 40%以上[4]。国内转炉普遍采用溅渣护炉技术，会对有效底吹强度有一定的抵消，在保证底吹寿命的前提下，提高高效率动力学效果，选择将底吹供气强度提高到 0.2Nm^3/t.min。

底吹长寿命维护技术：长寿命底吹效果是一个综合技术应用效果的体现，不仅需要有供气系统装备、供气元件作为保证，还需设计合理的底吹总体方案、吹炼工艺，辅之以日常的底吹长寿维护工艺[5]。

通过研究冶炼过程的成渣及冶炼规律，获得大型转炉的成渣、元素选择性高效氧化规律并在此基础上获得了大型转炉高效冶炼工艺；高强度顶底复合吹炼是为高效率冶炼提供动力学保证，供气强度提高到 3.5~3.8Nm^3/(t.min)，底吹供气强度达到 0.2 Nm^3/(t.min)，可保证高效率冶炼；炉底渣层厚度控制在 100mm 以内，冶炼终点碳氧积平均稳定控制在 0.0015 以内，炉渣氧化性大幅降低；高效率、长寿命顶底复吹吹炼技术的关键是获得转炉冶炼规律、满足动力学条件和氧化性控制，通过 10 年的研发和持续应用，形成高效率冶炼技术体系，冶炼效率大幅提升。

Highly Efficiency and Long Life Combined Blowing Technology of Big Converter

Yang Libin[1], Deng Yong[2], Jiang Xiaofang[3], Tian Yong[4]

(1. Central Iron and Steel Research Institute, Beijing 100081, China; 2. Masteel, Ma'anshan 243000, China; 3. Baosteel, Shanghai 201999, China; 4. Ansteel, Anshan 114000, China)

低成本新双渣法转炉炼钢工艺开发

杨 健[1]，杨文魁[1]，张润灏[1]，孙 晗[1]，鲁欣武[2]，刘万善[2]

（1. 上海大学材料科学与工程学院，上海 200444；2. 宁波钢铁有限公司，浙江宁波 315000）

摘 要：本文在220 t转炉中进行了高效低成本新双渣法转炉炼钢工艺的开发，结果表明，在较低碱度0.9～2.6范围的脱磷阶段，随着脱磷渣碱度的增加，磷分配比LP和脱磷率均增大。脱磷渣中含有深灰色富磷相1、浅灰色液态渣相2和白色富铁相3。随着碱度的增加，脱磷渣中各相的形貌显著变化，富磷相的面积分数和P_2O_5含量增加。采用新双渣法转炉炼钢工艺，脱碳后试验炉次的平均脱磷率为90%，钢水中的磷含量平均为0.0142%，整个转炉吹炼阶段的石灰和菱镁矿的平均单耗分别为25.3，10.4kg/吨钢，对比传统工艺分别下降23.6%和24.1%。实现了降低石灰和菱镁矿等辅料消耗的目的。

关键词：新双渣法；转炉炼钢；脱磷；碱度；富磷相

Development of New Double Slag Converter Steelmaking Process with Low Cost

Yang Jian[1], Yang Wenkui[1], Zhang Runhao[1], Sun Han[1], Lu Xinwu[2], Liu Wanshan[2]

(School of Materials Science and Engineering, Shanghai University, Shanghai 200444, China; 2. Ningbo Iron and Steel Co., Ltd., Ningbo 315000, China)

Abstract: In this paper, a new double-slag converter steelmaking process with high efficiency and low cost has been developed in a 220t converter. The results show that both phosphorus distribution ratio LP and dephosphorization ratio increase with increasing basicity of dephosphorization slag in the lower basicity range of 0.9-2.6. The dephosphorization slag contains dark gray P-rich phase 1, light gray liquid slag phase 2 and white Fe-rich phase 3. With increasing basicity, the morphology of each phase in dephosphorization slag changes greatly, and the area fraction and P_2O_5 content of P-rich phase increase. After decarbonization, the average dephosphorization ratio is 90% and the average phosphorus content in molten steel is 0.0142%. The average consumption of lime and magnesite per ton steel in the whole converter blowing stage is 25.3kg/ton steel and 10.4kg/ton steel, respectively, which is 23.6% and 24.1% lower than those of the conventional process. The purpose of reducing the consumption of lime and magnesite is realized.

Key words: new double-slag; converter steelmaking; dephosphorization; basicity; P-rich phase

炼钢及精炼过程中热力学动力学软件开发

张立峰

（燕山大学，河北秦皇岛　066004）

摘　要：先进的冶金热力学和动力学数据、理论和方法可以用于指导钢铁材料洁净化生产。作者团队开发了一款具有我国自主知识产权的洁净钢冶金热力学和动力学相关计算软件平台。热力学平台功能为计算冶金反应过程主要由输入钢液成分、渣相成分和温度的初始条件。动力学平台功能为针对精炼、中间包和结晶器，通过输入成分、反应器尺寸和操作条件等冶炼参数，实现精炼和连铸过程中不同操作工艺条件下每一时刻钢液、夹杂物和精炼渣的在线预测，为我国钢铁工业的洁净化和精准化水平提升坚实基础。

关键词：热力学；动力学；精炼；连铸

Development of Thermodynamic and Kinetic Software in Steelmaking and Refining Processes

Zhang Lifeng

(Yanshan University, Qinhuangdao 066004, China)

Abstract: Advanced metallurgical thermodynamic and kinetic data, theories and methods can be used to guide the clean production of high quality steels. The author's team developed a software platform for calculation of metallurgical thermodynamics and dynamics of clean steel with independent intellectual property rights. The thermodynamic platform was used to calculate metallurgical reactions based on initial conditions of the steel composition, slag composition, and temperature. The kinetic platform was used to online predict for composition evolutions of steel, inclusions, and refining slag with time. Effects of various operating parameters such as composition, reactor size during the refining, tundish, and mold processes were calculated, which provided a fundamental for the improvement of the level of cleanliness and precision of Chinese steel industry.

Key words: thermodynamics; kinetics; steelmaking; refining

铝镇静钢中夹杂物塑性化的工艺研究与实践

沈　昶[1,2]，陆　强[1,2]，郭俊波[1,2]，杨　峥[1,2]

（1. 马鞍山钢铁股份有限公司技术中心，安徽马鞍山　243000；
2. 安徽省高性能轨道交通新材料及安全控制重点实验室，安徽马鞍山　243000）

摘　要：利用中高碳钢的成分特点，研究开发了中高碳铝镇静钢中 MnS 以 Al_2O_3 为形核质点的非均质形核工艺，将钢中 Al_2O_3 脆性夹杂用塑性 MnS 包裹，解决了疲劳应力钢因脆性非金属夹杂引起的疲劳断裂问题。本文通过对微细、弥散 Al_2O_3 夹杂生成条件、MnS 非均质形核析出热力学条件的研究，开展了钢中关键元素的成分设计、精炼及连铸集成工艺的设计与开发。工业实践表明：(1)低活度氧条件下进行铝终脱氧可以形成 3-5μm 微细弥散的 Al_2O_3 夹杂，并作为非均质形核的核心在二次枝晶晶间的凝固末端析出弥散、细小的粒状 MnS；(2) 成品 T.O 含量平均为 6.18ppm，较原工艺的 7.39ppm 降低了 16%；(3) 成品的夹杂物中 MnS 及 MnS 包裹 Al_2O_3 夹杂占比＞96%；(4) MnS 塑性夹杂工艺可明显提高材料的疲劳性能，成品的平均断裂韧性为 83.47MPa.m$^{1/2}$，较原工艺的 67.31MPa.m$^{1/2}$ 提高了 24%。

关键词： 非金属夹杂物；MnS 塑性夹杂；非均质形核；MnS 包裹 Al_2O_3

Research and Practice of Inclusion Plasticity in Al Killed Steel

Shen Chang[1,2], Lu Qiang[1,2], Guo Junbo[1,2], Yang Zheng[1,2]

(1. Technical Center of Masteel, Ma'anshan 243000, China; 2. Anhui Province Key Laboratory of High-performance Rail Transportation New Materials and Safety Control, Ma'anshan 243000, China)

Abstract: Based on the composition characteristics of medium and high carbon steel, the process which the heterogeneous nucleation of MnS in steel takes Al_2O_3 as nucleation particle is developed, and the brittle inclusions of Al_2O_3 in steel is wrapped by plastic MnS, the problem of fatigue fracture caused by brittle nonmetallic inclusion is solved. In this paper, the composition design of key elements in steel, the design and development of refining and continuous casting integrated process were carried out by studying the formation conditions of fine and dispersed Al_2O_3 inclusions and the thermodynamic conditions of MnS heterogeneous nucleation and precipitation. Industrial practice shows that:(1) The final deoxidation by aluminum with low activity oxygen can form 3-5 μ m fine dispersed Al_2O_3 inclusions, which act as the core of heterogeneous nucleation and precipitate dispersed and fine granular MnS at the solidification end of secondary dendrites; (2) The average T.O. content of the finished product is 6.18ppm, which is 16% lower than 7.39ppm of the original process;(3) The proportion of MnS and MnS wrapped Al_2O_3 inclusions is more than 96% of the total inclusions in the finished product; (4) The average fracture toughness of the finished product is 83.47 MPa.m$^{1/2}$, which is 24% higher than 67.31MPa.m$^{1/2}$ of the original process.

Key words: non-metallic inclusion; MnS plastic inclusion; heterogeneous nucleation; MnS wrapped Al_2O_3

高拉速板坯连铸凝固传热行为及其均匀性控制

朱苗勇，蔡兆镇

（东北大学冶金学院，辽宁沈阳 110819）

摘 要： 高速连铸是发展新一代高效连铸的主题，是实现直轧、铸轧的前提保障，是实现钢铁制造高效、绿色的具体体现。目前我国板坯的实际工作拉速基本在 1.8 m/min 以下，包晶钢拉速大都为 1.2-1.4m/min。拉速提升，影响结晶器凝固传热的不利因素更加凸显，热通量增加，保护渣消耗量降低，凝固坯壳与结晶器铜壁间的润滑变得越来越差，高温凝固坯壳承受各种应力应变的能力变得越来越弱，漏钢和裂纹成为最大挑战。如何确保高拉速条件下结晶器内凝固坯壳的均匀性与安全性以及铸坯质量，是实现高速连铸必须要面对和解决的技术难题。本文从分析包晶钢高速连铸结晶器凝固传热行为为特征入手，阐述高效传热结晶器凝固均匀性控制技术，为高速连铸技术开发和生产提供借鉴。

关键词： 高拉速连铸；均匀凝固；高效传热结晶器

Heat Transfer Behavior and Homogenous Solidification Control for High-speed Slab Continuous Casting of Steel

Zhu Miaoyong, Cai Zhaozhen

(School of Metallurgy, Northeastern University, Shenyang 110819, China)

Abstract: High-speed continuous casting is the theme for developing new generation of high-efficiency continuous casting technology and the premise to make direct rolling or continuous casting and rolling come true, that is the embody for realizing high-efficiency and green steelmaking production line. Presently, the actual casting speed for slab in China is no more than 1.8 m/min and it is in the range of 1.2-1.4m/min for continuous casting of peritectic steel. With increasing casting speed, the negative factors affecting the solidification in continuous casting mold show more obvious, and the lubrication between solidifying shell and mold copper plate becomes worse and worse due to the increase of heat flux and decrease of mold flux consumption, therefore the capability of solidifying shell for resisting the all kinds of stress and strain during casting becomes weaker and weaker, and the occurrence of breakout and cracks with high frequency has been the most challenge. How to ensure homogenous growth and safety of solidifying shell and slab quality, the technical issues, should be solved for high-speed casting. The behavior of heat transfer and solidification in mold with high-speed casting was analyzed and the control technology for mold with high-efficiency heat transfer and homogenous solidification was presented and discussed in this paper.

Key words: high-speed continuous casting; homogenous solidification; continuous casting mold with high-efficiency heat transfer

连铸坯中非金属夹杂物数量、尺寸和成分空间分布的定量预报

张立峰[1]，任　英[2]，张月鑫[2]，王举金[2]，陈　威[1]

（1. 燕山大学，河北秦皇岛　066004；2. 北京科技大学，北京　100083）

摘　要：本研究建立了用以预报非金属夹杂物数量、尺寸、成分在连铸坯全断面上空间分布的数学模型。该模型耦合了结晶器和液相穴凝固全长度上的三维流动、传热、凝固和元素偏析，夹杂物成分转变的热力学和动力学，基于离散相模型的夹杂物在钢液中的运动和在凝固前沿的捕获。定量预报和讨论了管线钢连铸坯断面上、重轨钢大方坯断面上、帘线钢小方坯断面上夹杂物的空间分布。结果表明，为了准确预报连铸坯断面上夹杂物的数量和成分的空间分布，模型中必须考虑钢中溶质元素包括酸溶铝、钙、镁、氧和硫等元素的偏析和扩散行为。

关键词：非金属夹杂物；空间分布；热力学；动力学；元素偏析；捕获；连铸坯

Prediction on the Three Dimensional Distribution of the Number Density, Size and Composition of Non-metallic Inclusions in Continuous Casting Products

Zhang Lifeng[1], Ren Ying[2], Zhang Yuexin[2], Wang Jujin[2], Chen Wei[1]

(1. Yanshan University, Qinhuangdao 066004, China;
2. University of Science and Technology Beijing, Beijing100083, China)

Abstract: In the current study, a comprehensive mathematical model was established to quantitatively predict the spatial distribution of the number density, size and composition of non-metallic inclusions in continuous casting products. The model coupled the three dimensional fluid flow, heat transfer, solidification and element segregation, the thermodynamics and kinetics for the composition transformation of inclusions, and the motion and entrapment of inclusions by the Discrete Phase Model in the full length of the casting strand. A special criterion for the entrapment of inclusions at the solidification front was used in the current model. The distribution of inclusions in the continuous casting pipeline steel slab, the heavy rail steel bloom and the tire cord steel billets was predicted and quantitatively discussed. The result indicated that the segregation and the kinetic diffusion of the dissolved elements such as aluminum, calcium, magnesium, oxygen, sulfur must be included in the model in order to correctly predict the distribution of the composition and the number density of inclusions in continuous casting products.

Key words: non-metallic inclusions; spatial distribution; thermodynamics; kinetics; element segregation; entrapment; continuous casting product

高品质特殊钢长材生产连铸工艺研究与新认识

张家泉

（北京科技大学，北京 100083）

摘 要：高品质特殊钢应兼具稳定一致的优良加工性能与服役性能，对材料设计、装备条件、生产工艺及其协同控制均具有较高的要求。研究表明，铸态组织与轧材产品质量具有强关联性，基于连铸源头及其工序界面的研究与合理控制是提升特殊钢长材质量稳定性与一致性的关键。以连铸冶金反应器研究为例，介绍了对钢包、中间包、结晶器水口与铸流装备等冶金功能要求的新认识；以高端特殊钢长材生产研究为例，介绍了当前以铸代锻生产轨道交通用齿轮钢、车轴钢、轴承钢等铸坯质量控制技术特点。最后，基于抗硫管用微合金高强钢服役性能要求，重点讨论基于铸态组织调控改善产品带状缺陷与服役性能的成效。

关键词：连铸；特殊钢；质量与控制

Study and Understanding of Continuous Casting Practices for an Improved High-quality Special Steel Long Products

Zhang Jiaquan

(University of Science and Technology Beijing, Beijing 100083, China)

Abstract: High-quality special steel should have the characteristics of identical superior process ability and service properties, which has a higher demand to materials design, equipment, practices of production and their coordination control in mass production. New research shows that the final as-rolled steel quality is closely related to a selective control to its as-cast solidification morphology, and both the understanding and reasonable application of overall continuous casting practices play a key role to upgrade steel quality stability for same lot production. In terms of our recent study on each individual reactor, such as ladle, tundish, mold nozzle and the instrumentation of casting strands in continuous casting processes, additional demands to their conventional metallurgical functions have been presented. For industrial study on some high end special steel long productslike rail way gear, axle, and bearing steels,unique CC practice has been carried outto obtain their adaptive as-cast products. Additionally, an effective control to the popular band defects in hot rolled products has been introduced from casting operations, together with a much improved HIC resistant for oil pipe steels.

Key words: continuous casting; special steels; quality & control

喷淋水量分布对轴承钢大方坯凝固传热的影响

王慧胜[1]，刘　青[1]，王　超[2]，李　明[3]，董文清[3]

（1. 北京科技大学钢铁冶金新技术国家重点实验室，北京　100083；
2. 北京科技大学高等工程师学院，北京　100083；
3. 南京钢铁股份有限公司，江苏南京　210035）

摘　要：针对某钢厂连铸 250mm×300mm GCr15 轴承钢工艺，建立了基于喷嘴实际水量分布的凝固传热数学模型，分析了二冷区各段不同的喷淋水量分布对铸坯表面温度分布的影响。结果表明：现有生产工艺条件下，二冷 1 段喷淋水量分布不合理，二冷 2~4 段喷淋水量分布较好；在二冷 1 段末，铸坯表面横向温度波动最大，在铸坯内弧和侧弧表面横向温度波动最大值分别为 4.9℃/mm、7.5℃/mm，在二冷 2 段末，铸坯内弧表面横向温度波动最小，为 1.6℃/mm；在二冷 1 段末，铸坯内弧和侧弧表面中心纵向回温较大，分别为 125.6℃/m、172.1℃/m，在矫直区铸坯角部温度部分落入其第三脆性温度区间；通过喷嘴选型和喷淋高度调节，可改善铸坯表面温度分布，降低裂纹缺陷的发生几率。

关键词：GCr15 轴承钢；大方坯；凝固传热；水量分布；二次冷却

Influence of Spray Water Distribution on Solidification Heat Transfer of Bearing Steel Bloom

Wang Huisheng[1], Liu Qing[1], Wang Chao[2], Li Ming[3], Dong Wenqing[3]

(1. State Key Laboratory of Advanced Metallurgy, University of Science and Technology Beijing, Beijing 100083, China; 2. School of Advanced Engineers, University of Science and Technology Beijing, Beijing 100083, China; 3. Nanjing Iron & Steel Co., Ltd., Nanjing 210035, China)

Abstract: Aiming at the continuous casting process parameters of a 250 mm×300 mm GCr15 bearing steel in a steel plant, a mathematical model of solidification heat transfer was established based on the actual spray water distribution of the nozzles, and the influence of different spray water distribution in different stages of secondary cooling on the surface temperature value of bloom was analyzed. The results show that the spray water distribution in the first stage of secondary cooling is not reasonable, and the spray water distribution in the second to fourth stages of secondary cooling is well. At the end of the first stage of secondary cooling, the transverse temperature value of the bloom surface has the largest fluctuation, and the maximum transverse temperature fluctuation of the inner arc and side arc surface of the bloom is 4.9℃/mm and 7.5℃/mm, respectively. At the end of the second stage, the transverse temperature fluctuation of the inner arc surface of the bloom has little change, which was 1.6℃/mm. At the end of the first stage of secondary cooling, the inner arc and side arc surface center of the bloom have a large longitudinal temperature reheating, which is 125.6℃/m and 172.1℃/m, respectively. In the straightening zone, a part of corner temperature of the bloom falls into the third brittle temperature range. The surface temperature distribution of the bloom can be improved and the probability of crack defect can be reduced by nozzle selection and spray height adjustment.

Key words: GCr15 bearing steel; bloom; solidification heat transfer; water distribution; secondary cooling

厚板连铸机末端高温全连续淬火技术研发与应用

蔡兆镇，王少波，刘志远，匡泓旭，董嘉宁，朱苗勇

（东北大学冶金学院，辽宁沈阳 110819）

摘 要：热模拟研究了含铌桥梁钢在不同温度及冷速条件下的组织结构转变与碳氮化物析出特点，确定了高温钢组织高塑化最佳淬火控冷起始温度与冷速。结合某钢厂厚板坯连铸生产实际，建立其厚板坯连铸三维温度场计算模型，分析获得了铸坯表面最佳淬火位置。在此基础上，设计开发了厚板连铸机铸坯表面在线全连续淬火系统并获得了应用。结果表明，厚板坯连铸机末端实施超强全连续淬火控冷，可高塑化铸坯表层20mm范围内的组织，满足连铸坯热送轧制需求。

关键词：连铸；铸坯表面淬火；组织转变；裂纹

Development and Application of High Temperature Continuously Quenching Technology at the End of Thick Slab Continuous Caster

Cai Zhaozhen, Wang Shaobo, Liu Zhiyuan, Kuang Hongxu, Dong Jianing, Zhu Miaoyong

(School of Metallurgy, Northeastern University, Shenyang 110819, China)

Abstract: The evolution of the microstructure and the precipitation characteristic of carbonitride of Nb containing bridge steel were thermally simulated by Gleeble simulator under different quenching temperatures and cooling rate, and the optimum quenching temperature and the optimum cooling rate were obtained. Moreover, a 3D heat transfer model was established according to the practical continuous casting conditions of thick slab in a plant. The optimum position of slab surface quenching was determined by the slab heat transfer simulation. Based on these, the quenching online equipment was developed and applied. The results show that the application of the continuously quenching process can greatly improve the plasticity of the micro-structure of the thick slab surface to meet the requirement of hot charging for slab.

Key words: continuous casting; slab surface quenching; microstructure transformation; crack

结晶器内多相流动和卷渣定量化统计的大涡模拟研究

陈　威[1]，张立峰[2]

（1. 燕山大学机械工程学院，河北秦皇岛　066044；2. 燕山大学亚稳材料制备技术与科学国家重点实验室，河北秦皇岛　066044）

摘　要：连铸结晶器内卷渣的发生对连铸坯表面质量有重要影响，因此通过研究结晶器内多相流动对卷渣发生几率的影响，对减少连铸坯夹杂缺陷有重要意义。本研究通过耦合 LES（Large Eddy Simulation, LES）湍流模型、VOF（Volume of Fraction, VOF）多相流模型以及 DPM 离散相模型，建立三维结晶器钢液-渣相-空气相-氩气泡瞬态四相流动模型，对实际连铸过程结晶器内多相流动和卷渣进行研究。并通过用户自定义程序（User Define Function, UDF）定量化研究了卷入渣滴的平均直径、速度、空间分布以及弯月面处卷渣发生位置等信息。结果表明卷入渣滴直径主要在 2～4 mm 之间，且主要发生在结晶器宽度 1/4 处。

关键词：多相流；卷渣；渣滴分布；大涡模拟；结晶器

Large Eddy Simulation on Multiphase Flow and Quantitative Statistics of Slag Entrainment in a Continuous Casting Mold

ChenWei[1], Zhang Lifeng[2]

(1. School of Mechanical Engineering, Yanshan University, Qinhuangdao 066044, China; 2. State Key Lab of Metastable Materials Science and Technology, Yanshan University, Qinhuangdao 066044, China)

Abstract: Theslag entrapment in the continuous casting (CC) mold has an important impact on the surface quality of the CC slab. Therefore, it is of great significance to reduce the surface defect of the CC slab to study the influence of multiphase flow on the probability of the slag entrapment. In the current study, a three dimensional mathematical model, coupled with the large eddy simulation (LES), VOF model, and DPM model, was established to investigate the steel-slag-air-argon bubble four phases flow and slag entrainment during the CC process. The average diameter, speed, spatial distribution of the entrained slag droplet, and the location of the slag entrainment on the meniscus werequantitatively studied using a userdefined function (UDF).The result shows that the diameter of the entrained slag droplet was mainly between 2-4 mm, and it mainly occurred at 1/4 width of the CC mold.

Key words: multiphase flow; slag entrainment; distribution of slag droplets; large eddy simulation; mold

直弧型 450mm 特厚板坯连铸机的技术装备及生产实践

郝瑞朝，陈卫强，陈　杰，王　颖

（中冶京诚工程技术有限公司，北京　100176）

摘　要：主要介绍了由中冶京诚自主研发设计的直弧型特厚板坯连铸机的开发背景、技术特点、核心装备及现场应用效果，阐述了保证特厚连铸板坯质量所采用的技术措施。450mm 特厚板坯的生产实践表明，中冶京诚自主设计研发的直弧型特厚板坯连铸机主体设备运行良好，铸坯的表面质量和内部质量优良，连铸机各项关键技术和质量控制能力达到了国际领先水平，有力推动了我国高端装备制造和先进钢铁材料的发展。

关键词：特厚板坯；直弧型连铸机；技术装备；生产实践

Technical Equipment and Production Practice of Straight Arc 450mm Extra-thick Slab Caster

Hao Ruichao, Chen Weiqiang, Chen Jie, Wang Ying

(Capital Engineering & Research Incorporation Limitied, Beijing 100176, China)

Abstract: This paper mainly introduces the development back ground, technical features, core equipment and field application effect of the straight arc extra-thick slab caster, which is independently developed and designed by CERI. The technical measures used to ensure the quality of extra-thick continuous casting slab are described. The production practice of 450mm extra-thick slab show that the main equipment of the straight arc extra-thick slab caster in dependently developed and designed by CERI runs well, the surface quality and internal quality of the slab was excellent, and the key technologies and quality control ability of the caster reach the international leading level. It has effectively promoted the development of high-end equipment manufacturing and advanced steel materials in China.

Key words: extra-thick slab; straight arc caster; technical equipment; production practice

高端冶金产品技术变革与自主创新

黄庆学[1,2]，王 涛[1,2]

（1. 太原理工大学 先进金属复合材料成形技术与装备教育部工程研究中心，山西太原 030024；
2. 太原理工大学 先进成形与智能装备研究院，山西太原 030024）

摘 要：金属复合材料广泛应用于航空航天、海洋工程、核电装备等重点领域，是国民经济的重要支柱之一。随着冶金产品生产规模的逐渐扩大，传统生产方式已向着追求极限制造、提高产品附加值、节约能源消耗和降低生产成本的方向发展。由于国外核心技术的封锁和自身基础理论研究的薄弱，在复合轧制、极限轧制、柔性轧制等重点领域关键技术储备仍然不足。本报告以复合板轧制、极薄带轧制以及装备智能化设计开发为案例，阐述"材料-工艺-装备"全流程研发体系，剖析金属材料发展新方向，为推动材料和装备创新、支撑产业升级、建设制造强国提供新动能。

关键词：先进轧制；复合板；极薄带；极限制造；智能化装备

Technological Transformation and Independent Innovation of High-end Metallurgical Products

Huang Qingxue[1,2], Wang Tao[1,2]

(1. Engineering Research Center of Advanced Metal Composites Forming Technology and Equipment, Ministry of Education, Taiyuan University of Technology, Taiyuan 030024, China; 2. Institute of Advanced Forming and Intelligent Equipment, Taiyuan University of Technology, Taiyuan 030024, China)

Abstract: As one of the important pillars of the national economy, metal composites are widely used in key fields such as aerospace, marine engineering, nuclear power equipment, etc. With the gradual scale expansion of metallurgical products, traditional production methods have developed towards pursuing extreme manufacturing, increasing additional value, saving energy and reducing costs. Due to the blockade of foreign core technologies and the weakness of our basic theoretical research, the key technology reserves in momentous areas such as clad plate rolling, ultra-thin strip rolling and flexible rolling are still insufficient. Taking clad plate rolling, ultra-thin strip rolling as well as intelligent equipment design and development as examples, this report expounds the whole process design and development system of "material-process-equipment", analyzes the new development direction of metal materials, and provides new kinetic energy for promoting material and equipment innovation, supporting industrial upgrading and building a world manufacturing power.

Key words: advanced rolling; clad plate; ultra-thin strip; extreme manufacturing; intelligent equipment

面向双线双智控的宝武马钢热轧智慧工厂建设实践

邵 健[1]，何安瑞[1]，陈雨来[1]，杨 荃[1]，毛学庆[2]，司小明[2]，闻成才[2]

（1. 北京科技大学工程技术研究院，北京 100083；
2. 马鞍山钢铁股份有限公司第四钢轧总厂，安徽马鞍山 243000）

摘 要：经历以自动化、信息化为主要特征的工业3.0阶段后，以高效、高精度、低成本为目标的热轧智慧工厂"双线双智控"模式先行示范具有重要意义。双线双智控指项目构建的热轧智慧工厂覆盖两条产线，并包含操维集控平台和协同智慧平台，实现操维和业务双智控。操维集控平台将板坯库、加热炉、轧线、磨辊间、能介、远程运维六大单元进行集中操控，打破区域管理边界，实现颠覆式的集约化生产。协同智慧平台以多级KPI指标为牵引，将热轧车间运行过程生产、质量、设备、能源、成本、交付、人员、安环八大任务进行串接，开发44个智能化模型，形成数字钢卷、实时诊断、智能助力、绩效导航四大主题，并在物理车间和虚拟车间迭代运行，形成生产在线优化和管控精益求精的运行新模式。热轧智慧工厂建设实现了主线人员效率提升35%，产能提升15%，成本下降5%，质量损失率降低10%。

关键词：热轧；智慧工厂；操作集控；业务协同；大数据中心

宽厚板坯连铸大压下及超低压缩比特厚板轧制技术

康永林，朱国明，姜 敏

（北京科技大学，北京 10083）

摘 要：首先，简要介绍了宽厚板坯连铸凝固末端及凝固后大辊径大压下技术的原理方法、变形渗透数值模拟分析和研究开发的大压下技术装备的特点。其次，介绍分析了针对厚度400mm、300mm宽厚板坯连铸过程实施大压下的板坯内部组织、缩孔、疏松变化情况，以及1.9~2.4超低压缩比轧制160~200mm特厚板的工艺与组织性能情况。工业性试验结果表明，研究开发的宽厚板坯连铸大压下工艺及装备充分利用了连铸后期铸坯从表面至芯部由凝固冷却自然形成的≥400~500℃的大梯度温度场以及由此产生的表层硬、芯部软的物理特征，实施大压下可以显著改善铸坯中心的缩孔、疏松和偏析缺陷，为后期以超低压缩比轧制高质量特厚板建立了关键技术基础。

Heavy Reduction and Ultra-low Compression Bit Thick Plate Rolling Technology for Wide and Thick Slab Continuous Casting

Kang Yonglin, Zhu Guoming, Jiang Min

(University of Science and Technology Beijing, Beijing 100083, China)

Abstract: Firstly, the principle and method of large roll diameter and large reduction technology at solidification end and after solidification of wide and heavy plate continuous casting, numerical simulation analysis of deformation infiltration and characteristics of large reduction technology equipment are briefly introduced. Secondly, the changes of internal structure, shrinkage cavity and porosity of wide and heavy plate with thickness of 400 mm and 300 mm during large reduction in continuous casting process were introduced and analyzed, as well as the process, microstructure and properties ofultra-heavy plate rolling with ultra-low compression ratio of 1.9~2.4 and thickness of 160~200 mm. The industrial test results show that the developed continuous casting large reduction process and equipment of wide and heavy plate make full use of the large gradient temperature field≥400~500℃ in casting blank naturally formed by solidification and cooling from surface to core at the later stage of continuous casting, as well as the resulting physical characteristics of hard surface and soft core. The implementation of large reduction can significantly improve the shrinkage cavity, porosity and segregation defects in the center of casting blank, as well as establish the key technology foundation for the later rolling of high-quality ultra-heavy plate with ultra-low compression ratio.

Comparing Rolling and Additive Technologies, Challenges and Opportunities

Pedro Rivera-Diaz-del-Castillo, Hossein Eskandari Sabzi

(Department of Engineering, Lancaster University, LA1 4YW, United Kingdom)

Abstract: Metal additive manufacturing (MAM) is a rapidly growing industry: projections show that the £2bn global market is to increase to ~£6bn by 2024, i.e. a Compound Annual Growth Rate (CAGR) of 25%, without parallel in any other manufacturing technology. In comparison, rolling of advanced metals is a well-established technology dominating the majority of advanced alloys production, but with moderate growth compared to MAM. MAM can deliver enhanced products but at a higher cost owing to its unit production costs. This presentation reviews both technologies from the point of view of the microstructures they can deliver and their resulting properties. The conditions under which MAM can prevail over rolling are discussed in terms of properties, geometric constraints, sustainability and unit costs. A pathway to combine MAM and rolling technologies is suggested.

Key words: metal; rolling; additive manufacturing; microstructures; properties

冶金装备-工艺-产品控制技术融合发展

彭 艳

（燕山大学机械学院，河北唐山 066000）

摘 要： 随着学科发展与第四次工业革命智能制造技术的到来，机械装备所具备的平台和载体属性越来越显现，提高装备-工艺-产品之间匹配度是永恒课题。以冶金装备中的关键部件，轧辊的制造、使用和再制造等环节为背景，总结介绍轧辊全生命周期关键技术群及其新的发展趋势，阐释装备-工艺和产品控制技术融合发展的重要性。

Metallurgical Equipment-Process-Product Control Technology Integration Development

Peng Yan

(School of Mechanics, Yanshan University, Tangshan 066000, China)

Abstract: With the development of disciplines and the arrival of intelligent manufacturing technology in the fourth industrial Revolution, the platform and carrier properties of mechanical equipment are becoming more and more obvious. It is an eternal subject to improve the matching degree between equipment, process and products. Based on the manufacturing, use and remanufacturing of key components of metallurgical equipment and roll, this paper summarizes and introduces the key technology groups and their new development trend in the whole life cycle of roll, and explains the importance of the integrated development of equipment-process and product control technology

超高强度耐久型桥梁缆索成套技术研究与应用

张剑锋

（江阴兴澄特种钢铁有限公司，江苏无锡 214000）

摘 要：桥梁缆索的高强化和轻量化要求，对于缆索用钢制造提出了更高层次的要求，兴澄特钢从材料的力学性能实现原理出发，设计了 2000 MPa 级超高强度缆索钢的成分、组织目标及其配套的生产工艺路径。应用浇注钢水物性控制、搅拌-冷却-轻压下配合技术，降低铸坯偏析指数至 0.95~1.05，基于钢种的应变-相转变及温度-相变等材料基础研究，优化盘条控制轧制，应用自主研发的在线水浴技术（XDWP-Xingcheng Direct Water Patenting），轧后 890±10℃水浴至 550~680℃，绿色高效地实现盘条在相变区索氏体化等温韧化处理，结合优化设计的冷拉减面率位错密度控制技术，得到成分均匀、组织均匀、力学性能良好的盘条，实现盘条索氏体含量≥90%，索氏体片层间距 100~150 nm，断面收缩率>25%。通过材料、工艺设计及生产过程优化控制，直径 φ7.0 mm 2000 MPa 级超高强度桥梁缆索钢丝强度、抗扭转性能稳定达到大型桥梁建设要求，应用于沪苏通跨长江公铁两用斜拉大桥。

关键词：绿色低碳; 高效; 超高强度; 桥梁缆索线材

高性能颗粒增强钛基复合材料制备技术研究

路 新，潘 宇

（北京科技大学工程技术研究院，北京 100000）

摘 要：不规则氢化脱氢（HDH）钛粉易受间隙氧污染，导致其粉末冶金制件力学性能大幅受限。为此基于有机覆膜阻氧原理，抑制氧在钛基体中的扩散固溶并原位生成陶瓷相颗粒以实现粉末钛制件的增强增韧。本研究以低氧高活性 HDH 钛粉和陶瓷先驱体聚合物聚碳硅烷（PCS）为原料，通过粉末冶金工艺制备原位自生 TiC 颗粒增强钛基复合材料，探究了 HDH 钛粉的自钝化行为，PCS 的引入对材料控氧效果、组织演变规律和力学性能强韧化的作用机制。研究表明：采用湿混包覆工艺可以将 PCS 包覆于 Ti 粉表面，有效控制材料制备过程中的氧增，制件的氧含量可以控制在 0.21~0.24 wt.%。在烧结过程中，PCS 受热分解并与 Ti 基体原位反应生成 TiC 颗粒，弥散分布于基体中，而 Si 元素则固溶于 Ti 基体起到固溶强化作用。PCS 的引入对钛基体的强度和塑性具有显著的改善作用，同时结合后续的热变形工艺，可以进一步实现强度塑性的大幅提升。本研究将为新型低成本高强高韧钛基复合材料的研发提供理论和技术依据。

关键词：钛基复合材料；粉末冶金；聚碳硅烷；力学性能

Application of Artificial Intelligence for Efficient Development, Manufacturing and Implementation of Steel

Hedström Peter[1,2], Mu Wangzhong[1], Serrano Ismael García[2], Kolli Satish[2], Chowdary Kailash[2], Holmström Claes[2]

(1. KTH Royal Institute of Technology, Department of Materials Science and Engineering, Brinellvägen 23, SE-100 44 Stockholm, Sweden; 2. Ferritico AB, Valhallavägen 79, SE-114 28 Stockholm, Sweden)

Abstract: Data-driven methods, based on artificial intelligence (AI) or machine learning, contribute to a paradigmatic shift in materials science and engineering at present. The conventional approach of materials development, manufacturing and implementation, requires extensive experimental testing, and by replacing certain parts, or ultimately most parts, to AI-based methods this process can be made significantly more efficient. We will present the underlying methodology which has been developed over the past years in collaboration between Materials Science at KTH Royal Institute of Technology and the company Ferritico AB. We will give application examples and an outlook on AI for steels. The work aims at supporting the implementation of the concept Steel Industry 4.0.

Key words: artificial intelligence; machine learning; manufacturing; material design; steels; industry 4.0

热轧钢铁材料绿色加工技术与应用

袁　国[1]，康　健[1]，张元祥[1]，窦为学[2]，王国栋[1]

（1. 东北大学轧制技术及连轧自动化国家重点实验室，辽宁沈阳　110819；
2. 敬业钢铁有限公司，石家庄　050400）

摘　要：在“双碳目标”的大背景下，钢铁企业持续不断的进行在线处理以及近终成形等短流程技术实践，从而达到产品高质化、生产绿色化的最终目标。本报告以热轧无缝钢管在线组织性能调控工艺与技术、薄带铸轧短流程技术为例，介绍了相关工艺技术的自主创新与应用实践进展。其一，在提出针对圆形断面特征的非对称射流冲击冷却换热机制及控制方法基础上，开发出具有内外壁快速均匀冷却和直接淬火功能的热轧无缝钢管在线控制冷却技术与装备，以及工业线的生产应用情况。其二，分析了薄带连铸工艺亚快速凝固及短流程特点与电工钢组织与织构控制要求的契合性，介绍了电工钢薄带连铸浇注策略、液位控制以及磁性能调控等工业化技术的应用进展。

关键词：节能减排；短流程；在线冷却；薄带连铸

Green Processing Technology and Application of Hot-Rolled Steel

Yuan Guo[1], Kang Jian[1], Zhang Yuanxiang[1], Dou Weixue[2], Wang Guodong[1]

(1. State Key Laboratory of Rolling and Automation, Northeastern University, Shenyang 110819, China;
2. Hebei Jingye Group Co., Ltd., Shijiazhuang 050400, China)

Abstract: Under the background of "emissions peak and carbon neutrality", steel industries carry out online processing, near net shaping and other short process technology practice so as to produce high-quality and green products. The report takes the online microstructure and property regulation technology of hot-rolled seamless steel tube and the short process technology of strip casting for examples and introduces independent innovation and application of related process technology. Firstly, the online controlled cooling technology and equipment including the functions of rapid uniform cooling in inner and outer walls and direct quenching of hot-rolled seamless steel tube have been developed on the basis of the heat transfer mechanism and control method of asymmetric jet impingement cooling according to the characteristics of circular section. Secondly, it's analyzed that the agreement between the characteristics of sub-rapid solidification and short process of strip-cast process and the requirements of microstructure and texture of electrical steel. And introducing the application progress of industrial technology such as pouring strategy, liquid level control and magnetic properties regulation.

Key words: energy saving and emission reduction; short process; online cooling; strip casting

板带钢全生命周期数字化技术及应用

朱国明，康永林

（北京科技大学材料科学与工程学院，北京　100083）

摘　要：数字化制造技术是十四五期间先进基础材料发展的重点之一。其侧重于产品全生命周期数字化技术的应用，是实现智能制造的基础，而数值模拟技术是数字化的中坚。文章总结了材料加工数值模拟技术的发展方向，并围绕板带钢从连铸到应用服役，开展了板坯连铸一维、二维到整台铸机连铸全程的三维热力耦合模拟，在支持板坯大直径辊大压下工艺装备的落地实施的同时开发了应变渗透、压下负荷等相关预测系统；针对热连轧、宽厚板轧制的核心工艺环节，从加热、定宽、往复轧制、展宽轧制、热连轧及控制冷却等，完成全轧程三维热力耦合数值模拟及组织演变、性能预测；建立了卷取、板卷冷却、开卷矫直过程数值计算模型并完成模拟分析；对板带轧制中遇到的常见问题：翘头、浪型、边部折叠、开卷过程的“龟背”现象、支撑辊应力集中以及特厚板轧后平面形状及边部鼓形建立了针对性的模型并进行模拟分析；开发了热连轧、宽厚板轧制数字化系统平台；针对冷轧连退核心工艺环节，建立了 6 辊弹性辊轧制数值计算模型，得到了显式化浪型结果；完成连退炉内“W”形辐射管内燃烧模拟，并建立了连退过程辐射加热带钢的数值计算模型，并得到了连退炉中走带瓢曲、边浪、跑偏的模拟结果；完成了 CVC 平整机弹性辊二次压下过程模拟，得到了边降、轧制纵纹等计算结果；在板带钢产品成形及应用服役方面，模拟了多种成形工艺，完成了极端服役条件及选材相关的模拟研究。

关键词：板带钢；全生命周期；数字化技术；数值模拟

Digital Technology and Application of Whole Life Cycle of Plate and Strip Steel

Zhu Guoming, Kang Yonglin

(School of Materials Science and Engineering, University of Science and Technology Beijing, Beijing 100083, China)

Abstract: Digital manufacturing technology is one of the key points in the development of advanced basic materials during the 14th Five Year Plan period.It focuses on the application of digital technology in the whole product life cycle, which is the basis of intelligent manufacturing, and numerical simulation technology is the backbone of digitization.This paper summarizes the development direction of material processing numerical simulation technology, and introduce author's related research work around the plate and strip steel from continuous casting to application.For the wholecontinuous casting processfrom 1D and 2D to entire strand's thermal mechanical coupled3Dmodel are built and simulated; For supporting the implementation of slab casting largediameter roll large reduction process equipment, relevant prediction systems such as strain infiltration and reduction load are developed;Aiming at the core process links of hot continuous rolling and plate rolling, 3Dthermal mechanical coupled numerical simulation, microstructure evolution and performance prediction of the whole rolling process are completed from heating, sizing press, reversing rolling, spread rolling, hot continuous rolling and controlled cooling;The numerical model of coiling, coil cooling and uncoiling straightening process is established and simulated;The common problems encountered in plate and strip rolling: ski-up, wave shape, edge folding, C warp phenomenon in uncoiling process, stress concentration of backup roll, plane shape and edge double drumshape ofultra-heavy plate rolling are established and simulated;The digital system platform of hot continuousrolling and plate rolling is developed;Aiming at the core process of cold rolling continuous annealing, the 6 high elastic roll numerical model is established and the explicit wave shape results are obtained;The combustion simulation in the "W" shaped radiant tube in the continuous annealing furnace is completed, the numericalmodel of radiant heating strip steel in the continuous annealing process is established, and the simulation results ofstripbuckling, edge wave and running deviation in the continuous annealing furnace are obtained;The secondary reduction process of CVC temper mill elastic roll is simulated, and the calculation results of the edge drop and rolling longitudinal waveare obtained;In terms of sheet and strip product forming and application service, a variety of forming processes are simulated, and the simulation research related to extreme service conditions and material selection is completed.

Key words: plate and strip steel; the whole product life cycle; digital technology; numerical simulation

厚板轧制绿色智慧新技术

焦四海，丁建华

（宝山钢铁股份有限公司中央研究院，上海　201900）

摘　要：厚板产品对制造大国重器、促进社会降碳减排起到关键作用。报告从通过复合轧制技术和厚板数字化与智能制造平台两个方面介绍了厚板轧制绿色智慧新技术。通过复合轧制技术，可以增强型钢铁材料功能性，解决碳钢材料易锈蚀的本质痛点，从而提升大国重器的本质安全、可靠性，延长其使用寿命，从而在全生命周期上实现降碳、降本；基于业务需求不断开发、优化和应用数字化与智能制造平台，可以实现产品质量、生产效率、生产效能不断提升，还可以带来知识沉淀与传承的等有益效果。实践表明，轧制复合技术已经打开了广阔的产品空间；基于业务需求开发和应用工厂级数字中心和智慧应用平台，可以成为流程型制造业数字化转型的一种模式，但是要实现真正的技术自主，还需要大量基础工作。

关键词：厚板；复合轧制；数字化转型

Green and Smart Technologies for Plate Steel Rolling

Jiao Sihai, Ding Jianhua

(Central Research Institute, Baoshan Iron & Steel Co., Ltd., Shanghai 201900, China)

Abstract: Heavy plate steel products play an important key role in manufacturing key equipment and infrastructural facilities. In this report, two category new technologies for green and smart plate rolling are presented as hot roll bond and the work-shop level data center and intelligent manufacturing platform. The corrosion weakness of carbon steel could be

solved by enhancing its functionality with hot roll bond technology, leading to the improvement of the intrinsic safety and reliability of equipment, prolongation of its service life, and the reduction of life-cycle carbon emission and cost. The continuous development, optimization and application of the work-shop level data center and intelligent manufacturing platform driven by business requirements can promote digital transformation of plate mill, realizing the continuous improvement of product quality, production efficiency and effectiveness, and knowledge accumulation, learned and applied, and furthermore could be a effective way of digital transformation of process manufacturing.
Key words: plate steel; hot roll bond; digital transformation

首钢冷轧高强钢轧制技术的实践

于 孟[1]，林海海[1]，王永强[1]，张晓峰[2]，任新意[3]，王春海[1]

（1. 首钢集团有限公司技术研究院，北京 100043；2. 首钢京唐钢铁联合有限责任公司冷轧部，河北唐山 063200；3. 首钢京唐钢铁联合有限责任公司技术中心，河北唐山 063200）

摘 要：先进高强钢产品在冷轧生产过程中轧制负荷高、板形与尺寸精度控制难度大。本文介绍了首钢冷轧产线配置及主要产品情况，比较了首钢京唐 2230mm 冷连轧机与 1750mm 十八辊单机架轧机的设备与工艺特点，分析了两条产线在先进高强钢生产中板形控制、厚度控制、轧制负荷等方面的优劣。
关键词：冷轧；高强钢；轧制负荷；板形；尺寸精度

Practice of Cold Rolling High Strength Steel in Shougang

Yu Meng[1], Lin Haihai[1], Wang Yongqiang[1], Zhang Xiaofeng[2], Ren Xinyi[3], Wang Chunhai[1]

(1. Research Institute of Technology, Shougang Group Co., Ltd., Beijing 100043, China;
2. Cold Rolling Department, Shougang Jingtang United Iron & Steel Co., Ltd., Tangshan 063200, China;
3. Technology Center, Shougang Jingtang Iron and Steel Co., Ltd., Tangshan 063200, China)

Abstract: In the cold rolling process of advanced high strength steel products, the rolling load is high, and it is difficult to control the shape and dimensional accuracy. This paper introduces the configuration and main products of Shougang cold rolling line, compares the equipment and process characteristics of Jingtang 2230mm tandem cold mill and 1750mm 18 high single stand mill, and analyzes the advantages and disadvantages of the two production lines in shape control, thickness control, rolling load, rolling efficiency and yield in the production of advanced high strength steel.
Key words: cold rolling; high strength steel; rolling load; flatness; dimensional accuracy

热轧钢材组织性能演变的智能数字解析与工艺逆向控制

刘振宇

（东北大学）

摘　要： 我国钢产量世界第一，是国民经济建设的支柱性产业，但产品性能波动大、稳定性差等质量问题，仍是困扰我国钢铁工业的重中之重。成分、工艺、组织和结构与性能之间相关关系极为复杂，且存在各种交互作用，开发钢材生产全流程组织与综合性能的精准数字解析，实现最优工艺在线决策与控制，是钢铁生产方式的一次重大转变，将引领钢铁生产由传统自动化与信息化向智能化转变。然而，钢材质量控制水平取决于生产中能否对组织演变进程进行动态最优控制。但当前生产以“工艺-性能”控制为主，组织演变过程处于黑箱状态，导致生产控制目标模糊，是当前钢铁生产急需解决的核心问题。

为此开发出以组织性能预测与优化技术。通过开发描述析出、再结晶、相变等组织演变的物理冶金学方法，结合工业大数据驱动和机器学习，实现了全流程组织演变行为的数字解析，构建起工艺-组织-性能的数字孪生体，既准确描述生产实际中物理冶金学规律，又大幅提升了其对工业生产过程的适应性，将热轧过程复杂物理过程高精度映射为数字信息。此外，针对生产过程多变量、强耦合的特点，产品性能指标多为分散单独控制，难以实现全局动态优化和柔性化生产的问题，开发出了基于多目标优化算法的工艺反向决策系统，解决复杂系统多维度、多目标优化的准确性与高效性相统一问题，实现热轧过程最优工艺在线决策与控制，有效解决产品质量稳定性等重大问题。

相关技术已在包括鞍钢、河钢、首钢、涟钢等多家钢铁企业的十余条生产线进行了推广应用，实现了热轧的集约化、绿色化生产。

核工程用钢板技术

王　勇

（鞍钢集团钢铁研究院，辽宁鞍山　114009）

摘　要： 介绍了核电做为安全、清洁、稳定的能源，现已成为世界经济发展的重要支柱之一，以及我国核电三步走战略，对我国在建、在运的核电机组类型及材料现状进行分析，总结提出了核工程用钢板特点及要求。之后重点介绍了鞍钢核工程钢板主要生产装备及取得的业绩，核岛关键设备系列用钢、核反应堆安全壳系列用钢、核电常规岛设备系列用钢、核电不锈钢及异质复合钢板四大系列核电用钢典型钢种及用途，阐述了核工程用钢板新技术，同时提出建立我国核工程钢板从设计、研发、生产、应用一体化运行机制，引领全球行业发展。

关键词： 核工程；钢板；典型钢种；新技术

Steel Plate Technology for Nuclear Engineering

Wang Yong

(Iron and Steel Research Institute of Angang Group, Anshan 114009, China)

Abstract: As a safe, clean and stable energy, nuclear power has become one of the important pillars of world economic development. This paper introduces the three-step strategy of nuclear power in China, and analyzes the types and material status of nuclear power units under construction and operation in China, which summarizes and puts forward the characteristics and requirements of steel plates for nuclear engineering. The manuscript focuses on the main production equipment and achievements of Angang's nuclear engineering steel plate, the typical steel types and applications of four series of nuclear power steel, including steel for nuclear island key equipment series, nuclear reactor containment series, nuclear power conventional island equipment series, nuclear power stainless steel and heterogeneous composite steel plate, and expounds the new technology of nuclear engineering steel plate, Furthermore, it is proposed to establish an integrated operation mechanism from design, research and development, production and application of China's nuclear engineering steel plate, which lead the development of the global industry.

Key words: nuclear engineering; steel plate; typical steel grade; new technique

极限规格特种板带钢连续淬火装备技术与应用

付天亮，王昭东，邓想涛，王国栋

（东北大学轧制技术及连轧自动化国家重点实验室，辽宁沈阳 110819）

摘 要：极限规格特种钢板是重大装备制造关键材料，热处理是其制备的瓶颈之一。本文基于高强均匀射流换热理论，提出单相倾斜射流壁面换热机制和极薄板准瞬态相变控制技术路线，获得梯度组织调控方法，发明约束淬火、非对称淬火等新技术，研发极薄 3mm 级、超宽 5m 级、特厚 300mm 级极限规格钢板热处理核心装备与示范产线，解决了极限规格钢板热处理核心技术装备缺失、复杂窄窗口工艺难以实现等瓶颈难题，极薄板淬火不平度≤4mm/m，特厚板心部冷速比传统浸入式淬火提高 1 倍，实现极限规格高强钢板高质量、高效率生产。

关键词：极限规格特种钢板；热处理；淬火；冷却路径；组织调控

Technology and Application of Continuous Quenching Equipment for Extreme Specification Special Plate

Fu Tianliang, Wang Zhaodong, Deng Xiangtao, Wang Guodong

(State Key Laboratory of Rolling and Automation, Northeastern University, Shenyang 110819, China)

Abstract: Extreme specification special steel plate is the key material for major equipment manufacturing, and heat treatment is one of the bottlenecks in its preparation. Based on the high-strength uniform jet heat transfer theory, this paper puts forward the wall heat transfer mechanism of single-phase inclined jet and the quasi transient phase transformation control technical route of very thin plate, obtains the gradient structure control method, invents new technologies such as constrained quenching and asymmetric quenching, and develops the core equipment and demonstration production line for heat treatment of extremely thin 3mm, ultra wide 5m and extra thick 300mm extreme specification steel plates, The bottleneck problems such as the lack of core technical equipment for heat treatment of limit specification steel plate and the difficulty of realizing complex narrow window process are solved. The quenching flatness of ultra thin plate is ≤4mm/m, and the cooling speed of the center of ultra thick plate is twice higher than that of traditional immersion quenching, so as to realize the high-quality and high-efficiency production of limit specification high-strength steel plate.

Key words: extreme specification special steel plate; heat treatment; quenching; cooling path; microstructure structure control

低碳环境下优特钢长材生产技术与装备

王京瑶

（中冶京诚工程技术有限公司）

摘　要： 针对目前钢铁行业存在的问题，在十四五钢铁高质量发展、碳达峰绿色发展双重目标下，对外应对国际市场贸易战，对内满足国内市场需求的环境下，总结近几年长材技术和发展，分别介绍在优特钢性能、特种高附加值材料、生产装备三个领域的应用进展。

关键词： 长材技术；优特钢；特种材料；生产装备

氮钒原子比对高强度抗震钢筋性能的影响

王卫卫，李光瀛

（钢铁研究总院冶金工艺研究所，北京　100081）

摘　要： 本文通过对含氮合金的物理化学性质及其在钢水中的溶解特征、合金加入方法对收得率的影响等理论分析与试验研究，提出了在 VN 合金基础上适当添加氮化硅锰合金的增氮新技术，同时研究了不同 V 和 N 元素含量及配比对高强度钢筋力学性能的影响。结果表明，钒氮的最佳配比不仅可以提高力学性能，而且可以节约 V 合金，提高 V 的利用率。

关键词： 钒氮合金；氮化硅锰；增氮新技术；力学性能；VN 最佳配比

The Effect of the Optimum ratio of Vanadium and Nitrogen on the Performance of High Strength Aseismic Hot Rolled Bars

Wang Weiwei, Li Guangying

(Metallurgical Technology Institute of Central Iron & Steel ResearchInstitute, Beijing 100081, China)

Abstract: The physical and chemical properties of nitrogen containing alloy and its dissolution characteristics, in molten steel alloys adding method of theoretical analysis and experimental study, new nitrogen adding technologyby added nitride silicon manganese alloy on the basis of vanadium nitrogen alloy were given in this paper. At the same time in different V and N content and ratio the influence on themechanic performance of high strength reinforced bars were also researched. The results show that the optimum ratio of vanadium and nitrogen can not only improve mechanical property, but also save the V alloy and improve the utilization of V.

Key words: vanadiumnitrogen alloy; nitrid silicon manganese alloy; new nitrogen adding technology; mechanical property; optimum ratio of vanadium and nitrogen

氢加剧腐蚀的研究以及高强韧抗氢钢的开发

石荣建[1,2]，乔利杰[1,2]，庞晓露[1,3]

（1. 北京科技大学北京材料基因工程高精尖创新中心，北京　100083；2. 北京科技大学腐蚀与防护中心，北京　100083；3. 北京科技大学材料科学与工程学院，北京　100083）

摘　要： 氢是金属材料中普遍存在的元素，然而氢加剧腐蚀却并未引起人们的关注。应力腐蚀过程中氢能进入材料内部并在裂纹尖端富集，使阳极溶解速度增加 1-2 个数量级，并降低应力腐蚀开裂门槛值；氢会增大金属点蚀敏感性，由此开启了氢对腐蚀影响研究的新领域。另一方面，高强钢的氢脆是限制其工业应用的重要瓶颈，而碳化铌（NbC）纳米析出相在提升强度的同时也能提高抗氢脆性能。本研究利用高分辨透射电镜（HRTEM）直接观察到马氏体钢中 NbC 与 Fe 基体的半共格界面处存在大量失配位错，在回火态样品中含有大量的尺寸在 5-10nm 的半共格 NbC 纳米析出相，通过 HRTEM 原子级观察得到 NbC（平均直径为 10.0 ± 3.3 nm）与 Fe 基体之间为 K-S 半共格取向关系，并且其半共格界面处存在高密度的两套失配位错。为了揭示 NbC 深氢陷阱的本质，即界面处高密度的失配位错是否是深氢陷阱的捕获位点，我们对 NbC 与 α-Fe 的 B-N 界面、K-S 半共格界面进行第一性原理计算（DFT）建模与计算，计算其与氢的结合能并观察氢的捕获位点。在 B-N 的共格界面处，氢处在 Fe 的弹性应变场，结合能为 0.06 eV（约为 6kJ/mol）；对于含有界面碳空位的 B-N 共格界面，氢处在碳空位中心，结合能为 0.33 eV，即 32kJ/mol；在透射电镜下得到的 K-S 半共格界面中，氢处在失配位错核心，结合能为 0.80 eV (77 kJ/mol)。采用热脱附谱法（TDS）实验得到的激活能为 81.8 kJ/mol。结合 DFT 和 TDS 证实了这些失配位错核心是 NbC 作为深氢陷阱的根源，并基于这种理念得到具有优异抗氢脆性能的高强马氏体钢。

关键词：氢；腐蚀；氢脆；高强韧钢；纳米相

Fundamental Principles of Hydrogen Exacerbated Metal Corrosion and Atomic-scale Investigation of Hydrogen Embrittlement Resistant High-strength Steels

Shi Rongjian[1,2], Qiao Lijie[1,2], Pang Xiaolu[1,3]

(1. Beijing Advanced Innovation Center for Materials Genome Engineering, University of Science and Technology Beijing, Beijing 100083, China; 2. Corrosion and Protection Center, University of Science and Technology Beijing, Beijing 100083, China; 3. School of Materials Science and Engineering, University of Science and Technology Beijing, Beijing 100083, China)

Abstract: Hydrogen is a common element in metals, but the aggravation of corrosion by hydrogen has not attracted attention. Pits are observed on the hydrogen-charged specimen after 6 days of immersion in 6% $FeCl_3$ solution, while no pits on the uncharged specimen even after more than 30 days of immersion, which indicates that hydrogen promotes pitting initiation and pit growth. A critical shortcoming for high-strength martensitic steels is the catastrophic hydrogen embrittlement (HE), which is not fully understood for more than a century. The precipitation of niobium carbide (NbC) is a superior approach to mitigating HE. The role of the semi-coherent interface between NbC and α-Fe on hydrogen trapping and HE resistance in high-strength tempered martensitic steel was investigated in this study. High-resolution transmission electron microscopy observations are performed to reveal the atomic-scale crystallographic orientation relationship, atomic arrangements, and associated crystalline defects in the NbC/α-Fe semi-coherent interface. We observed the Kurdjumov–Sachs orientation relationship between the NbC and α-Fe phases. Noticeably, two sets of misfit dislocations would be the deep hydrogen trapping sites, were characterized in the NbC/α-Fe semi-coherent diffuse interface. In addition, density functional theory-based first-principles calculations revealed that the deep binding energy between the NbC/α-Fe semi-coherent interface and hydrogen is 0.80 eV, which well matches the hydrogen desorption activation energy of 81.8 kJ/mol determined via thermal desorption spectroscopy experiments. These demonstrate that the nature of the deep hydrogen trapping sites of the NbC/α-Fe semi-coherent interface is the misfit dislocation core. Distinguished HE resistance was obtained and ascribed to the deep hydrogen trapping of uniformly dispersed NbC nanoprecipitates with an average diameter of 10.0 ± 3.3 nm. The strategy of deep hydrogen trapping in the NbC/α-Fe semi-coherent interface is beneficial for designing HE-resistant steels.

Key words: hydrogen; corrosion; hydrogen embrittlement; high strength steel; precipitates

彩涂产品定制化方案应用实践

郭丽涛

（山东冠洲股份有限公司）

摘　要：针对工业环境中，不同的环境特点造成涂层失效的原因进行分析，并提出定制化解决方案，延长彩涂产品的使用寿命。

彩涂产品广泛的应用于军工、家电、家装幕墙、公共建筑、大型钢铁集团、热电、化工、畜牧养殖、玻璃陶瓷、造纸、医药、电子、冷链物流等，但不同的行业所面对的环境有较大的差异。

特别是在工业建筑中，彩涂产品面对的环境也不能仅仅依靠 C1~CX 的环境级别去进行区分，工业环境中彩涂产品所面临的环境要远严苛与 CX 环境，在面对高温、高湿、酸碱性的腐蚀环境更严重，因此在 C3 环境及以下可使用 10 年寿命的彩涂产品，在此类环境中，寿命缩短到 5 年以内，特别是以化工生产行业为代表的环境中，彩涂产品的使用寿命有些在 1 年左右。

彩涂产品的失效方式基本体现在涂层的失效和基板的失效，涂层的失效主要体现在光照造成的褪色、粉化，使用一段时间后漆膜的完整性也无法保证；基板的失效当前较多的体现在切口的腐蚀、红锈的产生。但有些失效案例分析中，较多的存在涂层封闭能力和基板耐腐蚀能力的双重体现。

针对涂层、基板、涂层和基板的共同作用，山东冠洲股份有限公针对工业环境做了大量调研，对造成彩涂产品失效方式进行分析，并针对不同的环境形成了不同的方案，便于工业环境中彩涂产品的寿命得到保证。

针对彩涂产品寿命的延长，冠洲从涂层的封闭性能提升、不锈钢彩涂产品设计进行积极研究攻关，推出了冠洲全连续高封闭涂层、冠洲不锈钢彩涂系列产品。

冠洲全连续高封闭涂层在封闭性能上，以耐酸碱指标性能为典型数据，对比普通涂层的封闭能力达到 5~15 倍，广泛的应用于工业建筑领域，大大提高了建筑寿命。

冠洲不锈钢彩涂产品，形成了轧制--退火--彩涂的全流程控制模式，保证涂层与基板的附着力，主要在附着力、盐雾实验、耐酸碱实验、紫外线加速老化测试进行对比测试，不锈钢彩涂各项指标均得到大幅度的提升。并通过对比测试，不锈钢彩涂产品对比镀层彩涂产品，更需要保证附着力及涂层封闭性能，以高的产品质量推动不锈钢彩涂产品广泛应用。

关键词：彩涂；定制化

超薄镀锡板高效制造技术与应用

方　圆[1]，文　杰[1]，王永强[1]，莫志英[2]，徐海卫[3]，于　孟[1]

（1. 首钢集团有限公司技术研究院，北京　100043；2. 首钢京唐钢铁联合有限责任公司镀锡板事业部，河北唐山　063200；3. 首钢京唐钢铁联合有限责任公司技术中心，河北唐山　063200）

摘　要：为了实现镀锡板一次冷轧和二次冷轧产品厚度减薄及生产效率提升，研究了超薄镀锡板冶炼、连铸、酸连轧、连退在线二次冷轧与离线二次冷轧等关键工艺控制技术，攻克了镀锡板连铸高拉速与微小夹杂物控制、超薄带酸连轧高速卷取穿带、连退在线二次冷轧与离线二次冷轧极薄规格稳定轧制等诸多难题，最终实现了超薄厚度镀锡板产品高效生产，连铸拉速提高到 1.7m/min，夹杂物尺寸控制在 20μm 以内，酸连轧实现最薄厚度 0.11mm 高速稳定轧制，连退实现 0.14mm DR 材在线高效稳定生产，离线 DCR 最薄厚度减薄到 0.08mm。

关键词： 镀锡板；超薄；高效；连铸；一次冷轧；二次冷轧

High-efficiency Manufacturing Technology and Application of Ultra-thin Tinplate

Fang Yuan[1], Wen Jie[1], Wang Yongqiang[1], Mo Zhiying[2], Xu Haiwei[3], Yu Meng[1]

(1. Research Institute of Technology, Shougang Group Co., Ltd., Beijing 100043, China;
2. Tinplate Department, Shougang Jingtang United Iron & Steel Co., Ltd., Tangshan 063200, China;
3. Technology Center, Shougang Jingtang Iron and Steel Co., Ltd., Tangshan 063200, China)

Abstract: In order to reduce the thickness of the single and double cold-rolled tinplate products and increase the production efficiency, the smelting, continuous casting, acid continuous rolling, and continuous retreat on-line double cold-rolling and offline secondary are studied. Sub-cold rolling and other key process control technologies have overcome the high-speed drawing speed and micro-inclusion control of tin-plated continuous casting, high-speed coiling and threading of ultra-thin strip acid continuous rolling, continuous retreat online double cold rolling and offline double cold rolling. Many problems such as stable rolling of extremely thin gauges have finally realized the efficient production of ultra-thin thickness tin plate products. The continuous casting speed increased to 1.7m/min, the inclusion size was controlled within 20μm, and the acid continuous rolling achieved the thinnest thickness of 0.11mm, continuous retreat realizes the efficient and stable online production of 0.14mm DR material, and the thinnest thickness of offline DCR is reduced to 0.08mm.

Key words: tinplate; ultra-thin; high efficiency; continuous casting; single cold rolling; double cold rolling

钢板连续涂镀装备方面的最新进展

仲海峰

（中国钢研科技集团钢研工程设计有限公司，北京 100081）

摘　要：（1）高效喷淋电解装置：一种节能、清洗效果好、占地小、维护简单，更换电机不需停机的高新装备。（2）炉气净化装置：抑制锌灰和镁灰在炉鼻子内的产生，关系到合金镀层产品表面质量的关键技术之一。（3）镀后高速移动风冷装置：采用新型吊轨移动方式，用于镀后快速降低合金镀层表面温度，防止镁系合金镀层液态条件下快速氧化，形成小锌花结构，改善表面质量。（4）镀后高效冷却装置：用于合金化镀层与厚规格产品的快速降温至镀后冷却塔顶转向辊所需的温度，同时最大程度上减少了钢板的抖动。（5）新型整体移动卧式辊涂机：四辊水平涂层机是全自动化涂层机，具有很高操作灵活性，并能够在带钢的一侧或两侧精确计量涂料量。可以在前向和反向程序模式下工作，具有大涂层厚度范围功能，带有两个涂机的涂层段允许一台涂机在工作位置，而另一台处于维修状态；每个涂机可以不切断带钢而离线维修。（6）新型排渣炉鼻子：新型抽锌锌鼻子它带有漏斗及抽锌装置，能将锌鼻子内大部分的锌渣排出。（7）平整机高压水清洗装置：广泛应用在热镀锌、连退家电、汽车板生产线上，具有自主知识产权的国际领先水平。高压水枪往复横移强力清洗辊面,洁净压力无波动的清洗介质能在较高轧制力状态洗净辊面粘锌色差,提升毛化工作辊粗糙度寿命。（8）辊涂机恒温控制：辊涂机涂料恒温控制设备，该设备可高精度控制辊涂机涂料温度，大幅度降低厂家稀释剂用量，为用户节约大量资金，实现减排环保。（9）表面缺陷在线检测系统:表面缺陷在线检测系统，它广泛用于钢铁、造纸、塑料薄膜、等行业，能够实现 100%的范围对各种高速、连续生产产品的表面质量检测，为提高工厂自动化和确保质量控制提供有效的解决方案。

酒钢锌铝镁镀层板带研发进展

王　瑾

（甘肃酒钢集团，甘肃嘉峪关　735100）

摘　要：本文对酒钢锌铝镁合金镀层板带（SCS）的研发进展、质量特性、应用等方面进行了综合阐述。酒钢SCS的镀层合金成分为11%Al-3%Mg-Si-Ni-Re类型，其显微组织为多相组织，由富铝相、$MgZn_2$、Zn-Al-$MgZn_2$三元共晶组织、Mg2Si等构成。在腐蚀环境下，除Mg、Al、Zn对基体金属的牺牲阳极保护作用外，Mg的添加使切口耐腐蚀性能更加优越，Si、Ni、稀土的添加，优化了镀层的附着性和耐蚀性能。与镀锌、镀铝锌镀层相比，酒钢SCS在弯折、切口等工艺状态下具有更加优异的耐蚀性能，各种试验结果显示，SCS在中性、碱性盐雾下的腐蚀速度是镀锌的1/10，镀层硬度约是镀锌硬度的2-3倍，摩擦损失约是镀锌失重的1/10，抗风沙能力约是镀锌的4倍。作为预镀锌产品，镀层质量138g/m^{2}加工构件的盐雾耐蚀数据优于热浸700g/m^{2}后镀锌处理的加工构件。因酒钢SCS产品具有优异表面耐蚀性和切口耐蚀性，可以增加构建服役寿命，减少后期维护成本、降低构件因腐蚀带来的失效几率，从而提高构件使用安全性能，并且节约制造成本，环境友好，绿色环保。未来酒钢SCS产品在替代厚镀锌热浸锌产品，替代部分不锈钢、铝合金材料的外观或内饰件等方向具有极大的市场潜力。

关键词：连续热镀锌；锌铝镁镀层；切边耐蚀性

Research and Development of Zinc-Aluminum-Magnesium Alloy-Coated Steel Strip of JISCO

Wang Jin

(JISCO, Jiayuguan 735100, China)

Abstract: In this paper, the development, quality characteristics and applications of Zn-Al-Mg alloy-coated strip (SCS) of JISCO are reviewed. The coating alloy composition of JISCO®SCS is 11%Al-3%Mg-Si-Ni-Re type, the micro-structure of the coating is multiphase structure, which is composed of aluminum rich phase, $MgZn_2$, Zn-Al-$MgZn_2$ ternary eutectic structure, Mg_2Si, etc. In the corrosive environment, In addition to sacrificial anode protection of substrate by Mg, Al and Zn in corrosive environment, the addition of Mg makes the notch corrosion resistance better, Si, Ni and RE can optimize the adhesion and corrosion resistance of the coating. Compared with Zinc coating and Aluminum-Zinc coating, JISCO SCS coating has better corrosion resistance under bending and cutting conditions. Various tests show that the corrosion rate of SCS coating under neutral and alkaline salt fog is 1/10 of that of zinc, the hardness of SCS is about 2-3 times of that of zinc coating, and the friction loss is about 1/10 of the weight loss of zinc.The sand resistance is about 4 times that of zinc coating. As a pre-galvanized product, the salt spray corrosion resistance data of SCS with coating quality of 138g/m^2 is better than that of galvanized after hot-dip 700g/m^2. JISCO®SCS products have excellent surface corrosion resistance and notch corrosion resistance, which can increase the service life of construction, reduce the later maintenance cost input, reduce the failure probability of components due to corrosion, improve the safety performance of components, and save manufacturing costs. JISCO®SCS products are environmentally friendly, in the future, JISCO®SCS will have great market potential in replacing thick galvanized hot dip zinc products and replacing part of stainless steel and aluminum alloy material appearance or interior parts.

Key words: continuous hot dip galvanizing; Zn-Al-Mg alloy-coated strip; notch corrosion resistance

金属制品行业发展概况

毛海波

（中钢集团郑州金属制品研究院有限公司，河南 郑州 450001）

摘 要：金属制品是线材深加工产品，其应用领域十分广泛，普遍用于建筑、交通、冶金、矿山、机械、化工、航空航天、国计民生。本报告一方面将按照产品类型介绍主要金属制品产品的发展概况，一方面介绍金属制品的原材料线材/盘条的发展概况。

关键词：金属制品；线材；发展概况

超高强钢辊弯成形回弹机理及控制研究

韩 飞，尚伟勇，孟伊帆，牛丽丽，李荣健，操召兵

（北方工业大学机械与材料工程学院，北京 100144）

摘 要：辊弯成形技术能够生产高强度、性能佳、高精度的型钢产品，并且具有高效、节能、省材等特点，在许多领域得到广泛运用。但由于超高强钢成形难度大，易发生回弹，是影响辊弯成形精度的重要原因。本文采用试验及有限元仿真的方法，对回弹规律产生机理及控制进行了研究，首先，对Q&P980超高强钢板材进行1步及2步循环拉伸实验，对其循环加载性能进行准确表征，研究发现，实验屈服轨迹呈外凸性，部分屈服轨迹不对称，屈服轨迹随变形程度的增加向外扩大，弹性模量随应变的增加而降低，降低到一定程度后趋于平缓；其次，设计了三组道次数不同的V型件辊弯成形试验，研究发现，在单道次辊弯成形中，随着弯曲角度地增加，回弹角度呈先升后降的趋势，对于多道次辊弯成形，增多道次数有利于提高塑性变形量的积累、减小弹性应变，增大弯角增量使板材弯角处厚度减薄，促进角部材料流动，有效减小了回弹量；并对V型件弯曲部位进行了微观组织分析，发现随着弯曲角度的增加，残余奥氏体在马氏体板条边界处发生相变，使板条状马氏体及块状马氏体含量增多，通过相变强化机制改善了Q&P高强钢板材的力学性能；然后设计了两种轧辊设计方案，分析不同设计方案下板材在成形过程中的等效塑性应变和等效应力的分布情况，提出控制回弹的成形工艺方法，轧辊优化设计后的成形件边部纵向应变值降低并趋于平缓，可有效控制板材的回弹问题；最后将闭环反馈控制的概念引入到辊弯成形回弹控制过程中，建立了辊弯成形回弹闭环控制模型，可根据补偿算法预测最优轧辊形状，实验结果表明，所提出的补偿算法可有效控制辊弯成形回弹问题。

关键词：超高强钢；辊弯成形；回弹控制；有限元仿真

Research on Springback Mechanism and Control of Ultra High Strength Steel in Roll Forming Process

Han Fei, Shang Weiyong, Meng Yifan, Niu Lili, Li Rongjian, Cao Zhaobing

(School of Mechanical and Materials Engineering,
North China University of Technology, Beijing 100144, China)

Abstract: Roll forming technology can obtain better products which have high strength, better performance and high precision and has been widely used in many fields. However, because of the difficulty of forming ultra-high strength steel, it is prone to springback which is an important factor influencing the roll forming accuracy. In this paper, the generation mechanism and control of springback law are studied by using the method of experiment and finite element simulation. Firstly, the one-step and two-step cyclic tensile tests of Q&P980 ultra-high strength steel plate were carried out to accurately describe its cyclic loading performance. It was found that the experimental yield trajectory was convex and part of the yield trajectory was asymmetric. The yield trajectory expanded outward with the increase of deformation degree, and the elastic modulus decreased with the increase of strain, and tended to be gentle after decreasing to a certain degree. Secondly, three groups of v-shaped parts roll forming tests with different number of passes were designed, the results showed that for single stand roll forming, the springback angle increased first and then decreased with increasing forming angle, for multiple stands roll forming, increasing the number of stands was conducive to increasing the accumulation of plastic deformation and reducing the elastic strain, corner increment that sheet thickness was thinned, corner place promoted the material flow at the corner, effectively reduced the amount of springback. The microstructure of the forming part of the v-shaped steel was analyzed. It was found that with the increase of forming angle, the residual austenite phase transformation occurred at the boundary of martensite lath, which increased the content of lath martensite and massive martensite. The mechanical properties of Q&P high strength steel sheet were improved by the mechanism of phase transformation strengthening. Then designed two kinds of design of roll, sheet under different design analysis in the process of forming the equivalent plastic strain and equivalent stress distribution, put forward the method to control the springback of forming technology, roll forming a longitudinal strain of edge after optimization design values to reduce and flatten out, sheet could effectively control the springback problem. Finally, the concept of closed-loop feedback control was introduced into the springback control process of roll forming, and the closed-loop control model of roll forming springback was established, which could predict the optimal roll shape according to the compensation algorithm. Experimental results showed that the proposed compensation algorithm could effectively control the springback problem of roll forming.

Key words: ultra high strength steel; roll forming; springback control; finite element simulation

真空制坯轧制 TC4 钛合金/EH690 海工钢复合板的界面组织及力学性能研究

王丰睿，郭　胜，骆宗安，谢广明

（东北大学　轧制技术及连轧自动化国家重点实验室，辽宁沈阳　110819）

摘　要： 钛/钢复合板兼具高耐蚀、低成本及高强韧性的优点，在海工领域应用十分广泛，但目前关于钛合金复合板的研究报道较少。本文利用真空轧制复合法制备 TC4 钛合金/EH690 海工钢复合板，并研究了轧制温度及金属铌中间层对复合界面相组成、元素扩散和力学性能的影响。结果表明，随着轧制温度升高，界面两侧 Ti、Fe、C 等元素相互扩散加剧，界面生成了明显的 TiC 层，在 950℃轧制温度下界面出现 Fe_2Ti 化合物，极大影响了界面的结合强度，865℃时界面剪切强度最高，也仅为 155MPa。为改善界面结合性能，在界面引入铌中间层。结果表明，铌中间层有效地隔绝了 Ti、Fe 和 C 元素的扩散，抑制了 Ti-Fe 脆性相和 TiC 相的生成，在界面形成了 Nb-Ti 固溶体，界面结合性能优异，断口具有塑性断裂特征，平均剪切强度高达 287MPa。

关键词： 钛/钢复合板；真空轧制复合；扩散；界面；中间层

Study on Interfacial Microstructure and Mechanical Properties of TC4 Titanium Alloy/EH690 Marine Steel Clad Plate by Vacuum Rolling

Wang Fengrui, Guo Sheng, Luo Zongan, Xie Guangming

(State Key Laboratory of Rolling and Automation, Northeastern University, Shenyang 110819, China)

Abstract: Titanium/steel clad plates have the advantages of high corrosion resistance, low cost and high strength and toughness, which are widely used in the field of marine engineering. However, there are few reports about titanium alloy clad plates. In this paper, TC4 titanium alloy /EH690 marine steel clad plates were prepared by vacuum rolling cladding. The effects of rolling temperature and Nb interlayer on the interfacial phase composition, element diffusion and mechanical properties of the clad plates were studied. The results show that with the increase of rolling temperature, Ti, Fe, C and other elements on both sides of the interface diffusion intensifies, and an obvious TiC layer is formed at the interface. Fe_2Ti compound appears at the interface at 950℃, which greatly affects the bonding strength of the interface. The shear strength of the interface at 865℃ is the highest, which is only 155MPa. In order to improve the interfacial bonding performance, Nb interlayer was introduced into the titanium steel interface. The results show that the Nb interlayer effectively isolates the diffusion of Ti, Fe and C elements, inhibits the formation of Ti-Fe brittle phase and TiC phase, and forms Nb-Ti solid solution at the interface. The interface bonding property is excellent, and the fracture has the characteristics of plastic fracture, the average shear strength is up to 287MPa.

Key words: titaniumclad steel plate; vacuum rolling cladding; diffusion; interface; interlayer

双碳背景下钢材深加工产业的减碳路径

张光明，董馨浍，陈　剑，于治民，杨梅梅

（冶金工业信息标准研究院冶金信息研究所，北京　100006）

摘　要：在“碳达峰”和“碳中和”的双碳背景下，我国钢铁工业面临巨大的减碳压力，钢材深加工产业作为钢铁工业与制造业间的界面产业，做好钢材深加工产业的减碳工作对我国钢铁工业和制造业的低碳转型升级具有重要意义。本文结合我国钢材深加工产业发展的现状，论述了双碳背景下钢材深加工产业面临的机遇及挑战，并指出科技创新、产业协同发展、减量化设计、信息化和智能化建设、低碳标准体系等因素是钢材深加工产业减碳的主要路径；钢材深加工产业应发挥界面产业的优势，做好与钢铁产业和制造业之间的产业链接，实现产业低碳转型升级，助力我国双碳目标的实现。

关键词：双碳；钢材深加工；减碳路径；碳减排

Carbon Reduction Path for Steel Deep Processing Industry Under the Background of Carbon Emission Peak and Carbon Neutrality

Zhang Guangming, Dong Xinhui, Chen Jian, Yu Zhimin, Yang Meimei

(China Metallurgical Information and Standardization Institute, Metallurgical Information Research Department, Beijing 100006, China)

Abstract: Under the background of "carbon peak" and "carbon neutral", carbon reduction is a huge pressure for iron and steel industry in China. As the interface industry between the iron and steel industry and manufacturing industry, it's of great significance to low carbon transformation and upgrading of iron and steel industry and manufacturing industry in China if we do a good job of carbon reduction in steel deep processing industry. The present situation of steel deep processing industry development in China, opportunities and challenges of steel deep processing industry facing are discussed in this paper, while points out that factors such as scientific and technological innovation, the coordinated development of industry, reduction design, the construction of informatization and intelligent, low carbon standard system are the main path of carbon reduction for steel deep processing industry. To realize transformation and upgrading of low carbon, Steel deep processing industry should give a full play to the superiority of the interface industry and link the iron and steel industry and manufacturing industry well, then it will help achieve the goal of "carbon peak" and "carbon neutral" in China.

Key words: carbon peak and neutrality; steel deep processing; carbon reduction path; carbon reduction

汽车用超高强钢板先进成形技术

米振莉，苏　岚，唐　荻

（北京科技大学，北京　100083）

摘　要：中国汽车工业正处于高速发展时期，虽然产销量进入瓶颈期，但仍有较大的发展空间，使用超高强钢替代传统钢材将是今后汽车板发展的主流。发展汽车用超高强钢板深加工技术要结合下游汽车行业的需求，解决钢铁材料在应用层面的问题，从而促进钢铁生产流程的质量提升。本研究通过介绍热金属气胀成形技术、超高强钢辊弯成形技术、汽车零部件的辊冲技术及高强钢管件的液压胀形技术等先进成形技术，打破传统超高强钢研究只针对汽车用轻量化材料的研发和设计理念，通过材料的强度结合成形后的性能组合设计实现零部件优化设计的新思路。零部件的几何形状、力学性能分布等都会影响其使用性能，因此，将零件的力学行为、微观组织及其分布、几何形状等综合考虑，从而设计出所需要零件的形状和力学性能。此外，报告还将针对超高强钢板深加工过程的成形回弹、延迟开裂、焊接等相关研究成果进行阐述和发表。

关键词：汽车用超高强钢；深加工技术；热金属气胀成形；辊弯成形；辊冲成形；液压胀形

Advanced Forming Technology of Ultra-high-strength Steel Sheet for Automobiles

Mi Zhenli, Su Lan, Tang Di

(University of Science and Technology Beijing, Beijing 100083, China)

Abstract: China's automobile industry is developing rapidly. Although the production and sales of automobiles have reached a bottleneck, its development can not to be underestimated. The use of ultra-high-strength steel to replace traditional steel will be the mainstream for the development of automotive sheets in the future. The development of deep processing techniques of ultra-high-strength steel plates for automobiles must be combined with the needs of the downstream automotive industry. The application problems of ultra-high-strength steel plates should be solved in order to further promote the quality improvement of the steel production process. The traditional ultra-high-strength steel research is only aimed at the R&D and design concept of lightweight materials for automobiles. In comparison with the traditional research of lightweight materials, the current research focuses on advanced forming technologies such as hot metal gas forming technology, ultra-high-strength steel roll forming technology, automotive parts roll-stamping technology, and high-strength steel tube hydroforming technology. A new idea of optimal design of parts is realized through the combined design of the strength of the material and the performance after forming. The shape of parts and the local mechanical properties will affect the service performance of the whole part. Therefore, the mechanical behavior, microstructure distribution, geometric shape of the parts are comprehensively considered to design the parts. In addition, the report will also introduce related research results on forming springback, delayed cracking, and welding during the deep processing procedure of ultra-high-strength steel plates.

Key words: ultra-high-strength steel for automobiles; deep processing technique; hot metal gas forming; roll forming; roll-stamping; hydroforming

先进钢铁材料深加工过程的夹杂物冶金及智能化设计

牟望重[1]，杨永刚[1,2]，王 勇[1]，Park Joo Hyun[1,3]，米振莉[2]

（1. 瑞典皇家工学院，材料科学与工程系，斯德哥尔摩瑞典 SE10044；2. 北京科技大学，工程技术研究院，北京 100083；3. 汉阳大学，材料学科与化学工程系，韩国 15588）

摘 要：先进钢铁材料的深加工过程涉及拉拔、冲压、冷弯等传统深加工工艺，以及激光拼焊、液压成形、气胀成形、热成形、3D 打印等先进深加工工艺技术，对钢铁材料最终产品的性能有着极其重要的影响。此工作对在瑞典皇家工学院开展的先进钢铁材料的夹杂物冶金以及智能化设计两方面内容进行系统的总结。夹杂物冶金方面，以中锰钢热轧过程中的不同种类的夹杂物的变形为例，阐明氧化物、硫化物、复合夹杂物等的变形规律及机理研究。同时简要介绍如不锈钢焊接、3D 打印过程夹杂物的演变规律等。智能设计部分介绍应用机器学习模型预测不同钢铁材料的组织演变，如形变诱导马氏体转变温度及组织分数等。通过此报告旨在推广夹杂物冶金及智能化材料设计在金属材料深加工过程的应用。

关键词：先进钢铁材料；金属深加工；夹杂物冶金；机器学习；智能设计

Inclusion Metallurgy and Intelligent Material Design of Processing of Advanced Steels

Mu Wangzhong[1], Yang Yonggang[1,2], Wang Yong[2], Park Joo Hyun[1,3], Mi Zhenli[2]

(1. Department of Materials Science and Engineering, KTH Royal Institute of Technology, Stockholm, Sweden SE-100 44; 2. Institute of Engineering Technology, University of Science and Technology Beijing, Beijing 100083, China; 3. Department of Materials Science and Chemical Engineering, Hanyang University, Ansan, Korea 15588)

Abstract: Control the processing of advanced steels play an important role on the mechanical properties of final products. The processing includes the conventional manufacturing methods, e.g. drawing, stamping, cold bending, etc. as well as the state-of-the-art process, e.g. laser welding, hydro-forming, inflatable forming, thermal forming, and additive manufacturing (3D print), etc. Here, a brief summary of research activities focusing on i) inclusion metallurgy and ii) intelligent material design was presented. For the case of inclusion metallurgy, one detailed example of inclusion deformation behaviors in the medium Mn steels during hot rolling is presented, the deformation behaviors of oxide, sulfide and complex inclusions is described and the related mechanisms are discussed. Besides, other topics focusing on inclusion transformations during welding and additive manufacturing will be shorted summarized. Finally, the intelligent material design activities utilizing the machine learning method is presented. An example of predicting deformation martensite transformations in austenitic stainless steels is given.

Key words: advanced steels; material processing; inclusion metallurgy; machine learning; intelligent design

第三代先进高强钢的机会与挑战

王　利

（宝钢中央研究院，汽车用钢开发与应用技术国家重点实验室，上海　201000）

摘　要：先进高强度钢，特别是兼有高强度和高延伸率的第三代先进高强钢是近年来的发展方向。第三代先进高强钢的出现，为汽车复杂结构零件的设计提供了除了热冲压外另外一种全新的技术解决方案。亚稳奥氏体微观组织的精细调控，是第三代先进高强度钢的关键技术。淬火延性钢（QP）是第三代先进高强钢中产业化最早的钢种，已经成功地应用于汽车工业中近 10 年了。本文简要概述了第三代先进高强钢不同技术路线的进展，结合“双碳”战略，通过具体案例分享了第三代先进高强度钢特别是 QP 钢在汽车上的应用。结合工业生产和应用情况，分析了第三代先进高强度钢面临的技术挑战。

关键词：先进高强度钢；第三代；亚稳奥氏体；挑战

Opportunities and Challenges for Third-generation Advanced High-strength Steels

Wang Li

(Baosteel Central Research Institute, State Key Laboratory of Automotive Steel Development and Application Technology, Shanghai 201000, China)

Abstract: Advanced high-strength steels, especially the third generation of advanced high-strength steels with both high strength and excellent elongation, have been the focus of the development direction in recent years. The third generation advanced high-strength steels provide a new technical solution for the design of complex structural components of automobiles, in addition to the hot press forming steels. Fine tuning of meta-stable austenite microstructure is the key technology for the third generation of advanced high-strength steels. Quench and partitioning steel (QP) is the earliest industrialized steel grade in the third generation of advanced high-strength steels and has been successfully used in the automotive industry for nearly 10 years. This paper briefly outlines the progress of different technological routes of third-generation advanced high-strength steels and shares the application of third-generation advanced high-strength steels, especially QP steels, in automotive applications through specific cases in the context of the "double carbon" strategy. The technical challenges of third-generation advanced high-strength steels will be discussed in the context of industrial production and applications.

Key words: advanced high strength steel; third generation; meta-stable austenite; technical challenges

异构中锰 TRIP 钢制备及变形行为的研究

张 宇，丁 桦

（东北大学材料科学与工程学院，辽宁沈阳 110819）

摘 要：基于异构组织设计的思想，对 Fe-7Mn-1Al-0.2C（wt.%）冷轧实验钢进行部分奥氏体化逆相变退火（partial austenite reversed transformation，PART），制备出同时具有等轴和片层状铁素体/奥氏体的异构组织。与具有均匀组织的两相区退火（intercritical annealing，IA）以及奥氏体逆相变（austenite reversed transformation，ART）实验钢相比，PART 实验钢在保持优异伸长率的前提下，呈现出更高的屈服强度和抗拉强度，且同时具有较高且持续的应变硬化，这可归因为实验钢的组织异构和成分异构。PART 实验钢的高屈服强度主要源于体积分数约为 80%的超细晶等轴组织。准原位 EBSD 观察表明，随着应变的增加，不同形貌的组织间出现了应变配分，片层状组织比等轴组织承担了更多的变形，这种由组织异构引起的力学不相容性导致 PART 实验钢产生了很强的异构变形诱发（hetero-deformation induced, HDI）硬化效果，因此 PART 实验钢在拥有高抗拉强度的同时也保持了优异的伸长率。此外，经 PART 处理的实验钢由于扩散动力学过程的差异，其片层状奥氏体的 Mn 元素浓度高于等轴奥氏体，因此 PART 实验钢中存在成分异构。具有不同稳定性的奥氏体在实验钢不同的变形阶段持续发生马氏体相变而产生 TRIP 效应，为实验钢提供更为持久均匀的应变硬化，从而有助于提高其力学性能。

关键词：中锰钢；HDI 硬化；异构组织；应变配分

An Investigation in the Preparation of a Heterostructured Medium Mn TRIP Steel and Its Deformation Behaviors

Zhang Yu, Ding Hua

(School of Materials Science and Engineering, Northeastern University, Shenyang 110819, China)

Abstract: A mixture of UFG granular and lath-typed ferrite/austenite microstructure was prepared by partial austenite reversed transformation (PART) in the cold rolled medium Mn steel with a composition of Fe-7Mn-1Al-0.2C (wt.%). Compared with samples prepared by intercritical annealing (IA) and austenite reversed transformation (ART), the PART sample presented higher yield strength and ultimate tensile strength as well as sustainable strain hardening ability without sacrificing the ductility, which is attributed to the microstructural and compositional heterogeneity. The higher yield strength was originated from the UFG granular grains with a predominate volume fraction of about 80%. Ex-situ EBSD observation showed that strain partitioning occurred between the microstructures with different morphology with strain increasing since the lath-typed microstructure accommodated more plastic deformation in comparison to the granular microstructure. Such mechanical incompatibility induced by microstructural heterogeneity produced significant hetero-deformation induced hardening (HDI hardening), leading to higher ultimate tensile strength as well as adequate total elongation. The compositional heterogeneity of Mn concentration was observed in PART sample. Mn enrichment in lath-typed austenite was higher than the one in granular austenite due to the different diffusive kinetics of Mn element in the PART process. The compositional heterogeneity of Mn element gave rise to the formation of austenite with different stability and thus a sustainable martensite transformation in different stages of plastic deformation, leading to an improvement in mechanical properties of the material.

Key words: medium Mn steels; HDI hardening; heterostructure; strain partitioning

M3 组织钢变形与断裂行为研究

王存宇[1]，韩　硕[1,2]，周峰峦[1]，常　颖[2]，曹文全[1]，董　瀚[1,3]

（1. 钢铁研究总院 特殊钢研究院，北京　100081；2. 大连理工大学汽车工程学院，
辽宁大连　116024；3. 上海大学 材料科学与工程学院，上海　200444）

摘　要：具有多相、亚稳、多尺度的 M3 组织钢，如 Q&P 钢和逆相变中锰钢，利用高强度基体和亚稳奥氏体的 TRIP 效应可获得高强度高塑性的力学性能，提升材料的碰撞安全性和成形能力，是汽车钢重要发展方向。M3 组织钢在微观组织上与第一代汽车钢（如 DP 钢）具有显著差别，尤其是亚稳奥氏体相的大量存在，在加工、成形和服役过程中的性能表现以及组织性能的演变具有重要的研究价值。本文研究了 M3 组织钢在剪切变形、拉伸变形和疲劳载荷条件下的变形和断裂行为，结果表明，亚稳奥氏体转变行为导致剪切边加工硬化程度更大，塑性损失受落料间隙影响大；拉伸变形条件下，M3 组织钢以界面开裂为主，晶界大尺寸奥氏体的转变是变形初期奥氏体转变量较大的主要原因；低应力水平下，循环载荷对 M3 组织钢力学性能影响不明显，而高应力条件下将发生大量奥氏体转变行为。

关键词：亚稳奥氏体；剪切；变形；断裂；组织演变

Study on Deformation and Fracture Behavior of M3 Structure Steel

Wang Cunyu[1], Han Shuo[1,2], Zhou Fengluan[1],
Chang Ying[2], Cao Wenquan[1], Dong Han[1,3]

(1. Special Steel Institute, Central Iron & Steel Research Institute, Beijing 100081, China;
2. School of Automotive Engineering, Dalian University of Technology, Dalian 116024, China;
3. School of Material Science and Engineering, Shanghai University, Shanghai 200444, China)

Abstract: Multi-phase, meta stable, multi-scale M3 structure steels, such as Q&P steel and ATR-annealing medium manganese steel, use high-strength matrix and the TRIP effect of meta-stable austenite to obtain mechanical properties of high-strength and good ductility, and improve the collision safety and forming ability of the steel, they are important development direction of automotive steel. The microstructure of M3 steel is significantly different from the first generation of automotive steels (such as DP steel), especially the exist of many metastable austenite phases. The performance during processing, forming and service, and the evolution of microstructure and properties are worth studying. This paper studies the deformation and fracture behavior of M3 structure steel under cutting deformation, tensile deformation, and fatigue loading. The results show that the metastable austenite transformation behavior leads to a greater degree of work hardening at the cutting edge, and the ductility loss is greatly affected by the clearance; It 's suggesting that the main reason for a massive volume fraction decrease of austenite during the early stages of deformation lies in the rapid phase transformation of austenite near the grain boundaries, As the simulations suggest, the preferential crack initiation site is the interface between phases. Along with the initiation of a new crack is the butterfly-shaped distributed stress zone that favors the martensite transformation on the lateral sides instead of the crack front. The mechanical properties of the steel remain stable under the cyclic loading of median fatigue limit, and a large amount of austenite transformation behavior will occur under high stress conditions.

Key words: metastable austenite; cutting; deformation; fracture; phase transformation

首钢汽车钢研发进展

滕华湘，刘华赛，蒋光锐，韩 赟

（首钢集团有限公司技术研究院，北京 100043）

摘 要：在“碳达峰、碳中和”的“双碳”驱动下，首钢汽车用钢着力于在轻量化、资源减量化、产品长寿化等方面进行产品的开发和研究工作。为适应汽车制造环节低碳节能发展的需求，汽车车身涂装工艺趋向于更加环保的免中涂工艺，基于此，首钢成功开发了免中涂汽车外板的全流程工艺控制及表面结构控制技术，并批量应用。在车身轻量化用材发展方面，首钢开发了增强成形性双相钢（DH 钢）系列产品，DH 钢通过在钢基体中引入一定比例的残余奥氏体，延伸率得以明显提升，全局成形能力得到改善。针对局部成形特性需求和更好的碰撞吸能效果，首钢开发了 980MPa 和 1180MPa 级别的高扩孔高强钢，扩孔率在 45%以上。对于先进高强钢的涂镀性能改善方面，首钢采用了镀前的纳米涂层技术，更好地保障了高强钢的耐蚀性能。此外，首钢正在开发低铝低镁系列锌铝镁涂层外板及高强钢，在提升汽车用钢的耐蚀性能的同时，减少锌资源的消耗。本报告对上述高强钢产品的设计原理，工艺流程及服役性能做出具体阐述。

关键词：汽车钢；研发进展

The Development of High-performance Automobile Steels in Shougang Group

Teng Huaxiang, Liu Huasai, Jiang Guangrui, Han yun

(Research Institute of Shougang Group, Beijing 100043, China)

Abstract: Driven by the “carbon neutrality” and “emission peak” control policy, Shougang Group focuses on the research and development in the fields of light weight, resource reduction and long duration of the steel products. In order to meet the demand of low-carbon and energy-saving development during automobile manufacturing, the auto body coating process tends to be more environment friendly. Shougang Group has successfully developed the whole process control and surface structure control technology for the middle coat free outer plate and realized the batch production. In the development of lightweight body materials, Shougang Group has developed a series of formability enhanced dual-phase steel (DH steel) products. By introducing a certain percentage of residual austenite into the steel matrix, the elongation of DH steel is significantly raised, and the global formability is improved. For local forming characteristics and better collision energy absorption, Shougang Group has developed 980 MPa and 1180 MP grade AHSS with hole expansion rate above 45%. For the improvement of coating performance of AHSS, Shougang Group has adopted the nano-coating technology before plating to improve the corrosion resistance. In addition, a low-aluminum and low-magnesium series of zinc-aluminum-magnesium coated outer plates from AHSS have been developed to improve the corrosion resistance and reduce zinc consumption. The product features and the process flow will be introduced during the talk.

Key words: automotive steels; research and development

汽车钢板断裂行为的多尺度研究

郭晓菲[1,2]，张 梅[1]，赵洪山[1]，Bleck Wolfgang[2]，董 瀚[1]

（1. 上海大学材料科学与工程学院，上海 200444；
2. 亚琛工业大学钢铁研究所，德国 D-52072）

摘 要：汽车用钢的研发沿着高强韧轻量化路径不断前进，在近二十年的历程中，第二、三代高强汽车钢板在生产工艺及力学性能的优化方面日趋成熟。新一代高强汽车板的组织结构多通过细晶化，多相化设计提高强度，并通过引入不稳态奥氏体结构的 TRIP 效应，以及孪晶硬化的 TWIP 效应改进材料塑性。但多相及 TRIP/TWIP 效应的引入也带来一系列服役方面的问题，材料的服役性能，如滞后断裂，疲劳，远低于材料表观力学性能，影响产品的使用安全。因而，通过多尺度表征与测试解析新材料断裂机制与断裂行为，从而进一步改进材料组织结构设计，并设定材料安全应用边界，对高强韧新型汽车钢的推广有重要的指导意义。本研究以高锰奥氏体钢、中锰钢为例，通过对材料断裂行为从微观，介观到宏观的多尺度表征，揭示材料在不同尺度的断裂行为。在微观表征上采用扫描隧道显微（ECCI）和电子背散射衍射（EBSD）相结合的方式，表征高锰奥氏体钢中位错和孪晶行为对裂纹萌生的影响。在介观到宏观尺度表征上，采用结合 GOM 数字图像关联技术（DIC）及断口区域的电子背散射衍射组织分析，探讨局部形变对组织结构及裂纹萌生的影响。进而探索建立材料组织结构与材料断裂和服役安全性能的联系，为现行高强汽车钢板的应用及服役性能评价提供参考。

关键词：高强汽车钢；微观组织；断裂行为

Multi-scale Characterization of the Fracture Behaviors in Automotive Sheet Steels

Guo Xiaofei[1,2], Zhang Mei[1], Zhao Hongshan[1], Bleck Wolfgang[2], Dong Han[1]

(1. School of Material Science and Engineering, Shanghai University, Shanghai 200444, China;
2. Steel Institute, RWTH Aachen University, Aachen, Germany D-52072)

Abstract: The research and development of automotive steels advance along the routine of high strength and light weight design. In the past decades, the second and third generation high strength automotive sheet steels have experienced great optimization in production process and mechanical properties. The multiphase, ultrafine microstructure design generally improves the material strength, while the introduction of TRIP/TWIP effect of meta-unstable austenite phase further improves the material ductility. However, the introduction of secondary phases or microstructure defects also induces a series of in-service problems, such as delayed fracture, fatigue failure. The in-service mechanical properties are much lower than their apparent mechanical properties, which affects their in-service safety. Therefore, it is important to understand the fracture mechanism and fracture behavior of the new generation of materials through multi-scale characterization and testing. It helps to improve the material microstructure design from one side, and from another side, to set the boundary condition for the safe application of new automotive steels. The current study takes high manganese austenitic steels and medium manganese steels as examples and reveals the fracture behaviors at different scales through multi-scale characterization of fracture behaviors. A combination of electron channeling contrast imaging (ECCI) analysis and electron backscatter diffraction (EBSD) analysis is used in the microscopic level to characterize the effect of dislocation and twinning behaviors on crack initiation in high manganese austenitic steels. In the mesoscopic to macroscopic scale level, a combination of GOM digital image correlation (DIC) and EBSD analysis of the fractured region is used to investigate the effect of local deformation on the crack initiation. The connections between microstructure, fracture behavior and in-service performance will be discussed. It tries to provide references for the evaluation of in-service properties of current high-strength automotive sheet steels.

Key words: high strength steels; microstructure; fracture behaviors

超高强度汽车钢复杂加载路径试验与成形仿真

常 颖[1]，王宝堂[1]，詹 华[2]，张 军[2]，陈新力[2]，李晓东[1]，余树洲[1]

（1. 大连理工大学汽车工程学院，辽宁大连 116024；
2. 马鞍山钢铁股份有限公司技术中心，安徽马鞍山 243000）

摘 要：超高强度汽车钢的工业化应用是实现汽车轻量化的核心技术，当前，冷冲压汽车钢板的强度水平不断刷新纪录，但强度级别超过 780MPa（如 980MPa 和 1180MPa）的冷冲压汽车钢仅占实际应用比例的 5%左右，应用技术的研究明显落后于钢板研发和生产，尤其是钢板成形过程中成形精度问题，成为困扰各大汽车厂、零部件厂和钢厂的瓶颈，是阻碍超高强度钢板推广的重要原因之一。在解决冲压回弹的过程中，最为棘手的问题是难以实现回弹的准确预测。本项目组首先从材料试验和宏观力学性能出发，创新性地开展了综合考虑材料各向异性，弹性模量衰减和包辛格行为的材料模型开发和回弹预测研究，使得 DP 钢和 QP 钢等主流超高强度汽车钢满足一般零件回弹预测误差不高于 5%，复杂零件回弹预测误差不高于 15%的工业要求，为模具的优化和补偿提供了技术支持；同时，通过控制成形工艺参数和拉延筋的设置，开发了回弹的控制策略。本项目组的研究成果为推动超高强度汽车钢的顺利应用具有重要的意义。

关键词：汽车钢；超高强度；回弹预测；包辛格行为；弹性模量衰减

Complex Loading Path Experiment and Forming Simulation of Ultra-high Strength Automobile Steels

Chang Ying[1], Wang Baotang[1], Zhan Hua[2], Zhang Jun[2],
Chen Xinli[2], Li Xiaodong[1], Yu Shuzhou[1]

（1. School of Automotive Engineering, Dalian University of Technology, Dalian 116024, China;
2. Technology center, Magang (Group) Holding Co., Ltd., Maanshan 243003, China）

Abstract: The industrial application of ultra-high strength automobile steel is one of the core technologies to realize automobile lightweight. At present, the strength level of cold-stamping automobile steel plate is constantly breaking the record, but the cold-stamping automobile steel with a strength level of more than 780MPa (such as 980MPa and 1180MPa) accounts for only about 5% of the actual application proportion. It indicates that the research on application technology obviously lags behind the research, development and production of steel plate. In particular, the forming accuracy problem in the forming process of steel plate has become a bottleneck in the automobile factories, parts factories and steel factories, which is one of the important reasons hindering the promotion of ultra-high strength steel plate. Springback prediction is the most difficult topic in the process of solving stamping springback. In our study, starting from the material experiment and macro mechanical properties, the development of material model and springback prediction research are conducted by considering material anisotropy, elastic modulus attenuation and Bauschinger behavior innovatively. The springback prediction error of mainstream ultra-high strength automobile steel parts such as DP steel and QP steel can meet the industrial requirements of no more than 5% for common parts and no more than 15% for complex parts. It provides technical support for die optimization and compensation. At the same time, the springback control strategy is developed by controlling the forming process parameters and the drawbead setting. The research results of the project team are of great significance to promote the further application of ultra-high strength automobile steels.

Key words: automobile steels; ultra-high strength; springback prediction; bauschinger behavior; elastic modulus attenuation

高品质抗湿硫化氢腐蚀管线钢板的生产实践

孙宪进

（中信特钢集团江阴兴澄特钢钢铁有限公司特板研究所，江苏江阴 214400）

摘 要：抗湿硫化氢腐蚀管线钢板主要用于输送含有硫化氢腐蚀介质的油气管道工程中，目前主要市场需求在中东地区，对材料的纯净度、质量稳定性要求很高。目前我国很多钢铁企业均能够生产，但我国在国际市场上的份额较小，认可度较低，主要市场仍旧被国外钢厂占有。兴澄特钢中厚板产线自 2010 年投产以来，坚持以特种钢板为发展目标，在设备配置、产品研发到市场开拓方面进行不断探索，完成 38mm 厚度抗 HIC 海底管线钢的开发和产业化，抗 HIC 管线钢应用客户覆盖中东和印度等主要国际知名钢管企业，累计出口抗 HIC 钢板 10 万吨以上，本文重点分享该类钢板的生产实践，供大家参考。

在湿硫化氢腐蚀环境下，目前实验评价的主要失效形式有三种，氢致开裂 Hydrogen Induced Cracking（HIC）、抗硫化物应力开裂 Sulfide Stress Corrosion Cracking（SSCC）和应力导向氢致开裂 Stress Oriented Hydrogen Induced Cracking（SOHIC），其 SSCC 和 SOHIC 均在应力加载状态下进行，SSCC 失效位置主要在焊接接头处，SOHIC 主要在环焊缝位置处，与钢管的焊接工艺相关性更大，HIC 在母材及焊缝处均有发生，母材相关性更大。因此，从钢板生产实践的角度，我们重点讨论与 HIC 质量稳定性相关的控制问题，通过对典型失效案例和成功案例的分析，探讨提高性能稳定性的控制方法，同时对标国际同行，查找自身存在的不足以及未来的改进方向。

关键词：抗湿硫化氢腐蚀；管线钢板；氢致开裂

钒微合金化钢的技术进展与应用

杨才福

（钢铁研究总院，北京 100081）

摘 要：介绍了钒微合金化技术的最新进展以及钒钢的开发与应用情况。氮是含钒钢中有效的合金元素，含钒钢中增氮，优化了钒在钢中的析出，显著提高沉淀强化效果。采用钒氮微合金化设计，配合适当的轧制工艺，促进 V(C,N) 在奥氏体中析出，起到了晶内铁素体形核核心作用，实现了含钒钢的晶粒细化。最新的研究成果表明钒微合金化可以提高双相钢、贝氏体钢、相变诱导塑性钢、孪晶诱导塑性钢、热成型马氏体钢等汽车用先进高强度钢的强度并改善使用性能，显示出良好的应用前景。钒氮微合金化技术在中国高强度钢筋、高强度型钢、非调质钢、薄板坯连铸连轧高强度带钢等产品中获得广泛应用，大大促进了中国钒微合金化钢的发展。

关键词：钒微合金化；析出强化；晶内铁素体；先进高强度钢；高强度钢筋

Recent Development and Applications of Vanadium Microalloying Technology

Yang Caifu

(Central Iron and Steel Research Institute, Beijing 100081, China)

Abstract: The recent developments of V microalloying technology and its applications in HSLA steels were reviewed. Enhanced-N in V microalloyed steels promote precipitation of fine V(C,N) particles, and improves markedly precipitation strengthening effectiveness of vanadium in steel. V-N process can be used effectively to refine ferrite grain size by the nucleation of intragranular ferrite promoted by V(C,N) precipitates in austenite. The latest research results show the potential benefits of V microalloying technology to advanced high strength steels (AHSS) for the automotive sector including DP steels, bainitic steels, TRIP steels and TWIP steels, etc. The wide applications of V-N process in high strength rebars, section steels, forging steels and thin slab direct rolling strips, etc., had greatly promoted the rapid development of V microalloyed steels in China.

Key words: V microalloying; precipitation strengthening; intragranular ferrite; advanced high strength steel (AHSS); high strength(HS) rebar

高性能止裂钢组织调控技术与工业实践

王 华[1,3]，尚成嘉[2,3]，严 玲[1,3]，韩 鹏[1]，李秀程[2]

（1. 鞍钢股份有限公司钢铁研究院，辽宁鞍山 114009；
2. 北京科技大学钢铁共性技术协同创新中心，北京 100083；
3. 海洋装备用金属材料及其应用国家重点实验室，辽宁鞍山 114009）

摘 要：根据国际船级社协会的要求，超大型集装箱船的建造必须使用止裂钢板。超高强度特厚钢板的止裂性源于材料的显微结构设计。本文将从多相组织塑性变形行为出发，介绍由铁素体+贝氏体组成的多相组织低合金钢中软硬相在变形过程中位错塞积、微孔洞形核及其在不同组织中的演变特征等方面的基础研究，论述多相组织提高止裂韧性的原理，并且通过工业实践结果验证利用多相组织设计大幅度提升厚板钢止裂性能的技术路线。

关键词：止裂钢；多相组织；位错滑移；微孔洞；止裂韧性

Multi-phase Microstructure Design of High-performance Crack Arresting Heavy Plate Steel and Its Industry Practices

Wang Hua[1,3], Shang Chengjia[2,3], Yan Ling[1,3], Han Peng[1], Li Xiucheng[2]

(1. Iron & Steel Research Institutes of ANSTEEL Group Corporation, Anshan 114009, China;
2. Collaborative Innovation Center of Steel Technology of USTB, Beijing 100083, China;
3. State Key Laboratory of Metal Materials for Marine Equipment and Application, Anshan 114009, China)

Abstract: According to the requirements of the International Association of Classification Society, crack arrest plate steel must be used in the construction of super large container ships. The crack arrest of ultra-high strength heavy plate steel plate is derived from the microstructure design of the material. Based on the plastic deformation behavior of multi-phase structure, this paper introduces the basic researches on the dislocation accumulation in soft and hard phases, micropore nucleation and evolution characteristics of different phases during deformation in multi-phase low alloy steel composed of ferrite and bainite, and discusses the principle of improving crack arrest toughness by multi-phase microstructure design. The technical route of using multi-phase design to greatly improve the crack arrest performance of heavy plate steel is also verified by industrial practice results.

Key words: crack arrest steel; multi-phase microstructure; dislocation slip; micropore; crack arrest toughness

极薄取向电工钢制造与应用

刘宝志

（威丰稀土电磁材料，内蒙古包头　014000）

摘　要：极薄取向电工钢是支撑我国现代化建设的关键性材料，是电力、电子工业中一种重要的软磁材料，也是国家重要的战略资源，属国家紧缺类产品。主要依赖进口，价格高且限制国防应用。包头威丰通过解决无底层原料制备问题、性能一致性控制困难、涂液及涂敷工艺控制等工艺难题，掌握了 0.05mm、0.08mm、0.10mm 的取向电工钢极薄带产品全套工艺技术，工艺、设备及原辅材料技术，搭建了产品特殊服役工况性能测试平台，研制的极薄取向电工钢产品已经成功应用于高新技术装备及雷达等特殊领域。

关键词：材料；极薄取向电工钢

Manufacturing and Application of Extremely Thin Orientation Electrical Steel

Liu Baozhi

(Weifeng Rare Earth Electromagnetic, Baotou 014000, China)

Abstract: Extremely thin oriented electrical steel is the key material supporting China's modernization construction, is an important soft magnetic material in power and electronics industry, is also an important strategic resource of the country, is a national shortage of products. It relies heavily on imports, is expensive and limited to national defense applications. Baotou Weifeng has mastered the whole set of technology, process, equipment and raw and auxiliary material technology of 0.05mm, 0.08mm and 0.10mm oriented electrical steel thin strip products by solving the problems of raw material preparation without substrate, difficulty in performance consistency control, coating solution and coating process control, and built a performance test platform for products in special service conditions. The extremely thin oriented electrical steel products developed have been successfully applied in special fields such as high-tech equipment and radar.

Key words: materials; extremely thin oriented electrical steel

“双碳”背景下中国取向硅钢需求趋势分析

吴树建

（宝钢股份，上海　200000）

摘　要：讲述中国取向硅钢自主化对国家电力建设的支撑成果，并从十四五国家电力规划分析取向硅钢用材结构及相对于“十三五”的变化。

关键词：国家电力规划；取向硅钢

Analysis of the Demand Trend of Oriented Silicon Steel in China under “Carbon Neutrality” and “Peak Carbon Dioxide Emissions” Background

Wu Shujian

(Baoshan Iron & Steel, Shanghai 200000, China)

Abstract: This paper describes the supporting results of China's independent oriented silicon steel for national electric power construction, and analyzes the material structure of oriented silicon steel from the 14th five-year national electric power planning and the changes compared with the 13th five-year plan.

Key words: national power planning; oriented silicon steel

不锈钢产品及工艺技术发展的思考

李建民

（太原钢铁（集团）有限公司，山西太原　030003）

摘　要：近年来，全球不锈钢市场仍呈增长态势。1980~2019 年，全球不锈钢复合增长率 5.33%，同时期全球碳钢复合增长率仅为 2.49%。同时中国不锈钢产业的地位日益重要。2014 年后中国不锈钢产量、消费量的全球占比稳定在 50%以上。2010~2020 年中国不锈钢产量复合增长率 10.35%。但中国的铬、镍等战略资源自然禀赋不足，上游原料高度依赖进口，需要引导企业海外布局。对比发达经济体，尤其是第二产业与我国相近的韩国，其不锈钢人均消费量 33kg，中国不锈钢人均消费量 18kg 仍有提升空间。“十四五”中国不锈钢表观消费量预测平均增速为 8%。

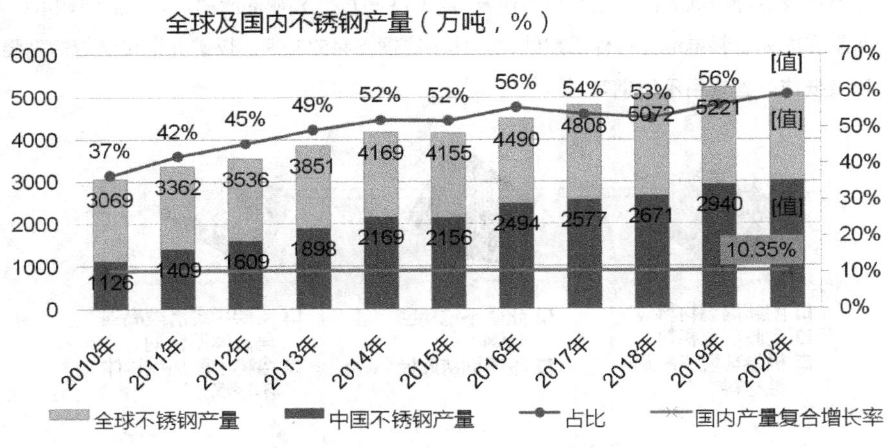

图 1　全球及国内不锈钢产量

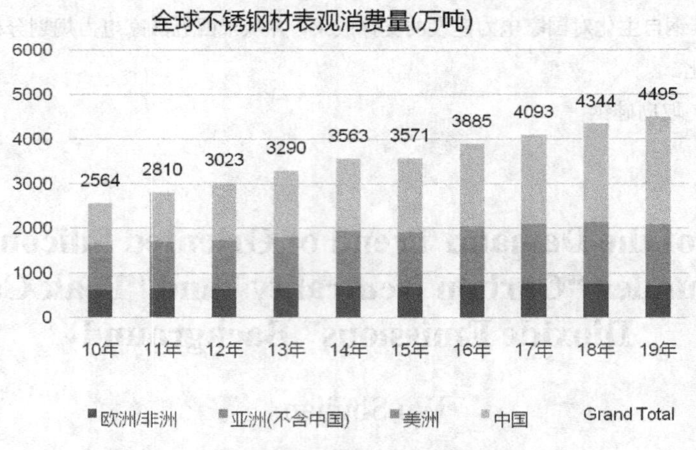

图 2　全球不锈钢表观消费量

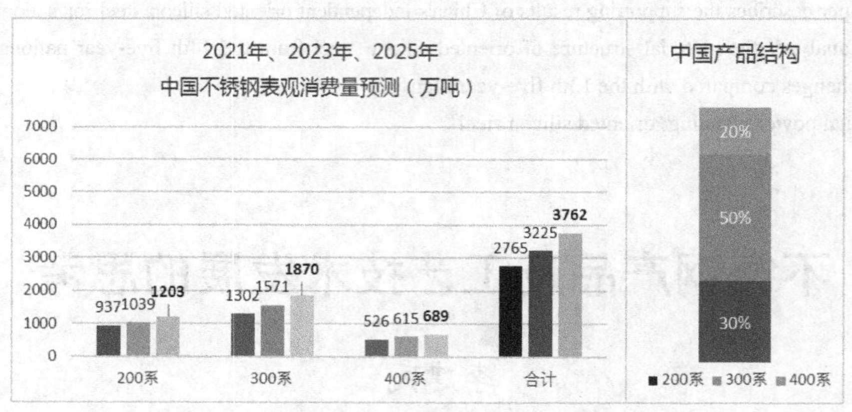

图 3　中国不锈钢表观消费量的预测

经过一百多年的发展，不锈钢已经成为特殊钢中第一大钢种。未来不锈钢产品将围绕以下三个方面开展品种研发。（1）极端苛刻环境下的材料研发与应用，超高温环境下如超超临界锅炉管、光伏多晶硅反应器用铁镍基特厚板、国际热核聚变实验堆（ITER 项目）计划用不锈钢等，超低温环境下如–196℃低温空气动力试验装置用不锈钢、储氢容器用不锈钢板材等，以及高耐蚀材料如超级双相不锈钢、超级奥氏体不锈钢以及镍基耐蚀合金等。（2）制造难度较大的品种研发，如宽幅超薄不锈精密带钢和第四代核电快堆用 316H 特厚板等极限规格产品，和含稀土宽钢带铁铬铝合金板带材以及装饰用高表面质量材料的研发等。（3）更广泛领域的应用，如新能源领域的氢能源电堆双极板用不锈精密带钢和固体燃料电池连接件用超纯铁素体不锈钢 S44537 等，煤矿井下传送带的高强支架 TSZ410 代替传统碳钢，以及低成本建筑用不锈钢等。

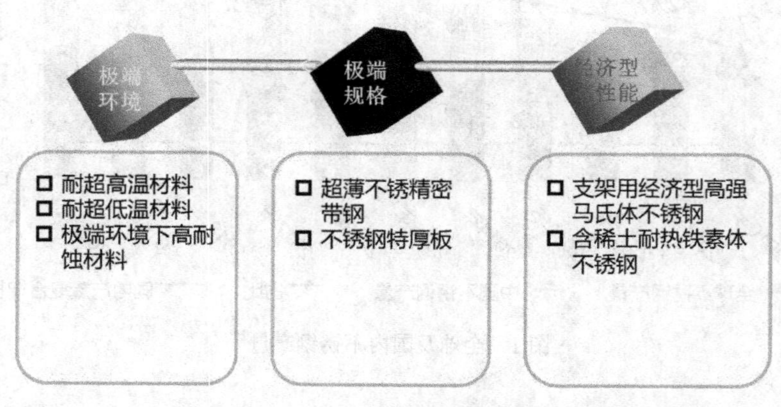

图 4　未来不锈钢品种发展的趋势

围绕未来不锈钢产品研发的趋势，不锈钢工艺技术的发展将围绕低成本原料制备、高洁净度和极限元素、高表面质量、酸洗与高效轧制、低碳绿色发展等五个方面持续推进。

图5 不锈钢工艺技术的发展趋势

我国超级双相不锈钢的研发进展

宋志刚

（钢铁研究总院特殊钢研究院，北京 100081）

摘 要：结合“十三五”国家重点研发计划、海洋油气开采领域材料研发相关工作，重点介绍分享我国超级双相不锈钢材料的分类、需求、技术现状和近年来超级双相不锈钢板、管产品研发的进展。

关键词：超级双相不锈钢；海洋油气；钢板；钢管

石化新能源用钢技术发展展望

张国信

（中石化广州工程有限公司，广东广州 510620）

摘 要：主要从石化行业领域用氢能、LNG、关键装备用材料未来技术发展，技术要求，开发方向等方面进行分析阐述，以期带动国产钢铁、装备制造行业的产业发展和技术进步。

关键词：能源石化；钢铁材料；不锈钢；装备制造

不锈钢的组织性能调控及质量提升技术研究

宋仁伯

（北京科技大学材料科学与工程学院，北京 100083）

摘 要：本文主要介绍了高品质不锈钢钢丝的质量控制及工艺技术研究，以及低成本不锈钢的开发及应用基础研究。

关键词：不锈钢；钢丝；组织调控；低成本

给夹杂物穿上"铌铠甲"—利用铌微合金化改善双相不锈钢耐蚀性的新方法

姜周华

（东北大学，辽宁沈阳　110167）

摘　要： 针对夹杂物易引起腐蚀失效的问题，本研究创新性提出了给夹杂物穿上"铌铠甲"的方法，即将铌微合金化思想应用于双相不锈钢，促使富CrMo高耐蚀含铌Z相以夹杂物为核心析出，如同"铠甲"的含铌相将夹杂物严密包裹，有效抵御了苛刻环境的腐蚀。该方法在系列双相不锈钢中均适用，有望成为该类材料腐蚀防护领域一项关键共性技术。

关键词： 双相不锈钢；夹杂物；铌微合金化；腐蚀

酒钢超纯铁素体不锈钢产品研发及应用

李具仓

（酒钢钢研院，甘肃嘉峪关　735100）

摘　要： 超纯铁素体克服了传统铁素体不锈钢的脆性转变温度高、室温低温韧性差、缺口敏感性高、耐晶间腐蚀性能差、加工性能差、焊接性能差等缺点，同时在生产成本、耐应力腐蚀、氯离子腐蚀、晶间腐蚀等方面优于奥氏体不锈钢，是奥氏体不锈钢的良好替代材料；酒钢成功解决了超纯铁素体不锈钢水口结瘤、铸坯横裂、热轧黏辊、表面色差、山形压痕、砂金等关键技术难题和不同钢种在不同工序的其他质量问题，开发出优质的409L/M、436L/D、439、441、442D、443、444/444E、445J1/J2等超纯铁素体不锈钢产品,广泛应用于汽车排气系统、汽车装饰条、电梯、建筑装饰、厨具、家电和五金制品等领域。

关键词： 超纯铁素体不锈钢；成型性能；耐腐蚀性能；应用领域

Research and Application of JISCO Ultra-pure Ferritic Stainless Steel Products

Li Jucang

(Research Institute, JISCO, Jiayuguan 735100, China)

Abstract: Ultra-pure ferritic stainless steel overcomes high brittle transition temperature, poor room temperature, low temperature toughness, high gap sensitivity, poor intercrystal corrosion resistance, poor processing performance, poor welding performance of traditional ferritic stainless steel, and is superior to austenite stainless steel in production cost, stress corrosion resistance, resistance to chloride corrosion and Intercrystal corrosion, which is a good alternative to austenitic stainless steel; JISCO has successfully solved the key technical problems such as ultra-pure ferritic stainless steel water outlet nodules, cast billet transverse crack, hot rolling adhesive roll, surface color difference, mountain indentation, sand gold and other quality

problems of different steel types in different processes. Developed high quality 409L/M, 436L/D, 439, 441, 442D, 443, 444/444E, 445J1/J2 and other ultra-pure ferritic stainless steel products, widely used in the automotive exhaust system, automotive trim strips, elevators, architectural decoration, kitchenware, home appliances, hardware products and other fields.
Key words: ultra-pure ferritic stainless steel; forming performance; corrosion resistance; application field

不锈钢冷轧制新工艺及技术发展

张子强，李旭东

（中冶南方工程技术有限公司，湖北武汉 430023）

摘 要：本文通过对不锈钢冷轧制原料、工艺特点和我国目前冷轧制工艺现状分析，在传统不锈钢黑卷轧制、白卷连轧以及可逆轧制工艺基础上，提出了不锈钢黑卷强力连轧+白卷强力连轧+直接可逆轧制（仅超薄带材使用）新工艺，并进行了轧制 0.15mm（常规热轧料）、0.175mm（炉卷热轧料）工业性实践和分析，得出了在该新工艺下，在不进行额外中间退火的基础上，可以实现现代不锈钢冷轧市场产品需求全覆盖，为做强、做大我国不锈钢冷轧建设及生产提供了新建议、新思路。
关键词：不锈钢；冷轧制；新工艺

New Process and Technology Development of Stainless Steel Cold Rolling

Zhang Ziqiang, Li Xudong

(WISDRI Engineering & Research Incorporation Limited, Wuhan 430023, Hubei, China)

Abstract: This essay analyzes the raw material of stainless steel cold rolling, process characteristics and domestic process status in China. Based on the process of the conventional stainless steel cold rolling using hot-rolled strips, tandem rolling with strips after annealing and pickling and reversible rolling, a new process has been put forward which contains “high deformation tandem rolling of hot-rolled strips” + “high deformation tandem rolling of strips after annealing and pickling” + “immediate reversible rolling (only for ultrathin strips)”. The industrial practice and analysis of strips reduced to 0.15mm (using conventional hot rolled strips) and strips reduced to 0.175mm (using hot rolled strips by Steckel mill) are also carried out in this essay. And this indicates that all ranges of products demanded by markets will be covered by following this novel production process, without extra intermediate annealing, which provides a new method for improvement of stainless steel cold rolling in China.
Key words: stainless steel; cold rolling; new process

合金元素 Mo 在超级奥氏体不锈钢界面偏析行为及 B 的影响

韩培德

（太原理工大学，山西太原 030024）

摘　要：超级奥氏体不锈钢具有优异的机械和耐蚀性能，但σ相析出敏感，导致热塑性差、热加工难度大等问题，是制约其生产的关键。本文分析硼对超奥钢σ相回溶、析出及耐蚀性能的影响。具体包括：B调控奥氏体不锈钢析出相的溶解、析出行为；合金元素在奥氏体不锈钢晶界占位偏析倾向，及B对Mo等元素不锈钢晶界偏析的影响；合金元素在奥氏体不锈钢界面占位倾向，及B、稀土对Cr、Mo等元素在表-界面体系偏析的影响。
关键词：超级奥氏体不锈钢；第二相；钼；硼；界面偏析

面向绿色环保的交通用先进高强不锈钢开发及应用

毕洪运

（中国宝武中央研究院，上海　200126）

摘　要：介绍了中国宝武面向汽车交通领域广泛应用的低生命周期成本高强不锈钢产品设计及应用。着重介绍氮合金化系列先进高强不锈钢的性能、组织特征、耐腐蚀性等，分享了高强不锈钢在汽车交通领域的典型应用案例。
关键词：汽车交通；高强不锈钢；氮合金化；组织；应用

Fe基非晶/纳米晶软磁复合材料的制备与性能

惠希东[1]，沈宁宁[1]，李育洛[1]，李福山[2]

（1. 新金属材料国家重点实验室，北京科技大学，北京　100000；
2. 材料科学与工程学院，郑州大学，河南郑州　450001）

摘　要：本文系统研究了利用水氧化法制备具有磁性（Fe_2O_3和Fe_3O_4）包覆层的Fe-P-C-B-Si-Mo非晶态与FINEMET纳米晶态的软磁复合材料。分别采用气雾化和球磨法制备了Fe-P-C-B-Si-Mo非晶球形粉末和FINEMET片状粉末，并用冷压法制成了的磁环。将非晶与纳米晶态的软磁复合材料分别在不同温度下真空热处理以消除内应力，提高磁环的软磁性能。研究发现，水氧化法制备的软磁复合材料具有优异的高频性能，对磁导率和磁饱和强度的减弱作用小。在B_m=0.05T，f=100kHz条件下，对气雾化制备的非晶软磁复合材料与采用球磨与水氧化结合制备的FINEMET软磁复合材料的损耗进行了测试，所有软磁复合材料的磁导率均在300kHz内保持恒定。本研究采用的水氧化法具有无污染，工艺简单等优点，适合于工业生产，具有广阔的应用前景。
关键词：软磁复合材料；Fe基非晶/纳米晶合金；水氧化；磁损耗

新型软磁性钴基块体非晶合金的制备及其性能研究

张　伟

（大连理工大学材料科学与工程学院，辽宁大连　116024）

摘　要：软磁性Co基非晶合金具有低矫顽力（H_c）、高磁导率、低铁损、接近于零的磁致伸缩系数等特性，尤其是其高频软磁性能极佳，在电力、电子产业及 5G 通信技术领域显示出越来越重要的应用价值。但 Co 基合金的玻璃形成能力（GFA）相对较低，且饱和磁感应强度（B_s）不高，限制了它们的应用范围。通过添加前过渡金属元素（ETM）合金化，可有效提高 Co 基合金的 GFA，研发出了一系列 Co 基块体非晶合金（BMGs）体系。但由于这些 BMGs 含有较多量 ETM 而使它们 B_s 明显降低，甚至失去磁性。最近，我们制备出了高 B_s Co 基 Co-Fe-B-Si-P BMGs，其 B_s、H_c 和有效磁导率分别在 1.02-1.24T、0.8-4.6A/m 和 12700-18500 之间，并具有优异的力学性能。此外，我们还研发出的具有优异软磁性能的不含 ETM 的 Co 基(Co, Fe)-RE-B BMGs，其 B_s 值达到 1.17T；并探讨合金成分与 GFA 及磁性能的相关性。

关键词： Co 基非晶合金；合金化；玻璃形成能力；软磁性能；饱和磁感应强度

Synthesis and Properties of New Soft Magnetic Co-based Bulk Metallic Glasses

Zhang Wei

(School of Materials Science and Engineering, Dalian University of Technology, Dalian 116024, China)

Abstract: Co-based metallic glasses possess high permeability and low core loss in high frequency as well as near-zero magnetostriction, which exhibit good promise for applications in the high-frequency fields. However, the glass-forming ability (GFA) and saturation magnetic flux density (B_s) of Co-based alloys are inferior to that of Fe-based alloys, which increases the production difficulty and limits the form and size of the products. Enhancing GFA of the Co-based alloys is highly needed for preparing bulk metallic glasses (BMGs) or glassy powders by atomization. Most soft magnetic Co-based BMGs contain a quantity of nonmagnetic early transition metals (ETMs), which play an important role in the enhancement of GFA, whereas they severely deteriorate the B_s at the same time. Recently, we successfully developed Co-based (Co, Fe)-B-Si-P BMGs, which possess outstanding soft magnetic properties with high B_s of 1.02-1.24T and good mechanical properties. In addition, we also developed new soft magnetic Co-based (Co, Fe)-Y-B BMGs without ETMs, which exhibit high B_s of 1.17T. The correlation of the alloy composition with GFA and magnetic properties was discussed.

Key words: Co-based metallic glasses; alloying; Glass-forming ability; soft magnetic property; saturation magnetic flux density

$Fe_{78}Si_9B_{13}$ 非晶合金熔体的原子扩散研究

张 博，胡金亮

（松山湖材料实验室，广东东莞 523808）

摘 要： 液态金属的原子扩散系数是理解液体金属动力学行为以及动力学和结构之间的关联的关键参数。同时，准确可靠的扩散数据也是液态金属凝固过程模拟的必备参数。然而，由于实验测量上的困难，可靠的实验数据非常缺乏。对于高熔点的金属熔体，扩散数据更加稀少，几乎是空白。采用自主研发的多层滑动剪切技术，实验获得了 $Fe_{80}Si_{20}$ 和 $Fe_{80}B_{20}$ 二元合金熔体的互扩散数据。三元及三元以上多组元互扩散系数的求解也是公认的难题。基于 Onsager 关系，我们自主提出了一种高效的求解多组元扩散系数的新算法。采用该算法，利用多层滑动剪切技术的实验数据，我们首次实验获得了典型的非晶合金 $Fe_{78}Si_9B_{13}$ 熔体的互扩散系数。二元和三元合金熔体扩散实验结果揭示，B 和 Si 组元的添加降低了合金熔体的原子互扩散，这和 $Fe_{78}Si_9B_{13}$ 体系的良好非晶形成能力相吻合。

关键词： 液态金属；原子扩散；Fe 基非晶合金

基于掺杂选区激光熔化技术制备金属构件性能研究

王　丽，王胜海

（山东大学（威海）机电与信息工程学院，山东威海　264209）

摘　要：选区激光熔化技术（Selective laser melting，SLM）作为一种极具前景的增材制造技术，备受国内外学者和世界知名企业的关注。随着打印机设备技术的飞速发展和不断完善，打印原材料金属粉末反而成为当前制约 3D 打印技术进一步发展的主要障碍。目前，用于 3D 打印的金属粉末主要是由国内外制粉商提供的通用粉末，与传统铸造合金相比较，品种稀少，且金属粉末性能局限性大。具备特定性能的金属粉末定制困难且价格昂贵。在此，我们基于合金掺杂改性原理，通过掺杂合理调控合金粉末成分，采用 SLM 工艺制备出高强高韧的高温合金和高熵合金、高延展性的钛合金、耐腐蚀性优异和生物相容性良好的不锈钢、高强度高耐磨铝合金等，同时采用超声辅助电化学钻削技术对 3D 打印构件进行处理，来满足航空航天、海洋生物、石油化工等不同应用领域的特殊性能需求。

关键词：选区激光熔化；耐腐蚀；高温合金；高熵合金

The Investigation of Metal Fabricated by Selective Laser Melting based on Doping Principle

Wang Li, Wang Shenghai

(School of Mechanical and Electrical Engineering,
Shandong University (Weihai), Weihai 264209, China)

Abstract: As a promising additive manufacturing technology, selective laser melting (SLM) has attracted the attention of scholars and world-famous enterprises. With the rapid development and continuous improvement of printer equipment technology, the raw metal powder has become the main obstacle restricting the further development of 3D printing technology. At present, the metal powder used for 3D printing is mainly provided by powder manufacturers. Compared with traditional casting alloys, the category of printing metal power is few, andthe properties of metal powderarelimited. The metal powder with special components is difficult to customize and expensive. Here, based on the principle of alloy doping modification, we use SLM process to prepare high-strength and high toughness superalloys and high entropy alloys, high ductility titanium alloys, stainless steel with excellent corrosion resistance and biocompatibility, high-strength and high wear-resistant aluminum alloys, etc. to meet the requirements of aerospace, marine organisms and petrochemical industry Special performance requirements in different application fields. At the same time, the post processing technology such as ultrasonic assisted electrochemical drilling technology has beenused to treat the SLM components.

Key words: selective laser melting; corrosion resistance; superalloy; high entropy alloy

快速退火对 Fe-Si-B-P-Cu 纳米晶合金薄带的磁性能与弯折韧性的影响

常春涛[1]，孟　洋[2]，赵成亮[1]，白雪垠[2]，逄淑杰[2]，张　涛[2]

（1. 东莞理工学院，广东 东莞 523808；2. 北京航空航天大学，北京 100191）

摘　要：由 Fe 基非晶合金前驱体带材退火后形成纳米晶合金可提高其饱和磁感应强度、优化其软磁性能，但会导致显著的脆性，不利于其在电力电子器件等领域的广泛应用。因此，发展可获得兼具高饱和磁感应强度（Bs）、低矫顽力（Hc）及良好韧性的 Fe 基纳米晶合金的热处理工艺具有重要意义。已有研究表明，通过对 Fe 基非晶合金带材进行快速退火热处理，可以使非晶基体中析出均匀细小的纳米晶，从而改善 Fe 基纳米晶合金的软磁性能，但其对于纳米晶合金韧性的影响及其机理尚不明确。本工作针对快速退火及常规退火热处理对 Fe-Si-B-P-Cu 非晶合金带材的微观结构、磁性能、弯折韧性的影响进行了系统的对比研究，并探讨了相关机理。研究结果表明，与常规退火相比，快速退火的最佳热处理温度区间显著向高温区移动，其较高的退火温度及较短的退火时间有利于在非晶基体上析出更多、更均匀细小的 a-Fe 纳米晶，在提高其磁性能的同时，改善了纳米晶合金薄带的弯折韧性。对 Fe-Si-B-P-Cu 非晶合金带材在 520℃快速退火 30 秒获得的纳米晶合金兼具高 Bs（1.82T）、低 Hc（6.2A/m）及良好的韧性，具有良好的应用前景。

关键词：铁基非晶合金；纳米晶合金；弯折韧性；软磁性能

Nanocrystalline $Fe_{83}Si_4B_{10}P_2Cu_1$ Ribbons with Improved Soft Magnetic Properties and Bendability Prepared Via Rapid Annealing of the Amorphous Precursor

Chang Chuntao[1], Meng Yang[2], Zhao Chengliang Zhao[1], Bai Xueyin[2], Pang Shujie[2], Zhang Tao[2]

（1. Dongguan, 523808, China; 2. Beijing 100191, China）

Abstract: In this study, the microstructure, soft magnetic properties and bendability of nanocrystalline $Fe_{83}Si_4B_{10}P_2Cu_1$ ribbons acquired via rapid annealing of the amorphous precursor were investigated in comparison with those of the ribbons prepared through the conventional annealing. It was found that the appropriate rapid annealing with higher temperatures and shorter heating time is beneficial to the higher crystallinity and the formation of the α-Fe nanocrystallites with higher size-uniformity in the resultant nanocrystalline alloys, leading to the good magnetic properties and ribbon bendability simultaneously. The $Fe_{83}Si_4B_{10}P_2Cu_1$ nanocrystalline ribbon with a high saturation magnetization of 1.82 T and low coercivity of 6.2 A/m in combination with improved bend1ability, which are desirable for the applications in the electronic components with complex geometries, was successfully fabricated by rapidly annealing the amorphous precursor at 520 ℃ for 30 s.

Key words: Fe-based alloy; nanocrystalline alloy; bendability; soft magnetic property

T_A 处金属玻璃熔体动力学转折的内在特征

胡丽娜，管鹏飞

(1. 山东大学材料科学与工程学院，山东济南 250061；2. 北京科学计算研究中心，北京 100083)

摘 要：金属玻璃熔体中在 T_A 处普遍存在 non-Arrhenius 转折。实验证实，T_A 约等于 2Tg,说明玻璃转变的某些内在信息在 T_A 处就有所体现。对 12 种金属玻璃体系从高温熔体到玻璃转变附近较宽温度区间内的动力学性质进行了模拟研究，发现了弛豫时间与温度关系的普适定量描述，确认了金属玻璃液体中弛豫时间与动力学不均匀性的普遍关联。T_A 明确为不同金属玻璃液体中动力学不均匀性大小相等的温度，对应 $\alpha_{2,\max} \approx 0.2$，且特征慢原子的平均空间连接性 k 在 T_A 接近 3。后者体现了 T_A 处的结构拓扑特征，T_A 以下的 non-Arrhenius 归因于 k>3 的原子通过发展稳定的拓扑网络进行的协同运动。上述结果说明了从高温源头预测金属玻璃液体动力学性质的重要意义：只要测得 T_A 及其对应的弛豫时间 τ_A 可预测从高温熔体至接近玻璃转变这一降温过程中弛豫时间的演变情况；而对动力学不均匀性，仅需确定 T_A（或 τ_A）即可预测其演变情况。

关键词：金属玻璃熔体；动力学不均匀性；弛豫

基于非晶合金实现可见光全波谱范围的调控

陈 娜

（清华大学材料学院）

摘 要：非晶合金具有类似于液体的短程有序、长程无序的原子密堆结构，通常也被称为液态金属。作为一种结构独特的金属材料，非晶合金表现出高强度、高硬度、耐腐蚀和高耐磨等优异性能。另一方面，由于非晶合金本质上是金属材料，因此并不像传统的氧化物玻璃具有透光特性，而是呈现出特有的金属光泽。如果可以把不透明的非晶合金变得像氧化物玻璃一样透明，那么非晶合金将会变身成集光学、力学、化学、物理性能于一体的“超能”材料。

我们通过在不透明的 Co-Fe-Ta-B 非晶合金中掺入氧诱导金属-半导体转变制备出一系列成分、结构与光学性能精准可调的非晶合金衍生物，其电学性能涵盖了金属、半导体和绝缘体的导电特性。随氧的不断加入，非晶合金逐渐变得透明，结构从单相非晶合金过渡到纳米非晶和纳米非晶氧化物的双相纳米玻璃，最后转变为完全透明的单相非晶氧化物。大范围精准调控非晶合金的光学性能是非晶合金面向光电薄膜器件等潜在应用亟需解决的关键瓶颈之一。复合高反射率的非晶合金与高透过率的非晶氧化物薄膜形成了双层膜结构，通过调控透光层非晶氧化物薄膜的厚度，利用薄膜干涉效应，实现了该复合膜在可见光波段的全色谱可调。该复合彩色功能膜同时具有高达 9GPa 的硬度，可镀于纳玻璃、石英玻璃、柔性基体 PET 或不锈钢基材上，可分别呈现出彩虹色或高亮度的可见光单色。此外，还可直接将透光层非晶氧化物镀在本征硅、n 型或 p 型硅片上实现可见光波段任意颜色的调控。由于该非晶氧化物为 p 型半导体，与 n 型硅或 p 型硅集成后可分别表现出 p-n 异质结开关功能或电阻功能。

该研究结果为研制全色谱可调且具有电学、磁学性能的大面积彩色多功能薄膜提供了一条新的途径。研究提出的设计理念可能适用于很多其他的非晶合金体系，通过这种氧调控制备出涵盖金属、半导体和绝缘体所有类型的非晶材料，形成被氧“点亮”的多彩非晶合金及其衍生物的材料家族，用于光电薄膜器件的研制

相分离 Zr-Cu-Fe-Al 大块非晶合金晶化动力学的原位中子和同步辐射衍射研究

兰　司，刘思楠

（南京理工大学材料科学与工程学院/格莱特研究院，江苏南京　210000）

摘　要：非晶合金的结构性能调控是领域热点难题。通过向 Zr-Cu-Al 三元大块非晶合金中引入微量 Fe 元素，可以诱导非晶纳米尺度相分离结构的产生。相分离 Zr-Cu-Fe-Al 非晶合金独特的结构异质性为调控其性能提供了可能。通过中子和同步辐射衍射的原位无损检测手段研究了相分离的 Zr-Cu-Fe-Al 块状金属玻璃的热稳定性和结晶动力学，利用中子和同步辐射衍射元素衬度差异，发现在相分离非晶合金在两个晶化阶段中，富铜相和富铁相析出的先后顺序存在差别。具有优异玻璃形成能力的相分离 Zr-Cu-Fe-Al 非晶合金在晶化过程中具有明显的两步结晶过程，晶化产物及生成顺序比均质 Zr-Cu-Al 非晶合金更复杂。随着温度的升高，结晶产物从纳米级面心立方的 Zr2Cu 转变为体心四方 Zr2Cu 相，而基体转变为正交 Zr3Fe 相。与铸态的 Zr-Cu-Fe-Al 块体金属玻璃相比，含有立方 Zr2Cu 相和非晶基体的 Zr-Cu-Fe-Al 复合材料的强度增加，塑性也提高。本研究所揭示相分离非晶合金的晶化顺序将对未来性能可调控的纳米异质结构玻璃提供参考并有助于合成具有高强度和良好塑性的大块玻璃纳米晶体复合材料。

关键词：大块非晶合金；液态相分离；中子与同步辐射；非晶晶化；结构异质性

高温高压下超高密度玻璃态的探索

曾桥石，楼鸿波

（北京高压科学研究中心，北京　100005）

摘　要：金属玻璃具有优异的性能。对金属玻璃性能的调控在过去的研究中往往是通过合成后样品的机械、热处理来实现的。在实践中发现，这些方法虽然能有效的改变玻璃的性质，但是玻璃的整体结构却变化很小，在高精度的 X 射线衍射和电镜分析下几乎不可区分。这一现象告诉我们，要么玻璃的结构确实不能显著调控，要么说明我们对性质的调控还在一个很小的区间内，远没有实现更大范围伴随结构显著改变的调控。在本报告中，我将介绍我们课题组利用高温高压为手段，结合原位的同步辐射 X 射线技术，通过熔体的快速冷却，把不同的高温高压条件的液体结构冷冻下来，发现了在结构上显著不同的金属玻璃态。这些高密度态在一定程度能保留到常温常压下来，通过高分辨电镜，3DAP，DSC 证实 Lee 高密度玻璃态样品均匀而无序的玻璃结构和玻璃转变现象。不同的玻璃态之间也存在温度和压力诱导的转变现象。这一现象在多种典型金属玻璃体系中都存在，说明在更广阔的温度压力空间里，玻璃可能普遍存在丰富的以前没有被发现的多形态现象，为理解玻璃材料和调控玻璃结构和性质打开了新的大门。

非晶合金的微结构和力学性能调控

王　刚

（上海大学，上海 200444）

摘　要：非晶合金作为一种新型金属材料，与传统晶态合金材料相比，展现出优异的性能，是近几十年来材料领域的研究重点和热点。然而，由于变形极易形成局域化剪切带，多数非晶合金在室温环境下表现出差的延展性，这成为非晶合金大规模工程应用的主要瓶颈。考虑到非晶合金的亚稳态特性和结构非均匀特性，制备工艺和外场条件将会对其微观结构产生极大的作用，进而影响非晶合金的宏观力学性能。为此，本研究采用先进的材料结构和性能表征手段，观察制备工艺以及外场（温度、辐照）刺激下非晶合金的微结构响应特性，探索结构演化对非晶合金力学性能的作用机理。研究发现，制备条件和外场环境能够有效调控非晶合金的微观结构，使得结构的非均匀程度增加，从而改善非晶合金的力学性能。本研究不仅为深刻认识非晶合金结构-性能调控提供有效的实验和理论基础，而且对推进非晶合金这一先进材料的工程化应用具有重要的理论与实际意义。

关键词：微结构；力学性能；外场；非晶合金

Manipulation of Microstructures and Mechanical Properties in Metallic Glasses

Wang Gang

(Shanghai University, Shanghai 20044, China)

Abstract: As a new type of metal material, metallic glasses (MGs) have been the research focus and hot topic in the field of materials science in recent decades because of their superior properties as compared with their traditional crystalline counterparts. However, localized shear bands can be easily formed during deformation, most MGs exhibit poor ductility at room temperature, which makes it being the main bottleneck for large-scale engineering applications of such alloys. Taking into account of the metastable characteristics and the nature of structural heterogeneity of MGs, preparation process and external fields will essentially have a significant effect on the microstructures, and then influence their macroscopic mechanical properties. To this end, this research will utilize advanced material structure and performance characterization methods to observe the microstructural responses of MGs under the stimulation of preparation processes and external fields such as temperature and ion irradiation, and get insights into the role of structural evolution in dominating their mechanical properties. It is found that the preparation conditions and the external fields can effectively manipulate the microstructures of MGs and increase the degree of structural heterogeneities and, eventually, improve their mechanical properties. These findings not only provide an experimental and fundamental basis for the deep understanding of structure-performance regulation of MGs, but also has crucial theoretical and practical significance for engineering applications of the advanced materials.

Key words: microstructures; mechanical properties; external fields; metallic glasses

高熵非晶合金的纯净化及其工艺进展

张 勇[1,2,3]，闫薛卉[1]，吴亚奇[1]，蔡永森[1]，刘芳菲[1]

（1. 北京科技大学，北京 100083；2. 青海大学，青海西宁 810016；
3. 北京科技大学顺德研究生院，广东佛山 528399）

摘 要：高性能的材料一般需要超高的洁净度，非晶也不例外，大体积的非晶合金形成一般需要高纯度的原料（>99.9%），合金中的杂质极易诱发异质形核，从而降低合金的玻璃形成能力。合金的纯净话一般有如下几种技术：（1）多孔陶瓷过滤或者陶瓷坩埚；（2）重熔，熔体包覆或电渣重熔或电子束重熔等；（3）微量稀土元素合金化。一般认为第三种是简单易行的工艺，已经在非晶的科研和生产过程中广泛应用。本报告将综述高熵非晶领域的最新工艺进展，并对非晶合金的未来进行展望。

关键词：非晶合金；纯净化；微合金化；稀土元素；高熵非晶；非晶薄膜

Purification of High-entropy Amorphous Alloys and Its Technological Progress

Zhang Yong[1,2,3], Yan Xuehui[1], Wu Yaqi[1], Cai Yongsen[1], Liu Fangfei[1]

1. University of Science and Technology Beijing, Beijing 100083, China;
2. Qinghai University, Xining 810016, China; 3. Shunde Graduate School,
University of Science and Technology Beijing, Foshan 528399, China)

Abstract: High-performance materials generally require ultra-high cleanliness, as well as amorphous alloys. The formation of large-volume amorphous alloys generally requires high-purity raw materials (> 99.9%), and impurities in the alloys can easily induce heterogeneous nucleation, thus reducing the glass forming ability of the alloys. The high-purity of alloy can be obtained generally by the following technologies: 1. By using porous ceramic filtration and/or ceramic crucibles; 2 remelting, fluxing by B_2O_3, electroslag remelting or electron beam remelting; 3 alloying with trace rare earth elements. Generally, the third technology is simple and easy, which has been widely used in the research and production of amorphous materials. This report will summarize the latest technological progress in the field of high-entropy amorphous alloys, and look forward to the future of amorphous alloys.

Key words: amorphous alloy; purifying; microalloying; rare earth elements; high entropy amorphous; amorphous films

铁基非晶软磁复合材料的研究进展

鲁书含[1,2]，王明罡[1,2]，赵占奎[1,2]

（1. 先进结构材料教育部重点实验室（长春工业大学），吉林长春 130012；
2. 长春工业大学材料科学与工程学院，吉林长春 130012）

摘 要：铁基非晶软磁合金是利用现代快速凝固冶金技术合成的一种新型功能材料，具有普通金属和玻璃的优良力学、物理和化学性能。它具有短程有序和长程无序的非晶结构。在众多种类的非晶合金中，铁基非晶软磁合金是应用最广泛的，具有制备和应用的双绿色特点，在碳达峰碳中和的双碳目标背景下具有重要的研究意义。尽管铁基非晶合金原子外层一些自由电子由于近程续结构呈局域态，其电阻率仍为 10^2~10^3 μΩ·cm 数量级的良导体。为避免高频下大的涡流损耗，应采用超薄层压或粉末复合铁芯。采用粉末冶金工艺生产铁基软磁材料时需对非晶合金软磁颗粒进行绝缘包覆和成型。本文综述了铁基非晶软磁复合材料的发展历史，以及铁基非晶软磁复合材料的种类、绝缘包覆和成型制备方法，展望了铁基非晶软磁复合材料未来的研究方向和应用前景。

关键词：铁基非晶合金；软磁复合材料；绝缘包覆技术；成型制备技术；现状与展望

Research Status and Development Trend of Fe-based Amorphous Soft Magnetic Composites

Lu Shuhan, Wang Minggang, Zhao Zhankui

（Key Laboratory of Advanced Structural Materials, Ministry of Education, School of Material Science and Engineering, Changchun University of Technology, Changchun130012, China）

Abstract: Amorphous alloy is a new type of alloy material which is synthesized by modern rapid solidification metallurgical technology and has excellent mechanical, physical and chemical properties of common metal and glass. It has short range ordered and long range disordered amorphous structure. In many different kinds of amorphous alloys, magnetic amorphous alloys are the most widely used. Although the disordered atomic structure causes some electrons to be in the local state, the resistivity of the amorphous is higher than that of corresponding crystalline material. Most of the free electrons in the outer layer are not affected by the periodic loss and the amorphous is still a good conductor whose resistivity is in the order of 10^2~10^3 μΩ·cm. To avoid large eddy current loss (Pe), ultra-thin laminating or powder composite core should be used. Amorphous soft magnetic particles are insulated and coating during the production of Fe-based soft magnetic materials using a powder metallurgy process. In this paper, the development history of Fe-based amorphous soft magnetic composites are reviewed, and the classes, preparation methods, insulation coating technology, forming preparation technology and application fields of Fe-based amorphous soft magnetic composites which are mainly studied in the domestic and overseas are prospected after the introduced.

Key words: Fe-base amorphous alloy; soft magnetic composite material; insulation coating technology; forming preparation technology; present situation and prospect

高强韧非晶复合材料设计与制备

朱正旺[1,2]，林诗峰[1,2]，张　龙[1,2]，张海峰[1,2]

（1. 中国科学院金属研究所师昌绪先进材料创新中心，辽宁沈阳　110016；
2. 中国科学院核用材料与安全评价重点实验室，辽宁沈阳　110016）

摘　要：非晶复合材料是一类兼具晶态材料和非晶态材料优点的先进材料，已在相关领域获得重要应用。如何进行复合材料材料成分设计、探讨宏微观结构与力学性能的关系，仍然是该领域的研究重点。报告以 TiZr 基非晶复合材料为研究对象，系统地研究了该类材料的凝固特性，揭示了合金成分分布-析出相形态和体积分数之间关系，提出了内生非晶复合材料可控制备新思路，获得了系列大尺寸、高性能内生非晶复合材料。在此基础上，探索了一种新的复合材料制备方法，该方法结合了内生和外加两种复合材料制备方法优点，实现了复合材料两相界面可控和宏微观结构设计，为高强高韧非晶复合材料设计、制备和应用奠定了坚实基础。

关键词：非晶合金；非晶复合材料；力学性能

Design and Preparation of Novel Bulk Metallic Glass Composites with High Strength and Toughness

Zhu Zhengwang[1,2], Lin Shifeng[1,2], Zhang Long[1,2], Zhang Haifeng[1,2]

(1. Shi-changxu Innovation Center for Advanced Materials, Institute of Metal Research, Chinese Academy of Sciences, Shenyang 110016, China; 2. CAS Key Laboratory of Nuclear Materials and Safety Assessment, Institute of Metal Research, Chinese Academy of Sciences, Shenyang 110016, China)

Abstract: Bulk metallic glass composites (BMGCs) with high strength and high toughness have significant application prospects in the many fields. How to design the composition, and interpret the relationship between mechanical properties and macro- and micro-scale structure are the key issues to promote the widespread applications. This presentation will introduce the solidification behaviors and microstructure control of TiZr-based BMGCs. The relationship of composition and dendrite morphologies including size and volume fraction, *etc.*, will be disclosed. As a result, the large-size and high performance *in-situ* BMGCs have been prepared successfully. Meanwhile, a novel preparation method will be proposed to obtain the *in-situ* interface and structural control. These findings will promote and benefit the practical application of BMGCs.

Key words: bulk metallic glass; bulk metallic glass composite; mechanical property

非晶合金的多维度制造

马 将

（深圳大学，广东深圳 518060）

摘 要：非晶合金是一种长程无序的新型金属材料，在力学、物理和化学等方面具有优异的性质。然而，材料尺寸问题及成分-性能调控问题在科学及应用上制约了该材料的发展。本文通过利用先进制造的方法，在材料尺寸及成分调控等多个维度对非晶合金进行改善，获得了大尺寸和性能可调控的非晶合金新材料，为其应用有一定的意义。

关键词：非晶合金；成分调控；微纳制造；功能应用

铁磁性非晶合金结构调控与相关性能研究

沈宝龙

（东南大学 材料科学与工程学院，江苏南京 211189）

摘 要：本文通过冷热循环、等温吸氢处理及磁场热处理等方法调控铁磁性非晶合金微观结构，系统研究结构演化与铁磁性非晶合金力学性能、磁热性能以及软磁性能的关联，探索制备了系列高性能铁磁性非晶合金。（1）在力学性能方面，利用冷热循环实现结构回春和 1nm 左右类晶体有序结构增加，显著增加非晶合金结构不均匀性，提高铁磁性块体非晶合金塑性应变能力，获得 4060 MPa 屈服强度和 6.1%压缩塑性的$[(Fe_{0.5}Co_{0.5})_{0.75}B_{0.2}Si_{0.05}]_{96}Nb_4$非晶合金，此外，0.1 at.% Cu 元素添加可降低冷热循环温度并进一步提升其力学性能，$\{[(Fe_{0.5}Co_{0.5})_{0.75}B_{0.2}Si_{0.05}]_{96}Nb_4\}_{99.9}Cu_{0.1}$块体非晶合金表现出 7.4%压缩塑性，4350 MPa 屈服强度及 5050 MPa 断裂强度的优异力学性能，合金中类液区增加带来剪切转变区的形核和渗逾，增加结构不均匀性，促进剪切稳定性，从而提高塑性变形能力。(2) 在磁热性能方面，通过等温吸氢处理，在 GdTbDyCoAl 高熵非晶合金基体上析出 7.5nm 左右纳米晶，主要包括稀土二氢化物和少量稀土-铝/钴相，提高结构不均匀性，改善低温磁热性能。最大磁熵变由 8.8 提高到 13.6 J $kg^{-1}K^{-1}$，磁转变温度由 59 降低到 8 K，同时磁滞损耗可忽略，是一种潜在低温磁制冷材料。(3) 在软磁性能方面，通过磁场热处理促进形成规则平面磁畴，大幅降低钉扎作用，获得了饱和磁感应强度达 1.86T，矫顽力低至 1.2A/m，有效磁导率达 16300 具有优异综合软磁性能的 $Fe_{66.65}Co_{16}Si_2B_{14}Cu_{1.35}$ 非晶合金。外加磁场抑制中程序团簇有序化，使得合金局域结构更加均匀，显著提升软磁性能。本工作澄清了结构不均匀性对铁磁性非晶合金力学、磁热以及软磁性能的影响机制，对于开发高强高韧、大磁熵变低温磁制冷、优异软磁性能铁磁性非晶合金材料具有指导意义。

关键词： 铁磁性非晶合金；结构不均匀性；力学性能；磁热性能；软磁性能

非晶合金先进激光与机械加工成形研究进展

王成勇，丁　峰，唐梓敏，胡治宇，张智雷

（广东工业大学，广东广州　510006）

摘　要： 大尺寸、精密、复杂的非晶合金零件离不开先进的加工制造技术。本报告首先探讨了非晶合金超快激光微织构加工理论与工艺方法，及其在大尺寸非晶合金超塑性成形中的应用；介绍了团队在非晶合金高效精密切削、磨削加工理论、加工工具与工艺方法的最新进展；最后列举了先进激光与机械加工技术在超锋利非晶合金手术刀和非晶血管夹制造中的应用。报告内容为大尺寸非晶合金的制造成形提供了新方向；为非晶合金零部件的高效精密生产提供了新的解决思路。

关键词： 大块非晶合金；超快激光；切削与磨削加工

非晶合金的非局域结构特征及其与性能的关联性

李茂枝

（中国人民大学物理系，北京　100872）

摘　要： 非晶合金的结构长程无序，但是具有短程序，甚至中程序，非常复杂，现代微观结构分析和表征技术对其结构的分析能力非常有限，也无法用现代晶体学对其认识和理解，所以建立非晶合金的微观结构与宏观性能的关系一直是科学难题。到目前为止，非晶中局域结构对性能的确切作用仍存在争议，而越来越多的研究表明，原子尺度水平以外的非局域结构对非晶合金及合金液体的动力学、力学等性质起着更重要的作用。本报告将简要介绍短程序的空间连接的重要性及其定量表征。通过引入图论、网络等方法表征非晶结构的非局域关联性质，建立了短程序的连接与非晶合金动力学和力学性能之间的定量关系，揭示了非晶合金的非局域结构关联特性是影响其宏观性能的更本质因素，对非晶的结构和结构性能关系提供了新的认识。

关键词： 非晶合金；非局域结构；结构性能关系

非晶合金的高通量制备与表征

李明星，孙奕韬，王　超，汪卫华，柳延辉

（中国科学院物理研究所，北京　100190）

摘　要：非晶合金组成元素的多样性和复杂性给性能优化带来巨大挑战。材料基因工程是最近发展起来的材料研发新理念，高通量实验是材料基因工程的主要组成部分。通过高通量制备和表征，不仅可以加快非晶合金新材料探索效率，而且在高通量表征中所获得大量实验数据还可以帮助理解非晶合金中的科学问题。本报告将介绍采用材料基因工程理念和高通量实验方法在高性能非晶合金的成分设计和探索中应用，以高通量方法研制出的高温高强非晶合金材料新体系为例，证明材料基因工程在新材料探索中的有效性和高效率。

关键词：非晶合金；非晶形成能力

非晶合金晶化行为的调控及其对磁热性能的影响

霍军涛，王军强

（中国科学院宁波材料技术与工程研究所，浙江宁波　315201）

摘　要：晶化是非晶合金的固有特征，其对调控非晶合金的性能至关重要。而怎么去控制非晶合金的晶化行为是关键难题。本工作中，通过建立非晶合金升降温过程中熔体、玻璃相和晶化相的相图，我们发现快冷速制备的非晶合金更难以晶化，这主要是由于快冷速制备的非晶合金中 β 弛豫模式含量较高，而 β 弛豫模式对晶化具有抑制作用。另外，基于控制晶化的思想，通过优化合金成分和制备工艺，我们设计了一种 Gd 基非晶纳米晶复合纤维材料。这种复合纤维材料兼具了 Gd 基非晶合金和 Gd 纳米晶的优点，在大于 200K 的温度区间内都具有较大的磁熵变值，其磁制冷能力参数（refrigerant capacity，RC）可以达到 985 Jkg^{-1}. 通过控制合金成分和制备工艺，可以实现对非晶合金晶化行为的控制，进而调控合适的非晶相和晶化相比例，这是一种获得优异力学性能和功能特性的有效方法。

关键词：非晶合金；弛豫；晶化行为；复合材料；磁热性能

Regulation of Metallic Glasses' Crystallization Behavior and Its Effect on Magnetocaloric Properties

Huo Juntao, Wang Junqiang

(Ningbo Institute of Materials Technology and Engineering, Chinese Academy of Sciences, Ningbo 315201, China)

Abstract: Crystallization is an intrinsic characteristic that derive from the metastable nature of metallic glasses, which is critical for regulating their functional properties. How to control the crystallization of metallic glasses is a key issue. In this work, diagrams composed of liquid, glass and crystal are established to illustrate the phase transitions upon heating and cooling a metallic glass-forming material. We find that the fast-cooled metallic glass is more difficult to crystallize, which is attributed to the activation of β relaxation that can depress the crystal nucleation. Based on the crystallization controlling theory, Gd-based amorphous/nanocrystalline composite fibers are designed by optimizing components and melt extraction methods. The designed composite fibers exhibit a good combination of the advantages of amorphous and crystalline magnetocaloric materials, which possess large magnetic entropy changes over a much wider temperature range (> 200 K) and giant refrigerant capacity (985 Jkg^{-1}). Our work suggests that designing composites of amorphous and crystalline phases by controlling alloy compositions and preparation methods is a superior method not only for achieving advanced mechanical properties but also for obtaining advanced magnetic functional properties.

Key words: metallic glass; relaxation; crystallization; composite; magnetocaloric properties

FeCuNbSiB 纳米晶软磁合金的感生各向异性及机理研究

刘天成，潘　贇

（安泰科技股份有限公司）

摘　要：非晶纳米晶合金中由于双相结构的存在，往往表现各向同性。但在有外界势场存在（如磁场，应力场等）时，势场影响相变过程，使合金结构出现感生各向异性，进而影响其性能。实验主要针对非晶纳米晶合金感生各向异性，探究了磁场退火、张力退火等工艺对 FeCuNbSiB 纳米晶合金感生各向异性及磁性能的影响，分析了磁畴结构与感生各向异性之间的作用机理。

实验系统地研究了张力退火过程中退火张力、退火温度、保温时间、加热速率和冷却速率等因素对 FeCuNbSiB 纳米晶合金组织及磁性能的影响规律。通过定义直流偏置场 H_{98} 表征合金感生各向异性，直流偏置场 H_{98} 为合金磁导率 μ_e 随磁场强度增加降低至初始值的 98 %时所对应的磁场强度值。实验表明利用恒温静张力退火处理后 FeCuNbSiB 纳米晶合金直流偏置场 H_{98} 可以超过 3500 A/m，各向异性明显。同时发现张力退火后 FeCuNbSiB 纳米晶合金内部形成垂直于带轴和磁化方向的条形畴结构，随退火张力增加磁畴宽度减小且趋于一致；在恒温静张力退火过程中保温时间达到 900 s 时，磁畴结构开始从垂直于磁化方向的条形畴转变为平行于磁化方向的波纹形畴。

综合分析退火过程中磁场和力场对非晶纳米晶合金磁各向异性的影响可知，存在力场作用时，晶体相会产生晶格畸变，畸变的晶格产生磁晶各向异性，同时由于晶体相和非晶相之间存在磁弹性各向异性，所以张力退火后 FeCuNbSiB 纳米晶合金会产生明显的各向异性进而表现抗直流恒导磁性能。而力场对原子团簇作用较弱，故非晶状态下施加力场无法获得各向异性；存在磁场作用时，非晶相中原子团簇的磁各向异性会沿磁场方向进行排列，最终使退火后的合金具有磁场感生各向异性。实验结果解释了非晶纳米晶软磁合金各向异性与组织结构的联系，验证了不同工艺对非晶纳米晶软磁合金感生各向异性的影响，阐明了各向异性非晶纳米晶合金的工艺路线。

关键词：非晶纳米晶合金；张力退火；磁畴结构；各向异性

通过热机械固结铁与其他元素混合粉制备优质钢材

张德良，张有鋆，赵晓丽，Valladares Luis De L.S.

（东北大学，先进粉末冶金材料与技术实验室，辽宁沈阳　110819）

摘　要：采用低成本还原铁粉和其他需要的合金元素粉末，通过热机械粉末固结直接制备钢材具有低能耗，短流程，高效率，成分与组织均匀细化等优势，是将来低碳物理冶金工艺重要发展途径之一。本团队采用该工艺成功制备 Q235, HRB335 和 Fe-20Mn（wt%）等棒材，并研究了样品的微观组织和拉伸性能，发现 Q235 和 Fe-20Mn 棒材具有优异的力学性能。由于加入的冶金渣料带进的大尺寸硬质颗粒的影响，HRB335 棒材的强度很高，但拉伸塑性较差。本报告将展示研究结果并讨论以粉末冶金为基础的低碳固态冶金发展前途。

关键词：固态冶金；粉末冶金；钢；热机械粉末固结；力学性能

Fabrication of High Quality Steel by Thermomechanical Consolidation of Blends of Fe and other Elemental Powders

Zhang Deliang, Zhang Youyun, Zhao Xiaoli, Valladares Luis De L.S.

(APM-Lab, School of Materials Science and Engineering, Northeastern University, Shenyang 110819, China)

Abstract: Direct fabrication of steel by thermomechanical consolidation of blends of Fe powder and other elemental and/or master alloy powders needed for alloying has many advantages. They include low energy consumption, shortened processing route, high efficiency, uniform composition and fine microstructure. This process route is believed to be one of the important low carbon-emission physical metallurgy process development routes. Using this technology, we have successfully fabricated Q235, HRB335 and Fe-20Mn (wt%) alloy rods, and studied their microstructures and tensile properties. The Q235 and Fe-20Mn alloy rods have good tensile properties. The HRB335 rod has a high strength, but poor tensile ductility due to the large sized hard particles brought in by the added slag. This report will present the findings and discuss the future development of low carbon-emission solid state metallurgy.

Key words: solid state metallurgy; powder metallurgy; steel; thermomechanical powder consolidation; mechanical properties

粉末冶金 FeCrNi 中熵合金的微观组织与性能研究

付　遨，刘　彬，李伟华

（中南大学，湖南长沙　410000）

摘　要：基于 3d 过渡金属元素形成的中熵合金具有高的强韧性、优异的耐蚀性以及良好抗辐照损伤性能等，在航空航天、海洋、核能等领域具有广泛的应用前景，已成为金属材料研究领域的热点。本文基于 Fe、Cr、Ni 等 3d 过渡金属元素，采用粉末冶金方法制备了高 Cr 含量的单相 FCC 结构的新型 FeCrNi 中熵合金，系统地研究了该合金的微观组织、准静态/动态力学性能以及耐腐蚀性能。研究结果表明：FeCrNi 中熵合金具有优异的室温拉伸性能和抗绝热剪切能力，变形过程中高度激活的多重变形行为（如孪生、微带、层错等）造成的“动态霍尔-佩奇”效应是该合金优异强度-韧性组合和抗绝热剪切性能的主要原因。同时，FeCrNi 中熵合金在 NaCl 溶液中相较于传统 304 和 316 不锈钢表现出更优异的耐蚀性能，其中合金中较高的 Cr 含量能极大的改善合金的钝化能力，有利于合金表面富 Cr_2O_3/$Cr(OH)_3$ 钝化膜的形成，进而使得合金具有更加优异的耐蚀性能。

Research on Microstructure and Properties of Entropy Alloy in Powder Metallurgy FeCrNi

Fu Ao, Liu Bin, Li Weihua

(China Central South University, Changsha 410083, Hunan)

Abstract: The medium entropy alloy formed based on 3d transition metal elements has high strength and toughness, excellent corrosion resistance and good resistance to radiation damage. It has a wide range of application prospects in aerospace, marine, nuclear energy and other fields, and has become a metal Hot spots in the field of materials research. In this paper, based on Fe, Cr, Ni and other 3d transition metal elements, a new type of FeCrNi medium-entropy alloy with a single-phase FCC structure with high Cr content was prepared by powder metallurgy, and the microstructure and quasi-static/dynamic mechanical properties of the alloy were systematically studied. And corrosion resistance. The research results show that the FeCrNi medium-entropy alloy has excellent room temperature tensile properties and resistance to adiabatic shear. The highly activated multiple deformation behaviors (such as twinning, microstrips, stacking faults, etc.) The "odd" effect is the main reason for the alloy's excellent strength-toughness combination and adiabatic shear resistance. At the same time, FeCrNi medium-entropy alloy shows better corrosion resistance in NaCl solution than traditional 304 and 316 stainless steel. The higher Cr content in the alloy can greatly improve the passivation ability of the alloy and is beneficial to the surface of the alloy. The formation of Cr_2O_3/$Cr(OH)_3$-rich passivation film makes the alloy have more excellent corrosion resistance.

粉床增材制造 H13 模具钢研究进展

刘世锋

（西安建筑科技大学，陕西西安　710055）

摘　要：H13 钢是国际上广泛使用的一种热作模具钢，主要应用于热镦锻、挤压和压铸模具的制造。近年来随着消费级电子产品及工业品的升级换代，对模具复杂度提出了更高的需求，金属增材制造由于其离散堆积的成形特点，成为制备异性模具最有竞争力的技术之一。金属增材制造具有微小熔池、快速熔凝和反复热处理等冶金特点，使得制备构件表现出与传统工艺完全不同的显微组织。本文系统研究选区激光熔化（Selective laser melting，SLM）和选区电子束熔化（Selective Electron Beam Melting，SEBM）两种粉床增材 H13 钢的组织性能演变，以及相应的后处理调控工艺。结果表明，SEBM 制备的 H13 钢中无粗大的碳化物析出（碳化物尺寸~100nm）。由于自回火效应，可显著降低残余奥氏体含量和残余应力。深冷后处理可以改变 SLM 制备 H13 钢残余奥氏体应力状态，显著提高强度和塑性。

关键词：增材制造；H13 模具钢；组织演变

Progress in Additive Manufacturing on H13 Die Steel

Liu Shifeng

(School of Metallurgical Engineering, Xi'an University of Architecture and Technology, Xi'an 710055, China)

Abstract: As a kind of hot work die steel, H13 steel is widely used in hot heading forging, extrusion and die casting die manufacturing industry. In recent years, with the upgrading of consumer electronic products and industrial products, the complexity of die has been put forward a higher demand. Metal additive manufacturing has become one of the most competitive technologies for the preparation of heterosexual die because of its discrete accumulation forming characteristics. and repeated heat treatment, which makes the fabricated components show completely different microstructure from the traditional process.

This paper systematically studies the microstructure evolution, mechanical properties and the corresponding post-treatment control technology of H13 steel, which were prepared by selective laser melting (SLM) and selective electron beam melting (SEBM). The results show that there is no coarse carbide precipitation (carbide size ~100 nm) in the H13 steel prepared by SEBM. Residual austenite content and residual stress can be significantly reduced due to the self-tempering effect. Furthermore, cryogenic treatment can change the residual austenite stress state of H13 steel prepared by SLM, and improve the strength and plasticity significantly.

Key words: additive manufacturing; H13 die steel; microstructural evolution

热爆反应结合脱合金制备微纳多孔铜及电催化性能研究

冯培忠

（中国矿业大学，材料与物理学院/现代分析与计算中心，江苏徐州 221000）

摘　要：纳米多孔金铜具有高比表面积、低密度、高通透性和结构灵活可调等特点，有望在传感、催化、分离、能源等领域得到广泛应用。针对现有的纳米多孔铜制备方法能耗高、时间长、工艺繁琐的特点，本研究使用Cu粉和Al粉为原料，通过压片成型、热爆反应与脱合金法相结合的工艺制备微纳多孔铜（MNPC）粉末。该方法充分利用热爆反应自放热的特征，具有工艺简便、耗时短的优点，产物结构均匀，性质稳定，韧带及通道尺寸可灵活调控。研究了原子比（Cu:Al=40:60、30:70、20:80）、脱合金时间（0.5h、1h、2h）、腐蚀温度（25℃和75℃）和腐蚀液浓度（5%，10%，15%，20wt% NaOH）对产物物相组成、微观结构和腐蚀过程的影响，另外，制备了前驱体成分为Cu18Ni2Al80（at.%）的MNPC，发现掺杂Ni使微纳多孔铜电催化性能有明显的提升。

关键词：粉末冶金；热爆反应；脱合金；多孔铜

Thermal Explosion Reaction Combined with Dealloying to Prepare Micro-nano Porous Copper and Its Electrocatalytic Properties

Feng Peizhong

(School of Materials and Physics, China University of Mining and Technology, Xuzhou 221009, China)

Abstract: Nano-porous gold copper has the characteristics of high specific surface area, low density, high permeability and flexible and adjustable structure. It is expected to be widely used in sensing, catalysis, separation, energy and other fields. In view of the high energy consumption, long time and cumbersome process characteristics of the existing nanoporous copper preparation methods, this study uses Cu powder and Al powder as raw materials to prepare micro-nano materials through a combination of sheet molding, thermal explosion reaction and dealloying method. Porous copper (MNPC) powder. The method makes full use of the self-exothermic characteristics of the thermal explosion reaction, has the advantages of simple process, short time-consuming, uniform product structure, stable properties, and flexible control of ligament and channel size. The atomic ratio (Cu:Al=40:60, 30:70, 20:80), dealloying time (0.5h, 1h, 2h), corrosion temperature (25℃ and 75℃) and corrosion solution concentration (5%, 10%, 15%, 20wt% NaOH) on the phase composition, microstructure and corrosion process of the product. In addition, MNPC with a precursor composition of Cu18Ni2Al80 (at.%) was prepared, and it was found that doping with Ni makes micro-nano porous The electrocatalytic performance of copper has been significantly improved.

Key words: powder metallurgy; thermal explosion reaction; dealloying; porous copper

增材制造高强度铝合金成分设计、制备及应用

李瑞迪[1]，周科朝[1]，袁铁锤[1]，刘　咏[1]，史玉升[2]，祝弘滨[3]

（1. 中南大学 粉末冶金国家重点实验室，湖南长沙　410083；2. 华中科技大学 材料成形与模具技术国家重点实验室，湖北武汉　430074；3. 中车工业研究院有限公司，北京　100015）

摘　要：以探月、火星探测、空间站建设等为代表我国航天事业正蓬勃发展。铝合金激光增材制造 (Selective Laser Melting, SLM) 较好地契合了航天领域对材质轻量化、结构镂空轻量化的迫切需求，有望成为下一代运载火箭、卫星等核心零部件成形的关键技术。因此，开发出新一代增材制造高强铝合金，已经成为当前航天增材领域亟待完成的一个重要基础研究任务。在国家重点研发计划、国家自然科学基金的支持下，与中车开展合作，针对现有增材制造商用铝合金性能不足难题，提出基于层错能效应发展增材制造铝镁合金新型强韧化机理。针对易裂问题，从 SLM 铝镁合金的热裂机理出发，揭示了合金成分与凝固应力敏感性因子、层错能及力学性能的关系，发现了激光增材制造高镁含量铝合金中存在 9R 相。通过抑裂机制-层错能效应强韧化机制-成分设计-疲劳性能-构件质量控制的系统研究，制备了屈服强度 520MPa，拉伸强度 570MPa，延伸率 12%的增材制造铝合金。研发的铝合金粉末已成形出 200×200mm 的复杂零件，且通过疲劳性能测试，在中车工业获得应用验证。

关键词：激光增材制造；选区激光熔化；铝合金

Composition Design, Preparation and Application of Additive Manufacturing High-strength Aluminum Alloy

Li Ruidi[1], Zhou Kechao[1], Yuan Tiechui[1], Liu Yong[1], Shi Yusheng[2], Zhu Hongbin[3]

(1. State Key Laboratory of Powder Metallurgy, Central South University, Changsha 410083, China; 2. State Key Laboratory of Material Forming and Mould Technology, Huazhong University of Science and Technology, Wuhan 430074, China; 3.CRRC Research Institute Co., Ltd., Beijing 100015, China)

Abstract: China's aerospace industry is developing vigorously, represented by lunar exploration, Mars exploration, and space station construction. Aluminum alloy laser additive manufacturing (SLM) meets the urgent needs of the aerospace industry for lightweight materials and hollow structures, and is expected to become a key technology for the formation of core components such as next-generation launch vehicles and satellites. Therefore, the development of a new generation of additive manufacturing high-strength aluminum alloys has become an important basic research task urgently to be completed in the current aerospace additive field. With the support of the National Key R&D Program and the National Natural Science Foundation of China, it cooperated with CRRC to address the problem of insufficient performance of the existing additive manufacturing commercial aluminum alloys, and proposed the development of new strengthening and toughening of additive manufacturing aluminum-magnesium alloys based on the stacking fault energy effect. mechanism. Aiming at the problem of easy cracking, starting from the hot cracking mechanism of SLM aluminum-magnesium alloy, the relationship between alloy composition and solidification stress sensitivity factor, stacking fault energy and mechanical properties was revealed, and 9R was found in laser additive manufacturing of high-magnesium aluminum alloy. Mutually. Through the systematic research of crack suppression mechanism-stacking fault energy effect strengthening and toughening mechanism-composition design-fatigue performance-component quality control, an additively manufactured aluminum alloy with a yield strength of 520MPa, a tensile strength of 570MPa and an elongation of 12% was prepared. The aluminum alloy powder developed has been formed into complex parts with a size of 200×200mm, and has passed the fatigue performance test, and has obtained application verification in CRRC.

Key words: laser additive manufacturing; selective laser melting; aluminum alloy

核壳结构氧化物/钨复合纳米粉体的制备及其烧结特性研究

董　智，马宗青，刘永长

（天津大学材料科学与工程学院，天津　300072）

摘　要：与相应的合金基体相比，氧化物弥散强化(ODS)的合金具有显著提升的强度、组织稳定性和蠕变抵抗力，这使得它们在许多关键的领域具有很大的应用潜力。但由于添加的氧化物趋于在金属基体晶界处团聚长大，不能很好地弥散分布，这大大削弱了氧化物添加对合金的改善效果。基于此，我们以 ODS-W 合金为实验对象，首先结合水热和冷冻干燥法制备了具有钨包覆氧化物核壳结构的复合纳米粉末。在这种独特复合粉末中，纳米尺寸的氧化物(约 2~5 nm)被均匀地包覆到钨晶粒内部，同时它们与周围的钨基体呈现出完全共格的界面特征。以该复合粉末为先驱粉，经低温烧结和高能锻加工以后，获得了强度和韧性匹配良好的 ODS 钨基合金。力学性能测试的结果表明：该 ODS 钨基合金在室温下延伸率为 2.5%，抗拉强度为 1390MPa，这表明该钨合金的韧脆转变温度降到了室温以下，打破了传统 ODS-W 合金在低温下呈脆性的局限性。600℃下强度为 720MPa，延伸率为 14%，具有优异的强韧性。综上，这种通过核壳结构前驱粉末来原位引入超细、共格的弥散强化相的方式能够显著改善合金的强韧性，并且有望应用于其他 ODS 合金体系。

Preparation and Sintering Characteristics of Core-shell Structure Oxide/Tungsten Composite Nano-powder

Dong Zhi, Ma Zongqing, Liu Yongchang

(School of Materials Science and Engineering, Tianjin University, Tianjin 300072, China)

Abstract: Compared with the corresponding alloy matrix, oxide dispersion strengthened (ODS) alloys have significantly improved strength, structure stability and creep resistance, which makes them have great application potential in many key areas. However, since the added oxide tends to agglomerate and grow at the grain boundary of the metal matrix, it cannot be dispersed well, which greatly weakens the improvement effect of the oxide addition on the alloy. Based on this, we took ODS-W alloy as the experimental object, and firstly combined hydrothermal and freeze-drying methods to prepare composite nanopowders with a core-shell structure of tungsten-coated oxide. In this unique composite powder, nano-sized oxides (approximately 2-5 nm) are uniformly coated inside the tungsten crystal grains, and at the same time they and the surrounding tungsten matrix exhibit completely coherent interface characteristics. Using the composite powder as the precursor powder, after low-temperature sintering and high-energy forging processing, an ODS tungsten-based alloy with well matched strength and toughness is obtained. The results of the mechanical properties test show that the elongation rate of the ODS tungsten alloy at room temperature is 2.5%, and the tensile strength is 1390MPa, which indicates that the ductile brittle transition temperature of the tungsten alloy has dropped below room temperature, breaking the traditional ODS-W alloy. The limitation of brittleness at low temperature. The strength at 600℃ is 720MPa, the elongation is 14%, and it has excellent strength and toughness. In summary, this method of introducing ultrafine and coherent dispersion strengthening phases in situ through the core-shell structure precursor powder can significantly improve the strength and toughness of the alloy, and is expected to be applied to other ODS alloy systems.

Key words: metallurgy; materials

一种特殊“壳核”结构的包覆型粉体

杨亚锋，李少夫

（中国科学院过程工程研究所，北京　100190）

摘　要：基于粉体颗粒的粉末冶金和 3D 打印近净成形是实现高性能复杂构件低成本制造的重要技术。高质量粉体颗粒的研发是重中之重。针对当前市场上存在的粉体种类单一、复合粉体缺乏以及功能粉体制备困难等问题，本报告将重点报道一种特殊“壳核”结构包覆型粉体的制备原理及宏量化制备装置开发技术，并介绍该特色粉体在大尺寸、复杂结构粉末冶金制品的控形控性，高性能金属基复合材料制备，复合材料、高反射率金属以及超细晶材料的 3D 打印等方面的应用实践。

关键词：粉体改性；粉末冶金；增材制造；组织；复合材料

A Type of Coated Powder with Special Core-shell Structure

Yang Yafeng, Li Shaofu

(Institute of Process Engineering, Chinese Academy of Sciences, Beijing 100190, China)

Abstract: Powder metallurgy and 3D printing of powders are two key near-net-shaping and low-cost manufacturing technologies for fabricating components and parts with complex structure and high performance. Developing high-quality powder are significantly important for the both technologies. This report aims at the typical issues of current powders, including of less diversity in the type of powder, the lack of composite powder, and the difficulty in making the powder with specific function. According to the urgent need of high-quality powder, it introduces a type of coated powder with core-shell structure and reveals the corresponding preparation mechanism. Meanwhile, the macro-scaled fabrication technology and corresponding equipment of the composite powder are also reported here. As another focus of this study, the engineering applications of the special powder in the aspects of (i) controlling the shape and property of large-size and complex parts made by powder metallurgy, (ii) the fabrication of high-performance metal matrix composite, (iii) 3D printing of composite materials, the metals with high laser reflectivity, and the materials with ultrafine microstructure are systemically introduced.

Key words: powder modification; powder metallurgy; additive manufacturing; microstructure; composite

铼的形变组织及力学性能研究

章 林，曲选辉，秦明礼，魏子晨，李星宇，阙忠游

（北京科技大学新材料技术研究院，北京 100083）

摘 要： 铼具有高熔点、力学性能优异等特性，是航空、电真空、大型变温炉等领域的关键材料。制备显微组织致密均匀、性能高的铼产品至关重要；杂质元素的引入，也会对铼的变形加工行为产生影响。本文研究了纯铼冷轧中的组织演变规律，主要分为两个阶段：原始晶粒破碎和后期晶粒长大。在轧制过程中，{11$\bar{2}$1}拉伸孪生很容易激活，孪生的面积分数随着变形量的增大先增加后减小，小变形量时，滑移和孪生共同作用，变形量大于70%时以滑移为主要变形方式；基面织构类型并无明显变化，织构强度逐渐增强并转向 TD 方向倾斜 30°的双峰分布。研究了不同晶粒尺寸纯铼的拉伸行为，非基面滑移和{11$\bar{2}$1} <$\bar{1}\bar{1}$26>拉伸孪晶是大晶粒尺寸样品中的主要变形模式，晶粒尺寸的减小导致非基面滑移被抑制，使延展性和强度降低。烧结态的 C、O 以团簇的形式存在于晶粒内 N、Na、W 等游离式地分散，轧制后由于可动位错的迁移等，使得 C、O、N、Na、W 偏聚于晶粒边界处。杂质元素主要影响是延缓再结晶或抑制非基面织构的产生；在等时退火过程中杂质元素的主要影响是减弱晶粒粗化行为和引起硬化。

关键词： 铼；组织均匀性；烧结；冷轧；孪晶；滑移；杂质元素

Study on the Deformed Structure and Mechanical Properties of Rhenium

Zhang Lin, Qu Xuanhui, Qin Mingli, Wei Zichen, Li Xingyu, Que Zhongyou

(Institute of New Materials Technology, Beijing University of Science and Technology, Beijing 100083, China)

Abstract: Rhenium has the characteristics of high melting point and excellent mechanical properties, and is a key material in the fields of aviation, electric vacuum, and large-scale temperature-changing furnaces. It is very important to prepare rhenium products with dense and uniform microstructures and high performance; the introduction of impurity elements will also affect the deformation processing behavior of rhenium. This paper studies the microstructure evolution law of pure rhenium cold rolling, which is mainly divided into two stages: original grain crushing and later grain growth. In the rolling process, $\{11\bar{2}1\}$ tensile twins are easily activated. The area fraction of twins first increases and then decreases with the increase of deformation. When the deformation is small, slip and twins work together, and the deformation is greater than 70%. When slipping is the main deformation mode; the base surface texture type has not changed significantly, and the texture strength gradually increases and turns to a bimodal distribution inclined 30° in the TD direction. The tensile behavior of pure rhenium with different grain sizes is studied. Non-basal surface slip and $\{11\bar{2}1\}$ $\langle\bar{1}\bar{1}26\rangle$ stretching twins are the main deformation modes in samples with large grain sizes. Surface slip is suppressed, reducing ductility and strength. The sintered C and O exist in the form of clusters in the crystal grains. N, Na, W and other freely dispersed, after rolling due to the migration of movable dislocations, etc., the C, O, N, Na, W deviation Gather at the grain boundary. The main effect of impurity elements is to delay recrystallization or inhibit the generation of non-basal surface texture; the main effect of impurity elements in the isochronous annealing process is to weaken the grain coarsening behavior and cause hardening.
Key words: rhenium; uniformity of structure; sintering; cold rolling; twinning; slip; impurity elements

钢铁行业系统节能和减碳技术

杜　涛[1,2]

（1. 东北大学 冶金学院，辽宁 沈阳 110819；
2. 东北大学 国家环境保护生态工业重点实验室，辽宁 沈阳 110819）

摘　要： 面对能源约束的持续加剧和“双碳”目标的加快推进，作为高能耗、高污染的钢铁行业须进一步推进节能减碳工作。基于系统节能理论广泛开展物质流、能量流网络优化等方面工作在实现钢铁生产过程节能降耗上取得巨大成效，为钢铁行业减碳奠定了基础。为实现“双碳”目标须优先加快结构调整、淘汰落后产能，优化产业布局。同时在现有资源和技术条件下源头上采取低碳能源替代技术，调整能源结构；过程上强化余热余能回收技术，优化企业工艺流程，提升生产过程能源效率；末端上开展碳捕集和利用技术，实现碳资源化利用。
关键词： 钢铁行业；结构调整；系统节能；碳捕集与利用；碳减排

System Energy Conservation and CO_2 Reduction Technology in the Iron and Steel Industry

Du Tao[1,2]

(1. School of Metallurgy, Northeastern University, Shenyang 110819, China;
2. SEP Key Laboratory of Eco-industry, Northeastern University, Shenyang 110819, China)

Abstract: Facing the continuous intensification of energy constraints and the accelerated implementation of the "double carbon" goal, the iron and steel industry with high energy consumption and high pollution must further promote energy conservation and CO_2 reduction. Based on the system energy conservation theory, extensive work has been carried out on the optimization of material flow and energy flow network, which has achieved great results in realizing energy conservation and consumption reduction, and laid a foundation for CO_2 reduction in iron and steel industry. To achieve the goal of "double carbon", speeding up structural adjustment, eliminating backward production capacity and optimizing industrial layout are priority measures. Besides, under the existing resource and technical conditions, the following measures should be taken: adjusting energy structure, adopting low-carbon energy substitution technology at the source; strengthening the waste heat and energy recovery technology, optimizing the enterprise process flow, and improving the energy efficiency in the process; and carrying out carbon capture and utilization technology at the end.

Key words: iron and steel industry; structural adjustment; system energy conservation; carbon capture and utilization; CO_2 reduction

双碳形势下对冶金工程的思考

谢国威

（中钢集团鞍山热能研究院有限公司，辽宁鞍山 114004）

摘 要：双碳形势下如何有效降低冶金生产过程中CO_2，实现冶金生产降碳工作是行业的热点，各类方案层出不穷。本研究在典型钢铁企业能源结构、CO_2形成及排放特点分析基础上，结合冶金反应装置综合性能比较，在高碳储能材料和低碳共性技术方面进行研讨，探讨“源头减碳化—产储调优化—回收资源化”立体式减碳方案。

关键词：高碳储能；低碳技术；减碳化；资源化

Thoughts on Metallurgical Engineering to Achieve Carbon Peak and Neutrality Goals

Xie Guowei

(Sinosteel Anshan Research Institute of Thermo-Energy Co., Ltd., Anshan 114004, China)

Abstract: To achieve carbon peak and neutrality goals, it is urgent to effectively reduce the CO_2 emissions of metallurgical production processes. In this work, the characteristics of energy structure, CO_2 formation and emission of typical steelworks are analyzed, the comprehensive performance of metallurgical reaction devices is compared, high-carbon energy storage materials and low-carbon common technologies are investigated, and a comprehensive carbon-reducing solution, consisting of carbon reduction at emitting sources, optimization of production and storage, and resource recovery and valorization, is discussed.

Key words: high-carbon energy storage; low-carbon technology; carbon reduction; resource valorization

冶金重污染厂房高效低耗控烟技术及应用

朱晓华

（钢铁工业环境保护国家重点实验室，北京 100088）

摘 要：通过统计数据分析冶金重污染厂房烟尘控制的能耗水平，详细阐述烧结、焦化、高炉、转炉等冶金主要工序重污染厂房的大气污染现状。结合超低排放改造及工业建筑环境保障的迫切需要，重点介绍了高炉出铁场、烧结机尾、转炉一次二次等主要高温烟尘的高效低耗控烟技术，以及在湛江钢铁环境除尘 BOO 运营中的实际应用效果。最后对冶金重污染厂房控烟技术未来绿色低碳和智能化的发展方向进行展望。

关键词：冶金；重污染；高效低耗；控烟

High Efficiency and Low Consumption Smoke Dust Control Technology and Application in Metallurgical Heavy Polluted Plants

Zhu Xiaohua

(State Key Laboratory of Iron and Steel Industry Environmental Protection, Beijing 100088, China)

Abstract: The energy consumption level of smoke dust control in metallurgical heavily polluted plants is analyzed through statistical data, and the air pollution status of heavily polluted plants in main metallurgical processes such as sintering, coking, blast furnace and converter is described in detail. Combined with the urgent needs of ultra-low emission transformation and environmental protection of industrial buildings, this paper focuses on the high-efficiency and low-consumption smoke dust control technology of main high-temperature smoke dust in blast furnace tapping yard, sintering tail and primary and secondary converter, as well as the practical application effect of environmental smoke dust removal Building-Owning-Operation in Zhanjiang iron and steel. Besides, the future development direction of green, low-carbon and intelligent smoke dust control technology in metallurgical heavy pollution plant is prospected.

Key words: metallurgy; heavy pollution; high efficiency and low consumption; smoke dust control

碳中和愿景下钢铁行业超低排放技术与绿色低碳发展展望

邢　奕[1,2]

（1. 北京科技大学能源与环境工程学院，北京　100083；
2. 工业典型污染物资源化处理北京市重点实验室，北京　100083）

摘　要：近年来，钢铁行业各项主要污染物排放量已超过电力行业，成为工业领域最大的排放源，实施钢铁行业超低排放改造是打赢蓝天保卫战的重要措施。同时，钢铁行业作为重度碳素燃料使用行业，其碳排放量占比全国碳排放的 18%以上，中国提出在 2030 年和 2060 年分别实现“碳达峰”和“碳中和”。要实现钢铁行业的“超低排放”与“碳中和”的减排目标，需围绕半干法、湿法、干法全覆盖，统筹兼顾污染物治理、节能、副产物资源化进行技术创新与研究。本次报告聚焦钢铁的绿色低碳发展：钢铁以煤基化石燃料为主，能源-污染物-碳排放具有同源性，高效低耗多功能耦合超低排放技术是污染物治理的科学之道；在现有常规污染物减排的同时要逐步关注非常规污染物减排如预烟气脱氯和精脱硫，实现全面超低排放是碳捕集的基础，并对未来钢铁行业减污降碳和碳减排路线进行了探索和展望。

关键词：钢铁行业；超低排放；碳中和

Ultra-low Emission Technology and Green and Low-carbon Development Prospect of Steel Industry under the Background of Carbon Neutrality

Xing Yi[1,2]

(1. School of Energy and Environmental Engineering, University of Science and Technology Beijing, Beijing 100083, China; 2. Beijing Key Laboratory of Source-based Treatment of Industrial Typical Pollutants, Beijing 100083, China)

Abstract: In recent years, the emissions of major pollutants in the iron and steel industry have exceeded that of the electric power industry and become the largest emission source in the industrial field. The implementation of ultra-low emission transformation in the iron and steel industry is an important measure to win the battle against blue sky. At the same time, the steel industry as a heavy carbon fuel use industry, its carbon emissions accounted for more than 18% of the national carbon emissions, China proposed to achieve "carbon peak" and "carbon neutral" by 2030 and 2060 respectively. In order to achieve the emission reduction goals of "ultra-low emissions" and "carbon neutrality" in the steel industry, it is necessary to focus on the full coverage of semi-dry process, wet process and dry process, and overall consideration of pollutant treatment, energy saving and resource recovery of by-products to carry out technological innovation and research. This report focuses on the green and low-carbon development of iron and steel: iron and steel is dominated by coal-based fossil fuels. Energy-pollutant-carbon emissions have homology. In addition to the reduction of conventional pollutants, we should gradually pay attention to the reduction of unconventional pollutants, such as pre-flue gas dechlorination and fine desulphurization, to achieve comprehensive ultra-low emissions is the basis of carbon capture, and the future iron and steel industry pollution reduction and carbon reduction routes are explored and prospected.

Key words: steel industry; ultra-low emissions; carbon neutral

焦炉荒煤气强化重整脱焦提质的研究

谢华清，张津宁，于震宇

（东北大学冶金学院，辽宁沈阳　110819）

摘　要：提出一种吸附强化焦炉荒煤气蒸汽重整制氢的工艺概念，利用焦炉荒煤气自身尚未回收利用的高温显热在CO_2吸附剂的参与下强化其焦油等大分子有机组分蒸汽重整制氢进程，在实现焦炉煤气余热利用、在线焦油脱除的同时，达到焦炉煤气增质增量化制氢的目的，以推动氢冶金技术的发展以及钢铁行业绿色低碳转型升级。通过热力学分析和实验研究对该工艺重整制氢过程进行探究。热力学结果表明，焦炉荒煤气经过吸附强化蒸汽重整反应后，其焦油组分能够完全裂解、重整为H_2等小分子气体，氢气放大倍数接近理论最大值 4.38，氢气浓度超过 95%，同时反应体系中的碳主要以CO_2被收集，吸附率可达 98%，体现出显著的CO_2减排效益。制备了吸附-重整双功能催化剂（Ni-CeO_2-CaO），应用于焦炉荒煤气蒸汽重整实验中，结果表明在温度 800℃，水碳比 15，质量空速（WHSV）0.0847h^{-1}时，氢产量达到 83.80%，H_2产量以及重整后净 COG 的量分别放大 3.86 倍和 2.52 倍。结果有效验证了该工艺的可行性和先进性。

关键词：焦炉荒煤气；蒸汽重整；吸附强化；氢气

Study on Raw Coke Oven Gas Enhanced Reforming with Tar Removal

Xie Huaqing, Zhang Jinning, Yu Zhenyu

(Northeastern University School of Metallurgy, Shenyang 110819, China)

Abstract: The concept of sorption-enhanced steam reforming of raw coke oven gas for hydrogen production is proposed. In this process, the tar and other macromolecular organic components were reformed to produce hydrogen within situ CO_2 adsorption, and the heat supplying such an endothermic process was from the sensible heat of coke oven gas which was wasted before. Thus, the novel process can not only realize the hydrogen enrichment also the heat waste recovery, and then promoting the development of hydrogen metallurgy technology and the low-carbon transformation of the steel industry. For this process, the thermodynamic analysis and experimental research were conducted. Thermodynamic results showed that the tar component of raw coke oven gas can be cracked completely and reformed into H_2 and other small molecular gases after sorption-enhanced steam reforming reactions, with the hydrogen amplification ratio close to the theoretical maximum of 4.38 and the hydrogen concentration over 95%. Meanwhile, the carbon in the reaction system was mainly collected as CO_2, and the capture rate can reach 98%, showing the significant CO_2 emission reduction benefits. The sorption-reforming dual-function catalyst (Ni-CeO_2-CAO) was prepared and applied to the steam reforming experiment of raw coke oven gas. The results showed that the hydrogen yield reached 83.80% at the temperature of 800℃, the water carbon ratio of 15, the weight hourly space velocity (WHSV) of 0.0847h^{-1}. H_2 production and the amount of COG after reforming were amplified 3.86 and 2.52 times, respectively. The results effectively proved the feasibility and advance of the novel process.

Key words: raw coke oven gas; steam reforming; adsorption-enhanced; hydrogen

提高炼钢转炉二次烟气捕集效率的技术研究与应用

刘昌健，黄艳秋，易勇兵

（1. 中冶南方工程技术有限公司，湖北武汉　430223；2. 西安建筑科技大学，陕西西安　710055）

摘　要：传统的炼钢转炉二次烟气捕集效率较低，相应产生的粉尘逃逸成为影响钢铁企业炼钢车间环境质量和工人工作环境的主要问题。本文通过系统的分析和研究，提出了多种提高转炉二次烟气捕集效率的技术措施，经过理论分析及工程实践，表明应用这些技术措施后，二次烟气捕集效率可以大幅度提高，达到实际工程中目测二次烟气没有明显外逸的效果。

关键词：捕集效率；CFD 模拟；排烟罩

Research and Application of Improveing the Collecting Efficiency of Converter Secondary Fume

Liu Changjian, Huang Yanqiu, Yi Yongbing

(1. WISDRI Engineering & Research Incorporation Limited, Wuhan, 430223, China;
2. Xi'an University of Architecture and Technology, Xi'an, 710055, China)

Abstract: Traditional converter secondary fume collecting system's efficiency is low, the severely fume 's escape become the major issues affecting the steelmaking workshop environment quality and working condtion.Through analysis and research, the paper worked out a variety of technical measures to improve the collection efficiency. Theoretical analysis and engineering practice shows that after applying these measures, secondary flue collection efficiency can be greatly improved to achieve the effect that no obvious fume escape in practical engineering.

Key words: collecting efficiency; CFD simulation; exhaust hood

环保抑尘处理技术在轧钢生产线的发展

徐言东，王兆辉，韩 爽，王占坡，尹浩彬

（1. 北京科技大学工程技术研究院，北京 100083；2. 冶金自动化研究设计院，北京 100071；3. 首钢股份公司迁安钢铁公司，河北唐山 063200）

摘　要：轧钢生产线在钢铁生产中是实现成材的最后一环，高线、棒材、板带热轧、冷轧酸洗生产线的生产过程会伴随着诸多氧化粉尘、油脂烟气、废酸气体的生成，这些产物除了会对生产质量、操作人员职业健康和设备造成损害，还会污染外部环境。随着国家环保政策标准的提高，对企业的环保治理实行一票否决，国内外相关研究机构和企业加强了研究投入，经过近年来的实践，发现美国某品牌喷雾降尘系统，尤其是日本某著名品牌干雾环保抑尘处理系统等技术的应用取得了较好的效果，引领了环保抑尘技术的发展方向。

关键词：轧钢生产线；环保抑尘；氧化粉尘；油脂烟气；废酸气体

Development of Environmental Protection and Dust Suppression Technology in Steel Rolling Production Line

Xu Yandong, Wang Zhaohui, Han Shuang, Wang Zhanpo, Yin Haobin

(1. Engineering and Technology Research Institute, University of Science and Technology Beijing, Beijing 100083, China; 2. Automation Research and Design Institute of Metallurgical Industry Beijing, Beijing 100071, China; 3. Beijing Shougang Qianan Steel Co., Ltd., Tangshan 063200, China)

Abstract: Steel rolling line is the last link in steel production. The production process of high-speed wire, bar, strip hot rolling, cold rolling and pickling line will be accompanied by the generation of a lot of oxidation dust, grease smoke and waste acid gas. These products will not only damage the production quality, occupational health of operators and equipment, but also pollute the external environment. With the improvement of the national environmental protection policy standards, one vote veto on environmental protection of enterprises has been carried out. Relevant research institutes and enterprises at home and abroad have intensified their research investment. After years of practice, it has been found that the application of spray dust reduction system of a brand in the United States, especially the dry fog environmental protection and dust suppression treatment system of a famous brand in Japan, has achieved good results. Leading the development direction of environmental protection and dust suppression technology.

Key words: steel rolling line; environmental protection and dust suppression; oxidized dust; oil smoke; waste acid gas

熔融钢渣高效罐式有压热闷及发电技术与装备

郝以党，吴　龙

（中冶建筑研究总院有限公司，北京　100088）

摘　要：中冶建筑研究总院长期致力于钢铁渣处理利用技术研发推广工作，20 余年来自主创新成功研发一至四代钢渣热闷技术，并在国内外开展了大量的工程应用实践。近年来，新研发熔融钢渣高效罐式有压热闷技术与装备成功应用于首钢京唐、宝钢湛江、武钢、河钢乐亭、沙钢等 53 家钢铁企业，并出口至一带一路沿线马来西亚联合钢铁（大马）集团，累计建设钢渣处理生产线 109 条,设计钢渣处理能力 3866 万吨，服务钢铁产能超过 3.0 亿吨。该技术采用装备自动化工艺，处理适用性强，热闷时间缩短至 1.5-3h；处理生产过程烟气排放浓度<10mg/Nm^3；处理后钢渣粉化率（粒径<20mm）>70%，f-CaO 含量<2.5%，浸水膨胀率<1.5%；同时，处理过程产生大量水蒸气，具备了余热发电利用的条件。该技术总体实现了钢渣处理清洁化生产，改变了钢渣处理现状，促进了钢渣的资源化利用，实现了钢渣处理模式跨越式发展。

关键词：钢渣；罐式有压热闷；发电

Molten Steel Slag High Efficiency Tank Type Pressured Hot-disintegrating and Power Generation Technology and Equipment

Hao Yidang, Wu Long

（1. State Key Laboratory of Iron and Steel Industry Environmental Protection, Beijing 100088, China;
2. Energy Conservation and Environment Protection Co., Ltd., MCC Group, Beijing 100088, China;
3. Central Research Institute of Building and Construction Co., Ltd., MCC Group, Beijing 100088, China）

Abstract: Central Research Institute of Building and Construction contributes to iron and steel slag treatment and utilization technology research and promotion in a long term. The first to the fourth generation of steel slag heat-treatment technology has been successfully developed through independent innovation over the past 20 years. Moreover, a large number of engineering applications have been carried out at home and abroad. Recent years, Molten steel slag high efficiency tank type pressured hot-disintegrating technology has been successfully applied in 53 steel enterprises such as Jingtang Shougang, Baosteel, Wisco, Leting Hesteel and Shagang. Furthermore, the technology was exported to United Iron and Steel (Malaysia) Group along the Belt and Road.109 steel slag processing production lines has been built and the designed steel slag processing capacity was up to 38.66 million tons. The served steel production capacity is higher than 300 million tons. The technology adopts equipment automation process and the processing applicability is strong. The treatment time is shortened to 1.5-3h and flue gas emission concentration can be controlled lower than 10mg/Nm^3 during treatment production process. After treating, the pulverization rate (particle size <20mm) is higher than 70%. The f-CaO content and soaking expansion rate are lower than 2.5% and 1.5%, respectively. At the same time, a large amount of water vapor is generated during the treatment process, which has the conditions for power generation and utilization of waste heat. In general, this technology has realized the clean production of steel slag treatment, which changing the status and promoting the resource utilization of steel slag. It realizes the leapfrog development of steel slag treatment mode.

Key words: steel slag; tank type pressured hot-disintegrating; power generation

精炼铸余渣熔态处理新技术的研究及应用

孙 健，吴 桐

（中冶节能环保有限责任公司）

摘 要：我国精炼铸余渣年产量超过 1500 万吨，含铁量超过 50%，主要采用热泼法处理。热泼后形成的渣钢砣，单重一般在 20~60 吨左右，不符合循环利用的单件尺寸和重量要求，钢铁企业需设置独立车间对渣钢砣进行落锤处理或者氧气切割，力求实现金属铁资源的全量回收。该方法工艺流程长、自动化程度低、环境污染大、人员工作环境差且安全隐患大，急需技术升级。武钢、攀钢等钢铁企业尝试对铸余渣进行熔态改性后，作为 LF 炉脱硫精炼渣直接回用，因对 LF 工艺影响较大，仅停留在试验阶段。宝钢尝试使用滚筒法处理铸余渣，因设备运行维护成本高，没有大范围推广。首钢、宝钢采用的格珊法，效果较好，但冶金渣混凝土格栅为一次性使用，且预制成本较高，超过 1800 元/个。本技术根据铸余渣含铁量高的特点，采用模铸法处理铸余渣。模铸试验结果表明，该方法实现了铸余渣熔态分割，并具有尾渣易脱模，渣铁易分离等优点。以模铸法为核心开发的整套工艺装备，实现了铸余渣处理的自动化、装备化和清洁化，并在新余完成了试验生产线建设及试生产，效果良好。

关键词：铸余渣；热泼法；格珊法；模铸法

Research and Application of New Technology of Refining Casting Residue Treatment

Sun Jian, Wu Tong

(Energy Conservation and Environmental Protection Co., Ltd., MCC Group)

Abstract: The annual output of refining casting residue in China is more than15 million tons, and its iron content is more than 50%, which is mainly treated by hot splashing method. The single weight of the steel chunk in slag formed after the hot splashingis generally about 20~60tons, which does not meet the recycling requirements of the size and weight of the single piece.Therefore iron and steel enterprises need to set up independent workshops to hammer and oxygen cut the steel chunk into pieces, and strive to achieve the full recovery of metal iron resources. This method has long process flow, low degree of automation, high environmental pollution, poor working environment and great security risks, which is in urgent need of technical upgrading.Wisco, Pangang and other iron and steel enterprises try to modify the cast residue in the molten state and reuse it directly as the desulphurization refining slag of LF furnace. Because of the great influence on LF process, the method only stays in the experimental stage. Baosteel tried to use rotary cylinder to treat residual slag, but it was not widely promoted due to the high cost of equipment operation and maintenance.Shougang, Baosteel used grating method, which got a good effect, but the metallurgical slag concrete grating is disposable, and the prefabrication cost is higherthan1800 yuan per grating.According to the characteristics of high iron content of casting residue, the die casting method is used.The results of die casting tests show that this method can achieve the molten separation of casting residue, and has the advantages of easy demoulding of tailings and easy separation of slag and iron.With the core of die casting method,the whole set of process equipment developed which realized the automation, equipment and cleaning of casting residue treatment, and the construction of test production line and trial production in Xinyu have been completed with good results.

Key words: casting residue; hot splashing method; gratingmethod; die castingmethod

烧结烟气脱硝废弃SCR催化剂资源化利用途径分析

龙红明，钱立新，丁　龙，余正伟

（1. 安徽工业大学冶金工程学院，安徽马鞍山　243032；
2. 冶金减排与资源综合利用教育部重点实验室（安徽工业大学），安徽马鞍山　243002）

摘　要：选择性催化还原法（SCR）已成为烧结烟气脱硝主流技术之一，导致钒钨钛系废催化剂（危废）产生量逐年增加。随着环保要求日益严厉，加强对这类危废的有效处置利用已成为行业急需解决的关键共性难题。介绍了烧结烟气脱硝废催化剂的失活机理及超低排放背景下未来3-5年废催化剂的产生数量，从活性再生、资源化利用、无害化处置等方面综述了国内外关于废催化剂处置的最新研究进展及挑战。结合钢铁企业“固废不出厂”的新发展理念，分析了废催化剂作为含钛资源在钢铁生产流程中资源化利用的相关研究思路，并对废催化剂进一步钢铁企业内部绿色清洁利用进行了展望。

关键词：烧结烟气脱硝；废弃催化剂；资源化利用；含钛炉料；高炉护炉

Analysis on Resource Utilization of Waste SCR Catalyst from Sintering Flue Gas Denitrification

Long Hongming, Qian Lixin, Ding Long, Yu Zhengwei

（1. School of Metallurgical Engineering, Anhui University of Technology, Ma’anshan 243032, China;
2. Key Laboratory of Metallurgical Emission Reduction & Resources Recycling (Anhui University of Technology), Ministry of Education, Ma’anshan 243002, China）

Abstract: Selective catalytic reduction (SCR) has become one of the mainstream technologies for sintering flue gas denitration, which leads to the increasing production of vanadium, tungsten, and titanium waste catalyst (hazardous waste) year by year. With the increasingly stringent environmental protection requirements, strengthening the effective utilization of this solid waste has become a key common problem urgently needed to be solved in the industry. The deactivation mechanism of spent catalyst from sintering flue gas denitrification and the production quantity in the next 3-5 years under the background of ultra-low emission were introduced. The latest research progress and challenges of spent catalyst disposal at home and abroad were reviewed from the aspects of catalyst regeneration, resource utilization, and harmless disposal. Combined with the new development concept of “solid waste not leaving the factory” in iron and steel enterprises, the research ideas of spent catalyst as titanium-containing resource utilization in iron and steel production process were analyzed. Finally, the further green and clean utilization approaches of spent catalyst in iron and steel enterprises were prospected.

Key words: sintering flue gas denitrification; spent catalyst; resource utilization; titanium-bearing charge; blast furnace lining protection

河钢集团固废协同优化处置及资源化利用

孙宇佳，田京雷

（河钢集团有限公司，河北石家庄 050023）

摘 要：报告阐述近年来河钢集团在钢铁行业固废协同处置方面的实践与发展，重点介绍高炉水渣、钢渣、脱硫脱硝副产物、冶金尘泥以及废旧 SCR 脱硝催化剂这几类典型固废的处置及研究，旨在与下游建筑、建材、交通、环境治理等产品应用领域深度融合，打通部门间、行业间堵点和痛点。并结合近期循环经济的政策法规和规划的方向，为我国钢铁工业固废综合利用产业与上下游产业结合构成循环经济发展，以及促成我国双碳目标达成、节污降碳提供思路。

关键词：固废；协同处置；循环经济；节污降碳

Collaborative Optimal Disposal and Resource Utilization of Solid Waste in HBIS Group

Sun Yujia, Tian Jinglei

(HBIS Group Co., Ltd., Shijiazhuang 050023, China)

Abstract: The report describes the practice and development of HBIS Group in the collaborative disposal of solid wastes in the iron and steel industry in recent years, focusing on the disposal and research of typical solid wastes such as blast furnace slag, steel slag, desulfurization and denitration by-products, metallurgical dust and sludge and waste SCR denitration catalyst, in order to deeply integrate with the application fields of downstream construction, building materials, transportation, environmental treatment and other products, and open up inter departmental blocking points and pain points among industries. Combined with the recent policies, regulations and planning direction of circular economy, it provides ideas for the combination of solid waste comprehensive utilization industry of China’s iron and steel industry and upstream and downstream industries to form the development of circular economy, and promote the achievement of China's double carbon goal, pollution saving and carbon reduction.

Key words: solid waste; collaborative disposal; circular economy; saving pollution and reducing carbon

配碳还原氧化铁皮制备多孔不锈钢的机理研究

张 芳，赵立杰，彭 军，明守禄

（内蒙古科技大学材料与冶金学院，内蒙古包头 014010）

摘 要：为了充分发挥氧化铁皮含铁品位高、杂质元素含量低、产生量大的优势和特点，并开发高附加值的金属制品，本文采用还原烧结的方法制备了 316 多孔不锈钢。研究过程中采用高温真空管式炉还原烧结制备了多孔不锈钢试样；采用直读光谱、氧氢氮联合检测仪、XRD 衍射分析仪、场发射 SEM-EDS 等设备对试样的化学成分、物相组成和微观形貌进行了检测。研究结果表明，以氧化铁皮为含铁原料，配入适量的石墨及合金粉末，在 10^{-3}Pa 真空度、1200℃下保温 3h 可以制备出 316 多孔不锈钢，在此条件下 Mn 合金粉末的收率最低，为 61%；制备出的不锈钢试样的组织为单一奥氏体，少数局部位置存在铬元素不均匀的情况；同时在晶内有 CrC 析出，晶界附近存在 σ 相；本论文制备的 316 多孔不锈钢在加入造孔剂之前的孔隙率为 37.6%，每增加 10%质量比的碳酸氢铵，试样的孔隙率增加 9.3%。

关键词：氧化铁皮；316 多孔不锈钢；还原烧结；孔隙率

Mechanism of Preparing Porous Stainless Steel by Reducing Iron Scale with Carbon

Zhang Fang, Zhao Lijie, Peng Jun, Ming Shoulu

Abstract: In order to give full play to the advantages and characteristics of iron scale with high iron grade, low impuritycontent and large production, and develop high added value metal products, 316 porous stainless steel was prepared by reduction sintering in our work. The porous stainless steel samples were prepared in high temperature vacuum tube furnace. The chemical compositions, phase compositions and microstructure of the samples were detected by direct reading spectroscopy, O-H-N combined detector, XRD diffraction analyzer and field emission SEM-EDS. The results show that 316 porous stainless steel can be prepared by adding appropriate graphite and alloy powder with iron oxide as raw material at 10^{-3}Pa vacuum and 1200℃ for 3h. Under this condition, the yield of Mn alloy powder is the lowest, which is 61%. The microstructure of the prepared stainless steel sample is single austenite, and chromium element is not uniform in a few of positions. CrC precipitates austenite in the grain and σ phase exists near the grain boundary. The porosity of 316 porous stainless steel prepared in this paper is 37.6% before adding pore-forming agent, and the porosity of the sample increases by 9.3% with an increase of 10% mass ratio of ammonium bicarbonate.

Key words: iron oxide; 316 porous stainless steel; reduction sintering; porosity

ISO 水回用技术标准创新发展与启示

梁思懿，陈　卓，巫寅虎，胡洪营

（1. 中冶京诚工程技术有限公司，北京　100176；2. 清华大学环境学院，北京　100084）

摘　要：污水再生利用是保障用水安全、改善水环境质量和建设生态文明的国家重大需求。国际标准化组织水回用技术委员会（ISO/TC282）自 2017 年起陆续发布了《污水再生处理反渗透反渗透脱盐系统设计》（ISO23070）、《再生水安全性评价与方法》（ISO20761）等多项水回用领域国际标准，规定了污水回用系统设计原则和方法，阐释了水源、处理、储存、输配、监测等水回用关键环节的设计要求和关键水质指标，明确了处理工艺，管网输配和利用途径之间的关系。在清华大学等单位联合编制的 5 项国际标准中，创新性提出了“污水特质评价”、“反渗透膜污堵潜势评级”等评价指标体系和方法，相应催生了污水再生利用的新理念和新工艺，对中国钢铁行业废水高效资源化利用具有很强的指导作用。

关键词：水回用系统；评价和质量管理；安全高效利用

Innovation and Development of ISO International Standards on Water Reuse Technology and Its Enlightenment

Liang Siyi, Chen Zhuo, Wu Yinhu, Hu Hongying Hu

(1. MCC Capital Engineering and Research Incorporation Ltd., Beijing 100176, China;
2. School of Environment, Tsinghua University, Beijing 100084, China)

Abstract: Water reuse is a major national demand to ensure water safety, improve water environment quality and build ecological civilization. International Organization for Standardization Technical Committee on Water Reuse (ISO/TC282) issued a number of international standards in the field of water reuse since 2017, including *Design principle of RO System of municipal wastewater (ISO23070)*, *Guideline for water reuse safety evaluation: assessment parameters and methods (ISO20761)*, etc. This forms a basis in water reuse field which proposes design principles and methods of water reuse system considering the key aspects of source, treatment, storage, transmission and monitoring, clarifies the relationship among treatment technology, transmission and distribution and end uses. Tsinghua University and other institutions have compiled five ISO international standards. The evaluation methods and index system such as "evaluation of water feature" and "RO fouling potential" were innovatively put forward. Accordingly, the new concept and technology on water reuse are produced. It can play a strong guiding role in the efficient resource utilization of reclaimed water in China in iron and steel industry.

Key words: water reuse system; evaluation and quality management; safe and efficient use

碳达峰背景下的钢铁工业水系统优化

吕光辉

（太原钢铁（集团）有限公司能源环保部，山西太原 030003）

摘 要：2000 年以来，我国钢铁企业节水减排已取得了显著的成效，各钢铁企业逐步形成了完整的水系统模式，吨钢耗新水、吨钢排水量均有大幅度的下降，外排水 COD、氨氮等部分指标达到国际先进水平。在当前碳达峰、碳中和目标的背景下，钢铁企业水系统如何通过系统优化、技术创新协同实现减污降碳、提高水资源循环率等多目标值，是当前摆在我们面前迫切需要解决的问题。本文分析了减污降碳、提高水循环利用率、水污染环境风险防控等多方面协同增效的潜力，构建了钢铁企业高效水效水系统模式，提出了水处理技术创新的方向。

关键词：钢铁工业；水系统优化；减污降碳；原级资源化

Optimization of Water System in Iron and Steel Industry under the Background of Carbon peak

Lü Guanghui

(Taiyuan Iron and Steel Group Co., Ltd., Taiyuan 030003, China)

Abstract: Since 2000, China's iron and steel enterprises have made remarkable achievements in water saving and emission reduction, and various iron and steel enterprises have gradually formed a complete water system model, some indexes such as COD and ammonia nitrogen of the external drainage reach the international advanced level. In the context of the current carbon peak and carbon neutral targets, how the water system of iron and steel enterprises can achieve multi-objective values such as reducing pollution, reducing carbon and increasing water resource recycling rate through system optimization and technological innovation, and take the lead in achieving the mid-term target of 30% carbon reduction, it is an urgent problem that needs to be solved. This paper analyzes the potential of synergetic effect in reducing pollution and carbon, improving water cycle utilization rate and preventing and controlling water pollution environmental risk, constructs the mode of high efficient water efficiency water system in iron and steel enterprises, and puts forward the direction of innovation of water treatment technology.

Key words: iron and steel industry; optimization of water system; reduce pollution and carbon; primary resourcing

钢铁渣固废的堆存场地风险管控和资源高效循环利用发展方向

岳昌盛，卢光华

（钢铁工业环境保护国家重点实验室，北京　100088）

摘　要：我国钢铁渣年产量数以亿吨计，其中钢渣综合利用率较低，钢铁渣固废堆存量大，不仅占用土地，而且会对土壤、地下水和大气带来一定的环境危害。从中短期来看，钢铁渣固废堆场的风险管控是减少环境危害的重要手段，包括基于固废堆场三维立体污染物扩散构建的垂直阻隔和水平覆盖组合方式可有效避免污染物在土壤、地下水和大气中的污染扩散，同时结合场地地下水抽提技术解决已渗漏污染地下水的环境问题；从中长期来看，钢铁渣固废的资源高效循环利用是解决固废产生和堆存的较好渠道，除应用于路基材料、水泥、建材制品等传统建材技术外，钢渣在矿山充填、土壤改良、人工渔礁等方面的发展已经得到了应用，前景良好；另外，钢铁渣在碳捕获与封存（carbon capture and storage, CCS）方面具有较好的发展前景，其矿物碳酸化方式将有助于为我国“双碳”发展贡献力量。

关键词：钢渣；风险管控；利用

Risk Control of Steel Slag Solid Waste Storage Sites and Development Direction of Efficient Recycling of Steel Slag Resources

Yue Changsheng, Lu Guanghua

（1. State Key Laboratory of Iron and Steel Industry Environmental Protection, Beijing 100088, China;
2. Energy Conservation and Environment Protection Co., Ltd., MCC Group, Beijing 100088, China;
3. Central Research Institute of Building and Construction Co., Ltd., MCC Group, Beijing 100088, China）

Abstract: The annual output of steel slag in China was hundreds of millions of tons, of which the comprehensive utilization rate of steel slag was low and the stock of steel slag solid waste pile was large, which will not only occupied land, but also bring certain environmental hazards to soil, groundwater and atmosphere. In the medium and short term, the risk control of steel slag solid waste yard was an important means to reduce environmental hazards, including the combination of vertical barrier and horizontal coverage based on the three-dimensional pollutant diffusion of solid waste yard, which could effectively avoid the pollution diffusion of pollutants in soil, groundwater and atmosphere, At the same time, combined with the site groundwater extraction technology to solve the environmental problems of groundwater polluted by leakage; In the medium and long term, the efficient recycling of iron and steel slag solid waste resources was a good channel to solve the generation and stockpiling of solid waste. In addition to the traditional building materials technologies such as subgrade materials, cement and building materials products, the development of steel slag in mine filling, soil improvement and artificial reef has been applied and had a good prospect; In addition, iron and steel slag had a good development prospect in carbon capture and storage (CCS), and its mineral carbonation method would contribute to the "double carbon" development in China.

Key words: slag; risk control; utilize

深化超低排放治理 持续推进绿色发展

——唐山东华钢铁企业集团有限公司超低排放改造评估检测工作实践

梁志敏，赵小宇

（唐山东华钢铁企业集团有限公司）

摘 要：唐山东华钢铁企业集团有限公司按照生态环境部、省、市关于环境整治、超低排放等要求，持续开展治污减排、环保提升改造工作。截至 2019 年底，投入超低排放改造资金约 25 亿元，涵盖了从原料储存，烧结、炼铁、炼钢、轧钢到清洁运输的全工序改造；先后完成了环保料库建设、烧结机头烟气脱硫脱硝和烟气循环处理、高炉均压煤气放散全回收、高效布袋除尘器和轧钢塑烧板除尘器新建及改造、加热炉换向煤气反吹回收等项目。为进一步深化超低排放改造工作，2021 年上半年又投资 7.5 亿元，实施热风炉脱硫、轧钢加热炉脱硫脱硝等深度减排处理措施。唐山东华钢铁始终贯彻落实绿色发展和精细化管理理念，全面推进评估检测工作。目前，超低排改造项目已达标验收并稳定高效运行，企业环境面貌焕然一新，经济效益大幅提升。

关键词：超低排放改造；评估监测；绿色发展

Deepen Ultra-low Emission Control and Continue to Promote Green Development

Ultra-low Emissiontechnical Transformation Evaluation and Testing Practice of Tangshan Donghua Iron and Steel Enterprise Group Limited Liability Company

Liang Zhimin, Zhao Xiaoyu

(Tangshan Donghua Iron and Steel Enterprise Group Limited Liability Company)

Abstract:In accordance with the requirements ofthe Ministryof Ecology and Environment of People's Republic of China, the province and the city on environmental remediation and ultra-low emissions, TangShanDonghua iron and steel enterprise group Co., Ltd. has continuously carried out pollution control and emission reduction, environmental protection upgrading and transformation.By the end of 2019, about 2.5 billion yuan has been invested inultra-low emission technical transformation, including the whole processfrom raw material storage, sintering, iron making, steel making, steel rolling to clean transportation. It has successivelyfinished the construction of environmental protection warehouse, the cleaning and circulationtreatment of flue gasin sintering machine front section, fullrecovery of blast furnace equalized pressure releasing gas, new construction and transformation of efficient bag filter and steel rolling plastic burning plate filter, the blowback recoveryof heating furnace reversing gas and other projects. For further deepening the technicaltransformation of ultra-low emission, Donghuahas continued to invest 750 million yuan to carry out in-depth emission reduction measures in projects such as desulfurization of hot blast stoves, desulfurization and denitrification of rolling steel reheating stoves and other environmentalfriendly technology methods in the first half of 2021. In addition, Donghua always keepsthe concept of green development and fine management to advancethe assessment and testing work. At present, the ultra-low emission project has been achieved withstable and efficient operation, which makes enterprise take a new look and economic benefits have been significantly improved.

Key words: ultra-low emission technical transformation; assessment and monitoring; green development

关于钢铁企业超低排放与碳减排的浅显认识与思考

刘恩辉

（首钢京唐钢铁联合有限责任公司，河北唐山　063200）

摘　要：结合钢铁企业超低排放改造，就实施过程中的组织、技术、管理等方面进行了分析，对实施过程中存在的问题、加强企业日常环保管理，提出了改进建议，同时结合当前正在看进行的碳减排工作，如何加强与环保工作的协同、系统实施，进行了阐述，从企业视角进行思考，提出了一些看法和意见。

关键词：超低排放；碳减排

Plain Understanding and Thinking on Ultra-low Emission and Carbon Reduction of Iron and Steel Enterprises

Liu Enhui

（Shougang Jingtang United Iron & Steel Co., Ltd., Tangshan 063200, China）

Abstract: Combined with the ultra-low emission transformation of iron and steel enterprises, we analyzed the organization, technology and management in the implementation process. In addition, we put forward suggestions for improving the problems existing in the implementation process and strengthening the daily environmental management of enterprises. At the same time, combined with the ongoing carbon emission reduction work, we have described how to strengthen the coordination and systematic implementation with environmental protection. Finally, we put forward some views and opinions from the perspective of enterprises.

Key words: ultra-low emissions; carbon emission reduction

棒线材无头焊接轧制系统及经济效益

洪荣勇

（福建三钢闽光股份有限公司，福建三明　365000）

摘　要：当今棒线材的生产机仍主要属于单坯轧制，间断的坯料进给不仅给设备额外增加了许多载荷，还会造成堆钢现象，进而影响产能和管理。无头焊接轧制系统将两根坯料进行头尾焊接，可实现产品的无间断轧制生产，大幅度降低了成本。因此，本文就无头焊接轧制系统的设备组成及其创造的经济效益进行了阐述，因经济效益明显，具有推广价值

关键词：无头焊接轧制系统；移动焊机；经济效益

Structure and Economic Benefit of Endless Rod and Wire Welding Rolling System

Hong Rongyong

(Fujian Sangang Minguang Co., Ltd., Sanming 365000, China)

Abstract: Nowadays, The production mechanism of rod and wire still mainly belong to single billet rolling. The intermittent billet feeding not only adds a lot of additional load to the equipment, but also causes the phenomenon of steel piling, which in turn affects production capacity and management. The endless welding rolling system can realize the uninterrupted rolling production of products by welding the head and tail of the two blanks, which greatly reduces the cost. Therefore, this paper expounds the equipment composition of endless welding rolling system and its economic benefits. Because of the obvious economic benefits, it has popularization value.

Key words: endless welding rolling; moving welder; economic benefit

金属 3D 打印的模型校准、验证和预测控制

奚志敏

（工业和系统工程，罗格斯大学 – 新布朗斯维克，美国）

摘　要：激光粉末床融合过程是一种流行的通过一层一层的金属粉末融化和固化过程来制造金属部件的增材制造技术。该技术目前有很多成功的产品原型。然而，对于其产品缺乏在质量和长期可靠性上的自信可能是一个阻碍该技术被工业界广泛应用的主要原因。在众多原因中，缺乏精确和高效的模型来仿真这个过程可能是最主要的原因之一因为可靠性和质量分析需要精确的模型，否则就需要大量的实验数据来确保可靠性和质量。另外，目前的激光粉末床融合过程主要是一个开环的控制过程并具有一些实时监测能力。所以如果过程中有任何的干扰因素就不能保证制造质量和可靠性。这个报告讲解如何实现有限元模型和数据挖掘模型的整合，并通过利用有限的实验数据进行系统的激光粉末床融合过程的模型校准和验证。基于一个已经验证过的模型，我们继续研究了 PID 和模型预测控制的有效性。

关键词：模型验证；激光粉末床融合；模型预测控制；质量和可靠性

Model Calibration, Validation and Predictive Control for Metal 3D Printing

Xi Zhimin

(Department of Industrial and Systems Engineering, Rutgers University – New Brunswick, USA)

Abstract: Laser power bed fusion (LPBF) process is one of popular additive manufacturing (AM) techniques for building metal parts through the layer-by-layer melting and solidification process. To date, there are plenty of successful product prototypes manufactured by the LPBF process. However, the lack of confidence in its quality and long-term reliability could be one of the major reasons prevent the LPBF process from being widely adopted in industry. Among many reasons, the lack of accurate and efficient models to simulate the process could be the most important one because reliability and quality quantification relies on accurate models; otherwise, large amount of experiments should be conducted for reliability and quality assurance. In addition, the existing LPBF process is an open loop control system with some in-situ monitoring capability. Hence, manufacturing quality and long-term reliability of the part cannot be guaranteed if there is any disturbance during the process. This talk presents the integration of finite element and data-driven modeling with systematic calibration and validation framework for the LPBF process based on limited experiment data. I will also present the study on control effectiveness using the PID control and the model predictive control (MPC) for the LPBF process based on a validated model.

Key words: model validation; laser power bed fusion; model predictive control; quality and reliability

深冷轧制工艺开发与发展

喻海良

（中南大学轻合金研究院，湖南长沙　410083）

摘　要：深冷轧制是近年新兴的一种轧制工艺。本报告介绍钛合金材料、铝合金材料、铜合金材料在深冷环境中的力学性能变化情况。基于他们在深冷环境下的变形能力，重点介绍上述三类材料在深冷轧制过程中的变形特点。对三种典型的材料在深冷轧制中的变形特点和解决的问题进行了介绍。对于 Ti-6Al-4V 合金材料，深冷轧制可以细化第二相尺寸，进而，提升材料性能。对于 Al-Cu-Li 合金材料，深冷轧制可以大幅降低材料的时效处理温度，并实现材料性能提升。对于 Cu-Ni-Sn 材料，深冷轧制可以抑制材料内部偏析行为，解决热轧和冷轧出现裂纹的问题。

关键词：深冷轧制；钛合金；铝合金；铜合金

Development of Cryorolling Technique

Yu Hailiang

(Institute of Light Metals, Central South University, Changsha 410083, China)

Abstract: Cryorolling is a new rolling process in recent years. This report introduces the changes in mechanical properties of titanium alloys, aluminum alloys, and copper alloys in a cryogenic environment. Based on their deformability in a cryogenic environment, the deformation behavior of the above three types of materials during cryorolling is introduced. Finally, the deformation characteristics of three typical materials subjected to cryorolling are introduced. For Ti-6Al-4V alloy materials, cryorolling can refine the size of the second phase, thereby improving material properties. For Al-Cu-Li alloy materials, cryorolling can greatly reduce the material's aging treatment temperature and improve material performance. For Cu-Ni-Sn materials, cryorolling can suppress the internal segregation behavior of the material and solve the problem of cracks in hot rolling and cold rolling.

Key words: cryorolling; titanium alloy; aluminum alloy; copper alloy

一种有监督的适用于缺失数据的高维数据降维方法及其在设备寿命预测中的应用

方晓磊

（北卡罗来纳州立大学，罗利，北卡　27695，美国）

摘　要：多数用于设备状态监测的统计及机器学习的模型包含两步：特征提取（从高维数据中提取低维特征）和特征建模（利用提取的低维特征进行回归，分类，聚类等分析）。此类两步方法的缺点是第一步提取的特征不一定最适用于第二步的特征建模。为了解决这个问题，本报告提出一种新的有监督的数据降维方法将前述两步的数据分析方法合并为一步。本方法构建一个分为两部分的优化模型，第一部分为能够从高维的、不完整的数据里提取低维特征，第二部分利用第一部分提取的特征进行回归分析。通过同时优化前述的两个部分，第一部分提取的特征能够更有效的用于第二部分的建模。为了估计模型参数，我们提出一种基于区块坐标下降的优化方法并证明了其收敛性质。我们用仿真的和 NASA 的飞机退化数据来验证我们提出的方法的有效性。

关键词：退化建模；矩阵填充；数据融合

A Supervised Dimension Reduction-based Prognostics Model for Applications with Incomplete Signals

Fang Xiaolei

Abstract: In prognostic analysis, the high-dimensionality and high-incompleteness of degradation signals as well as the censoring of historical failure times pose significant challenges for residual useful life prediction. To address the foregoing challenges, this paper develops a novel supervised dimension reduction-based prognostics methodology. The methodology builds an optimization problem that combines a feature extraction term and a regression term. The first term is capable of extracting low-dimensional features from high-dimensional incomplete degradation signals, while the second term regresses the features against censored failure times. By simultaneously optimizing the two terms, the extracted features are guaranteed to be most informative for predicting failure times. To solve the optimization problem, a Block Prox-Linear Coordinate Descent algorithm with a global convergence property is developed.

Key words: degradation; matrix completion; data fusion

基于机器视觉的轧钢无人驾驶精准控制技术

徐　冬，杨　荃，王晓晨，何海楠，刘　洋

（1. 北京科技大学工程技术研究院，北京　100083；

2. 北京科技大学设计研究院有限公司，北京　100083）

摘　要：在热轧过程中，中间坯镰刀弯、翘扣头，精轧带钢跑偏、甩尾等非对称缺陷，不仅会造成成品质量问题，而且还会影响整个轧制过程的连续性与稳定性，是实现轧钢过程无人驾驶的重要瓶颈。本文针对不同场景，基于机器视觉技术设计开发了翘扣头、镰刀弯、跑偏等检测系统，实现了粗轧阶段中间坯三维形状检测与精轧阶段高频跑偏数据检测，并基于检测结果构建了机理、数据联合驱动的雪橇系数、轧机调平量在线控制系统。应用结果表明：测控系统运行稳定，能够显著减低非对称缺陷的产生，提高了产品形状精度和生产稳定性，有效减少人工操作，为热轧生产的无人化奠定了基础。

关键词：热轧；镰刀弯；翘扣头；跑偏；机器视觉；精准控制

汽车轻量化发展及用材策略

高永生

（首钢股份有限公司，北京　100043）

摘　要：汽车工业是我国国民经济重要支柱产业之一，汽车用材技术直接体现了我国钢铁冶金工艺技术和材料应用技术水平。本世纪以来，汽车轻量化概念逐渐被汽车主机厂应用并且在汽车结构设计和选材策略方面取得了长足的进展。但是近年来汽车智能驾驶技术以及国家双碳战略的布局，给汽车用材提供了新的研究方向，也给汽车安全法规的制定和执行提供了新的变化。因此，依据国家的政策和战略，钢企和汽车主机厂都对汽车用材进行的重新思考和布局，提质、增效、工艺降成本已经成为减碳和绿色可持续发展的主要举措，这也为钢企的材料品种开发提出了方向。

关键词：汽车；轻量化；减碳

The Development of Auto Lightweighting Technologies and Material Application Strategy

Gao Yongsheng

(Beijing Shougang Co., Ltd., Beijing 100043, China)

Abstract: Automobile industry is one of the most important pillar industries of China's national economy. Automobile material technology directly reflects the level of iron and steel metallurgical process technology and material application technology. Since this century, the concept of automobile lightweighting has been gradually applied by automobile OEMs, and great progress has been made in automobile structure design and material selection strategy. However, in recent years, the intelligent driving technology and the layout of the national "dual carbon" strategy have not only provided a new research direction for automotive materials, but also provided new changes for the formulation and implementation of automotive safety regulations. Therefore, according to the national policies and strategies, both of steel and automotive makers rethink and layout of steel catalogues and grades of auto materials. Improving quality, increasing efficiency and reducing process cost have become the main measures for carbon reduction and green sustainable development, which also puts forward the direction for the development of material varieties of steel enterprises.

Key words: automotive; lightweighting; carbon reduction

Inclusions Control in High-quality Bearing Steels based on the Prediction of Fatigue Life

Gu Chao, Lian Junhe, Münstermann Sebastian

(1. State Key Laboratory of Advanced Metallurgy, University of Science and Technology Beijing, Beijing 100083, China; 2. Department of Mechanical Engineering, Aalto University, Espoo 02150, Finland; 3. Steel Institute, RWTH Aachen University, Intzestraße 1, Aachen 52074, Germany)

Abstract: Bearing steel is an important manufacturing material of basic parts, which is widely used in aerospace, high-speed railway, shipbuilding and other fields. Improving the fatigue life of bearing steel under special service conditions is research hotspot. Accurate prediction in fatigue limit is crucial to the service safety. Traditional characterization method in fatigue properties of materials is acceptable for low cycle fatigue test; however, it is difficult to determine the fatigue limit within the effective test cycle for the high or ultra-high cycle fatigue-resistant and damage-tolerant materials. In this work, we select martensitic bearing steel as the research steel grade, using representative volume element (RVE) model, data process software, such as Matlab, Python and Origin, and finite element method to develop the multi-scale fatigue life prediction model on the basis of micro-structure, including the most common inclusion in bearing steel, to realize the prediction of the fatigue life of bearing steel under different inclusion conditions. Additionally, based on the multi-scale model, the production conditions can be optimized to meet the micro-inclusion and the fatigue life of materials design. The research work provides the basis for improving the high-quality bearing steel fatigue life and satisfying the customized design thoughts.

Key words: fatigue life prediction; micro-inclusion; multi-scale model; bearing steel

Mechanics and Practice of Novel Nanolubrication in Hot Steel Rolling

Wu Hui, Huang Shuiquan, Xing Zhao, Jiao Sihai, Huang Han Jiang Zhengyi

(1. School of Mechanical, Materials, Mechatronic and Biomedical Engineering, University of Wollongong, Wollongong, NSW 2522, Australia; 2. School of Mechanical and Mining Engineering, The University of Queensland, Brisbane, QLD 4072, Australia; 3. Baosteel Research Institute (R&D Centre), Baoshan Iron & Steel Co., Ltd., Shanghai 200431, China)

Abstract: Lubrication applied in hot rolling of steels has been a long-standing research topic as it can significantly improve both the roll service life and the strip surface quality, and also reduces the energy consumption by decreasing rolling force and torque. Conventional oils and oil-in-water emulsions have been used extensively in hot steel rolling over the past decades. The use of these oil-containing lubricants, however, inevitably brings environmental issues when burned and discharged, and regular maintenance of oil pipes and nozzles is usually a demanding job. This study is to develop eco-friendly and recyclable water-based nanolubricants in large quantity that are applicable in hot steel rolling production line. Fundamental research on novel water-based nanolubricants containing TiO_2 (~20nm in diameter) was first conducted to examine their tribological performance at ambient and elevated temperatures up to 500℃ using an innovative ball-on-disk tribometer. The use of 4wt% nanolubricant decreased the coefficient of friction (COF) and wear of ball (WOB) over 40% and 80%, respectively, at all temperatures. Hot rolling tests were then carried out at rolling temperatures of 850 and 950℃ with a rolling reduction of 30% and a rolling speed of 0.35 m/s using 4wt% nanolubricant. This enabled a decrease in rolling force over 6% and a decrease in oxide scale thickness over 40%, compared to those obtained using pure water. To further enhance the lubrication performance, the formulation of developed water-based nanolubricants was optimised using sodium dodecyl benzene sulfonate (SDBS). The COF and WOB obtained by use of pure water were decreased over 70% and 80%, respectively, when using the nanolubricant containing 4wt% TiO_2 and 0.4wt% SDBS. The hot rolling results showed a decrease in rolling force above 8% at a rolling temperature of 850℃ with a rolling reduction of 30% and a rolling speed of 0.35m/s. The lubrication mechanisms using such nano-sized TiO_2 (~20nm) in water were contributed by ball bearing, mending, polishing and lubricating film of the nanoparticles (NPs). For the purpose of accelerating the application and popularisation of water-based nanolubricant in industrial-scale hot steel rolling, the synthesis and operation costs of applied lubricants were greatly reduced using relatively coarse TiO_2 NPs with the aids of glycerol, SDBS and Snailcool (a novel extreme pressure agent) that are low-cost aqueous additives. The results indicated that the water-based nanolubricants consisting of 4wt% TiO_2, 10wt% glycerol, 0.2wt% SDBS and 1wt% Snailcool exhibited the best lubrication performance by reducing the rolling force, surface roughness and oxide scale thickness of rolled strip up to 8.1%, 53.7% and 50%, respectively, at a rolling temperature of 850℃ with a rolling reduction of ~30% and a rolling speed of 0.35m/s. The lubrication mechanisms of applied water-based nanolubricants were ascribed to the synergistic effect of ball bearing, laminae and mending of TiO_2 NPs. The pilot-scale hot rolling tests were subsequently conducted at Baosteel, indicating a decrease in rolling force above 5.5% at a rolling temperature of 850℃ with varying reductions of 20%-40%.

Keywords: eco-friendly; water-based nanolubricant; tribological performance; hot steel rolling

基于演化博弈和收益分配的物流资源共享主体决策研究

孔继利，陈梓语

（北京邮电大学）

摘　要：随着物流业的快速发展，城市地区的“最后一公里配送”越来越受到重视。物流共享作为一种节能、高效且可实行的方法受到了物流业的欢迎，但在实际应用中效果不佳。因此，本文聚焦城市末端的物流共享问题。建立快递企业间非对称演化博弈模型，探讨各因素的影响路径。应用 MATLAB 和 K-means 算法进行仿真研究，验证模型有效性。在此基础上，研究物流资源共享主体的收益分配问题，构建了收益分配模型。根据模型结论和数值结果，最后提出不同情况下构建物流共享联盟的策略。

Petri 网建模求解混流装配生产线调度问题

高慕云

（北京科技大学物流工程系）

摘　要：针对混流装配生产线缩短生产工期的需求，本文首先应用赋时变迁 Petri 网（Timed Transition Petri Nets, TTPN）仿真模型建模描述装配生产线工序间的复杂关系，以装配生产工期最短为优化目标，建立装配生产线调度问题的优化模型，最后使用改进的遗传算法求解优化模型。为了实现灵活输入加工产品数据的目标，本文在工序编码方式的基础上提供一种分区间解码的思路，实现对同类产品不同数量的灵活调整，最终得出产品最短生产周期及其对应的加工调度方案。通过应用实例计算结果对比，有效缩减了产品的装配生产周期。

基于 SIR 模型的应急医疗物资仓库选址

姜肖依

（北京科技大学物流工程系）

摘　要：应急医疗设施对于人民的生命财产安全具有重要意义，一个合理的选址策略将有效提高国家对疫情的预防与控制水平。本文以应对某种疫情扩散为背景，基于复合种群的传染病模型以及 P 中值问题模型，构建了综合考虑人口及交通因素的应急医疗物资仓库选址模型。同时本文基于改进的遗传算法，使用真实运输距离数据，对选址问题进行求解。实验结果验证了本文提出的模型和算法的有效性。

冶金学的时代命题——打通流程，沟通层次，开新说

殷瑞钰

摘　要：经过近百年的探索、研究发展进程，当代冶金学（冶金科学与工程）已逐步构成了由三个层次的知识集成构建而成的框架体系，即（1）原子/分子层次上的微观基础冶金学；（2）工序/装置层次上的专业工艺冶金学；（3）全流程/过程群层次上的宏观动态冶金学。冶金流程工程学是一个新的冶金学分支，其定位是总体集成的冶金学，顶层设计的冶金学，宏观动态运行的冶金学，工程科学层次上的冶金学。

关键词：绿色化；智能化；冶金流程工程学；宏观动态运行；冶金学新分支

The Topic of the Times of Metallurgy——Get through the Process, Communicate Different Levels and Open up a New Theory

Yin Ruiyu

Abstract: After nearly one hundred years of exploration, recent metallurgy (metallurgical science and engineering) has gradually formed a framework system constructed by the integration of three levels of knowledge, namely (1) micro-metallurgy at the atomic/molecular level; (2) process metallurgy at the procedure/device level; (3) macro-dynamic metallurgy at the manufacturing process/process group level.Metallurgical process engineering is a new branch of metallurgy. And it is the overall integrated metallurgy, top-level designed metallurgy, macro-dynamic operated metallurgy, engineering science level metallurgy.

Keywords: greenization; intelligence; metallurgical process engineering; macro-dynamic operation; new branch of metallurgy

多模式薄板坯连铸连轧生产实践

朱国森，邓小旋，刘　珂，季晨曦，李海波

（首钢技术研究院，北京　100043）

摘　要：MCCR（Multi-mode Continuous Casting & Rolling）是首钢京唐公司与达涅利公司合作开发的多模式薄板坯连铸连轧产线，具备单坯轧制、半无头和全无头轧制三种模式。板坯厚度为 110mm、123mm（主要规格：110mm），宽度 900～1600mm，设计最大拉速为 6.0m/min。自 2019 年 4 月热试以来，先后攻克了板坯高拉速连铸工艺技术、表面冶金缺陷控制技术等工艺技术难题，有力支撑了产线达产达效。
为了满足全无头轧制的需求，首先必须攻克高拉速连铸技术。以全无头轧制 1.0mm×1200mm 带钢为例，连铸拉速必须提高至 5.0m/min 以上。但拉速提高后保护渣耗量降低，出结晶器坯壳厚度约 10mm，漏钢风险大幅度提升。国内某企业建设的全无头轧制生产线经过三年的攻关，漏钢次数才控制在平均约 1 次/月。为此，本研究开发了新型保护渣，大幅度提高保护渣中 Li_2O 含量至 1.6%，粘度从 0.08Pa.s 降低至 0.04Pa.s，熔点从 1160℃降至 1020℃。与原有保护渣相比，新型保护渣消耗量提高了 28%。主要钢种的工作拉速达到了 5.4m/min，产线漏钢发生率降至 0.5 次/月以下。

拉速提高后的第二个难题是液面波动大导致的卷渣问题，特别是辊间非稳态鼓肚（对应频率为 0.55Hz）与结晶器流场控制不合理导致液面波动超过了 10mm，卷渣发生率高达 12.5%。为此，本研究在对结晶器内钢水流动和传热开展大量数值模拟的基础上优化了浸入式水口的结构、EMBR 工艺参数和 1~4 扇形段二冷水量。结晶器液面波动控制在 6mm 以下，卷渣发生率大幅度降低至 0.3%以下。

在攻克了高拉速连铸技术后，全无头轧制的比例稳定提高到 95%。但是 MCCR 钢卷表面仍然存在线状缺陷。大量 SEM 分析表明：此类缺陷成分主要为钙、铝、氧。与钢中夹杂物的成分类似，推测其形成原因是钢中内生的大型夹杂物或者是钢中小型的内生夹杂物和二次氧化产物碰撞后形成的夹杂物未充分上浮去除，在结晶器中被凝固坯壳捕获。本研究开发了中间包流场控制技术与中间包预吹氩技术。优化了中间包内部坝、堰和湍流抑制器的位置及尺寸，与防止二次氧化技术相结合，稳定浇铸时中间包钢水的总氧含量稳定控制在 0.0020%以下。采用以上技术后，钙铝酸盐夹杂物导致的线状缺陷发生率大幅度降低。

在上述工艺技术的支撑下，MCCR 产线成功实现了超薄规格高强钢的批量生产，月产量达到 18.9 万吨，产品规格≤1.5mm 的比例超过 50%，成功开发了 980MPa 热轧超高强汽车用钢，1.5mm×1500mm 薄宽规格 700MPa 高强钢及 1.2mm 薄规格热成型钢 SHR1500HS 等产品；耐候钢实现了最薄 1.42mm 的薄规格产品开发。“以热代冷”产品同板性能波动在 20MPa 以内，厚度精度±30μm 命中率控制在 99%以上。

Production Practice of Multi-mode Continuous Casting & Rolling Line

Zhu Guosen, Deng Xiaoxuan, Liu Ke, Ji Chenxi, Li Haibo

（Shougang Research Institute of Technology, Beijing 100043, China）

智能工厂建设实践与思考

罗 禹，孙小东

（中冶赛迪重庆信息技术有限公司，重庆 400031）

摘 要：随着钢铁行业智能制造探索的不断深入，新一代信息技术加速赋能钢铁智能工厂。中冶赛迪围绕安全、效率、质量、成本及人力资源缺口等行业问题，基于 60 余年的钢铁领域知识，颠覆了传统信息化架构，提出了基于 HCPS 的新一代钢铁智能工厂体系架构，创新采用工业互联网平台+智能应用的模式，推出了智能料场、智能炼铁、智慧热轧、智慧能源等一批行业首创的核心技术与产品，有效解决了工序协同不足、凭经验操作、现场安全环境等难题，成果已应用至宝武韶钢、宝武湛江、宝武马钢、永锋钢铁、南京钢铁等多个大型钢铁企业，并于 2020 年 12 月获得第二届中国工业互联网大赛全国第一名。未来，中冶赛迪将携手业内同行，持续探索钢铁智能工厂，助力行业数字化转型发展。

关键词：智能工厂；HCPS；数字化转型

Practice and Thinking of Smart Factory Construction

Luo Yu, Sun Xiaodong

(CISDI Chongqing Information Technology Co., Ltd., Chongqing 400031, China)

Abstract: As the exploration of smart manufacturing in the steel industry continues to deepen, a new generation of information technology is accelerating the empowerment of smart steel factories. Focusing on industry issues such as safety, efficiency, quality, cost and human resource gaps, based on more than 60 years of knowledge in the field of iron and steel, China Metallurgical CCID has overturned the traditional information architecture, and proposed a new generation of HCPS-based intelligent steel factory system architecture, innovatively adopted The industrial Internet platform + smart application model has launched a number of industry-first core technologies and products such as smart stockyard, smart ironmaking, smart hot rolling, smart energy, etc., which effectively solves the lack of process coordination, empirical operation, and on-site safety environment The results have been applied to many large steel companies such as Baowu Shaogang, Baowu Zhanjiang, Baowu Maanshan Iron and Steel, Yongfeng Iron and Steel, Nanjing Iron and Steel, etc., and won the first place in the second China Industrial Internet Competition in December 2020. In the future, MCC CCID will join hands with peers in the industry to continue to explore smart steel factories to help the industry's digital transformation and development.

Key words: smart factory; HCPS; digital transformation

SAR 雷达高炉料面形态场与流场可视化研究进展

陈先中，侯庆文

（北京科技大学自动化学院，北京 100083）

摘 要：钢铁工业是高耗能、高污染、资源型产业。现代大型高炉属一类耗能巨大的“黑箱”设备，操作人员希望获取高炉内部冶炼实时状态，炉况数字化，可控制，可预测，高效完成碳氧高温还原反应，实现节能减排和稳定顺行目标。研究了粗糙料面多参数、多维度电磁特征的获取、认知、解释的理论问题，研究了封闭空间和多场多相复杂环境下，高性能雷达探测以及电磁点云图像深度学习，通过料面电磁散射特征逆问题求解，映射出炉顶上部空间真实的密度和速度场分布的物理状态。研制了恶劣环境 SAR 雷达扫描成像装置，可以掌握炉内料形变化趋势，下料速度，煤气流分布等关键参数，初步实现了高温、高压、高粉尘、强黏附环境下的料层可视化。项目成果在宝武、攀枝花、首钢和南钢等投入使用，最长免维护周期达到 18 个月，料线测量精度在 1%以内。

关键词：高炉；SAR 雷达；料线测量；料层可视化；煤气流分布

Advances in the Visualization of the Morphological Field and Flow Field of the BF Surface by SAR Radar

Chen Xianzhong, Hou Qingwen

(School of Automation and Electrical Engineering, University of Science and Technology Beijing, Beijing 100083, China)

Abstract: The iron and steel industry is a high-energy, high-pollution, and resource-based industry. Modern large-scale blast furnace(BF) is a kind of "black box" equipment with huge energy consumption. The operator hopes to obtain the real-time status of the internal smelting of the BF, realize the digitalization of the BF condition, controllable, predictable, efficient completion of the high-temperature reduction reaction of carbon and oxygen, and achieve the goals of energy saving, emission reduction and stable direct travel. The theoretical problems of the acquisition, cognition and interpretation of the multi-parameter and multi-dimensional electromagnetic characteristics of the rough BF surface are studied. Research on high-performance radar detection and deep learning methods of electromagnetic point cloud images in closed spaces and multi-field and multi-phase complex environments. By solving the inverse problem of electromagnetic scattering characteristics of the surface, the physical state of the real density and velocity field distribution in the upper space of the BF is mapped out. Developed a harsh environment SAR radar scanning imaging device, which can obtain the change of the burden surface, the feeding speed, the gas flow distribution and other key parameters, and initially realized the visualization of the material layer under the environment of high temperature, high pressure, high dust, and strong adhesion. The Radar is applied in companies such as BaoWu, Panzhihua, ShouGang and NanGang Iron & Steel company. The longest maintenance-free time of the Radar reaches 18 months, and the measuring accuracy of the burden line is within 1%.

Key words: blast furnace; SAR radar; burden line measurement; burden layer visualization; gas flow distribution

中冶京诚工业大数据应用

李　胜，薛颖健

（中冶京诚工程技术有限公司，北京　100000）

摘　要：当前正值我国国民经济和社会发展的“十四五”规划期，大数据作为制造业的数字化转型、智能化升级的重要技术支撑，在钢铁行业得到越来越多的应用。随着我国钢铁行业向高质量发展迈进，生产特点由传统大规模、批量生产向多品种、柔性定制转变，这些都对质量提出了更高的要求。传统质量管控分析系统辅助生产，还存在如下问题：仅关注本工序的质量数据，缺乏对全流程的关注，没有解决工序间遗传性导致的产品质量问题；生产的不确定性导致传统质量模型无法有效进行产品质量预测，使用效果不理想等等；随着云计算、大数据、物联网等新技术的发展，行业内迫切需要基于工业大数据构建全流程质量管控平台辅助企业进行质量管控。中冶京诚质量大数据分析管控系统针对上述钢铁行业产品质量管控方面及技术手段局限性等问题，将大数据及人工智能技术与钢铁生产过程相融合，突破钢铁行业智能工厂应用共性关键技术，构建产品全流程质量分析与管控平台，实现企业生产工艺整体优化、实现生产高效化与绿色化，提高钢铁产品质量档次和稳定性，从而既可推动钢铁行业由大变强，也可充分利用智能工厂技术提升政府治理能力、促进经济转型升级。

关键词： 大数据；质量管控；全流程；人工智能

Industrial Big Data Application of CERI

Li Sheng, Xue Yingjian

(CERI, Beijing 100000, China)

Abstract: At present, it is in the 14th Five Year Plan period of China's national economic and social development. As an important technical support for the digital transformation and intelligent upgrading of the manufacturing industry, big data has been more and more applied in the iron and steel industry. As China's iron and steel industry moves towards high-quality development, the production characteristics have changed from traditional large-scale and batch production to multi variety and flexible customization, which put forward higher requirements for quality. The traditional quality control analysis system can assist producting, but there are also the following problems: Only paying attention to the quality data of one process, lack of attention to the whole process, and it's hard to solve the product quality problems caused by heredity between processes; Because of the uncertainty of production, the traditional simple model can't effectively predict the product quality, and the application effect is not good and so on. With the development of new technologies such as cloud computing, big data and Internet of things, there is an urgent need to build a whole process quality control platform based on industrial big data to assist enterprises in quality control. For the above problems Quality big data analysis and control system of CERI integrates big data and artificial intelligence technology with the iron and steel production process, breaks through the application of common key technologies in intelligent factories in the iron and steel industry, constructs a product whole process quality analysis and control platform, and realizes the overall optimization of enterprise production process To achieve efficient and green production and improve the quality, grade and stability of iron and steel products can not only promote the iron and steel industry from large to strong, but also make full use of intelligent factory technology to improve government governance capacity and promote economic transformation and upgrading.

Key words: industrial big data; quality management and control ; whole process; artificial intelligence(AI)

基于工业互联网的钢铁企业智慧物流架构研究与实践

王　奕[1]，黄港明[2]，姜玉河[3]

（1. 上海宝信软件股份有限公司工业互联网研究院，上海　201900；2. 宝钢湛江钢铁有限公司制造管理部，广东湛江　524033；3. 宝钢湛江钢铁有限公司物流部，广东湛江　524033）

摘　要： 钢铁企业的物流管理是一个复杂的体系，需要在业务层面倡导基于整体供应链的物流管理理念，在系统层面基于新一代信息技术构建信息系统，并融合两者以提升物流效率。在调研相关技术进展与分析钢铁企业物流管理业务和系统的基础上，结合自身在钢铁企业的工作经验，融合基于供应链的物流管理理念和工业互联网体系架构，提出了基于工业互联网的钢铁企业智慧物流系统架构，并在宝钢湛江钢铁有限公司进行实践，为钢铁企业物流的数字化、智慧化发展提供了新思路。

关键词： 智慧物流管理；系统架构；工业互联网；人工智能；物联网

Research and Practice of Iron and Steel Enterprise Smart Logistics Architecture based on Industrial Internet

Wang Yi[1], Huang Gangming[2], Jiang Yuhe[2]

(1. Industrial Internet Research Institute, Shanghai Baosight Software Co., Ltd., Shanghai 201900, China; 2. Products and Technique Management Department, BaoSteel Zhanjiang Iron and Steel Co., Ltd., Zhanjiang 524033, China; 3. Logistics Department, BaoSteel Zhanjiang Iron and Steel Co., Ltd., Zhanjiang 524033, China)

Abstract: Logistics management in modern iron and steel enterprises involves a complex architecture. It is the integration of the supply chain-based logistics management concept at the business level, and information systems exploiting the new information technology at the infrastructure level. By investigating the related technological progress and analyzing the logistics management business and systems in steel enterprises, combined with my own work experience in steel enterprises, a smart logistics system architecture is proposed for iron and steel enterprises based on the industrial Internet by integrating the supply chain-based logistics management concept and the industrial Internet system architecture. The practice of this new system architecture in Zhanjiang Iron and Steel has provided new ideas for the digital and intelligent development of steel enterprise logistics.

Keywords: smart logistics management; system architechture; industrial Internet; artificial intelligence (AI); IoT

基于“supOS”的钢铁厂数字化解决方案

褚　健，杨溪林

（浙江中控技术股份有限公司，浙江杭州　310053）

摘　要：“supOS”工业操作系统是以工厂全信息集成为突破口，实现生产控制、生产管理、企业经营等多维、多元数据的融合应用，提供对象模型建模、大数据分析 DIY、智能 APP 组态开发、智慧决策和分析服务，以集成化、数字化、智能化手段解决生产控制、生产管理和企业经营的综合问题，打造服务于企业、赋能于工业的“智慧大脑”。本文介绍了基于“supOS”构建的钢铁厂数字化转型升级解决方案。

关键词： 冶金；数字化；智能制造

Digital Solution of Steel Plant based on "supOS"

Chu Jian, Yang Xilin

(Zhejiang Supcon Technology Co., Ltd., Hangzhou 310053, China)

Abstract: "supOS" industrial operating system takes the full information set of the factory as the breakthrough, realizes the integration and application of multidimensional and multivariate data such as production control, production management and enterprise operation, provides object model modeling, big data analysis DIY, intelligent APP configuration development, intelligent decision-making and analysis services, and solves the comprehensive problems of production control, production management and enterprise operation, and creates a "smart brain" that serves enterprises and industries. This article introduces the digital transformation and upgrading solution of steel plant based on "supOS".

Key words: metallurgy; digitization; intelligent manufacturing

工业互联网平台助力冶金行业数字化转型

吴志强

（北京国信会视科技有限公司，北京 100000）

摘 要：介绍工业互联网平台在轨道行业的应用，探讨基于工业互联网平台，实现冶金行业设备预测性维修。通过设备运营数据采集、多方知识融合和深度关联分析，实现设备状态监控、故障预测、故障诊断、健康评估和维修决策。降低设备的故障率，提高设备利用率和企业生产效率。

（1）设备运维应用的多源数据融合处理技术。研究数据融合处理技术，开展多源异构数据的融合与集成处理，解决由于采集信息数据类型、格式及产生方式不同而难以统一处理等问题，实现标准化的设备运维数据采集与融合。

（2）设备故障预测与健康管理。利用设备运行数据和运维数据进行数据挖掘分析，实现设备的状态监测、故障诊断、故障预测等功能，并提供故障处理知识和决策信息，提升设备安全运营保障能力，助力设备维修模式从计划维修向预测性维修转变。

（3）设备关键零部件全生命周期管理。构建设备构型管理，开展设备关键零部件全生命周期管理能力建设，开展设备和关键零部件履历信息管理，为设备运维提供基础数据。

（4）设备故障知识库管理。建立设备故障处理层级结构和故障知识库，以规范和关联相关维修活动，搭建故障代码体系并提供故障标准化管理流程，实现故障快速诊断和处理，提升精细化故障修能力。

关键词：工业互联网；冶金；设备；运维

A Way to Digital Transformation of Metallurgy Industry: Internet of Things

Wu Zhiqiang

(Beijing Halosee Technology Co., Ltd., Beijing 100000, China)

Abstract: Introducing the application of the industrial Internet platform in the rail industry, and explore the feasibility of predictive maintenance of equipment in the metallurgical industry based on the industrial Internet platform. Equipment status monitoring, fault prediction, fault diagnosis, health assessment and maintenance decision-making can be achieved through equipment operation data collection, multi-dimensional knowledge integration and in-depth correlation analysis. As a consequence, reducing equipment failure rate, improving equipment utilization and enterprise production efficiency.

1. Multi-source data integration processing technology for equipment operation and maintenance. Research on data integration processing technology, carry out the integrated processing of multi-source heterogeneous data, solve the

difficulties in unifying data due to different types, formats of collected information, and finally standardize equipment operation and maintenance data collection and integration

2. Equipment failure prediction and health management. Use equipment operation data and operation & maintenance data for data mining analysis to achieve equipment status monitoring, fault diagnosis, fault prediction and other functions, and provide fault handling knowledge and decision massage, improve capabilities of equipment safety operation, and hence help equipment maintenance transforming from planned maintenance to predictive maintenance.

3. Full life cycle management of equipment key parts. Construct equipment configuration management. Provide basic data for equipment operation and maintenance by carrying out the full life cycle management of equipment key parts and building equipment key parts history information management.

4. Knowledge base management of equipment failure. Establish a hierarchical structure of equipment fault handling and a fault knowledge base to standardize and correlate related maintenance activities, build a fault code system and provide a standardized process for faults, achieve fast fault diagnosis and processing, and improve the ability of refined repair.

Key words: internet of things; metallurgy; equipment; maintenance

钢铁产业碳中和技术途径

储满生，王国栋

（东北大学矿冶学科群）

摘　要：钢铁行业是资源、能源密集型产业。2019 年我国粗钢产量 9.96 亿吨，占全世界 53.3%左右，产生碳排放约 15 亿吨，约占我国碳总排放 16%。钢铁产业低碳绿色发展是实现国家双碳战略目标的重大需求。碳中和愿景下的钢铁低碳技术路径建议：第一阶段减排过渡期（2025-2035），在工艺优化、强化冶炼、余热和二次资源高效循环利用、超低排放改造、系统节能等基础上，研发应用低碳高炉、高效连铸、铸轧一体化等低碳冶炼技术，同时开发全流程信息物理系统，实现智能化冶炼，提高能源利用效率，实现碳减排 30%以上，为碳中和奠定基础；第二阶段快速减排期（2035-2050），实现废钢在钢铁生产流程的增量利用，同时基于碳捕集利用，研发应用钢铁-化工-氢能一体化网络集成 CCU 技术，实现碳循环利用，为我国高炉-转炉长流程为主的钢铁产业实现碳净零排放提供合理解决方案；第三阶段中和达成期（2050-2060），以氢能替代化石能源，应用氢基竖炉-电炉短流程新工艺技术，实现钢铁工艺流程革新和能源结构优化，为低碳或无碳钢铁生产提供新途径；贯穿全程，加强氢能廉价绿色制备技术研发应用，为社会贡献低碳减排的绿色产品。

Steel Industry Carbon Neutralization Technical Approaches

Chu Mansheng, Wang Guodong

(Group of Mining and Metallurgy, Northeastern University)

Abstract: Steel industry is a resource and energy intensive industry. In 2019, China's crude steel output was 996 million tons, accounting for about 53.3 % of the world, while carbon emissions from steel industry were about 1.5 billion tons, accounting for about 16 % of China's total carbon emissions. The low-carbon and green development of steel industry is a major demand for realizing the national dual-carbon strategic goal. Suggestions on low carbon technology of steel industry under the vision of carbon neutralization are as follows. Phase 1: Emission Reduction Transition Period (2025-2035). On the basis of process optimization, strengthening smelting, efficient recycling of waste heat and secondary resources, ultra-low emission transformation and system energy saving, low carbon smelting technologies suchaslow-carbon blast furnace,high efficiency continuous casting and integrated casting and rolling, will be developed and applied. At the same time, the cyber physical system of whole process will be developed to realize intelligent smelting. All above measures will raise energy utilization efficiency and achieve a carbon reduction of more than 30%, which lay the foundation for carbon neutralization. Phase 2: Rapid Emission Reduction Period (2035-2050). The incremental utilization of scrap steel in the steel production process will be realized. And based on carbon capture utilization, the innovative technology of Steel-Chemicals-Energy Networking Integration CCU will be developed and applied to realize carbon recycling, which provides a reasonable solution for long routesteel production byBF-BOF to achieve net-zero carbon emission.Phase 3: Carbon Neutralization Achievement Period (2050-2060). Fossil energy will be replaced by hydrogen energy. And theshortroute new technology by hydrogen-based shaft furnace-electric furnace will be applied, which realizes steel process innovation and energy structure optimization, and provides a creative way for low-carbon or non-carbon steel production. Throughout the whole phases, it should strengthen the application of cheap-green hydrogen energy production technologies, and contribute green products with low carbon emissionsto the society.

科技创新助力河钢集团低碳绿色发展路径

田京雷

（河钢集团有限公司，河北石家庄　050000）

摘　要：本报告从三个方面进行汇报，首先是低碳绿色研究背景，当前，应对气候变化是全球面临的共同课题，低碳发展是中国应对气候变化的自身需求，而中国钢铁行业是资源消耗密集型产业和典型的高碳排放行业，是碳达峰、碳中和目标实现的重点领域和责任主体。第二部分是河钢集团低碳绿色发展路径，基于现有碳排放强度及未来产业规划，河钢集团制定了面向“碳达峰、碳中和”的低碳绿色发展行动目标，并详细制定了六大路径十五项技术方案助力实现双碳目标，第三部分是低碳绿色技术创新，重点阐述钢铁行业要真正实现“碳减排”必须遵循的 3 大创新路径，即能源结构创新、工艺结构创新和材料技术创新。

关键词：冶金；低碳

Technological Innovation Helps HBIS Group’s Low-carbon and Green Development Path

Tian Jinglei

(HBIS Group, Shijiazhuang 050000, China)

Abstract: This report reports from three aspects. The first is the background of low-carbon green research.At present, addressing climate change is a common issue facing the world. Low-carbon development is China’s own demand for addressing climate change. China’s steel industry is a resource-intensive industry and a typical high-carbon emission industry.It is the key area and main body of responsibility for achieving the carbon peak and achieving carbon neutral goals.The second part is the low-carbon and green development path of HBIS. Based on the current carbon emission intensity and future industrial planning, HBIS has formulated a low-carbon and green development action goal oriented toward "carbon peak and carbon neutrality". And it has formulated in detail six major paths and fifteen technical solutions to help achieve the dual-carbon goal.The third part is low-carbon green technology innovation, focusing on the three major innovation paths that the steel industry must follow to truly achieve "carbon emission reduction", namely, energy structure innovation, process structure innovation and material technology innovation.

Key words: metallurgy; low carbon

钢铁行业碳达峰、碳中和实施路径研究

上官方钦[1]，刘正东[2]，殷瑞钰[2]

（1. 中国钢研科技集团有限公司钢铁绿色化智能化技术中心，北京　100081；
2. 钢铁研究总院，北京　100081）

摘　要：概括了中国钢铁行业绿色化发展进程，即节能、减排、脱碳三个阶段；分析了中国钢铁行业的 CO2 排放现状及实现碳达峰、碳中和的关键时间节点、思路、切入口和技术路线图等；指出钢铁行业越早实现碳达峰，越有利于后续碳“下坡”和碳中和的实现；削减粗钢产出总量和流程结构调整发展全废钢电炉短流程钢厂是中国钢铁行业实现碳中和过程中的两大可行的抓手。

关键词：碳达峰；碳中和；钢铁；实施路径；绿色化发展

Study on the Implementation Path of Carbon Peak and Carbon Neutrality in the Steel Industry in China

Shangguan Fangqin[1], Liu Zhengdong[2], Yin Ruiyu[2]

(1. Steel Industry Green and Intelligent Manufacturing Technology Center, China Iron and Steel Research Institute Group, Beijing 100081, China; 2.Central Iron and Steel Research Institute, Beijing 100081, China)

Abstract: Green development process of the steel industry in China was summarized, which included three stages: energy saving, emission reduction and decarbonization. And the current situation of CO_2 emission in the steel industry in China and the key time nodes, ideas, cut-off points and technical roadmap to achieve carbon peak and carbon neutrality were analyzed in this paper. It is pointed out that the earlier the steel industry reaches the carbon peak, the more favorable it is for the subsequent ‘carbon downhill’ and carbon neutrality. However, reducing the total output of crude steel and adjusting the process structure are two feasible ways to realize carbon neutrality in the steel industry in China.

Key words: carbon peak; carbon neutrality; steel; implementation path; green development

生命周期评价方法在钢铁企业低碳发展规划中的应用

刘颖昊

（宝山钢铁股份有限公司中央研究院，上海　201999）

摘　要：在当前“碳达峰”“碳中和”的背景下，很多钢铁企业在制订低碳发展规划的过程中，对于减碳措施的减碳潜力没有数字化的概念，缺乏量化的评价手段和科学的数据支撑。为此，提出应用生命周期评价（LCA）进行钢铁企业低碳发展规划的方法，建立产品碳足迹与组织层级碳核算的关联。在此基础上，通过建立覆盖全公司的产品生命周期评价模型，以量化评估新技术新工艺应用、产品结构变化、能源结构变化、废钢利用率提升、节能减排改进、供应链优化等因素的对于企业组织层面的碳减排绩效，可实现数字化碳减排路线图的描绘。给出了钢铁企业主要措施与策略碳减排潜力的评价方法与案例，可为钢铁企业的低碳发展规划提供参考。

关键词： 碳达峰、碳中和；低碳发展规划；生命周期评价；减排潜力；数字化

Application of Life Cycle Assessment in Low-carbon Planning of Steel Company

Liu Yinghao

(The Central Research Institute of Baoshan Iron and Steel Co., Ltd., Shanghai 201999, China)

Abstract: In the current context of "carbon peak", and "carbon neutrality", many steel companies are formulating low-carbon plans. However, they have no digital concept on the potential of carbon reduction measures, and lack of quantitative evaluation methods and scientific data support. A method for applying life cycle assessment (LCA) to low-carbon planning for steel enterprises was proposed, and the relationship between product carbon footprint and organization-level carbon accounting was established. On this basis, through the company-wide product life cycle assessment model, it was possible to quantitatively evaluate the carbon emission reduction performance of the strategies such as the application of new technologies and new processes, product structure changes, energy structure changes, increased utilization of scrap steel, energy conservation and emission reduction, and supply chain optimization at the company organization level. In this way, the digital carbon emission reduction roadmap could be described. Examples of how to apply LCA to access the carbon emission reduction potential of various measures were listed, which provide reference for the low-carbon planning of steel companies.

Key words: carbon peak and carbon neutrality; low-carbon planning; life cycle assessment; emission reduction potential; digitalization

碳循环-富氢低碳高炉炼铁技术

左海滨，王静松，薛庆国

（北京科技大学钢铁冶金新技术国家重点实验室，北京 100083）

摘 要： 在国家“双碳”目标驱动下，钢铁工业势必要加快其低碳化进程，一是为促进我国钢铁工业可持续发展及提高未来钢铁产品的国际竞争力，二是要为他新兴产业提供一定的碳排放容量。钢铁工业的低碳化是涉及技术、经济、环境、社会影响等多个方面的重要课题，尤其是超过世界50%产量的中国钢铁工业的低碳化必将为世界瞩目。本文基于中国钢铁工业长流程为主的特点，分析了我国钢铁工业低碳化的潜在路径，重点剖析了碳循环高炉、富氢高炉及碳循环耦合富氢高炉的节碳潜力。针对不同工艺流程的关键技术问题，如煤气加热、炉内煤气流分布、多相喷吹风口设计、原燃料适应性等进行了系统研究，为基于高炉的炼铁低碳化发展奠定基础。同时本研究也将为中国钢铁工业未来低碳化路径的选择提供理论指导与借鉴。

关键词： CO_2减排；炼铁；碳循环高炉；富氢高炉

BF Ironmaking Technologies with Top Gas Recycling and Hydrogen Enriched Injection for Mitigation of CO_2

Zuo Haibin, Wang Jingsong, Xue Qingguo

(State Key Laboratory of Advanced Metallurgy, University of
cience and Technology Beijing, Beijing 100083, China)

Abstract: Under the drive of national double carbon target, iron and steel industry will inevitablyaccelerate the process of promoting low carbon transition, not only promoting the sustainable development of iron and steel industry and improving the worldwide competitiveness of steel products, but also providing CO2 emission space foremerging industries. Low carbon development of iron and steel industry involves technologies, economy, environment and social influence, especially the low carbon development in China with steel production over 50% of the world will catch the eyes of the whole world. In this paper, considering the dominant position of long process flow in China, the potential route for low carbon development was analyzed, emphatically focusing on the top gas recycling blast furnace, hydrogen enriched injection and coupled process with above two methods. Aiming at the key technologies for different processes, the systematical studies were carried out, such as gas heating, gas distribution in blast furnace, tuyere structure design with multiple substances injection and adaption of materials and fuels et al. This research will establish the basis for low carbon development of ironmaking based on blast furnace, meanwhile, providing the theoretical direction and suggestion for route choose of low carbon development of iron and steel industry.

Key words: carbon dioxide reduction; ironmaking; blast furnace with top gas recycling; blast furnace with hydrogen injection

低温还原炼铁新工艺

郭培民

（钢铁研究总院先进钢铁流程及材料国家重点实验室，北京 100081）

摘　要：钢铁研究总院开发的低温还原炼铁新工艺是将铁精矿粉和煤粉按照一定比例混匀、冷压成型，然后在低温还原新装备中加热并完成快速还原，还原后的金属化球团热送、热装电炉熔分生产高纯生铁或优质钢水。新工艺不需要烧结、焦化、高炉等传统高炉炼铁工序，是颠覆性的钢铁冶炼新工艺。新工艺采用国内铁精矿粉、有色行业产生的铁红为主原料，保证低温快速还原进行：物料平均温度 900～1000℃，将高温还原所需碳耗降低 20%;反应热降低 20%；物理热降低 35%。新工艺采用上下间接加热的新型低温还原炉，创造还原气氛，确保高的还原率高（大于 90%）和快速还原（炉内停留时间 20~30min）。低温还原炉自产煤气简单、高效低成本返回使用回收，实现工艺"零化学热"煤气外排。新工艺从源头大幅度减少硫氧化物、氮氧化物等排放（直接达到钢铁超低排放要求，无需末端治理）。吨钢煤粉消耗量仅 300kg，为高炉-转炉系统（含烧结、焦化等）的工艺的 42.8%，吨钢碳排放从高炉-转炉的 2t 水平降低到 1t 以下水平，下降幅度超过 50%，节能、减碳、降本和绿色效果显著。本工艺已完成 5 万吨/年的示范，正在建设更大规模的示范线或生产线。

关键词：低温还原；低碳排放；炼铁新工艺

New Ironmaking Process of Iron Ore Reduction at Low Temperature

Guo Peimin

(State Key Laboratory of Advance Steel Process and Products,
Central Iron and Steel Research Institute, Beijing100081, China)

Abstract: The new low-temperature reduction ironmaking process developed by China Central Iron and Steel Research Institute.Iron concentrate powder and pulverized coal are mixed according to a certain proportion, and formed at room temperature, then heated in the new low-temperature reduction equipment and rapid reduction is realized thereby. High-purity pig iron or high-quality molten steel is produced by hot delivery and hot charging of reduced metallized pellets in electric furnace. The new process does not need traditional blast furnace ironmaking processes such as sintering, coking and blast furnace. It is a subversive new iron and steel smelting process. The new process uses domestic iron concentrate powder and iron red produced by nonferrous industry as the main raw materials to ensure rapid reduction at low temperature: the average material temperature is 900 ~ 1000 ℃, reducing the carbon consumption required for high temperature reduction by 20%; The reaction heat is reduced by 20%; Physical heat reduced by 35%. The new process adopts a new low-temperature reduction furnace indirectly heated up and down to create a reduction atmosphere and ensure high reduction rate (greater than 90%) and rapid reduction (residence time in the furnace is 20 ~ 30min). The self-produced gas of low-temperature reduction furnace is simple, efficient, and low-cost, and can be recycled to realize the "Zero chemical heat" gas discharge of the process.The new technology greatly reduces the emissions of sulfur oxides and nitrogen oxides from the source (directly meeting the ultra-low emission requirements of iron and steel without end treatment). The consumption of pulverized coal per ton of steel is only 300kg, which is 42.8% of the process of blast furnace converter system (including sintering, coking, etc.). The carbon emission per ton of steel is reduced from 2 tons of blast furnace converter to <1 ton, a decrease of >50%. The energy-saving, carbon reduction, cost reduction and green effects are remarkable.The 50000 T/a demonstration of this process has been completed, and a larger demonstration line or production line is under construction.

Keywords: low temperature reduction; low carbon emission; new ironmaking process

面向低碳高效的铁矿石烧结工艺数学模型

吕学伟

（重庆大学）

摘　要：铁矿石烧结是典型的依赖于液相作用产生粘接效果的高温物理化学反应过程。料层温度随着燃烧前沿的推进先升高后下降，随着温度变化先后发生固相反应、固液同化和液相结晶等，其中同化过程是烧结工艺的核心。本文基于冶金热力学和动力学原理，建立了烧结高温物理化学过程的数学模型，利用该模型探讨了化学成分、矿石粒度等因素对初始铁酸钙、液相量、液相成分和结晶产物的影响规律。模型的建立为实现烧结过程的智能控制和高效低耗生产提供了支撑。

Mathematical Models of Iron Ore Sintering Process for High Efficiency and Low Energy Consumption

Lü Xuewei

(Chongqing University)

Abstract: The iron ore sintering process is a high temperaturephysical chemistryprocess which dependsonliquid phase formation to bond the ore particles together. In sintering process, the temperature firstlyincreasesthen decreases, accompany with carbon combustion, solid-state reaction, liquid phase formation, solid-liquid reaction, as well as crystallization. The assimilation is the key step for sintering process. In this paper, a model of high temperature reaction in iron ore sintering process was established considering the solid-state reaction, solid-liquid dissolution and crystallizationbehaviors. The influence of chemical composition, particle size of iron ore fines and sintering temperature on initial calcium ferrite, liquid phase amount, liquid phase composition and phase composition of sinter were investigated by present model. It was found that the amountofinitial calcium ferriteincreased with the increasing of temperature and basicity, theamountof initial calcium ferrite decreased with the increasing of particle size. The liquid phase amount increased with increase of temperature, theliquid phase amount first increased then slight decreased with the increasing of basicity, while the liquid phase amount seriouslydecreased with the particle size increased. This model provides a new method to calculate and predict the reactions at high temperature in sintering process.

氢气竖炉内气固热质传递行为数值模拟研究

邵　磊，曲迎霞，邹宗树

（东北大学，辽宁沈阳　110819）

摘　要：随着对钢铁行业绿色低碳发展要求的日益迫切，氢气竖炉已成为目前涉及氢冶金工艺的研发焦点。由于 H2 还原铁氧化物为强吸热反应，氢气竖炉的供气强度主要由还原反应和加热固相炉料对应的物理能需求决定，因此造成炉内物理能与化学能的利用严重不匹配。鉴于此，建立了氢气竖炉 CFD 模型，并利用其定量研究氢气竖炉内复杂的气固两相热质传递行为。研究结果可为氢气竖炉的设计优化提供一定理论指导。

关键词：氢冶金；竖炉；直接还原铁；热质传递

Numerical Simulation of Gas-solid Heat and Mass Transfer Behavior in H2 Shaft Furnace

Shao Lei, Qu Yingxia, Zou Zongshu

(Northeastern University, Shenyang 110819, China)

Abstract: With the increasing demand of green and low-carbon development of iron and steel industry, the H2 shaft furnace has become the focus of research and development regarding hydrogen metallurgy processes. Since the reduction of iron oxides by H2 is strongly endothermic, the gas feed rate for a H2 shaft furnace is mainly determined by the demand of physical energy for the endothermic reaction and the heating of solid phase, thus leading to a serious mismatch between the utilization of physical energy and chemical energy in the furnace. Therefore, a CFD model of H2 shaft furnace was built to quantitatively study the complicated in-furnace gas-solid heat and mass transfer behavior. The findings of this work may serve as guidelines for future design of H2 shaft furnace.

Key words: hydrogen metallurgy; shaft furnace; direct reduced iron; heat and mass transfer

富氢煤气在竖炉工艺中的生产能力计算与分析

姜 鑫，周宇露，安海玮，高强健，郑海燕，沈峰满

（东北大学，辽宁沈阳 110819）

摘 要：本文计算选择的裂解介质为 O_2，首先对理想条件下富氢煤气在竖炉还原工艺中的生产能力进行了热力学计算。通过热力学计算可知：竖炉生产过程中，作为热源的富氢煤气消耗量大于作为还原剂的消耗量。补充竖炉内热需求量的方法有两种，不同的热量补偿方式，金属铁产量的变化趋势不同。（1）直接补充重整富氢煤气，金属铁的产量随生成温度的升高而降低。120 万吨焦炉产生的焦炉煤气在金属铁生成温度为 850℃和 900℃时，相应的竖炉生产能力分别为 50.72 万吨/年和 49.90 万吨/年。（2）先采用煤气自重整技术再补充额外富氢煤气的情况下，随着金属铁生成温度的升高，金属铁的产量先增大后减小，出现一个峰值。120 万吨焦炉产生的焦炉煤气在金属铁生成温度为 845℃和 900℃时，相应的竖炉生产能力分别为 56.69 万吨/年（最大值）和 55.68 万吨/年。本文研究内容及结果旨在为实际的富氢竖炉工艺选择合理的操作参数提供评估方法和理论依据。

关键词：富氢煤气；竖炉；直接还原；热力学计算；金属铁产量

Calculation and Analysis on Production Capacity of Hydrogen-enriched Gas in Shaft Furnace Process

Jiang Xin, Zhou Yulu, An Haiwei, Gao Qiangjian, Zheng Haiyan, Shen Fengman

(Northeastern University, Shenyang 110819, China)

Abstract: In this paper, the cracking medium is O_2. First, the production capacity of hydrogen-enriched gas in direct reduction process of shaft furnace under ideal conditions was calculated. The thermodynamic calculation results show that, in shaft furnace process, the hydrogen-enriched gas consumption as heat source is more than that as reducing agent. There are two methods to supply the heat gap. The tendency of metallic Fe production is different with two different methods. (1) In the case of directly supplying reformed hydrogen-enriched gas, the yield of metallic Fe decreases with increasing formation temperature of metallic Fe. For a coke oven with capacity of 1200000 tons, as the formation temperature are 850℃ and 900℃, the corresponding annual yields of shaft furnace are 507.2×103 tons/year and 499.0×103 tons/year. (2) In the case of ZR technology followed by supplying extra hydrogen-enriched gas, the yield of metallic Fe first increases and then decreases with increasing formation temperature of metallic Fe, and there is a peak value. For a coke oven with capacity of 1200000 tons, as the formation temperature are 845℃and 900℃, the corresponding annual yields of shaft furnace are 566.9×103 tons/year (maximum value) and 556.8×103 tons/year. The aim of present work is to provide evaluation method and fundamentals for choosing optimal parameters for an actual hydrogen-enriched shaft furnace process.

Key words: hydrogen-enriched gas; shaft furnace; direct reduction; thermodynamic calculation; yield of metallic Fe

高炉富氢冶炼的基础研究及展望

刘　然，吕　庆，张淑会，兰臣臣

（华北理工大学 冶金与能源学院，河北唐山　063210）

摘　要：我国现阶段仍以高炉炼铁工艺为主，高炉冶炼过程也是节能减排的主要环节，以H2替代部分碳进行还原是节能减排的有效手段。采用数值模拟的方法对高炉富氢后的运行状态进行研究，得到炉内煤气流场和速度的分布、压力场分布、温度场分布、矿石还原度的分布以及产量的变化；采用高温模拟实验和热力学计算的方法对矿石富氢还原的热动力学特点进行研究，得到H2含量对FeO的还原、煤气利用率、还原耗热、直接还原度、矿石低温还原粉化率、矿石还原过程、金属化率等参数的影响；采用高温模拟实验研究了高炉富氢后渣铁的形成过程，得到了高炉富氢后初渣形成、金属铁渗碳以及软熔带的变化规律。同时，提出了高炉实现富氢冶炼仍需深入研究的方向。

关键词：高炉；富氢冶炼；运行状态；软熔带；渣铁形成

Basic Research and Prospect of Blast Furnace Hydrogen-rich Smelting

Liu Ran, Lv Qing, Zhang Shuhui, Lan Chenchen

(College of Metallurgy & Energy, North China University of Science and Technology, Tangshan 063210, China)

Abstract: At present, China still focuses on the blast furnace ironmaking process, and the blast furnace smelting process is also the main link of energy conservation and emission reduction. H2 instead of partial carbon reduction is an effective means of energy conservation and emission reduction. The operation state of blast furnace after hydrogen-rich is studied by numerical simulation, and the distribution of gas flow field and velocity, pressure field, temperature field, ore reduction degree and output are obtained. The thermodynamic and kinetic characteristics of ore hydrogen-rich reduction are studied by means of high temperature simulation experiment and thermodynamic calculation. The effects of H2 content on FeO reduction, gas utilization, reduction heat consumption, direct reduction degree, ore low temperature reduction pulverization rate, ore reduction process, metallization rate and other parameters are obtained. The formation processes of slag and iron after hydrogen-rich are studied by high temperature simulation experiment, and the changes of initial slag formation, metal iron carburization and softening-melting zone are obtained. Concurrently, the further research directions of realizing hydrogen rich smelting in blast furnace are put forward.

Key words: blast furnace; hydrogen-rich smelting; operation status; soften-melting zone; slag iron formation

基于全流程系统优化理念的焦化行业发展模式探讨

石岩峰

（中国炼焦行业协会，北京　100120）

摘　要：为正确认识新常态、促进新发展，系统分析了我国焦化行业现状及所面临的主要发展瓶颈，阐述了国家对焦化行业环保的总体要求及其对整个行业发展所带来的影响，指出节能、环保、供给侧结构性改革是新时代焦化行业的重点发展方向。在此基础上，提出全流程系统优化理念下的焦化行业发展模式，分析了焦化行业全流程系统优化发展的意义，介绍全流程系统优化的工作方法。同时，基于实际调研及企业反馈，提出当前我国焦化行业所存在的共性技术问题及初步优化方向。

关键词： 焦化行业；化解过剩产能；共性技术问题；全流程系统优化；发展模式

Discussion on the Development Model of Coking Industry based on the Concept of Whole Process System Optimization

Shi Yanfeng

(China Coking Industry Association (CCIA), Beijing 100120, China)

Abstract: To correctly comprehend the new normal and promote new development in the new era, this paper thoroughly analyzes the current situation and main development bottlenecks of China's coking industry, discusses the overall national requirements for environmental protection of coking industry and its impact on the development of the whole industry, and points out that energy conservation, environmental protection and supply side structural reform are the key development direction of coking industry in the new era. On this basis, the development model of coking industry under the concept of whole process system optimization is put forward, the significance of whole process system optimization development of coking industry is analyzed, and the working methods of whole process system optimization are introduced. Meanwhile, based on the actual investigation and enterprise feedback, this paper also puts forward the technical issues in common existing in China's coking industry and the preliminary optimizing orientation.

Key words: coking industry; tackling overcapacity; technical issues in common; whole process system optimization; development model

钢铁制造流程炼铁区段耗散结构研究

张福明[1,4]，颉建新[2,4]，殷瑞钰[3]

（1. 首钢集团有限公司，北京 100041；2. 北京首钢国际工程技术有限公司战略技术部，北京 100043；3. 钢铁研究总院，北京 100081；4. 北京市冶金三维仿真设计工程技术研究中心，北京 100043）

摘　要： 由料场、焦化、烧结、球团、高炉等工序所组成的炼铁区段，不仅是钢铁制造流程中重要的物质/能源转换中心，也是全流程动态有序、协同连续运行的关键和基础环节。炼铁区段的物理本质是铁素物质流在碳素能量的驱动和作用下，经过一系列热量、质量和动量传输以及复杂的物理化学冶金反应工程，将铁矿石转换/转化成为高温液态生铁的过程。这一复杂的工艺过程需要多工序协同耦合、耗散结构优化、动态有序运行。分析了耗散结构理论的内涵及耗散结构形成与维持的基本条件，指出钢铁制造流程炼铁区段是一个典型的耗散结构。阐述了炼铁区段主要工序功能解析优化和工程演化的历程，指出焦化、烧结、球团等单元工序的形成和发展，是现代高炉功能解析优化、集成优化和重构优化的演化结果和工程效应。基于耗散结构理论，提出了炼铁区段主要单元功能解析、耗散结构特征、耗散结构体系建构及优化的理论与方法，论述了单元工序功能的对比分析和选择确定，设备/装置的能力和数量等参数的选择等。以耗散结构理论为指导，构建了首钢京唐钢铁厂炼铁区段概念设计、顶层设计、动态精准设计体系。生产实践证实，其耗散结构合理、物质和能量耗散低、生产运行指标先进、经济效益和社会效益显著，具有重大的理论研究和推广应用价值。

关键词： 钢铁制造流程；炼铁区段；耗散结构；工程设计；流程优化

洁净钢冶金热力学和动力学计算平台

张立峰

（燕山大学亚稳材料制备技术与科学国家重点实验室，河北秦皇岛　066044）

摘　要： 本文介绍了一种自主研发的用于钢铁冶炼过程的洁净钢冶金热力学和动力学计算平台及其应用。平台包括热力学计算、热力学数据库、动力学计算和动力学数据库四个模块，热力学模块可应用于钢液精准钙处理和热力学平衡的计算，动力学模块可应用于冶炼全流程钢、渣、夹杂物成分演变的预报，以及连铸过程连铸坯全断面夹杂物的成分分布预报。本研究完善了洁净钢冶炼过程相关的热力学和动力学，从理论层面建立了较为完善的具有自主知识产权的热力学、动力学计算平台。

关键词： 洁净钢；热力学；动力学；智能平台

考虑物质流-能量流协同优化的钢厂调度排程方法

郑　忠[1]，连小圆[1]，高小强[2]

（1. 重庆大学材料科学与工程学院，重庆　400044；
2. 重庆大学经济与工商管理学院，重庆　400044）

摘　要： 钢厂的物质流具有多因子性质（温度、成分、形态等），“流”随时间空间的变化具有矢量特征。调度方案作为物质流的运行程序，既与能量流的保障作用相关，也决定了物质流在流程网络中的运行方式和状态。因此，基于钢铁制造流程动态运行特征，以钢厂铁素物质流为对象，提出考虑物质流-能量流协同优化的钢厂非完全并行机的调度排程方法，提高调度排程可执行基础上的运行效果。该方法将物质流性质与调度问题的工艺路径规划、设备选择、时间安排进行协调，将传统调度主要关注物质流时间参数的调控，拓展到引入匹配度参数来度量物质流的温度和成分等性质与工序设备的匹配以及前后工序的衔接关系；并且考虑物质流-能量流的协同效应，改善调度排程来优化物质流的运行，进而提高生产运行的能源管控保障能力。基于某钢厂的生产实绩进行方法测试，结果表明：该方法能够有效实现物质流性质参数的合理衔接和优化，获得合理优化的钢厂调度方案，并且能够优化能源消耗、减少氧气需求的波动。

关键词： 钢铁制造流程；物质流；生产调度；运行优化

关于洁净钢生产新流程实践问题的讨论

吴双平，徐安军

（北京科技大学冶金与生态工程学院，北京 100083）

摘 要：首钢京唐炼钢厂积极进行洁净钢生产新流程的实践，积极应用“全三脱”冶炼工艺、“一罐到底”等技术。在实践的过程中主要有废钢熔化、生产计划的编制和“全三脱”工艺的比例等三个问题。基于此，本文从三个方面出发，阐述目前的实际现状，结合实验室已有的相关研究，对解决三个问题提出相应的改进方案，以期能够为更好地实施洁净钢生产新流程提供一定的指导意见。

关键词：洁净钢；“全三脱”；废钢；生产计划

钢铁企业煤气-蒸汽-电力转换系统运行调度优化

胡正彪，贺东风

（北京科技大学冶金与生态工程学院，北京 100083）

摘 要：钢铁企业实际生产过程中多工况变化不但影响能源转换设备的效率，甚至造成能源转换设备的启停，从而影响能源的利用和能耗。本文以能源运行成本最低和能源转换设备启停费用最小为目标函数，建立了针对钢铁联合企业煤气-蒸汽-电力转换系统运行调度优化模型。结果表明：以能源运行成本最小为单目标优化后，煤气柜起到充分存储作用，系统总费用比优化前降低 4.70%。综合考虑能源运行成本和能源转换设备启停费用的多目标优化后，在相同能源供需下，只需投入 7 台设备即可满足能源需求（1#35t 锅炉停运），且其余蒸汽锅炉效率相比单目标优化的结果有所提高，总费用比优化前降低了 5.20%，

关键词：钢铁企业；能源系统；运行负荷；设备启停；优化调度

炼钢–连铸区段多工序协同运行水平的量化评价

杨建平[1]，刘 青[1,2]

（1. 北京科技大学钢铁冶金新技术国家重点实验室，北京 100083；2. 北京科技大学钢铁生产制造执行系统教育部工程研究中心，北京 100083）

摘要：本文基于冶金流程工程学，构建炼钢–连铸区段层流运行水平、工序匹配水平和调度模型可用性三个评价模型，提出层流运行评价指数、工序匹配度、调度模型可用性评价指数等多工序协同运行水平的评价参数。以国内 A、B 两家钢厂为研究对象，首先，应用层流运行水平和工序匹配水平评价模型分析两厂 2019 年 4 月–7 月期间炼钢–连铸区段的生产运行状况；其次，针对运行状况欠佳的 A 厂，提出基于炉–机对应模式优化的调度模型，并应用调

度模型可用性评价模型对该模型进行评估。研究表明：2019 年 4 月–7 月期间 A 厂炼钢–连铸区段的层流运行程度低于 B 厂，两家钢厂的系统层流运行评价指数的月平均值分别为 0.638 和 1，工序匹配度的月平均值分别为 0.610 和 0.759；基于炉–机对应模式优化的调度模型可用性评价指数为 0.910，将该模型应用于 A 厂后，其炉–机对应关系得到显著改善，工序匹配度提高了 4.6%。

关键词：炼钢–连铸；多工序协同运行；量化评价模型；层流运行；工序匹配；调度模型

用数字钢铁驱动冶金流程界面

李立勋，杨晓江，康书广

（河钢集团唐钢公司，河北唐山　063000）

摘　要：钢铁产品是社会经济的重要基础，钢铁工业是制造业的重要分枝也是制造业进行数字化、网络化、智能化升级的主战场。数字钢铁是重要突破口，钢铁企业的全面数字化、冶金工艺高端模型化、企业深度信息化与多工序网络协调制造是新工业时代的三大工程。实现钢铁生产的操作自动化、工序集成化、流程紧凑化、管理数字化促进物质流、能量流、信息流高效协同精密运转，从而对钢铁生产经营活动实现流程与系统的全局统筹优化与网络协同制造，提高钢铁企业竞争力促进其可持续发展。

关键词：钢铁；数字化；数字钢铁；冶金工艺高端模型；网络协同制造；企业竞争力

鞍钢 11 号高炉热风炉的附加燃烧炉预热系统改造探讨

张维巍，战　奇，李世明，赵晓峰

（鞍钢集团工程技术有限公司炼铁室）

摘　要：针对鞍钢 11 号高炉热风炉采用的带用附加燃烧炉双预热系统迫切需要改造等问题。本文在详细分析了旧有系统改造的要点，并对下一步的施工提出建议。同时也对用风温评价预热系统的观点提出了想法。

关键词：热风炉；预热系统

1　矿业工程

精益生产管理在矿山企业的应用

慕园园

（内蒙古包钢钢联股份有限公司巴润矿业分公司，内蒙古包头 014080）

摘 要：由于行业形势的变化，矿山企业在生产组织中面临诸多困难。精益生产管理能为企业在经营中获取较大的利益，相较于传统的生产管理措施，精益生产管理具有明显的优点。构建企业精益生产管理体系，并结合生产实际，以创新为动力，以项目为抓手，在实际应用中利用精益生产管理的方法，提高生产效率，提升设备台效，降低消耗。运用多种创新方法、工具开展精益生产管理，消除过程存在的瓶颈制约、库存、设备故障、信息不畅通等问题，构建精益生产管理体系，助力企业健康发展。

关键词：精益生产；创新；实践

Mining Enterprises Lean Production Management System of Innovation and Practice

Mu Yuanyuan

(Barun Mining Company of Inner Mongolia Baotou Steel Union Co., Ltd., Baotou 014080, China)

Abstract: Due to the changes of the industry situation, mining enterprises face many difficulties in the production organization. Constructing enterprise's lean production management system, and connecting with the actual production, driven by innovation, with project as the gripper, in practice by using the method of lean production management, improve the production efficiency, improve equipment efficiency, reduce the consumption. Using a variety of innovative methods and tools to carry out the lean production management, to eliminate bottlenecks, inventory process exists, the problem such as equipment failure, the information is not clear, build lean production management system, boost the healthy development of enterprises.

Key words: lean production; innovation; practice

活性石灰旋窑燃烧智能化研究

余东晓，许 丽，李 伟，肖 玲，王泽健，姚腾宇

（武钢资源集团乌龙泉矿业有限公司，湖北武汉 430213）

摘 要：活性石灰是炼钢生产中最重要的造渣与脱磷脱硫熔剂，其质量是炼钢生产高效优质高产的重要影响因素之一。武钢资源集团乌龙泉矿业有限公司是我国最早引进旋窑生产活性石灰的企业，为了提高活性石灰生产效率和产品质量，针对活性石灰旋窑的燃烧过程各参数之间存在强相互耦合且系统具有时变、滞后、大惯性、非线性等特性无法建立准确的控制模型的难题，提出了采用大数据分析+神经网络+滑模变结构控制策略的解决方案，利用大数据分析对旋窑主要控制参数——石灰活性度、喷煤量、下料量、风量、转速、压力及温度等变量的历史数据进行了匹

配筛选，在此基础上利用神经网络算法构建了多变量关系模型，再通过滑模变结构控制模型实现了活性石灰旋窑燃烧智能化，并取得了满意的效果。

关键词： 活性灰；旋窑；大数据分析；神经网络解耦；滑模变结构控制

Study on Intelligent Combustion of Active Lime Rotary Kiln

Yu Dongxiao, Xu Li, Li Wei, Xiao Ling, Wang Zejian, Yao Tengyu

(Wulongquan Mining Co., Ltd. of WISCO Resources Group, Wuhan 430213, China)

Abstract: Active lime is the most important flux for slag making, dephosphorization and desulfurization in steelmaking production, and its quality is one of the important factors affecting high efficiency, high quality and high yield in steelmaking production. Wulongquan Mining Co., Ltd. of WISCO resources group is the first enterprise to introduce rotary kiln to produce active lime in China. In order to improve the production efficiency and product quality of active lime, it is difficult to establish an accurate control model due to the strong coupling among various parameters in the combustion process of active lime rotary kiln and the time-varying, lag, large inertia and nonlinear characteristics of the system, A solution of big data analysis + neural network + sliding mode variable structure control strategy is proposed. By using big data analysis, the historical data of main control parameters of rotary kiln, such as lime activity, coal injection quantity, cutting quantity, air volume, speed, pressure and temperature, are matched and screened. On this basis, a multivariable decoupling model is constructed by using neural network algorithm, Then, the intelligent combustion of active lime rotary kiln is realized by sliding mode variable structure control model, and satisfactory results are obtained.

Key words: activated lime; rotary kiln; big data analysis; neural network decoupling; sliding mode variable structure control

某金矿上向水平分层充填采矿参数优化研究

夏文浩[1,2]，宋卫东[1,2]

（1. 北京科技大学土木与资源工程学院，北京 100083；
2. 北京科技大学金属矿山高效开采与安全教育部重点实验室，北京 100083）

摘　要： 某金矿主要采用上向水平进路充填采矿法进行回采，由于回采断面较小，存在生产效率过低、年产量难以完成等问题。因此，考虑在保证安全的前提下，采用断面相对较大的上向水平分层充填法进行回采。根据同类型矿山经验，共设计了 9 种采场结构参数方案和 3 种回采顺序方案，并使用 FLAC^{3D} 软件进行了数值模拟，并从应力、位移和塑性区分布三方面对模拟结果进行了定量分析，最终确定最优采场结构参数为矿房跨度 8m，矿柱跨度 8m，分层采高 3.5m，最佳回采顺序为矿房超前两分层回采，研究结果对同类矿山采场结构参数选取有一定的借鉴意义。

关键词： 充填采矿法；结构参数优化；回采顺序优化；数值模拟

Optimization of Mining Parameters for Upward Horizontal Slicing and Filling in a Gold Mine

Xia Wenhao[1,2], Song Weidong[1,2]

(1. School of Civil and Resources Engineering, University of Science and Technology Beijing, Beijing 100083, China; 2. Key Laboratory of High-Efficient Mining and Safety of Metal Mines, Ministry of Education, University of Science and Technology Beijing, Beijing 100083, China)

Abstract: A gold Mine mainly adopts upward horizontal drift cemented filling mining method for deep mining. Due to the small mining section, the mining efficiency is too low and it is difficult to complete the annual production task. Therefore, on the premise of ensuring safety, the upward horizontal slicing and filling method with relatively large mining section is considered. A total of 9 kinds of stope structure parameters and 3 kinds of stope sequence schemes are designed, and the numerical simulation is carried out by using FLAC^{3D} software. The simulation results are quantitatively analyzed from three aspects of stress, displacement and plastic zone distribution. Finally, the optimal stope structure parameters of the gold mine are determined as follows: room span of 8m, pillar span of 8m, layered mining height of 3.5m, The best mining sequence is two layers ahead of the room. The research results have certain reference significance for the selection of stope structure parameters of similar mines.

Key words: filling mining method; structural parameter optimization; mining sequence optimization; numerical simulation

某金矿下向进路充填采矿参数优化研究

黄　坤[1,2]，宋卫东[1,2]

（1. 北京科技大学土木与资源工程学院，北京　100083；
2. 北京科技大学金属矿山高效开采与安全教育部重点实验室，北京　100083）

摘　要：采用三维有限元分析软件 FLAC^{3D}，对某金矿浅部采用下向进路充填采矿法的采场的回采进路进行参数优化，通过分析不同结构参数下顶板的位移、应力及塑性区分布规律，得出采场稳定性状况，并进而优选出最佳结构参数与进路布置形式。结果表明：最佳进路断面尺寸为 3.0m×3.0m，进路布置形式采用“品”字形布置时，可以改善假顶的受力状态，提高人工假顶的承载能力，从而保证顶板的稳定，保障作业人员的生命安全。

关键词：充填采矿法；数值模拟；结构参数优化；采场稳定性

Optimization of Mining Parameters for the Underhand Cut and Fill Mining Method in Agold Mine

Huang Kun[1,2], Song Weidong[1,2]

(1. School of Civil and Resources Engineering, University of Science and Technology Beijing, Beijing 100083, China; 2. Key Laboratory of High-Efficient Mining and Safety of Metal Mines, Ministry of Education, University of Science and Technology Beijing, Beijing 100083, China)

Abstract: Using the three-dimensional finite element analysis software FLAC^{3D}, this paper optimizes the parameters of the stope of a gold mine, which adopts the underhand cut and fill mining method, analyzes the displacement and stress changes of the stope under different stope structural parameters, obtains the stope stability under different parameters, and selects the best structural parameters and the layout form of the stope. The results show that the optimal section size of the access is 3.0m × 3.0m, and the "品" layout is adopted for the access layout, the stress state of the artificial roof can be improved, and the bearing capacity of the artificial roof can be improved, so as to ensure the stability of the roof and ensure the life safety of the operators.
Key words: filling mining method; numerical simulation; structural parameter optimization; stope stability

采矿汽车运输设备点检管理创新研究与实践

刘振陆

（鞍钢集团矿业有限公司齐大山分公司，辽宁鞍山　114043）

摘　要： 针对鞍钢集团矿业有限公司齐大山分公司采矿运输设备，大型矿用电动轮自卸汽车设备点检管理上出现的难点问题，通过管理创新，解决了影响矿用电动轮自卸汽车备件和材料成本消耗和运输效率等难题。通过研究和实践证明，点检管理创新是实现采矿汽车运输设备降本增效、保障生产任务完成行之有效的管理方法。
关键词： 采矿运输设备；点检管理创新；研究与实践

Research and Practice on Innovation of Spot Inspection Management of Mining Truck Transportation Equipment

Liu Zhenlu

(Anshan Group Mining Co., Ltd. Qidashan Branch, Anshan 114043, China)

Abstract: In view of the difficult problems in the spot inspection management of the mining and transportation equipment of Qidashan Branch of Anshan Iron and Steel Group Mining Co., Ltd., the problems affecting the spare parts and material cost consumption and transportation efficiency of the mine electric wheel dump truck were solved through management innovation. Through research and practice, it has been proved that spot inspection management innovation is an effective management method to reduce cost and increase efficiency of mining truck transportation equipment and ensure the completion of production tasks.
Key words: mining and transportation equipment; spot inspection management innovation; research and practice

长距离铁精矿矿浆管道除垢关键技术研究

刘亚峰，陶志宾

（内蒙古包钢钢联股份有限公司巴润矿业分公司，内蒙古包头　014080）

摘　要：固体物料管道水力输送技术是一种以液体介质作为载体，通过密闭管道输送固体物料的运输技术，是一种清洁、环保的生产方式，现被矿山行业广泛应用，也包头钢铁公司铁精矿输送采用此种技术。包头钢铁公司矿浆管道经多年运行，管道内部结垢越来越严重，造成主泵出口压力逐年升高，严重影响生产和产能提升。为清除管道垢质，恢复和提升管道输送能力，开展了清管技术研究，创新机械清管工艺，发明割刀、铣刀清管器，研发新型组合式机械清管器，有效完成清管除垢工作。

关键词：清管除垢；矿浆管道；铁精矿；机械清管器

Study on Key Technology of Descaling in Long-distance Iron Concentrate Slurry Pipeline

Liu Yafeng, Tao Zhibin

(Barun Mining Company of Inner Mougolia Baotou Steel Union Co., Ltd., Baotou 014080, China)

Abstract: Solid material pipeline transportation technology is a kind of transportation technology which uses liquid medium as carrier and transports solid materials through closed pipeline. This technology is a clean and environment-friendly production method, which is now widely used in mining industry. This technology is used in transportation of iron concentrate in Baotou Iron and Steel Company. After years of operation, the scaling in the slurry pipeline of Baotou Iron and Steel Company is becoming more and more serious, which causes the outlet pressure of the main pump to increase year by year, which seriously affects the production and productivity improvement. In order to remove pipeline scale, restore and improve pipeline transportation capacity, pigging technology research was carried out, mechanical pigging technology was innovated, cutters and milling cutters were invented, and new combined mechanical pigs were developed to effectively complete pigging and descaling work.

Key words: pigging and descaling; slurry pipeline; iron concentrate; mechanical pig

振动筛技术改造和应用实践

慕园园

（内蒙古包钢钢联股份有限公司巴润矿业分公司，内蒙古包头　014080）

摘　要：直线振动筛是目前应用很广泛的筛分机械之一。由于振动筛在工作过程中，要长期承受交变载荷的作用，其筛网与拉筋接触部位经常容易发生断裂等故障，这严重影响了振动筛的无故障运行时间和使用寿命。振动筛筛网的寿命长短直接关系到企业的生产效率和经济效益[1]，目前在钢铁行业爬坡过坎的关键时期，降低生产成本成了至关重要的措施。所以对振动筛的主要部位进行长寿化改造，提高振动筛的运行稳定性，同时节约生产成本，成为当下亟待解决的问题。本文通过介绍振动筛的几种常见问题，并结合生产实际对故障点进行改造，以此改善振动筛的运行状态，对提高筛网的寿命和筛机的生产效率有重要的意义[2]。

关键词：振动筛；技术；改造；应用

Vibrating Screen Technology Reform Practice and Application

Mu Yuanyuan

(Barun Mining Company of Inner Mongolia Baotou Steel Union Co., Ltd., Baotou 014080, China)

Abstract: Linear vibrating screen is currently one of the application of a wide range of screening machine. Because of the vibrating screen in the process of work, should carry alternating load for a long time, the screen mesh and brace contact parts often prone to fracture failure, the serious influence the trouble-free operation time and the service life of the vibrating screen. Vibrating screen mesh lifespan is directly related to enterprise's production efficiency and economic benefits, at present in the steel industry grade, critical period, reduce the production cost has become very important measures. So the main part of the vibrating screen for long life, improve the operation stability of the vibrating screen, save the cost of production at the same time, become a problem urgently to be solved in the present. In this paper, through the introduction of the several common problems of vibrating screen, and connecting with the actual production to the point of failure, in order to improve the running state of vibrating screen, to improve the life of the screen and screen production efficiency has important significance.

Key words: vibrating screen; technology; modification; application

管道输送系统精准计量技术研究

刘亚峰，裴　斌，陶志宾

（内蒙古包钢钢联股份有限公司巴润分公司，内蒙古包头　014080）

摘　要：在浆体管道输送系统中首次应用隔膜泵作为计量结算单元。通过分析隔膜泵入口压力、输矿浓度对成品隔膜泵流量的影响程度，提出改善措施；并设计应用快速自动标定系统定期对隔膜泵流量进行标定，最终保证浆体输送管道计量结算精准高效。

关键词：浆体管道；隔膜泵；计量结算；标定

Research on Accurate Measurement Technology of Pipeline Transportation System

Liu Yafeng, Pei Bin, Tao Zhibin

(Barun Mining Company of Inner Mougolia Baotou Steel Union Co., Ltd., Baotou 014080, China)

Abstract: Diaphragm pump is first used as metering and settlement unit in slurry pipeline conveying system. By analyzing the influence of inlet pressure and ore concentration of diaphragm pump on the flow of finished diaphragm pump, the improvement measures are put forward. The rapid automatic calibration system is designed and applied to calibrate the flow of diaphragm pump regularly to ensure the accuracy and efficiency of slurry delivery pipeline measurement and settlement.

Key words: slurry pipeline; diaphragm pump; measurement and settlement; calibration

ZNJQ-1150 型智能加球机在大红山铁矿的运用

唐国栋，杨雪莹，崔　宁，邓维亮

（玉溪大红山矿业有限公司，云南玉溪　653405）

摘　要：针对 MZS8848 半自磨机衬板出现延展、碎裂的现象，分析是否可以通过优化补加钢球方式来延长磨机衬板使用寿命，借助 ZNJQ-1150 型智能加球机的运用，通过“分时均量”修正“一次全量”的半自磨机钢球添加方式，使 MZS8848 半自磨机衬板延展、碎裂，衬板螺栓断裂的情况有所缓解，使用寿命得以延伸，同时提高了半自磨机的工艺技术指标，优化了选厂的选矿成本，并创造了一定的经济效益。

关键词：加球机；“分时均量”；补加球；增量；提效

Application of ZNJQ-1150 Intelligent Pelletizer in Dahongshan Iron Mine

Tang Guodong, Yang Xueying, Cui Ning, Deng Weiliang

(Yuxi Dahongshan Mining Co., Ltd., Yuxi 653405, China)

Abstract: Aiming at the extension and fragmentation of MZS8848 semi autogenous mill liner, this paper analyzes whether the service life of mill liner can be extended by optimizing the way of adding steel balls. With the help of ZNJQ-1150 intelligent ball adding machine, through the “time sharing average” correction of “one-time full” semi autogenous mill steel ball adding method, the MZS8848 semi autogenous mill liner can be extended and fragmented, and the liner bolt is broken At the same time, the technical indexes of the semi autogenous mill are improved, the beneficiation cost of the concentrator is optimized, and certain economic benefits are created.

Key words: adding ball machine; “time sharing average quantity”; adding ball; increment; improve efficiency

低品位赤铁矿湿式预选工艺研究及设计

于克旭，米红军，金　敏

（鞍钢集团矿业设计研究院工艺设计研究室，辽宁鞍山　114002）

摘　要：本文通过对磨前低品位赤铁矿采用预选技术进行选矿工艺研究，制定合理的工艺流程，在原有选矿厂入磨前新建磨前预选车间,根据选定的工艺进行现场的实际应用，并对相关选矿设备进行了详细的介绍。设计中选用了选矿常规设备解决了处理低品位赤铁矿的难题，为钢铁工业提供了新的矿产资源，提高矿产资源利用率，创造了显著的经济效益。

关键词：赤铁矿；预选；工艺；流程；设计

Study and Design of Wet Pre-concentration Process for Hematite

Yu Kexu, Mi Hongjun, Jin Min

(The Research Institute of Anshan Mining Company, Anshan 114002, China)

Abstract: This paper systematically introduces the process flow and design of wet pre-concentration of hematite, and introduces the ore dressing equipment in detail. In the design, the conventional equipment used for mineral processing is used to solve the problem of treating low grade hematite, which provides new mineral resources for the iron and steel industry, improves the utilization rate of mineral resources, and creates remarkable economic benefits.
Key words: hematite; pre-concentration; process; flow; design

八钢矿山自产铁精矿指标分析及改善措施

王振刚

（新疆八钢矿业资源有限公司，新疆乌鲁木齐　830022）

摘　要：近年来八钢各矿山自产铁精矿技术经济指标波动较大，部分指标不能满足八钢烧结、球团及高炉工序的需求，各矿山的总体经济效益不高。本文通过对八钢各矿山单位自产铁精矿指标进行分析，发现存在的问题并探索改善措施。
关键词：铁精矿；指标分析；改善措施

Eight Steel Mineral Mountain Self-producting Iron Jing Mineral Index Sign Analysis and Improvement Measure

Wang Zhengang

(Xinjiang Eight The Steel Mineral Industry Resources Limited Company, Urumqi 830022, China)

Abstract: Eight each mineral of steel mountains self product economic index sign of iron Jing mineral technique to undulate a little bit greatly in recent years, parts of index signs can not satisfy the need that eight steels burn knot, ball regiment and blast furnace work preface, and the macro economy efficiency of each mineral mountain isn't high. This text carries on analysis through self-producting to eight each mineral of steel mountain units iron Jing mineral index sign and discovers an existent problem and investigate to improve measure.
Key words: the mineral of iron Jing; index sign analyzes; improve measure

燃烧碘量法测定硫常见故障分析及应对措施

王　浩，崔俊豪，马博慧，路丽丽，刘亚敏

（中国宝武新疆八一钢铁股份有限公司，新疆乌鲁木齐　830022）

摘 要：燃烧碘量法测定硫是一种经典的硫分析方法，广泛应用于工企矿山等中小型化验室，为了提升燃烧碘量法测定硫的生产效率，有效降低故障率，应加强对燃烧碘量法测硫仪的维修保养，定期检修维护。基于此，本文针对燃烧碘量法测硫仪常见故障进行了详细分析，提出了相应的应对措施，以供相关人员参考。

关键词：测硫仪；常见故障；应对措施

Analysis and Countermeasures of the Common Failures in the Determination of Sulfur by Combustion Iodimetry

Wang Hao, Cui Junhao, Ma Bohui, Lu Lili, Liu Yamin

(China Baowu Xinjiang Bayi Iron & Steel Co., Ltd., Urumqi 830022, China)

Abstract: The burning iodimetric determination sulfur is one classical sulfur analysis method, widely applies in the labor business mine and so on the middle and small scale laboratory, in order to promote the burning iodimetric determination sulfur the production efficiency, reduces the failure rate effectively, should strengthen to the burning iodimetric measures the sulfur meter the service maintenance, preventive maintenance maintenance. Based on this, this article in view of the burning iodimetric measured the sulfur meter common breakdown has carried on the multianalysis, proposed corresponding should to the measure, for the related personnel to refer.

Key words: test sulfur scanner; a common problem; response measures

2 炼铁与原燃料

2.1　炼焦化学

7.63m 焦炉四大机车控制系统总体设计思路浅析

王　宁
（首钢京唐钢铁联合有限责任公司，河北唐山　063200）

摘　要：提高炼焦行业焦炉四大机车（推焦车、拦焦车、熄焦车和装煤车）的自动化控制水平对安全可靠的生产有重要意义。四大机车相互间的通信、定位、连锁，自动走行以及计算机的管控协调等一系列技术问题有待深入研究。由于生产现场环境恶劣，高温、多尘、多干扰以及车辆的频繁移动和工艺上要求的精确定位等因素，凸显了对自动化技术的迫切需求。本文仅对四大机车控制系统的总体设计思路进行简单介绍。

关键词：四大机车；总体设计；控制系统

焦炉荒煤气放散点火系统应用浅析

王　宁
（首钢京唐钢铁联合有限责任公司，河北唐山　063200）

摘　要：本文主要介绍京唐钢铁公司焦化作业部 7.63m 焦炉所使用的荒煤气点火放散系统的特点和功能，并对其工艺原理、系统整体结构和网络通讯、自动控制的实现以及 HMI 监控画面进行简要阐述。

关键词：荒煤气点火放散；通讯；自动控制；HMI

焦炉脱硫脱硝系统中 PLC 技术应用浅析

王　宁
（唐山首钢京唐西山焦化有限责任公司，河北唐山　063200）

摘　要：本文着重介绍 PLC 技术在焦化脱硫脱硝工段中的应用，并简要阐述其工艺原理、整个控制系统的软件、硬件结构、主要控制、HMI 等方面。

关键词：PLC；脱硫脱硝；控制系统；焦化；技术

影响中间相炭微球质量的机理研究

贾楠楠[1]，芦　参[2]，张馨予[1]，刘海丰[1]，何　莹[1]

（1. 鞍钢化学科技有限公司，辽宁鞍山　114021；
2. 鞍钢股份有限公司炼焦总厂，辽宁鞍山　114021）

摘　要：为了满足市场对中间相炭微球质量的要求，本文从沥青中间相的生成机理入手，研究了反应温度、反应时间、溶剂选择和热处理对中间相炭微球质量的影响。同时，利用扫描电镜检测中间相炭微球的形貌，确定不同反应条件对中间相炭微球质量的影响。结果表明，中间相炭微球的收率随着反应温度的升高先增加后降低，在反应温度430℃时达到最大值；随着反应时间的增加中间相炭微球的粒径逐渐增大；热处理能有效提高中间相炭微球表面的光洁度；利用 DMF 替代甲苯可以提高工艺安全性。在最有利条件下可制备出粒径 8~15μm，比容量大于 330mAh/g 的中间相炭微球。

关键词：中间相炭微球；反应温度；反应时间

Research on Quality of Control of MCMBs

Abstract: To meet the requirement of MesoCarbon MircoBead (MCMBs) from the market, the influence of react temperature, react time, solvent selection and heat treatment on the content of the mesophase in modified pitch are studied in this paper. The research above is based on the SEM about the MCMB. Through the reflected light microscope to ensure the content of the MCMB, the impressions of different react conditions on it were got. The results indicated that the yield of MCMBs increased when the temperature rised at first, and became the maximum at 430℃. However, it decreased with the temperature went on increasing. The particle size of MCMBs increased with the adding react time and the smoothness of the surface on MCMBs could be improved through heat treatment. And the process safety could be improved when DMF replaced toluene. Under the best conditions, MCMBs with 8-15μm particle size and specific capacity more than 330mAh/g could be got.

Key words: mesocarbon mircobeade; react temperature; react time

煤系中间相沥青的性能及制备探究

何　莹[1]，刘海丰[1]，张大奎[1]，薛占强[1]，王晓楠[1]，宋克东[2]，郑慧娜[3]

（1. 鞍钢化学科技有限公司，辽宁鞍山　114000；2. 大连理工大学，辽宁大连　116000；
3. 郑州中科新兴产业技术研究院，河南郑州　450000）

摘　要：中间相沥青是新型碳材料的前驱体，在高新材料领域有广阔的应用前景。本文以来源丰富、价廉易得的煤系沥青为原料，对原料组成结构进行评价分析，通过净化工艺得到精制煤沥青，采用四氢萘作为供氢试剂对精制煤沥青进行氢化改性，再对氢化煤沥青进行热聚合反应得到性能优异的煤系中间相沥青。

关键词：煤沥青；中间相沥青；氢化改性；制备

Research on Performance and Preparation of Coal based Mesophase Pitch

He Ying[1], Liu Haifeng[1], Zhang Dakui[1], Xue Zhanqiang[1], Wang Xiaonan[1], Song Kedong[2], Jia Huina[3]

(1. Angang Chemical Technology Co., Ltd., Anshan 114000, China;
2. Dalian University of Technology, Dalian 116000, China;
3. Zhengzhou Institute of Emerging Industry Technology, Zhengzhou 450000, China)

Abstract: Mesophase pitch is the precursor of new carbon materials, which has broad application prospects in the field of high-tech materials. In this paper, the coal tar pitch with rich source, low price and easy availability is used as raw material, the composition and structure of raw material are evaluated and analyzed, the refined coal tar pitch is obtained by purification process, tetrahydronaphthalene is used as hydrogen donor to hydrogenate the refined coal tar pitch, and then the hydrogenated coal tar pitch is thermally polymerized to obtain coal mesophase pitch with excellent performance.

Key words: coal tar pitch; mesophase pitch; hydrogenation modification; preparation

针状焦理化指标与微观结构分析评价浅析

刘海丰，何　莹，张大奎，薛占强，姚　君，王晓楠

（鞍钢化学科技有限公司，辽宁鞍山　114000）

摘　要：通过分析几种针状焦的理化指标及微观结构组成，测定其石墨化后热膨胀系数和石墨化度，得出结论，针状焦的微观结构组成与真密度、粉末电阻率、石墨化度及热膨胀系数存在一定联系。通过分析针状焦的真密度和粉末电阻率，可评判针状焦微晶排列的规整度和石墨化度，最终实现对针状焦性能的快速分析评价。

关键词：针状焦；理化指标；微观结构；分析

Analysis and Evaluation of Physical and Chemical Indexes and Microstructure of Needle Coke

Liu Haifeng, He Ying, Zhang Dakui, Xue Zhanqiang, Yao Jun, Wang Xiaonan

(Angang Chemical Technology Co., Ltd., Anshan 114000, China)

Abstract: By analyzing the physicochemical indexes and microstructure composition of several needle coke, the coefficient of thermal expansion and degree of graphitization of needle coke were determined. It was concluded that the microstructure composition of needle coke was related to true density, powder resistivity, degree of graphitization and coefficient of thermal expansion. By analyzing the true density and powder resistivity of needle coke, the regularity and graphitization degree of needle coke microcrystalline arrangement can be evaluated, and finally the rapid analysis and evaluation of needle coke performance can be realized.

Key words: needle coke; physicochemical indexes; microstructure; analysis

煤焦油深加工产品及发展潜能概述

张立伟，张　旭，王政强，王海涛，王　鹏，刘庆佩，梁　丰

（鞍钢化学科技有限公司，辽宁鞍山　114000）

摘　要：我国焦炭产量巨大，高温煤焦油是焦炭生产过程中的主要副产物，由近万种化合物组成。目前，煤焦油加工企业主要通过高温蒸馏分离得到轻油、酚油、萘油、洗油、蒽油及煤沥青等化工产品。这些产品是合成塑料、纤维、农药、染料、医药、涂料等精细化工产品的主要基础原料，也是冶金、纺织、交通等行业的重要原料。煤焦油下游深加工产品主要包括苯酚、甲酚、苯酐、2-萘酚、喹啉、甲基萘、吲哚、联苯、咔唑、泡沫炭、活性炭和针状焦等。煤焦油深加工产品应用广泛，市场前景广阔。因此，开展煤焦油深加工综合利用，加大下游产品开发力度，提高煤焦油的附加值是煤焦油加工企业创效的一个重要途径。

关键词：煤焦油；深加工；下游产品；煤沥青；炭材料

An Overview of Deep Processing Products and Development Potential of Coal Tar

Zhang Liwei, Zhang Xu, Wang Zhengqiang, Wang Haitao, Wang Peng, Liu Qingpei, Liang Feng

(Angang Chemical Technology Co., Ltd, Anshan 114000, China)

Abstract: Chinese coke production is huge, and high-temperature coal tar is the main by-product in the process of coke production. High-temperature coal tar is composed of nearly 10,000 compounds. At present, coal tar processing enterprises obtain chemical products such as light oil, phenol oil, naphthalene oil, washing oil, anthracene oil and coal pitch through high-temperature distillation, which are the basic raw materials for fine chemical products such as synthetic plastics, synthetic fibers, pesticides, dyes, medicines, coatings, etc., as well as important raw materials for metallurgy, textile, transportation and other industries. The downstream deep processing products of coal tar mainly include phenol, cresol, phthalic anhydride, 2-naphthol, quinoline, methyl naphthalene, indole, biphenyl, carbazole, foamed carbon, activated carbon and needle coke. Deep processing products of coal tar are widely used and have broad market prospects. Therefore, to carry out the comprehensive utilization of coal tar deep processing, increase the development of downstream products and the value of coal tar are important ways for coal tar processing enterprises to create benefits.

Key words: coal tar; deep processing; downstream products; coal pitch; carbon material

7.63m 焦炉拦焦车操作时间优化研究

冯敏超[1]，王　宁[2]

（1. 河北省唐山市唐山科技职业技术学院，河北唐山　063000；
2. 唐山首钢京唐西山焦化有限责任公司，河北唐山　063000）

摘　要：本文介绍了 7.63m 焦炉熄焦车的控制流程及特点，分析了熄焦车在单体控制和系统控制方面的缺陷，并通过改进这些缺陷，减少了熄焦车操作时间，提高了自动化控制水平。

关键词：7.63m 焦炉；熄焦车；操作时间；优化

7.63m 焦炉装煤车操作时间优化研究

刘永国，王　宁

（唐山首钢京唐西山焦化有限责任公司，河北唐山　063000）

摘　要：本文介绍了 7.63m 焦炉装煤车的控制流程及特点，分析了装煤车在单体控制和系统控制方面的缺陷，并通过改进这些缺陷，减少了装煤车操作时间，提高了自动化控制水平。

关键词：7.63m 焦炉；装煤车；操作时间；优化

7.63m 焦炉熄焦车操作时间优化研究

刘永国，王　宁

（唐山首钢京唐西山焦化有限责任公司，河北唐山　063000）

摘　要：本文介绍了 7.63m 焦炉熄焦车的控制流程及特点，分析了熄焦车在单体控制和系统控制方面的缺陷，并通过改进这些缺陷，减少了熄焦车操作时间，提高了自动化控制水平。

关键词：7.63m 焦炉；熄焦车；操作时间；优化

关于干熄焦冷却室温度控制的思路和方法

李梦年

（临涣焦化股份有限公司，安徽淮北　235000）

摘　要：本文介绍了干熄焦在生产运行时影响冷却室温度的各类因素，通过控制焦炭下落速度和优化调风装置两项主要举措，实现冷却室温度平衡的安全生产要求，并以此来提高干熄率。

关键词：干熄焦；温度平衡器；调风装置

干熄焦余热锅炉过热器爆管的研究与分析

李梦年

摘　要：通过对两台干熄焦余热锅炉过热器爆管的工艺运行、设备缺陷、材质分析等方面进行调查研究，总结出爆管的原因及改进措施。

关键词：过热器；爆管；干熄焦；余热锅炉

Abstract: Based on the investigation and Research on the process operation, equipment defects and material analysis of two CDQ waste heat boiler superheaters, the causes and improvement measures of tube explosion are summarized.

Key words: superheater; tube explosion; CDQ; waste heat boiler

提高配合煤堆比重试验研究

陈　健

（重庆钢铁制造管理部，重庆　401220）

摘　要：本文通过对重庆钢铁现有配煤结构下配合煤细度进行优化调整，研究配合煤堆比重的变化规律，同时开展小焦炉试验，研究焦炭 CSR 与配合煤堆比重之间的线性关系，得出如下结论：(1)随着配合煤细度降低，配合煤堆比重逐步提高，当配合煤细度处于 72%左右时，对应堆密度 0.7531t/m3 最佳;(2)堆比重提高焦炭 CSR 改善，当配合煤细度与堆比重处于最佳条件时，焦炭 CSR 最优。

关键词：配煤结构；细度；堆密度；CSR

浅谈宝钢湛江钢铁焦炉的少人化

李明岩，周利鹏，袁浩杰，张　帆

（宝钢湛江钢铁有限公司，广东湛江　524000）

摘　要：炼铁厂三期为新建 2×65 孔碳化室高 7m，宽 530mm 的复热式顶装焦炉，机车无人操作少人值守模式，应用机械数字孪生，无烟装煤车，煤口堵塞检测，主动防碰撞系统，故障诊断系统，辅助功能拓展等技术应用，实现焦炉机车自动化。结合焦炉本体的液压交换机，炉底巡检机器人，上升管余热利用及自动开闭和火落判定技术等功能相互配合，完成焦炉少人作业，文章同时分析了焦炉机车无人驾驶的可行性。

关键词：无人驾驶；数字孪生；机车防碰撞；无烟装煤；炉底巡检；炭化室控制

Talking about the Unmanned Coke Oven of Baosteel Zhanjiang Iron and Steel

Li Mingyan, Zhou Lipeng, Yuan Haojie, Zhang Fan

(Baosteel Zhanjiang Steel Co., Ltd., Zhanjiang 524000, China)

Abstract: The third phase of the ironmaking plant is a newly built 2×65 hole carbonization chamber with a height of 7 meters and a width of 530mm. The reheating top loading coke oven is operated by no one and few people are on duty. The automation of coke oven locomotive is realized by the application of mechanical digital twinning, smokeless coal loading car, coal opening blockig detection, active anti-collision system, fault diagnosis system, auxiliary function expansion and other technologies. Combined with the coke oven body's hydraulic switch, furnace bottom inspection robot, rising tube waste heat utilization and automatic opening and closing and fire fall determination technology and other functions to cooperate with each other to complete the coke oven unmanned operation, the paper also analyzes the feasibility of coke oven locomotive unmanned driving.

Key words: unmanned; digital twin; locomotive collision avoidance; smokeless coal; the furnace bottom inspection; carbonization chamber control

干熄炉长寿化技术应用实践

杨邵鸿

（本钢板材股份有限公司，辽宁本溪 117017）

摘 要：随着世界性能源短缺加剧以及环保法规日趋严格，干熄焦技术作为焦化行业重大节能环保技术已成为炼焦行业配套的重要环保、节能、增效装备,其长寿化是关键技术之一。因此，为减少干熄炉检修时间，提高综合干熄率，提升企业经济效益，研究干熄炉长寿化技术势在必行。本钢北营通过与中冶焦耐设计专家深入分析该厂 4 套干熄炉检修内容及频次，特别是 1 号干熄炉内环墙破损原因，探讨炼焦、运焦及干熄焦生产组织优化方案，采取一系列改进措施，特别是引进新一代干熄焦炉长寿化耐火材料、优化砖型设计、改良砌筑工法、规范生产操作及新型分割斜道技术等干熄炉长寿化技术，增加单砖体间咬合能力，提高内环墙结构的整体性，实现了干熄焦装置四年一中修目标，有效延长了干熄炉检修周期。

关键词：干熄炉；长寿化技术；应用实践

Application Practice of Longevity Technology in CDQ Furnace

Yang Shaohong

(Bengang Steel Plates Co., Ltd., Benxi 117017, China)

Abstract: As the world's energy shortage intensifies and the environmental protection laws and regulations become more and more strict, the dry quenching technology, as a major energy-saving and environmental protection technology in the coking industry, has become an important environmental protection, energy-saving and efficiency-enhancing equipment for the coking industry, its longevity is one of the key technologies. Therefore, in order to reduce the overhaul time of CDQ, improve the comprehensive CDQ rate and enhance the economic benefits of enterprises, it is imperative to study the technology of CDQ. Through an in-depth analysis of the overhaul contents and frequency of the four cdq furnaces in the plant, especially the reason for the breakage of the inner ring wall of the No. 1 CDQ Furnace, Bx Steel Beiying and experts in coke resistance design at mcc have discussed the optimization scheme of the production organization of coking, coke transportation and CDQ, a series of improvement measures have been taken, in particular, the introduction of new generation of Refractory, optimization of brick design, improvement of masonry methods, standardization of production operations and new technology of dividing ramps, etc., to increase the occlusive ability of single brick body and improve the integrity of Inner Ring Wall structure, the intermediate repair goal of cdq device is achieved, and the repair period of cdq furnace is extended effectively.

Key words: CDQ furnace; ongevity technology; application practice

大型焦炉热平衡测试与节能分析

甘秀石[1]，郝　博[1]，李卫东[1]，边子峰[2]，王　磊[2]

（1. 鞍钢集团钢铁研究院，辽宁鞍山　114009；2. 鞍钢股份炼焦总厂，辽宁鞍山　114021）

摘　要：总结分析了常规大型焦炉热平衡测试和计算结果，定量掌握焦炉能量消耗和分配情况，结合焦炉自身情况找出炼焦工序节能降耗的途径。

关键词：焦炉；热平衡；节能降耗

Energy Consumption Analysis on Coke Oven of Ansteel

Gan Xiushi[1], Hao Bo[1], Li Weidong[1], Bian Zifeng[2], Wang Lei[2]

(1. Iron & Steel Institutes of Ansteel Group Corporation, Anshan 114009, China;
2. General Coking Plant of Angang Steel Co., Ltd., Anshan 114021, China)

Abstract: It is summaried and analysed the results of heat balance determination and calculation on large coke oven. According to the results of measurement and calculation and the present situations of coke oven, the ways of reducing the heat loss and achieving the saving of coke oven energy in the coking process were pointed.

Key words: coke oven; heat balance; energy conservation

一体化锤头的改进及对配合煤细度的影响研究

赵振兴[1]，朱庆庙[1]，王　旭[1]，朱亚光[2]，杜少彬[2]

（1. 鞍钢集团钢铁研究院，辽宁鞍山　114009；2. 鞍钢股份炼焦总厂，辽宁鞍山　114021）

摘 要：为了解决反击式粉碎机分体式锤头频繁脱落弊端，配合煤细度产生较大波动问题。对比分析了相同质量下一体式锤头与分体式锤头的生产参数，对分体式锤头进行一体式改进，达到了降低粉碎细度 0.819%，过细颗粒减少 20%，提高粉碎效率，锤头更换周期由 6~7 个月延长至 10~12 个月，单孔装煤量增加 0.3~0.4t。

关键词：备煤车间；粉碎机；一体式锤头；配合煤细度

Study on Improvement of Integrated Hammer Head and Its Influence on Fineness of Blending Coal

Zhao Zhenxing[1], Zhu Qingmiao[1], Wang Xu[1], Zhu Yaguang[2], Du Shaobin[2]

(1. Iron & Steel Institutes of Ansteel Group Corporation, Anshan 114009, China;
2. General Coking Plant of Angang Steel Co., Ltd., Anshan 114021, China)

Abstract: In order to solve the problem of frequent falling off of the split hammer head of the impact pulverizer, the problem of large fluctuation in coal fineness is produced. The production parameters of the integrated hammer head and the split hammer head with the same quality were compared and analyzed. The integrated improvement of the split hammer head achieved a reduction of the crushing fineness by 0.819%, a reduction of the fine particles by 20%, and an improvement of the crushing efficiency. The replacement cycle of the hammer head was extended from 6-7months to 10-12months, and the coal loading of the single hole was increased by 0.3-0.4tons.

Key words: coal preparation workshop; mill; integrated hammer head; with coal fineness

鞍钢焦炉在线自动测温技术实践

王 超[1]，韩树国[2]，甘秀石[1]，朱庆庙[1]，肖泽坚[2]，刘福军[1]

（1. 鞍钢集团钢铁研究院，辽宁鞍山 114009；
2. 鞍钢股份炼焦总厂，辽宁鞍山 114021）

摘 要：针对人工测量焦炉直行温度调节煤气流量和烟道吸力过程中存在的的测温误差大、调节反馈滞后、浪费资源等焦炉热工管理关键问题，鞍钢股份炼焦总厂在全厂焦炉应用焦炉温度自动测量及自动控制系统。实现焦炉温度变化趋势在线监测，消除人工测温误差干扰，降低人工劳动强度；实现全自动焦炉加热控制，通过快速调节加热用煤气流量，减少炉温波动；实现焦炉安定系数≥0.85，节省加热用高炉煤气 1.5%以上。

关键词：焦炉；在线测温；自动加热

Practice of on Line Automatic Temperature Measurement Technology for Coke Oven of Ansteel

Wang Chao[1], Han Shuguo[2], Gan Xiushi[1], Zhu Qingmiao[1], Xiao Zejian[2], Liu Fujun[1]

(1. Iron & Steel Institutes of Ansteel Group Corporation, Anshan 114009, China;
2. General Coking Plant of Angang Steel Co., Ltd., Anshan 114021, China)

Abstract: Aiming at the key problems of coke oven thermal management in the process of manual measurement of coke oven longgitudinal temperature, adjustment of gas flow and flue suction, such as large temperature measurement error, adjustment feedback lag, waste of resources, etc., the coke oven temperature automatic measurement and automatic control system is applied in the General Coking Plant of Angang Steel Co., Ltd. It can realize on-line monitoring of coke oven temperature change trend, eliminate the interference of manual temperature measurement error and reduce the labor intensity; The automatic coke oven heating control can be realized, and the furnace temperature fluctuation can be reduced by rapidly adjusting the gas flow for heating; The coke oven stability coefficient is greater than to 0.85, and the blast furnace gas for heating is saved by more than 1.5%.

Key words: coke oven; online temperature measurement; automatic heating

干熄焦一次除尘系统故障分析及控制优化研究

朱庆庙[1]，甘秀石[1]，王　超[1]，武　吉[1]，侯士彬[2]，王　旭[1]，赵振兴[1]

（1. 鞍钢集团钢铁研究院，辽宁鞍山　114009；
2. 鞍钢股份有限公司炼焦总厂，辽宁鞍山　114021）

摘　要：分析干熄焦一次除尘的重力除尘原理及除尘设备的选择；分析一次除尘故障损坏部位及根本原因；从年修方式上改进干熄焦一次除尘的维护方式；改进干熄焦一次除尘的自动控制方法；优化干熄焦一次除尘的操作方法，避免干熄焦一次除尘系统排空吸入空气造成的岔型溜槽板结问题，保证干熄焦系统的连续稳定生产，实现干熄焦年修周期2年的目标。

关键词：一次除尘；故障；分析；控制；优化

Study on Fault Analysis and Control Optimization of Primary Dust Removal System for CDQ

Zhu Qingmiao[1], Gan Xiushi[1], Wang Chao[1], Wu Ji[1],
Hou Shibin[2], Wang Xu[1], Zhao Zhenxing[1]

(1. Technology Center of Angang Steel Co., Ltd., Anshan 114009, China;
2. Ansteel Stock Company Coking Plant, Anshan 114021, China)

Abstract: This paper analyzes the principle of gravity dust removal and the selection of dust removal equipment for CDQ primary dust removal, analyzes the damaged parts of primary dust removal and the basic reasons, and improves the maintenance mode of primary dust removal for CDQ from the annual maintenance mode To improve the automatic control method of once dust removal in CDQ, to optimize the operation method of once dust removal in CDQ, to avoid the problem of the forked slot plate sticking caused by the air sucked into the primary dust removal system of CDQ, the continuous and stable production of CDQ system is guaranteed, and the annual repair period of CDQ IS 2 years.

Key words: primary dedusting; malfunctions; analysis; control; optimizes

固废离子交换树脂在炼焦技术中的应用

甘秀石，王　旭，朱庆庙，王　超，武　吉，赵振兴，刘福军

（鞍钢集团钢铁研究院，辽宁鞍山　114009）

摘　要：利用配合煤与固废的共炭化原理，在分析固废离子交换树脂技术指标的基础上，本文通过固废树脂配煤炼焦试验，检验焦炭质量变化，探讨固废离子交换树脂再利用的技术。结果表明，固废离子交换树脂中会对焦炭质量起到劣化作用；使用固废离子交换树脂回配炼焦，回配比例为0.1%时，焦炭质量下滑不明显，但回配比例超过0.1%，对焦炭质量影响加大，焦炭质量明显下降。通过控制固废离子交换树脂的配入量，可以达到稳定焦炭质量，解决固废离子交换树脂不易处理的问题。

关键词：固废离子交换树脂；配煤炼焦；焦炭冷强度；焦炭热态性能

Research on Utilization of Solid Waste Resin

Gan Xiushi, Wang Xu, Zhu Qingmiao, Wang Chao,
Wu Ji, Zhao Zhenxing, Liu Fujun

(Anshan Iron & Steel Research Institules, Anshan 114009, China)

Abstract: Using the principle of co-carbonization of coal and solid waste, on the basis of analyzing the technical index of solid waste ion exchange resin, through analysis solid waste technology index of ion exchange resin, and solid waste resin blending coking test, test coke quality change, explore the recycling technology development and application of ion exchange resin.The results show that the solid waste ion exchange resin will degrade the coke quality. When solid waste ion-exchange resin was used to backmix coke, the coke quality declined not obviously when the backmix ratio was 0.1%, but the backmix ratio exceeded 0.1%, which increased the influence on the coke quality and significantly decreased the coke quality. By controlling the amount of solid waste ion-exchange resin, the coke quality can be stabilized and the environmental protection problem that the solid waste ion-exchange resin is not easy to be treated can be solved.

Key words: solid waste ion exchange resin; coal blending coking; coke cold strength; thermal properties of coke

煤质分析在炼焦配煤中的应用

王晓光，祝开宇，刘晓桃，杜　彬，张秋菊，宋文杰

（鞍钢集团朝阳钢铁有限公司，辽宁朝阳　122000）

摘　要：随着我国钢铁工业的快速发展，对焦炭质量和产量要求也越来越高，造成我国优质炼焦煤资源严重短缺价格偏高，直接影响炼铁成本。因此，在满足高炉生产需要的前提下，合理利用具有价格优势的煤资源，寻求降低配煤成本稳定焦炭质量的配煤方案，已成为企业迫切需要解决的问题。为了提高企业竞争优势，降低生产成本，本文

将炼焦煤的煤质煤性分析结合大数据分析与炼焦操作指标等生产实际参数相结合，研究通过煤质检验对炼焦成本的影响，切实降低炼焦成本的研究。

关键词：配煤炼焦；煤质分析；煤岩学；大数据分析

Application of Coal Quality Analysis in Coking and Coal Blending

Wang Xiaoguang, Zhu Kaiyu, Liu Xiaotao, Du Bin, Zhang Qiuju, Song Wenjie

(Chaoyang Iron and Steel Company Ansteel Group, Chaoyang 122000, China)

Abstract: With the rapid development of chinese iron and steel industry, the demand for coke quality And output is also increasingly high，resulting in a serious shortage of high-quality coking coal resources in China，the price is on the high side, directly affecting the cost of iron making.Therefore, on the premise of satisfying the producting needs of blast furnase, it has become an urgent problem for enterprises to rationally ulilize the coal resources with price advantages and seek a coal blending a stabilize the coke quality. In order to improve the competitive advantage of enterprises and reduce the production cost, this paper combines the coal quality, and coal property analysis of coking with big data analysis and coking operation index and actual production parameters to study the impact of coal property inspection on the coking cost and affectively reduce the coking cost.

Key words: blending coal coking; coal quality analysis; anthracology; big data analysis

6m 焦炉单炭化室压力调节技术应用

戚惠杰，关云鹏，苏　宁，董　莆，韦庆志

摘　要：焦炉炭化室压力调节系统，在自动调节水封阀盘开度的基础上，集成了上升管水封盖自动开关、高/低压氨水自动切换功能。该 CPS 系统不仅可以实现炭化室压力的单独调节，CPS 开度无级调节的气缸进行全行程自动调节，该气缸一端连接在水封阀盘现有搬杆上（现有搬杆需要局部改造），另一端底座固定在现有集气管操作台上，环境温度~80℃，以气缸往复动作不同行程，实现水封阀盘的不同开度。在集气管操作台上增设 CPS 气控柜，并设置计算机控制设备

Abstract: Pressure regulating system for coke oven carbonization chamber, On the basis of automatically adjusting the opening of the water seal valve, Integrated rise pipe water sealing automatic switch high andomatic switch high and low pressure, ammonia water automatic switch function. The CPS system not only enables the independent regulation of the chamber pressure, CPS opening infinitely adjustable cylinder for full automatic adjustment, One end of the cylinder is connected to the water seal valve disc on the existing single rod, The other end the base is fixed on the existing machine tube operating table the ambient temperature is 80 degrees.Different opening of water seal valve disc can be realized by cylinder reciprocating action with different stroke. Set up CPS air control cabinet on machine tube operating table, and set up computer control equipment.

挂釉预制炉门衬砖在安钢焦化厂的应用

向　宇，张　俊

（安钢股份有限公司焦化厂，河南安阳　455001）

摘　要：介绍了挂釉预制炉门衬砖在安钢焦化厂 6m、7m 焦炉的使用情况，对原炉门衬砖石墨厚、炉头偏生、易冒烟等现象进行了阐述，通过试用挂釉预制炉门衬砖，对提高单炉焦炭产量、提升环保、降低劳动强度、节省高炉煤气消耗有明显效果。

关键词：焦炉；炉门；挂釉预制衬砖

Application of Preglazed Furnace Door Lining Brick in Coking Plant of Anyang Iron and Steel Co.

Xiang Yu, Zhang Jun

(The Coking Plant of Anyang Iron and Steel Stock Co., Ltd., Anyang 455001, China)

Abstract: This paper introduces the application of preglazed furnace door lining brick in 6 m, 7 m coke oven, expounds the phenomenon that the preglazed furnace door lining brick is prone to smoke and so on, and through the trial of preglazed furnace door lining brick, it has obvious effect on improving single furnace coke output, improving environmental protection, reducing labor intensity and saving blast furnace gas consumption.

Key words: coke-oven; furnace door; precast lining brick

熔硫釜改进与使用

赵国玉

（安阳钢铁股份有限公司，河南安阳　455001）

摘　要：对安钢焦化精脱硫工序熔硫釜的使用现状进行分析，进一步介绍了对焦化厂精脱硫熔硫釜安全阀，直侧排改进之后，新设备的使用情况。

关键词：熔硫釜；改进使用；设备优化

Abstract: This paper analyzes the current situation of the use of the sulfur melting kettle in the fine desulfurization process of Anyang Iron and Steel Co., Ltd., and further introduces the use of the new equipment after the improvement of the safety valve of the fine desulfurization sulfur melting kettle in the coking plant.

Key words: sulfur melting kettle; improved use; equipment optimization

煤岩参数在配煤炼焦生产中的深入开发与实际应用

庞文娟，刘凤娥，王晓峻，封伟政

（内蒙古包钢钢联股份有限公司煤焦化工分公司，内蒙古包头　014010）

摘　要：本文主要介绍了煤岩学的配煤原理及煤岩参数在实际配煤炼焦生产中的应用，并提出煤岩学更为广阔的应用领域和思路，为今后配煤炼焦生产提供有效的解决途径和方法。

关键词：煤岩学；配煤炼焦；镜质组最大反射率

Deep Development and Practical Application of Coal Rock Parameters in Coking Production of Coal Blending

Pang Wenjuan, Liu Feng'e, Wang Xiaojun, Feng Weizheng

(Coal Coking Chemical Branch of Inner Mongolia Baogang Steel Union Co., Ltd., Baotou 014010, China)

Abstract: This paper mainly introduces the coal blending principle of coal petrology and the application of coal rock parameters in the actual coal blending coking production, and puts forward that coal petrology is more. It provides an effective solution for the wide application fields and ideas and for the coal blending coking production in the future.

Key words: coal petrology; coking with coal blending; maximum reflectance of vitrinite

浅析蒙古 1/3 焦煤在包钢焦化的研究与应用

庞文娟[1]，李晓炅[2]，史学军[1]，付利俊[2]

（1. 内蒙古包钢钢联股份有限公司煤焦化工分公司，内蒙古包头　014000；
2. 内蒙古包钢钢联股份有限公司技术中心，内蒙古包头　014000）

摘　要：目前包钢用煤区域包括乌海、包头等地区，基本上都属于高灰、高硫区域，山西地区高硫煤也较多，使得包钢所生产的焦炭与同行业对比其灰分、硫分相对较高，影响高炉相关经济技术指标，因此在用煤结构上需要进行适当调整。蒙古煤资源丰富，是世界上品种齐全，储量最丰富的地区之一。据统计其资源储量达 1624 亿吨，其中探明储量为 176 亿吨，约占总储量的 10.8%。该煤种具有灰分、硫分较低、挥发分较高、G 值衰减较快，结焦性较差的特点。能够起到降低包钢焦炭灰分和硫分的作用，且符合包钢目前实际情况，值得深入研究。因此，开发蒙古地区的 1/3 焦煤对于包钢焦化的长远发展具有较好的工艺质量优势和经济效益优势。

关键词：蒙古煤；蒙古 1/3 焦煤；配煤炼焦；焦炭质量；煤种性质研究

The Study and Application of Mongolia 1/3 Coking Coal in the Coking Process of Baotou Steel Co.

Pang Wenjuan[1], Li Xiaojiong[2], Shi Xuejun[1], Fu Lijun[2]

(1. Coal and Coke Chemical Branch, Inner Mongolia Baotou Steel Union Co., Ltd., Baotou 014000, China;
2. Technology Center of Inner Mongolia Baotou Steel Link Co., Ltd., Baotou 014000, China)

Abstract: At present baotou coal area including the wuhai, baotou and other regions, basically belongs to the high ash, high sulfur area, shanxi district high sulfur coal is more makes steel production coke industry compared with the relatively high ash content, sulfur content, the influence of blast furnace related economic and technical indicators, therefore, need to be adjusted in the use of coal structure. Mongolia is rich in coal resources, is the world's variety is complete, one of the most abundant storage areas. According to statistics, its resource reserves reach 162.4 billion tons, of which 17.6 billion tons are proved reserves, accounting for about 10.8% of the total reserves. This kind of coal is characterized by low ash and sulfur content, high volatile content, fast attenuation of G value and poor coking. It can play a role in reducing the ash and sulfur content of the coke of Baotou Iron and Steel, and is in line with the actual situation of Baotou Iron and Steel, and is worthy of further study. Therefore, the development of 1/3 coking coal in Mongolia has better technological quality and economic benefit advantages for the long-term development of coking in Baotou Steel Group.

Key words: Mongolian coal; 1/3 coking coal of Mongolia; coking with coal blending; coke quality; study on coal properties

焦炉上升管余热利用的研究与应用——取代煤气净化管式炉

郭　飞，王永亮

（青岛特殊钢铁有限公司焦化厂，山东青岛　266400）

摘　要：结合焦炉工况下上升管内荒煤气的特点，就上升管余热回收利用的难点进行了分析与阐述，对焦炉上升管余热利用核心设备上升管蒸发器的几种常见形式行了简要对比，提出了将荒煤气余热用于煤气净化系统取代管式炉的方案，并就方案实施时的注意事项进行了说明。

关键词：焦炉；上升管；余热利用；管式炉

Research and Application of Waste Heat Utilization of Coke Oven Uprising–tube–Replacing Coke Oven Gas Purification Tubular Furnace

Guo Fei, Wang Yongliang

(Qingdao Special Steel Co., Ltd. Coking Plant, Qingdao 266400, China)

Abstract: combined with the characteristics of raw gas in the Uprising-tube under coke oven working conditions, this paper analyzes and expounds the difficulties in the recovery and utilization of waste heat in the Uprising-tube, briefly compares several common forms of Uprising-tube evaporator, the core equipment for the utilization of waste heat in the Uprising-tube of coke oven, and puts forward a scheme to use the waste heat of raw gas in the gas purification system to replace the tubular furnace, The matters needing attention in the implementation of the scheme are explained.

Key words: coke oven; uprising-tube; waste heat utilization; tubular furnace

干熄焦余热锅炉高温区爆管分析及控制措施

梁　波，孙尚华，谷安彤，李　勇，侯士彬，张维强，高　薇

（鞍钢股份有限公司，辽宁鞍山　114000）

摘　要：本文介绍某焦化厂一干熄焦余热锅炉高温区吊挂管爆管，严重影响干熄焦系统的安全稳定运行。经过对爆管的炉管进行取样研究，分析是由于该余热锅炉高温区域的吊挂管内部发生高温氧化，结垢后垢下腐蚀影响传热，炉管过热蠕胀加速，造成持久强度下降继而爆管，针对这一危害，本文详细分析其产生机理、影响因素，同时制定了相应控制措施。

关键词：余热锅炉；高温氧化；爆管

Analysis and Control Measures of Tube Burst in High Temperature Zone of Cdq Waste Heat Boiler

Liang Bo, Sun Shanghua, Gu Antong, Li Yong,
Hou Shibin, Zhang Weiqiang, Gao Wei

(Angang Steel Company Limited, Anshan 114000, China)

Abstract: In this paper, a set of dry quenching waste heat boiler high temperature hanging tube severe detonation tube, the tube of the furnace tube sampling research, analysis is due to the hanging pipe internal high temperature oxidation, corrosion under impact scale heat transfer, the furnace tube overheating creep bulge acceleration, then falling enduring strength tube, for this damage detailed analysis of its mechanism, influence factors, formulate the corresponding control measures.

Key words: waste heat boiler; high temperature oxidation; the pipe burst

利用循环氨水余热制冷的荒煤气初冷工艺优化

李昊阳[1]，张　丹[1]，梁　峰[1]，刘明军[2]

（1. 中冶焦耐（大连）工程技术有限公司，辽宁大连　116085；
2. 松下制冷（大连）有限公司，辽宁大连　116600）

摘　要：本文在对现有利用循环氨水余热制冷的荒煤气初冷工艺存在的问题分析基础上，通过对循环氨水的分级处

理，优化荒煤气初冷工艺，提高循环氨水进入制冷机组的温度，从而实现提高制冷机组效率，降低制冷机组造价，避免循环氨水喷洒喷头堵塞以及避免焦油在集气管内淤积堵塞的目标，保障了焦化生产过程的安全稳定顺行。

关键词：循环氨水余热利用；循环氨水制冷；循环氨水生产低温水；ASPEN PLUS 计算

Optimization of Raw Gas Primary Cooling by Utilizing Waste Heat from Flushing Liquor for Chilling

Li Haoyang[1], Zhang Dan[1], Liang Feng[1], Liu Mingjun[2]

(1. ACRE Coking & Refractory Engineering Consulting Corporation (Dalian), MCC, Dalian 116085, China;
2. Panasonic Appliances Air-conditioning and Refrigeration (Dalian) Co., Ltd., Dalian 116600, China)

Abstract: Analysis is made to the problems in raw gas primary cooling by utilizing waste heat from flushing liquor for current chilling. After taking measures, like treatment for the flushing liquor by stages, optimization for raw gas primary cooling process and increasing the temperature of the flushing liquor into the refrigerator, satisfactory results can be obtained such as higher efficiency of refrigerator unit, lower cost of the unit, and avoidance of blockage in flushing liquor nozzle or tar jamming inside the GCM so that safe and stable operation can be ensured.

Key words: waste heat utilization; flushing liquor chilling; chilled water; ASPEN PLUS calculation

鞍钢 7m 焦炉检测废气砣杆与废气拉板工作状态方法的研究

边子峰[1]，李 勇[1]，肖泽坚[1]，侯士彬[1]，梁 波[1]，王 超[2]，高 薇[1]

（1. 鞍钢股份炼焦总厂，辽宁鞍山 114021；2. 鞍钢集团钢铁研究院，辽宁鞍山 114009）

摘 要：本文主要介绍了鞍钢 7 米焦炉在检测废气砣杆与废气拉板等交换传动设备工作运行状态的方法。对目前焦炉废气拉板是否断裂无法检测，焦炉废气盘废气砣杆是否下落到位监测难度大，监测准确性低，部分异常工作状态无法准确监测的问题进行了研究。

关键词：焦炉；废气盘；砣杆；废气拉板；立火道；自动测温

Study on the Method of Detecting the Working State of Waste Gas Lead Rod and Waste Gas Pull Plate in 7m Coke Oven of Ansteel

Bian Zifeng[1], Li Yong[1], Xiao Zejian[1], Hou Shibin[1], Liang Bo[1], Wang Chao[2], Gao Wei[1]

(1. Coking Plant of Ansteel Co., Ltd., Anshan 114021, China;
2. Iron & Steel Institutes of Ansteel Group Corporation, Anshan 114009, China)

Abstract: The method of testing the working state of the transmission equipment was introduced, such as waste gas mound bar and waste gas pulling plate, in Angang's 7m coke oven. It is impossible to detect whether the drawplate of coke oven exhaust gas was broken, and it is difficult to monitor whether the coke oven waste gas plate waste gas lead rod falls in place, and the monitoring accuracy was low. the problem that some abnormal working conditions can not be accurately monitored was studied.

Key words: coke oven; exhaust disk; lead screw; exhaust pull board; vertical flue; online measurement

炼焦生产全流程烟尘超低排放技术与措施

孙刚森，霍延中

（中冶焦耐（大连）工程技术有限公司通风室，辽宁大连　116085）

摘　要：论述了炼焦生产全流程烟尘排放的位置以及排放特点，提出了相应的治理技术与措施。经过有效治理，污染物排放可以达到超低排放要求。

关键词：炼焦生产；烟尘治理；污染物排放

Ultra-low Emission Technology in Entire Cokemaking Process

Sun Gangsen, Huo Yanzhong

(ACRE Coking & Refractory Engineering Consulting Corporation, MCC, Dalian 116085, China)

Abstract: In this paper, emission points and features are discussed for the entire cokemaking process. Relevant technologies and measures are put forwarded. After effective treatment, the pollutant discharge can meet the requirement of ultra-low emission.

Key words: cokemaking; iontreatment; pollutant discharge

智能加热控制系统在焦炉生产上的应用

郭天胜[1]，李　勇[1]，王慧璐[2]，王君敏[1]，赵　锋[1]，甘秀石[3]，侯士彬[1]

（1. 鞍钢股份有限公司炼焦总厂，辽宁鞍山　114021；2. 中冶焦耐自动化有限公司，辽宁大连　116000；3. 鞍钢集团钢铁研究院，辽宁鞍山　114009）

摘　要：鞍钢股份有限公司炼焦总厂焦炉加热控制系统原设计为离散控制系统，通过人工手动定时测温来调节焦炉加热指导生产操作，存在劳动强度高、检测数据失真不连贯、能源浪费现象。为响应鞍钢集团公司“信息化、智能化”项目建设的要求，通过在线测温设备及 DCS 控制系统新设备新技术的应用，实现对焦炉温度的连续化检测与精控调节，有效的提升了焦炉的自动化检测与控制水平，同时改善焦炉温度的安定性、降低人工测温劳动强度、节省焦炉加热用煤气量，为焦炉全面实现减员增效、节能减排、安全生产的管理目标奠定了基础。

关键词：焦炉加热；精控调节；减员节能

Application of Intelligent Heating Control System in Coke Oven Production

Guo Tiansheng[1], Li Yong[1], Wang Huilu[2], Wang Junmin[1], Zhao Feng[1], Gan Xiushi[3], Hou Shibin[1]

(1. Coking Plant of Ansteel Co., Ltd., Anshan 114021, China; 2. MCC Coking Automation Co., Ltd., Dalian 116000, China; 3. Ansteel Iron & Steel Research Institutes, Anshan 114009, China)

Abstract: The coke oven heating control system of Angang Steel Company Limited coking plant was originally designed as a discrete control system, which regulates the coke oven heating to guide the production operation through manual and timed temperature measurement. There are some problems such as high labor intensity, distortion and incoherence of detection data, and energy waste. In order to respond to the requirements of "information and intelligent" project construction of Anshan Iron and Steel Group Corporation, through the application of online temperature measuring equipment and new equipment and new technology of DCS control system, the continuous detection and precision control of coke oven temperature is realized, which effectively improves the level of automatic detection and control of coke oven. At the same time, improve the stability of coke oven temperature, reduce the labor intensity of manual temperature measurement, save the amount of gas used for coke oven heating, and lay the foundation for the management goal of reducing personnel and increasing efficiency, energy saving and emission reduction, and safe production in the coke oven.

Key words: coke oven heating; fine control regulation; reducing staff and saving energy

焦炉煤气氨硫回收新工艺的实践与应用

胡 林，刘 麟，陈章翔，鲍淑春

（首钢水城钢铁（集团）有限责任公司，贵州六盘水 553028）

摘 要：水钢100万吨焦炭配套的煤气净化系统，在采用“剩余氨水解吸—HPF法脱硫—软水循环洗氨—焦炉烟道气余热负压蒸氨”流程组合后，确保了负压蒸氨运行的连续性、稳定性，焦炉煤气中氨、硫物质的洗涤净化得有效控制，在进行焦炉煤气烟气排放源头治理的同时还兼顾了余热的回收利用，实现了焦化行业的节能与环保治理齐头并进。

关键词：焦炉煤气；洗氨；脱硫；负压蒸氨；浓氨水

焦炉废气回配技术仿真计算与试验研究

康 婷，王进先，郝传松

（中冶焦耐（大连）工程技术有限公司，辽宁大连 116085）

摘 要：通过数值模拟计算和现场工业试验相结合的方法，研究了废气回配技术对焦炉燃烧室燃烧过程的影响机理，并对比了不同废气回配量下立火道内 NO 生成情况的变化。废气回配技术可降低助燃气体中 O_2 的浓度，降低高温区内 NO 的绝对生成量。当废气回配量达到 40%时，能使废气中 NO 含量降低 70%左右。

关键词：废气回配；模拟计算；工业试验；氮氧化物

Computer Simulation and Pilot Tests for Waste Gas Recycling Technology

Kang Ting, Wang Jinxian, Hao Chuansong

(ACRE Coking and Refractory Engineering Consulting Corporation (Dalian), MCC, Dalian 116085, China)

Abstract: Through computer simulation plus pilot tests, study is conducted in this paper for the mechanism of waste gas recycling technology influencing combustion process in the combustion chamber. Comparison is also made for the changes of NO produced in vertical flues under different amounts of waste gas to be added. This technology could help reduce the O_2 content in the combustion air and minimize the NO formation at high-temperature position in the flues, i.e. when the amount of waste gas to be mixed reaches up to 40%, the NO in the waste gas could be cut by about 70%.

Key words: waste gas recycling technology; simulation; pilot test; nitrogen oxide (NO_x)

低阶粉煤分质利用新技术——中冶焦耐 JNWFG 炉工艺

刘庆达，蔡承祐，李　超，刘洪春，刘承智

（中冶焦耐（大连）工程技术有限公司，辽宁大连　116085）

摘　要：低阶粉煤的分质利用对于提升煤炭资源价值、提高利用效率具有重要意义，JNWFG 炉以低阶粉煤为原料，获得高品质的焦油和煤气，半焦产品质量稳定。工业示范装置的稳定运行标志着 JNWFG 炉工艺已经实现了工业化生产，为低阶粉煤分质利用产业链打通了最为关键的环节。

关键词：低阶粉煤；分质利用；外热式直立炉

New Technology of Sorting Utilization for Low-rank Fine Coal—JNWFG Furnace Process by ACRE

Liu Qingda,Cai Chengyou, Li Chao, Liu Hongchun, Liu Chengzhi

(ACRE Coking & Refractory Engineering Consulting Corporation (Dalian), MCC, Dalian 116085, China)

Abstract: The sorting utilization for low-rank fine coal plays a significant role in upgrading coal resource value and increasing its utilization efficiency. This paper introduced JNWFG furnace, which uses the low-rank coal as raw material to produce high quality coal tar and gas as well as stable quality of semi coke. The stable operation of this demonstration plant marks the realization of JNWFG industrial production, which has moved the most critical step for the sorting utilization chain of low-rank fine coal.

Key words: low-rank fine coal; sorting utilization; external heat vertical furnace

对现有焦炉实现炭化室压力单独调节技术的改造分析

史瑛迪，洪志勇，周春蛟，张奎爽

（中冶焦耐（大连）工程技术有限公司，辽宁大连 116085）

摘 要：介绍了 3 种炭化室压力调节技术（德国 RPOven 技术、意大利 SOPRECO 技术及我国 CPS 技术）的基本原理。综合分析对比，在现有焦炉上进行炭化室压力调节技术改造，CPS 技术更具优势，值得推广。

关键词：炭化室压力调节技术；RPOven 技术；SOPRECO 技术；CPS 技术；不停产改造

Technological Upgrading of Individual Regulation for Oven Pressure of the Existing Coke Ovens

Shi Yingdi, Hong Zhiyong, Zhou Chunjiao, Zhang Kuishuang

(ACRE Coking & Refractory Engineering Consulting Corporation (Dalian), MCC, Dalian 116085, China)

Abstract: The basic working principles for three technologies of oven pressure regulation are introduced in this paper, i.e. German PROven, Italian SOPRECO and Chinese CPS. Through analysis and comparison, the CPS enjoys more advantages in upgrading of individual regulation for oven pressure of the existing coke ovens, which is worth popularizing and applying.

Key words: oven pressure regulating technology; PROven technology; SOPRECO technology; CPS technology; upgrading without shutdown

单孔炭化室压力稳定系统（CPS 系统）在新泰正大 6.78m 捣固焦炉上的应用

张洪波，崔义平，刘 宏，吕文才

（新泰正大焦化有限公司，山东新泰 271212）

摘 要：新泰正大焦化有限公司新建 6.78m 捣固焦炉是首个采用炭化室压力稳定系统（CPS 系统）的超大型捣固焦炉。实际应用效果表明，CPS 系统可以有效解决捣固焦炉装煤冒烟、炉门烟尘外逸等污染物治理问题，并能保证焦炉炭化室底部压力稳定，延长焦炉使用寿命。

关键词：CPS 系统；6.78m 捣固焦炉；单孔炭化室压力调节

Application of CPS System in 6.78m Stamp-charge Battery for Xintai Zhengda Coking Plant

Zhang Hongbo, Cui Yiping, Liu Hong, Lü Wencai

(Xintai Zhengda Co., Ltd., Xintai 271212, China)

Abstract: The newly-built 6.78m stamp-charge battery of Xintai Zhengda Co., Ltd. is an ultra-high capacity coke oven that first adopts CPS system in the world. The practical experience shows that the CPS system can effectively solve the problems such as charging emission escaping, oven door emission leakage, etc. and ensure stable pressure at the bottom of coking chamber so as to extend the service life of coke oven battery.

Key words: CPS (Chamber Pressure Stabilization) system; 6.78m stamp-charge battery; pressure regulation for individual coking chamber

"浆液进料、富氧焚烧"制酸工艺技术特点及其应用

刘　宏[1]，刘元德[2]，张素利[2]，白　玮[2]

（1. 山东新泰正大焦化有限公司，山东泰安　271200；
2. 中冶焦耐（大连）工程技术有限公司，辽宁大连　116085）

摘　要： 介绍了"浆液进料、富氧焚烧"脱硫废液制酸工艺技术特点及实际应用。该制酸工艺的突出特点是高效、稳定、安全、环保、自动化水平高。现该制酸工艺已在国内多家焦化企业投产使用，运行状况良好，有效解决了国内以 HPF、PDS 等为催化剂的焦炉煤气氨法湿式催化氧化脱硫工艺存在的低纯硫磺产品滞销、资源浪费以及脱硫废液环境污染技术难题，为实现焦化企业长期可持续发展提供了有力的技术保障。

关键词： 浆液进料；富氧燃烧；脱硫制酸；技术特点；应用

Features of Sulfuric Acid Making Process with "Sulfur Slurry Feeding" and "Combustion with Rich Oxygen Air" and Its Application

Liu Hong[1], Liu Yuande[2], Zhang Suli[2], Bai Wei[2]

(1. Shandong Xintai Zhengda Co., Ltd., Taian 271200, China; 2. ACRE Coking & Refractory Engineering Consulting Corporation (Dalian), MCC, Dalian 116085, China)

Abstract: In this paper, introduction is made for technical features and cases in sulfuric acid making process from desulfurization wastewater, which is characterized with "sulfur slurry feeding" and "combustion with rich oxygen air". This process has advantages of high efficiency, stable operation, safety, environment-friendliness and high automation level. The process has been put into operation in large-scale coking plants in China, which has effectively tackled the problems, such as the unmarketable low purity sulfur, poor utilization of sulfur resource and desulfurization wastewater pollution due to wet catalytic oxidation process with HPF & PDS as catalyst and with ammonia as alkali source. It is proved to be a solution for

sustainable development of coking industry in the long run.
Key words: sulfur slurry feeding; combustion with rich oxygen air; desulfurization and acid making; technical features; application

活性炭脱硫脱硝两段式吸附塔喷氨结构仿真模拟研究

李　超[1]，李旭东[1]，尹　华[1]，孙刚森[1]，霍延中[1]，张平存[2]

（1. 中冶焦耐（大连）工程技术有限公司，辽宁大连　116085；
2. 唐山钢联焦化有限公司，河北唐山　063000）

摘　要：以活性炭法脱硫脱硝两段式吸附塔喷氨结构为研究对象，依据小试实验及中试实验研究成果，采用 CFD 数值模拟方法，研究大差异流量-浓度焦炉烟道气与氨空混合气均匀混合及气流均布技术，通过调节氨空混合气供气方式及喷氨孔排布优化，提高喷氨管上方氨气分布的均匀性。
关键词：活性炭法；两段式吸附塔；喷氨结构；仿真模拟

Study on Simulation of Ammonia Spray Structure in Two-stage Absorber for $DeSO_x$ and $DeNO_x$ of Waste Flue Gas by Active Carbon Process

Li Chao[1], Li Xudong[1], Yin Hua[1], Sun Gangsen[1], Huo Yanzhong[1], Zhang Pingcun[2]

(1. ACRE Coking and Refractory Engineering Consulting Corporation (Dalian), MCC, Dalian 116085, China; 2. Tangsteel Meijin (Tangshan) Coal Chemical Industry Company Limited, Tangshan 063000, China)

Abstract: In this paper, a study is conducted for simulation of ammonia spray structure in two-stage absorber for $DeSO_x$ and $DeNO_x$ of waste flue gas by active carbon process. Based on the results from pilot tests and by adopting CFD simulation, a new streamline technology is studied, i.e. a big range of flow rate-concentration waste flue gas is uniformly mixed with air-NH_4 mass fraction and evenly distributed by regulating mixed fraction supply and optimizing ammonia nozzle arrangement so as to improve the uniformity of ammonia distribution above the nozzles.
Key words: active carbon process; two-stage absorber; ammonia spray structure; simulation

安全智能管控系统在鞍钢炼焦总厂的开发与应用

赵　锋，李昌胤，张允东，吴家珍，赵　明

（鞍钢股份有限公司炼焦总厂，辽宁鞍山　114021）

摘　要：介绍了安全智能管控系统在鞍钢炼焦总厂的开发与应用，投运后实现了操作牌电子化和信息化管理，对检修设备的停送电实施监控管理，避免了传统实牌形式存在的诸多危险，实现了检修和生产作业的本质安全。

关键词：智能管控系统；操作牌；本质安全

Development and Application of Intelligent Type Management and Control System

Zhao Feng, Li Changyin, Zhang Yundong, Wu Jiazhen, Zhao Ming

(General Coking Plant of Ansteel Co., Ltd., Anshan 114021, China)

Abstract: Introduction is made to the development and application of intelligent type management and control system in General Coking Plant of Ansteel, including informatization management for the operator ID card after the system put into operation, surveillance management for energization/de-energization of the equipment in maintenance so that potential risks of the conventional ID cards can be avoided and intrinsic safety in operation and maintenance can be realized.

Key words: intelligent type management and control system; operator ID cards; intrinsic safety

精细化智能配煤系统的应用及思考

姚　田，王　娟

（山西阳光焦化集团股份有限公司煤研中心，山西河津　043300）

摘　要：介绍了精细化智能配煤系统，结合实际应用情况可知，精细化智能配煤系统提供的配煤在一定程度上打破了经验配煤的模式，丰富了配煤工作者的配煤思路。但同时，精细化智能配煤系统也存在一些局限性，相信以大数据下的智能化配煤系统等“互联网+焦化”模式，智能配煤系统将会得到良好的改善。

关键词：智能配煤；专家经验；焦化生产

浅谈型煤炼焦技术

姚　田，王　娟

（山西阳光焦化集团股份有限公司煤研中心，山西河津　043300）

摘　要：本文说明了新时代煤炭的新特点及型煤是我国洁净煤技术中的重要技术，简述了国内外型煤技术的发展，重点介绍了型煤技术的工艺流程、影响因素，同时对型煤粘结剂做了简单的说明，指出了型煤炼焦过程中遇到的问题及挑战，希望能为焦化行业的进步以及今后型煤应用于配煤炼焦生产实际提供有效的参考。

关键词：型煤技术；工艺流程；影响因素；粘结剂；问题

焦化废水及污泥绿色处理技术

庞 江，成雪松，郭有林

（河钢宣钢焦化厂，河北宣化 075100）

摘 要：河钢宣钢焦化厂开发焦化废水及污泥绿色处理技术。首次将超导磁分离技术应用到焦化废水处理领域，并且对废水分离产物进行资源化处理，实现了绿色可持续发展，取得了非常好的效果。

关键词：焦化废水；污泥；绿色处理技术；超导磁分离技术

Coking Wastewater and Sludge Green Circular Treatment Technology

Pang Jiang, Cheng Xuesong, Guo Youlin

(HBIS Group Xuansteel Coking Plant, Xuanhua 075100, China)

Abstract: HBIS Group Xuansteel Coking Plant developed green treatment technology for coking wastewater and sludge. For the first time, superconducting magnetic separation technology was applied to the field of coking wastewater treatment, and the wastewater separation product was treated as a resource, which achieved green sustainable development and achieved very good results.

Key words: coking wastewater; sludge; green circular treatment technology; superconducting magnetic separation technology

全新绿色环保智能的现代焦化冷鼓新工艺

李 亮

（竣云环保科技工程（上海）有限公司，上海 200000）

摘 要：近年来，随着国家对于焦化行业环保的要求越来越高、目前新建焦化厂的规模也不断扩大，对焦油的处理量也日益增大，您还在准备上传统的机械化刮渣槽工艺吗？

目前，焦油压榨泵工艺是全球焦化冷鼓焦油氨水分离单元中最先进最经典的环保工艺。自2006年山东兖矿国际焦化引进二手焦油压榨泵等成套设备以来，这套工艺在中国发展近15年。这15年以来，我们公司在德国这套成熟工艺的基础上，根据中国的焦化现状，结合中国的煤质、焦炉的操作温度等多种综合因素，与德国MAVEG、GEA、SID制造厂商一起自主研发、完善、优化出了适合中国焦化发展的新型焦油压榨泵、离心机及焦油渣输送泵工艺装备。这三个核心关键设备，全部来自于德国欧洲制造，使整个工艺更稳定、更畅通；部分易损件经国产化改造后，后期运营成本更低。

压榨泵工艺较传统的刮渣槽+离心机工艺相比，占地面积小很多，对于企业目前工业用地紧张的情况下，有重要意义；压榨泵工艺是在密闭情况下对焦油渣进行破碎、分离及输送，整个工艺只有离心机一个出渣口，直接用管道输送到配煤指定位置，并且将处理完的焦油渣直接输送到配煤塔，全程自动化操作，无须增加人工及叉车等运力成本，能大大提高焦化厂的焦油回收率，焦油的产量及质量也能相应提高，帮助焦化厂彻底摆脱粉尘、异味、油污的困扰。既解决VOC治理问题，又解决了焦化厂的焦油渣危固废封闭式输送问题，有利于传统的焦化企业，打造全新绿色环保智能的现代化焦化企业。而传统的刮渣槽+离心机工艺，整个焦油氨水分离过程中出渣口的点位比较多，如果后期要进行封闭管道输送改造，难度较大，会增加投资费用。从投资角度来看，前期投资新工艺略大于传统工艺，但从长期运行及五年内工艺无需升级改造的层面看，选择焦油压榨泵的工艺是首选，投资反而是最省的。

近年来中国大多数新建焦化厂普遍使用传统的立式刮渣槽+离心机工艺，在日益严峻的环保问题面前，这种工艺肯定会面临多个刮渣口的危固废焦油渣的密闭处理输送问题，以及离心机出渣口的密闭管道输送改造问题。

从整个工艺建设方面看：中国新建的焦化厂产能规模不断扩大、焦油处理量及焦油渣的含固量也在不断增大，如果采用传统的（刮渣槽+离心机）组合工艺，不仅无法处理大渣块、齿轮链条遇到大渣块时也极易断裂，增加维修成本。无法处理回收大渣块中含有的焦油及水分。

我们的压榨泵工艺，是目前全世界焦化行业中环保的经典工艺，采用压榨泵+离心机的工艺组合，就可完全替代传统的刮渣槽工艺。流程简单，全流程只有离心机一个出渣口（刮渣槽工艺就有N个出渣口）。整个工艺占地面积小（20m×30m），结构方式简单、有利于油分、水分的回收利用，增加焦油回收率。

2013年，我们德国焦油压榨泵技术团队与上海宝钢化工研发了全新的大焦油压榨泵（36立方的大压榨泵可以破碎100mm以下的焦油渣；18立方的小压榨泵可以破碎50mm以下的焦油渣），我们对老款18立方，功率30kW的压榨泵改进行了改进，共同研发创新设计了带防护装置的36立方的大压榨泵。

解决了当前中国市场上压榨泵出现管道堵塞的问题，可以保证用户在7×24小时不间断工作，寿命达到4年以上不用更换易损件，金属异物进入泵体也不会损坏泵核心件。通过压榨泵破碎研磨输送后，管道中的焦油渣粒径基本保持在 3~4mm，有利于后期的配煤。同时，焦油压榨泵对后续离心机也是一种保护，离心机不容易堵，分离效率更好、并且可以将大块焦油渣及石墨块进行破碎研磨，较传统工艺来说能够大大提高焦油的产量。

如果后期进行智能化输送系统（焦油渣输送泵）改造，压榨泵工艺只需要在离心机点位上增加焦油渣输送泵就可以，相对来说难度低。如果是刮渣槽+离心机工艺就比较麻烦，除了考虑要在离心机点位增加焦油渣输送泵装置外，刮渣槽那边还要考虑 N 个刮渣口的焦油渣的统一收集及密闭输送处理。而且刮渣槽出来的焦油渣还含有一些油及水，如果不经过一台离心机处理，就会损失这部分油及水的回收。

如果要进行离心机回收，就要新增一台离心机处理，这种方式性价比不高。而对于渣量大的焦化企业来说，更应该权衡考虑一下。所以，如果您当下投资新建焦化厂，那从总体性价比上来说，我们的焦油压榨泵工艺最适合不过，整体工艺流程简洁，安全高效环保，5~10年一直保持工艺的先进性。经济效益、社会效益可观。

两种工艺性能对比表：

	对比项	传统立式刮渣槽	焦油压榨泵工艺
经济效益	土地成本	占地面积大	压榨泵体积小，整体布置简洁，结构紧凑，节省生产空间占地面积小、电机负荷小，使工程达到最大使用率
	处理量	2-3 台刮渣槽	1 台压榨泵
	清渣费用	槽子底部焦油渣清理	不需要清渣
	后期维护费用	故障率高，维修量大	几乎不需要维护
	维修费用	机械化澄清槽槽体结构刚性差，易变形，刮板机链条强度不足，极易发生事故。	压榨泵采用全进口特殊钢材质设计加工而成
		由于焦化生产的特殊性，事故发生后，无法排渣，易导致被迫停产。	可连续 7×24 小时运转，使用寿命长，几乎不需要维修，维修简易方便。
		难以清除槽内淤积焦油渣，维修十分困难，成本高，经济损失严重。	
		遇到大焦油渣及石墨块等固体，链条易发生断裂。	

两种工艺后期改造图对比表：

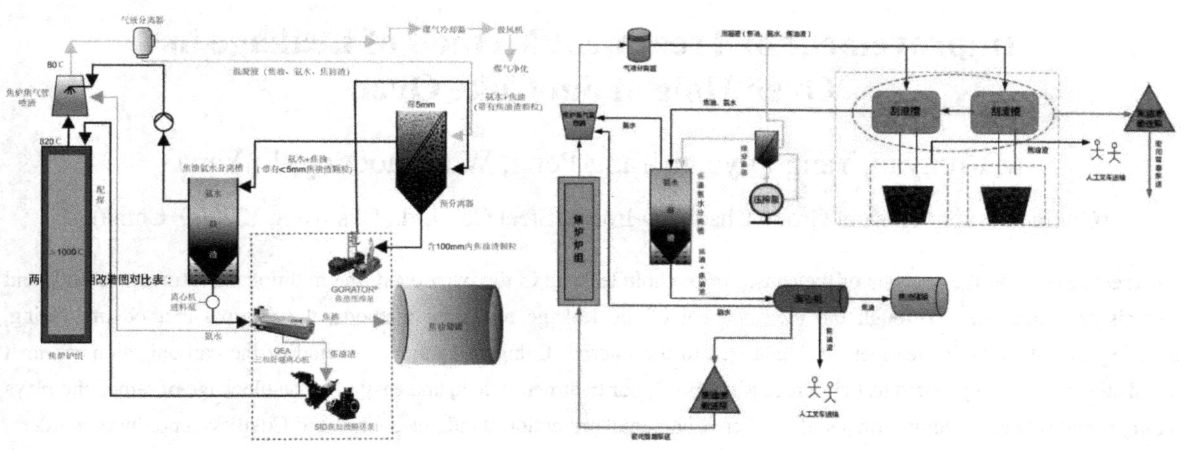

压榨泵工艺流程图　　　　刮渣槽+离心机工艺流程图

后期危固废改造方案参考：

单套刮渣槽的密闭式焦油渣输处理改造方案	
方案一（红色部分）	2 台焦油渣输送泵+焦油渣集中收集罐（难度大）
方案二（绿色部分）	1 台输送泵+1 台压榨泵（刮渣槽需有空地改造）

如果新建焦化项目一开始就采用我们的焦油压榨泵工艺，前期建设时只需焦油压榨泵和离心机，项目投产后，如果后期需要危固废封闭处理改造时，只需在离心机在一个点位使用输送泵，彻底解决您的后期改造所带来的损失。

考虑到业主前期投资受限，本方案也可以分两步走，一是确定压榨泵与离心机组合，二是项目投产后视效益情况先改造增加一台焦油渣输送泵，或同时改造两台，这样可以减少后期改造难度，也可减少投资压力。

所以综上所述，采用压榨泵工艺可以在源头上节省不必要的人力成本及设备成本、极大减少占地面积、增加焦油回收率、减少后期的改造费用，提高企业经济效益，为您打造全新绿色环保智能的现代化焦化企业！这也是目前只要用过压榨泵工艺的厂家，基本不会再考虑用其他传统工艺的原因。

6m 焦炉烘炉孔窜漏处理方法的改进

马俊尧，杨富元，唐　鹏，吴晓东，李亚娜

（鞍钢集团朝阳钢铁有限公司焦化厂，辽宁朝阳　122000）

摘　要：针对焦炉烘炉孔窜漏的原因进行分析，对传统的处理方法及危害进行介绍，通过窜漏处理方法的改进，由原来的放荒、灌浆、拆扒处理法改为炭化室内部抹补或喷补法，避免了传统处理法处理时间长、处理效果不好、容易造成斜道口堵塞等缺点，对环保达标控制、焦炉生产秩序的长期稳定起到重要作用。

关键词：炭化室；石墨；焦炉；烘炉孔

Improvement of Treatment Method of Leakage in Oven Hole of 6m Coke Oven

Ma Junyao, Yang Fuyuan, Tang Peng, Wu Xiaodong, Li Yana

(Coking Plant of Ansteel Group Chaoyang Iron & Steel Co., Ltd., Chaoyang 122000, China)

Abstract: Based on the analysis of the causes of the hole leakage in the oven oven, the traditional treatment methods and hazards are introduced. Through the improvement of the leakage treatment methods, the original method of grazing, grouting, and demolition treatment was changed to the internal filling or spraying method of the carbonization room. It avoids the disadvantages such as long processing time, poor treatment effect, and easy to cause blockage of ramp, and plays an important role in the long-term stability of environmental protection standards control and Coke oven production order.

Key words: carbonization room; graphite; coking furnace; oven hole

浅谈焦炉烟气末端治理

穆应东

（广东韶钢松山股份有限公司，广东韶关　512000）

摘　要：本文介绍焦化行业的末端烟气脱硫脱销除尘工艺技术，并分析优缺点，结合自己的工作经历，谈谈如何选择最优适合自己的焦炉烟气治理方案。

关键词：脱硫脱硝除尘；最优；治理方案

梅钢筒仓煤焦取制样检测系统在线应用

何志明[1]，薛　莹[1]，徐新华[1]，韩义峰[2]

（1.上海梅山钢铁股份有限公司制造管理部，江苏南京　210039；
2.山东济南中意维尔科技有限公司，山东济南　250000）

摘　要：本文介绍了梅钢筒仓煤焦取制样检测系统的设计方案及在线应用情况，该系统实现了煤和焦的自动取样、焦炭不同粒级自动检测、焦炭机械强度检测、自动收集热反应试样、自动收集大于 60mm 焦样、自动清理余料、现场无人值守和远程控制。该系统中焦炭机械强度测定过程中的配鼓精度较好，配鼓平均值为 49.9kg，标准偏差为 0.2kg。该系统为头部取样，取样代表性好，检测效率高，给外购焦炭质量控制及时提供了检测数据，同时改变了传统人工操作代表性差、劳动效率低等问题。

关键词：筒仓；煤；焦炭；采样；制样；粒级；机械强度

Online Application of Coal and Coke Sampling and Preparation Detection System in Silo of Meisteel

He Zhiming[1], Xue Ying[1], Xu Xinhua[1], Han Yifeng[2]

(1. Manufacture Department of Meishan Iron & steel Co., Nanjing 210039, China;
2. Shandong Jinan Zhongyiweier Technology Co., Ltd., Jinan 250000, China)

Abstract: This paper introduces the design scheme and on-line application of the coal coke sampling and preparation detection system in the silo of Meigang.The system realizes the automatic sampling of coal and coke, the automatic detection of different particle sizes of coke，the detection of coke mechanical strength, the automatic collection of thermal reaction samples, the automatic collection of coke samples greater than 60mm, the automatic cleaning of surplus materials, on-site unattended and remote control. In the system, the drum matching accuracy in the determination of coke mechanical strength is good. The average drum matching value is 49.9kg and the standard deviation is 0.2kg. The system is head sampling, with good sampling representativeness and high detection efficiency. It provides timely detection data for the quality control of purchased coke, and changes the problems of poor representativeness and low labor efficiency of traditional manual operation.

Key words: silo; coal; coke; sampling; sample preparation; detection; particle size, mechanical strength

焦炉热回收上升管传热和散热分析

王思维[1]，宋灿阳[1]，杜先奎[1]，李 平[2]，吴启富[2]

（1. 马钢技术中心，安徽马鞍山 243000；2. 马钢炼焦总厂，安徽马鞍山 243000）

摘 要：本文通过传热学公式计算焦炉荒煤气与上升管之间的对流和辐射传热量及上升管的散热量，采用热平衡原理测算炭化室红焦面对上升管辐射传热量。经过定量计算分析可知，荒煤气对上升管辐射传热约占总传热量 54%，荒煤气对流传热约占 33%，红焦面辐射对上升管传热量约占 13%，辐射传热是上升管得热的主要方式。

关键词：焦炉；上升管；热回收；传热

Analysis of Heat Transfer and Heat Dissipation in Gas Uptake of Coke Oven

Wang Siwei[1], Song Canyang[1], Du Xiankui[1], Li Ping[2], Wu Qifu[2]

(1. Tech. Center of Masteel, Maanshan 243000, China;
2. Coking Works of Masteel, Maanshan 243000, China)

Abstract: In this paper, the convection and radiative heat transfer capacity between raw gas and gas uptake and the heat dissipating capacity of gas uptake are calculated by heat transfer formula, and the radiative heat transfer of coking surface to gas uptake is calculated by the principle of heat balance. It can be seen that radiative heat transfer of raw gas accounts for 54% of total heat transfer, convective heat transfer of raw gas accounts for 33%, radiative heat transfer of coking surface to gas uptake accounts for 13%. Therefore, radiative heat transfer is the main way of heat gain in gas uptake.

Key words: coke oven; gas uptake; heat recovery; heat transfer

焦化厂废水深度处理达标回用技术研究

王 军，尚建芳，董海涛

（河钢集团邯郸分公司焦化厂，河北邯郸 056015）

摘 要：邯钢焦化厂废水深度处理回用技术研究，主要针对常规过滤器、陶瓷膜过滤器、活性炭吸附、化学氧化、臭氧催化氧化、膜脱盐处理技术进行研究，有效控制水质目标的情况下，减少改造工程量，节约项目投资成本。采取多介质过滤处理、非均相臭氧催化氧化、膜脱盐处理工艺相结合的方法，满足废水回用要求。

关键词：焦化废水；多介质过滤；臭氧催化氧化；反渗透

Research on the Advanced Treatment and Reuse Technology of Coking Wastewater

Wang Jun, Shang Jianfang, Dong Haitao

(Coking Plant of HBIS Group Hansteel Company, Handan 056015, China)

Abstract: Handan Iron and Steel Coking Plant wastewater advanced treatment and reuse technology research, mainly for conventional filters, ceramic membrane filters, activated carbon adsorption, chemical oxidation, ozone catalytic oxidation, membrane desalination treatment technology, under the condition of effective control of water quality goals, Reduce the amount of renovation projects and save project investment costs. A combination of multi-media filtration treatment, heterogeneous ozone catalytic oxidation, and membrane desalination treatment processes are adopted to meet the requirements of wastewater reuse.

Key words: coking wastewater; multi-media filtration; ozone catalytic oxidation; reverse osmosis

韶钢 6 号、7 号焦炉烟气脱硫脱硝项目超低排放达产实践

邱 旭，穆应东，郑水传，张 翅

（宝武集团中南钢铁有限公司，广东韶关 512000）

摘 要：韶钢为满足国家日益苛刻的烟气排放标准，进而降低烟气排放的环保风险，确保达到国家超低排放指标限值要求：焦炉烟气出口 SO_2≤30mg/Nm^3，NO_x≤150mg/Nm^3，粉尘浓度≤10mg/Nm^3，在 6 号、7 号焦炉烟气排放出口增设烟气脱硫脱硝项目，助力韶钢环保质量转型升级。

关键词：烟气；脱硫脱硝；超低排放

基于马钢焦炉煤气条件下的精脱硫技术方案比选分析

曹欣川洲，邱全山，刘自民，唐嘉瑞，饶 磊，周劲军
（马鞍山钢铁股份有限公司技术中心，安徽马鞍山 243000）

摘 要：本文介绍了马钢典型焦炉煤气的组成，焦炉煤气粗脱硫后煤气中硫化物的存在形式和焦炉煤气精脱硫的必要性，并重点介绍了目前行业内已投产的三种焦炉煤气精脱硫工艺（微晶吸附法、水解转化法和 DDS 生化法）的原理、优缺点以及现场运行情况，在此基础上，结合马钢焦炉煤气的条件进行了精脱硫工艺的比选。

关键词：焦炉煤气；精脱硫；微晶法；水解法；DDS 法

Comparison and Analysis of Technical Schemes for Fine Desulfurization Based on Masteel Coke Oven Gas

Cao Xinchuanzhou, Qiu Quanshan, Liu Zimin, Tang Jiarui, Rao Lei, Zhou Jinjun
(Maanshan Iron & Steel Co., Ltd. Technology Center, Maanshan 243000, China)

Abstract: The composition of typical coke oven gas of Masteel, the existence form of sulfide and the necessity of fine desulfurization of coke oven gas has been introduced in this paper. And the principle, advantages and disadvantages as well as operation conditions of three kinds of coke oven gas fine desulfurization processes (microcrystalline adsorption method, hydrolytic conversion method and DDS biochemical method) which have been put into operation in the industry are introduced emphatically. On this basis, combined with the conditions of coke oven gas at Masteel, the fine desulfurization process has been compared and selected.

Key words: coke oven gas; fine desulfurization; microcrystalline adsorption; hydrolysis catalysis; DDS biochemical method

浅谈炼焦煤粒度与煤、焦性质间的关系

武 吉[1,3]，来 威[2]，侯士彬[2]，未福宇[2]，王 旭[1]，刘福军[1]，陈立哲[2]，赵振兴[1]
（1. 鞍钢集团钢铁研究院，辽宁鞍山 114009；2. 鞍钢股份有限公司炼焦总厂，辽宁鞍山 114021；3. 东南大学能源与环境学院，江苏南京 210096）

摘 要：面对优质炼焦煤资源日趋匮乏的资源性问题，优质炼焦煤资源高效利用成为重要的研究方向。本文综述分析炼焦备煤工序中粒度选取的主要工艺流程及国内典型钢铁联合企业备煤工艺现状。对比分析了炼焦煤粒度与炼焦煤的黏结性、膨胀性、煤岩特性及焦炭质量间的相关性，并提出工业生产中应尽量降低煤粉过细或者过粗的粒度分

布，综合考虑炼焦煤粒度对炼焦煤的灰分、硫、结焦性能、煤岩分布及焦炭质量的影响规律，最终实现精准配煤与高效炼焦。

关键词： 粒度；黏结性；膨胀性；焦炭质量；备煤工艺

Discussion to the Correlation between Coal Particle Size and Properties of Coal and Coke

Wu Ji[1,3], Lai Wei[2], Hou Shibin[2], Wei Fuyu[2],Wang Xu[1], Liu Fujun[1], Chen Lizhe[2], Zhao Zhenxing[1]

(1. Ansteel Iron & Steel Research Institutes, Anshan 114009, China; 2. General Coking Plant of Angang Steel Co., Ltd., Anshan 114021, China; 3. School of Energy and Environment, Southeast University, Nanjing 210096, China)

Abstract: With the shortage of high-quality coking coal resources, it is an important research method toutilize high-quality coking coal resources efficiently. The main process flow of particle size selection in coking coal preparation process is summarized as well as comparative analysisof the correlation between particle size and coal char properties, including of cohesiveness, expansiveness, coal petrographic characteristics and coke quality. In addition to control the particle size distribution of coking coal, it is suggested that enterprises should take the influenceof coking coal particle size distribution on ash, sulfur, coking property, coal petrography and coke quality into account, and finally achieve accurate coal blending and efficient coking.

Key words: particle size; cohesiveness; expansiveness; coke quality; coal preparation process

鞍钢焦炉炉门衬砖特性及侵蚀行为研究

武　吉[1,3]，侯士彬[2]，赵　锋[2]，甘秀石[1]，周晓锋[2]，朱庆庙[1]，王　超[1]，张其峰[2]

（1. 鞍钢集团钢铁研究院，辽宁鞍山　114009；2. 鞍钢股份有限公司炼焦总厂，辽宁鞍山　114021；3. 东南大学能源与环境学院，江苏南京　210096）

摘　要： 对比分析焦炉生产服役过程中 3 种炉门衬砖物相成分与耐火性能，结合焦炉加热特点及煤结焦特性，分析不同炉门衬砖在炼焦周期内表面温度变化，最后对比分析堇青石炉门衬砖和莫来石-堇青石挂釉炉门衬砖的碳侵蚀行为。利用 XRD、SEM-DSC 等手段对不同阶段使用过的炉门砖进行了表征，含有堇青石（$Mg_2Al_4SiO_{18}$）的炉门衬砖使得焦炉加热的前期炉门保温效果得到改善，但炉门砖过度减薄，焦炉加热后期炉门散热损失增大。挂釉炉门砖中的钙长石（$CaAl_2Si_2O_8$）玻璃体充填于坯体的莫来石晶粒之间，使坯体致密而减少空隙，使得挂釉砖耐受侵蚀。堇青石衬砖裂缝内部和堇青石-莫来石挂釉炉门砖釉面微孔均发现 C 类物质，认为炼焦过程中产生的煤气或煤焦油中的炭类物质侵蚀至炉门衬砖中，进而造成炉门衬砖的损坏。

关键词： 焦炉炉门砖；堇青石-莫来石；炉门砖侵蚀；物相特性

Study on Characteristics and Corrosion Behavi of Coke Oven Door Lining Brick in Ansteel

Wu Ji[1,3], Hou Shibin[2], Zhao Feng[2], Gan Xiushi[1], Zhou Xiaofeng[2], Zhu Qingmiao[1], Wang Chao[1], Zhang Qifeng[2]

(1. Iron and Steel Research Institute of Angang Group, Anshan 114009, China; 2. Coking Plant of Angang Co., Ltd., Anshan 114021, China; 3. School of Energy and Environment, Southeast University, Nanjing 210096, China)

Abstract: The phase composition and fire resistance of three kinds of furnace door lining bricks in the production and service process of coke oven are compared and analyzed. Combined with the heating characteristics and coal coking characteristics of coke oven, the surface temperature changes of different furnace door lining bricks in the coking cycle are analyzed. Finally, the carbon erosion behavior of cordierite furnace door lining bricks and mullite cordierite glaze furnace door lining bricks is compared and analyzed. The furnace door bricks used in different stages were characterized by XRD, SEM-DSC. The furnace door lining brick containing cordierite ($Mg_2Al_4SiO_{18}$) improved the insulation effect of the furnace door in the early stage of coke oven heating, but the excessive thinning of the furnace door brick increased the heat loss of the furnace door in the later stage of coke oven heating. The glass body of calcium feldspar ($CaAl_2Si_2O_8$) in the glazed furnace door brick is filled between the mullite grains of the body, making the body compact and reducing the gap, making the glazed brick resistant to erosion. Class C substances are found in the cracks of cordierite lining brick and the micropores of cordierite mullite glaze furnace door brick. It is considered that the carbon substances in the gas or coal tar produced in the coking process corrode into the furnace door lining brick, and then cause the damage of the furnace door lining brick.

Key words: coke oven door brick; cordierite-mullite; erosion; phase

风化氧化对烟煤煤化度指标影响的研究

程启国，梁开慧，刘克辉

（宝钢集团广东韶关钢铁有限公司，广东韶关 512123）

摘　要：烟煤的煤化度指标挥发分和镜质组平均最大反射率是用来评价煤质是否适合炼焦的重要指标之一，煤的低温氧化对煤的挥发分和镜质组平均最大反射率均有重要影响，从而影响到煤质结焦性和生成焦炭的质量。为正确全面揭示低温氧化对烟煤挥发分和镜质组反射率影响规律，用于试验研究的煤样涵括了中国炼焦煤分类中所有代表性煤种，并有近年来国内普遍使用的几种进口烟煤。研究方法按照煤样常温室内混匀堆放、定期取样分析的方式开展。研究结果证明烟煤的挥发分和镜质组平均最大反射率受风化影响的变化规律与其煤化度有关，同时两者间也具备一定的线性相关。研究结果可用于正确鉴别煤质是否受到常温氧化及氧化判断对煤炭质量特别是用于炼焦时的结焦性能影响程度。

关键词：风化；烟煤；煤化度；挥发分；反射率

Study on the Influence of Weathering on the Coalification Degree Index of Bituminous Coal

Cheng Qiguo, Liang Kaihui, Liu Kehui

(Baosteel Group Guangdong Shaoguan Iron and Steel Co., Ltd., Shaoguan, 512123, China)

Abstract: The average maximum reflectance of volatile matter and vitrinite of bituminous coal is one of the important indexes to evaluate whether the coal quality is suitable for coking. The low temperature oxidation of coal has an important influence on the average maximum reflectivity of volatile matter and vitrinite of coal, thus affecting the coking property of coal and the quality of coke. In order to correctly and comprehensively reveal the influence of low temperature oxidation on volatile matter and vitrinite reflectance of bituminous coal, the coal samples used for experimental study include all representative coals in China's coking coal classification, and several imported bituminous coals are widely used in China in recent years. The research method is carried out in accordance with the normal temperature indoor mixing and stacking, regular sampling and analysis. The results show that the variation law of volatile matter and vitrinite average maximum reflectance of bituminous coal affected by weathering is related to its coal degree, and there is a certain linear correlation between them. The research results can be used to correctly identify whether the coal quality is oxidized at room temperature and to judge the influence of oxidation on coal quality, especially on coking performance.

Key words: weathering; bituminous coal; coalification degree; volatile matter; reflectivity

焦炭质量综合评价方法研究

李　铁，高云祥

（武钢集团昆明钢铁股份限公司技术中心，云南昆明　650302）

摘　要：依据钢铁企业的生产需求，运用数学方法 1-9 标度法、两两比较法初步建立了焦炭质量评价体系，以及决定焦炭质量优劣各因子评分标准，对武钢集团昆明钢铁股份有限公司（以下简称武昆股份）外购焦炭质量进行了客观、公正的评价。

关键词：焦炭；质量评价；1-9 标度法；两两比较法

Research on Comprehensive Evaluation Method of Coke Quality

Li Yi, Gao Yunxiang

(Wuhan Iron and Steel Group Kunming Iron & Steel Co., Ltd., Kunming 650302, China)

Abstract: According to the production demand of iron and steel enterprises, the evaluation system of coke quality was preliminarily established by using 1-9 scale method and two-by-two comparison method, and the evaluation criteria of each factor determining the quality of coke were also used to evaluate the quality of Wuhan Iron and Steel Group Kunming Iron and Steel Co., Ltd. (hereinafter referred to as wukun stock).

Key words: coke; quality evaluation; 1-9 scale method; pairwise comparison

7.63m 焦炉火落管理技术的开发与应用

殷喜和

（山西太钢不锈钢股份有限公司焦化厂，山西太原 030003）

摘 要：基于炼焦过程中的“火落”特性，借助先进的自动检测和分析、集成技术，建立数控模型，掌握结焦过程中发生“火落”现象的时刻，结合 7.63m 焦炉的自动控制系统，开发 7.63m 焦炉炼焦生产管理系统，本文介绍了该系统的原理、组成和功能，总结了应用效果和经验。

关键词：7.63m 焦炉；火落管理技术；炼焦生产管理系统；开发与应用

Development and Application of the Fire Fall Management Technology in 7.63m Coke Oven

Yin Xihe

(Shanxi Taigang Stainless Steel Co., Ltd., Coking Plant, Taiyuan 030003, China)

Abstract: Based on the characteristics of “fire fall”in coking process, with the help of advanced automatic detection, analysis and integration technology, a numerical control model was established to master the moment of “fire fall” phenomenon in the coking process, with the automatic control system of 7.63m coke oven, development of 7.63m set coking furnace prodction management system. this paper introduces the principle, composition and function of the system, and summarizes the application effect and experience.

Key words: 7.63m coke oven; fire fall management technology; coking production management system; development and application

焦煤全自动汽车采样系统的安全性

陈 刚，宋 娜，王振飞，赵攀峰

（首钢长治钢铁有限公司，山西长治 046000）

摘 要：首钢长治钢铁有限公司原传统的人工采样劳动强度大，既浪费时间，样品的代表性差，人为因素影响大，煤质情况难以及时、准确地掌握，严重影响了机组的安全稳定运行。随着新发展理念的引领，用科技转型的新步伐踏出坚实的一步，将科技力量注入传统作业，2020 年新上了进厂煤采制样全自动检验智能化项目，新增两套焦煤全自动汽车采样系统，至此煤质采样便不再是低效的代名词，也不是脏乱和危险的存在，实现了采、制、化的系统智能化，为“全方位、全时段、全过程”的质量安全生产奠定了坚实的基础。

关键词：焦煤；采样；自动采样系统；安全

Safety of Automatic Automobile Sampling System for Coking Coal

Chen Gang, Song Na, Wang Zhenfei, Zhao Panfeng

(Shougang Changzhi Iron ＆ Steel Co., Ltd., Changzhi 046000, China)

Abstract: The traditional manual sampling in Shougang Changzhi Iron and Steel Co., Ltd was labor-intensive, a waste of time, poor sample representativeness, and greatly affected by human factors. It was difficult to grasp the coal quality situation timely and accurately, which seriously affected the safe and stable operation of the unit. As the lead of new development concept, with new steps step solid step in the transformation of science and technology, the power of science and technology into traditional freight, in 2020 the new coal sampling automatic inspection into the factory on the intelligent project, add two sets of coking coal automatic sampling system, thus the coal sampling and is no longer synonymous with inefficient, nor is it a dirty and dangerous, It has realized the intelligent system of production, production and transformation, which has laid a solid foundation for the quality and safety production of "all-directional, whole-time and whole-process".

Key words: coking coal; sampling; automatic sampling system; safety

延长环境除尘提升阀使用时间

唐瑞晏，肖　超

（重庆钢铁股份有限公司，重庆　401220）

摘　要：环境除尘中提升阀作用是，当某个清灰室需要反吹清灰时，提升阀关闭，阻止气流通过，待清灰室内滤袋全部反吹后，提升阀重新打开。但实际使用中，提升阀经常故障，导致滤袋反吹效果差，各粉尘收集点扬尘四溢，不仅污染环境，还严重威胁职工健康状况。

关键词：环境除尘；提升阀；问题；措施

Life Environment Poppet Valve Use Time

Tang Ruiyan, Xiao Chao

(Chongqing Iron and Steel Co., Ltd., Chongqing 401220, China)

Abstract: The function of the poppet valve in environmental dust removal is that when a certain cleaning room needs to be blown back to clean the ash, the poppet valve is closed to prevent the air flow from passing through. After the filter bag in the cleaning room is all back blown, the poppet valve reopens. However, in actual use, the poppet valve often fails, resulting in poor back-blowing effect of the filter bag, and dust overflowing from various dust collection points, which not only pollutes the environment, but also seriously threatens the health of employees.

Key words: environmental dust removal; poppet valve; problem; measure

干熄焦炭烧损影响因素及烧损机理研究

张文成[1,2]，任学延[4]，张小勇[3]，郑明东[1]

（1. 安徽工业大学冶金学院，安徽马鞍山 243002；2. 宝钢研究院梅钢技术中心，江苏南京 210039；3. 安徽工业大学化学与化工学院，安徽马鞍山 243002；4. 上海梅山钢铁公司制造部，江苏南京 210039）

摘 要：本文通过对干熄焦生产数据分析及焦炭低浓度反应性试验，研究了干熄焦烧损率影响因素及烧损机理。研究结果表明保持合适的干熄焦处理量，焦炭烧损率较低；焦炭烧损随着风料比的增加而增加，随着排焦温度增加而增加，保持干熄焦稳定操作是控制焦炭烧损的重要条件之一；随着循环气体成分氢气和氧气含量降低，焦炭的烧损则随之增加，随着 CO_2 的增加而烧损率增加；在循环气体从斜道口进入循环烟道后，主要与导入的空气发生部分燃烧，同时粉焦继续发生碳素溶损；CO_2 在较高温度下与焦炭的溶碳反应是干熄炉内焦炭的烧损主要途径，控制循环气体中 CO_2 浓度成为控制烧损的关键。

关键词：焦炭；干熄焦；烧损；反应性；低浓度

Study on Influencing Factors and Mechanism of Coke Loss

Zhang Wencheng[1,2], Ren Xueyan[4], Zhang Xiaoyong[3], Zheng Mingdong[1]

(1. School of Metallurgy, Anhui University of Technology, Maanshan 243002, China; 2. Meishan Iron and Steel Corporation Technology Center, Nanjing 210039, China; 3. School of Chemistry and Chemical Engineering, Anhui University of Technology, Maanshan 243002, China; 4. Meishan Iron and Steel Corporation Manufacturing Management Department, Nanjing 210039, China)

Abstract: Through the analysis of dry-out coke production data and the low-concentration reactive test of coke, this paper studies the influencing factors and burn-out mechanism of dry-out scorching loss rate. The results show that maintaining a suitable dry-out coke treatment, the coke burn loss rate is low, the coke burn loss increases with the increase of wind ratio, and with the increase of coke temperature, it is one of the important conditions to keep the dry-out coke stable operation to control coke burn loss, and with the hydrogen component of circulating gas When the gas and oxygen content decreases, the coke burn loss increases, and the burn rate increases with the increase of CO_2, and after the circulating gas enters the circulating flue from the ramp, it is mainly partially burned with the imported air, while the carbon solubility of the powder coke continues The carbon-soluble reaction of CO_2 with coke at higher temperatures is the main way of coke burning in the dry-out furnace, and controlling the CO_2 concentration in circulating gas becomes the key to controlling the burn-out.

Key words: coke; CDQ; burn-out; reactive; low concentration

干熄炉斜道用复合碳化硅砖的开发与应用

钱 晶，董为纲，蔡 云，陆亚群，张军杰

（江苏诺明高温材料股份有限公司，江苏宜兴 214267）

摘　要：为适应干熄炉长寿化发展要求，研制开发了干熄炉斜道用复合碳化硅砖。研制的复合碳化硅砖具有高强度、高热震性和低膨胀系数的特点，其热态抗折强度（1100℃×0.5h）大于 30MPa，热震稳定性（1100℃，水冷）达 100 次，在干熄炉斜道区使用寿命达 4~5 年，取得了长寿化的应用效果。

关键词：干熄炉；斜道区；碳化硅；复合材料；长寿命

Development and Application of Composite Silicon Carbide Brick in Inclined Duct of Coke Dry Quenching

Qian Jing, Dong Weigang, Cai Yun, Lu Yaqun, Zhang Junjie

(Jiangsu Nuoming High Temperature Materials Co., Ltd., Yixing 214267, China)

Abstract: In order to meet the longevity requirements of CDQ development, composite silicon carbide brick for inclined flue area of CDQ. The composite silicon carbide brick has the characteristics of high strength, high thermal shock and low expansion coefficient. The thermal flexural strength (1100℃×0.5h) is greater than 30MPa, the thermal shock stability (1100℃, water cooling) is up to 100 times, the service life in inclined flue area of CDQ is up to 4-5 years, and the application effect of longevity is achieved.

Key words: CDQ; inclined duct; silicon carbide; compound material; longevity

2.2　烧结、球团

烧结点火制度的高效低耗化研究

周继良[1,2]，潘　文[1,2]，赵俊花[3]，张晓晨[1,2]，陈绍国[1,2]

（1. 首钢集团有限公司技术研究院，北京　100043；2. 绿色可循环钢铁流程北京市重点实验室，北京　100043；3. 首钢股份有限公司，河北迁安　064404）

摘　要：从成本、环保、烧结指标等方面考虑，对烧结点火系统的高效低耗和烧结机风箱结构进行了研究。首先利用多因素相关分析法找出了点火温度的主要影响因素；其次根据分析结果采用仿真手段对提高空燃比、使用低负压等进行研究；第三对现场烧结机风箱实际结构进行仿真研究，提高烧结料层压力均匀性；最后将上述研究最终方案应用到烧结机实际生产中，运行结果表明，空燃比最高增幅超过 40%，煤气消耗降低约 24.39%。

关键词：烧结点火系统；相关分析；空燃比；低负压；风箱

Study on High Efficiency and Low Consumption of Sintering Ignition System

Zhou Jiliang[1,2], Pan Wen[1,2], Zhao Junhua[3], Zhang Xiaochen[1,2], Chen Shaoguo[1,2]

(1. Shougang Research Institute of Technology, Beijing 100043, China; 2. Beijing Key Laboratory of Green Recyclable Process for Iron & Steel Production Technology, Beijing 100043, China; 3. Shougang Co., Ltd., Qian’an 064404, China)

Abstract: Consideration of cost and environmental protection and sinter index, the high efficiency and low consumption of sintering ignition system was studied. First of all, multivariate correlation analysis was used to find out the main indexes for temperature. Secondly, some simulation studies, based on indexes got from correlation analysis, were carried on, such as increase air-fuel ratio and use low negative pressure. Third, some simulation were taken to improve the uniformity of sintering bed based on wind boxes structure changes. Finally, some modifies were taken in sinter machine and as running shows that air-fuel ratio increased over 40% and fuel rate dropped about 24.39%.

Key words: sintering machine igniter system; correlation analysis; air fuel ratio; low negative pressure; wind boxes

湛江钢铁煤场扩建方案分析

汪　勇，夏光勇，谷显革

（中冶赛迪工程技术股份有限公司，重庆　401122）

摘　要：为解决进口煤商检贮存的问题，减少其中转倒驳进厂物流成本和焦煤混料，同时为增加配煤品种，降低配煤成本，湛钢有必要对已有煤场进行扩建。通过对新增 1 个和 2 个 D 型煤场方案进行对比，分析其对于煤的品种数、堆位数和贮量的影响，对进口煤比例的影响等。分析表明：建设 2 个 D 型煤场虽一次性投资高，但可减少倒驳约 128 万吨/年，增加 3 个低价小品种煤，进口煤比例可达 22.21%，能产生较好的经济效益。

关键词：商检；D 型煤场；进口煤

Scheme Analysis of Coal Yard Expansion for Zhanjiang Steel

Wang Yong, Xia Guangyong, Gu Xiange

(CISDI Engineering Co., Ltd., Chongqing 401122, China)

Abstract: In order to solve the problems of imported coal commodity inspection and storage, reduce logistics cost and coking coal blending, increase the varieties of coal for saving cost of blending coal, it is necessary for Zhanjiang steel to expand the existing coal yard. By comparing expanding 1 and 2 D-type coal yard, this paper analyses the effects on coal varieties, pile numbers, coal yard capacity, and imported coal ratio, etc. The results shows that the investment of expanding 2 D-type coal yard is higher, but this scheme can reduce about 1.28 million t each year of coal transportation required to offload to other port, increase three kinds of low-cost coal, the imported coal ratio can reach 22.21%, these measures can produce good economic benefit.

Key words: commodity inspection; D-Type coal yard; imported coal

海砂矿应用于鞍钢球团生产的试验研究

张　辉[1]，周明顺[1]，唐继忠[2]，马贤国[2]，翟立委[1]，刘　杰[1]，徐礼兵[1]

（1. 鞍钢集团钢铁研究院，辽宁鞍山　114009；

2. 鞍钢股份有限公司鲅鱼圈分公司，辽宁营口　115007）

摘 要：为了利用海砂矿生产低钛球团，进行了海砂矿用于球团生产的试验研究，结果表明：配加海砂矿后，生球质量变差；随海砂矿配比提高，粘结剂的配加量需要适当增加；海砂矿配比 14%的粘结剂配加量比海砂矿配比 8%的高 0.1~0.2 个百分点。配加海砂矿生产低钛球团的适宜焙烧制度为预热温度 950~975℃，焙烧温度 1220~1240℃；球团还原性随海砂矿配比的增加而变差，海砂矿配比 11%的还原度比基准降低 1.55 个百分点；球团膨胀率随海砂矿配比增加而略微改善。工业试验表明，配加 8%海砂矿所生产的低钛球团，其冷强度和高温冶金性能可以满足高炉冶炼要求；利用海砂矿低钛球团进行护炉冶炼，经济效益巨大。

关键词：海砂矿；低钛球团；抗压强度；冶金性能

Experimental Study on Marine Placer Applied in the Pellet Production of Ansteel

Zhang Hui[1], Zhou Mingshun[1], Tang Jizhong[2], Ma Xianguo[2], Zhai Liwei[1], Liu Jie[1], Xu Libing[1]

(1. Iron and Steel Research Institute, Anshan Iron and Steel Group Corporation, Anshan 114009, China; 2. Bayuquan Branch, Angang Steel Company Limited, Yingkou 115007, China)

Abstract: In order to produce low titanium pellets by marine placer, a series of experimental study on marine placer applied in pellet production was carried out. The results show that the quality of green pellets becomes worse after adding marine placer. With the increase of the proportion of marine placer, the dosage of bentonite needs to be increased appropriately. The adding amount of binder with the proportion of marine placer of 14% is 0.1~0.2 percentage points more than that with the proportion of marine placer of 8%. The suitable roasting system for the production of low-titanium pellets with marine placer is that the preheating temperature is between 950℃ and 975℃, the roasting temperature is between 1220℃ and 1240℃. The low temperature reduction pulverization and reducibility of pellets becomes worse as the proportion of marine placer increases. The reduction degree of pellet with 11% ratio of marine placer decreased by 1.55 percentage points compared with the benchmark. The expansion rate were slightly improved with the increase proportion of marine placer. The industrial test shows that the cold strength and high temperature metallurgical performance of the low titanium pellet produced by adding 8% marine placer can meet the smelting requirements of blast furnace. It is of great economic benefit to make use of low titanium pellet of marine placer to protect furnace smelting.

Key words: marine placer; low titanium pellets; compressive strength; metallurgical properties

回转窑煅烧不同特性石灰石工艺控制的生产实践

丁春辉，刘华建，刘世昌，赵恒起

（鞍山钢铁集团耐火材料有限公司，辽宁鞍山 114021）

摘 要：本文主要介绍了鞍山钢铁集团耐火材料有限公司利用两座日产 600t KM 型回转窑煅烧两种不同特性的原料石灰石，研究了煤气单耗、风煤比、石灰石在窑内停留时间等三项因素对冶金石灰产品质量的影响，同时优化回转窑生产工艺过程控制，通过大量工业试生产总结出两套分别适用于两种原料石灰石的煅烧工艺方案，最终均生产出满足钢厂需求的优质冶金石灰。

关键词：回转窑；石灰石；煅烧工艺；冶金石灰

Production Practice of Process Control for Calcining Limestone with Different Characteristics in Rotary Kiln

Ding Chunhui, Liu Huajian, Liu Shichang, Zhao Hengqi

(Anshan Iron and Steel Group Refractory Co., Ltd., Anshan 114021, China)

Abstract: This paper mainly introduces the use of Anshan Iron and Steel Group Refractory Co., Ltd., which use two 600-ton KM rotary kilns to calcinate two different raw material quicklime stones with different characteristics. The effects of gas consumption, air coal ratio and residence time of limestone in kiln on the quality of metallurgical lime were studied. At the same time, the process control of the rotary kiln production was optimized, and through a large number of industrial trial production, two sets of calcination process schemes suitable for two kinds of quicklime stone were summarized. Finally, it produces high-quality metallurgical lime that meets the needs of steel plants.

Key words: rotary kiln; limestone; calcination process; metallurgical lime

转炉渣制备高硅铁酸钙并用于烧结的试验研究

任 伟，王小强，张 伟，王 亮，国泉峰，刘沛江

（鞍钢股份有限公司技术中心，鞍山辽宁 114021）

摘 要：为探索转炉渣用于烧结的可行性，用矿热炉将转炉渣和巴西粉矿熔融制备 CaO/Fe_2O_3 为 0.17 到 0.7 的高硅铁酸钙，并以 0~10%比例用于烧结，结果表明：在高硅铁酸钙的外配比为 0~10%范围内，随着高硅铁酸钙的增加，转鼓强度从 58.4%提高到 60%左右，烧结利用系数稍稍增加，但不明显，当添加量为 2%时，利用系数可达 $1.7t/m^2h$，而添加量为 5%时，利用系数又下降到 $1.37t/m^2h$ 左右，烧结中<5mm 颗粒的质量增加，从 10%升至 15%，烧结固体燃耗最多降低 5kg/t 左右，从经济效益来看，以鞍钢为例，每生产 1t 高硅铁酸钙需亏损 49 元。

关键词：转炉渣；烧结；铁酸钙；成本

The Experimental Study on Sintering with High Silica Calcium Ferrites

Ren Wei, Wang Xiaoqiang, Zhang Wei, Wang Liang, Guo Quanfeng, Liu Peijiang

(Angang Steel Company Limited, Anshan 114021, China)

Abstract: An experimental study on sintering with from 0 to10 percent high silica calcium ferrites which was produced by LF slag and Brazil fines to obtain the utilization feasibility of LF slag, and it was shown that the drum index of sinter increased from 58.4% to 60% with a growing high silica calcium ferrites, and the utilization coefficient has a few increase but not impressive, and the max value can reach 1.7 ton per square meter hour while the high silica calcium ferrites addition was 2 percent, as for the solid fuel consumption can be reduced by 5 kilo grams per ton sinter. After the economic calculation, to produce one ton high silica calcium ferrites may spend RMB 49.

Key words: LF slag; sintering; calcium ferrite; cost

影响球团矿膨胀率因素正交试验分析

段祥光，韩 峰，张亚群，贾 熊，刘景权，郑 凯

（内蒙古包钢稀土钢炼铁厂，内蒙古包头 014010）

摘 要：本文根据包钢钢联股份有限公司球团矿的实际用料情况，研究了焙烧温度、焙烧时间、皂土配比及预热温度等四个因素对球团矿还原膨胀率的影响。采用了四因素四水平正交试验法。试验结果发现，影响球团矿还原膨胀的因素由强到弱为：焙烧温度、皂土配比、焙烧时间、预热温度；其中，最优水平为：焙烧温度 1280℃，皂土配比 4.0%，焙烧时间 10min，预热温度 900℃；最差水平为：焙烧温度 1160℃，皂土配比 1.0%，焙烧时间 8min，预热温度 600℃。微观结构和矿物组成分析表明，球团矿的膨胀与气孔率、液相比例有关。可以通过优化焙烧温度、皂土配比等参数，改变其显微结构，从而改变其性能。

关键词：球团矿；膨胀；正交试验；参数

Analysis of Factors Affecting the Expansion Rate of Pellets by Orthogonal Experiment

Duan Xiangguang, Han Feng, Zhang Yaqun, Jia Xiong, Liu Jingquan, Zheng Kai

(Dept. of Rare Earth Steel Iron-making Plate of Baotou Steel (Group) Corp., Baotou 014010, China)

Abstract: The object of this article is to explore how the roasting temperature, roasting time, proportion of bentonite and preheating temperature influence the rate of expansion, on the condition of fixed raw material. We adopt the orthogonal experiment of four factors at four different levels. The result is that the key factor is roasting temperature, the important factor is proportion of bentonite, and the unimportant factors are roasting time and preheating temperature; the best factors are roasting temperature 1280℃, proportion of bentonite 4%, roasting time 10 minutes, preheating temperature 900℃; the worst factors are sintering temperature 1160℃, proportion of bentonite 1%, roasting time 8 minutes, preheating temperature 600℃.The microstructure and mineral composition suggest that the expansion of pellets relates to porosity and liquid proportion. So optimizing the parameters of roasting temperature and proportion of bentonite can change the microstructure, leading to the change of property.

Key words: pellet; expansion; orthogonal test; parameter

DMAIC 方法在烧结生产提高台时产量中的应用

朱付涛

（武钢炼铁厂，湖北武汉 430080）

摘 要：本文运用六西格玛方法对烧结台时产量指标进行改进，利用 FMEA、DOE 分析等方法进行分析优化，以

提高过程能力控制水平。其中对混合料水分、生石灰配比、料层厚度、焦粉配比等关键因子参数优化。对比实验前后，烧结矿的台时产量均值明显提高、波动幅度明显改善，项目改善效果显著，过程控制稳定。

关键词：六西格玛；DMAIC；台时产量

烧结生产中 NO_x 的形成及过程减排措施

张江鸣

（武钢炼铁厂，湖北武汉 430080）

摘 要：随着国家将氮氧化物列为污染物总量控制的约束性指标，脱硝被认为是下一步钢铁烧结烟气治理的主要方向之一。本文根据武钢烧结厂的生产实际，分析了烧结过程中 NO_x 的形成机理、影响因素，主要从原料使用及生产过程方面实施对 NO_x 的减排措施。

关键词：氮氧化物；形成机理

Formation of NO_x in Sintering Process and Measures for Reducing NO_x Emission

Zhang Jiangming

(Ironmaking Plant of Wisco, Wuhan 430080, China)

Abstract: As the state has listed nitrogen oxide as a binding target for the total amount control of pollutants, denitration is considered to be one of the main directions for the next step in the control of flue gas from iron and steel sintering. According to the production practice in sintering plant of Wisco, the formation mechanism and influencing factors of NO_x in sintering process are analyzed.

Key words: the nitrogen oxide; the mechanism

稳定混匀矿质量的生产实践

梅 奇，柯昌华，芦 川

（武钢炼铁厂，湖北武汉 430080）

摘 要：为稳定混匀矿质量、提升关键质量指标，通过对混匀建堆的全过程跟踪，运用质量分析工具，采取堆料工艺调整、计划成分实时调整、配料计划优化、减少皮带机运行故障、配料设备缺陷整改等措施，促进二混匀作业区 TFe/SiO_2 合格率和标准偏差指标的完成。

关键词：工艺调整；设备缺陷；计划优化；成分调整

360m² 烧结机节能技术研究与实践

袁平刚

（河钢股份有限公司承德分公司，河北承德　067002）

摘　要：本文通过研究改善 360m² 烧结机烧结过程粒度、料温、透气性等参数，研发混合机增加在线自动清料和燃料破碎皮带平整料装置，引进耐磨输灰管道等综合节能技术，有效降低固体燃料和空压风风消耗；烧结机工序能耗明显降低。对发挥钒钛磁铁精粉资源高产、低耗及综合利用具有较强的实际意义，为国家钒钛资源高效、综合利用提供强有力地保证，同时对于烧结工序减少碳排放等指标，改善区域大气环境发挥积极作用。

关键词：烧结机；节能；碳排放

Research and Practice on Energy-saving Technology of 360m² Sintering Machine

Yuan Pinggang

(Chengde Branch of Hebei Iron and Steel Co., Ltd., Chengde 067002, China)

Abstract: This paper studies and improves the parameters of the sintering process of 360m² sintering machine such as particle size, material temperature, air permeability, etc., develops the mixer to increase the online automatic cleaning and fuel crushing belt leveling device, and introduces comprehensive energy-saving technologies such as wear-resistant ash conveying pipelines to effectively reduce Solid fuel and compressed air consumption; energy consumption of sintering machine process is obviously reduced. It has strong practical significance for the high-yield, low-consumption and comprehensive utilization of vanadium-titanium magnet powder resources, provides a strong guarantee for the efficient and comprehensive utilization of national vanadium-titanium resources, and at the same time reduces carbon emissions and other indicators for the sintering process, and improves the regional atmosphere The environment plays a positive role.

Key words: sintering machine; energy saving; carbon emission

富氧对烧结过程和烧结矿质量的影响

张亚鹏[1,2]，季　斌[1,2]，张晓臣[1,2]，潘　文[1,2]，陈绍国[1,2]，赵志星[1,2]

（1. 首钢集团有限公司技术研究院钢铁技术研究所，北京　100043；
2. 绿色可循环钢铁流程北京市重点实验室，北京　100043）

摘　要：本文重点研究了富氧率和富氧阶段对烧结过程、成品矿物理指标和冶金性能等方面的影响；同时，着重分析了富氧烧结对烧结烟气成分变化的影响。研究表明：随着富氧浓度增加，垂直烧结速度增大，烧结利用系数提高，烧结成品率由基准方案的 82.9%增加至 85.21%，烧结固体燃耗由基准方案的 49.94kg/t 降低至 48.04kg/t；随着富氧时间段向后推移，垂直烧结速度和利用系数呈现出先增大后减小的趋势，成品率降低、固体燃耗升高；从有利于烧结的角度，富氧宜在烧结前中期，同时选择高富氧浓度；对于烧结烟气成分，在富氧烧结阶段 CO 浓度出现“凸台效应”，富氧开始时 CO 浓度曲线出现跃升，而当富氧结束时 CO 浓度曲线出现跃降。

关键词： 富氧烧结；复合铁酸钙；CO；烧结烟气

Effect of Oxygen Enrichment on Sintering Process and Sinter Quality

Zhang Yapeng[1,2], Ji Bin[1,2], Zhang Xiaochen[1,2], Pan Wen[1,2], Chen Shaoguo[1,2], Zhao Zhixing[1,2]

(1. Shougang Group Co., Ltd., Research Institute of Technology, Beijing 100043, China; 2. Beijing Key Laboratory of Green Recyclable Process for Iron & Steel Production Technology, Beijing 100043, China)

Abstract: The effects of oxygen enrichment on sintering process, physical index of finished ore and metallurgical properties were studied. At the same time, the influence of oxygen-enriched sintering on the composition change of sintering flue gas was analyzed emphatically. The results show that with the increase of oxygen enrichment, the vertical sintering speed and sintering utilization coefficient increased, the sintering yield increased from 82.9% to 85.21%, and the solid fuel consumption decreased from 49.94kg/t to 48.04kg/t. The vertical sintering speed and utilization coefficient increased first and then decreased with the oxygen enrichment period delayed, while the yield decreased and the solid fuel consumption increased. Oxygen enrichment should be in the early and middle stages of sintering with high concentration of oxygen enrichment, which was beneficial to sintering process and sintered ore quality. For the composition of sintering flue gas, the “hump effect” of CO concentration appeared in the oxygen-enriched sintering stage, and the CO concentration curve jumped to a higher level at the beginning of oxygen enrichment, and then jumped to a lower level at the end of oxygen enrichment.

Key words: oxygen enrichment sintering; SFCA; CO; flue gas of sintering

宝钢 C 型料场存在问题及改进实践

吴旺平

（宝山钢铁股份有限公司炼铁厂，上海　201900）

摘　要： C 型料场在宝钢投入使用后，出现了诸如品种切换不易、平料时间长、实际储量达不到设计等问题。针对这些问题，宝钢原料场采取了如下措施：（1）改进品种切换方式提高料场物料品种配置的灵活性；（2）通过智能化技术提升提高平料速度及效率；（3）加强 C 型料场堆积管理减少滑落的可能及水分大物料的输出。以上措施值得国内其他采用 C 型料场的企业参考。

关键词： C 型料场；品种切换；储量；平料时间

Existing Problems and Improvement Practice of Type-C Closed Stockyard in Baosteel

Wu Wangping

(Ironmaking Plant, Baoshan Iron & Steel Co., Ltd., Shanghai 201900, China)

Abstract: Since Type-C closed stockyard have been put into prodution in Baosteel, some problems have arisen,such as : hard to change varieties、long time to scrap the raw material、the actual reserves can not meet the design requirements. Aiming to solve above problems, some relevant measures have been taken in Baosteel. Such as: (1) Improving varieties switching make the variety allocation agile; (2) enhance the speed of scraping by intelligent technology; (3) Daily stacking management of C-type stockyard can aviod raw material sliding and ensure high moisture material output. All above measures will offer the reference for peer company.

Key words: Type-C closed stockyard; variety switching; actual storage; srcap time

提升球团生产过程中 NO_x 控制能力的探索与实践

王鹏飞，刘金英，杨金保，沈国良

（北京首钢股份有限公司，河北迁安 064404）

摘 要：本文主要通过对 SCR 脱硝技术、球团生产工艺热工参数控制、回转窑主燃烧器改型等，不断提升 120 万吨链篦机-回转窑氧化球生产线氮氧化物控制能力，确保外排氮氧化物排放满足外排控制标准。首钢球团厂始建于 1985 年，一系列生产线 2000 年经过截窑改造，是我国第一条采用链箅机-回转窑-环冷机工艺的现代化球团生产企业，该生产线已成为整个球团生产行业的典范，是首钢重要的高炉原料生产基地。2018 年 9 月完成了烟气脱硝项目，采用中低温 SCR 工艺，2020 年 4 月投入 SNCR 脱销工艺，2021 年 3 月使用回转窑低氮燃烧器后。通过实施一系列措施控制后，在排放指标满足唐山市特别排放限值：粉尘小于 5mg/m^3、二氧化硫小于 20mg/m^3、氮氧化物小于 30mg/m^3 要求的同时，SCR 脱硝使用的氨水消耗显著降低。

关键词：球团；氮氧化物；脱硝技术；燃烧器

Exploration and Practice of NO_x Control in the Process of Ball Production

Wang Pengfei, Liu Jinying, Yang Jinbao, Shen Guoliang

(Beijing Shougang Co., Ltd., Qian'an 064404, China)

Abstract: This paper mainly improves the nitrogen oxide control capacity of SCR denitration technology, the thermal parameter control of the rotary furnace, to ensure that the discharge nitrogen oxide emission meets the discharge control standard. Shougang pellet Factory was built in 1985. A series of production lines underwent kiln transformation in 2000. It is the first modern pellet production enterprise in China to adopt the grate machine-rotary kiln-ring cooler process. This production line has become a model of the whole pellet production industry and an important blast furnace raw material production base in Shougang.The flue gas denitration project was completed in September 2018, using medium and low temperature SCR process. SNCR cancellation process was put in April 2020, and low nitrogen burner of rotary kiln was used in March 2021.After the implementation of a series of measures, the ammonia water consumption used in SCR denitration is significantly reduced while the emission indicators meet the Tangshan special emission limit: dust is less than 5mg/m^3, sulfur dioxide is less than 20mg/m^3, and ammonia oxide is less than 30mg/m^3.

Key words: pellets; nitrogen oxides; offpin technology; burners1

球团矿还原膨胀的一种新机制

沈茂森[1]，吕志义[2]，康文革[2]，白晓光[2]，李玉柱[2]，孟文祥[1]

（1. 包钢集团公司矿山研究院，内蒙古包头　014030；
2. 包钢集团公司技术中心，内蒙古包头　014010）

摘　要：本文介绍了球团矿还原膨胀的一种新机制。由于铁矿石中富含易生成挥发分的包裹体，进而导致球团矿中富含挥发分。在还原过程中，球团矿气孔中挥发分的爆发，扩大了基体裂块间的空隙，加快了还原气体的扩散，增加了反应面积，使还原速度加快，从而导致了球团矿恶性膨胀。球团矿的“挥发分膨胀理论”是一种全新的理论，对抑制球团矿的恶性膨胀具有重要指导意义。

关键词：球团矿；挥发分；显微结构；还原膨胀；机制

A New Mechanism for Reduction Swelling of Pellets

Shen Maosen[1], Lü Zhiyi[2], Kang Wenge[2], Bai Xiaoguang[2],
Li Yuzhu[2], Meng Wenxiang[1]

(1. Mine Research Institute of Baotou Steel Group, Baotou 014030, China;
2. Technical Center of Baotou Steel Group, Baotou 014010, China)

Abstract: A new mechanism of reduction swelling of pellets is introduced in this paper. Volatiles are abundant in pellets because iron ore is rich in inclusions which are easy to generate volatiles. In the process of reduction, the eruption of volatiles in the pores of pellets expands the gap between the matrix crack blocks, accelerates the diffusion of reducing gas, increases the reaction area, and speeds up the reduction rate, thus leading to the malignant swelling of pellets. Volatile Swelling Theory of pellets is a new theory, which has important guiding significance to restrain the malign swelling of pellets.

Key words: pellet; volatiles; microstructure; reduction swelling; mechanism

玉钢 210m^2 烧结机降低烧结工序燃料单耗生产实践

王　冲，周　宾

（玉溪新兴钢铁有限公司烧结厂，云南玉溪　653100）

摘　要：烧结燃料单耗占烧结工序加工费很大的比重，降低烧结工序能耗首先要降低固体燃料消耗。如何降低燃料单耗是所有钢厂一直探索和追寻的目标。本文对国内外生产技术指标先进的钢铁企业的研究及生产实践进行总结，寻找出固体燃料消耗指标与国内外一些钢铁企业之间存在的差距，结合自身用矿结构、装备技术、环境等，对标挖潜摸索出适宜烧结厂降低烧结固体燃耗的措施。

关键词：燃料单耗；烧结矿成本；实践

Reducing Sintering Process of 210m² Sintering Machine in Yugang Production Practice of Unit Fuel Consumption

Wang Chong, Zhou Bin

(Yuxi Xinxing Iron and Steel Co., Ltd. Sintering Plant, Yuxi 653100, China)

Abstract: The unit consumption of sintering fuel accounts for a large proportion of the processing cost of sintering process. To reduce the energy consumption of sintering process, we must first reduce the consumption of solid fuel. How to reduce the unit fuel consumption is the goal that all steel mills have been exploring and pursuing. This paper summarizes the research and production practice of iron and steel enterprises with advanced production technical indicators at home and abroad, finds out the gap between solid fuel consumption indicators and some iron and steel enterprises at home and abroad, explores the potential against the standard in combination with their own ore structure, equipment technology and environment, and finds out the measures suitable for sintering plants to reduce sintering solid fuel consumption.

Key words: fuel consumption; sinter cost; practice

脉石成分对固相下赤铁矿烧结过程二次赤铁矿生成的影响

郭兴敏，刘洪波，赵洁婷

（北京科技大学冶金与生态工程学院，北京　100083）

摘　要：低温还原粉化指数（RDI）是衡量烧结矿质量的一个重要参数，烧结矿内二次赤铁矿存在显著影响 RDI 指标。本文，通过热重法和 XRD 法，跟踪了赤铁矿与脉石成分 Al_2O_3、SiO_2 和 MgO 混合试样升降温过程中二价铁与三价铁及其矿物间转变，定量地研究了固相下含脉石赤铁矿烧结过程中二次赤铁矿的生成规律乃至机理上差异。实验表明，SiO_2 成分增加，阻碍了赤铁矿分解后氧的扩散，可以抑制二次赤铁矿生成；Al_2O_3 成分增加，加速了赤铁矿向磁铁矿转变，可以促进二次赤铁矿生成。与以上两者不同，MgO 成分增加，可以提高磁铁矿（含镁铁矿）晶型低温下稳定性，在抑制二次赤铁矿生成上效果显著得多。

关键词：二次赤铁矿生成；升降温过程；脉石成分；赤铁矿烧结

Effect of Gangue Composition on Solid-state Formation of Secondary Hematite in Sintering Process of Hematite

Guo Xingmin, Liu Hongbo, Zhao Jieting

(State Key Laboratory of Advanced Metallurgy and School of Metallurgical and Ecological Engineering, University of Science and Technology Beijing, Beijing 100083, China)

Abstract: Reduction degradation index (RDI) at low-temperature is an important parameter to judge the quality of iron ore sinter, it is related to secondary hematite formation in the sinter. In this work, thermogravimetric and X-ray diffraction methods were used to follow transformations between ferric and ferrous ions, orthese minerals respectively in temperature-increasing and temperature-decreasing processes for investigating quantitativelyeffect of gangue composition on the solid-state formation of secondary hematite and it's different mechanismin sintering of hematite. It shows thatthe formation of secondary hematitedecreasedwith SiO_2 content due to hindering the diffusion of oxygen in sample whereas increasedwith Al_2O_3contentdue to promoting the transformation of hematite to magnetite. In addition, MgO was added into hematite, it increased the crystal structure stability of magnetite (or magnesioferrite), resulting in a superior effect to prevent formation of secondary hematite than SiO_2 and Al_2O_3.

Key words: secondary hematite formation; temperature-increasing and -decreasing processes; gangue composition; sintering of hematite

巴卡粉与超特粉互补配矿烧结试验

马怀营[1,2]，朱　旺[3]，辛　越[3]，潘　文[1,2]，高新洲[3]，安乃鲜[3]

（1. 首钢集团有限公司技术研究院，北京　100043；2. 绿色可循环钢铁流程北京市重点实验室，北京　100043；3. 北京首钢股份有限公司，河北迁安　064400）

摘　要：为扩充铁矿粉资源，优化烧结配矿，利用烧结杯研究了高比例磁精粉条件下巴卡粉与超特粉的互补配矿情况。结果表明，在保持烧结矿成分稳定前提下，提高巴卡粉与超特粉的总配比能够改善烧结混合料的粒度组成，提高烧结速度和烧结利用系数，但对烧结矿的质量呈负面影响，烧结矿转鼓指数、平均粒径及低温还原粉化指数均呈下降趋势，建议巴卡粉与超特粉的总配比控制在15.36%以内。

关键词：巴卡粉；超特粉；磁精粉；优化配矿；烧结

Sintering Experiment on Complementary Ore Proportioning of SFCJ and SS

Ma Huaiying[1,2], Zhu Wang[3], Xin Yue[3], Pan Wen[1,2], Gao Xinzhou[3], An Naixian[3]

(1. Shougang Group Corporation, Research Institute of Technology, Beijing 100043, China; 2. Beijing Key Laboratory of Green Recyclable Process for Iron & Steel Production Technology, Beijing 100043, China; 3. Beijing Shougang Co., Ltd., Qian'an 064400, China)

Abstract: In order to expand the resources of iron ores and optimize the ore blending, the complementary proportioning of SFCJ (sinter feed Carajas) and SS (super special fines) with high proportion magnetite concentrate powder was studied by the sinter pot. The results show that on the premise of keeping the composition of sinter stable, increasing the total ratio of SFCJ and SS can improve the particle size of sinter mix, increase the sintering speed and sintering productivity. However, it has a negative impact on the quality of sinter, that is, it deteriorates the sinter tumble index, average particle size and low temperature reduction degradation index. It is suggested that the total ratio of SFCJ and SS should be controlled within 15.36%.

Key words: SFCJ; SS; magnetite concentrate; ore blending optimization; sintering

烧结温度对烧结矿矿相及其显微力学性能的影响

陈绪亨[1,2]，王　炜[1,2]，杨代伟[1,2]，罗　杰[1,2]

（1. 湖北省冶金二次资源工程技术研究中心，湖北武汉　430081；
2. 武汉科技大学省部共建耐火材料与冶金国家重点实验室，湖北武汉　430081）

摘　要：本文基于铁矿粉粒度分布对铁矿粉进行粒级缩放，通过微型烧结实验研究了烧结温度对烧结矿矿相及其力学性能的影响，结果表明：随着烧结温度升高，烧结矿中赤铁矿、铁酸钙含量逐渐减少，磁铁矿和硅酸盐含量逐渐增加，气孔率呈先减小后增加的趋势；在烧结温度较低时烧结矿中铁酸钙主要是针状为主，提高烧结温度后逐渐出现条柱状；随着烧结温度的升高，烧结矿中赤铁矿和磁铁矿显微硬度先增加后降低，铁酸钙的显微硬度逐渐增加，硅酸盐显微硬度无明显变化。

关键词：烧结温度；矿相组成；矿相结构；显微硬度

Effect of Sintering Temperature on the Mineral Phase and Micromechanical Properties of Sinter Ore

Chen Xuheng[1,2], Wang Wei[1,2], Yang Daiwei[1,2], Luo Jie[1,2]

(1. Hubei Provincial Engineering Technology Research Center of Metallurgical Secondary Resources, Wuhan 430081, China; 2. State Key Laboratory of Refractories and Metallurgy, Wuhan University of Science and Technology, Wuhan 430081, China)

Abstract: Based on the particle size distribution of iron ore powder, this article scales the size of iron ore powder. The effect of sintering temperature on the ore phase and mechanical properties of sintered ore were studied by micro-sintering experiments. The results showed that with the sintering temperature increasing, the contents of hematite and calcium ferrite gradually decreased, and the contents of magnetite and silicate gradually increased, and the porosity decreased first and then increased. When the sintering temperature was low, the calcium ferrite in the sinter was mainly needle-shaped, but the columnar shape gradually appears with the sintering temperature increasing. With the sintering temperature increasing, the microhardness of hematite and magnetite first increased and then decreased. The microhardness of calcium ferrite gradually increased, while the microhardness of silicate did not change significantly.

Key words: sintering temperature; mineral phase composition; mineral phase structure; microhardness

烧结机侧部密封装置改进与优化数值模拟

任素波，李忠阳

（燕山大学机械设计及理论系，河北秦皇岛　066004）

摘　要：为进一步降低烧结设备在烧结过程中的漏风率，在烧结机侧部密封装置内增加了一种能够降低负压的结构，通过对改进装置入口速度的数值模拟优化密封装置内新增结构的尺寸参数，最后，尝试改变其边界条件降低入口速度，提高密封性。结果表明，密封腔内两凹槽的距离为16mm，入口流体速度最低，凹槽厚度变化对入口速度没有显著影响；而且入口流体速度随着出口负压减小而减小，密封腔内增加的凹槽具有降低压强和滞留流体的作用。
关键词： 烧结机；密封腔；凹槽；数值模拟

Numerical Simulation of Improvement and Optimization of Side Sealing Device of Sintering Machine

Ren Subo, Li Zhongyang

(Department of Mechanical Design and Theory, Yanshan University, Qinhuangdao 066004, China)

Abstract: In order to further reduce the air leakage rate of sintering equipment in the sintering process, a structure capable of reducing negative pressure was added in the side sealing device of sintering machine. The size parameters of the newly added structure in the sealing device were optimized by numerical simulation of the inlet speed of the improved device. Finally, the boundary conditions were changed to reduce the inlet speed and improve the sealing performance. The results show that the distance between two grooves in the sealed cavity is 16mm, and the inlet fluid velocity is the lowest, and the change of groove thickness has no significant effect on the inlet velocity Moreover, the inlet fluid velocity decreases with the decrease of outlet negative pressure, and the increased groove in the sealed cavity has the functions of reducing pressure and retaining fluid.
Key words: sintering machine; sealing cavity; groove; numerical simulation

昆钢新区金布巴粉烧结杯试验及研究

苏亚刚[1]，桂林峰[1]，赵　彧[1]，赵红全[2]，罗英杰[2]

（1. 昆钢新区烧结厂，云南昆明　650300；2. 昆钢技术中心，云南昆明　650300）

摘　要： 结合昆钢新区烧结厂生产实际，以五因素、四水平正交试验模型制定混匀矿配比方案，考察金布巴粉、燃料的配加比例及水分、负压等因素的变化对烧结矿转鼓指数、成品率及垂直烧结速度三个技术经济指标的影响情况。
关键词： 金布巴粉；正交试验；烧结矿；技术经济指标

Test and Study on Sinter Cup of Jinbuba Powder in New Area of Kunming Iron and Steel Co.

Su Yagang[1], Gui Linfeng[1], Zhao Yu[1], Zhao Hongquan[2], Luo Yingjie[2]

(1. Kunshan Steel New Area Sintering Plant, Kunming 650300, China;
2. Kunming Iron & Steel Technology Center, Kunming 650300, China)

Abstract: Based on the actual production of the sintering plant in Kunshan Iron and Steel Co., Ltd., the mixing ratio of gold buba powder and fuel, as well as the change of moisture and negative pressure on the three technical and economic indexes of sinter drum index, yield and vertical sintering speed were investigated by using the five-factor and four-level orthogonal experimental model.

Key words: jinbuba powder; orthogonal test; the sinter; technical and economic index

昆钢新区适宜的混匀矿原料结构评价及分析

赵 彧[1]，桂林峰[1]，苏亚刚[1]，罗英杰[2]，赵红全[2]

（1. 昆钢安宁公司新区烧结厂，云南昆明 650300；2. 昆钢技术中心，云南昆明 650300）

摘 要：对昆钢新区所用 125#~150#混匀矿堆进行实际使用过程中的数据进行汇总分析，剔除日均新料量低于 400t/h 的异常情况，对混匀矿烧结性能做综合评价，并进行优选。对优选出的混匀矿料堆进行汇总，并分析其原料结构情况、各单矿种适宜的配矿比例，为优化配矿工作提供参考依据。

关键词：混匀矿；原料结构；烧结性能；评价分析

Evaluation and Analysis of Suitable Raw Material Structure of Mixed Ore in New Area of Kunshan Iron and Steel

Zhao Yu[1], Gui Linfeng[1], Su Yagang[1], Luo Yingjie[2], Zhao Hongquan[2]

(1. New District Sintering Plant of Anningcompany, Kunming 650300, China;
2. Kunming Iron and Steel Technology Center, Kunming 650300, China)

Abstract: The data of 125#~150# mixed ore pile used in the new area of Kunming Iron and Steel Co., Ltd. were summarized and analyzed to eliminate the abnormal situation that the daily average fresh material quantity was less than 400t/h. The sintering performance of the mixed ore was comprehensively evaluated and optimized. The optimized mixed ore piles are summarized, and the structure of raw materials and the suitable ore blending proportion of each single ore are analyzed, so as to provide a reference for optimizing ore blending work.

Key words: blend ore; raw material structure; sintering performance; evaluation and analysis

X-荧光光谱法测定锰矿石的化学成分

马晓云，吴作立，郑 莉

（新疆八一钢铁有限公司制造管理部，新疆乌鲁木齐 830022）

摘 要：锰矿石在添加助熔剂后，于高温熔融成玻璃圆片，在降低元素效应的同时，也消除了矿物效应，从而实现 X-荧光光谱法对锰矿石中 TFe、TMn、P 的测定。利用具有浓度梯度标样绘制标准曲线，测定锰矿石中 TFe、TMn、P 的成分。其中精密度、准确度均符合国家标准。将方法用于测定锰矿石生产样，锰矿石的 X-荧光法测定结果与化学测定结果均在误差允许范围内。

关键词：锰矿石；X-荧光光谱法；化学成分；玻璃熔片法

Determination of Chemical Composition of Manganese Ore by X-fluorescence Spectrometry

Ma Xiaoyun, Wu Zuoli, Zheng Li

(Manufacturing Management Department of Xinjiang Bayi Iron and Steel Co., Ltd., Urumqi 830022, China)

Abstract: After adding flux, manganese ore is melted into glass discs at high temperature, which reduces the element effect and eliminates the mineral effect, thereby realizing the determination of TFe, TMn and P in manganese ore by X-fluorescence spectroscopy. A standard curve was drawn using a standard sample with a concentration gradient to determine the composition of TFe, TMn and P in manganese ore. The precision and accuracy are in line with national standards. The method is used to measure manganese ore production samples, and the results of X-fluorescence and chemical determination of manganese ore are within the allowable range of error.

Key words: manganese ore; X-fluorescence spectrometry; chemical composition; glass frit method

TGS680 石灰窑生产碱性球原料的实践与探讨

刘志浩[1]，毕 燕[2]，刘树钢[3]，王 刚[4]，张德国[4]，许贵宾[5]

（1. 唐山精研实业有限责任公司，河北唐山 063020；2. 唐山助纲炉料有限公司，河北唐山 063020；3. 唐山今实达科贸有限公司，河北唐山 063020；4. 北京首钢国际工程技术有限公司，北京 100043；5. 宣化正朴铁业有限责任公司，河北张家口 075100）

摘 要： 中国铁精粉 SiO_2 含量普遍较高，这加剧了熔剂性球团（或称碱性球团）生产和应用中存在的五大难题，成为进一步优化高炉炉料结构的关键问题。本文介绍了高炉煤气 TGS 石灰窑的技术特点和熔剂性球团熔剂的生产情况，再配合上裹皮高镁碱性球团（又称熔剂性复合球团）生产技术，可较好地解决上述难题，为利用高 SiO_2 铁精粉生产优质碱性球团探索了一条可行的工艺路线，并提出了碱性球团熔剂生产工艺和质量标准的意见。

关键词： 炉料结构；熔剂性球团；烧结矿；TGS 石灰窑；高炉

Practice and Discussion on the Production of the Raw Materials for Fluxed Pellets in TGS680 Lime Kiln with the BF Gas

Liu Zhihao[1], Bi Yan[2], Liu Shugang[3], Wang Gang[4], Zhang Deguo[4], Xu Guibin[5]

(1. Tangshan Jingyan Industry Co., Ltd., Tangshan 063020, China; 2.Tangshan Zhugang Burden Co., Ltd., Tangshan 063020, China; 3. Tangshan Jinshida Technology Trading Co., Ltd., Tangshan 063020, China; 4. Beijing Shougang International Engineering Co., Ltd., Beijing 100043, China; 5. Xuanhua Zhengpu Iron LLC, Zhangjiakou 075100, China)

Abstract: In China, the content of SiO_2 in iron concentrate is generally high. It aggravates five major problems in the production and application of fluxed pellets (alkaline pellets), which becomes the key difficulties to further optimize the burden structure of blast furnace. This paper introduces the technical characteristics of TGS lime kiln with the BF gas and the production situation of the flux for alkaline pellets. Combined with the production technology of the coated high magnesium alkaline pellets (also known as fluxed composite pellets), it can solve the above problems. It explores a feasible process route for the production of high quality fluxed pellets with high SiO_2 iron concentrate, and provides several suggestions for production process and quality standards of the flux for alkaline pellets.

Key words: burden structure; fluxed pellets; sinter; TGS lime kiln; blast furnace

粉矿种类及配比对双层烧结产质量的影响

翟立委[1]，周明顺[1]，刘沛江[2]，刘　杰[1]，徐礼兵[1]，张　辉[1]

（1. 鞍钢集团钢铁研究院，辽宁鞍山　114009；2. 鞍钢股份有限公司炼铁总厂，辽宁鞍山　114000）

摘　要：针对鞍钢采用的创新式新工艺双层烧结生产技术，通过烧结杯实验室实验研究，摸索出适用于双层烧结的粉矿品种及配比。研究结果表明：在 1000mm 双层烧结中，对于双层烧结的产质量，配加 28%的铁矿粉 B 为最佳粉矿品种与配比。而随着粉矿配比增加到 45%，对于烧结产质量指标，配加铁矿粉 B 仍为最优，铁矿粉 A 可配至 45%，铁矿粉 C 配比不超过 35%，铁矿粉 D 配比不可超过 45%。对于双层烧结矿粒度分布的影响，配加 28%的铁矿粉 A 为最优选择，铁矿粉 B 次之，铁矿粉 D 配比在 35%时烧结矿粒度分布较好，而铁矿粉 C 配比应尽量减少。

关键词：粉矿种类及配比；双层烧结；烧结产质量

Effect of Type and Proportion of Powder Ore on Production and Quality of Double-layer Sintering

Zhai Liwei[1], Zhou Mingshun[1], Liu Peijiang[2], Liu Jie[1], Xu Libing[1], Zhang hui[1]

(1. Iron-making Technology Institute of Ansteel Group Corporation Iron and Steel Research Institute, Anshan 114009, China; 2. Iron-making Plant, Angang Steel Co., Ltd., Anshan 114000, China)

Abstract: According to the innovative new double-layer sintering technology adopted by Anshan Iron and Steel Co., through the laboratory experiment of Sinter Cup, the kinds and proportion of powder ore suitable for double-layer sintering were found out. The results show that: for the production and quality of double-layer sintering in 1000mm double-layer sintering, adding 28% newman powder is the best variety and proportion of powder. As the proportion of powder increases to 45%, the addition of newman powder is still the best for sintering production and quality index, PB powder can be added to 45%, FMG powder can not exceed 35%, yangdi powder can not exceed 45%. For the influence of double-layer sinter particle size distribution, adding 28% PB powder is the best choice, newman powder is the second, Yangdi powder is the best when the proportion is 35%, and FMG powder should be reduced.

Key words: type and proportion of powder ore; double-layer sintering; sintering yield and quality

开发新炉料推广新技术，促进炼铁低碳清洁高效

周明顺[1]，李忠武[2]，赵东明[3]，赵正洪[3]，朱建伟[1]，李 仲[1]，翟立委[1]

（1. 鞍钢集团钢铁研究院炼铁技术研究所，辽宁鞍山 114009；2. 鞍钢股份有限公司，辽宁鞍山 114021；3. 鞍钢股份有限公司炼铁总厂，辽宁鞍山 114021）

摘 要：为提高鞍钢股份有限公司本部炼铁总厂高炉利用系数、降低燃料消耗和推动本部炼铁总厂的绿色发展，分析、梳理了目前本部炼铁总厂高炉原料工艺结构方面存在的问题与短板，针对问题与短板，从原料工艺结构创新、设备大型化、固废的利用、高炉炉料结构重新构建等入手，提出具体技术措施与解决方案。基于此，针对鞍钢本部高炉炉料结构的发展提出了提铁降硅、减少中型规模烧结产线、球团熔剂性球团关键工艺技术开发与产业化等 3 条战略判断。

关键词：高炉；利用系数；焦比；炉料结构；发展战略

Developing New Burden, Popularizing New Technology, Promoting Low Carbon, Clean and High Efficiency in Ansteel

Zhou Mingshun[1], Li Zhongwu[2], Zhao Dongming[3], Zhao Zhenghong[3], Zhu Jianwei[1], Li Zhong[1], Zhai Liwei[1]

(1.Iron-making Technology Institute of Ansteel Group Corporation Iron and Steel Research Institute, Anshan 114009, China; 2. Angang Steel Co., Ltd., Anshan 114021, China; 3. Iron-making Plant, Angang Steel Co., Ltd., Anshan 114021, China)

Abstract: In order to improve the utilization factor of blast furnace, reduce fuel consumption and promote the green development of the main ironmaking plant of AISC, this paper analyzes and sorts out the existing problems and short plates in the technological structure of the blast furnace raw materials in the main iron-making plant of the headquarters, aiming at the problems and short plates, starting from the innovation of raw material technological structure, the large-scale equipment, the utilization of solid waste and the reconstruction of blast furnace burden structure, the concrete technical measures and solutions are put forward. In view of the development of burden structure of blast furnace in AISC, three strategic judgments are put forward, such as raising iron and reducing silicon, reducing medium-scale sintering production line, developing key technology and industrialization of pellet flux pellet.

Key words: blast furnace; utilization coefficient; coke rate; charge structure; development strategy

气基还原用绿色球团工艺研究与工程应用

张 晨，杨木易

（中钢国际工程技术有限公司，北京 100000）

摘　要： 研究和开发用于气基直接还原的带式焙烧球团工艺，不仅可以推动我国带式焙烧球团的快速发展，对节能减排也有着极大的意义。要满足 Midrex 气基直接还原的生产要求，氧化球团必须具有较高的强度，均匀的粒度，适宜的还原强度。本文模拟带式焙烧机工艺进行焙烧实验，考察不同的热工制度对抗压强度的影响，在保证球团矿产量和质量的前提下提供适宜 Midrex 工艺的氧化球团热工参数，为带式焙烧机的设计提供理论基础。

关键词： 气基还原；节能减排；带式焙烧球团；抗压强度

Research & Engineering Application of Green Pellet Process for Gas-based Direct Reduction

Zhang Chen, Yang Muyi

(Sinosteel Engineering &Technology Co., Ltd., Beijing 100000, China)

Abstract: Research and development on traveling grate process applied to gas-based direct reduction not only can be a driving force to the acceleration of development for traveling grate process pellet production in China, but also is of great significance to energy conservation and emission reduction. High strength, homogeneous grain size and adequate reducibility are essential to oxide pellets in order to meet the requirement to Midrex gas-based direct reduction production. In this paper, Test can be conducted via simulation of pellet induration process in travelling grate as to evaluate different thermal effects on compressive strength, thereby obtaining thermal parameters suitable for oxide pellets in Midrex process on the premise of ensuring output and quality of pellets. A theoretical basis is thus given on working out the engineering of traveling grate induration machine.

Key words: gas-based direct reduction; energy conservation and emission reduction; traveling grate process pellet; compressive strength

“双碳”背景下烧结工序温室气体减排研究

臧疆文

（宝钢集团八钢公司碳中和办公室，新疆乌鲁木齐　830022）

摘　要： 通过对八钢烧结工序的现有装备、CO_2 排放量及构成的分析和国内外低碳烧结新技术的研究，提出烧结工序应持续完善电石渣利用技术，提高电石渣在烧结与脱硫中的用量，可以优先实施烧结机台车加宽技术、新型环冷机技术、变频技术，有选择性地实施喷吹富氢气体燃料技术、烟气循环技术、料面喷洒蒸汽技术，研究与跟踪微波烧结技术、生物质能烧结技术等前沿低碳烧结技术，减少烧结工序温室气体排放。

关键词： 温室气体；烧结；CO_2 排放强度；台车加宽；烟气循环；微波烧结；生物质

Study on Greenhouse Gas Emission Reduction in Sintering Process Under Dual Carbon Background

Zang Jiangwen

(Carbon Neutral Office, Bayi Iron & Steel Co., Baosteel Group, Urumqi 830022, China)

Abstract: Based on the analysis of the existing equipment, CO_2 emission and composition of sintering process in Bayi Steel and the research of new low CO_2 sintering technology at home and abroad. The utilization technology of carbide slag should be improved continuously in sintering process. Increasing the dosage of calcium carbide slag in sintering and desulfurization. It can give priority to the implementation of sintering machine trolley widening technology, new ring cooler technology and frequency conversion technology. Selective implementation of hydrogen rich gas fuel injection technology, flue gas circulation technology, surface spraying steam technology. Research and track the microwave sintering technology, biomass energy sintering technology and other cutting-edge low-carbon sintering technology. Reducing greenhouse gas emission in sintering process.

Key words: greenhouse gas; sinter; CO_2 emission intensity; trolley widening; flue gas recirculation; microwave sintering; biomass

添加炼钢除尘灰对赤磁混合铁精矿生球性能的影响

王代军[1,2]，贺万才[1,2]

（1. 北京首钢国际工程技术有限公司，北京　100043；
2. 北京市冶金三维仿真设计工程技术研究中心，北京　100043）

摘　要：为确定添加炼钢尘对赤磁混合铁精矿造球的合适工艺参数，研究用磁铁矿 1、2、4 和球团尘均属于弱成球性矿种，磁铁矿 3 属于中等成球性矿种，炼钢尘成球性能非常好，比较容易成球。依据球团生产原料供应情况及配矿方案，试验确定适宜的成球参数为：膨润土用量为 0.9%，混合料水分为 9.0%，造球过程中生球的水分为 8.5%，造球时间为 10min，造球盘转速为 22r/min，对应的生球落下强度、抗压强度及爆裂温度满足 GB50491—2009 标准要求。

关键词：炼钢尘；赤铁矿；落下强度；抗压强度；爆裂温度

Effect of Adding Steelmaking Dust on Green Pellets Properties of He-magnetic Mixed Iron Concentrate

Wang Daijun[1,2], He Wancai[1,2]

(1. Beijing ShouGang International Engineering Technology Co., Ltd., Beijing 100043, China;
2. Beijing Metallurgy Three Dimensional Simulation Design Engineering Technology Research center, Beijing 100043, China)

Abstract: In order to determine the appropriate process parameters of adding steel-making dust to he-magnetic mixed iron concentrate, the magnetite 1#, 2#, 4# and the pellet dust are belong to weak pelletizing minerals, the magnetite 3# belongs to medium pelletizing minerals. The pelletizing performance of steel-making dust is very good, and it is easy to pelletize. According to the raw material supply and the ore blending scheme of pelletizing production, the suitable pelletizing parameters are determined as follows: the dosage of bentonite is 0.9%, the moisture content of mixture is 9.0%, the moisture content of green pellet in pelletizing process is 8.5%, the pelletizing time is 10min, and the rotating speed of pelletizing plate is 22rpm. The drop strength, compressive strength and burst temperature corresponding green pellets are met the requirements of GB 50491—2009 standard.

Key words: steelmaking dust; hematite; drop strength; compressive strength; burst temperature

2.3 炼　铁

火花放电直读光谱法测定生铁样品分析时间的研究

李青青，陈　俊，沈　克

（武钢有限质检中心，湖北武汉　430080）

摘　要：采用火花放电直读光谱法分析炼钢生铁样品，研究各阶段时间对试样分析的影响，探讨直读光谱法分析生铁的分析时间，以此来更准确、快速、稳定地为钢厂生产服务。结果表明，当冲洗时间为 3s、第一次预燃时间为 4s、第二次预燃时间为 13s、曝光时间为 5s 时，光强曲线最稳定，在此分析时间下分析生铁样品的准确度和精密度良好，能很好地满足实际生产需要。

关键词：直读光谱法；生铁；分析时间

Study On Determination Of Analysis Time Of Iron Sample By Spark Discharge Direct Reading Spectrometry

Li Qingqing, Chen Jun, Shen Ke

(Quality Inspection Center of Wuhan Iron and Steel Group Company, Wuhan 430080, China)

Abstract: The influence of various stages of time on sample analysis and the analysis time of iron were analyzed by direct reading spectrometry, which was studied by Spark discharge direct reading spectrometry to provide more accurate, rapid and stable production services for steelworks. The results indicate that the most stable light intensity curve when the flushing time is 3 seconds, the first preburn time is 4 seconds, the second preburn time is 13 seconds and the exposure time is 5 seconds, which can meet the requiements of actual production.

Key words: direct reading spectrometry; iron; analysis time

Through the Looking Glass: Comparison of Blast Furnace Operational Practices in China and Europe

Geerdes Maarten[1], Sha Yongzhi[2]

(1. Geerdes Advies, The Netherlands; 2. China Iron and Steel Research Institute, Beijing 100081, China)

Abstract: A comparison is made of blast furnace operation practices in China and Europe and North America from benchmark data and site visits. Productivity in China is generally higher than in Europe, but in Europe furnaces are operated at lower coke rates. The data are analyzed from the perspective that the most critical step in blast furnace operation is the melting of the ferrous burden. Melting is more difficult at higher primary slag basicity (as in China) and can be optimized by using burden distribution tools as is done in Europe.

Key words: blast furnace; coke rate; coal injection; slag basicity; burden distribution; productivity; melting of ferrous burden

MnSiN 和 Si_3N_4 用于铁水增氮的实验研究

任　伟，国泉峰，王小强，张　伟，王　亮，韩子文，朱建伟

（鞍钢股份有限公司 技术中心，辽宁鞍山　114021）

摘　要：为探索铁水增氮的可行性，实验室范围内在 1400~1600℃向铁水中添加 0~10%增氮剂 Si_3N_4 和 MnSiN，结果表明：该 2 种增氮剂铁水增氮效果较差，铁水中[N]含量均不超过 0.01%，且表现为无规律性。分析认为氮在铁水中溶解度低，增氮剂比重低且动力学条件不充分是主要原因，不建议以炉料方式进入高炉并增氮。采用预埋的方式向铁水中添加 Si_3N_4 和 Fe_2Ti，当增氮剂 Si_3N_4 和 Fe_2Ti 的质量比为 1∶2，温度为 1500℃，在 1h 内可形成含 TiN 为 10%的高温不熔物。

关键词：增氮；铁水；钛铁；TiN

Experimental Study on Hot Metal Nitrogen Addition by MnSiN and Si_3N_4

Ren Wei, Guo Quanfeng, Wang Xiaoqiang, Zhang Wei, Wang Liang, Han Ziwen, Zhu Jianwei

(Angang Steel Company limited, Anshan 114021, China)

Abstract: In order to explore the feasibility of increasing nitrogen content of hot meal for blast furnace, the hot metal nitrogen addition by Si_3N_4 and MnSiN has been conducted at the temperature from 1400℃ to 1600℃ in lab scale, and the nitrogen content of hot metal sample was no more than 0.01 percent by this two nitrogen reagents, which also displayed no principle with time and reagents addition. It was thought that the low nitrogen solution and density played a important role after analyse, and the thermodynamic condition was another factor to influence the hot metal nitrogen addition, so it was not recommend to add these tow reagents directly. Finally, the TiN content can reach 10 percent from the solid powered resident by embedded Si_3N_4 and Fe_2Ti in hot metal at 1500℃ with weight proportion 1∶2, and this plan was supposed to be a good way to generate TiN.

Key words: nitrogen addition; hot metal; Fe_2Ti; TiN

鞍钢 7 号高炉 96h 休风恢复实践

范崇强，杨维元，王啸男，白天野

（鞍钢股份有限公司，辽宁鞍山　114000）

摘　要：鞍钢 7 号高炉是炉容为 2580m³ 的一座现代化高炉。2021 年 2 月 9 日接总厂指令，由于雾霾原因限产休风 96h。送风恢复顺利，参数快速恢复正常。但由于此次休风特殊原因，高炉无充足时间严格按照配料计算上完休风料，同时由于休风前冷却壁破损漏水、休风后焦炭质量变差等原因，炉缸热储备不足，炉况出现一段时间的波动。本文对此次休风过程进行总结，为以后的操作提供借鉴。

关键词：休风；复风；焦炭负荷

Blowing-on Practice of No.7 Blast Furnace in Ansteel

Fan Chongqiang, Yang Weiyuan, Wang Xiaonan, Bai Tianye

(Anshan Iron & Steel Co. of Angang Steel Co., Ltd., Anshan 114000, China)

Abstract: AnGang No.7 blast furnace is a modern blast furnace with capacity of 2580 m³. On February 9, 2021, it was ordered to limit production and blowing down for 96 hours. The air supply recovered smoothly and the parameters quickly returned to normal. However, due to the special reasons of damping, the blast furnace didn't have sufficient time to calculate the rest material strictly according to the batching. At the same time, due to the damage and leakage of the cooling stave before the damping and the poor quality of coke after the damping, the hearth hot reserve was insufficient, and the furnace condition fluctuated for a period of time. This paper summarizes the process of the damping, so as to provide reference for the future operation.

Key words: blowing-down; blowing-on; coke burden

鞍钢球团比例对料面形状的影响

姜　喆[1]，张　磊[2]，朱建伟[1]，赵长城[2]，曾　宇[2]，车玉满[1]

（1. 鞍钢股份有限公司技术中心，辽宁鞍山　114009；
2. 鞍钢股份有限公司炼铁总厂，辽宁鞍山　114021）

摘　要：本文结合鞍钢 7 号高炉炉料组成和布料制度，建立了无料钟高炉布料模型，并计算了球团比例增加后 7 号高炉炉喉处的料面形状和径向矿焦比。认为随着球团比例的增加，炉喉中心无矿区面积减小，而边缘 O/C 也相应减小。为保证炉内煤气流合理分布，应适当增加中心加焦比例或缩小批重。

关键词：高炉；布料模型；料面形状；中心加焦；批重

Effect of High Pellet Ratio on Burden Layer Profile

Jiang Zhe[1], Zhang Lei[2], Zhu Jianwei[1], Zhao Changcheng[2],
Zeng Yu[2], Che Yuman[1]

(1. Technology Center of Angang Steel Co., Ltd., Anshan 114009, China;
2. General Ironmaking Plant of Angang Steel Co., Ltd., Anshan 114021, China)

Abstract: Based on the burden composition and charging system of Ansteel, the charging model of bell-less top blast furnace was established, and Burden surface profile together with radial distribution of O/C of 7#BF at Ansteel was also calculated as the pellet ratio increasing. The results show that the central ore-free area and the O/C at the edge of throat increased as the pellet ratio going up. In order to ensure reasonable distribution of gas flow, measures such as increasing central coke and decreasing batch could be adopted.

Key words: BF; charging model; burden surface profile; central cook; batch

鞍钢高炉炉缸快速修复技术应用

符显斌，张延辉，李建军，赵长城

（鞍钢股份有限公司炼铁总厂，辽宁鞍山 114000）

摘 要：为彻底解决炉缸严重侵蚀带来的安全隐患，鞍钢7号高炉进行了炉缸浇注修复施工。修复施工包括放残铁、炉缸清理及炉缸浇注及修复后开炉送风恢复等环节。基于充分的可行性评估，施工前可靠的侵蚀计算和现场测量，以及成熟的炉缸浇注技术，此次炉缸浇注修复达到了预期目标。完成修复后，高炉各项技术经济指标迅速得到强化，达到了可靠、经济地延长高炉炉缸寿命的预期目标。

关键词：高炉；炉缸浇注；修复；放残铁

Practice of Hearth Pouring Repair of No. 7 Blast Furnace in Ansteel

Fu Xianbin, Zhang Yanhui, Li Jianjun, Zhao Changcheng

(Ansteel, Anshan 14000, China)

Abstract: In order to completely solve the potential safety hazard caused by serious erosion of hearth, the hearth pouring repair construction was carried out for No. 7 blast furnace of Ansteel. The repair construction includes discharging the residual iron, cleaning the hearth, pouring the hearth and restoring the air supply after opening the furnace after repair. Based on sufficient feasibility assessment, reliable erosion calculation and field measurement before construction, and mature hearth pouring technology, the hearth pouring repair has achieved the expected goal. After the repair, the technical and economic indexes of the blast furnace have been strengthened rapidly, and the expected goal of reliably and economically prolonging the service life of the blast furnace hearth has been achieved.

Key words: blast furnace; hearth pouring; repair; residual iron

鞍钢冶金粉尘短流程清洁再利用技术

刘德军

（鞍钢集团钢铁研究院，辽宁鞍山 114009）

摘 要：通过对当前冶金除尘灰各种再利用方式的特点进行详细分析与甄别，阐明了“除尘灰与粉煤同喷”这种与“熔融还原”巧妙结合利用方式的科学性；明确了冶金粉尘再资源化科学利用的重要性；同时就冶金粉尘高炉综合喷吹技术对冶金粉尘的种类及量等因素的普适性和最佳性、含铁除尘灰等冷料下料量的工程级精准配料、高炉喷吹含铁除尘灰回旋区热补偿实现手段、消除喷吹除尘灰输送管道易堵塞技术以及整个工艺过程无二次扬尘绿色技术等进行了详细的论述；分析了冶金粉尘高炉综合喷吹技术在鞍钢鲅鱼圈炼铁部的应用状况。

关键词： 冶金除尘灰；科学利用；再资源化；集成技术

Cleaning and Recycling Technology of Metallurgical Dust in Short Process at Ansteel

Liu Dejun

(Iron & Steel Research Institutes of Ansteel Group Corporation, Anshan 114009, China)

Abstract: It's stated that it's important to utilize scientifically metallurgical dusts recycling. After analyzing and distinguishing features of different kinds of metallurgical dusts, we concluded that it's scientificity of the technology that injecting metallurgical dust and pulverized coal into burst furnace together. At the same time, it's expounded on (1) universality and optimality, about the technology to the kinds and amount of dusts injected into BF, are confirmed; (2) the accurately mixture in engineering of the materials such as dusts with iron injected into BF; (3) thermal compensation of dusts with iron in the blast furnace raceway; (4) avoidance of blocking the tunnel as injecting into the BF and (5) the green environment technology without second generate dust in the whole process. This paper introduces the influence of the application of the metallurgical dust blast furnace comprehensive injection technology to the BF in the of Ansteel Ba Yu Quan Iron-making Department.

Key words: metallurgical dust; science utilization; recycling; integration technology

鞍钢单一低热值高炉煤气下的高风温技术

刘德军

（鞍钢集团钢铁研究院，辽宁鞍山 114009）

摘　要： 重点介绍了鞍钢高炉高风温技术的进步。核心就鞍钢热风炉长期使用单一低热值煤气烧炉的特点，介绍了鞍钢梯次实施的热风炉结构形式的改造，和热风炉自预热、前置炉的板换改造及辅助热风炉等根本性改造；继而开展了针对热风炉的板换替代管换实施双预热、送风换炉技术优化、富氧烧炉、复合涂料的使用、送风系统关键部位预制预警技术等多项综合节能技术的研究与应用，实现了热风温度的大幅提高和热风炉烧炉煤气消耗的大幅降低，取得了良好的效果，极大地推动了鞍钢高炉热风炉技术的进步。

关键词： 高炉；热风炉；技术；低成本

High Hot Air Temperature Technology under Single Low Calorific Value BF Gas at Anshan Steel

Liu Dejun

(Iron & Steel Research Institutes of Ansteel Group Corporation, Anshan 114009, China)

Abstract: The progress of high blast temperature and corresponding energy saving technology in Anshan iron and steel co.

In this paper, the features of low-calorific value gas fired furnace in Anshan iron and steel co, This paper introduces the structural transformation of hot blast furnace implemented, and self-preheating, front furnace and auxiliary heating furnace and other fundamental transformation by steps in Anshan iron and steel group co. Then conducted for hot blast stove plate for replacement tube in the implementation of double preheating, the air distribution in the furnace technology optimization, the rich oxygen burning furnace, the use of the composite coating, the air supply system key parts prefabrication early warning technology and so on many research and application of comprehensive energy saving technology, realize the large increase in the hot blast temperature and hot blast stove burning furnace gas consumption greatly reduced, and achieved good effect, greatly promoted the Anshan iron and steel blast furnace hot blast stove technology progress.

Key words: blast furnace; hot blast stove; technology; lowcost

鞍钢鲅鱼圈 4038m^3 高炉大修出残铁操作实践

李伟伟，蒋　益，姜彦冰，张　南，董建兴，滕雪亮

（鞍钢股份有限公司鲅鱼圈钢铁分公司，辽宁营口　115007）

摘　要：本文阐述了鞍钢鲅鱼圈 1#高炉（第一代）大修停炉出残铁操作的过程以及经验总结。本次出残铁操作，为鞍钢首次 4000m^3 高炉出残铁，从出残铁方案的审定，残铁口位置的确定，残铁沟与残铁坑的设计与制作均进行了严密的论证。从操作实践结果看，放残铁效果较好，炉缸残留渣铁较少，放残铁操作圆满成功，为高炉大修工程顺利进行奠定了基础。

关键词：出残体；大修；大高炉

Practice of Emptying the Residual Molten Iron before the Overhaul of Bayuquan No. 1 Blast Furnace of Anshan Steel

Li Weiwei, Jiang Yi, Jiang Yanbing, Zhang Nan, Dong Jianxing, Teng Xueliang

(Bayuquan Branch of Anshan Iron and Steel Co., Ltd., Yingkou 115007, China)

Abstract: This article described the process and summed up the experience of emptying the residual molten iron before the overhaul of Bayuquan No.1 Blast Furnace (First generation) of Anshan Steel. This work of emptying residual molten iron which is the first time of 4000m^3 blast furnace in Anshan Steel has been rigorously argued previously, including the plan approval, the determination of the position of residual iron mouth and the design and manufacture of residual iron ditch and pit. Practice showed that the work is so successful that there were little iron and slag left in the hearth after emptying, which laid a solid foundation for the safety and smooth of blast furnace overhaul.

Key words: emptying the residual molten iron; overhaul; large blast furnace

大型高炉用无水炮泥的制样方法及性能研究

钟　凯，崔园园，温太阳，杨　彬

（首钢集团有限公司技术研究院钢铁技术研究所，北京　100043）

摘　要： 为了较准确地测定无水炮泥的物理性能，设计开发了一套适用于无水炮泥试样的制备方法，并选取了某大型高炉使用中的两种无水炮泥，通过对炮泥进行加热和成型，获得带有模具的试样坯；将带模试样坯在-15℃的温度下冷冻 2h 时间后，进行脱模，获得炮泥试样。制备的炮泥试样外观尺寸稳定，无鼓胀问题。测量了两种炮泥试样在 300℃热处理后的体积密度、耐压抗折强度和线变化率等物理性能并探讨了炮泥的高温抗折强度和与高炉出铁口深度等性能的关系。

关键词： 无水炮泥；高炉；制样；高温抗折；铁口深度

Study on Sample Preparation Method and Properties of Anhydrous Taphole Mixes for Large Blast Furnace

Zhong Kai, Cui Yuanyuan, Wen Taiyang, Yang Bin

(Institute of Iron and Steel Technology, Technology Research Institute of Shougang Group Co., Ltd., Beijing 100043, China)

Abstract: In order to accurately determine the physical properties of anhydrous taphole mixes, a set of preparation methods for anhydrous taphole mixes samples were designed and developed. Two kinds of anhydrous taphole mixes used in a large blast furnace were selected, and the sample billet with mold was obtained by heating and forming the anhydrous taphole mixes; The sample with mould is placed in -15℃ for 2 hours, demoulding was carried out to obtain anhydrous taphole mixes samples. The appearance and size of the sample are stable without bulging. The bulk density, compressive strength and linear change rate of two kinds of anhydrous taphole mixes samples after heat treatment at 300℃ were measured, and the relationship between the high temperature flexural strength and the taphole depth of blast furnace was discussed.

Key words: anhydrous taphole mixes; blast furnace; sample preparation; high temperature bending strength; taphole depth

鞍钢朝阳钢铁高炉 2600m^3 高炉强化冶炼生产实践

胡德顺，赵正洪，李泽安，吕宝栋，刘金存，金　瑶

（鞍钢集团朝阳钢铁有限公司，辽宁朝阳　122000）

摘　要： 朝阳钢铁主要通过是优化高炉装料制度、提煤降焦、建立赶超目标、优化炉前出铁等提产降耗措施的实施，2019 年下半年高炉技经指标高炉燃料比稳定在 530kg/Tf 左右，高炉煤比实现 150kg/TF 以上，高炉利用系数稳定在 2.30t/dm^3 以上。

关键词： 装料制度；提煤降焦；利用系数

Practice of Increasing Production and Reducing Consumption of Blast Furnace in Chaoyang Iron and Steel

Hu Deshun, Zhao Zhenghong, Li Zean, Lü Baodong, Liu Jincun, Jin Yao

(Anshan Iron and Steel Group Chaoyang iron and Steel Co., Ltd., Chaoyang 122000, China)

Abstract: Chaoyang iron and steel mainly through the optimization of blast furnace charging system, raising coal and reducing coke, establishment of catch-up objectives, optimization of the front of the furnace iron and other measures to increase production and reduce consumption. In the second half of 2019, the technical and economic index of blast furnace fuel ratio is stable at about 530kg/TF, the coal ratio of blast furnace is more than 150kg/TFe, and the utilization coefficient of blast furnace is stable at more than 2.30t/dm^3.

Key words: charging system; coal raising and coke reduction; utilization coefficientt

鞍钢朝阳钢铁炼铁工序低成本理念生产实践

赵正洪，胡德顺，李泽安，吕宝栋，刘金存，金　瑶

（鞍钢集团朝阳钢铁有限公司，辽宁朝阳　122000）

摘　要：朝阳钢铁通过双跑赢理念在铁前低成本战略的实施，使各部门建立清晰的目标管理体系，明确了铁前低本运行战略的关键因素，有效地提高采购计划及时性和准确性，促进铁前工艺操作标准的实施及高炉原燃料底线的制定，推进铁前技经指标稳步提升，保证了低成本战略持续发扬光大。

关键词：铁前成本；双跑赢；控制；管理

The Control and Practice of Low Cost Concept in the Ironmaking Process of Chaoyang Iron and Steel

Zhao Zhenghong, Hu Deshun, Li Zean, Lü Baodong, Liu Jincun, Jin Yao

(Anshan Iron and Steel Group Chaoyang Iron and Steel Co., Ltd., Chaoyang 122000, China)

Abstract: Chaoyang iron and Steel Co., Ltd. has established a clear target management system through the implementation of the low-cost strategy in front of the iron, clarified the key factors of the low-cost operation strategy in front of the iron, effectively improved the timeliness and accuracy of the procurement plan, promoted the implementation of the pre iron process operation standard and the formulation of the bottom line of raw materials and fuels for the blast furnace, promoted the steady improvement of the pre iron technical and economic indicators, and ensured low success This strategy continues to be brilliant.

Key words: pre iron cost; double win; control; management

煤岩性质对煤粉燃烧行为的影响研究

张　华，张生富，况雨岑，锁广胜，温良英

（重庆大学材料科学与工程学院，重庆　400044）

摘　要：煤岩分析方法作为重要的煤质分析手段，已广泛应用于煤岩配煤炼焦，但在高炉喷吹煤粉的燃烧行为方面研究甚少。本文通过单种煤、混合煤的热分析实验，研究了煤岩显微组分及其燃烧行为的关系。结果表明，神华煤

的镜质组含量及平均最大反射率均低于永城煤，可燃性和燃尽性优于永城煤，说明镜质组含量高的煤燃烧性能更差；混煤燃烧性能随神华煤配比的增加而提高，失重速率及放热速率曲线由高温区向低温区移动，说明神华煤所含煤岩组分的燃烧可促进永城煤的燃烧；动力学结果表明煤粉燃烧前后期分别符合二级、一级反应机理，且混煤燃烧过程神华煤对永城煤的促进主要体现在燃烧反应后期。研究成果可为优化高炉喷煤种类和配比以及提升喷煤比提供理论参考。

关键词：喷吹煤粉；煤岩；热分析；燃烧行为；动力学

Study on the Impacts of Coal Petrography on the Combustion Behavior of Coal

Zhang Hua, Zhang Shengfu, Kuang Yucen, Suo Guangsheng, Wen Liangying

(College of Materials Science and Engineering, Chongqing University, Chongqing 400044, China)

Abstract: As a vital means of coal quality analysis, coal petrography analysis method has been widely used in coal blending coking, however, there are few research on the combustion behavior of pulverized coal injected (PCI) into blast furnace. In this paper, the relationship between coal petrographic composition and combustion behavior of PCI was studied by means of the thermal analysis experiment of single coal, mixed coal. The results show that the average maximum reflectance and content of vitrinite of shenhua coal are lower than those of yongcheng coal, the flammability and burnout of shenhua coal are better than that of yongcheng coal, which indicates that the combustion performance of coal with high vitrinite content is worse; The combustion performance of blended coal increases with the increase of shenhua coal ratio, and the curve of weight loss rate and heat release rate moves from high temperature zone to low temperature zone, indicating the combustion of coal petrographic components in shenhua coal can promote the combustion of coal macerals contained in yongcheng coal; The kinetic results show that the second-order and the first-order reaction mechanisms are followed in the early and late stages of pulverized coal combustion, the promotion of shenhua coal to yongcheng coal is mainly reflected in the later stage. The research results can provide a theoretical reference for optimizing the type and proportion of coal injection and improving the coal injection ratio.

Key words: pulverized coal injection; petrology; thermal analysis; combustion behavior; kinetic

鞍钢7号高炉炉缸浇注修复实践

谢明辉，车玉满，李　仲，郭天永，姜　喆，姚　硕，邵思维

（鞍钢股份有限公司，辽宁鞍山　114009）

摘　要：鞍钢7号高炉进入炉役末期以后，炉缸、炉底侵蚀严重。为了保证安全生产，高炉大幅度降低冶炼强度，并使用钛矿护炉，导致高炉效率降低、成本升高。针对此问题，利用高炉中修采用炉缸浇注技术修复炉缸内衬，消除安全隐患，使高炉效率提升、成本降低。

关键词：高炉；炉缸浇注

Practice of Pouring and Repairing Hearth of NO.7 BF in Ansteel

Xie Minghui, Che Yuman, Li Zhong, Guo Tianyong, Jiang Zhe, Yao Shuo, Shao Siwei

(Anshan Iron and Steel Co., Ltd., Anshan 114009, China)

Abstract: The No.7 BF at ANSTEEL is at the later stage of campaign and its hearth has been eroded seriously. Several measures, such as decreasing smelting intensity and using vanadium-titanomagnetite for furnace protection, have been introduced to ensure safe production, which also lead to reduction of smelting efficiency and increase of cost. In the light of the problem, hearth pouring has been carried out during medium pritod to repair hearth lining, eliminate the potential security risks, enhance smelting efficiency and decrease cost.

Key words: blast furnace; hearth pouring

焦炭抗碱度研究分析

孙宝芳，吴艺鹏，何 波，宋 辉

（青岛特殊钢铁有限公司，山东青岛 266200）

摘 要：焦炭各项性能指标的好坏对高炉顺行的影响是至关重要的，现阶段，热态性能已被认为是大高炉用焦炭最重要的考核指标之一。为此，本文研究了青钢焦炭分别在 1100℃和 1200℃下的气化反应特性。同时，由于碱金属在高炉内循环富集日益加剧，碱金属对焦炭气化反应有明显的催化作用，因此，当前十分有必要对青钢焦炭在富碱条件下的劣化行为进行研究。

关键词：焦炭；碱金属；高炉炼铁

Study on Alkali Resistance of Coke

Sun Baofang, Wu Yipeng, He Bo, Song Hui

(Qingdao Special Steel Co., Ltd., Qingdao 266200, Cjina)

Abstract: The influence of each performance index of coke on BF is very important. Now, Hot state performance has been considered as one of the most important indexes of coke used in large blast furnace. Therefore, the gasification characteristics of coke at 1100 and 1200 DEG C were studied.Besides as the alkali is enriched in the BF, the alkali play an important catalytic role in the gasification of coke.Therefore, it becomes necessary to study the degradation of coke under the condition of rich alkali.

Key words: coke; alkali; BF

红钢 3#高炉开炉达产实践

查海超，陈　礁，凡则松，杨永刚，洪佳勇，赵树逵

（红河钢铁有限公司，云南蒙自　661111）

摘　要： 红钢 3#高炉（$1350m^3$）开炉达产总结，通过开炉前精心周密的工作准备，合理恰当选择开炉用料配比及开炉参数，点火复产后及时组织渣铁排放，准确把握风口打开时机，准确选择合理负荷。避免持续高硅，快速喷煤富氧，实现了开炉后的快速达产。

关键词： 高炉；烘炉；开炉；达产

Practice of Hong gang 3# Blast Furnace's Start-up and Production

Zha Haichao, Chen Jiao, Fan Zesong,
Yang Yonggang, Hong Jiayong, Zhao Shukui

(Ironmaking plant of Honghe Iron and Steel Co., Ltd., Mengzi 661111, China)

Abstract: Summary of Hong gang 3# blast furnace startup and production, through carefully work preparation before opening the furnace, reasonable selection of material ratio and parameters for start-up, timely organize the discharge of slag iron after ignition and resumption of production, accurately grasp the opening time of blast furnace tuyere and selecting reasonable load. Avoiding continuous high silicon and rapid coal injection and oxygen enrichment, and finally, the rapid production after furnace opening is realized.

Key words: blast furnace; oven; blowing in; reach production

废钢和反应气氛对含铬型钒钛磁铁矿软熔滴落性能的影响

宋翰林，张金鹏，刘建兴，程功金，薛向欣

（东北大学冶金学院，辽宁省冶金资源循环科学重点实验室，东北大学钒钛产业技术创新研究院，辽西地区钒钛磁铁矿资源综合利用产业创新研究院，辽宁朝阳　122000）

摘　要： 废钢作为含铁量高杂质少的优质固废之一，在钢铁生产中的重新利用可以有效缓解资源紧张，降低对优质铁矿石的对外依存度。此外，由于含铬型钒钛磁铁矿含有过量的 TiO_2（>2%），在软熔带中被还原为具有高熔点的碳氮化钛，继而出现炉渣稠化、泡沫渣、渣铁难分等严重危害高炉顺行的问题。因此，通过配加废钢来代替过去配加优质铁矿石的方法以降低 TiO_2 的比例，从而改善钒钛高炉的生产状况。本研究通过改变废钢的配加比例和反应气氛组成，探索了不同反应气氛和废钢比对含铬型钒钛磁铁矿的软熔滴落性能和有价金属元素迁移的影响。随着废钢比从 2%增加到 10%，钒钛矿的软化温度 T_4 和 T_{40} 变化较为稳定，有利于改善块状区的透气性并保持气固反应的

稳定性。此外，当反应气氛中掺入CO_2时，也有助于改善钒钛矿的软熔滴落性能。根据实验后渣样的成分分析和微观图像，适当的废钢比可以降低渣中氮化物的比例，尤其是碳氮化钛的存在，这可以认为是改善炉渣性能和软熔滴落性能的原因之一。此外，由于CO_2具有一定的弱氧化性，改变了炉腹内的还原氧势，同时N_2的含量的降低，这也有利于抑制碳氮化钛的生成，从而进一步改善了炉渣性能和软熔滴落性能。

关键词：废钢；含铬型钒钛磁铁矿；软熔滴落性能；反应气氛；碳氮化钛

Effect of Scrap Steel and Reaction Atmosphere on Softening-elting-ripping Performance of Chromia-bearing Vanadia-Titania Magnetite

Song Hanlin, Zhang Jinpeng, Liu Jianxing, Cheng Gongjin, Xue Xiangxin

(School of Metallurgy, Northeastern University, Liaoning Key Laboratory of Recycling Science for Metallurgical Resources, Innovation Research Institute of Vanadium and Titanium Resource Industry Technology, Northeastern University, Innovation Research Institute of Comprehensive Utilization Technology for Vanadium-Titanium Magnetite Resources in Liaoxi District, Chaoyang 122000, China)

Abstract: As one of the high-quality solid wastes with high iron content and low impurities, scrap steel can be reused in steel production to effectively alleviate the shortage of resources and reduce the degree of dependence on high-quality iron ore. In addition, because the chromia-bearing vanadia-titania magnetite contains excessive TiO_2 (>2%), it is reduced to titanium carbonitride with a high melting point in the reflow zone, and then slag thickening, foaming slag, and slag iron are difficult to occur. In this study, by changing the proportion of scrap steel and the composition of the reaction atmosphere, the effects of different reaction atmosphere and scrap ratio on the softening-melting-dripping performance and the migration of valuable metal elements of chromia-bearing vanadia-titania magnetite were explored. As the scrap ratio increases from 2% to 10%, the changes in the softening temperature T_4 and T_{40} are relatively stable, which is conducive to improving the gas permeability. According to the composition analysis and microscopic images of the slag sample after the experiment, an appropriate scrap ratio can reduce the proportion of nitrides in the slag. In particular, the presence of titanium carbonitride can be considered as one of the reasons for the improvement of slag performance and softening-melting-dripping performance. In addition, CO_2 changes the reducing oxygen potential in the furnace belly. At the same time, the decrease in the content of N_2 is also beneficial to inhibit the formation of titanium carbonitride, thereby further improving the slag performance and softening-melting-dripping performance.

Key words: scrap steel; chromia-bearing vanadia-titania magnetite; softening-melting-dripping performance; reaction atmosphere; titanium carbonitride

块矿-烧结矿混合炉料冶金性能及温室气体排放研究

Doostmohammadi Hamid[1]，Hoque Mohammad[1]，Chibwe Deside[1]，刘新亮[2]，O'Dea Damien[2]，Honeyands Tom[1]

（1. 澳大利亚纽卡斯尔大学，新南威尔士州 2287；2. 必和必拓 铁矿石市场部，上海 200021）

摘 要：本文研究了0~40%块矿配比下块矿、烧结矿二者混合炉料的冶金性能，烧结矿碱度根据块矿比例调整以维持混合炉料碱度恒定；熔滴试验结果表明混合炉料软熔行为与具有相同碱度的单一烧结矿炉料较为相似。

同时，本文采用二维高炉模型模拟了不同炉料的温室气体排放，其结果表明，提升块矿比例、降低烧结矿使用量，有助于减少烧结过程二氧化碳排放总量，但由于高块矿比炉料烧结矿碱度升高，其吨矿烧结矿二氧化碳排放强度有所升高；此外，由于块矿还原性弱于烧结矿，高炉燃料比略有升高；综合而言，提高块矿比例至40%后，吨铁二氧化碳排放强度降低约 141kg CO_2。本文同时采用 DEM 模拟考察了烧结矿、块矿混合炉料布料模式优化，结果表明，在上料皮带上烧结矿和块矿分层布料、并采用并罐式布料方式可获得最有混匀效果。

关键词：块矿；混合炉料；温室气体；布料模式；交互作用

Lump-Sinter Mixed Burden Operation-Softening and Melting Results, Practical Mixing and Greenhouse Gas Emissions

Doostmohammadi Hamid [1], Hoque Mohammad[1], Chibwe Deside[1],
Liu Xinliang[2], O'Dea Damien[2], Honeyands Tom[1]

(1. University of Newcastle Australia, NSW 2287;
2. BHP, Iron Ore Marketing Strategy and Technical, Shanghai 200021, China)

Abstract: In this study, a range of sinter-lump mixed burdens were created ranging from 0 to 40% lump. In each case the basicity of the sinter was varied so that the overall mixed burden basicity was constant. Even at 40% lump ore, the mixed burden softening and melting results were found to be comparable to sinter only burden.

The overall greenhouse gas emissions from the process were estimated using a two stage heat and mass balance model. The primary result of increasing the lump percentage is to decrease the requirement for sinter, and the associated sintering greenhouse gas emissions. However, the sinter basicity increases as the lump percentage increases, leading to higher emission per tonne of sinter. The reducibility of the lump ore is also lower than sinter, affecting blast furnace fuel rates. The overall effect was to decrease greenhouse gas emissions by approximately 141kg/t hot metal as lump percentage increased. A discrete element method (DEM) model was used to assess practical methods to mix sinter and lump burdens prior to charging to the blast furnace. Layering of sinter on top of lump on the conveyor belt feeding a parallel hopper system was found to be the most effective at mixing the burden.

These results empower blast furnace operations to use more lump ore and promise more sustainable ironmaking operations through optimised burden design.

Key words: lump; mixed burden; greenhouse gas; charging; interaction

武钢 $2600m^3$ 高炉强化冶炼与节能降耗

帅　照，徐　伟

（武汉钢铁有限公司，湖北武汉　430080）

摘　要：本文对武钢 4 号高炉（$2600m^3$）强化冶炼与节能降耗工作进行了总结。通过提高原燃料质量、优化上下部制度、控制操作参数、炉前渣铁管理、强化设备检修维护等措施改善料柱透气性和炉缸活跃性、维护合理操作炉型，实现长周期稳定顺行，在提高利用系数、降低燃料比方面取得了较大的进步。2020 年 4 号高炉有效容积利用系数达 2.607t/(m^3·d)，燃料比降至 511.5kg/t，均创开炉以来最好水平。

关键词：高炉；操作制度；强化冶炼；节能降耗

Strengthening Smelting and Energy Saving of 2600m^3 Blast Furnace in WISCO

Shuai Zhao, Xu Wei

(Wuhan Iron & Steel Co., Ltd., WuHan 430080, China)

Abstract: The work on the strengthening smelting and saving energy and reducing consumption of No.4 blast furnace (2600m^3) in WISCO is summarized. Through improving the quality of raw materials, optimizing the operating regulations of upper and lower parts, controlling the smelting parameters, managing the slag and iron tapping regulations and strengthening the maintenance of equipment, so as to improve the permeability of the burden and the activity of the hearth. These measures help to maintain a reasonable operation status of blast furnace, realize a long term smooth smelting, and make a great progress in improving the capacity of blast furnace and reducing the fuel consumption. In 2020, the effective volume utilization coefficient of No.4 blast furnace reached 2.607 t/(m^3·d), and the fuel ratio decreased to 511.5kg/t, both of which were the best indexes since its running.

Key words: blast furnace; operating regulations; strengthening smelting; saving energy and reducing consumption

武钢 4 号高炉长期稳定顺行操作

仉翼鹏，杨亚魁

（武汉钢铁有限公司，湖北武汉 430083）

摘 要：本文对高炉长期稳定顺行及产量创新高的经验进行了总结：在原料管理方面，从“稳”和“净”上入手，对不同品种的焦炭进行定槽管理，减小焦炭品种波动对炉况的影响；在筛分环节，建立了新的原料检查机制，既有助于操作者第一时间了解焦炭情况，也最大限度地减少了焦粉入炉；把入炉锌负荷纳入日常管控，当烧结矿中锌含量超过 0.02%时进行预警并及时干预；下部调剂主要是采取缩小进风面积，并辅以精准的上部调剂，风量维持在 5200~5400Nm^3/min，煤气利用达到 47%左右，渣皮动态平衡，炉型相对稳定。

关键词：高炉；焦炭；稳定顺行；进风面积

Operation of Long-time Stable and Smooth in No. 4 Blast Furnace of WISCO

Zhang Yipeng, Yang Yakui

(Wuhan Iron and Steel Co., Ltd., Wuhan 430083, China)

Abstract: This article summarizes the experience of long-term stability and high output of blast furnace. Starting from ‘stable’ and ‘clean’ in terms of raw material management, the management of different types coke is carried out to reduce the fluctuation of coke varieties. In the screening process, a new raw material inspection mechanism has been established, which not only helps operators get the information of the coke situation in the first time, but also reduces the entrance of coke powder into the blast furnace. The load of zinc element entering the furnace is tested in daily control, necessary measures are taken when the zinc content in the sinter exceeds 0.02%. Though the the tuyere area is reduced, but the air volume of 4BF is maintained at 5200-5400Nm^3/min, and the gas utilization rate reaches 47% Up and down. At last, the slag skin is dynamically balanced, and the furnace type is relatively stable.

Key words: blast furnace; coke; stable and smooth; tuyere area

武钢 4 号高炉高产降耗实践

岳 锐，熊 伟

（宝钢股份武钢有限炼铁厂，湖北武汉 430080）

摘 要：对武钢 4 号高炉 2020 年高产降耗的经验及所取得的成就进行了总结，通过加强原燃料管理，优化高炉操作制度，活跃炉缸，控制合理操作炉型，加强炉前管理等措施，高炉达到高产状态。并对进一步提产的措施进行了探讨。

关键词：高炉；活跃炉缸；操作制度；高产；降耗

The Production Practice for High Yield Low Consumption of NO.4 BF in Bao steel

Yue Rui, Xiong Wei

(Ironmaking Plant of Baosteel Wuhan Iron and Steel Co., Ltd., Wuhan 430080, China)

Abstract: The production practice for high yield low consumption of NO.4 BF of Bao steel is analyzed and summarized. By strengthening the management of raw materials and fuels, optimizing the operation system of blast furnace, activating hearth, controlling suitable operation of furnace type, strengthening the management of furnace front, and reaching the condition of high yield.

Key words: blast furnace; active hearth; operational system; high yield; low consumption

高炉煤气干法除尘系统优化

尹作明，杜凤祥

（宝武股份武钢有限，湖北武汉 430072）

摘 要：针对武钢 5 号高炉煤气干法除尘运行中出现的故障，从温度、输灰与清灰周期等方面，分析、优化干法除尘系统运行方式，总结故障后处理经验。

关键词：高炉煤气；干法除尘；温度；输灰系统

Optimization of Dry Dedusting System for Blast Furnace Gas

Yin Zuoming, Du Fengxiang

(Baosteel Wuhan Iron and Steel Co., Ltd., Wuhan 430072, China)

Abstract: In view of the failures in the dry dedusting removal operation of No. 5BF gas in WISCO, the operation mode of the dry dedusting system was analyzed and optimized from the aspects of temperature, ash transport and dust removal cycle, and the experience of post-fault treatment was summarized.

Key words: blast furnace gas; dry dedusting system; temprature; ash conveying system.

2500m³钒钛矿高炉无干熄焦冶炼实践

牛西园，姜 汀，石云鹏，付国庆，邓小辉

（河钢股份有限公司承德分公司炼铁事业部，河北承德 067102）

摘 要：中滦一期焦炭于2019年10月末停产，造成河钢承钢某2500m³高炉长期不能供应中滦干熄焦。2020年6月高炉大修开炉后，完全停止配加干熄焦，该高炉通过加强原燃料管理、加强炉前管理、改进冷却设施、优化炉型、采取合理上下部调剂等措施，成功做到了无干熄焦高炉的稳定生产。

关键词：钒钛磁铁矿；高炉；停配；干熄焦；稳定生产

Practice of Smelting 2500m³ V-Ti Magnetite Blast Furnace without CDQ Coke

Niu Xiyuan, Jiang Ting, Shi Yunpeng, Fu Guoqing, Deng Xiaohui

(HBIS Cheng Steel, Chengde 067102, China)

Abstract: No.5 BF of HBIS Cheng Steel can not supply CDQ for a long time. In June 2020, after the blast furnace was overhauled and opened, the addition of CDQ was completely stopped. Through strengthening the management of raw materials and fuels, strengthening the management in front of the furnace, improving the cooling facilities, optimizing the furnace type, and taking reasonable upper and lower blending measures, the stable production of no CDQ blast furnace was success fully achieved.

Key words: vanadium titanomagnetite; blast furnace; stop adding; coke dry quenching; stable production

论高炉水力冲渣节能环保运行发展趋势

张 勇，张志江，刘利芳，赵永峰

（河钢宣钢，河北张家口 075100）

摘　要： 高炉渣是高炉冶炼得到的主要副产品。对高炉渣不同的处理方式后，均有不同的用途。综合考虑环保及高炉渣处理后的再利用，水力冲渣是综合效益较好的处理方式。在当前低碳绿色发展战略的大背景下，高炉渣处理要充分考虑工序能耗的合理性，特别要深挖降低水力冲渣粒化水低流量与脱水器低转速的潜力，综合考虑炉渣再利用、降低烟气排放，合理利用工业废水进行水力冲渣生产，不断挖掘高炉渣处理工艺及设备设施综合配套潜力，充分进行余热利用，实现高炉渣处理节能环保运行发展。

关键词： 高炉渣处理；水力冲渣；粒化水；节能；环保；废水利用

On the Development of Energy Saving and Environmental Protection in Hydraulic Flushing BF Slag System

Zhang Yong, Zhang Zhijiang, Liu Lifang, Zhao Yongfeng

(Xuanhua Steel Co., Ltd. of HBIS Group, Zhangjiakou 075100, China)

Abstract: BF slag is the main by-product of BF smelting. After different treatment of BF slag, there are different uses. Considering environmental protection and reuse of BF slag treatment, Hydraulic Flushing BF Slag is a better treatment method. Under the background of current low-carbon green development strategy, BF slag treatment should fully consider the rationality of process energy consumption, especially the potential of deep excavation to reduce the low flow rate of hydraulic slag granulation water and the low speed of dehydrator.

Key words: BF slag treatment; hydraulic flushing BF slag; hydraulic water; energy saving; environmental protection; waste water utilization

六西格玛在提高炉缸侧壁温度受控率中的应用

路　鹏，褚润林，吕志敏，闫　军

（河钢宣钢炼铁厂，河北宣化　075100）

摘　要： 通过对宣钢 1 号高炉生产数据的收集与整理，运用 6σ 管理模式和 Minitab 分析软件，对高炉炉缸侧壁温度受控率及所选参数进行 C&E 矩阵、FMEA 和多元回归分析，确定了影响高炉炉缸侧壁温度受控率的主要因素为入炉钛负荷、风口面积、风口长度和铁口深度。通过制定改进方案，实施控制计划，2020 年 11，12 月份，1 号高炉炉缸侧壁温度分别完成 90.6%和 90.8%，实现了炉缸侧壁温度受控率的提高，稳定了高炉生产。

关键词： 炉缸活跃性；6σ；Minitab 分析软件；多元回归分析；入炉钛负荷

Application of Six Sigma in Improving Temperature Control Rate of Hearth Sidewall

Lu Peng, Chu Runlin, LüZhimin, Yan Jun

(Ironworks, Hesteel Group Xuansteel Company, Xuanhua 075100, China)

Abstract: Through collecting and collating the production data of No.1 blast furnace of Xuanhua Steel, using 6 sigma management model and Minitab analysis software, C & E matrix, FMEA and multiple regression analysis were performed on the temperature control rate and selected parameters of blast furnace hearth sidewall. The main factors affecting the temperature control rate of hearth sidewall of of blast furnace are titanium load, area of tuyere, length and depth of iron hole. Though development of improvement plan, implementation of control plan, November, December 2020, The temperature of hearth sidewall of No.1 BF is 90.6% and 90.8%respectively, which increases the temperature control rate and stabilizes the production of BF.

Key words: hearth activity; six sigma; minitab analysis software; multiple regression analysis; loading of titanium into furnace

烧结矿还原性对炉料软熔滴落性能的影响

闫瑞军[1]，储满生[2]，柳政根[1]，刘培军[1]，李 峰[1]，唐 珏[1]

（1. 东北大学冶金学院，辽宁沈阳 110819；
2. 东北大学轧制技术及连轧自动化国家重点实验室，辽宁沈阳 110819）

摘 要：在现有高炉工艺基础上，优化炉料结构成为进一步节能降耗的有效途径。本文针对不同还原性烧结矿，配以两种低、高反应性焦炭，研究烧结矿不同还原性对炉料熔滴性能的影响。实验结果表明：在低反应性焦炭下，随烧结矿还原性（80%~85%~90%）的增加，炉料的软化开始温度（1150℃~1097℃~1046℃）、软化结束温度（1270℃~1228℃~1185℃）均降低，但软化区间逐渐变宽；熔化开始温度（1271℃~1240℃~1197℃）降低，滴落温度（1477℃~1497℃~1516℃）升高，熔化区间（软熔带）逐渐变宽。在高反应性焦炭下，随烧结矿还原性的增加，炉料的软化开始温度（1139℃~1124℃~1064℃）、软化结束温度（1256℃~1256℃~1192℃）均降低，软化区间逐渐变宽；熔化开始温度（1272℃~1257℃~1194℃）降低，滴落温度（1477℃~1497℃~1500℃）升高,熔化区间逐渐变宽。

关键词：烧结矿；还原性；软化性能；熔化性能

Effect of Sinter Reducibility on Softening-Melting-Dropping Properties of Burden

Yan Ruijun[1], Chu Mansheng[2], Liu Zhenggen[1], Liu Peijun[1], Li Feng[1], Tang Jue[1]

(1. School of Metallurgy, Northeastern University, Shenyang 110819, China;
2. State Key Laboratory of Rolling and Automation, Northeastern University, Shenyang 110819, China)

Abstract: On the basis of existing blast furnace technology, optimizing the burden structure becomes one of the effective ways to further save energy and reduce consumption. In this paper, three kinds of sinters with different reducibility and two kinds of cokes with low and high reactivity were used to study the influence of different reducibility of sinters on the softening-melting-dripping performance of burden. The experimental results show that the softening starting temperature (1150℃-1097℃-1046℃) and softening ending temperature (1270℃-1228℃-1185℃) of the burden all decreased with the increasing sinter reducibility (80%-85%-90%) under low reactivity coke, and the softening range gradually widens. The melting start temperature (1259℃-1240℃-1197℃) decreased, the dripping temperature (1477℃-1497℃-1516℃) increased, and the melting zone (soft melt zone) gradually widen. Under the condition of high reactivity coke, with the increasing sinter reducibility, the softening starting temperature (1139℃-1124℃-1064℃) and softening ending temperature (1256℃-1256℃-1192℃) of the burden all decreased, and the softening range gradually widen. The initial melting temperature (1272℃-1257℃-1194℃) decreased, while the dripping temperature (1477℃-1497℃-1500℃) increased. The melting zone gradually widen.

Key words: sinter; reducibility; softening property; melting property

河钢邯钢 8 号高炉激光料面仪应用实践

郭先燊，张志斌

（河钢集团邯钢公司，河北邯郸　056015）

摘　要：对河钢邯钢 8 高炉激光料面仪的应用实践进行了总结。该料面仪通过安装在炉喉密封盖上南北方向对称的两个激光照射扫描炉内料面，同时使用正东向的红外线摄像头进行拍摄录像，最后用配套软件将拍摄的照片与视频进行重叠处理，显示出激光轨迹图，即料面形状图。操作者可以直观地观测到炉内料面形状，也可以通过配套软件推算出料面深度、平台宽度等参数，实现高炉上部调剂的可视化。

关键词：高炉；激光；料面仪

Application Practice of Laser Material Surface Meter in No. 8 Blast Furnace of Handan Steel

Guo Xianshen, Zhang Zhibin

(Ironmaking Department, HBIS Group Handan Steel Co., Ltd., Handan 056015, China)

Abstract: The application practice of laser surface meter for blast furnace in Handan Iron and Steel is summarized. The material surface meter is used to scan the material surface in the furnace by two lasers symmetrical in the north and south direction installed on the sealing cover of the furnace throat. At the same time, an infrared camera facing east is used to take videos. Finally, the photos and videos are overlapped with the supporting software to display the laser trajectory diagram, namely the material surface shape diagram. The operator can visually observe the shape of the feed surface in the furnace, and can also calculate the depth of the feed surface, the width of the platform and other parameters through the supporting software to realize the visualization of the upper adjustment of the blast furnace.

Key words: blast furnace; laser material; surface meter

首钢股份公司炼铁冲渣水下轴头设备优化改进实践

齐立东，王海滨，胡　荣

（北京首钢股份有限公司设备部，河北迁安　064404）

摘　要：目前我国高炉炉渣处理工艺主要是水淬渣。首钢股份公司炼铁作业部冲渣设备即采用水力冲渣的处理工艺，搅笼机是系统中的核心设备。水下轴头是搅笼机下侧支撑运转的关键部件，由于工作环境比较恶劣，水下轴头频繁出现轴承故障，设备使用寿命周期较短，直接影响整个渣处理系统的正常生产。通过结合现场实际分析，从如何提高水下轴头轴承箱密封的技术性能入手，确保轴承箱的密封性，保障轴承的润滑，从而使设备达到长周期稳定运行。我们对水下轴头密封结构进行了改进，同时对润滑方式和润滑制度进行了调整，有效解决了以上难题。

关键词： 冲渣；水下轴头；润滑；密封

Practice of Optimization and Improvement of Underwater Shaft Head Equipment for Iron Making Slag Washing in Shougang Co. Ltd

Qi Lidong, Wang Haibin, Hu Rong

(Equipment Department, Beijing Shougang Co., Ltd., Qian'an 064404, China)

Abstract: At present, water quenching slag is the main processing technology of blast furnace slag in China. The slagging equipment of iron making operation department of Shougang Company adopts hydraulic slagging processing technology, and the mixing machine is the core equipment in the system. Underwater shaft head is a key component of the lower side support operation of the cage agiter. Due to the harsh working environment, the underwater shaft head frequently suffers bearing failures, and the service life cycle of the equipment is short, which directly affects the normal production of the whole slag treatment system. Combined with the actual field analysis, starting with how to improve the sealing technical performance of the underwater shaft head bearing box, to ensure the sealing of the bearing box, to ensure the lubrication of the bearing, so that the equipment to achieve a long period of stable operation. We improved the sealing structure of the underwater shaft head, and adjusted the lubrication mode and lubrication system, which effectively solved the above problems.

Key words: flushing slag; underwater shaft head; lubrication; seal

攀钢钒钛铁水配加废钢试验分析

蒋　胜[1]，陈　利[2]，朱凤湘[1]，王禹键[1]

（1. 攀钢集团研究院有限公司，四川攀枝花　617000；2. 西昌钢钒有限公司，四川西昌　615000）

摘　要： 为了研究加废钢对钒钛铁水的成分、温度和铁水罐皮重的影响规律，以在现有条件下不同加入方式进行了试验应用研究。结果表明：现有条件下在炉台上一次性加入废钢的方式较为合适，钒钛铁水中[V]、[C]、[S]元素和铁水温度都随废钢加入量增加而降低，铁水罐皮重则增重，粘结更加严重。并得出废钢加入量不宜超过30kg/t铁水。

关键词： 钒钛铁水；高炉；废钢；铁水罐

Analysis of Vanadium-Titanium Hot Metal with Scrap at Panzhihua Iron and Steel Co.

Jiang Sheng[1], Chen Li[2], Zhu Fengxiang[1], Wang Yujian[1]

(1. Pangang Group Research Institute Co., Ltd., Panzhihua 617000, China;
2. Xichang Steel & Vanadium Co., Ltd., Pangang Group, Xichang 615000, China)

Abstract: In order to study the effect of scrap on the composition, temperature and ladle of V-Ti molten iron, experiments were carried out with different adding methods. The results show that the way of adding scrap steel on the stove top is more suitable. The elements [V], [C], [S] and the temperature of V-Ti molten iron decrease with the increase of scrap steel addition, while the weight of ladle increases, and the adhesion becomes more serious. It is concluded that the amount of scrap should not exceed 30kg/t molten iron.

Key words: V-Ti hot molten iron; BF; scrap; ladle

太钢 4350m^3 高炉悬料机理的研究

刘文文

（山西太钢不锈钢股份有限公司技术中心，山西太原 030003）

摘 要：本文对太钢 4350m^3 高炉悬料的机理进行分析可得：此高炉不同高度的压差控制极值为 h_1<95kPa、h_2<62kPa、h_3<138kPa；当压差 180kPa 持续时间在 4h 或压差 185kPa 持续时间在 1h 以上时，调整矿焦比；边缘煤气流温度高于 150℃，并且多个方位气流过剩时，控制煤气流速，避免发生管道及悬料。

关键词：悬料；压差；煤气静压力；原燃料

Study on Suspension Mechanism in 4350 m^3 Blast Furnace of TISCO

Liu Wenwen

(Technology Center, Taiyuan Iron and Steel Group Co., Ltd., Taiyuan 030003, China)

Abstract: Based on the analysis of the mechanism of the suspension in the 4350m3 blast furnace of TISCO, the following results can be obtained: the extreme pressure difference control values of the blast furnace at different heights are h_1 < 95kPa, h_2 < 62kPa and h_3< 138kPa; When the pressure 180kPa is 4 hours or 185 kPa is more than 1 hour, the coke ratio is adjusted. When the temperature of the edge gas flow is higher than 150℃, and the gas flow in multiple directions is excessive, the gas flow rate should be controlled to avoid the occurrence of pipelines and suspension.

Key words: suspension; pressure different; gas pressure; original fuel

应对炉缸异常侵蚀条件下风口鼓风参数分布的模拟研究

张兴胜[1]，迟臣焕[1]，邹宗树[2]

（1. 本钢板材研发院，辽宁本溪 117000；2. 东北大学冶金学院，辽宁沈阳 110819）

摘 要：为解析应对高炉炉缸异常侵蚀条件下调整部分风口直径后的风量分布，利用数值模拟方法研究了热风管道及各风口的压力、速度、风量及鼓风动能分布。计算结果表明，在全部风口直径相同的条件下，鼓风参数并非完全均匀分布。随着热风向三岔口对向流动，各鼓风参数整体呈现逐渐增高趋势。为应对炉缸铁口区域异常侵蚀，减小对应风口的直径，风速增加，但风量降低，鼓风动能降低，符合保护炉缸、抑制侵蚀的目的。
关键词：高炉；炉缸异常侵蚀；鼓风分布；鼓风动能；数值模拟

Simulation Study on the Distribution of Tuyere Blast Parameters under the Condition of Hearth Abnormal Erosion

Zhang Xingsheng[1], Chi Chenhuan[1], Zou Zongshu[2]

(1. R&D Institute of Bengang Steel Plates Co., Ltd., Benxi 117000, China;
2. School of Metallurgy, Northeastern University, Shenyang 110819, China)

Abstract: In order to analyze the blast volume distribution under the condition of adjusting tuyere diameter to cope with the abnormal erosion of blast furnace hearth, numerical simulation was conducted to study the circumferential distribution of blast pressure, velocity, volume flow rate and kinetic energy in the hot blast duct and each tuyere. The calculation results show that, under the condition of uniform tuyere diameters, the blast parameters are not circumferentially uniform. As the hot air flows to the opposite side of the duct entry, the magnitudes of tuyere blast parameters show a gradual increasing trend. Under the condition of the tuyere diameter distribution in response to the hearth abnormal erosion, when the diameter the corresponding tuyere is decreased, its blast velocity increases, but the volume flow rate and the blast kinetic energy decreases, which meets the purpose of inhibiting local erosion and protecting the hearth.
Key words: blast furnace; abnormal hearth erosion; blast distribution; blast kinetic energy; numerical simulation

宁钢 2 号高炉提高煤比的措施

李　刚，顾尚领，冯小均，俞晓林，丁德刚

（宁波钢铁有限公司，浙江宁波　315807）

摘　要：由于高炉喷煤中的煤粉在燃烧带停留的时间极短（10~30s），实验室和实际高炉取样表明，煤粉的燃烧率在 75%左右，且随着煤比的提高，燃烧率下降。采用富氧、高风温、氧煤枪等技术可使燃烧率提高到 85%~95%，但进入炉内的未然煤粉绝对量仍很大，导致炉内未燃煤粉增加，恶化高炉的透气性，从而破坏炉况的顺行[1]。因此本文从制约提高煤比的原因分析入手，找出高炉提高喷煤比的具体方法。日常生产中主要通过优化炉料的精细化管理、调整上下部制度、提高炉温的管控标准，减少炉温波动，强化炉前出铁等一系列措施，使热负荷稳定性大幅度提高，高炉炉况稳定明显改善，实现了高炉煤比的大幅提升和燃料消耗的降低，为高炉高煤比，低消耗生产积累了成功的实践经验。
关键词：优化炉料结构；装料制度；煤比；炉温；热负荷

Measures to Increase Coal Ratio of No.2 BF in Ningbosteel

Li Gang, Gu Shangling, Feng Xiaojun, Yu Xiaolin, Ding Degang

(Ningbo iron and Steel Co., Ltd., Ningbo 315807, China)

Abstract: Due to the short time (10~30s) that the pulverized coal stays in the combustion zone in the blast furnace injection, the laboratory and actual blast furnace samples show that the combustion rate of pulverized coal is about 75%, and with the increase of coal ratio, the combustion rate decreases. The combustion rate can be increased to 85%~95% by using oxygen enriched, high air temperature, oxygen coal gun and other technologies, but the absolute amount of unburned pulverized coal entering the furnace is still large, which leads to the increase of unburned pulverized coal in the furnace, worsens the permeability of the furnace, and thus destroys the smooth operation of the furnace[1]. Therefore, this paper starts with the analysis of the reasons that restrict the increase of coal ratio, and finds out the specific methods to increase the coal injection ratio of blast furnace. In daily production, a series of measures such as optimizing the fine management of charge, adjusting the upper and lower system, improving the control standard of furnace temperature, reducing the fluctuation of furnace temperature, strengthening the tapping in front of furnace, etc. are taken to greatly improve the stability of heat load and the stability of furnace condition, so as to greatly increase the coal ratio of blast furnace and reduce the fuel consumption, so as to increase the production volume of high coal ratio and low consumption of blast furnace Tired of successful practical experience.

Key words: optimize charge structure; charging system; coal ratio; furnace temperature; thermal load

高炉铁水中硫加速炉缸炭砖侵蚀机理及对策

陈立达[1]，邓　勇[1]，刘　然[1]，陈艳波[2]

（1. 华北理工大学冶金与能源学院，河北唐山　063210；
2. 首钢京唐钢铁联合有限责任公司技术中心，河北唐山　063200）

摘　要：为了延缓炉缸炭砖侵蚀，分析了炉缸铁水硫含量变化趋势，研究了硫元素加速炉缸炭砖侵蚀机理，提出了现代大型高炉脱硫技术措施。结果表明：高炉-铁水预处理联合脱硫、使用高比例球团是炉缸铁水硫含量升高的主要原因；炉缸炭砖与碳含量欠饱和的铁水接触是炭砖侵蚀的直接原因，硫含量升高使铁水表面张力下降、黏度下降，提高了界面反应速率、增大了铁水中碳的传质系数，加速了炭砖侵蚀。在低渣比条件下，采用控制炉渣成分和铁水成分的协同脱硫技术，是现代大型高炉脱硫的有效措施。

关键词：硫含量；炭砖侵蚀；铁水性能；炉渣脱硫；低渣比

Erosion Mechanism of Sulfur Accelerated Hearth Carbon Brick in Molten Iron of Blast Furnace and Its Countermeasures

Chen Lida[1], Deng Yong[1], Liu Ran[1], Chen Yanbo[2]

(1. College of Metallurgy and Energy, North China University of Science and Technology, Tangshan 063210, China; 2. Technique Center, Shougang Jingtang United Iron and Steel Co., Ltd., Tangshan 063200, China)

Abstract: In order to delay the erosion of carbon brick in hearth, the variation trend of sulfur content of molten iron in hearth was analyzed, the erosion mechanism of sulfur accelerating hearth carbon brick was studied, and the desulfurization technology of modern large blast furnace was put forward. The results show that the main reasons for the increase of sulfur content of molten iron are the combined desulfurization of blast furnace and hot metal pretreatment and the use of high proportion pellets. The direct cause of carbon brick erosion is the contact between the hearth carbon brick and the molten iron with less saturated carbon content. The increase of sulfur content reduces the surface tension and viscosity of molten iron, improves the interface reaction rate, increases the mass transfer coefficient of molten iron, and accelerates the erosion of carbon brick. Under the condition of low slag ratio, the technology of co-desulfurization of slag composition and molten iron composition is an effective measure for desulfurization of modern large blast furnace.

Key words: sulfur conten; carbon brick erosion; properties of molten iron; desulphurization of slag; low slag ratio

韶钢 2200m^3 高炉提高铁水硅偏差生产实践

周凌云，陈生利，陈国忠，陈开泉

（宝武集团广东韶关钢铁有限公司，广东韶关　512123）

摘　要：铁水硅偏差（铁水[Si]含量控制在 0.25%~0.55%范围）是衡量高炉铁水质量的重要指标。2018 年韶钢 7 号高炉硅偏差命中率仅为 52.78%，高炉铁水硅含量平均达到 0.56%，通过开展提高铁水硅偏差稳定率系列攻关措施，现实铁水硅偏差稳定率达到 80%，焦比 345kg/t 良好的技术经济指标。

关键词：高炉；铁水；硅偏差

Production Practice of Increasing Silicon Deviation in Hot Metalof 2200m^3 Blast Furnace at SISG

Zhou Lingyun, Chen Shengli, Chen Guozhong, Chen Kaiquan

(Baowu Group Guangdong Shaoguan Steel Co., Ltd., Shaoguan 512123, China)

Abstract: The silicon deviation of hot metal (the content of [Si] in hot metal is controlled in the range of 0.25%-0.55%) is an important index to measure the quality of hot metal. In 2018, the silicon deviation hit rate of No. 7 BF at SISG was only 52.78%, and the average silicon content of hot metal reached 0.56%. Through a series of measures to improve the silicon deviation stability rate of hot metal, the actual silicon deviation stability rate of hot metal reached 80%, good technical and economic index of coke ratio 345kg/t.

Key words: blast furnace; molten iron; silicon deviation

红钢炼铁厂喷煤系统整合优化设计

李晓芹，李碧中

（昆钢集团设计院有限公司，云南昆明　650302）

摘　要：根据红钢炼铁产量提升工作计划要求，红钢 3#高炉年喷煤量相应要求提升，针对现有 3#高炉的 2#喷煤系统制粉量产量不足，难以满足 3#高炉年喷煤量相应提升要求，在保 3#高炉喷煤量提升，平稳喷煤，保炉况稳定顺行，对红钢炼铁厂 1#、2#喷煤系统进行整合，优化工艺体系设计，达到弥补现有 2#喷煤系统制粉量产量不足，满足 3#高炉产量提升喷煤量增加的需求，通过提煤降焦，降低综合燃料消耗效率，使生产经营经济效益，技术指标优化，有效的提量降本，达成提高生产效益的目的。

关键词：炼铁；喷煤；整合；优化设计

Integrated and Optimized Design of Coal Injection System in Honggang Iroworks

Li Xiaoqin, Li Bizhong

(Kunming Iron and Steel Group Design Institute Co, Ltd, Kunming 650302, China)

Abstract: According to the request of red steel plans to raise the ironmaking production work, red steel quantity of 3 # blast furnace coal injection in the corresponding requirements, in view of the existing control of 3 # of 2 # blast furnace injection coal powder quantity production shortfalls, difficult to meet the quantity of 3 # blast furnace coal injection in the corresponding requirements, in the 3 # blast furnace coal injection volume increase, smooth coal injection, a stable furnace condition along the line, the red steel iron 1 #, 2 # coal injection system integration, optimization of process system design, to make up for the existing control of 2 # injection coal powder quantity production shortfalls, meet the 3 # blast furnace production coal injection quantity increase in demand, by coal reducing tar and reduce fuel consumption in the integrated efficiency, make the production operation and economic benefit, Optimize the technical index, effectively raise the quantity and reduce the cost, and achieve the purpose of improving the production efficiency.

Key words: ironmaking; coal injection; integration; the optimization design

红钢 1350m^3 高炉低品位高有害元素条件下强化冶炼实践

陈元富，卢俊旭

（红河钢铁有限公司，云南蒙自　661100）

摘　要：对近年红钢 1350m^3 高炉（以下称 3#高炉）在现有装备及原燃料条件下的强化冶炼生产实践进行总结和分析，通过优化高炉炉料结构、优化操作制度、管理创新、全员全方位对标挖潜、提高生产组织效率及操作技术水平等措施，2021 年 4 月份以来 3#高炉实现了长周期、安全、稳定、高水平、低耗的生产，产能不断提高，各项技术经济指标不断优化，其中一些技术经济指标达到 3#高炉开炉以来历史最好水平。

关键词：高炉；产量；强化冶炼

Under the Condition of Low Grade and High Harmful Elements in 1350 m^3 Blast Furnace of Honggang Strengthen Smelting Practice

Chen Yuanfu, Lu Junxu

(Kunming Iron & Steel Co., Ltd. Honghe Iron & Steel Co., Ltd., Mengzi 661100, China)

Abstract: This paper summarizes and analyzes the intensive smelting production practice of 1350 m3 blast furnace (hereinafter referred to as 3# blast furnace) in Honggang in recent years under the existing equipment and raw fuel

conditions. By optimizing the burden structure of blast furnace, optimizing the operation system, innovating management, tapping the potential of all employees, improving the production organization efficiency and operating technology level, etc., 3# blast furnace has achieved long-term, safe, stable, high-level and low-consumption production since April 2021.
Key words: blast furnace; output; strengthening smelting

某高炉炉缸浇注炉身喷涂技术实践

马小青，孙赛阳，喇校帅，胡大伟，杜尚斌

（北京联合荣大工程材料股份有限公司，北京　101400）

摘　要：高炉在运行一个阶段后，由于侵蚀冲涮的原因，炉内耐材会有不同程度的减薄，当减薄到一定程度后，就会对高炉的正常运行产生影响，带来一定的安全隐患。针对某钢厂高炉出现炉缸侧壁温度升高的情况，对该高炉进行高炉炉型修复施工。整个修复过程包括放残铁、炉缸清理、炉缸浇注和炉身喷涂等步骤，通过施工前的跟踪和炉缸清理过程中现场测量，得到炉内具体的侵蚀情况，应用成熟的炉缸浇注技术，使得这次高炉中修达到了预期目标，延长了高炉的使用寿命，修复完成后迅速达产，减少了钢铁企业的经济损失。
关键词：高炉；炉缸浇注；高炉喷注；炉缸清理

Practice of Spraying Technology for Casting Furnace Body of a Blast Furnace Hearth

Ma Xiaoqing, Sun Saiyang, La Xiaoshuai, Hu Dawei, Du Shangbing

(Beijing AlliedRongda Engineering Materials Co., Ltd., Beijing 101400, China)

Abstract: After a period of operation of the blast furnace, due to erosion and flushing, the refractory material in the furnace will be thinned to a certain degree. When the thickness is reduced to a certain level, it will affect the normal operation of the blast furnace and bring certain safety. Hidden dangers. In response to the temperature rise of the hearth side wall of a blast furnace in a steel plant, the blast furnace was repaired. The entire repair process includes the steps of discharging residual iron, hearth cleaning, hearth pouring and furnace body spraying. Through the tracking before construction and on-site measurement during the hearth cleaning process, the specific corrosion situation in the furnace is obtained, and mature hearth pouring is used. Technology has enabled the blast furnace to achieve the expected goal during the mid-repair, prolong the service life of the blast furnace, and quickly reach production after the repair is completed, reducing the economic losses of the iron and steel enterprises.
Key words: blast furnace; hearth pouring; blast furnace injection; hearth cleaning

玉钢 3#高炉钒钛矿冶炼炉况波动的原因及对策

何飞宏，薛　锋，黄德才

（玉溪新兴钢铁有限公司炼铁厂，云南玉溪　653100）

摘　要：对玉钢3#1080m^{3}高炉钒钛矿冶炼中频繁出现憋风、低炉温、生铁排Zn进行分析和总结，找出影响炉况的主要因素，通过改善和稳定原燃料质量，优化工艺操作参数和操作制度，炉况的稳定性大幅提升，各技术经济指标得到改善。

关键词：高炉；钒钛矿；二级焦；稳定

Causes and Countermeasures of Furnace Condition Fluctuation in Vanadium-Titanium ore Smelting of No.3 Blast Furnace in Yugang

He Feihong, Xue Feng, Huang Decai

(Yuxi Xinxing Iron and Steel Co., Ltd. Ironmaking Plant, Yuxi 653100, China)

Abstract: This paper analyzes and summarizes the frequent occurrence of blast holding, low furnace temperature and pig iron Zn in the smelting of vanadium titanium ore in the 3 # 1080m^3 blast furnace of Yugang, and finds out the main factors affecting the furnace condition. By improving and stabilizing the quality of raw materials and fuels, optimizing the process operation parameters and operation system, the stability of the furnace condition is greatly improved, and the technical and economic indicators are improved.

Key words: last furnace; vanadium-titanium ore; secondary coke; stabilization

玉钢1080m^{3}高炉2021年技术进步

冯光进，李裕华

（玉溪新兴钢铁有限公司炼铁厂，云南玉溪　653100）

摘　要：2021年是玉钢重大转折的一年，也是炼铁厂1080m^{3}高炉冶炼技术取得突破性进步的一年。利用系数突破4.0t/m^3·d，6月19日达到4.014t/m^3·d，保持了长周期稳定顺行高产。尤其是自4月以来，坚持以高炉为中心，坚持精料入炉的方针、优化高炉操作参数等技术手段，5月日均产量达到4143.48t；在稳产高产的前提下不断优化指标，焦比燃料比逐月下降（5月完成焦比417kg/t，燃料比536.71kg/t）。

关键词：高炉；利用系数；精料；稳定顺行

Technical Progress of 1080m^3 Blast Furnace in Yugang in 2021

Feng Guangjin, Li Yuhua

(YuxiXinxing Iron and Steel Co., Ltd. Ironmaking Plant, Yuxi 653100, China)

Abstract: The year of 2021 is a major turning point in Yugang and a breakthrough in the smelting technology of 1080m^3 blast furnace in ironmaking plant. The utilization coefficient exceeded 4.0t/m^3·d, and reached 4.014 t/m^3·d on June 19, maintaining long-term stable and high yield. Especially since April, adhering to the blast furnace as the center, adhering to the policy of charging concentrate, optimizing the operation parameters of blast furnace and other technical means, the average daily production in May reached 4143.48t. On the premise of stable production and high yield, the coke ratio fuel ratio decreased month by month (417kg/t coke ratio and 536.71kg/t fuel ratio were completed in May).

Key words: blast furnace; utilization factor; concentrated feeding stuff; stable smooth operation

前苏联环形燃烧室内燃式热风炉数值模拟研究

陈浩宇，程树森，程晓曼
（北京科技大学冶金与生态工程学院，北京 100083）

摘 要：建立了前苏联环形燃烧室内燃式热风炉的流动、传热及燃烧三维数学模型，对煤气在燃烧室内的燃烧进行了数值计算，对空气与煤气在进行混合燃烧的过程中环形内燃室出口，球形拱顶内的流场、温度场、浓度场、火焰形状进行了分析讨论。结果表明：环形内燃室出口处形成涡流，其火焰形状沿涡流向下聚集并延伸至蓄热室内；环形内燃室出口截面形成沿圆周分布均匀的温度、速度、浓度梯度区，最大温差为 379k，最小最大速度比为 0.57，CO 最高摩尔浓度为 0.22。

关键词：环形内燃室热风炉；燃烧；数值模拟；流场

Numerical Simulation of Internal Combustion Hot Blast Stove with Annular Combustion Chamber in Former Soviet Union

Chen Haoyu, Cheng Shusen, Cheng Xiaoman
(College of Metallurgical and Ecological Engineering University of Science and Technology Beijing, Beijing 100083, China)

Abstract: A three-dimensional mathematical model of flow, heat transfer and combustion in a Soviet annular combustion chamber was established. The numerical calculation of gas combustion in the combustion chamber was carried out. The flow field, temperature field, concentration field and flame shape at the exit of the annular combustion chamber and in the spherical arch during the mixed combustion of air and gas were analyzed. The results show that vortex is formed at the exit of annular combustion chamber, and the flame shape gathers downward along the vortex and extends to the regenerative chamber. The outlet section of the annular combustion chamber forms a gradient area of temperature, velocity and concentration evenly distributed along the circumference. The maximum temperature difference is 379K, the minimum maximum velocity ratio is 0.57, and the maximum molar concentration of CO is 0.22.

Key words: hot blast stove with annular internal combustion chamber; combuston; numerical simulation; flow field structure

基于混煤耦合燃烧的高炉高效煤粉喷吹技术研究

马利科[1]，庞海清[2]，赵鸿波[1]
（1. 本钢板材股份有限公司研发院，辽宁本溪 117021；
2.辽宁科技大学冶金学院，辽宁鞍山 114051）

摘 要：通过对本钢用煤的官能团机构分析与燃烧性能的检验，阐明了烟煤能否助燃无烟煤和助燃程度多少的原因，就是混煤是否具有耦合性，达到耦合燃烧，选择具有良好的助燃作用的烟煤和无烟煤进行优化搭配，能达到烟煤助燃无烟煤，提高高炉喷吹效率的作用；高炉煤粉喷吹生产应用试验证明：取得了提高喷吹效率 4.2%，降低校正焦比 15kg/t 的效果。

关键词：煤粉喷吹；耦合燃烧；官能团；燃烧性能；喷吹效率；焦比

Research on High Efficiency Pulverized Coal Injection Technology for Blast Furnace Based on Mixed Coal Coupling Combustion

Ma Like[1], Pang Haiqing[2], Zhao Hongbo[1]

(1. Research and Development Institute of Benxi Steel Plate Co., Ltd., Benxi 117021, China;
2. School of Metallurgy, Liaoning University of Science and technology, Anshan 114051, China)

Abstract: Through the analysis of the functional group mechanism of the coal used in Benxi Iron and Steel Group Co. and the test of the combustion performance, this paper expounds the reasons for whether the bituminous coal can be used for combustion of anthracite and how much it can be used for combustion, that is, whether the blended coal has the coupling property and achieves the coupling combustion, the optimum combination of bituminous coal and anthracite with good combustion-supporting function can achieve the function of combustion-supporting anthracite coal and increase the injection efficiency of blast furnace, the effect of reducing the corrected coke ratio by 15kg/t.

Key words: pulverized coal injection; coupling combustion; functional group; combustion performance; injection efficiency; coke ratio

宝钢 2 高炉（二代）空料线停炉操作实践

居勤章，王 超

（宝山钢铁股份有限公司炼铁厂，上海 200941）

摘 要：宝钢 2 高炉 2020 年 8 月 29 日开始降料线操作，降料线休风过程共耗时 17.5h，降料线过程基本按照停炉方案计划节点进行，受环缝入口温度偏高的影响，累计打水量较大。空料线过程中煤气放散时间较短，基本实现全过程煤气回收，最大程度实现安全环保，确保顺利停炉。

关键词：高炉；顶温；打水；停炉

Practice of Lowering Stock Level Method Shutdown in Baosteel No.2 BF

Ju Qinzhang, Wang Chao

(Baoshan Iron & Steel Co., Ltd., Shanghai 200941, China)

Abstract: Operation of lowering stock level was started in 29 Aug. 2020, it cost 17.5 hours till damping down. The whole process was mostly based on operation planning, Due to high inlet temperature of Bischoff scrubbing system, Cumulative consumption of spray water was high. gas discharge time was short during the lowering stock level process, almost all the gas generated in the whole process was recycled, safety and environmental protection was achieved to an extreme, operation process of shutdown was smooth and safe.

Key words: blast furnace; top temperature; water spray; shutdown

高炉主皮带机滚筒轴向窜动分析与改进

李晓农，何建峰，赵 靖

（北京首钢股份有限公司设备管理室，河北迁安 064400）

摘 要：滚筒作为高炉上料主皮带机的关键部件，使用寿命直接影响高炉的正常生产。本文首先对滚筒轴向窜动故障进行了描述，分析了滚筒轴向窜动的原因，改进滚筒轴承座结构，在保留剖开式轴承座的基础上增加轴肩，轴承紧定套与轴肩之间制作特殊的轴向定位套，定位套与透盖组合成密宫式密封，保证轴承的良好润滑；轴承座端盖由固定式改为在线可拆式端盖，在紧急情况下可以在不停机的情况下检查滚筒，为早期处理相关故障提供的保障，同时针对轮毂与轴之间胀套，根据滚筒的受力情况，选取相对应的胀套，以上的改进经过实际使用证明是可行、合理、可靠和实用的，产生了较高的经济效益，为生产的顺畅和效益提供了保障。

关键词：滚筒；轴承座；窜轴；胀套；改进

Analysis and Improvement of Axial Movement of Main Belt Conveyor Drum of Blast Furnace

Li Xiaonong, He Jianfeng, Zhao Jing

(Equipment Management Office of Beijing Shougang, Co. Ltd, Qian'an, Hebei 064400, China)

Abstract: The main feeding belt conveyor of blast furnace is one of the most critical equipment in modern blast furnace production. As the key component of the main feeding belt conveyor of blast furnace, the service life of the roller directly affects the normal production of blast furnace. This paper first describes the fault of the axial movement of the drum, analyzes the causes of the axial movement of the drum, improves the structure of the drum bearing seat, increases the shaft shoulder on the basis of retaining the split bearing seat, makes a special axial positioning sleeve between the bearing tightening sleeve and the shaft shoulder, and combines the positioning sleeve and the through cover to form a palace seal to ensure the good lubrication of the bearing; The end cover of bearing pedestal is changed from fixed type to on-line detachable type. In case of emergency, the roller can be checked without stopping the machine, so as to provide guarantee for early treatment of relevant faults. At the same time, for the expansion sleeve between hub and shaft, the corresponding expansion sleeve is selected according to the stress condition of roller. The above improvement has been proved to be feasible, reasonable, reliable and practical by actual use, It has produced higher economic benefits and provided guarantee for smooth production and benefit.

Key words: roller; bearing pedestal; axial displacement; expansion sleeve; improvement

富氢燃料喷吹高炉 CO_2 排放能力的理论分析

李海峰[1]，徐万仁[2]，张晓辉[1]，邹宗树[1]

（1. 东北大学冶金学院，辽宁沈阳 110890；2. 宝山钢铁股份有限公司，上海 201900）

摘　要：基于焦炭置换比数学模型，获得了煤粉、天然气、焦炉煤气对焦炭的置换比。同时，基于二氧化碳排放水平数学模型，获得了不同喷吹燃料工序下煤气利用率和炉顶煤气温度的变化，分析了不同工序下二氧化碳排放水平，还考察了喷吹气体量、煤粉喷吹量、富氧率等重要操作参数对二氧化碳排放水平的影响，模型计算结果为高炉节能减排寻求更为合理的喷吹工艺提供理论依据。

关键词：高炉；富氢燃料；焦炭置换比；二氧化碳排放水平

Theoretical Analysis and Practice of CO_2 Emission Capability under Hydrogen-rich Fuel Injection in Blast Furnace

Li Haifeng[1], Xu Wanren[2], Zhang Xiaohui[1], Zou Zongshu[1]

(1. School of Metallurgy, Northeastern University, Shenyang 110819, China;
2. Baoshan Iron and Steel Co., Ltd., Shanghai 201900, China)

Abstract: Based on the mathematical model of coke replacement ratio, the coke replacement ratios of pulverized coal, natural gas and coke oven gas are obtained. At the same time, based on the mathematical model of carbon dioxide emission level, the changes in gas utilization rate and top gas temperature under different fuel injection processes are obtained, and the carbon dioxide emission level under different processes is analyzed. The influence of important operating parameters such as the volume of injected gas, the mass of pulverized coal, and the oxygen enrichment rate on the level of carbon dioxide emissions are also investigated. The model calculation results provide a theoretical basis for seeking a more reasonable injection process for energy saving and emission reduction of blast furnace.

Key words: blast furnace: hydrogen-rich fuel; coke replacement ratio: CO_2emission level

昆钢新区 2500m^3 高炉排锌事故处理

麻德铭，马杰全，张品贵，李　淼

（武昆股份安宁公司炼铁厂，云南昆明　650300）

摘　要：本文针对昆钢新区 2500m^3 高炉 2020 年 6 月 26 日高炉排锌造成炉凉的事故进行了总结归纳，分析了此次排锌事故发生的原因，总结了短时间内炉况恢复调整到位的有效措施，并依据查找到的原因，制定了预防措施，为今后预防高炉集中排锌及炉凉事故的处理提供参考依据。

关键词：高炉；炉凉；排锌；处理

Treatment of Zinc Discharge Accident of 2500m^3 Blast Furnace in New District of Kunming Iron and Steel Co., Ltd

Ma Deming, Ma Jiequan, Zhang Pingui, Li Miao

(Ironmaking Plant of Anning Company, Kunming 650300, China)

Abstract: This paper summarizes the accident of blast furnace cooling caused by zinc discharge of 2500m^3 blast furnace in New District of Kunming Iron and Steel Co., Ltd. on June 26, 2020, analyzes the causes of the accident, summarizes the effective measures for furnace condition recovery and adjustment in a short time, and formulates the preventive measures according to the found reasons, It provides a reference for the prevention of centralized zinc discharge and cooling accident of blast furnace in the future.

Key words: blast furnace; cooling; zinc discharge; treatment

高炉富氧喷吹天然气的数值分析

彭　星，王静松，左海滨，王　广，薛庆国

（北京科技大学钢铁冶金新技术国家重点实验，北京　100083）

摘　要：基于物料平衡与热平衡的高炉数学模型，计算了高炉富氧喷吹天然气对高炉主要经济指标的影响，并通过高炉风口回旋区三维燃烧数值模型，分析了天然气以及煤粉燃烧行为。结果表明：以维持理论燃烧温度为 2000℃为标准，随着鼓风氧气含量增加，可接受喷吹的天然气量增加，直接还原度、焦比、焦炭置换比下降，CO_2减排效果增加。当鼓风氧气含量为 29%时，可接受喷吹的天然气量为 105Nm^3/tHM，节焦 78kg/tHM，CO_2 减排量为 192.3kg/tHM，CO_2减排率达 10.5%，天然气在回旋区的转化率为 74.8%，而煤粉燃尽率下降 5%。

关键词：天然气；高炉；节焦；CO_2 排放；煤燃烧；数值分析

Numerical Analysis of Injection of Natural Gas in an Oxygen-enriched Blast Furnace

Peng Xing, Wang Jingsong, Zuo Haibin, Wang Guang, Xue Qingguo

(State Key Laboratory of Advanced Metallurgy, University of Science and Technology Beijing, Beijing 100083, China)

Abstract: Based on the blast furnace mathematical model of material and heat balance, the influence of natural gas injection in oxygen-enriched blast furnace on the main economic indicators of the blast furnace is calculated, and the combustion behavior of natural gas and pulverized coal is analyzed through the three-dimensional combustion numerical model of the tuyere-raceway zone of the blast furnace. The results show that with the adiabatic flame temperature of 2000℃ as the standard, with the increase of blast oxygen content, the acceptable natural gas injection increases, the direct reduction degree, coke rate and coke replacement ratio decrease, and the CO_2 emission reduction effect increase. When the blast oxygen content is 29%, the acceptable amount of natural gas injected is 105Nm^3/tHM, the coke saving is 78kg/tHM, CO_2 emission reduction is 192.3kg/tHM, and the CO_2 emission reduction rate is 10.5%. The conversion rate of natural gas in the raceway zone is 74.8%, while the pulverized coal burnout drops by 5%.

Key words: natural gas; blast furnace; coke saving; CO_2 emission; coal combustion; numerical analysis

中国宝武富氢碳循环高炉工艺路线研究与实践

袁万能，田宝山

（新疆八一钢铁有限公司炼铁厂，新疆乌鲁木齐　830022）

摘　要：温室效应导致的全球变暖已成了引起世人关注的焦点问题。燃烧煤、石油等化石能源所产生的二氧化碳是导致全球变暖的罪魁祸首，全球气候变化对自然生态系统和社会经济的影响正在加速，本方介绍了炼铁工艺围绕降低碳排放所采取的碳循环及富氢冶金工艺路线研究与实践，通过试验探索，为从源头上减少炼铁冶炼工艺的碳排放提供实践和理论依据。

关键词：氧气高炉；富氢；碳循环；富氧率；碳排放

Research and Practice on Process Route of Baowu Hydrogen Rich Carbon Cycle Blast Furnace in China

Yuan Wanneng, Tian Baoshan

(Ironmaking Plant of Xinjiang Bayi Iron and Steel Co., Ltd., Urumqi 830022, China)

Abstract: The global warming caused by the greenhouse effect has become the focus of the world. The carbon dioxide produced by burning fossil energy such as coal and oil is the main cause of global warming. The impact of global climate change on natural ecosystem and social economy is accelerating. This paper introduces the research and practice of carbon cycle and hydrogen rich metallurgy process adopted by ironmaking process to reduce carbon emissions, It provides practical and theoretical basis for reducing carbon emission of ironmaking process from the source.

Key words: oxygen blast furnace; rich in hydrogen; carbon cycle; oxygen enrichment rate; carbon emission

国内外氢冶金发展现状及需要研究解决的主要问题

徐万仁[1]，朱仁良[1]，毛晓明[1]，储满生[2]，唐　珏[2]

（1. 中国宝武钢铁集团有限公司，上海　201999；2. 东北大学冶金学院，辽宁沈阳　110819）

摘　要：本文介绍了国外氢冶金工艺研发和实施进展，以及我国氢冶金工艺研发应用的现状。对富氢还原高炉和富氢气基竖炉直接还原工艺的 CO_2 减排潜力进行了计算分析，提出了我国氢冶金的适宜工艺路线，分析了我国应用氢冶金技术需要研究解决的主要问题，并对我国氢冶金技术的未来应用前景作了展望。

关键词：CO_2 减排；氢冶金；高炉；气基竖炉直接还原；制氢；储氢

Current Status and Main Problems of Hydrogen Metallurgy

Xu Wanren[1], Zhu Renliang[1], Mao Xiaoming[1], Chu Mansheng[2], Tang Jue[2]

(1. China Baowu Iron & Steel Group Co., Shanghai 201999, China;
2: School of Metallurgy, Northeastern University, Shenyang, Liaoning 110819, China)

Abstract: The progress of research and application of hydrogen metallurgy process in the foreign countries and China was introduced in this work. Meanwhile, the CO_2 reduction potential of hydrogen-rich reduction blast furnace and hydrogen-rich gas-based shaft furnace direct reduction is calculated and analyzed, and the suitable process route of hydrogen metallurgy in China is proposed. Moreover, the main problems that need to be researched and solved for the application of hydrogen metallurgy technology in China are discussed. And then, the future prospects for the application of hydrogen metallurgy technology in China are presented.

Key words: CO_2 reduction; hydrogen metallurgy; blast furnace; gas-based shaft furnace direct reduction; hydrogen production; hydrogen storage

闪速还原炼铁技术基础研究

郭 磊，鲍其鹏，杨逸如，郭占成

（北京科技大学钢铁冶金新技术国家重点实验室铁矿资源高效利用研究所，北京 100083）

摘 要：闪速还原炼铁技术由于具有还原速度快、直接使用粉状原料，可以摆脱烧结、焦化工序等技术优势近年来受到了国内外专家学者的关注。本文对闪速还原这一方法的衍生过程进行了剖析。对近年来闪速还原炼铁工艺在国内外的发展进行了介绍。重点介绍了作者所在研究团队在闪速还原炼铁方面所进行的实验室研究工作，并列举了闪速还原炼铁工艺工业化上有可能面临的主要问题。

关键词：闪速还原；直接还原；铁矿粉；非高炉炼铁；闪速炼铁

Fundamental Research on Flash Reduction Ironmaking Technology

Guo Lei, Bao Qipeng, Yang Yiru, Guo Zhancheng

(State Key Laboratory of Advanced Metallurgy, University of Science and Technology Beijing, Beijing 100083, China)

Abstract: Because of its fast reduction speed and direct use of pulverized raw materials, flash reduction ironmaking technology can get rid of technical advantages such as sintering and coking processes, which has attracted the attention of domestic and foreign experts and scholars in recent years. This article analyzes the derivation process of the flash reduction method. The development of flash reduction ironmaking process at home and abroad in recent years is introduced. It focuses on the laboratory research work of the author's research team in flash reduction ironmaking, and enumerates the main problems that may be faced in the industrialization of flash reduction ironmaking process.

Key words: flash reduction; direct reduction; iron ore fines; non-blast furnace; flash ironmaking

CaO-SiO_2-CaF_2-La_2O_3-(P_2O_5)渣中稀土结晶行为研究

郭文涛[1]，丁子琦[1]，吴 珺[1]，王金明[1]，刘玉宝[2]，赵增武[1]

（1. 内蒙古自治区白云鄂博矿多金属资源综合利用重点实验室，材料与冶金学院，内蒙古科技大学，内蒙古包头 014010；2. 白云鄂博稀土资源研究与综合利用国家重点实验室，包头稀土研究院，内蒙古包头 014030）

摘 要：针对含稀土冶金渣中稀土资源利用问题，研究不同结晶温度和冷却速率条件下 P_2O_5 含量对 CaO-SiO_2-CaF_2-La_2O_3 渣中稀土结晶相形貌和结构的影响。发现添加 P_2O_5 后渣中稀土结晶相数量增加，未添加 P_2O_5 时稀土以

$Ca_2La_8(SiO_4)_6O_2$形式结晶，当渣中添加P_2O_5后稀土以$Ca_3La_2[(Si,P)O_4]_3F$形式结晶。降低结晶温度和冷却速率均能促进的$Ca_2La_8(SiO_4)_6O_2$和$Ca_3La_2[(Si,P)O_4]_3F$晶体的长大。

关键词： 稀土；结晶相；P_2O_5；温度；冷却速率

Crystallization Behavior of Rare Earth in CaO-SiO_2-CaF_2-La_2O_3-(P_2O_5) Slag

Guo Wentao[1], Ding Ziqi[1], Wu Jun[1], Wang Jinming[1], Liu Yubao[2], Zhao Zengwu[1]

(1. Key Laboratory of Integrated Exploitation of Bayan-Obo Multi-Metal Resources, Inner Mongolia University of Science &Technology, Baotou 014010, China; 2. Baotou Research Institute of Rare Earths, State Key Laboratory of Baiyunobo Rare Earth Resource Researches and Comprehensive Utilization, Baotou 014030, China)

Abstract: to promote the utilization of rare earth resources in metallurgical rare earth containing slag，the influences of P_2O_5 content on the morphology and strcture of rare earth crystal phase in CaO-SiO_2-CaF_2-La_2O_3 slag were investigated under different crystallization temperature and cooling rate conditions. With the added P_2O_5, the number of rare arth crystal phase in slag increased. In the absence of P_2O_5, rare earth crystallizes in the form of $Ca_2La_8(SiO_4)_6O_2$, while when P_2O_5 is added, it crystallizes into $Ca_3La_2[(Si,P)O_4]_3F$. The decreased crystallization temperature and cooling rate facilitate the growth of $Ca_2La_8(SiO_4)_6O_2$ and $Ca_3La_2[(Si,P)O_4]_3F$ crystals. Crystallization behavior of Rare earth in CaO-SiO_2-CaF_2-La_2O_3-(P_2O_5) Slag.

Key words: rare earth; crystalline phase; P_2O_5; temperature; cooling rate

3　炼钢与连铸

大会特邀报告

分会场特邀报告

矿业工程

炼铁与原燃料

★ 炼钢与连铸

轧制与热处理

表面与涂镀

金属材料深加工

先进钢铁材料及其应用

粉末冶金

能源、环保与资源利用

冶金设备与工程技术

冶金自动化与智能化

冶金物流运输

其他

3.1 炼 钢

复合粉剂包芯线应用研究

康 伟[1,2]，栗 红[1,2]，曹 东[1,2]，常桂华[1,2]，廖相巍[1,2]

（1. 海洋装备金属材料及应用国家重点实验室，辽宁鞍山 114009；
2. 鞍钢钢铁集团研究院炼钢技术研究所，辽宁鞍山 114009）

摘 要：将制备的 Fe-Si+CaO 复合粉剂包芯线加入钢液，研究其对去除夹杂物的作用，结果表明：复合粉剂包芯线能够加入到钢液，其中粉剂可以扩散分布到钢液，与钢液中夹杂物形成硅钙铝复合夹杂物，外观呈圆形，容易聚集上浮排出钢液，工业应用中，复合粉剂包芯线在不影响钢种成分控制的情况下，能够起到去除夹杂净化钢液的作用，应用复合粉剂的罐次铸坯 T[O]、夹杂物面积含量、夹杂物数密度显著下降，残留在铸坯中的夹杂物尺寸绝大多数小于 4μm，未出现大于 10μm 夹杂物。

关键词：复合粉剂包芯线；钢液纯净化；钢液去夹杂；复合夹杂物

Study on Application of Composite Powder Cored Wire

Kang Wei[1,2], Li Hong[1,2], Cao Dong[1,2], Chang Guihua[1,2], Liao Xiangwei[1,2]

(1.Key Laboratory of Metal Materials for Marine Equipment and Application, Anshan 114009, China;
2. Ansteel Group Iron and Steel Research Institute, Anshan 114009, China)

Abstract: Fe Si + CaO composite powder cored wire was added into liquid steel for study its effect on removing inclusions. The results show that the composite powder cored wire can be added into liquid steel, the composite powder diffuse into liquid steel and form silicon calcium aluminum composite inclusions, which is round in appearance and easy to gather and float up . In industrial trial show that composite powder cored wire using remove inclusions and purify liquid steel without affecting the control of steel grade composition, the T [O] in slab, area content and number density of inclusions decrease significantly, The inclusions size remaining in the slab are mostly less than 4μm and no inclusions larger than 10μm.

Key words: composite powder cored wire; liquid steel purification; liquid steel removing inclusions; composite inclusions

电渣重熔高温合金的熔渣物化性质研究

侯 栋[1]，王德永[1]，姜周华[2]，李花兵[2]

（1. 苏州大学 钢铁学院，江苏苏州 215000；2. 东北大学 冶金学院，辽宁沈阳 110819）

摘 要：高温合金中 Al、Ti 是 $Ni_3(Al,Ti)$相形成元素，具有强化基体作用。电渣重熔初期升温段渣-金反应易引起

Al、Ti 成分不均匀，严重致使电渣锭报废。当熔渣物性不合理时极易引起电渣锭表面缺陷。为此，本文测定了电渣重熔含 Al、Ti 高温合金 GH660 的基础物性，为优选渣系和改善电渣锭表面质量提供基础。通过对不同渣系的半球温度的实验分析可得，当 TiO_2 的含量从 4%增加到 10%，熔点先降低后急剧增加；通过 Factsage7.3 热力学软件进一步研究了上述渣系的熔化特性，将渣系成分点标在相图内，发现 12 组渣系的熔点范围在 1250~1400℃，这与实验所得的结果趋势一致，且熔点最低的 IVS3 渣系最符合冶炼所需。通过热力学计算得到了渣中 $\lg(X^3_{TiO2}/X^2_{Al2O3})$ 随钢液中 $\lg([Ti]^3/[Al]^4)$的变化情况，这与实验结果基本相一致。因此，在给定目标钢种时的铝钛含量时，可根据所得到的关系在确定目标渣系下 Al_2O_3 含量后，可以计算得到渣中 TiO_2 的添加量。

关键词：高温合金；电渣重熔；Factsage；热力学；熔渣基础物性

Physical and Chemical Properties of Slag Used for Electroslag Remelting of Superalloy

Abstract: In the superalloy, Al and Ti are Ni3(Al,Ti) phase forming elements, which have the function of strengthening matrix. The slag-gold reaction in the early stage of electroslag remelting is easy to cause the uneven composition of Al and Ti, which seriously results in the scrap of electroslag ingot. The surface defects of electroslag ingot can be easily caused when the physical properties of molten slag are unreasonable. Therefore, the basic physical properties of the Al and Ti superalloy GH660 in ESR were determined in this paper, which provides a basis for optimizing the slag system and improving the surface quality of ESR ingots. Through the experimental analysis of the hemispheric temperature of different slag systems, it can be found that when the content of TiO2 increases from 4% to 10%, the melting point decreases first and then increases sharply. Factsage7.3 thermodynamics software was used to further study the melting characteristics of the above slag systems. The composition of the slag systems was mapped in the phase diagram, and the melting points of 12 groups of slag systems were found to be in the range of 1250~1400℃. This trend is consistent with the experimental results, and the IVS3 slag system with the lowest melting point is the most suitable for smelting. The variation of $\lg(X^3_{TiO2}/X^2_{Al2O3})$ in slag with that of $\lg([Ti]^3/[Al]^4)$ in molten steel is obtained by thermodynamic calculation, which is basically in agreement with the experimental results. Therefore, when the content of Al and Ti in the target steel is given, the addition amount of TiO_2 in the slag can be calculated after the content of Al_2O_3 in the target slag system is determined according to the obtained relationship.

Key words: superalloy; electroslag remelting; factsage; thermodynamics; physical properties of slag

BOF-CC 流程生产高品质低合金钢

夏金魁，曹龙琼，刘复兴

（宝武集团鄂城钢铁有限公司炼钢厂，湖北鄂州　436002）

摘　要：文介绍了 BOF-CC 工艺生产高品质 Q355B 低合金钢的生产工艺，通过成分钛微合金化技术、钢包全程加盖技术及合金烘烤等措施有效降低转炉出钢温度，为缩短工艺流程提供了温度支撑；通过保碳出钢、出钢前期弱脱氧、钛微合金化前铝脱氧、氩站钙处理等措施解决了含钛钢浇铸结瘤问题，经生产试验对比，相对 BOF-LF-RH-CC 生产工艺，BOF-CC 流程生产 Q355B 低合金钢炼钢工序附加成本降低 100 元/吨以上，而其轧制钢板探伤合格率达到 98.5%以上，符合中国钢铁工业低碳、绿色高质量发展要求。

关键词：温降；微合金化；保碳出钢；低成本；高品质

Compact Production Process of High Quality Low Alloy Steel

Xia Jinkui, Cao Longqiong, Liu Fuxing

(Steelmaking Plant of Ecsteel, China Baowu Steel Group, Ezhou 436002, Hubei, China)

Abstract: The production process based on BOF-CC for high quality Q355B is introduced. By means of micro-alloying, ladle capping, alloy preheating operation, the tapping temperature of BOF is effectively reduced, which provides a temperature support for shortening the process flow. By means of carbon holding tapping, semi- deoxidization in the early stage of tapping, aluminum deoxidization before micro-alloying, calcium treatment in argon station, problem of nozzle blocking in continuous casting process is solved. Compared with BOF-LF-RH-CC tradition alprocess, the additional cost of BOF-CC process for Q355B production process is reduced by more than 100 yuan per ton, while the qualified rate of rolled steel plate flaw detection is over 98.5%, which meets the requirements of low carbon, green and high quality development of China's iron and steel metallurgy industry.

Key words: temperature drop; micro-alloying; holding carbon of steel tapping; low cost; high quality

260t 转炉高硅铁水单渣生产实践

李 超[1]，金 辉[2]，陈 晨[1]

（1. 鞍钢股份鲅鱼圈钢铁分公司，辽宁营口 115007；
2. 鞍钢集团鞍钢教培中心，辽宁鞍山 114000）

摘 要： 本文通过对高硅铁水操作相关理论与实践研究，结合本厂实际情况对温度制度、造渣制度、终点控制与供氧制度进行了优化改进。采用生白云石压渣、冷料计算与加入时机、规范三批次加料、终点控制原则与枪位控制要求等改进手段，基本实现高硅铁水无喷溅，同时，高硅铁水熔时降低 7min。

关键词： 温度；造渣；喷溅；枪位

260t Converter Production Practice of Once Making Slag for High Silicate Iron

Li Chao[1], Jin Hui[2], Chen Chen[1]

(1. Bayuquan Iron&Steel Subsidiary of Angang Steel Co., Ltd., Yingkou 115007, China;
2. Angang Group Angang Education and Training Center, Anshan 114000, China)

Abstract: This paper studies the theory and practice of high silicon iron operatio, according of our factory situation, it improves the temperature system and slag-making and the end controlling and the oxygen supply system. Using improvement methods such as raw dolomite suppressing slag, cold material calculation and addition time, standard three-batch addition, end controlling principle and lance controlling requirements, basically realize high iron silicon iron without splashing, while high iron silicon iron producing time is reduced by 7min.

Key words: temperature; slag-making; splashing; oxygen lance position

260t 转炉溅渣工艺优化实践

李　超，陈　晨，崔福祥，王富亮，李海峰，刘　博

（鞍钢股份鲅鱼圈钢铁分公司，辽宁营口　115007）

摘　要：本文通过对溅渣相关理论与实践研究，结合本厂实际情况对终渣、溅渣操作进行了优化改进。通过确定溅渣基本原则、保证溅渣条件、针对不同阶段与炉况如何溅渣、保证溅渣时间、溅渣与底吹协调等手段，实现平均每炉次减少补炉翻料时间 1.3min、炉衬喷补维护时间减少 2.1min，转炉的作业率提高 0.7%。

关键词：溅渣；溅渣条件；溅渣时间；底吹

Optimization Practice of Slag Splashing in 260t Converter

Li Chao, Chen Chen, Cui Fuxiang, Wang Fuliang, Li Haifeng, Liu Bo

(Bayuquan Iron & Steel Subsidiary of Angang Steel Co., Ltd., Liaoning Yingkou 115007, China)

Abstract: This paper optimizes and improves the operation of final slag and slag splashing by studying the theory and practice of slag splashing. Through determining the basic principle of slag splashing, ensuring the prerequisite of slag splashing, slag splashing aiming at different stages and furnace conditions, ensuring the time of slag splashing, coordinating the slag splashing and bottom blowing, the average feeding time per furnace is reduced by 1.3min, the lining spraying maintenance time is reduced by 2.1min, and the operating rate of converter is increased by 0.7%.

Key words: slag splashing; the prerequisite of slag splashing; the time of slag splashing; bottom blowing

顶底复吹转炉用镁碳砖的损毁机理

崔园园[1]，高　攀[1]，温太阳[1]，钟　凯[1]，邵俊宁[2]

（1. 首钢集团有限公司技术研究院，北京　100043；

2. 北京首钢股份有限公司，河北迁安　064404）

摘　要：为解决顶底复吹转炉用镁碳砖侵蚀严重问题，进行镁碳砖的残砖、抗渣和喷吹试验研究。结果表明：镁碳砖残砖分为挂渣层、反应层和原质层。挂渣层物相以 $CaO-SiO_2$ 和 FeO 为主，少量 MgO-FeO 固溶体；反应层物相主要是 MgO 和 MgO-FeO 固溶体；随着渣侵深入，Ca、Si 和 Fe 含量逐渐减少；镁碳砖的抗侵蚀和抗冲刷能力受砖本身材质和制砖工艺、石墨氧化、炉渣和气氛等综合影响；随着温度或气体流量增大，镁碳砖的磨损量、喷吹深度和喷吹直径增大，抗冲刷能力逐渐减弱。在实际使用过程中，结合所喷气体种类、底吹强度对转炉炉底所用耐材进行优化选材。

关键词：转炉；镁碳砖；抗渣侵；抗冲刷

Damage Mechanism of Magnesia Carbon Brick for Top-bottom Combined Blowing Converter

Cui Yuanyuan[1], Gao Pan[1], Wen Taiyang[1], Zhong Kai[1], Shao Junning[2]

(1. Research Institute of Technology of Shougang Group Co., Ltd., Beijing 100043, China;
2. Beijing Shougang Co., Ltd., Qian'an 064404, China)

Abstract: In order to solve the serious erosion problem of magnesia-carbon bricks for top-bottom combined blowing converters, the residual bricks, slag resistance and injection tests of magnesia-carbon bricks were studied. It is preliminarily believed that the residual magnesia-carbon bricks are divided into a slag layer, a reaction layer and an original layer. The phases of the slag layer are mainly CaO-SiO_2 and FeO, with a small amout of MgO-FeO solid solution. The phases of the reaction layer are mainly MgO and MgO-FeO solid solution. As the slag penetrates deeper, the content of Ca, Si and Fe gradually decreases. The anti-erosion and anti-scouring ability of magnesia-carbon bricks are affected by material and manufacturing of the brick, graphite oxidation, slag and atmosphere. As the temperature or gas flow increases, the wear volume, the injection depth and the injection diameter of the magnesia-carbon bricks increase and the erosion resistance gradually weakens. In the actual use process, the refractory materials used for bottom of the converter are optimized in combination with the type of gas injected and the strength of bottom blowing.

Key words: converter; magnesia-carbon brick; anti-erosion; anti-scouring

转炉钢水利用氮气增氮工艺实践

李洪涛[1]，蒋晓放[2]，马志刚[1]，吴亚明[1]

（1. 宝山钢铁股份公司炼钢厂，上海 200941；
2. 宝山钢铁股份有限公司中央研究院，上海 201999）

摘 要：本文从理论上分析了钢液增氮的热力学和动力学影响因素，研究了300吨转炉全程底吹氮气、不同底吹风口状态、不同底吹氮气供气强度、转炉停吹镇静时间、出钢合金化以及出钢过程钢包底吹氮气对钢水增氮的影响，结果表明转炉底吹风口模糊、氮气供气强度提高、出钢过程充分脱氧以及钢包底吹氮气有利于提高转炉钢包钢水氮成分。300t转炉工业试验表明利用转炉全程吹氮气和出钢钢包底吹氮气，能够将转炉钢水氮含量稳定控制在0.0100%以上。

关键词：300t转炉；底吹系统；氮气供气强度；钢包底吹氮气

Practice of Increasing Nitrogen in Converter Molten Steel by Using Nitrogen

Li Hongtao[1], Jiang Xiaofang[2], Ma zhigang[1], Wu Yaming[1]

(1. Steelmaking Plant, Baoshan Iron & Steel Co., Ltd., Shanghai 200941, China;
2. Research institute, Baoshan Iron & Steel Co., Ltd., Shanghai 201999, China)

Abstract: In this paper, the thermodynamic and kinetic factors of nitrogen increase in molten steel are analyzed theoretically.

The effects of bottom blowing nitrogen in the whole process of 300 t converter, different bottom blowing tuyere conditions, different bottom blowing nitrogen supply intensity, converter blowing stop and sedation time, tapping alloying and bottom blowing nitrogen gas in ladle on nitrogen increase in molten steel are studied Sufficient deoxidation during tapping and bottom blowing of nitrogen in ladle are beneficial to increase the nitrogen content of molten steel in converter ladle. The industrial test of 300 t converter shows that the nitrogen content in molten steel can be controlled above 0.0100% by using the whole process of blowing nitrogen in converter and bottom blowing nitrogen in ladle.

Key words: 300 t converter; bottom blowing system; nitrogen supply intensity; bottom blowing nitrogen in ladle

精炼含 S 钢 A 类夹杂物工艺技术的研究

付谦惠，黄继利

（宝武集团韶关钢铁有限公司炼钢厂，广东韶关 512123）

摘 要： 随着汽车工业的发展，含 S、A1 钢（S：0．010%~0．070%；A1：0.010%~0.050%）的需求不断增长，使得连铸具备生产含 S、A1 钢的能力，本文摸索出一种稳定增硫、可浇性良好含 S 钢精炼工艺。通过研究含 S 钢中 A 类夹杂物的关键控制技术，提升含 S 钢 A 类夹杂物控制能力，提高钢水可浇性，实现连铸连浇最高炉数 5 炉以上，通过改善含 S 钢中 A 类夹杂物，使含 S 钢 A 类夹杂物控制合格率≥80%。

关键词： 含 S、A1 钢；连铸；精炼工艺；A 类夹杂物；合格率

Study on Refining Technology of CLass A Inclusions in Steel Containing S

Fu Qianhui, Huang Jili

(Shaoguan Iron & Steel Co., Ltd., Baowu Group, Shaoguan 512123, China)

Abstract: With the development of automobile industry, steel containing S, A1 (S: 0.010%~0.070%; A1: 0.010%~ 0.050%), so that continuous casting has the conditions to produce steel containing S and A1. In this paper, a kind of steel-containing S refining process with stable increasing sulfur and good pouring ability has been found out. By studying the key control technology of Class A inclusions in steel containing S, the control ability of Class A inclusions in steel containing S is improved, the pouring ability of molten steel is improved, and the maximum number of blast furnaces for continuous casting and continuous casting is more than 5 furnaces. By improving Class A inclusions in steel containing S, the qualified rate of Class A inclusions in steel containing S is greater than 80%.

Key words: steel containing S and A1; continuous casting; refining process; class A inclusions; percent of pass

合金杂质含量分析及合金化工艺优化实践

黎均红[1]，刘晓峰[1]，王少波[1,2]，张 波[1,3]，郝 苏[1]

（1. 重庆钢铁股份有限公司炼钢厂，重庆 401258；2. 宝山钢铁股份有限公司炼钢厂，上海 201900；3. 广东韶钢松山股份有限公司炼钢厂，广东韶关 512100）

摘 要：没有洁净的铁合金，就无法生产高质量的洁净钢。通过对重庆钢铁精炼工序使用的硅铁合金、高碳锰铁、金属锰铁、钛铁和硅钙包芯线进行检测分析，采取完善内控标准、同类型合金替代、合金化工艺优化等改进措施，精炼工序的主要技术经济指标改善明显，取得了良好的实践效果。

关键词：铁合金；氧含量；氮含量；标准；合金化

Analysis of Alloy Impurity Content and Optimization of Alloying Process

Li Junhong[1], Liu Xiaofeng[1], Wang Shaobo[1,2], Zhang Bo[1,3], Hao Su[1]

(1. Steelmaking Plant of Chongqing Iron and Steel Co., Ltd., Chongqing 401258, China;
2. Steelmaking Plant of Baoshan Iron and Steel Co., Ltd., Shanghai 201900, China;
3. Steel Making Plant of Songshan Iron and Steel Co., Ltd., Shaoguan 512100, China)

Abstract: It is impossible to produce high quality clean steel without clean ferroalloy. Based on the detection and analysis of ferrosilicon alloy, high carbon ferromanganese, metallic ferromanganese, ferrotitanium and calcium silicon cored wires used in the refining process of Chongqing Iron and Steel Co., Ltd., the main technical and economic indexes of the refining process have been significantly improved by taking improvement measures such as improving internal control standards, substituting the same type of alloy and optimizing the alloying process.

Key words: ferroalloy; oxygen content; nitrogen content; standard; alloying

120t 转炉氧枪喷头设计与应用

宋 健，孙 波，王 勇，吴发达，解文中，牛金印

（马鞍山钢铁股份有限公司第一钢轧总厂，安徽马鞍山 243000）

摘 要：针对某厂 120t 转炉炉型特点和冶炼工艺情况，对原有氧枪 4 孔喷头参数进行了优化设计。对氧枪喷孔数、喷孔夹角、工况压力、喉口直径、出口直径和扩张段长度等都进行了适当调整，优化后的氧枪喷头马赫数由 1.98 提高到 2.05。结合该厂生产实践表明，优化后的氧枪 5 孔喷头可增加转炉供氧强度，优化转炉内动力学反应条件，吹氧时间平均缩短 128s，脱磷率提高 3.79%，终渣 w(TFe)量降低 2.75%，铁水消耗降低 22 kg/t，不仅有利于转炉提高废钢比，还可提高金属收得率。

关键词：转炉；氧枪；喷头；参数优化；供氧强度；脱磷率

Design and Application of 120 t Converter Oxygen Lance Nozzle

Song Jian, Sun Bo, Wang Yong, Wu Fada, Xie Wenzhong, Niu Jinyin

(Steel Making and Rolling General Plant of Ma’anshan Iron and Steel Co., Ltd., Ma'anshan 243000, China)

Abstract: Aiming at the characteristics of the 120t converter and smelting process of a certain plant. The parameters of the original four-hole nozzle of the oxygen lance were optimized. The number of nozzle holes, nozzle angle, operating pressure, throat diameter, outlet diameter and expansion length have all been adjusted appropriately. The optimized oxygen lance nozzle Mach number has been increased from 1.98 to 2.05. Combined with the production practice of the plant, it shows that the optimized 5-hole nozzle of oxygen lance can increase the oxygen supply intensity of converter, optimize the dynamic reaction conditions in converter, shorten the oxygen blowing time by 128s，increase the dephosphorization rate by 3.79%, decrease the w(TFe)offinal slag by 2.75%, and decrease molten iron consumption by 22 kg/t, which is not only conducive to improvingthe scrapsteel ratio of converter, but also to improving the metal yield.

Key words: converter; oxygen lance; nozzle; parameter optimization; oxygen supply intensity; dephosphorization rate

120t 转炉冶炼过程硫含量控制与分析

孙　波，解养国，王　勇，吴发达，宋　健，牛金印

（马鞍山钢铁股份有限公司第一钢轧总厂，安徽马鞍山　243000）

摘　要：通过对渣钢间脱硫反应热力学计算，在结合某厂生产实践的基础上，分析了该厂 120t 转炉冶炼低硫洁净钢种(w([S])≤0.0060%)时相关增硫因素，提出减少 KR 脱硫渣、废钢和造渣料等入炉原辅料硫质量分数和优化转炉冶炼工艺参数等控制措施，为转炉冶炼低硫钢水和加强终点硫控制提供参考借鉴。结果表明，控制入炉原辅料钢水增硫质量分数 Δw([S])≤0.0082%，转炉终点温度为 1640~1680℃，炉渣碱度为 3.0~4.0，渣中 w((FeO))为 10%~18%，渣钢硫分配比 LS 为 5.11~8.16，转炉终点硫质量分数合格率由 80%提升至 98%。

关键词：转炉；低硫钢；增硫；热力学；脱硫；控制

Analysis and Control on Sulfur Content in the Process of 120t Converter Steelmaking

Sun Bo, Xie Yangguo, WangYong, Wu Fada, Song Jian, Niu Jinyin

(Steel Making and Rolling General Plant of Ma'anshan Iron and Steel Co., Ltd., Ma'anshan 243000, China)

Abstract: Based on the thermodynamic calculation of the desulfurization reaction between slag and steel and the production practice of a certain plant, the relevant sulfur increase factors when smelting low-sulfur clean steel (w([S])≤0.0060%) in 120t converter of the plant are analyzed. Control measures such as reducing the sulfur content of raw and auxiliary materials such as KR desulfurization slag, scrap and slagging materials and optimizing converter smelting process parameters are proposed. It provides a reference for converter smelting low-sulfur molten steel and strengthening end-point sulfur control. The results show that the sulfur-increasingcontent of raw and auxiliary materials entering the furnaceless than0.0082%, the converter end temperature is 1640~1680℃, the slag basicity is 3.0~4.0, and the w((FeO)) in the slag is 10%~ 18%, the slag steel sulfur distribution ratio LS is 5.11~8.16, and the final sulfur contentpass rate of the converter is increased from 80% to 98%.

Key words: converter; low sulfur content steel; resulfurization; thermodynamics; desulfurization; control

KR 新型脱硫剂的研究

宋吉锁[1]，曹　祥[1]，王一名[1]，吴跃鹏[1]，曹　琳[1]，曾　涛[2]

（1.鞍钢股份炼钢总厂，鞍山辽宁　114021，2.鞍钢工程技术公司，辽宁鞍山　114069）

摘　要：本文对KR脱硫机理进行简单的阐述，研究用铝矾土代替萤石制作脱硫剂，在保证脱硫剂中Al_2O_3含量在5%~10%，石灰活性度在280mL以上的新型脱硫剂能达到普通的脱硫剂效果。

关键词：铁水预处理；KR法

Research of New Desulfurizer for KR

Song Jisuo[1], Cao Xiang[1], Wang Yiming[1], Wu Yuepeng[1], Cao Lin[1], Zeng Tao[2]

(1. General Steelmaking Plant of Angang Steel Co., Ltd., Anshan 114021, China;
2. Ansteel Engineering Technology Co., Ltd., Anshan 114069, China)

Abstract: In this ariticle, the desulfurizing mechanism was simply discussed. In this research, bauxite was used as desulfurizer to replace fluorite. Under the condition of ensuring Al_2O_3 content between 5%-10%, new desulfurizer achieved the same desulfurizing effects of normal kind, with the reactivity of quick-lime above 280mL.

Key words: hot metal pretreatment; KR desulfurization

声纳化渣结合煤气分析仪应用实践

尚世震[1]，李　旭[1]，张　帅[1]，马　超[2]

（1.鞍钢股份有限公司炼钢总厂，辽宁鞍山　114021；
2.鞍钢集团钢铁研究院，辽宁鞍山　114009）

摘　要：本文介绍了炼钢总厂一分厂3座90t顶吹转炉，通过声纳化渣曲线与转炉冶炼过程实际化渣情况对比，得到了冶炼过程工艺参数与化渣曲线的相关性，并且通过煤气分析仪的CO变化数值辅助分析冶炼过程化渣操作，改变了原有落后的经验判断的准确率低，随意性强的现象，使转炉的跑渣率下降2.5%，一拉出钢率提高5%，化渣合格率提高35%，平均冶炼周期降低2min/炉。

关键词：声纳化渣；煤气分析仪；顶吹转炉

高锰铁水双渣法冶炼余锰含量控制生产实践

李玉德，李叶忠，陈志威，齐志宇，张立辉

（鞍钢股份有限公司，鞍山　114000）

摘　要：鞍钢260t转炉高锰铁水生产成品锰≤0.10%类钢种，采用双渣法冶炼模式，确定合理的双渣时机，降低双渣前炉渣碱度，优化转炉吹炼过程枪位、拉碳枪位及拉碳时间，控制转炉终点温度等措施，转炉一拉终点锰≤0.10%合格率可到达95%以上，可满足生产低锰类钢种成分要求。

关键词：高锰铁水；双渣法；余锰；低锰钢种

Double Slag Method Production Practice of Controlling Remaining Manganese Content in Smelting High Manganese Hot Metal

Li Yude, Li Yezhong, Chen Zhiwei, Qi Zhiyu, Zhang Lihui

(Ansteel Co., Ltd., Anshan 114000, China)

Abstract: Ansteel 260t converter high manganese hot metal production finished manganese less 0.10% steel grade, using double slag smelting mode, Determine the reasonable time of double slag, reducing basicity of slag before double slag, optimization the position of oxygen lance and carbon drawing time, control of converter end point temperature, the qualified rate of finished manganese less 0.10% is over 95%, It can meet the composition requirements of low manganese steel.

Key words: high manganese hot metal; double slag method; residual manganese; low manganese steel

不同脱氧合金对 LF 无氟精炼渣脱硫影响的工业实践

杜　林，孙　群，刘　磊

（鞍钢股份有限公司炼钢总厂，辽宁鞍山　114021）

摘　要： LF 冶炼采用不同的脱氧合金会对顶渣的渣系造成影响，在不采用含氟的精炼渣精炼情况下，顶渣的流动性在不同脱氧合金下区别较大。在 $CaO\text{-}SiO_2\text{-}Al_2O_3$ 三元渣系中调整助熔渣的加入量同样可以获得较低熔点区的渣系，对比两不同脱氧合金使用下 LF 炉处理过程，结果表明：硅预脱氧相比于铝脱氧而言成本更低，但整体渣系流动性偏差，脱硫效果不好。

关键词： 脱氧合金；LF 精炼；$CaO\text{-}SiO_2\text{-}Al_2O_3$ 渣系；脱硫

Industrial Practice of the Effect of Different Deoxidizing Alloys on LF Desulphurization Using Fluoride Free Refining Slag

Du Lin, Sun Qun, Liu Lei

(Steelmaking Plant of Ansteel Company Limited, Anshan 114021, China)

Abstract: LF refining using different deoxy alloy will affect the slag system, in the case of not using fluorine-containing refining, the fluidity of slag in different deoxy alloy is different. In $CaO\text{-}SiO_2\text{-}Al_2O_3$, the slag system with lower melting point can also be obtained by adjusting the amount of fluxing slag in the system. Comparing the treatment process of two different deoxidized alloys in LF furnace, the results show that the cost of Si deoxidization is lower than that of Al deoxidization, but the fluidity deviation of the whole slag system results in poor desulfurization effect.

Key words: deoxidized alloy; LF refining; $CaO\text{-}SiO_2\text{-}Al_2O_3$ slag system; desulfurization

废钢智能识别系统的应用及展望

林 海，李国军，董淑华

（鞍钢集团朝阳钢铁有限公司，辽宁朝阳 122000）

摘 要：由于废钢判级“目视”判别的特殊性，人为因素影响较大，国内各钢厂虽然采取了各种预防措施，效果仍不理想。目前国内已开发出智能废钢验质系统，在多家钢铁厂推广应用。一定程度上解决了当前人工废钢评级带来的识别不准，客观性无法保证等问题。本文就当前智能验质系统发展现状，存在问题，及发展前景进行分析、论述。

关键词：废钢；智能；验质；判级

Application and Prospect of Scrap Intelligent Identification System

Lin Hai, Li Guojun, Dong Shuhua

(Anshan Iron and Steel Group Chaoyang Iron and Steel Co., Ltd., Chaoyang 122000, China)

Abstract: Due to the particularity of “visual” discrimination of scrap grade, the influence of human factors is great. Although various preventive measures have been taken by domestic steel mills, the effect is still not ideal. At present, the intelligent scrap quality inspection system has been developed in China and applied in many steel works. To a certain extent, it solves the problems of identification inaccuracy and objectivity which are caused by the current manual scrap grade. This paper analyzes and discusses the current development status, existing problems and development prospects of intelligent quality testing system.

Key words: scrap steel; intelligent; check quality; sentence level

RH 前后钙处理对高钛钢夹杂物的影响

舒宏富，熊华报，霍 俊

（马鞍山钢铁股份有限公司技术中心，安徽马鞍山 243000）

摘 要：针对在“LF→RH→钙处理”路径生产高钛钢时存在钛含量控制不稳定的问题，本文开展将钙处理调整至LF结束RH前的试验，并与原工艺进行对比以考察夹杂物的变化。结果表明：（1）RH前喂钙线炉次的钛损平均比RH后喂钙线炉次的少0.0041%；（2）RH前喂钙线的夹杂物数量和尺寸都小于RH破空后喂线，且夹杂物的类型更接近于钙铝酸盐，对热轧高强钢成品的力学性能和冲击功影响不大。

关键词：钙处理；夹杂物；高钛含量高强钢；钢洁净度

Effect of Calcium Treatment before and after RH on Inclusions in the Steel with High Titanium Content

Shu Hongfu, Xiong Huabao, Huo Jun

(Maanshan Iron and Steel Co., Ltd., Ma'anshan 243000, China)

Abstract: Aiming at the problem of unstable control of titanium content in the production of high titanium steel by "LF→ RH→ Ca treatment" path, the experiment of adjusting calcium treatment to the end of LF was carried out, and the change of inclusions was investigated by comparing with the original process. The results show that: (1) The titanium loss of pre-RH calcium was 0.0041% less than that of post-RH calcium treatment. (2)The number and size of inclusions in the pre-RH calcium treatment are smaller than those in the post-RH, and the type of inclusions is more similar to calcium aluminate, which has little effect on the mechanical properties and impact energy of hot-rolled HSS.

Key words: calcium treatment; inclusions; high titanium content high strength steel; steel cleanliness

钢渣尾渣在转炉造渣的初探

韩东亚

（新疆八一钢铁股份有限公司炼钢厂，新疆乌鲁木齐 830022）

摘 要： 炼钢钢渣尾渣内含有一定量的 CaO、FeO，通过其替代部分渣料，改善转炉造渣条件及达到少渣冶炼目的，实现了炼钢钢渣尾渣的有效利用。

关键词： 钢渣尾渣；造渣；有效利用

Preliminary Study on Slag Formation of Steel-making in Converter

Han Dongya

(Xinjiang Ba Yi Iron & Steel Co., Ltd., Urumqi 830022, China)

Abstract: There are some CaO and FeO in steel-making slag tail slag. By replacing part of slag, the slag-making condition of converter is improved and less slag smelting is achieved, the effective utilization of steel-making slag tail slag is realized.

Key words: steel slag; slag-making; effective utilization

120t 转炉双联脱硅法脱硅炉造渣制度的分析与探讨

孙学刚

（新疆八一钢铁股份有限公司第一炼钢厂，新疆乌鲁木齐 830022）

摘　要：为应对欧冶炉高硅铁水 120t 转炉参考双联脱磷工艺开发出了双联脱硅的工艺方法，在脱硅炉内铁水[Si]含量从正常的不到 1%提高到了 5%，硅氧化过程中必然会释放出大量热，由于废钢配比不足，脱硅工艺最大困难就是保持热平衡，通过研究脱硅炉次的渣料加入量、渣成分、渣碱度、渣矿相，推导出最优的转炉双联脱硅造渣制度。

关键词：高硅铁水；双联脱硅；造渣制度；矿相；碱度

Analysis and Discussion on Slag-making System of 120t Converter Double Desiliconization

Sun Xuegang

(No.1 Steel-making Plant, Xinjiang Bayi Iron & Steel Co., Ltd., Urumqi 830022, China)

Abstract: In order to cope with 120t converter with high silicon hot metal in COREX to double-link dephosphorization process, a double-link desiliconization process has been developed. The content of hot metal [Si] in the desiliconization furnace has been increased from less than 1% to 5% in normal condition.A lot of heat will be released during the silica oxidation process. Due to the insufficient ratio of scrap steel, the biggest difficulty in the desiliconization process is to maintain the thermal balance. By studying the amount of slag added, the composition of slag, the alkalinity of slag and the phase of slag in the desiliconization furnace, the optimal system of double desiliconization slagging in converter is deduced.

Key words: high silicon hot metal; duplex desilication; slagging system; mineralogical phase; alkalinity

高氮复合合金在 HRB400E 钢上的应用

马进国

（新疆八一钢铁股份有限公司，新疆乌鲁木齐　830022）

摘　要：在八钢热轧带肋钢生产中，通过添加一种高氮复合合金替代贵重钒氮合金试验，取得了一定的成效，在满足产品成分及性能合格的同时实现合金成本的进一步降低。

关键词：热轧带肋钢；高氮复合合金；替代；合金成本

汽车大梁钢（B510L）精炼过程钢中全氧分析探讨

李立民，刘军威

（宝武集团新疆八一钢铁股份有限公司，新疆乌鲁木齐　830022）

摘　要：生产汽车大梁钢时，通过取样分析 LF 精炼炉冶炼过程不同阶段钢中全氧含量，研究了 $CaO-Al_2O_3-SiO_2-MgO-CaF_2$ 五元渣系下组分的不同、钢中铝含量、钢中[Ca]/[Al]、软吹时间对钢中全氧含量的影响，结合生产实际，优化渣系中[CaO]/[Al_2O_3]比值在 1.5~2.0，钢中铝含量为 0.020%~0.030%，钢中[Ca]/[Al]比值≥0.1，软吹时间控制在 14~20min，精炼冶炼结束钢中全氧含量≤25×10^{-6}。

关键词：LF 精炼炉；全氧含量；大梁钢

Discussion on Total Oxygen Analysis in Automobile Beam Steel (B510L) Refining Process

Li Limin, Liu Junwei

(Baowu Xinjiang Bayi Iron &Steel Stock Co., Ltd., Urumqi 830022, China)

Abstract: as the automobile big beam steel production, different stages through the sampling analysis of LF refining process total oxygen content in the steel, and studied the $CaO\text{-}A1_2O_3\text{-}SiO_2\text{-}MgO\text{-}CaF_2$ five yuan slag series of components under different, steel aluminum content, [Ca]/[Al] in steel, soft blowing time effect on the total oxygen content in steel, combined with the actual production, optimization of slag ratio of $[CaO]/[A1_2O_3]$ in 1.5~2.0, steel aluminum content is 0.020%~0.030%, [Ca]/[Al] in steel ratio of 0.1 or higher, soft blowing time control in 14~20min, The total oxygen content in the steel after refining is no more than 25×10^{-6}.

Key words: LF refining furnace;total oxygen content;big beam steel

120t 转炉复吹工艺优化的物理模拟试验

张　岭[1,3]，于　波[2]，钟良才[1,3]，王立新[2]，贺龙龙[1,3]，翁　莉[2]

（1. 东北大学低碳钢铁前沿技术研究院，辽宁沈阳　110819；2. 建龙集团抚顺新钢铁炼钢厂，辽宁抚顺　113001；3. 东北大学冶金学院，辽宁沈阳　110819）

摘　要：以 120t 顶底复吹转炉为原型，依据相似原理在实验室开展了几何相似比为 1:10 的顶底复吹转炉复吹工艺物理模拟试验，研究了顶吹气体流量、底吹气体流量以及氧枪枪位对于转炉熔池搅拌混匀的影响，结果表明：当顶吹流量为 $122m^3/h$，底吹流量为 $1.074m^3/h$，氧枪枪位为 200mm 时，熔池的搅拌效果最好，混匀时间最短，冲击深度约为熔池深度的一半，在合理范围之内。本试验条件下认为复吹时枪位在 200mm 可获得较短的混匀时间，后搅时底吹流量为 $1.608m^3/h$ 时搅拌效果好，混匀时间较低。

关键词：复吹转炉；熔池；搅拌混匀；冲击深度；物理模拟

Physical Modeling Experiment on Combined Blown Process Optimization in 120t Converter

Zhang Ling[1,3], Yu Bo[2], Zhong Liangcai[1,3], Wang Lixin[2], He Longlong[1,3], Weng Li[2]

(1. Advanced Research Institute of Low Carbon Steel, Northeastern University, Shenyang 110819, China; 2. Jianlong Fushun New Iron and Steel Plant, Fushun 113001, China; 3. School of Metallurgy, Northeastern University, Shenyang 110819, China)

Abstract: Physical experiment in a model with a geometric similarity ratio of 1:10 as a prototype based on the similarity theory was conducted in the laboratory for a 120 t top-bottom combined blown converter. The effects of top gas flow rate, bottom gas flowrate and top lance height on the bath mixing were studied. The results show that when the top gas flow rate was 122 m^3/h, the bottom blowing flow rate was 1.074 m^3/h, and the top lance height was 200 mm, the mixing effect in the converter bath was the best and the mixing time was the shortest. The impact depth is about half of the converter bath depth, which was within a reasonable range. Under the conditions in the study, it was considered that 200 mm top lance height in the top and bottom combination blowing can obtain a shorter bath mixing time. In the post-stirring, the bath stirring effect is batter, and the mixing time is lower with the bottom gas flow rate of 1.608 m^3/h.

Key words: combined blown converter; bath; stirring and mixing; impact depth; physical modeling

基于随机森林算法的转炉石灰加入量模型

杨仕存[1,2]，钟良才[1,2]，赵 阳[3]，张 岭[1,2]，贺龙龙[1,2]

（1. 东北大学低碳钢铁前沿技术研究院，辽宁沈阳 110819；2. 东北大学冶金学院，辽宁沈阳 110819；3. 建龙集团抚顺新钢铁科技处，辽宁抚顺 113001）

摘 要：针对转炉吹炼石灰加入量，基于 100t 转炉炼钢实际生产数据，通过对数据预处理和特征选择，提出随机森林石灰加入量预测模型。采用 960 炉的实际生产数据对模型进行训练，240 炉的数据用于验证模型的预测效果。实际生产数据仿真结果显示预测误差范围在±400kg、±300kg 和±200kg 的命中率分别为 93.33%，81.25%和 62.67%。相较于支持向量机模型，随机森林模型有着更高的预测精度。

关键词：石灰加入量；转炉炼钢；随机森林模型；命中率

Lime Addition Model of Converter based on Random Forest Algorithm

Yang Shicun[1,2], Zhong Liangcai[1,2], Zhao Yang[3], Zhang Ling[1,2], He Longlong[1,2]

(1. Advanced Technology Research Institute of Low Carbon Steel, Northeastern University, Shenyang 110819, China; 2. School of Metallurgy, Northeastern University, Shenyang 110819, China; 3. Science and Technology Division, Jianlong Fushun New Iron and Steel, Fushun 113001, China)

Abstract: Aiming at the lime addition in converter blowing, a prediction model of lime addition with random forest was proposed by data preprocessing and feature selection from the actual production data of 100t converter steelmaking in a steel plant. The actual production data of 960 heats were used to train the model, and another 240 heats were used to verify the accuracy of the model. The simulation results of actual production data showed that the hit rate of prediction error in the range of ±400kg, ±300kg and ±200kg is 93.33%, 81.25% and 62.67%, respectively. Compared with the support vector machine, the random forest model has higher prediction accuracy.

Key words: amount of lime addition; converter steelmaking; random forest model; hit rate

红钢 50t 转炉终渣成分分析与探讨

和 浩，邱 肖，杨冠龙，程 蛟，茶维杰，杨昌文

（红河钢铁有限公司炼钢厂，云南蒙自 661100）

摘　要：采用热力学平衡模型对红钢终渣成分进行计算分析，结果表明，红钢冶炼终点渣—钢反应远未达到平衡，其主要原因是受操作工艺参数的影响，通过生产实践实验对红钢的工艺参数进行优化，使冶炼过程喷溅降低，进一步提高金属收得率、一到合格率。有效降低了钢铁料消耗。

关键词：转炉；吹炼终点；终渣成分

Analysis and Discussion on the Final Slag of Honggang 50t Converter

He Hao, Qiu Xiao, Yang Guanlong, Cheng Jiao, Cha Weijie, Yang Changwen

(Steel Mill of Honghe Iron and Steel Co., Ltd., Mengzi 661100, China)

Abstract: A thermodynamic equilibrium model calculation and analysis of red steel end slag composition, the results showed that the red steel slag smelting end-steel reaction is far from equilibrium, the main reason is because of the influence of operating parameters, through production practice test to optimize the process parameters for the red steel, reduce spillage of smelting process, further improve the metal yield, one to the percent of pass. Effectively reduce the consumption of steel material.

Key words: converter; blowing end; end slag composition

钢水夹杂物控制在 0.5 级的研究与实践

李军辉，张永亮，王克忠

（山东泰山钢铁集团有限公司，山东济南　271100）

摘　要：通过对转炉生产操作的改进与优化，采用滑板挡渣技术，出钢前期实施顶渣改质技术，出钢 1/3 加入脱氧剂，酸溶铝控制在 0.004%～0.007%以内，出钢 2/3 加入合金化合金，吹氩过程喂入 10～30m 铝线进行深脱氧，吹压站吹压 10～12min，连铸保护浇铸。既降低了合金的消耗，有增加了钢水纯净度，使夹杂物控制水平达到 0.5 级以下。

关键词：改进优化；前期加入渣洗剂；酸溶铝控制；吹氩过程深脱氧；吹氩时间

浅谈超级电弧炉技术

石秋强[1]，张豫川[2]，谈存真[2]，杨宁川[1]，黄其明[2]

（1. 中冶赛迪工程技术股份有限公司，重庆　401122；
2. 中冶赛迪技术研究中心有限公司，重庆　401122）

摘　要：通过对双石墨电极直流电弧炉技术、废钢连续加料及预热技术、IGBT 柔性直流电源技术的介绍和对比，说明了超级电弧炉技术的特点和显著优势。

关键词：直流电弧炉；双石墨电极；废钢预热；IGBT 柔性直流电源

Talking about Super Electric Arc Furnace Technology

Shi Qiuqiang[1], Zhang Yuchuan[2], Tan Cunzhen[2],
Yang Ningchuan[1], Huang Qiming[2]

(1. CISDI Engineering Co., Ltd., Chongqing 401122;
2. CISDI Research & Development Co., Ltd., Chongqing 401122)

Abstract: Through the introduction and comparison of dual graphite electrode DC electric arc furnace technology, continuous scrap feeding and preheating technology, and IGBT flexible DC power supply technology, the characteristics and significant advantages of super electric arc furnace technology are demonstrated.

Key words: DC EA; dual graphite electrodes; scrap preheating; IGBT flexible DC power supply

低锰钢冶炼工艺研究与应用

杨　峰[1]，罗海明[2]，张嘉华[2]，陈　胜[2]

（1. 内蒙古包钢钢联股份有限公司技术中心，内蒙古包头　014010；
2. 内蒙古包钢钢联股份有限公司稀土钢板材公司，内蒙古包头　014010）

摘　要：工业纯铁要求钢中锰含量控制在 0.005%以内，为了研究低锰钢冶炼工艺，对脱锰反应的热力学和动力学条件进行了分析。研究表明，钢水温度越低、炉渣氧化性越高、钢水氧含量越高、炉渣中氧化锰含量越低则锰的氧化反应越容易进行。当铁水锰含量在 0.04%~0.055%时，将转炉吹炼后钢水终点碳含量控制在 0.025%~0.035%，终点温度控制在 1620℃以下，可将钢水残锰含量降低到 0.04%~0.08%。钢水脱锰率随着转炉渣量的增加而升高。转炉出钢过程中不对钢水进行脱氧处理，钢包中的钢水和炉渣保持较高氧化性，在 LF 精炼处理过程中，可以使钢中锰含量平均降低 32%，降幅为 0.01%~0.04%。结果表明，采用转炉和 LF 炉复合脱锰工艺，可稳定的将钢中锰含量降低到 0.05%以内。

关键词：工业纯铁；脱锰反应；碳含量；温度；复合脱锰工艺

The Research and Application of Low Manganese Steel Smelting Process

Yang Feng[1], Luo Haiming[2], Zhang Jiahua[2], Chen Sheng[2]

(1. Baotou Iron and Steel Corporation Technology and Research Center, Baotou 014010, China;
2. Rare Earth Steel Plate Plant, Baotou 014010, China)

Abstract: Industrial pure iron requires the manganese content of steel should be within 0.005%. In order to study the smelting process of low manganese steel, the thermodynamic and dynamic conditions of demanganese reaction are analyzed. The research shows that the lower the liquid steel temperature, the higher the oxidation property of the slag, the higher oxygen content of the steel, and the lower the manganese oxide content in the slag, the easier the manganese oxidation reaction is. When manganese content of hot metal is 0.04%~0.05%, if the final carbon content of the steel in the converter after blowing is controlled to 0.025%~0.035%, and the final temperature is controlled below 1620℃, the residual manganese content in steel can be reduced to 0.04%~0.08%. The demanganese rate of liquid steel increases with the increase of converter slag volume. Without deoxidation treatment during tapping,the liquid steel in ladle and also the slag would maintain higher oxidizability. During the refining process in LF, the manganese content ofthe steel can be reduced by 32%, or 0.01%~0.04%. The results show that the composite manganese removal process of converter and LF can stably reduce the steel manganese content to within 0.05%.

Key words: industrial pure iron; demanganese reaction; carbon content; temperature; the composite manganese removal process

Quik-Tap 投弹副枪系统在 120t 转炉的实践与应用

杜秀峰，潘艳华，王春锋，罗焕松，刘东清，王书鹏

（武汉钢铁有限公司条材厂，湖北武汉　430083）

摘　要：本文阐述了 Quik-Tap 型投弹式副枪系统的装备结构、技术原理及其在武钢有限条材厂一炼钢分厂 120t 转炉上的应用效果。指出 Quik-Tap 终点控制系统可达到炉前人工测温检测控制的同等效果，并降低炉前测温取样安全风险，缩短转炉冶炼周期，　降低石灰消耗，减少过程温降，从而全面提高一炼钢分厂转炉的生产效率。

关键词：投弹副枪；转炉冶炼周期；效率提升；应用实践

Practice and Application of Quik-Tap Sub-gun System in 120 T Converter

Du Xiufeng, Pan Yanhua, Luo Huansong, Wang Chunfeng, Liu Dongqing, Wang Shupeng

(Plant of Long Product, Wuhan Iron & steel Co., Ltd., Wuhan 430083, China)

Abstract: In this paper, the equipment structure, technical principle and application effect of Quik-TAP projectile sub-gun system on 120t converter in No.1 steelmaking branch of WISCO Limited Bar Factory are described. It is pointed out that the Qui-TAP terminal control system can achieve the same effect as the control of manual temperature measurement in front of furnace, reduce the safety risk of temperature measurement and sampling in front of furnace, shorten the smelting cycle of converter, reduce the consumption of lime and reduce the temperature drop in the process, so as to comprehensively improve the production efficiency of converter in a steelmaking branch plant.

Key words: sub-gunsystem; converter smelting cycle; efficiency improvement; the application practice

转炉-RH 流程生产 GCr15 氧含量控制

段光豪，廖扬标，王彦林，张昌宁

（宝钢股份武汉钢铁有限公司条材厂，湖北 武汉　430083）

摘 要：本文研究了轴承钢氧含量的控制，对轴承钢中氧和铝的关系进行了分析，转炉出钢碳含量、LF 精炼加热时间、RH 处理过程的铝损耗以及浇注过程中的二次氧化，都影响钢中氧的控制。通过提高转炉终点碳，优化精炼调铝工艺，提高自浇率等措施，钢中的氧含量可稳定控制在 0.0006%左右。

关键词：轴承钢；氧含量；工艺优化

Oxygen content control of GCr15 Bearing Steel by Converter -RH Process

Duan Guanghao, Liao Yangbiao, Wang Yanlin, Zhang Changning

(Plant of Long Product , Wuhan Iron & steel Co., Ltd., Wuhan 430083, China)

Abstract: This paper studies the contol of oxygen content in the production process of bearing steel, The relationship between oxygen and aluminum in bearing steel is analyzed,The carbon content of converter tapping,heating time of LF refining,aluminum drop during RH treatment and seconday oxidation during pouring all affect the control of oxygen in steel.The oxygen content in the steel can be controlled at about 0.0006% by increasing the end-point carbon of converter, optimizing the refining and aluminum adjusting process and increasing the self pouring rate.

Key words: bearing steel; oxygen content; process optimization

精炼渣对夹杂物的冶金作用机理

任 英[1]，张立峰[2]，任昶宇[1]，成 功[1]

（1. 北京科技大学冶金与生态工程学院，北京 100083；
2. 燕山大学亚稳材料制备技术与科学国家重点实验室，河北秦皇岛 066044）

摘 要：精炼渣对钢中非金属夹杂物的有成分改性和吸附去除的作用。对于铝脱氧钢，当铝含量较低的情况下，精炼渣不能改性 Al_2O_3 夹杂物，其对夹杂物的作用主要为吸附去除。当钢含量较高时，钢中 Al_2O_3 夹杂物才会被完全改性为液态 $CaO-Al_2O_3-MgO$ 夹杂物。对于硅锰脱氧钢，精炼渣可以有效改性降低夹杂物中的 Al_2O_3 含量，精炼渣也对夹杂物有一定的吸附作用。

关键词：洁净钢；精炼渣；夹杂物

Metallurgical Effect Mechanism of Refining Slag on Non-metallic Inclusions in Steel

Ren Ying[1], Zhang Lifeng[2], Ren Changyu[1], Cheng Gong[1]

(1. School of Metallurgical and Ecological Engineering, University of Science and Technology Beijing, Beijing 100083, China; 2. State Key Lab of Metastable Materials Science and Technology, Yanshan University, Qinhuangdao 066044, China)

Abstract: The refining slag can modify the composition of non-metallic inclusions in steel and remove inclusions by slag adsorption. For Al-killed steel, the refining slag can hardly modify the composition of inclusions in low Al steels. The effect on inclusions was mainly removal by slag adsorption. When the steel content is high, Al_2O_3 inclusions in the steel were completely modified into liquid CaO-Al_2O_3-MgO inclusions. For Si-Mn-killed steel, refining slag effectively reduced the content of Al_2O_3 in inclusions, and refining slag also has an adsorption effect on inclusions in steel.

Key words: clean steel; refining slag; inclusions

LF 炉快速造渣工艺优化研究与实践

何　晴，王金星，黄　山，单红超，范英权

（河钢股份有限公司承德分公司，河北承德　067102）

摘　要：为提高 LF 炉生产效率，满足高效化生产的需要，对 LF 炉快速造渣进行了一系列的优化研究，采取提高转炉终点控制水平、提高挡渣效果、转炉出钢预成渣工艺、出钢后炉渣预脱氧、钢包浇余循环利用、优化化渣剂使用等一系列措施实现了 LF 炉的快速造渣。研究表明：在出钢过程中随合金加入 300-400kg 石灰，可有效的对精炼前炉渣进行预处理，降低精炼处理周期。通过出钢后使用硅铁粉及铝线段进行炉渣预脱氧工艺的实施，为快速造渣及保证生产顺行创造良好的条件。出钢结束后，在吹氩之前，对钢包内钢水进行折渣操作，折渣量控制在 1.5～3t，既能保证钢包净空足够精炼，又能保证循环炉渣的冶金效果。精炼过程中使用一部分包渣替代改质剂等高价造渣物料，既能够满足精炼化渣、造渣需要，又能降低生产成本。通过以上工艺的综合利用，能够实现 LF 造渣低成本、高效率生产。

关键词：快速造渣；预成渣；预脱氧；钢包浇余循环利用；包渣

Research and Practice of Rapid Slag Making Process

He Qing, Wang Jinxing, Huang Shan, Shan Hongchao, Fan Yingquan

(HBIS Company Limited Chengde Branch, Chengde 067102, China)

Abstract: In order to improve LF furnace production efficiency and meet high efficiency production, a series of optimization research on LF furnace, adopting a series of measures to improve the control level of converter end, slag retaining effect, slag, and optimization of slag agent. It show that 300 - 400kg lime with alloy can effectively preprocess the refining slag and reduce the refining processing cycle. slag pre - oxidation process is implemented using iron silicon powder and aluminum section, creating good conditions for rapid slag production and ensuring smooth production. After the steel exit, the steel water in the steel bag is controlled at 1.5 - 3 tons, which can ensure that the steel bag clearance is refined enough, but also the metallurgical effect of circulating slag. In the refining process, some slag bags are used to replace quality changing agents and other expensive slag making materials, which can not only meet the needs of refining slag and slag making, but also reduce the production cost. Through the comprehensive utilization of the above processes, LF slag production can be achieved at low cost and high efficiency.

Key words: rapid slag building; preformed slag; pre-deoxidation; recirculation of steel package; bag lag

炼钢氧枪喷头优化对转炉终点钢液洁净度的影响

张明博，陈树军，张　东，李彦军，康爱元

（河钢股份有限公司承德分公司，河北 承德　067102）

摘　要： 采用理论计算方法对 150t 转炉氧枪喷头结构进行优化，并实施工业对比试验验证优化效果。理论研究和工业试验表明：当氧枪喷头马赫数 2.02，喷孔夹角 14.5°，喷孔数 5 孔，喉口直径 41.3mm，出口直径 54.1mm，设计流量 28000～33000Nm^3/h，熔池具有良好的搅拌效果和冲击动能。较原氧枪喷头相比，优化后的氧枪喷头，可使转炉终点[O]含量、碳氧积、终渣 FeO 含量的平均值分别降低 147.7×10^{-6}、0.0002 和 7.48%，提高初始钢液洁净度的同时，降低渣中 FeO 含量，提高金属收得率。

关键词： 转炉炼钢；氧枪喷头优化；碳氧积；洁净度

Influence on the Cleanliness of Converter End-point Molten Steel by Optimizing the Structre of Oxygen Lance

Zhang Mingbo, Chen Shujun, Zhang Dong, Li Yanjun, Kang Aiyuan

(HBIS Company Limited Chengde Branch, Chengde 067102, China)

Abstract: The structure of Oxygen Lance nozzle of 150t converter was optimized by theoretical calculation method, and the optimization effect was verified by industrial contrast test. Theoretical research and industrial tests show that when the oxygen lance nozzle Mach number is 2.02, the nozzle angle is 14.5°, the nozzle number is 5 holes, the throat diameter is 41.3mm, the outlet diameter is 54.1mm, and the design flow rate is 28000-33000Nm^3/h, the molten pool has a good Stirring effect and impact kinetic energy.Compared with the original oxygen lance nozzle, the optimized oxygen lance nozzle can reduce the final [O] content, oxygen content of the converter, and FeO content of the final slag by an average of 147.7×10^{-6}, 0.0002 and 7.48% respectively.It can be seen that optimizing the nozzle of the oxygen lance not only improves the cleanliness of the initial molten steel, but also reduces the FeO content in the slag and improves the metal recovery rate.

Key words: BOF steelmaking; optimization of oxygen lance nozzle; carbon oxygen equilibrium; cleanliness

利用转炉终渣修补出钢面炉衬技术应用与实践

潘　军，吴　坚，杨应东，赵　斌

（马鞍山钢铁股份有限公司长材事业部，安徽 马鞍山　243000）

摘　要： 介绍了马钢股份公司长材事业部 65t 顶底复吹转炉利用转炉终渣修补出钢面炉衬技术，通过对转炉终渣成分的控制和利用转炉终渣修补出钢面炉衬过程操作的优化，有效提高了转炉出钢面炉衬耐侵蚀能力和转炉作业率。

与传统补炉砂补钢面技术相比，出钢面炉衬补炉频次由 25/月次下降至 16 次/月，年补炉砂吨钢用量下降了 20.33%，年节约补炉砂耐材成本 200 万元。

关键词： 出钢面炉衬；转炉终渣成分控制；过程操作优化；修补

Application and Practice of Repairing Steel Lining with the Final Slag of Converter

Pan Jun, Wu Jian, Yang Yingdong, Zhao Bin

(Excellent talent division of Ma'anshan Iron & Steel Co., Ltd., Ma'anshan 243000, China)

Abstract: Technology of repairing steel lining with converter slag in 65t top and bottom combined blowing converter of Masteel, optimizing operation of repairing steel lining bycontrolling composition of converter end slag and cooling converter end slag, while reducing the frequency of reinforcing steel surface and the cost of refractory, the erosion resistance of converter lining and converter operation rate are effectively improved.Compared with the traditional furnace sand filling process for steel surface, furnace lining frequency decreased from 25 times per month to 16 times per month, the amount of sand per ton steel used for recharging decreased by 20.33%, the 200 million yuan cost of refractory sand refractory was saved by per annum.

Key words: furnace lining; control the final slag of converter; optimization of operating process; repair

中国电炉冶炼普碳钢的趋势分析

余光光，李永忠，姚　远，武国平

（北京首钢国际工程技术有限公司，北京市冶金三维仿真设计工程技术研究中心，北京　100043）

摘　要： 电炉炼钢工艺比高炉转炉炼钢工艺温室气体排放量大幅度降低，它是绿色制造、循环经济的工业技术。发改委制定政策鼓励发展电炉炼钢，实现中国钢铁工业的工艺优化、转型升级。本文分析电炉冶炼普碳钢的影响因素，如废钢、环保政策等，建议中国钢铁企业可以借鉴美国电炉发展经验，发展冶炼普碳钢的短流程电炉炼钢，降低碳排放，规范废钢分类和保证社会废钢积累。

关键词： 电炉；绿色制造；碳排放；废钢

Analysis of the Trend of EAF for Producing Plant Carbon Steel in China

Yu Guangguang, Li Yongzhong, Yao Yuan, Wu Guoping

(Beijing Shougang International Engineering and Technology Co., Ltd., The Research Center of Engineering Technology of Three Dimensions Simulating Design of Beijing, Beijing 100043, China)

Abstract: CO_2 emission by EAF steelmaking route is much less than that by BF-BOF steelmaking route.The EAF route is the technology of green manufacturing and recycle eceonomy. NDRC is encouraging the ratio of EAF steel products in China, to optimize the production process.It is analyzed that scrap output, environment policy are key factors to the development of EAF route.With the experience of American EAF development, Chinese steel companies can adopt EAF minimill to

produce plain carbon steel, reducing CO_2 emission,standarizing the scrap classification and acculmulating the scrap resource.
Key words: EAF; green manufacturing; carbon emission; scrap

耐候防腐焊接用钢的研究与应用

张玉海

（河钢集团宣钢公司，河北宣化　075100）

摘　要：介绍了耐候焊接用钢 T55－G 盘圆生产过程，经过剖析各合金元素对耐候焊丝焊接功能的影响，结合耐大气腐蚀指数、抗裂纹敏感系数，确定化学成分范围，选择合理的冶炼、浇注、轧制工艺参数等措施，形成了量化生产耐候焊丝钢的工艺。独特的冷却工艺对盘圆进行冷却，避免有害组织出现，生产出强度高、耐腐蚀等综合性能优良的盘圆。
关键词：耐候焊丝；化学成分；热轧盘圆

Research and Application of Weathering and Anticorrosion Welding Steel

Zhang Yuhai

(HBIS Group Xuansteel Company, Xuanhua 075100, China)

Abstract: This paper introduces the production process of weather resistant welding steel T55-G wire rod. By analyzing the influence of various alloy elements on the welding function of weather resistant welding wire, combining with the atmospheric corrosion resistance index and anti crack sensitivity coefficient, the chemical composition range is determined, and the reasonable smelting, pouring and rolling process parameters are selected, the quantitative production process of weather resistant welding wire steel is formed. The unique cooling process is used to cool the wire rod to avoid the appearance of harmful structure. The wire rod with high strength and corrosion resistance is produced.
Key words: weather resistant welding wire; chemical composition; hot rolled wire rod

Formation and Characterization of Complex Inclusions in Gear Steels with Different Sulphur and Calcium Contents

Yu Sha L[1,3], Ahmad Haseeb[1], Ma Xiaodong[1], Huang Zongze[2], Xu Yingtie[2], Zhao Sixin[2], Zhao Baojun[1]

(1. Sustainable Minerals Institute, University of Queensland, 4072, Brisbane, Australia; 2. Baoshan Iron and Steel Co., Ltd., Shanghai 200122, China; 3. Department of Materials Science and Engineering, Southern University of Science and Technology, Shenzhen, 518055, China)

Abstract: Suitable MnS inclusions in gear steel can significantly improve the steel machinability and reduce the manufacturing costs. Two gear steel samples with different sulphur contents were prepared by aluminiumdeoxidation followed by calcium treatment. The shape, size, composition, and percentage distribution of the inclusions present in the steel samples have been analyzed by electron probe micro-analysis (EPMA) technique. The average diameter of MnS precipitated on oxide is less than 5μm. It was found that the steel with high sulphur contains a greater number of elongated MnS precipitates than low sulphur steel. Also, there are more oxide inclusions such as calcium-aluminates and spinel with a little amount of solid solution of (Ca,Mn)S in low content sulphur steel after calcium treatment, which indicates the modification of solid alumina inclusions into liquid aluminates. The typical inclusions generated in high sulphur steel are sulphide encapsulating oxide inclusions and some core oxides were observed as spinel. The formation mechanisms of complex inclusions with different sulphur and calcium contents were discussed. The results are in good agreement with thermodynamic calculations.

Key words: gear steel; inclusions characteristics; EPMA; sulphide modification

基于烟气分析的转炉自动炼钢生产实践

李学禹[1]，刘　钊[2]，程树森[2]

（1. 河钢唐钢，河北唐山　063000；2. 北京科技大学冶金与生态工程学院，北京　100083）

摘　要：介绍了唐钢新区炼钢车间烟气分析动态控制炼钢技术。以转炉内物理化学反应为基础，分析了烟气中 CO 浓度与碳-氧反应的关系。冶炼过程采用恒枪变压的模式，供氧流量随冶炼过程逐步升高。造渣剂的加入分为多个批次进行，能够保证冶炼的稳定进行。终点碳-氧浓度积控制在 0.0015~0.0025；脱磷率控制在 83.5%~96.8%；吹损控制在 12.9%以内。实现了高效平稳冶炼。

关键词：转炉；烟气分析；终点控制

Production Practice of Automatic Converter Steelmaking based on Flue Gas Analysis

Li Xueyu[1], Liu Zhao[2], Cheng Shusen[2]

(1. Tangshan Iron and Steel, HBIS Group Co., Ltd, Tangshan 063000, China; 2. School of Metallurgical and Ecological Engineering, University of Science and Technology Beijing, Beijing 100083, China)

Abstract: This paper introduces the dynamic control steelmaking technology of flue gas analysis in steelmaking workshop of TangSteel New district. Based on the physical and chemical reaction in converter, the relationship between CO concentration in flue and carbon oxygen reaction is analyzed. The mode of constant lance and variable pressure is adopted in the smelting process, and the oxygen flow rate increases gradually with the smelting process. The addition of slagging agent is divided into several batches, which can ensure the stability of smelting. The end-point carbon oxygen concentration product was controlled between 0.0015 and 0.0025. The dephosphorization rate was controlled at 83.5%~96.8%. The blowing loss should be controlled within 12.9%. High efficiency and stable smelting is realized.

Key words: converter; flue analyse; endpoint control

250t 转炉二次燃烧氧枪射流及多相流行为研究

郑淑国，刘　超，朱苗勇

（东北大学冶金学院，辽宁沈阳　110819）

摘　要：本文结合某钢厂250t转炉，运用Fluent软件建立三维模型，对比分析了周边5孔加中心1孔的普通六孔氧枪与单流道二次燃烧氧枪的射流特性，并通过VOF多相流模型对气-液-渣三相流行为进行了模拟研究。结果表明，相比于普通氧枪，因一侧副孔流股快速汇入中心孔射流而使二次燃烧氧枪有更大的径向横截面积，且其中心孔射流速度大，故二次燃烧氧枪有更大的冲击面积和冲击深度。通过气-液-渣三相模拟可知，与普通氧枪相比，二次燃烧氧枪射流对熔池表面的冲击力更大；二次燃烧氧枪的冲击面积和冲击深度分别是普通氧枪的1.30倍和1.19倍。

关键词：转炉；单流道二次燃烧氧枪；射流行为；多相流行为；数值模拟

Research on Jet and Multiphase Flow Characteristics of Post Combustion Oxygen Lance in a 250t Converter

Zheng Shuguo, Liu Chao, Zhu Miaoyong

(School of Metallurgy, Northeastern University, Shenyang 110819, China)

Abstract: The three-dimensional model of a 250t steelmaking converter was established by FLUENT software. The jet characteristics of a conventionalsix-holeoxygen lance with 5 holes around and one hole in the center and a post combustion (PC) oxygen lance with a single flow channel were compared and analyzed. The interaction between gas-liquid-slag three phases was simulated by VOF multiphase flow model.The research results show that, compared with the conventional oxygen lance, The post combustion oxygen lance has a larger radial cross-sectional area due to the rapid flow of one side secondary hole into the center hole jet, and the jet velocity of the center hole is large, so the post combustion oxygen lance has a larger impact area and impact depth.Through gas-liquid-slag three-phase simulation, compared with conventional oxygen, the impact force of post combustion oxygen lance jet on the surface of molten pool is greater.The impact area and impact depth of the post combustion oxygen lance are 1.30 times and 1.19 times that of the ordinary oxygen lance, respectively.

Key words: converter; single-channelpost combustion oxygen lance; jet characteristics; multiphase flow characteristics; numerical simulation

转炉提高废钢比工艺讨论

张　胤，云　霞，张怀军，刁望才

（内蒙古包钢钢联股份有限公司技术中心，内蒙古包头　014010）

摘　要：国家明确指出，炼钢生产逐步减少铁矿石比例和增加废钢消耗比重。伴随近几年社会废钢资源不断扩大，加大废钢资源的合理开发和利用，将有利于钢铁企业实现节能环保、降低生产成本、减少对以铁矿石为主的传统炼铁工艺的过度依赖。随吨钢铁耗降低，吨钢效益随之增加。

关键词：铁钢比；废钢比；热平衡

Discussion on Increaseingscrap Ratio in Converter

Zhang Yin, Yun Xia, Zhang Huaijun, Diao Wangcai

(Technical Center of Inner Mongolia Baotou Steel Union Co., Ltd, Baotou 014010, China)

Abstract: The state has clearly pointed out that the proportion of iron ore and the proportion of scrap steel consumption should be gradually reduced in steelmaking production. With the continuous expansion of scrap steel resources in recent years, increasing the rational development and utilization of scrap steel resources will be beneficial to iron and steel enterprises to achieve energy conservation and environmental protection, reduce production costs, and reduce the over-reliance on the traditional iron-making process, which is mainly based on iron ore. With the decrease of steel consumption perton, the benefit of steel per ton increases.

Key words: iron and steel ratio; scrap ratio; heat balance

超低碳冷轧搪瓷钢浇铸蓄流原因分析及改进

李应江，邓　勇，李宝庆，胡晓光

（马鞍山钢铁股份有限公司第四钢轧总厂，安徽马鞍山　243011）

摘　要：对超低碳冷轧搪瓷钢蓄流原因进行了分析，并通过工艺优化改善了钢水可浇性，得出如下结论：（1）内生的 TiOx-Al_2O_3 夹杂物高熔点夹杂物是引起 Ti-IF 冷轧搪瓷钢水口蓄流的主要原因，而夹杂物内部的金属冻结加重了水口堵塞；（2）控制 RH 终点钢包渣 TFe、禁止 RH 吹氧以及提高浇铸期钢水 Als 含量是提高 Ti-IF 冷轧搪瓷钢钢水可浇性的关键措施；（3）改进后的工艺可显著改善超低碳冷轧搪瓷钢钢水可浇性，连浇炉数由 2～3 炉提升至 5 炉。

关键词：超低碳冷轧搪瓷钢；水口蓄流；TiOx-Al_2O_3 夹杂物；炉渣 TFe

Analysis and Improvement on the SEN Clogging of Coldrolled Ti-IF Enamel Steel

Li Yingjiang, Deng Yong, Li Baoqing, Hu Xiaoguang

(No.4 Steelmaking and Rolling General Plant of
Maanshan Iron and Steel Co., Ltd., Ma'anshan 243011, China)

Abstract: SENcloggings of cold rolled Ti-IF enamel steelwere analyzed.And the castability of molten steel was improved by optimizing the process.The results showed that SEN clogging was mainly caused by endogenous TiOx-Al_2O_3 inclusions high melting point inclusions are the main cause of nodulation at the enamel steel mouth of Ti-IF cold rolling, and the metal freezing in the inclusions aggravates the blockage of the water mouth.Controlling ladle slag TFe, forbidding RH blowing oxygen and increasing the content of Als in molten steel during casting are the key measures to improve the pouring ability of Ti-IF cold rolling enamel steel.The improved process can significantly improve the steel pouring ability of Ti-IF cold rolling enamel steel, and the number of continuous pouring furnaces is increased from 2-3 to 5.

Key words: cold rolled Ti-IF enamel steel; SEN clogging; TiOx-Al_2O_3inclusions; Ladle slag TFe

铁钢比与效益最大化动态预测模型研究

胡文华，赵　滨，刘　威

（中国宝武马钢集团长材事业部，安徽马鞍山　243003）

摘 要： 随着降低碳排放压力的加大，钢铁企业为了追求效益最大化，纷纷增加废钢用量，降低铁钢比；而随着各种原材料价格以及产品边际利润的变化，并非铁钢比越低越好。本文通过建立模型，对于不同铁钢比组产时各条件变化进行分析，进而对效益进行预测，以帮助企业实现效益最大化。

关键词： 铁钢比；预测模型；效益最大化

Research on Dynamic Prediction Model of Iron-steel Ratio and Benefit Maximization

Hu Wenhua, Zhao Bin, Liu Wei

(Long Products Division of China Baowu Masteel Group, Ma'anshan 243003, China)

Abstract: With the pressure of reduce carbon emissions increasing, the amount of scrap was increased in many steel companies in order to maximize benefits,and the iron-steel ratio was reduced more and more.But with the changes in the prices of various raw materials and product margins, the lower iron-steel ratio does not mean the better. The various conditions was analyzed during different iron-steel ratio in this article,and the marginal benefit was predicted to get maximize benefit for steel industry.

Key words: iron-steel ratio; predictive model; maximize benefit

引流砂在钢厂的使用与管理

蔡常青

（福建三钢闽光股份有限责任公司炼钢厂，福建三明 365000）

摘 要： 福建三钢闽光股份有限责任公司炼钢厂自转炉大型化后一直用铬质引流砂作为钢包自开的导流材料，它具有自开率高，重复性好、可靠性高等优点而备受钢厂推崇，通过选择合适的引流砂和加强安装、施工和过程工艺管理，钢包自开率一直稳定 99.60%以上，也减少了钢水的污染。

关键词： 引流砂；自开率；管理

Application and Management of Sand Drainage in Steel Works

Cai Changqing

(Steelmaking Plant, Fujian Steel Minguang Co., Ltd., Sanming 365000, China)

Abstract: Since the large-scale of the converter, the chrome sand has been used as the diversion material for the ladle to open itself in Steelmaking plant, Fujian steel Minguang Co., Ltd,. It has the advantages of high self-opening rate, good and repeatability and high reliability and so on, and is highly praised by steel mills. By selecting suitable drainage sand and strengthening installation, construction and process management, translate into. The self-opening rate of ladle has been stable 99.60%, it also reduces the contamination of Molten Steel.

Key words: drainage sand; self opening rate; management

镁钙不同添加方式时钢中 Al_2O_3-MgO-CaO 系夹杂物演变

施利魏，周星志，王德永，屈天鹏，田 俊，侯 栋

（苏州大学沙钢钢铁学院，江苏苏州 215006）

摘 要：随着洁净钢技术的发展，钢中非金属夹杂物的控制变得越来越重要，Al_2O_3-MgO-CaO 系夹杂物是铝脱氧钢中最为常见的夹杂物之一。为此，本文在实验室研究了铝脱氧钢中先镁后钙、先钙后镁、镁钙同时添加三种方式对钢中夹杂物改性效果的影响，并探讨了夹杂物随时间的演变规律。结果表明：三种添加方式下夹杂物的演变规律截然不同，但最终都形成了以镁铝尖晶石夹杂物为核心，外层包括 CaO -Al_2O_3-MgO 的复合夹杂物。与此同时，采用热力学计算揭示了三种添加方式下钢液中镁铝尖晶石与铝酸钙之间的相互转化关系，热力学计算与试验结果具有良好的一致性。

关键词：夹杂物；镁钙复合处理；热力学计算；镁铝尖晶石；铝脱氧钢

Evolution Mechanisms of Al_2O_3-MgO-CaO Inclusions in Steel

Shi Liwei, Zhou Xingzhi, Wang Deyong, Qu Tianpeng, Tian Jun, Hou Dong

(School of Iron and Steel, Soochow University, Suzhou 215006, China)

Abstract: With the development of clean steel technology, the control of non-metallic inclusions in steel is of increasing importance. Al_2O_3-MgO-CaO Inclusion is one of the most common inclusions in aluminum killed steels. In view of this, how three addition methods (i.e. adding Mg before Ca, adding Mg after Ca, and adding Mg together with Ca) influenced the modification effect of inclusions in steel was experimentally studied, and how these inclusions evolved with time was discussed in this paper. The results demonstrated that despite the sharp difference in their inclusion evolution, composite inclusions with a magnesium aluminate spinel (MAS) core and an outer CaO -Al_2O_3-MgO layer were formed by all the three addition methods, with the average inclusion size of 1-2μm. Furthermore, thermodynamic calculation was adopted to reveal the transformation relationship between MAS and calcium aluminate in molten steel in each of the three addition methods. The thermodynamic calculation results agreed well with the experiment data.

Key words: inclusions; Mg-Ca treatment; thermodynamic calculation; magnesium aluminate spinel; Al - killed steel

提高大线能量焊接性能的 Mg 氧化物冶金技术开发

杨 健[1]，徐龙云[2]，潘晓倩[1]，张银辉[1]，刘德坤[1]

（1. 上海大学材料科学与工程学院，上海 200444；
2. 湖南华菱湘潭钢铁有限公司厚板厂，湖南湘潭 411104）

摘　要： 本文研究了 Mg 氧化物冶金技术对提高厚板焊接热影响区韧性的影响。随着钢中 Mg 含量由 0ppm 升高至 27ppm 再到 99ppm，夹杂物中 Mg 含量增加，夹杂物尺寸从 2.3μm 减小到 1.7μm，数量密度由 100 个/mm^2 增加到 200 个/mm^2 以上。焊接 HAZ 组织由 FSP(49.0%)+ GBF(22.8%)+Bu(21.6%)变成 IAF(55.4%)+GBF(35.0%)再变成 PF(95.2%)，PAG 尺寸由 220μm 减小至 206μm 再到 64μm。钢样 400 kJ/cm 大线能量焊接 HAZ -20°C 夏比冲击功分别为 27 J、179 J、199 J。利用 Mg 氧化物冶金技术开发的大线能量焊接用钢焊接 HAZ 性能优异，满足 400 kJ/cm 大线能量焊接的要求。

关键词： Mg 脱氧；氧化物冶金；焊接热影响区；韧性

Development of Mg Oxide Metallurgy Technology for Improving HAZ Toughness after High Heat Input Welding

Yang Jian[1], Xu Longyun[2], Pan Xiaoqian[1], Zhang Yinhui[1], Liu Dekun[1]

(1. School of Materials Science and Engineering, Shanghai University, Shanghai 200444, China;
2. Xiangtan Iron & Steel Co., Ltd., of Hunan Valin, Xiangtan 411104, China)

Abstract: In this paper, the effect of Mg oxide metallurgy on the toughness of HAZ of steel plate is studied. With increasing Mg content from 0ppm to 27ppm and then to 99ppm, the Mg content in inclusions increases and the size of inclusions decreases from 2.3ppm μm to 1.7μm. The number density increases from 100 /mm^2 to more than 200 /mm^2. The microstructure of HAZ changes from FSP (49.0%) + GBF (22.8%) + Bu (21.6%) to IAF (55.4%) + GBF (35.0%) to PF (95.2%), and the size of PAG changes from 220μm to 206μm to 64μm. After the welding with 400 kJ/cm high heat input, the Charpy impact energies of HAZ of steel samples are 27 J, 179 J and 199 J at -20°C, respectively. The HAZ performance of high heat input welding for the steel plate developed by Mg oxide metallurgy technology is excellent, meeting the requirements of 400 kJ / cm high heat input welding.

Key words: Mg deoxidation; oxide metallurgy; heat affected zone; toughness

低钛 H13 电渣锭坯料的生产实践

魏　巍，何建武，刘宪民，李　虹，张志强

（石钢京诚装备技术有限公司技术中心，辽宁营口　115100）

摘　要： 为了控制液析碳化物的数量及尺寸，控制 TiN 夹杂物的析出，优化 H13 成分，将 C、V、Ti 含量按下限控制。使用低 P、低 Ti 合金及物料，倒净转炉前期渣，清理钢包残钢残渣，严格控制转炉出钢下渣。LF 精炼前期调整 Cr 含量，保证 LF 总渣量 30kg/t 钢以上，控制精炼渣较低的碱度。LF 渣系成分范围（质量分数）CaO 50%～53%、Al_2O_3 24%～28%、SiO_2 13%～15%、MgO 6%～8%、(TFe+MnO)≤0.5%、R 3.5～4.0，LF 精炼时间 130～150min。控制模温 50～80℃，锭身浇注速度 0.4～0.5t/min，帽口浇注速度 0.2～0.3t/min，保证钢水液面平稳上升，做好氩气保护浇注。可实现 H13 电渣锭 P 含量≤0.012%、Ti 含量≤0.0020%，有效控制液析碳化物和 TiN 夹杂的数量及尺寸，确保冲击功≥300J。

关键词： H13；TiN；液析碳化物；含 Ti 复合夹杂；钢锭

Production Practice of Low Titanium H13 Electroslag Ingot

Wei Wei, He Jianwu, Liu Xianmin, Li Hong, Zhang Zhiqiang

(Shi Gang Jing Cheng Equipment Development and Manufacturing Co., Ltd.,
Technology Center, Yingkou 115100, China)

Abstract: In order to control the quantity and size of primary carbide，and to control the precipitation of TiN inclusions.The composition of H13 is optimized and the content of C, V and Ti is controlled according to the lower limit.Use low P and low Ti alloy and materials, the slag in the early stage of BOF is emptied, the slag in ladle is cleaned up, and the Slag carry-overof BOF tapping is strictly controlled. In the early stage of LF refining, the content of Cr should be adjusted.The total amount of LF slag is more than 30kg/t steel and use low basicity refining slag.Composition range of LF slag system（mass fraction）:CaO 50%～53%、Al_2O_3 24%～28%、SiO_2 13%～15%、MgO 6%～8%、(TFe+MnO)≤0.5%、R 3.5-4.0.LF refining time is 130-150min.The ingot mold temperature is controlled at 50-80 ℃,The pouring speed of ingot body is 0.4-0.5t/min,The pouring speed of cap mouth is 0.2-0.3t/min.The liquid steel level rises steadily, and the argon protection pouring should be done well. H13 electroslag ingot can be P≤0.012%、Ti≤0.0020%.The quantity and size of primary carbide and TiN inclusion can be effectively controlled.Impact energy≥300J.

Key words: H13; TiN; primary carbide; Ti containing composite inclusion; ingot

120t 复吹转炉“留渣-双渣”脱磷工艺试验

孟华栋，杨　勇，姚同路，贺　庆，倪　冰，林腾昌

（钢铁研究总院冶金工艺研究所，北京　100081）

摘　要： 在国内某 120t 转炉钢厂采用“留渣-双渣”工艺技术进行脱磷工艺试验。结果表明：随着转炉前期脱磷率不断升高，终点脱磷率不断提高。铁水硅含量对前期的脱磷率影响最大。根据铁水成分在冶炼前期适当降低供氧强度，降低气固氧比，加入适量石灰及烧结矿，均有利于前期脱磷率的提高。在一倒时加入 4~8kg/t 石灰，不影响出钢温度，可提高一倒-终点阶段脱磷率，同时提高了终点脱磷率。从终点的控制效果可知，终点炉渣碱度应保持≥3.0，炉渣中（FeO）含量 16%~20%，并适当降低终点出钢温度在 1610~1630℃有利于终点脱磷率的提高。通过加强熔池搅拌，促进钢渣反应趋于平衡有利于终点磷分配比提高，从而进一步提高了终点脱磷率。

关键词： 复吹转炉；“留渣-双渣”工艺；脱磷效果；工艺优化

Experiment on Dephosphorization Process of "Remaining Slag and Double Slag" in 120t Combined Blown Converter

Meng Huadong, Yang Yong, Yao Tonglu, He Qing, Ni Bing, Lin Tengchang

(Department of Metallurgical Technology Research,
Central Iron and Steel Research Institute, Beijing 100081, China)

Abstract: The dephosphorization process test by the “remaining slag and double slag” technology was carried out in a 120t domestic converter steel plant.The results show that with increasing the dephosphorization rates in the early stage of converter, the end-point dephosphorization rates are improved obviously. The silicon content of molten iron plays the most important role in the dephosphorization rate in the early stage.According to the chemical composition of the molten iron, reducing oxygen supply intensity and gas-solid oxygen ratio, adding proper amount of lime and sinter are beneficial to raising the early dephosphorization rate.Adding 4~8kg/t lime at the first turn-downstage can both increase the dephosphorization rate in the first turn-down to end-point stageand the end-point dephosphorization rate without affecting the tapping temperature. It can be seen from the control effect of the end-point that the basicity of slag should be kept more than 3.0 and the content of slag (FeO) should be in the range of 16% to 20%, and the end-point tapping temperature should be appropriately reduced to 1610~1630℃, which are all conducive to the improvement of the end-point dephosphorization rate.By strengthening the bath stirring, the steel-slagreaction tends to equilibrium, which is good for the increase of end-point phosphorus partition ratio, thus further improving the end-point dephosphorization rate.

Key words: combined blowing converter; "remaining slag and double slag" process; dephosphorization effect; process optimization

浅谈炼钢厂工艺安全管理实践

安彬海，王　焕

（武汉钢铁有限公司炼钢厂，湖北武汉　430080）

摘　要：工艺安全管理（PSM）原是石油化工企业应用较为广泛的一种安全管理方法，它通过对工艺过程中的危害和风险进行识别、分析、评价、处理，从而避免与工艺相关的伤害和事故。本文探讨如何创新性的将 PSM 应用在炼钢厂的的安全管理中，我们根据钢厂工艺特点，建立新的 HAZOP、JSA 安全模型进行工艺安全分析，用 LS 法和 LEC 法进行风险评估，根据分析结果、针对性的进行防控。在风险防控上，应用 LOTO、DNT/DNT 例外这两个安全管理工具。目的是通过这种管理人员、技术人员和基层员工全员参与的安全管理方法，优化炼钢厂的安全管理，使得钢厂安全管理更加完善。

关键词：工艺安全管理；HAZOP；JSA；LOTO；DNT/DNT 例外；危险源

Discussion on the Practice of Process Safety Management in Steelmaking Plant

An Binhai, Wang Huan

(Steel Works of Wuhan Iron and Steel Co., Ltd., Wuhan 430080, China)

Abstract: Process safety management used to be a widely used safety management method in petrochemical enterprises.It identifies, analyzes, evaluates and deals with the hazards and risks in the process.This paper discusses how to apply the process safety management to the safety management of steelmaking plant innovatively.According to the process characteristics of the steel plant. Establish a new HAZOP, JSA safety model for process safety analysis. Risk assessment with LS methods and LEC methods. According to the analysis results, targeted prevention and control. On risk prevention and control, apply LOTO, DNT/DNT NOT management tools. The purpose of this safety managers, technicians and grass-roots employees, optimize the safety management of steel plant. Make the safety management of steel plant more perfect.

Key words: process safety management; HAZOP; JSA; LOTO; DNT/DNT not; hazard sources

Fe-P、Mn-P 和 Fe-Mn-P 合金熔体的热力学优化

游志敏[1]，姜周华[1]，Jung In-Ho[2]

（1. 东北大学冶金学院，辽宁沈阳　110819；

2. 首尔国立大学先进材料研究所材料科学与工程系，首尔　08826）

摘　要：本研究利用修正的准化学模型（MQM），考虑了液相短程有序，对全浓度范围 Fe-P、Mn-P 和 Fe-Mn-P 合金熔体进行了热力学模拟。基于对实验数据的批判性评价，本研究优化了 Fe-P 和 Mn-P 液相各组元活度和混合焓等热力学性质，同时分别确定了 Fe 液和 Mn 液中[P]的亨利活度系数($\gamma^{\circ}_{\text{P in Fe(l)}}$，$\gamma^{\circ}_{\text{P in Mn(l)}}$)和 P_2(g)的溶解吉布斯自由能变($\Delta G^{\circ}_{P_2\text{ in Fe(l)}}$，$\Delta G^{\circ}_{P_2\text{ in Mn(l)}}$)。在模拟 Fe-Mn-P 熔体时，引入了以 P 为非对称元素的 Toop 类型几何内插技术。通过对比实验数据可知，本模拟无需任何三元模型参数可以准确预测各温度和成分的 Fe-Mn-P 合金熔体的相平衡和 Fe、Mn、P 的活度（a^{R}_{Fe}、a^{R}_{Mn}、a^{R}_{P}）。作为本研究的结果，所优化的热力学数据库可用于计算 Fe-Mn-P 液相未被研究的相平衡和热力学性质等各种应用。

关键词：修正的准化学模型；Fe-P；Mn-P；Fe-Mn-P；合金熔体；相平衡关系；热力学性质

Thermodynamic Optimization of Molten Fe-P, Mn-P and Fe-Mn-P Alloys

You Zhimin[1], Jiang Zhouhua[1], Jung In-Ho[2]

(1. School of Metallurgy, Northeastern University, Shenyang 110819, China;

2. Department of Materials Science and Engineering, and Research Institute of Advanced Materials (RIAM), Seoul National University, Seoul 08826, South Korea)

Abstract: Thermodynamic modeling of molten Fe-P, Mn-P and Fe-Mn-P alloys in the entire composition were performed presently using the Modified Quasichemical Model (MQM) accounting for the short-range ordering of the solutions. In this study, thermodynamic properties including the activity of each component and enthalpy of mixing of liquid Fe-P and Mn-P solutions were optimized based on critical evaluation of all available experimental data. Meanwhile, Henrian activity coefficients of P ($\gamma^{\circ}_{\text{P in Fe(l)}}$, $\gamma^{\circ}_{\text{P in Mn(l)}}$) and Gibbs energy changes of P_2(g) dissolution ($\Delta G^{\circ}_{P_2\text{ in Fe(l)}}$, $\Delta G^{\circ}_{P_2\text{ in Mn(l)}}$) in liquid Fe and Mn were determined, respectively. In the modeling of liquid Fe-Mn-P solution, the Toop-type geometric interpolation technique with P as the asymmetric component was introduced to describe this system. Phase equilibria and activities of Fe, Mn, P (a^{R}_{Fe}, a^{R}_{Mn}, a^{R}_{P}) of molten Fe-Mn-P alloys at various temperatures and compositions were accurately predicted without any ternary model parameters, compared to experimental data. As results of this study, the optimized thermodynamic database can be used to back-calculate unexplored phase equilibria and thermodynamic properties of liquid Fe-Mn-P solution for various applications.

Key words: modified quasichemical model; Fe-P; Mn-P; Fe-Mn-P; molten alloys; phase equilibria; thermodynamic properties

一流板坯中间包长寿命技术实践

董战春，王伟光，秦文鹏，李曼曼，邓乐锐

（北京联合荣大工程材料股份有限公司，北京　101400）

摘　要：随着现代钢铁企业逐渐大型化和生产的高效化，连续铸钢技术的重要性日益显现，中间包用耐火材料能否满足连浇炉数的要求，对于钢铁企业的正常生产秩序和企业总体经济效益影响巨大。针对河北某钢铁公司 40t 一机一流板坯中间包的生产需要，为了提高该板坯中间包的连浇使用寿命至 50h，对各耐火材料侵蚀情况进行了系统分析并确定了相应的提寿方案，主要为 50h 试验结构方案设计、渣线干式料的调整、加装护板、挡渣墙和稳流器的改进以及塞棒和水口的改进调整等措施，目前连浇使用寿命已达到 50h 以上，降低了吨钢耐火材料消耗成本，减少了中间包浇余和铸坯切头、去尾数量，大幅提高了连铸生产作业率。

关键词：连铸；中间包；耐火材料；长寿命

Practice of Long Life Technology for One Flow Slab Tundish

Dong Zhanchun, Wang Weiguang, Qin Wenpeng, Li Manman, Deng Lerui

(Beijing Allied Rongda Engineering Material Co., Ltd., Beijing 101400, China)

Abstract: With the large-scale and high-efficiency production of modern iron and steel enterprises, the importance of continuous casting steel technology is becoming more and more obvious. Whether the refractory materials for tundish can meet the requirements of continuous casting furnace number has a great impact on the normal production order and overall economic benefits of iron and steel enterprises. According to the production needs of a 40t first-class slab tundish in a Hebei Iron and steel company, in order to improve the continuous casting service life of the slab tundish to 50h, the corrosion situation of various refractories was analyzed systematically and the corresponding service life improvement scheme was determined. The scheme mainly includes the structural scheme design of 50h test, the adjustment of dry vibration material for slag line, adding guard plate, the improvement of Preshaped skimmer block and Current regulator, as well as the improvement and adjustment of stopper and nozzle. At present, the service life of continuous casting has reached more than 50h, which reduces the consumption cost of refractories per ton of steel, reduces the quantity of molten steel in tundish and slab cutting and slab tailing, and improves greatly the operation rate of continuous casting.

Key words: continuous casting; tundish; refractory materials; long life

铁水捞渣技术应用与实践

杨明清，何　腾，郝于平

（宁波钢铁有限公司炼钢厂，浙江宁波　315807）

摘　要：铁水脱硫是转炉炼钢工艺中的重要一环。铁水脱硫后，需要将铁水表面的脱硫渣扒除干净，以减少转炉冶炼中回硫；同时要求扒渣过程中少带铁，以提高金属料收得率，达到降本增效的目的。铁水捞渣技术作为一种新型的除渣技术，被越来越多的冶金企业所认同并应用于生产实践中，给企业带来了直观的经济效益。
关键词：脱硫；捞渣；铁损

Application and Practice of Removing Slag in Molten Iron

Yang Mingqing, He Teng, Hao Yuping

(Ningbo Iron and Steel Co., Ltd., Ningbo 315807, China)

Abstract: In the metallurgical industry，soot particles with mars which is easy to damage the filter bag of baghouse system is very common.Therefore,it will affect dust pelletizing effect and increase the operation and maintenance cost of the dust collector without effective treatment.With the development of science and echnology and people continuously explore,mars capture device as an effective means to solve the problem is developed and become one of the main components of the bag filter.
Key words: desulfurization; removing slag; iron loss

底吹氩钢包精炼过程气泡破碎与聚合行为模拟研究

娄文涛，王晓雨

（东北大学冶金学院，辽宁沈阳　110819）

摘　要： 通过建立物理水模型实验和数值模拟，对底吹氩钢包内的气泡聚合、破碎行为机理展开研究。研究发现气泡破碎行为对于气泡尺寸及分布的影响不大，气泡间湍流随机碰撞和气泡上浮速度差碰撞对气泡尺寸起主导作用，而气泡尾涡捕捉机制影响微弱。相较于狭缝型透气砖，弥散型透气砖的气泡尺寸明显较小，并结合因此分析，确立了当前底吹系统下狭缝型透气砖和弥散型透气砖气泡直径的经验公式。
关键词： 钢包；CFD-PBM；气泡；破碎聚合；尺寸分布

Study of Bubble Breakup and Coalescence in Gas-Stirred Ladle Refining Process

Lou Wentao, Wang Xiaoyu

(School of Metallurgy, Northeastern University, Shenyang 110819, China)

Abstract: The bubble breakup and coalescence behavior and mechanisms in gas-stirred ladle were studied by establishing physical model and mathematical model. The mechanisms of bubble breakup and coalescence in the gas-stirred ladle and their influence on bubble size distribution were revealed. The results show thatthe bubble breaking behavior has little effect on bubble size and distribution. The bubble aggregation due to turbulent random collisions and buoyancy collision dominate the bubble size, while the bubble tail vortex capture mechanism has a weak effect.Compared with the slit-type ventilated brick, the bubble size of the diffused ventilated brick is significantly smaller, and based on this analysis, an empirical formula for the bubble diameter of the slit-type ventilated brick and the diffused ventilated brick under the current bottom blowing system is established.
Key words: ladle; CFD-PBM; bubble; breakup and coalescence; bubble size distribution

底喷粉转炉炉体振动行为的物理模拟研究

张靖实，娄文涛，朱苗勇

（东北大学冶金学院，辽宁沈阳　110819）

摘　要：本文采用相似原理研究方法，建立了模拟转炉底喷粉过程的物理模型，研究了炉体振动特性，考察了底吹布置和粉气质量比的影响规律。结果表明，底喷吹参数与模式对炉体振动行为有明显影响。随着粉气质量比的增加，炉体振动强度增大。当底吹喷嘴对称分布于离中心 0.5R 的双布置时，炉体振动强度得到有效减弱。底喷粉转炉炉体振动属于低频振动，炉底振动强度大于炉顶、炉衬和和耳轴处。该实验为转炉底喷粉炼钢的工业应用奠定了基础。

关键词：转炉炼钢；底喷粉；炉体振动；物理模型

Physical Modelling of Furnace Vibration Behavior in Converter with Bottom Powder Injection

Zhang Jingshi, Lou Wentao, Zhu Miaoyong

(School of Metallurgy, Northeastern University, Shenyang 110819, China)

Abstract: A cold model of top-bottom blown converter was set up to study the vibration performance of bottom powder injection converter. The effects of bottom blowing tuyere arrangements and powder to gas mass ratios on furnace vibration were studied. The results showed that the intensity of furnace vibration increases with increase of powder to gas mass ratios. When the bottom blowing tuyeres arranged at double arrangement 0.5R distance between tuyere and center of bath bottom, the intensity of vibration performance is weakened. The vibration of bottom powder injection converter belongs to low frequency vibration. The vibration intensity of bottom is greater than that of top, lining and trunnion. This experiment has laid the industrial application foundation for bottom powder injection in converter steelmaking.

Key words: converter steelmaking; bottom powder injection; furnace vibration; physical modeling

120t 钢包多点底吹氩下渣料卷混行为的模拟研究

王泽宇，娄文涛

（东北大学冶金学院，辽宁沈阳　110819）

摘　要：针对钢包双孔底吹氩模式流动死区大，对钢液搅拌效果无法满足快速渣金反应，脱硫周期长的现状，本文以国内某钢厂 120t 钢包为原型，结合物理模拟方法研究了不同透气元件数目、排布、底吹气量和渣料粒度对渣料卷混行为的影响。结果表明：渣料卷混程度与底吹气量呈正比，与渣料粒度呈反比；采用相同粒度粉剂在底吹气量相同的条件下，透气元件数目越大，渣料卷混效果越好，且在多点底吹情况下，改变排布对渣料卷混行为的影响更大，其中 4 透气元件的 14 号底吹布置方式，在底吹气量为 20NL/min 时，效果最优。

关键词：钢包；多点底吹；模拟；渣料卷混

Simulation Study on Mixing Behavior of Slag with Multi-point Bottom Blowing Argon in a 120t Ladle

Wang Zeyu, Lou Wentao

(School of Metallurgy, Northeastern University, Shenyang 110819, China)

Abstract: In view of the large dead zone of ladle bottom blowing argon mode, the stirring effect of molten steel can not meet the rapid slag gold reaction and the desulfurization period is long. In this paper, 120t ladle of a domestic steel plant is taken as the prototype to study the influence of different number of breathable elements, arrangement of breathable elements, bottom air blowing rate and slag particle size on slag rolling and mixing behavior by physical simulation method. The results show that the mixing degree of slag material is directly proportional to the bottom blowing volume and inversely proportional to the particle size of slag material. Under the condition of the same particle size powder and the same bottom air blowing volume, the larger the number of breathable elements, the better the mixing effect of slag material. In the case of multi-point bottom blowing, the change of distribution has a greater effect on the mixing behavior of slag material. And the No.14 bottom blowing arrangement of 4 breathable elements has the best effect when the bottom blowing volume is 20 NL /min.

Key words: ladle; multi-point bottom blowing argon; simulation; mixing of slag

$CaCl_2$在炼钢脱磷过程的作用

邓志银，闫子文，朱苗勇

（东北大学冶金学院，辽宁沈阳 110819）

摘　要：本文考察了不同 $CaCl_2$ 添加量对转炉渣的熔化特性、黏度以及脱磷能力的影响规律。研究发现，随着 $CaCl_2$ 添加量的增加，转炉渣的熔点总体上是不断上升的，而黏度不断降低。$CaCl_2$ 对渣黏度的影响并不是 Cl^-的作用，而是 Ca^{2+}影响了硅酸盐的结构。$CaCl_2$ 虽不能降低炉渣的熔点，但仍然可以有效促进石灰的溶解，也可使转炉渣与碳饱和铁液之间的磷分配比不断增加。含 $CaCl_2$ 渣的脱磷能力增强，主要是因为渣中富磷相 C_2S-C_3P 和 $Ca_5Cl(PO_4)_3$ 增加。

关键词：$CaCl_2$；熔点；黏度；脱磷；磷分配比

Effect of $CaCl_2$ on Dephosphorization during Steelmaking

Deng Zhiyin, Yan Ziwen, Zhu Miaoyong

(School of Metallurgy, Northeastern University, Shenyang 110819, China)

Abstract: The effects of $CaCl_2$ addition on the melting point, viscosity and dephosphorization ability of BOF steelmaking slag were investigated. It is found that with the addition of $CaCl_2$, the melting point of the slags climbs in general, while the viscosity drops continuously. The influence of $CaCl_2$ on the viscosity of the slags is mainly based on the impact of Ca^{2+} on the structures of silicates, not the effect of Cl^-. Although $CaCl_2$ can hardly reduce the melting point of the slag system, it is still helpful for the dissolution of CaO. Due to the increase of P-rich phases *viz.* C_2S-C_3P and $Ca_5Cl(PO_4)_3$ in slag, $CaCl_2$ can also result in the rise of P distribution ratio between slag and C-saturated iron.

Key words: $CaCl_2$; melting point; viscosity; dephosphorization; P distribution ratio

Mg 对含 Ti 高强度螺纹钢中夹杂物的影响

吴　静，田　俊，王德永，屈天鹏，侯　栋，胡绍岩

（苏州大学沙钢钢铁学院，江苏苏州　215137）

摘　要：对含 Ti 高强度螺纹钢进行了镁处理试验，分析了镁处理前后试样中夹杂物的成分和尺寸的变化，并用 FactSage 软件对夹杂物的形成和析出进行了计算。得出以下结论：加 Mg 前试样中夹杂物主要是 Al_2O_3、含有 Al_2O_3 核心的 TiN 和 TiN，加 Mg 后试样中夹杂物主要是 Al_2O_3、$MgO·Al_2O_3$、含有 Al_2O_3 或 $MgO·Al_2O_3$ 核心的 TiN 和 TiN。钢中氧含量越高，越容易形成 Al_2O_3，氧含量越低，越容易形成 MgO；Al 含量越高，越不容易形成 MgO。TiN 在钢液凝固过程中可以以氧化物为核心形核析出或者单独析出。镁处理后试样中夹杂物尺寸减小。$MgTi_2O_4$ 主要是在 Mg 含量较高和温度较低的情况下形成，试样中 $MgTi_2O_4$ 主要是在凝固过程形成。

关键词：螺纹钢；镁处理；夹杂物；析出

Effect of Mg on Inclusion of High Strength Rebars Containing Ti

Wu Jing, Tian Jun, Wang Deyong, Qu Tianpeng, Hou Dong, Hu Shaoyan

(Shagang School of Iron and Steel, Soochow University, Suzhou 215137, China)

Abstract: The experimentofmagnesium treatment of high strength rebars containing Ti was carried outbyvertical resistance furnacein the laboratory.The changes of composition and size of inclusions of the samples before and after magnesium treatment were analyzed, and the formation and precipitation of inclusions were calculated using FactSage software. Theresults showthattheinclusionsofthesample before Mgaddition wereAl_2O_3、TiN-Al_2O_3andTiN, andafterMgtreatment, theinclusionswereAl_2O_3、$MgO·Al_2O_3$、TiN-Al_2O_3、TiN-$MgO·Al_2O_3$ and TiN. It's easy to form Al_2O_3in the steel with high oxygen content, on the contrary, the MgO was formed easily. And, It is difficult to form MgO inclusion in the steel containing higher Al content.TiNinclusion couldbeprecipitated ontheoxideorjust precipitatedduringsolidification of steel. The size of inclusion decreased after magnesium treatment. The $MgTi_2O_4$ inclusion was mainly formed under the condition of thesteelwithhigh Mg content and low temperature, andthiskindofinclusion in the sample was mainly precipitated during solidification.

Key words: rebars; Mg treatment; inclusion; precipitate

转炉投弹式副枪在韶钢 120t 转炉的实践和应用

陈文亮，马　欢，肖双林，张建平

（宝武集团韶关钢铁有限公司炼钢厂，广东韶关　512122）

摘　要：本文简要叙述了投弹式副枪技术在韶钢 120t 转炉的应用效果情况。主要论述了在使用过程中投弹式副枪技术对转炉缩短冶炼周期、终点命中率 、降低铁钢比、经济效益等方面的影响。

关键词：转炉；降低；缩短；效益

Practice and Application of the Converter Projectile Gun in 120t Converter of Shaogang

Abstract: This paper briefly describes the application effect of projectile sublance technology in 120t converter of SISG. This paper mainly discusses the influence of the bullet type sublance technology on the shortening of smelting period, terminal hit rate, reduction of iron steel ratio and economic benefit of converter.

Key words: converter; reduction; shortening; benefit

Ce 夹杂物对钢力学性能影响的理论分析及试验研究

刘香军，王　婷，范明洋，杨昌桥，杨吉春

（内蒙古科技大学材料与冶金学院，内蒙古包头　014010）

摘　要：以 Ce 处理后的洁净钢(IF 钢)为研究对象，采用第一性原理计算 Ce 夹杂物的晶格参数、弹性常数、力学性质以及热膨胀系数等基本物理参数，定量分析夹杂物的物理性质对钢基体力学性能的影响。TiN、Al_2O_3 和 $CeAlO_3$ 夹杂物的体积模量、剪切模量、杨氏模量以及维氏硬度较大，呈现出较大的刚性和硬度，表现为脆性特征，而 Ce_2O_3 和 Ce_2O_2S 夹杂物表现为韧性特征。与 Al_2O_3 和 TiN 相比，Ce_2O_3、$CeAlO_3$ 夹杂的热膨胀系数与铁基体接近，而 Ce_2O_2S 夹杂物的热膨胀系数比铁基体稍大。Ce 夹杂物与基体在不可压缩性、刚性、硬度、韧脆性及热膨胀性等方面的差异较小，钢基体塑性变形的一致性得到提升，有利于延缓微孔洞微裂纹的萌生，有助于钢材力学性能的提高。

关键词：稀土 Ce；夹杂物变性；第一性原理计算；热膨胀系数；微裂纹萌生

Theoretical Analysis and Experimental Research on the Influence of Ce Inclusions on the Mechanical Properties of Steel

Liu Xiangjun, Wang Ting, Fan Mingyang, Yang Changqiao, Yang Jichun

(School of Materials and Metallurgy, Inner Mongolia University of Science and Technology, Baotou 014010, China)

Abstract: Taking Ce treated clean steel (IF steel) as the research object, the basic physical parameters such as lattice parameters, elastic constants, mechanical properties and thermal expansion coefficient of Ce inclusions are calculated using first principles, and quantitatively analyze the influence of the physical properties of inclusions on the mechanical properties of the steel matrix. The bulk modulus, shear modulus, Young's modulus and Vickers hardness of TiN, Al_2O_3 and $CeAlO_3$ inclusions are relatively large, they exhibit greater rigidity and hardness, and are characterized by brittleness, while Ce_2O_3 and Ce_2O_2S inclusions are characterized by toughness. Compared with Al_2O_3 and TiN, the thermal expansion coefficient of Ce_2O_3 and $CeAlO_3$ inclusions is close to that of the iron matrix, while the thermal expansion coefficient of Ce_2O_2S inclusions is slightly larger than that of the iron matrix. The difference between Ce inclusions and the matrix in terms of incompressibility, rigidity, hardness, toughness and brittleness, and thermal expansion is small, and the consistency of plastic deformation of the steel matrix is improved, which is beneficial to delay the initiation of micro voids and microcracks, and helps improvement of the mechanical properties of steel.

Key words: rare earths Ce; inclusions modification; first-principles calculation; thermal expansion coefficient; micro crack initiation

微合金元素 Ce, Ti, V, Nb 对 γ-Fe 力学性能和电子结构的影响

刘香军，李安鑫，刘 鹏，杨昌桥，杨吉春

（内蒙古科技大学材料与冶金学院，内蒙古包头 014010）

摘 要：采用第一性原理计算方法研究了微合金化元素 M(M=Ce、Ti、V、Nb)在钢中的固溶行为，分析了 M 对掺杂体系力学性能和电子结构的影响。根据溶解能和形成焓的计算结果表明，Ce、Ti、V、Nb 均可固溶在 γ-Fe 中。弹性模量计算结果表明，掺杂 M 降低了体系的不可压缩性和刚性，但韧性和可加工性有所提高。结合 Bader 电荷和差分电荷密度分析，掺杂体系不可压缩性和刚性降低的主要原因是 Fe-M 体系中金属键强度较纯 Fe 体系弱；其韧性增加的主要原因是掺杂体系具有较高的电子云密度。

关键词：微合金钢；稀土；第一性原理计算；力学性能；电子结构

Effect of Microalloyed Elements Ce, Ti, V, Nb on Mechanical Propertiesand Electronic Structures of γ-Fe

Liu Xiangjun, Li Anxin, Liu Peng, Yang Changqiao, Yang Jichun

(School of Materials and Metallurgy, Inner Mongolia University of Science and Technology, Baotou 014010, China)

Abstract: In this paper, first-principles calculation was used to study the solid solution behavior ofmicroalloyed elements M(M=Ce, Ti, V, and Nb) in steel, and the effects of M on the mechanical properties and electronic structures of the doped systemswere analyzed.The calculated results of the solvationenergy and formationenthalpy show that Ce, Ti, V, and Nb can be solubilized in γ-Fe. The elastic modulus calculated results show that M doped reduces the incompressibility and rigidity and of the doped system, but the toughness and machinability are improved. A combination of Bader charge and differential charge density analysis show that the metallic bond strength of Fe-M system is weaker than that of the pure Fe system, which is the main reason for the decrease in incompressibility and rigidity of the doped system; the higher electron cloud density of the doped system is the main reason for its toughness increase.

Key words: microalloyed steel; rare earth; first-principles calculation; mechanical properties; electronic structures

固溶 Ce 对 γ-Fe 力学性能及电子结构影响机理研究

杨昌桥，刘香军，车治强，刘 鹏，李安鑫，杨吉春

（内蒙古科技大学材料与冶金学院，内蒙古包头 014010）

摘 要：利用第一性原理计算研究了钢中常见元素对稀土 Ce 在钢中固溶的影响，并分析 Ce 对掺杂体系力学性能及电子结构的影响。形成焓计算结果表明，Ce 可以固溶在 γ-Fe 中，Cr、Ni、Cu、Nb、Mo 和 W 对 Ce 的固溶有负面影响，而 Si、V、Ti、Al 和 Mn 对 Ce 的固溶有促进作用，Si 的影响最大，Mn 的影响最小。弹性模量计算结果表明，Ce 的掺杂降低了体系的不可压缩性、刚性和硬度，但提高了韧性和可加工性。态密度表明，Fe-Ce 和 Si-Ce 之间的相互作用很强，而 Mn-Ce 之间几乎没有相互作用。Bader 电荷和差分电荷密度分析表明，Fe-Ce 体系的金属键强度弱于纯 Fe 体系，这是掺杂体系不可压缩性、刚性和硬度降低的主要原因，掺杂体系较高的电子云密度是其韧性增加的主要原因。另外，Si 与 Ce 之间的强相互作用，同时 Si 可以减少因 Ce 固溶引起的晶格畸变程度，这是 Si 显著增加 Ce 固溶的两个主要原因。

关键词：稀土钢；固溶；弹性模量；电子结构；第一性原理计算

Study on the Mechanism of the Influence of Rare Earth Ce on the Mechanical Properties and Electronic Structure of γ-Fe

Yang Changqiao, LiuXiangjun, Che Zhiqiang, Liu Peng, Li Anxin, Yang Jichun

(School of Materials and Metallurgy, Inner Mongolia University of Science and Technology, Baotou 014010, China)

Abstract: In this work, first-principles calculations were used to investigate the effects of the common elements on the solid solution of rare earth Ce in steel, and the effects of Ce on the mechanical properties and electronic structure of the doped system were analyzed. The calculated results of the formation enthalpy show that Ce can be solubilized in γ-Fe, and Cr, Ni, Cu, Nb, Mo, and W have negative effects on Ce solubility, while Si, V, Ti, Al, and Mn promote Ce solubility with the strongest effects from Si and the weakest from Mn. The elastic modulus calculated results show that Ce doping reduces the incompressibility, rigidity and hardness of the system, but the toughness and machinability are improved. Density of states shows that the interaction between Fe-Ce and Si-Ce is strong, while there are almost no interactions in Mn-Ce. A combination of Bader charge and differential charge density analysis shows that the strength of metallic bond of Fe-Ce system is weaker than that of the pure Fe system, which is the main reason for the decrease in incompressibility, rigidity, and hardness of the doped system; the higher electron cloud density of the doped system is the main reason for its increase in toughness. Furthermore, with the strong interaction between Si and Ce, and Si can effectively reduce the lattice distortion caused by the solid solution of Ce, which are the two main reasons why Si significantly increases the solid solution of Ce.

Key words: rare earth steel; solid solution; elastic modulus; electronic structure; first-principles calculation

应用极值分析法预测桥梁缆索用钢夹杂物最大尺寸的验证试验

郭洛方[1]，高永彬[1]，陈殿清[1]，李建开[2]，徐 凯[1]，张文涛[1]

（1. 青岛特殊钢铁有限公司线材研究所，山东青岛 266043；
2. 青岛特殊钢铁有限公司试验检测所，山东青岛 266043）

摘 要：本文应用 ASTM E2283 中极值分析的标准化方法（最大似然估计法）对桥梁缆索用钢 RH 精炼 10min 时的夹杂物进行了统计分析，并采用 Aspex 扫描电镜对多个试样进行夹杂物尺寸和成分检测，以验证预测结果的有效性。

分析结果表明：RH 精炼 10min 后仍能发现尺寸 30~55μm 的大尺寸夹杂物，夹杂物主要为 CaO-CaS-Al_2O_3 系类型；极值分析 Gumbel 概率分布 δ 和 λ 参数估计的不同计算方法中采用 Newton 迭代计算的结果与实际夹杂物尺寸吻合度最好，建议采用 Newton 迭代法数值求解；通过大量取样检测可以发现大尺寸夹杂物，与极值法预测的夹杂物最大尺寸相差不大，可见通过极值分析法来预测钢中夹杂物最大尺寸具有一定的可靠性。

关键词：极值分析；桥梁缆索用钢；夹杂物；尺寸分布；最大尺寸

Verification Test for Predicting the Maximum Size of Steel Inclusions for Bridge Cables by Extreme Value Analysis Method

Guo Luofang[1], Gao Yongbin[1], Chen Dianqing[1],
Li Jiankai[2], Xu Kai[1], Zhang Wentao[1]

(1. Qingdao Special Steel Co., Ltd., Research Institute of Wire Materials, Qingdao 266043, China;
2. Qingdao Special Steel Co., Ltd., Testing Institute, Qingdao 266043, China)

Abstract: In this paper, the standard method of extreme value analysis (maximum likelihood estimation method) in ASTM E2283 was used to analyze the inclusions during RH refining of bridge cable steel for 10 min. The size and composition of the inclusions were detected by Aspex scanning electron microscope on several samples to verify the validity of the predicted results. The analysis results show that: large inclusions with the size of 30~55μm can still be found after RH refining for 10min, and the inclusions are mainly CaO-CaS-Al_2O_3 series; Among the different calculation methods for the estimation of δ and λ parameters of Gumbel probability distribution in extreme value analysis, the results obtained by Newton iterative calculation are in good agreement with the actual inclusion size, so it is suggested to use Newton iterative method for numerical solution; Large size inclusions can be found through a large number of sampling and detection, which is not different from the maximum size of inclusions predicted by the extreme value method. Therefore, the maximum size of inclusions in steel predicted by the extreme value method has certain reliability.

Key words: extreme value analysis; bridge cables steel; inclusion; size distribution; maximum size

包钢铁稀土共生矿高效绿色利用生产技术研究

智建国[1,4]，包燕平[2]，吴 伟[3]，王 皓[1]，林 路[3]

（1. 内蒙古包钢钢联股份有限公司总工室，内蒙古包头 014010；2. 北京科技大学钢铁冶金新技术国家重点实验室，北京 100083；3. 钢铁研究总院工艺所，北京 100081；4. 内蒙古自治区稀土钢产品研发企业重点实验室，内蒙古包头 014010）

摘 要：白云鄂博铁稀土共生矿生产出来的铁水中磷含量高（含量在 0.15%~0.18%之间）、硅含量高且波动大（含量在 0.4%~1.0%之间），这给汽车板生产要求的低磷高洁净度带来技术难题。本文通过实验室模拟、进行热力学动力学分析开展了高磷铁水高效脱磷生产技术研究，得到了最佳的顶吹、底吹以及造渣工艺，结果表明，铁水脱磷率达到 92%以上，终点钢水碳氧积为 0.0018~0.0021，满足汽车板规模化生产的要求。针对加磷强化汽车板，通过加入稀土（留存量在 0.0020%以上），对钢中夹杂物进行变性处理，使氧化铝与硫化锰夹杂物尺寸变小，分散分布，并含稀土实现球化。稀土促使{111}有利织构比例增加、阻碍 FeTiP 相的析出。稀土在汽车板中起到改善表面质量，提高冲击性能和成形性。以上研究结果促进了白云鄂博铁稀土共生矿的高效、绿色化应用。

关键词：高磷铁水；转炉冶炼；加磷强化汽车板；夹杂物变性；冲击性能

Baotou Steel's Production Technology for High-efficiency and Green Utilization of Iron and Rare Earth Symbiotic Ore

Zhi Jianguo[1,4], Bao Yanping[2], Wu Wei[3], Wang Hao[1], Lin Lu[3]

(1. General Office of Inner Mongolia Baotou Steel Union Co., Ltd., Baotou 014010, China; 2. State Key Laboratory of Advanced Metallurgy, University of Science and Technology Beijing (USTB), Beijing 100083, China; 3. Department of Metallurgical Technology Research, Central Iron and Steel Research Institute, Beijing 100081, China; 4. Inner Mongolia Enterprise Key Laboratory of Rare Earth Steel Products Research and Development, Baotou 014010, China)

Abstract: The hot metal produced by the Baiyun ebo iron-rare earth symbiosis ore has high phosphorus content, between 0.15 and 0.18% and high silicon content, between 0.4 and 1.0%, which brings technical difficulties to the low-phosphorus and high cleanliness required for the production of automotive plates. In this paper, through laboratory simulation and thermodynamic kinetic analysis, the study on high-phosphorus hot metal dephosphorization production technology was carried out, and the best top-blowing, bottom-blowing and slagging processes were obtained. The results showed that the dephosphorization rate of hot metal reached more than 92%. The end-point carbon and oxygen content of molten steel is 0.0018-0.0021, which meets the requirements of large-scale production of automotive plates. For the phosphorus-enhanced automobile plate, the inclusions in the steel are modified by adding rare earths (the remaining amount is above 0.0020%), so that the alumina and manganese sulfide inclusions are reduced in size, dispersed and distributed, and containing rare earths to achieve spheroidization. Rare earth promotes the increase of {111} favorable texture ratio and hinders the precipitation of FeTiP phase. Rare earths play a role in improving surface quality, impact performance and formability in automotive plates. The above research results promoted the efficient and green application of the Baiyun ebo iron and rare earth symbiosis ore.

Key words: high phosphorus hot metal; bof smelting; phosphorus reinforced automotive plate; inclusion denaturation; impact properties

42CrMoA 曲轴钢发纹成因分析及控制措施

刘　君[1]，李成斌[2]，蒋　鹏[1]

（1. 宝山钢铁股份有限公司炼钢厂，上海　201999；
2. 宝山钢铁股份有限公司中央研究院，上海　201999）

摘　要：采用光学显微镜、扫描电镜以及能谱仪等仪器分析手段对 42CrMoA 曲轴钢机加工后出现发纹的原因进行了分析。分析了曲轴钢生产过程中钢水氧含量、LF 造渣及吹氩工艺、RH 合金量及软吹氩搅拌操作、浇注过程氩气密封效果等因素对曲轴磁粉探伤发纹的影响，确认影响曲轴发纹缺陷的主要因素是钢水氧含量、LF 造渣及吹氩工艺、RH 合金加入量等。结合生产实际对工艺进行了改进，通过提高钢水停吹氧合格率由 48%上升到 76%，LF 造渣达到合适的成分，以及提高 RH 合金加入量合格率由 45.7%上升到 70%，加铝量合格率由 65%上升到 80%，可有效减少钢水夹杂缺陷的产生。轧材探伤合格率由 92.4%提升至 98%以上，用户探伤发纹缺陷率由 0.95%降至 0.48%以下，大幅提高了曲轴的质量。

关键词：曲轴钢；发纹；夹杂物

The Reasons Analysis and Controlling Measure on Hair Cracks of 42CrMoA Crankshaft Steel

Liu Jun[1], Li Chengbin[2], Jiang Peng[1]

(1. Steelmaking Plant, Baoshan Iron&Steel Co., Shanghai 201999, China;
2. Central Research Institute, Baoshan Iron&Steel Co., Shanghai 201999, China)

Abstract: The reasons for hair cracks appearance on Crankshaft Steel surface after machining were analysed by means of metallographic examination,scanning electron microscope analysis and energy dispersive spectroscopy analysis. The causes such as liquid steel oxygen content, LF slagging and argon blowing process, RH alloy addition and soft argon blowing stirring operation, argon sealing effect in pouring process reflecting on the hair cracks of crankshaft steel inspected by magnetic powders were analyzed. It was found out that the primary causes brought to hair cracks of crankshaft steel were oxygen content, LF slagging and argon blowing process, RH alloy addition. The process was improved in combination with the actual production, the qualified rate of stopping blowing oxygen in molten steel was increased from 48% to 76%, LF slagging reaches the appropriate composition, and the qualified rate of RH alloy addition was increased from 45.7% to 70%, and the qualified rate of aluminum addition was increased from 65% to 80%, which can effectively reduce the occurrence of molten steel inclusion defects. The qualified rate of steel flaw detection increased from 92.4% to more than 98%, the flaw detection rate of hair cracks decreased from 0.95% to less than 0.48%, the quality of crankshaft was greatly improved.

Key words: crankshaft steel; hair cracks; inclusions

120t 顶底复吹转炉溅渣护炉水模试验

贺龙龙[1,3]，于　波[2]，钟良才[1,3]，孙　庆[2]，高　博[1,3]，魏志强[2]

（1. 东北大学低碳钢铁前沿技术研究院，沈阳　110819；
2. 建龙集团抚顺新钢铁炼钢厂，辽宁抚顺　113001；3. 东北大学冶金学院，沈阳　110819）

摘　要：在实验室按照 1:10 的几何相似比，建立了 120t 顶底复吹转炉模型，通过保证原型和模型的修正弗鲁德准数相等，在该顶底复吹转炉模型进行了溅渣护炉水模试验，研究了喷头距炉底距离、顶吹流量、留渣量对炉身和炉帽溅渣密度的影响，得出最佳溅渣工艺操作参数是顶吹流量为 76.6Nm^3/h、渣量为 12%、最佳溅渣枪位为 390～420mm，为实际转炉溅渣护炉工艺提供理论参考。

关键词：顶底复吹转炉；溅渣护炉；水模试验；工艺参数；优化

Water Model Experiment of Slag Splashing in 120t Top-bottom Combined Blowing Converter

He Longlong[1,3], Yu Bo[2], Zhong Liangcai[1,3], Sun Qing[2], Gao Bo[1,3], Wei Zhiqiang[2]

(1. Advanced Research Institute of Low Carbon Steel, Northeastern University, Shenyang 110819, China;
2. Fushun New Iron and Steel Plant, Jianlong Group, Fushun 113001, China; 3. School of Metallurgy, Northeastern University, Shenyang 110819, China)

Abstract: According to the geometric similarity ratio of 1:10, a 120t top and bottom combined blown converter model was established in the laboratory. By ensuring that the modified Froude number of the prototype and the model are equal, the water model experiment of slag splashing process was carried out in the top and bottom combined blown converter model. The effects of the distance between the nozzle and the furnace bottom, the top blown flow rate and the slag amount on the slag splashing density of the furnace body and cap were studied, The results show that the best operation parameters of slag splashing process are top blowing flow rate of 76.6 Nm^3/h, slag amount of 12%, and the best slag splashing lance position of 390~420mm, which provide theoretical reference for the actual slag splashing process of the converter.

Key words: top and bottom compound blowing; slag splashing; water model experiment; process parameters; optimization

提高耐火泥浆粘接强度测试精确度探讨

刘 晖，赵 磊，尹衍成，陈 伟

（青岛特殊钢铁有限公司，山东青岛 266000）

摘 要：全面分析了常用耐火泥浆所用原材料、添加剂、粘接用耐火砖试块选取及粘接强度试验过程中各因素对粘接强度测试结果的影响，进一步优化了 YB/T 22459.4—2008《耐火泥浆常温抗折粘接强度试验方法》中一些操作细节，完善后的试验方法可以使耐火泥浆粘接强度检测结果具有更好稳定性和重现性，提高了检测结果的精确度。检测人员能够在试验过程中对一些异常数据进行正确判定，以正常测试数据值的平均值作为测试结果。

关键词：泥浆；粘接强度；泥浆灰缝；精确度

Discussion on Improve Accuracy of Testing Results of Bonding Strength of Refractory Mud

Liu Hui, Zhao Lei, Yin Yancheng, Chen Wei

(Qingdao Special Iron and Steel Co., Ltd., Qingdao 266000, China)

Abstract: The influence of the raw materials, additives, bonding refractory bricks choice and various factors in the process of bonding strength test on the test results of the bonding strength is comprehensively analyzed, and some operational details in YB/T 22459.4—2008 Test Method for the Bending and Bonding Strength of Refractory Mud at Normal Temperature are further optimized. The improved test method can improve the stability and reproducibility of the testing results of the refractory mud bonding strength, and improve the accuracy of the testing results. Detectors can correctly judge some abnormal data in the test process and take the average value of normal test data as the test result.

Key words: mud; bonding strength; mud cement joint; precision

电工钢异常增 Cr 的分析与控制

蒋兴平，唐中队，孔勇江，李慕耘，张 弛，张 军，陈国威

（武汉钢铁有限公司炼钢厂生产技术室，湖北武汉 430080）

摘 要：2020年以来，武钢有限炼钢厂电工钢因异常增Cr，每月造成成分保留率约为4%，造成较大的质量、成本浪费，严重影响了产品的交付。本文通过分析炼钢厂不同工序Cr元素的来源，采取防止增Cr相应控制措施，取得良好效果，月度异常增Cr电工钢的保留率由4%降低至0.5%以下。

关键词：增铬；保留率；控制措施

Source Analysis and Control of Chromium in Si Al Killed Steel

Jiang Xingping, Tang Zhongdui, Kong Yongjiang, Li Muyun, Zhang Chi, Zhang Jun, Chen Guowei

(Production Technology Department of Steelmaking Plant of Wuhan Iron and Steel Co., Ltd., Wuhan, 430080, China)

Abstract: Since 2020, the composition retention rate of electrical steel is about 4% every month due to the abnormal increase of Cr in the steelmaking plant of WISCO, resulting in a large waste of quality and cost, seriously affecting the delivery of products. This paper analyzes the source of Cr element in different processes of steelmaking plant, takes corresponding control measures to prevent the increase of Cr, and achieves good results. The retention rate of electrical steel with abnormal monthly increase of Cr is reduced from 4% to less than 0.5%.

Key words: increasing chromium; retention rate; control measures

齿轮钢精炼过程中钢液-夹杂物-渣的热力学分析

张 剑[1]，刘 崇[2]，尹卫江[3]，马静超[4]，
于长鸿[5]，肖春江[3]

（1. 河钢集团承钢公司钒钛工程技术研究中心，河北承德 067102；
2. 河钢集团钢研总院，河北石家庄 050023；3. 河钢集团舞钢公司科技部，河南舞钢 462500；
4. 河钢集团战略研究院，河北石家庄 050023；5. 河钢集团唐钢公司热轧部，河北唐山 063000）

摘 要：通过钢液-夹杂物-渣之间的热力学计算，讨论了 $CaO-SiO_2-Al_2O_3$ 三元系中各物质的活度变化情况、准平衡状态下三个物相的热力学计算，以及齿轮钢中 Al_2O_3 夹杂物钙处理后可能变性程度及CaS夹杂生成条件。结果表明：随着二元碱度 CaO/SiO_2 比值的增加，与钢液平衡的[O]含量明显降低，CaO活度逐渐增加，SiO_2 活度逐渐降低；随着 CaO/Al_2O_3 比值的增加，Al_2O_3 活度在逐渐降低；通过热力学计算得到的1873K时的Ca-Al平衡曲线表明，为了获得变性良好的夹杂物应将[Ca%]、[Al%]含量控制在 $C_{12}A_7$ 曲线附近。

关键词：齿轮钢；等活度线；热力学；钙处理

Thermodynamic Analysis of Liquid Steel Inclusion Slag in Gear Steel Refining Process

Zhang Jian[1], Liu Chong[2], Yin Weijiang[3], Ma Jingchao[4],
Yu Changhong[5], Xiao Chunjiang[3]

(1. V-Ti Research Center for Engineering and Technology, HBIS Chengsteel, Chengde 067102, China;
2. Central Iron and Steel Reserch Institute, HBIS Group, Shijiazhuang 050023, China;
3. HBIS Group Wusteel Company, Wugang 462500, China; 4. Strategic Studies Institute, HBIS Group, Shijiazhuang 050023, China; 5. HBIS Group Tangsteel Company, Tangshan 063000, China)

Abstract: The activity change of each component in the CaO-SiO_2-Al_2O_3 ternary system, thermodynamic computing of the three phases under a quasi-equilibrium condition, the extent of modification of Al_2O_3 inclusion in calcium-treated gear steels, and the generation of CaS impurity were discussed through the thermodynamic calculation of the steel-inclusions-slag system. The activity of oxygen in molten iron which is in equilibrium with slag and SiO_2 decreases with increasing ratio of CaO/SiO_2, whereas, the activity of CaO increases with the increase of the binary basicity of CaO/SiO_2. Furthermore, the increase of the ratio of CaO/Al_2O_3 lowers the activity of Al_2O_3. The Ca-Al equilibrium curve at 1873K shows that the concentrations of Ca and Al should be adjusted near the $C_{12}A_7$ curve, such that better modified inclusions could be formed.

Key words: gear steel; isoactivity line; thermodynamics; calcium treatment

含钒铁水冶炼低磷钢的研究与应用

房　超，徐宇明，何　晴，黄　山

（河钢股份有限公司承德分公司，河北承德　067002）

摘　要：炼钢转炉正常吹炼过程中不倒前期渣，不加脱磷剂，留渣操作，利用自产包渣、渣铁缩短化渣时间，同时加大废钢用量，控制吹炼过程中炉内温度。吹炼过程中加入石灰、白云石、补热剂，石灰、白云石用于调整炉内渣况去除杂质，补热剂增补钢水温度，保证终点满足出钢要求。吹炼后期压枪操作，控制钢水回磷。

关键词：含钒铁水；低磷钢；研究

Study and Application of Smelting Low Phosphorus Wteel with Hot Metal Containing Vanadium

Fang Chao, Xu Yuming, He Qing, Huang shan

(Vanadium and Titanium Engineering Research Center, Chenggang, Chengde 067002, China)

Abstract: During the normal blowing process of converter, no slag is poured, no dephosphorization agent is added and slag is retained. The slag-covering and slag-iron produced by the converter are used to shorten the time of slag transformation, and the scrap steel consumption is increased to control the temperature in the furnace. Lime, dolomite and reheating agent are added in the blowing process. Lime and dolomite are used to adjust the slag condition in the furnace to remove impurities.

During the later stage of blowing, the pressing gun is operated to control the return of phosphorus in molten steel.

Key words: Hot Metal containing vanadium; low phosphorus steel; study

基于“转炉-LF-VD”工艺高速重轨钢非金属夹杂物控制技术探讨

储焰平[1]，张立峰[2]，葛君生[1]，封伟华[1]

（1. 中冶南方工程技术有限公司炼钢分公司，湖北武汉 430223；
2. 燕山大学亚稳材料制备技术与科学国家重点实验室，河北秦皇岛 066044）

摘 要：基于“转炉-LF-VD”冶炼工艺，从液体钢和固体钢的角度，对重轨钢生产过程中非金属夹杂物的行为进行了分析，为重轨钢中夹杂物的成分和尺寸控制提供参考。结果表明，重轨钢冶炼过程中，夹杂物成分不断发生变化，由 SiO_2-MnO 向 SiO_2-MnO-Al_2O_3 及 CaO-SiO_2-Al_2O_3-MgO 不断转变，精炼后夹杂物主要成分为液态 CaO-SiO_2-Al_2O_3-MgO。钢液凝固冷却过程中，不仅会导致夹杂物化学成分的变化，也会导致夹杂物内部相的不均匀性，影响轧制过程中夹杂物的变形，从而对钢轨的性能产生影响。合金辅料等带入钢中的钙、铝等杂志元素是导致夹杂物行为的主要原因，因此，从钢液脱氧及夹杂控制的角度，需要科学合理使用脱氧合金，尽量减少钢中钙、铝、镁等残余元素含量。

关键词：重轨钢；非金属夹杂物；凝固冷却；轧制

Discussion on the Control Technology of Non-metallic Inclusions in High-speed Heavy Rail Steel based on “ BOF-LF-VD ” Process

Chu Yanping[1], Zhang Lifeng[2], Ge Junsheng[1], Feng Weihua[1]

(1. Steelmaking Branch, WISDRI Engineering & Research Incorporation Ltd., Wuhan 430223, China;
2. State Key Lab of Metastable Materials Science and Technology,
Yanshan University, Qinhuangdao 066044, China)

Abstract: The behavior of non-metallic inclusions in the production process of heavy rail steel is analyzed from the perspective of liquid steel and solid steel based on the "converter-LF-VD" smelting process, which provides reference for the composition and size control of inclusions in heavy rail steel. The results show that the composition of inclusions changes continuously during the smelting process of heavy rail steel, from SiO_2-MnO to SiO_2-MnO-Al_2O_3 and CaO-SiO_2-Al_2O_3-MgO. And the main composition of the inclusions after refining is liquid CaO-SiO_2-Al_2O_3-MgO. During the solidification and cooling process of molten steel, not only the chemical composition of the inclusions will be changed, but also the inhomogeneity of the internal phases of the inclusions will be caused, which will affect the deformation of the inclusions during the rolling process, thereby affecting the performance of the rail. Calcium, aluminum and other magazine elements brought into the steel by alloy accessories are the main reason for the behavior of inclusions. Therefore, it is necessary to use deoxidizing alloys scientifically and rationally from the perspective of molten steel deoxidation and inclusion control to minimize the content of residual elements such as calcium, aluminum, and magnesium in the steel.

Key words: heavy rail steel; non-metallic inclusions; solidification and cooling; roll

3.2　连　铸

一种新型割嘴在方坯连铸机的应用

张银洲，郑向东，胡励克，刘　峰

（鄂钢公司转炉炼钢厂，湖北鄂州　436000）

摘　要：对转炉炼钢厂方坯连铸机的丙烷气割嘴存在的缺陷进行了剖析，通过改进割嘴内部通路及结构，有效减小了割缝，提高了切割端面质量。

关键词：割嘴；结构尺寸；割缝宽度

The Application of a New Type of Cutting Nozzle in a Square Blank Caster

Zhang Yinzhou, Zheng Xiangdong, Hu Like, Liu Feng

(Egang Converter Steelmaking Plant, Ezhou 436000, China)

Abstract: The defects of propane gas cutting nozzles in the billet caster of the converter steel mill are analyzed, and the cutting surface quality is effectively reduced and the cutting end surface quality is improved by improving the internal access and structure of the cutting nozzles.

Key words: cut mouth; structure size; cut width

45 钢小方坯高拉速连铸生产实践

李　伟，韦泽洪，胡楠楠

（宝武集团鄂城钢铁有限公司炼钢厂，湖北鄂州　436000）

摘　要：某公司炼钢厂根据生产需要，在 160mm×160mm 断面小方坯连铸机生产优质碳素结构钢 45 钢时，将拉速由原 2.4~2.5m/min 提高到 2.7~2.8m/min，通过对结晶器保护渣、结晶器铜管倒锥度、结晶器一次冷却水、中间包钢水过热度、电磁搅拌、二冷水等参数进行优化和控制，铸坯实物满足相关标准及质量要求，大幅度提高了生产效率。

关键词：拉速；连铸；铸坯质量；生产效率

Production Practice for High Speed Casting of 45# Billet

Li Wei, Wei Zehong, Hu Nannan

(Plant of Egang Iron&Steel Co., Ltd., Ezhou 436000, China)

Abstract: order to improve the steel production efficiency, casting speed of 160mm×160mmm 45# billet was increased from 2.4~2.5m/min to 2.7~2.8m/min, Through optimization of mold powder, mold pipe tapper, the primary cooling, superheat, electromagnetic stirring, secondary cooling. not only the billet meets the quality requirements, but also it improve production efficiency greatly.

Key words: casting speed; continuous casting; billet quality; production efficiency

高锰高铝钢连铸冶炼的保护渣物化性能研究

陈立峰，刘 坤，李艺璇，孙 康

（辽宁科技大学材料与冶金学院，辽宁鞍山 114000）

摘　要：高锰高铝钢的研发主要针对高品质特殊钢冶炼的生产和工业应用，对于这类特殊钢冶炼的研发一直是各国研发的重点，应变强化用高锰高铝钢，在钢中含有钛，锰等奥氏体稳定元素，应变强化高锰高铝钢拉伸应变速率越低，强化压力下的形变量越大，应变速率对这类特殊钢力学性能的影响随形变量的增加而逐渐减弱，钢中高铝含量的特殊钢增加钢材塑性，因此这类特殊钢具有优良的力学性能而被广泛应用在军工钢、船舶用钢、航空金属材料等领域。以往冶金工艺生产高锰高铝钢主要依靠模铸生产工艺，传统模铸生产效率和产品合格率低等问题，限制了这类特殊钢的生产，因此各国冶金工作者们为了提高生产效率和产品质量逐渐探索应用连铸工艺生产实践，但是在连铸浇注高铝高锰及含钛奥氏体钢时，由于钢中高含量的铝和传统 $CaO\text{-}SiO_2$ 基保护渣反应导致保护渣恶化，从而引起铸坯表面质量缺陷及连铸漏钢事故，这主要是连铸浇注过程中高含量的铝与传统 $CaO\text{-}SiO_2$ 基保护渣反应剧烈无法满足连铸浇注工艺要求，为此课题研究低反应性 $CaO\text{-}Al_2O_3$ 基连铸保护渣物化特性，已到达连铸浇注工艺要求的目的。低反应性 $CaO\text{-}Al_2O_3$ 基连铸保护渣中 CaO/Al_2O_3 比率控制在 1~4，并且保护渣中助溶剂 SiO_2 含量控制在 7wt%以内。当 CaO/Al_2O_3 比率从 1 到 2 时降低了保护渣的黏度，保护渣中 CaO/Al_2O_3=2~3 比率时黏度降低，并且随着 CaO/Al_2O_3 比率进一步增加到 4，这种黏度降低趋势变得很明显，随着 CaO/Al_2O_3 比从 1 增加到 2，表观活化能增加，保护渣光学碱度增加，CaO/Al_2O_3 比率从 3 到 4 会降低表观活化。使用傅立叶变换分析和测试了 $CaO\text{-}Al_2O_3$ 基连铸保护渣，同时分析了助熔剂的分子结构及红外光谱，结果表明，CaO/Al_2O_3 的比率从 1 增加到 2 降低了熔渣结构复杂性，熔渣结构简单导致黏度降低，随着 CaO/Al_2O_3 比率从 2 增加到 4，两种固相沉淀物相和熔体简单结构的形成有助于黏度降低，结晶动力学参数为由 Ozawa 方程，Avrami-Ozawa 方程组合和微分等转换确定弗里德曼的方法，结果发现 Ozawa 方法未能描述非等温连铸保护渣助溶剂的结晶行为，在 $CaO\text{-}Al_2O_3$ 基连铸保护渣中结晶后期的反应控制了二维生长具有较高助溶剂 B_2O_3 含量的连铸保护渣组分，对于较低助溶剂 B_2O_3 的连铸保护渣含量（10.8%），结晶是体积成核和反应控制在结晶初级阶段进行二维生长，SEM 试验结果支持这些实验分析结果，模具熔剂结晶特性对铸造高铝 TRIP 钢传热和润滑性能的影响也进行了分析。结果表明 $CaO\text{-}Al_2O_3$ 基保护渣助熔剂的结晶温度远低于 $CaO\text{-}SiO_2$ 基保护渣助熔剂的结晶温度。增加 B_2O_3 加入抑制了 $CaO\text{-}Al_2O_3$ 基保护渣的结晶，而助溶剂 Na_2O 增加了结晶性能。基于量子力学建立分子结构计算得出稳定结构排序为：$BaAl_2O_4$>$NaAlSiO_4$>$Ca_2Al_2SiO_7$>$LiAlO_2$，在 $LiAlO_2$ 结构中[AlO_4]结构需要正电荷补偿，Li-O 结构分离出 Li^+ 离子与[AlO_4]结构紧密结合。

关键词：分子结构计算；特殊钢；量子力学；$CaO\text{-}Al_2O_3$ 基保护渣

Study on Physicochemical Properties of Mold Fluxes in Continuous Casting and Smelting of High-Mn and High-Al Steel

Chen Lifeng, Liu Kun, Li Yixuan, Sun Kang

(School of Materials and Metallurgy, University of Science and Technology Liaoning, Anshan 114000, China)

Abstract: The research and development of high-manganese and high-aluminum steel is mainly aimed at the production and industrial application of high-quality special steel smelting. The research and development of such special steel smelting has always been the focus of research and development in various countries. High-manganese and high-aluminum steel for strain strengthening is contained in steel. Austenitic stabilizing elements such as titanium and manganese. The lower the tensile strain rate of strain-strengthened high-manganese and high-aluminum steel, the greater the deformation under the strengthening pressure. The effect of the strain rate on the mechanical properties of such special steels increases with the increase of the deformation Gradually weakened, special steel with high aluminum content in steel increases plasticity of steel. Therefore, this kind of special steel has excellent mechanical properties and is widely used in military steel, ship steel, aviation metal materials and other fields. In the past, metallurgical processes used to produce high-manganese and high-aluminum steel mainly relied on die-casting production processes. The traditional die-casting production efficiency and low product qualification rate restricted the production of this type of special steel. Therefore, metallurgical workers in various countries in order to improve production efficiency and products Quality is gradually exploring the application of continuous casting process production practice, but when continuous casting high-aluminum, high-manganese and titanium-containing austenitic steel, the high content of aluminum in the steel reacts with the traditional CaO-SiO_2 based mold flux and the mold slag deteriorates. Surface quality defects of slabs and continuous casting breakout accidents, which are mainly caused by the high content of aluminum and traditional CaO-SiO_2-based mold slag in the continuous casting process, which react violently and cannot meet the requirements of continuous casting and casting process. For this subject, low reactivity CaO-Al_2O_3 The physical and chemical properties of CaO-Al_2O_3-based continuous casting mold slag have reached the goal of continuous casting casting process requirements. The ratio of CaO/Al_2O_3 in the low-reactivity CaO-Al_2O_3-based continuous casting mold slag is controlled within 1-4, and the content of cosolvent SiO_2 in the mold slag is controlled within 7wt%. When the ratio of CaO/Al_2O_3 is from 1 to 2, the viscosity of the mold powder is reduced. When the ratio of CaO/Al_2O_3=2-3 in the mold powder, the viscosity decreases, and as the ratio of CaO/Al_2O_3 further increases to 4, this viscosity decrease trend changes. Obviously, as the CaO/Al_2O_3 ratio increases from 1 to 2, the apparent activation energy increases and the optical alkalinity of the mold powder increases. The CaO/Al_2O_3 ratio from 3 to 4 will decrease the apparent activation. Fourier transform was used to analyze and test CaO-Al_2O_3 based continuous casting mold powder. The molecular structure and infrared spectrum of the flux were also analyzed. The results showed that increasing the ratio of CaO/Al_2O_3 from 1 to 2 reduced the complexity of the slag structure and melted. The simple structure of the slag leads to a decrease in viscosity. As the ratio of CaO/Al_2O_3 increases from 2 to 4, the formation of two solid phases and a simple structure of the melt contributes to the decrease in viscosity. The crystallization kinetic parameters are determined by the Ozawa equation . The combination of Ozawa equations and differential transformations determined Friedman's method. It was found that Ozawa's method failed to describe the crystallization behavior of non-isothermal continuous casting mold flux cosolvent, and the reaction in the late crystallization stage of CaO-Al_2O_3-based continuous casting mold slag was controlled. Two-dimensional growth of continuous casting mold powder components with higher co-solvent B_2O_3 content. For the continuous casting mold slag content of lower co-solvent B_2O_3 (10.8%), crystallization is volume nucleation and reaction control is carried out in the primary stage of crystallization. Growth, SEM test results support these experimental analysis results, and the influence of mold flux crystallization characteristics on the heat transfer and lubrication properties of cast high-aluminum TRIP steel is also analyzed. The results show that the crystallization temperature of CaO-Al_2O_3-based flux slag is much lower than that of CaO-SiO_2-based flux slag. Increasing the addition of B_2O_3 inhibited the crystallization of CaO-Al_2O_3 based mold flux, while the cosolvent Na_2O increased the crystallization performance. Based on quantum mechanics to establish molecular structure calculations, the stable structure order is: $BaAl_2O_4$>$NaAlSiO_4$>$Ca_2Al_2SiO_7$>$LiAlO_2$. In the $LiAlO_2$ structure, the [AlO_4] structure requires positive charge compensation, and the Li-O structure separates Li^+ ions and closely combines with the [AlO_4] structure.

Key words: molecular structure calculation; special steel; quantum mechanics; CaO-Al_2O_3 based protective slag

浸入式水口结构对轴承钢非金属夹杂物影响

白李国[1,2]，逯志方[1,2]，王富扬[3]，吴 艳[1]，郑佳星[1]，申同强[1]

（1. 邢台钢铁有限责任公司，河北邢台 054027；2. 河北省线材工程技术创新中心，河北邢台 054027；3. 洛阳鼎辉特钢制品股份有限公司，河南洛阳 471322）

摘 要：研究了直通浸入式水口与两孔侧开浸入式水口对轴承钢中非金属夹杂物形貌、成分、尺寸及数量分布的影响，研究结果表明两孔侧开水口在促进钢中非金属夹杂物上浮去除方面具有明显的优势，两孔侧开水口各尺寸段夹杂物去除效率为 13.1%~81%，总体上呈现随夹杂物尺寸增加，两孔侧开水口夹杂物脱除效率增加的趋势，直通水口复合夹杂物中 CaS 含量较高，其他组元成分无明显区别。

关键词：轴承钢；夹杂物；浸入式水口；去除

Effect of the Structure of Submersed Entry Nozzle on Non-metallic Inclusion in Bearing Steel

Bai Liguo[1,2], Lu Zhifang[1,2], Wang Fuyang[3], Wu Yan[1], Zheng Jiaxing[1], Shen Tongqiang[1]

(1. Xingtai Iron and Steel Co., Ltd., Xingtai 054027, China; 2. Wire Engineering Technology Innovation Center, Xingtai 054027, China; 3. Luoyang Dinghui Special Steel Co., Ltd., Luoyang, 471322, China)

Abstract: The effect of the structure of submersed entry nozzle (SEN) on the morphology, composition, size and number of non-metallic inclusion in bearing steel is researched. The results show that the SEN with two side ports has a clear advantage in removal of nonmetallic inclusions over the single port. The removal efficiency in different size distribution is from 13.1%-81%, and there is an increasing trend with the increasing size. The SEN with single port has no visible difference in composition from two side ports, but the higher content on CaS.

Key words: bearing steel; inclusion; submersed entry nozzle; removal

汽车用钢新缺陷成因及对策

宋 宇，高立超，杜 林，潘统领，满 锐，张立辉

（鞍钢股份有限公司炼钢总厂，辽宁鞍山 114021）

摘 要：鞍钢股份有限公司炼钢总厂四分厂主要生产汽车用钢，年产量在 140 万吨左右。随着汽车用户的高标准、

严要求，鞍钢对于供给客户的热卷、冷卷表面质量要求提出了更高的标准。炼钢厂为了提高铸坯质量减少铸坯表面和内部缺陷引进了很多先进的技术和设备，如钢包下渣检测、火焰清理机等。炼钢铸坯的缺陷无外乎夹杂、夹渣、气泡、分层，但随着新技术，新设备的使用，新型缺陷也随之产生。本文阐述了 2019 年冷轧厂在生产炼钢总厂四分厂汽车钢时出现的新型缺陷，针对新缺陷进行分析、成因排查及整改措施，有效的降低了日后类似缺陷的产生。
关键词： 边部缺陷；火焰清理机；冲渣粒化水；氧化物点

Causes of New Defects of Automobile Steel and Countermeasures

Song Yu, Gao Lichao, Du Lin, Pan Tongling, Man Ri, Zhang Lihui

(General Steelmaking Plant of Angang Steel Co., Anshan 114021, China)

Abstract: The fourth branch of the General steelmaking plant of Anshan Iron and Steel Co., Ltd. mainly produces automobile steel with an annual output of about 1.4 million tons. With the high standards and strict requirements of automobile users, AISC has put forward higher standards for the surface quality of hot coil and cold coil. In order to improve the quality of slab and reduce the surface and internal defects of slab, many advanced technologies and equipments have been introduced, such as ladle slag detection, flame scarfing machine, etc. The defects of steel-making billet are inclusion, slag inclusion, blister and lamination, but with the new technology and new equipment, new defects also appear. In this paper, the new defects appeared in 2019 in the production of automobile steel in the fourth branch of the general steel-making plant of the cold-rolling mill are described.
Key words: edge defect; automatic scarfing machine; slag flushing and granulating water; oxide point

重型异型坯铸机漏钢原因与解决对策

李静文，王金坤，吴耀光，刘建坤，朱卫群，张学森

（马鞍山钢铁股份有限公司第一钢轧总厂，安徽马鞍山 243000）

摘　要： 马钢于 2020 年正式投产当前世界上最大断面重型异型坯，但是在 1030mm×440mm×130mm 断面上在翼缘顶部位置频繁发生漏钢事故，严重影响了铸机的正常生产。为了研究该断面漏钢的主要原因，通过对保护渣理化性能以及结晶器锥度适应性进行研究，研究结果表明漏钢的主要原因是结晶器保护渣理化性能与结晶器锥度设计不合理。通过研究结晶器保护渣中 Na_2O 质量分数对保护渣粘度，熔点与钢之间润湿性能影响，将保护渣中 Na_2O 质量分数由 4%提高至 8%，并且将不适用于当前生产结晶器锥度进行了调整，将结晶器液面区的锥度由 2.4%/m 减小至 1.1%/m。通过以上措施实施解决了 1030mm×440mm×130mm 断面翼缘顶部漏钢问题。
关键词： 重型异型坯；翼缘顶部；漏钢；保护渣；锥度

The Measures and Practices of Breakout Prevention of Heavy Beam Blank

Li Jingwen, Wang Jinkun, Wu Yaoguang, Liu Jiankun, Zhu Weiqun, Zhang Xuesen

(Steelmaking and Rolling Plant of Number 1, Maanshan Iron and Steel Co., Ltd., Ma'anshan 243000, China)

Abstract: Ma'anshan Iron & Steel Co., Ltd., officially put the world largest beam blank caster into production in 2020. However, breakout accidents frequently occur at the top of flange on the 1030mm×440mm×130mm cross-section, which seriously affects the normal production of the production line.In order to study the main reason of breakout in this section, through the study on the physical and chemical properties of the mold powderand the adaptability of the mold taper, the research results show that the main reason of breakout is that the physical and chemical properties of the mold powder and the mold taper design are unreasonable.By studying the mass fraction of Na_2O of mold powder on the viscosity、melting point and the wettability between steel, increase mass fraction of Na_2O conten of mold powder from 4% to 8%, and modified the unsuitable mold taper which is not adapt to current production, decrease the mold taper from 2.4%/m to 1.1%/m around the level domain. Through the above improvement measures eventually solved the top flange breakout problem of 1030mm ×440mm×130mm section.

Key words: heavy beam blank; flange top; breakout; mold powder; mold taper

Q235B 钢小方坯内部裂纹的控制

黄正华

（新疆八一钢铁股份有限公司炼钢厂，新疆乌鲁木齐　830022）

摘　要：根据八钢第一炼钢厂 Q235B 钢的生产实践，分析了 Q235B 钢小方坯生产过程中内部裂纹的产生及形成机理，针对目前生产 Q235B 钢内部裂纹比较严重的现状，结合实际生产过程中裂纹漏钢事故多的情况，提出了相应的解决措施，通过实践取得了较好的效果，Q235B 钢的内部裂纹缺陷得到了有效的控制，连铸坯质量较好的满足了轧钢工序的轧制要求，同时在拉速提升的情况下漏钢事故率得到了下降。

关键词：小方坯；内部裂纹；形成机理；措施

Internal Crack Control of Q235B Steel Billet

Huang Zhenghua

(Steel Making Plant of Xinjiang Bayi Iron and Steel Co., Ltd., Urumqi 830022, China

Abstract: according to the production practice of Q235B steel in No.1 Steelmaking Plant of Bayi Iron and Steel Co., Ltd., this paper analyzes the generation and formation mechanism of internal cracks in the production process of Q235B steel billet. Aiming at the serious internal cracks in the production of Q235B steel at present, combined with the situation of more cracks and breakout accidents in the actual production process, the corresponding solutions are put forward, and good results are achieved through practice, The internal crack defect of Q235B steel has been effectively controlled, the quality of continuous casting slab meets the rolling requirements of rolling process, and the breakout accident rate has been reduced with the increase of casting speed.

Key words: small square blank; internal cracks; formation mechanism; measures

钢水罐水口滑板液压系统蓄能器组增设的开发和应用

曹王杰，王革礼，杨万康

（新疆八一钢铁股份有限公司炼钢厂，新疆乌鲁木齐　830022）

摘　要：本文针对八钢炼钢厂钢水罐水口滑板执行机构及液压控制系统出现的问题进行分析，通过一个常规的液压控制原理，说明由于使用条件和场合的不同，而变得不适用的事例。既八钢炼钢厂钢水罐滑动水口液压控制系统安全蓄能器组增设前后作比较，正确的选用及改进，该系统运行平稳使用维护方便，可以有效避免防止设备与人身伤亡事故的发生。

关键词：滑动水口液压系统；蓄能器；钢水罐

Development and Application of Skateboard Hydraulic Accumulator Group of Steel Water Tank

Cao Wangjie, Wang Geli, Yang Wankang

(Steelmaking Plant of Xinjiang Bayi Iron and Steel Co., Ltd., Urumqi 830022, China)

Abstract: This paper analyzes the problems of the skateboard actuator and hydraulic control system of the steel steel mill. Through a conventional hydraulic control principle, the examples are not applicable due to different use conditions and occasions.Comparing the safety accumulator group of hydraulic control system is added, and correctly selected and improved. The system runs smoothly and maintained, which can effectively prevent the occurrence of equipment and personal casualty accidents.

Key words: slide water hydraulic system; accumulator; steel water tank

Q355B 角部横裂纹产生原因及其控制

钟　鹏，毛　鸣，范海宁

（马鞍山钢铁股份有限公司制造管理部，安徽马鞍山　243011）

摘　要：结合某厂的生产实际，对 Q355B 铸坯样进行金相显微镜和扫描电镜分析，找到了 Q355B 钢种产生角部横裂纹的主要原因，并通过成分优化，在保证产品使用性能的前提下，避开铸坯所处的包晶区，基本消除了该钢种的角部横裂纹，实现了铸坯的不清理直接轧制，同时降低了该钢种的生产成本，也为其它钢厂解决角部横裂纹提供参考依据。

关键词：包晶钢；角部横裂纹；微观组织；成分优化；性能

Causes and Control of Corner Transverse Crack of Q355B

Zhong Peng, Mao Ming, Fan Haining

(The Manufacturing Department of Maanshan Iron & Steel Co., Ltd., Ma' anshan 243011, China)

Abstract: Combined with the production practice of a factory, the main cause of transverse corner cracks of Q355B steel was found by metallographic microscope and scanning electron microscope analysis of Q355B casting billet samples. Through composition optimization, the transverse corner cracks of Q355B steel was completely eliminated by avoiding peritectic zone of billet on the premise of ensuring the service performance of product customers, and the billet was directly rolled without cleaning, At the same time, the production cost of the steel is reduced, and the reference for other steel plants to solve the transverse corner cracks is provided.

Key words: peritectic steel; corner transverse crack; microstructure; composition optimization; performance

热镀锌面板线状夹杂物缺陷的研究与改善

黄 君，杨新泉

（宝钢股份武钢有限炼钢厂，湖北武汉 430080）

摘 要： IF 钢热镀锌面板连铸坯中的夹杂物对热镀锌的表面质量有很大的影响，针对炼钢生产的 IF 钢热镀锌面板坯轧制后表面缺陷形态、成分、数量及分布，结合生产实际，分析夹杂物产生原因，采取一系列的改进措施，提升质量。

关键词： IF 钢；热镀锌；夹杂物

Research Methods and Control Measuer on Inclusion of If Hot Dip Galvanized Steel

Huang Jun, Yang Xinquan

(Baosteel WISCO Steel Co., Ltd., Steelmaking Plant, Wuhan 430080, China)

Abstract: The inclusions in IF steel billet have great impact on the quality of surface for hot dip galvanizing coil, In view of surface for the number of inclusions and composition、shape and distribution, combining with the actual production process, analysis the reasons, quite a few of the optimal measures for improving the quality of billet,improved quality.

Key words: IF; hot dip galvanizing coil; inclusions

高钛钢与 TiN 界面润湿性研究

邓雅岑，闫笑坡，张 勇，王强强

（重庆大学材料科学与工程学院，钒钛冶金与先进材料重庆重点实验室，重庆 400044）

摘 要： 随着钢中钛含量的提高，其产品韧性、加工性能以及耐磨性均大幅度提升，因而具有较好的应用前景。本研究采用改进的座滴法，测量高钛含量下钢与 TiN 之间的接触角；并结合电子探针显微分析仪和 FactSage 热力学计算，研究了高钛钢与 TiN 的界面润湿行为。结果表明，高钛钢与 TiN 的润湿性较差，其界面仅存在物理相互作用，无反应产物生成，为高钛钢连铸过程中 TiN 夹杂的去除提供了理论基础。

关键词： 润湿性；接触角；高钛钢；TiN

Interfacial Wettability Between High-titanium Steel and TiN

Deng Yacen, Yan Xiaopo, Zhang Yong, Wang Qiangqiang

(College of Materials Science and Engineering, Chongqing Key Laboratory of Vanadium-Titanium Metallurgy and Advanced Materials, Chongqing University, Chongqing 400044, China)

Abstract: With increasing titanium content, the toughness, machinability and wear resistance of steel products could be greatly improved. Therefore, it has a good application prospect. In this study, the contact angle between TiN and steel with high titanium content was measured by a modified sessile drop method. Moreover, based on the electron probe microanalyzer and FactSage thermodynamic calculation, interface wetting mechanism was discussed，no distinct reaction product was formed and the interface at high temperature underwent physical interaction. The results show that the wettability between high titanium steel and TiN was poor, thus providing theoretical guidance for the removal of TiN inclusions in the continuous casting of high titanium steel.

Key words: wettability; contact angle; high titanium steel; TiN

中间包内衬冲蚀及夹杂物生成和演化的研究

王　强[1,2]，谭　憧[1,2]，贺　铸[1,2]，李光强[1,2]，李亚伟[1,2]，王　强[3]

（1. 武汉科技大学耐火材料与冶金省部共建国家重点实验室，湖北武汉　430081；
2. 武汉科技大学钢铁冶金及资源利用省部共建教育部重点实验室，湖北武汉　430081；
3. 东北大学材料电磁过程研究教育部重点实验室，辽宁沈阳　110819）

摘　要：为了探究高温钢液对中间包内衬耐火材料的冲刷及夹杂物的生成与去除规律，本文建立了三维非稳态流固耦合数学模型，解析钢液在中间包内的流动和传热，以及中间包内衬耐火材料的温度分布。利用壁面切应力和总压分布计算夹杂物的生成位置、粒径和质量流量，然后采用欧拉—拉格朗日方法追踪夹杂物在中间包内的运动轨迹。结果表明，钢液对中间包内衬耐火材料的物理损毁可以分为冲击和摩擦两种来源。对于中间包而言，由于钢液冲击作用造成的耐火材料损毁要比由于钢液摩擦作用造成的耐火材料损毁更为严重。在长水口内壁产生的新生夹杂物数量占比 38%，在湍流抑制器底部产生的新生夹杂物数量占比 49%，而其余部位只产生了 13%的新生夹杂物。

关键词：中间包；耐火材料损毁；夹杂物；流固耦合；数值模拟

Numerical Simulation on Refractory Flow-Induced Erosion and Inclusion Formation in Continuous Casting Tundish

Wang Qiang[1,2], Tan Chong[1,2], He Zhu[1,2], Li Guangqiang[1,2], Li Yawei[1,2], Wang Qiang[3]

(1. The State Key Laboratory of Refractories and Metallurgy, Wuhan University of Science and Technology, Wuhan 430081, China; 2. Key Laboratory for Ferrous Metallurgy and Resources Utilization of Ministry of Education, Wuhan University of Science and Technology, Wuhan 430081, China; 3. Key Laboratory of Electromagnetic Processing of Materials (Ministry of Education), Northeastern University, Shenyang 110819, China)

Abstract: In order to understand the flow-induced erosion on the refractory lining and the formation and removal of inclusion in a continuous casting tundish, an unsteady 3D comprehensive numerical model of the respective fluid-structure interaction. The flow and heat transfer of the molten steel, as well as the refractory temperature profile was numerically clarified. The formation position, initial diameter, and mass flow rate of the inclusion were then determined by the wall shear stress and total pressure. Euler-Lagrange approach was then adopted to estimate the detachment and motion of the inclusion. The flow-induced erosion on the refractory lining could be divided into impacting and washing effects. As for the tundish, the damage caused by the impacting effect is more serious than that caused by the washing effect. At a 1.2 m/min casting speed, 49% and 38% of inclusions are created at the turbulent inhibitor inner bottom and long nozzle inner wall, respectively. In contrast, only 13% of new inclusions are produced at all other inner walls.

Key words: tundish; refractory damage; inclusion; flow-structure coupled; CFD

五轮轮带式连铸机结晶腔内流场及液面波动模拟

高 鲲，彭 艳

（燕山大学国家冷轧板带装备及工艺工程技术研究中心，河北秦皇岛 066044）

摘 要：为了了解五轮轮带式连铸机生产铸坯过程中金属液流动状态，本文建立了轮带式连铸机结晶腔内的三维模型，对不同入射角度、浇铸速度下流场分布以及自由液面波动情况进行数值模拟。结果显示，当入射角度为40°结晶腔内流场分布合理和自由液面波动幅度相对较小，综合考虑实际生产条件，当浇铸速度为12m/min时，可以得到质量较高的铸坯并满足产量的要求。

关键词：五轮轮带式连铸机；数值模拟；流场；自由液面

Simulation of Flow Field and Liquid Surface Fluctuation in the Mold of Five-wheel Belt Continuous Caster

Gao Kun, Peng Yan

(National Engineering Research Center for Equipment and Technology of Cold Rolling Strip, Yanshan University, Qinhuangdao 066044, China)

Abstract: In order to understand the state of molten metal flow during the production of cast slabs by the five-wheel belt continuous caster, this paper established a three-dimensional model of the mold of the five-wheel belt continuous caster, and analyze the flow field at different incident angles and casting speeds. The distribution and the fluctuation of the free surface are numerically simulated. The results show that when the incident angle is 40°, the flow field distribution in the crystal cavity is reasonable and the fluctuation of the free liquid level is relatively small. Considering the actual production conditions, when the casting speed is 12m/min, higher quality cast slabs can be obtained and the requirements of output can be met.

Key words: five-wheel belt continuous caster; numerical simulation; flow fluid; free surface

武钢 CSP 连铸生产 SPA-H 铸中黏连漏钢分析

王红军，许颖敏，叶　飞

（武汉钢铁有限公司条材厂 CSP 分厂，湖北武汉　430083）

摘　要： 以武钢 CSP 连铸生产 SPA-H 为背景，阐述了黏连漏钢的机理， SPA-H 的特性，分析了保护渣的性质、SPA-H 的高温力学性能、结晶器振动类型、冷却强度对 SPA-H 黏连的影响，在此基础了提出 5 条防止黏连漏钢的措施，CSP 箱板漏钢得到了很好的控制。

关键词： 连铸；黏连；SPA-H

Analysis of Sticking Breakout in SPA-H Casting in Wuhan Iron and Steel Co., Ltd.,

Wang Hongjun, Xu Yingmin, Ye Fei

(Plant of Long Product, Wuhan Iron＆steel Co., Ltd., Wuhan 430083, China)

Abstract: Based on the production of SPA-H by CSP continuous casting in Wuhan Iron and steel Co., Ltd., the mechanism of sticking and breakout and the characteristics of SPA-H described，the effects of the properties of the protective slag，the mechanical properties of SPA-H at the high temperature，the types of mold vibration and cooling intensity on the adhesion of SPA-H were analyzed. On this basis，five measures are put forward to prevent sticking leakage，and the leakage of CSP breakout was controlled well.

Key words: continuous casting; adhesion; SPA-H

硬线钢拉拔断丝机理及控制措施

杨　文[1]，张彦辉[1]，张立峰[2]，江金东[3]，伍从应[3]

（1. 北京科技大学冶金与生态工程学院，北京　100083；

2. 燕山大学亚稳材料制备技术与科学国家重点实验室，河北秦皇岛　066044；

3. 首钢水城钢铁（集团）有限责任公司，贵州六盘水　553028）

摘　要： 首先对硬线钢拉拔断丝试样进行了分析，通过对断裂试样的断口和剖面分析，结合连铸坯内部质量检测，得出硬线钢拉拔断丝主要是源于连铸坯严重的中心缩孔和中心偏析缺陷。然后通过施加连铸凝固末端电磁搅拌、控制钢水浇铸过热度在 30℃以内等措施，明显改善了硬线钢连铸坯中心缩孔和中心偏析，显著降低了 82B 硬线钢拉拔断丝率。

关键词： 硬线钢；拉拔断丝；内部质量；连铸参数

Mechanism and Control of Drawing Fracture of Hard Wire Steel

Yang Wen[1], Zhang Yanhui[1], Zhang Lifeng[2], Jiang Jindong[3], Wu Congying[3]

(1. School of Metallurgical and Ecological Engineering, University of Science and Technology Beijing, Beijing 100083, China; 2. State Key Lab of Metastable Materials Science and Technology, Yanshan University, Qinhuangdao 066044, China; 3. Shougang Shuicheng Iron and Steel (Group) Co., Ltd., Liupanshui 553028, China)

Abstract: The drawing fractured wire sample of hard wire steel was analyzed firstly. Through the analysis on the fractograph and section of the fracture specimen, combined with the internal quality inspection of continuous casting billet, it was concluded that the drawing fracture of hard wire steel was mainly caused by the serious central shrinkage and central segregation defects in the billet. Then, by applying final electromagnetic stirring (F-EMS) at the solidification end of continuous casting and controlling the superheat of molten steel during casting below 30℃, the central shrinkage and central segregation of hard wire steel billet were obviously restrained, significantly reducing the drawing fracture ratio of hard wire steel.

Key words: hard wire steel; drawing fracture; internal quality; continuous casting parameter

非反应性高铝钢连铸保护渣结构与性能研究

陈　阳，潘伟杰，王强强，何生平

（重庆大学材料科学与工程学院，钒钛冶金及新材料重庆市重点实验室，重庆　400044）

摘　要： 传统 CaO-SiO_2 系保护渣在浇铸高铝系钢种(w[Al]> 2.5%)时，渣中 SiO_2、B_2O_3、TiO_2 等组分易被钢液中的铝大量还原，产生强烈的渣-金界面反应，因此研究建立新型“非反应性保护渣”是重要方向。本文结合分子动力学模拟、旋转黏度计测试、XRD 检测分析及核磁共振波谱（NMR）实验，研究非反应性渣系 BaO–CaO–Al_2O_3–CaF_2–Li_2O 的微观结构与宏观性能之间的联系。结果表明，BaO–CaO–Al_2O_3–CaF_2–Li_2O 非反应性渣系形成了稳定的$[AlO_4]^-$四面体结构，熔体结构多以链状或层状结构单元为主；黏度实验表明保护渣在 1300 °C 时的黏度为 0.284 Pa·s，转折温度(T_{br})为 1120℃，符合非反应性保护渣设计要求；MD 模拟和 NMR 波谱实验结果表明，熔体中大部分 Al_2O_3 作为网络形成体，充当网络结构骨架。

关键词： 非反应性；保护渣；微观结构；核磁共振

Study on Structure and Properties of Non-reactive High-aluminum Steel Continuous Casting Slag

Chen Yang, Pan Weijie, Wang Qiangqiang, He Shengping

(College of Materials Science and Engineering, and Chongqing Key Laboratory of Vanadium–Titanium Metallurgy and Advanced Materials, Chongqing University, Chongqing 400044, China)

Abstract: During the continuous casting process of high Al Steel (w[Al] > 0.5%), SiO_2、B_2O_3、TiO_2 and other components in slag could be reduced by aluminum in molten steel, resulting in the slag-metal interface reaction seriously. Therefore, it is an important direction to research and establish the new "non-reactive protective slag". Based on molecular dynamics simulation, rotating viscometer test, XRD analysis and nuclear magnetic resonance spectroscopy (NMR) experiment, this paper investigates the relationship between microstructure and macroscopic properties of $BaO-CaO-Al_2O_3-CaF_2-Li_2O$ non-reactive mold flux. The results showed that the non-reactive slag system of $BaO-CaO-Al_2O_3-CaF_2-Li_2O$ formed stable structural units of $[AlO_4]^-$ tetrahedron, and the melt structure is dominated by chain or layered structure units. The results of viscosity test as follows: viscosity (1300℃): 0.284 Pa·s and break temperature: 1120℃, which meets the design requirements of non-reactive mold flux. The results of MD simulation and NMR spectrum experiments showed that the majority of Al_2O_3 in the melt acts as the network former and the framework of the network structure.

Key words: non-reactive; mold flux; microstructure; NMR

八钢板坯连铸机二冷控制优化实践

吴　军

（新疆乌鲁木齐，新疆八一钢铁股份有限公司炼钢厂）

摘　要：本文结合八钢板坯连铸机二次冷却系统现状，对板坯连铸机非稳态生产时不均匀的铸坯凝固速度导致裂纹、中心偏析和中心缩孔等缺陷形成的主要原因进行分析，通过对二次冷却系统进行优化，铸坯二冷分布更加均匀，保证了铸坯质量的稳定提升，现场质量改善效果明显，适应全钢种、钢水温度、拉速的变化。

关键词：二次冷却；裂纹；中心偏析；凝固；优化

Practice of Optimizing Secondary Cooling Control of Slab Caster in Xinjiang Bayi Iron and Steel Co., Ltd.,

Wu Jun

Abstract: Based on the current situation of the secondary cooling system of the slab caster in Xinjiang Bayi Iron and Steel Co., Ltd., the main reasons for the formation of defects such as cracks, central segregation and central shrinkage cavity caused by the uneven solidification rate of slab caster during unsteady production were analyzed in this paper. Through the optimization of the secondary cooling system, the secondary cooling distribution is more uniform and the quality of the cast billet is guaranteed. The optimized effect is obvious, suitable for the whole steel under different temperature, drawing speed control.

Key words: Secondary cooling; Crack; Central segregation; Coagulation; Optimal control

方坯连铸拉矫机冷却和铸坯保温优化措施

张翼斌

（河钢承钢工程技术有限公司，河北承德　067102）

摘　要：拉矫机是连铸机的重要设备之一，拉矫机液压缸使用喷水冷却时水直接流淌到铸坯上影响铸坯质量，同时造成铸坯温度快速下降。为保证铸坯质量和温度取消了拉矫机液压缸喷淋冷却，生产中液压缸活塞杆密封经常因高温炙烤损坏漏油，影响拉矫机的正常工作，增加了维修成本。而拉矫机前后立柱两侧的侧孔及中部横梁间的间隙为铸坯通过时的主要温降影响部位。因此对拉矫机进行了设备冷却和铸坯保温措施研究，提出了改进措施，并取得了成效，提高了连铸机作业率，保证了铸坯的高温直送率。

关键词：方坯连铸机；拉矫机；冷却；保温；优化

Optimization of Cooling and Slab Insulation for Billet Continuous Casting and Tensioning Machine

Zhang Yibin

(HBIS Chengsteel Maintain & Repair Center, Chengde, 067102, China)

Abstract: The tension leveler is one of the important equipments of the continuous casting machine. When the hydraulic cylinder of the tension leveler is sprayed with water, the water directly flows to the slab to affect the quality of the slab, and at the same time, the temperature of the slab is rapidly decreased. In order to ensure the quality and temperature of the slab, the hydraulic cylinder of the tensioning machine is sprayed and cooled. The piston rod seal of the hydraulic cylinder in production often damages the oil due to high temperature roasting, which affects the normal operation of the tension leveler and increases the maintenance cost. The gap between the side holes on both sides of the front and rear columns of the tension leveler and the middle beam is the main temperature drop influence point when the casting blank passes. Therefore, research on equipment cooling and slab insulation measures was carried out on the tension leveler, and improvement measures were put forward, and the results were achieved. The operation rate of the continuous casting machine was improved and the high temperature direct delivery rate of the slab was ensured.

Key words: billet continuous casting machine; straightening machine; cooling; insulation; optimization

小方坯连铸机高效化研究与实践

胡铁军，房志琦，刘宏春，刘　辉

（河钢集团承钢分公司棒材事业部，河北承德　067002）

摘　要：近年来，冶金行业的迅速发展，企业为提升效益，控制成本，提出了提产增效的目标，很多钢铁企业通过提升铸机拉速，提升产能，减少耐材成本及钢铁料消耗，达到提高产品利润的目的，我部通过对铸机进行工艺优化，提高铸机拉速，达到提产增效的目的。

关键词：拉速；工艺优化；提产增效

Research and Practice on High Efficiency of Billet Caster

Hu Tiejun, Fang Zhiqi, Liu Hongchun, Liu Hui

(Hegang Group Chenggang Branch Bar Business Department, Chengde 067002, China)

Abstract: In recent years, with the rapid development of metallurgical industry, enterprises have put forward the goal of increasing production and increasing efficiency in order to increase benefit and control cost. Many iron and steel enterprises have increased casting speed, increased production capacity, and reduced the cost of refractories and consumption of iron and steel, in order to increase the profit of the products, our department optimizes the technological parameters of the casting machine by studying the technology of the casting machine, improves the casting speed, and achieves the goal of increasing production and efficiency.

Key words: casting machine; process optimization; increase production and efficiency

600MPa 级螺纹钢筋表面裂纹控制工艺研究

胡春林，杨应东，付振宇，宋传文

（马钢股份长材事业部，安徽马鞍山　243011）

摘　要：针对 600MPa 级螺纹钢筋在试制过程中发现的表面裂纹问题。轧材裂纹金相分析、高温塑性试验，确定造成裂纹的主要原因为铸坯角部表面横裂。通过保护渣、二次冷却以及炼钢、连铸工艺调整，有效地改善了这一问题。

关键词：高强螺纹钢；角部裂纹；保护渣；二次冷却

Study on the Control of Surface Cracks of 600MPa Grade Rebar

Hu Chunlin, Yang Yingdong, Fu Zhengyu, Song Chuanwen

(Long Products Business Division of Maanshan Iron and Steel Co., Ltd., Ma'anshan, 243011, China)

Abstract: In order to solve the problem of surface cracks of 600MPa grade rebar. By the metallographic and high temperature plastic analysis, The results indicate that, the cause of surface crack of rebar was corner cracks on billets. By adjusted the mold powder, reduced billet's secondary cooling strength, steelmaking and continuous casting process adjustment, the problem was effectively improved.

Key words: high strength rebar; corner crack; mold powder; secondary cooling

B/C 级角钢成分优化及铸坯质量控制

张利江

（河钢集团宣钢公司，河北张家口　075100）

摘　要：本文分析了影响角钢铸坯生产成本因素，通过采取优化成分控制、稳定炼钢及连铸生产工艺，达到降低角钢铸坯生产成本、稳定铸坯质量目的。

关键词：角钢；成分优化；铸坯质量

Production Practice of Low Cost Angle Steel Billet in Xuanhua Steel

Zhang Lijiang
(HBIS Group Xuansteel Company, Zhangjiakou 075100, China)

Abstract: This paper analyzes the factors that affect the production cost of angle steel billet, and by adopting the optimized composition control, stabilizing the steel-making and continuous casting production process, the purpose of reducing the production cost and stabilizing the quality of angle steel billet is achieved.

Key words: angle steel; composition optimization; billet quality

斜极式结晶器电磁搅拌器对钢液磁场影响的数值模拟

张 静[1]，赵 震[1]，孟纯涛[1]，张立峰[2,3]

（1. 燕山大学车辆与能源学院，河北秦皇岛 066044；2. 燕山大学亚稳材料制备技术与科学国家重点实验室，河北秦皇岛 066044；3. 燕山大学机械工程学院，河北秦皇岛 066044）

摘 要：本文以设计发明的斜极式结晶器电磁搅拌器为研究对象，研究了不同铁芯偏斜角度和电磁参数对结晶器内钢液磁感应强度、电磁力分布的影响。结果表明，对斜极式电磁搅拌器线圈加顺时针方向电流，电磁力分布较好；磁感应强度和电磁力最大位置位于结晶器出口处；铁芯偏斜角度5°时磁场分布较优，钢液中心的最大磁感应强度值较传统电磁搅拌器提高了59.5 Gs，电磁力提高了98 N/m^3。

关键词：斜极式结晶器电磁搅拌器；铁芯偏斜角度；磁感应强度；电磁力

Numerical Simulation of the Influence of Deflected-pole Mold Electromagnetic Stirrer on Molten Steel Magnetic Field

ZhangJing[1], Zhao Zhen[1], Meng Chuntao[1], Zhang Lifeng[2,3]

(1. School of Vehicle and Energy, Yanshan University, Qinhuangdao 066044, China; 2. State Key Lab of Metastable Materials Science and Technology , Yanshan University, Qinhuangdao 066044, China; 3. School of Mechanical Engineering, Yanshan University, Qinhuangdao 066044, China)

Abstract: Aimed at deflected-pole electromagnetic stirrer, this article investigates the influence of different iron core deflection angles and electromagnetic parameters on the magnetic induction intensity and electromagnetic force distribution of molten steel in the mold. The results show that the electromagnetic force distribution is better when the clockwise current is added to the coil of the deflected-pole electromagnetic stirrer. The maximum magnetic induction intensity and electromagnetic force locate at the mold outlet. When the core deflection angle is 5°, the magnetic field distribution is better, the maximum magnetic induction intensity in the center of the molten steel is increased by 59.5Gs, which higher than that of the conventional electromagnetic stirrer, and the electromagnetic force is increased by 98 N/m^3.

Key words: deflected-pole mold electromagnetic stirrer; iron core deflection angle; magnetic induction intensity; electromagnetic force

包晶钢连铸保护渣的技术特征

李　刚，何生平，王强强

（重庆大学材料科学与工程学院，钒钛冶金与先进材料重庆重点实验室，重庆　400044）

摘　要：本研究是通过优化保护渣性能的方法，解决包晶钢板坯在连铸生产过程中表面纵裂纹控制不稳定、裂纹率较高的问题。主要是在保证润滑性能的前提下，通过提高保护渣碱度和 F 含量促使渣膜中析出的枪晶石含量增加，降低保护渣的熔点和凝固温度，提高其冷凝断口晶体比例，以此加强保护渣对传热性能的控制。现场试验表明，使用 S-1 保护渣能有效减少铸坯表面纵裂纹缺陷，优化思路正确。

关键词：包晶钢；保护渣；纵裂纹；碱度

Influence Technical Features of Peritectic Steel Continuous Casting Mold Flux

Li Gang, He Shengping, Wang Qiangqiang

(College of Materials Science and Engineering, Chongqing University, Chongqing 400044, China)

Abstract: This research is to solve the problem of unstable control of surface longitudinal cracks and high crack rate in the continuous casting process of peritectic steel slab by optimizing the performance of mold flux. Mainly on the premise of ensuring lubrication performance, increasing the alkalinity and F content of the flux promotes the increase of the content of gun spar in the slag film. And reduce the melting point and solidification temperature of the flux, and increase the ratio of the crystals of the condensation fracture, so as to strengthen the control of the mold powder on the heat transfer performance. Field tests show that the use of S-1 mold flux can effectively reduce the defects of longitudinal cracks on the surface of the cast slab, and the optimization idea is correct.

Key words: peritectic steel; mold flux; longitudinal crack; alkalinity

连铸坯中非金属夹杂物数量、尺寸和成分空间分布的数值模拟仿真

张立峰[1]，陈　威[2]，王举金[3]，张月鑫[3]

（1. 燕山大学亚稳材料制备技术与科学国家重点实验室，河北秦皇岛　066044；2. 燕山大学机械工程学院，河北秦皇岛　066044；3. 北京科技大学冶金与生态工程学院，北京　100083）

摘　要：连铸坯中非金属夹杂物的数量、尺寸和成分空间分布是影响钢产品质量的重要因素，因此实现夹杂物数量、尺寸和成分空间分布预测对提高钢产品质量有重要意义。本研究通过耦合大涡模拟模型（LES model）、传热、凝固模型、溶质传输模型及离散型模型（DPM model），实现了夹杂物数量、尺寸和空间分布分布的预测，并与连铸坯

全断面上夹杂物检测结果进行了对比验证。在此基础上耦合传热、凝固、夹杂物成分随温度变化的热力学转化和溶质元素在钢中的动力学扩散，建立夹杂物成分分布预测模型，实现了夹杂物在整个连铸坯断面上成分空间分布的预测。

关键词：夹杂物；空间分布；热力学；动力学；数值模拟

Mathematical Simulation of Number, Size, Composition, and Spatial Distribution of Non-metallic Inclusions in a Continuous Casting Slab

Zhang Lifeng[1], Chen Wei[2], Wang Jujin[3], Zhang Yuexin[3]

(1. State Key Lab of Metastable Materials Science and Technology, Yanshan University, Qinhuangdao 066044, China; 2. School of Mechanical Engineering, Yanshan University, Qinhuangdao 066044, China; 3. School of Metallurgical and Ecological Engineering, University of Science and Technology Beijing, Beijing 100083, China)

Abstract: The number, size, composition, and spatial distribution of non-metallic inclusions in continuous casting (CC) slabs are important factors that affect the quality of steel products. Therefore, realizing the prediction of the number, size, composition, and spatial distribution of inclusions is of great significance for improving the quality of steel products. In the current study, a mathematical model coupled with the large eddy simulation, heat transfer, solidification, solute transport, and DPM model, was established to predict the number, size, and spatial distribution of inclusions. The distribution of inclusions on the entail cross section of a CC slab was detected to validate the mathematical model. An integrated model coupled with the heat transfer, solidification, thermodynamic transformation of inclusion composition and diffusion of dissolved elements in steel was established to predict the composition distribution of inclusions in the CC billet.

Key words: inclusion; spatial distribution; thermodynamics; kinetics; mathematical simulation

板坯亚包晶钢高拉速工艺研究与实践

刘启龙，曹成虎，郑 晴，张 敏，罗 霄

（马鞍山钢铁股份有限公司第四钢轧总厂，安徽马鞍山 243011）

摘 要：为解决亚包晶钢浇注炉机周期匹配问题，提高生产效率，通过对亚包晶钢结晶器冷却、振动模式及结晶器保护渣等高拉速工艺技术进行优化，实现了亚包晶钢高拉速的稳定生产，最高工作拉速为1.6m/min；铸坯表面纵裂纹发生率小于0.3%，中心偏析质量稳定。

关键词：亚包晶钢；高拉速；铸坯质量

Study and Practice of Hypo-peritectic Steel Slab High Speed Technology

Liu Qilong, Cao Chenghu, Zheng Qing, Zhang Min, Luo Xiao

(No.4 Steelmaking and Rolling General Plant of Maanshan Iron and Steel Co., Ltd., Ma’anshan 243011, China)

Abstract: For solving the problem of period matching of hypo-peritectic steel casting between BOF and CC, improving production efficiency, the stable production of high speed of hypo-peritectic steel is realized by optimizing the technology of high speed such as mold cooling, oscillation mode and mold powder, the maximum speed is 1.6 m/min; the incidence of longitudinal cracks on the surface of slab is less than 0.3 percent and the quality of central segregation is stable.

Key words: hypo-peritectic steel; high speed; slab quality

控制板坯连铸机漏钢的探索与实践

蔡常青[1]，王龙飞[2]

（福建三钢闽光股份有限公司炼钢厂，福建三明 365000）

摘 要：福建三钢闽光股份有限公司炼钢厂板坯连铸机投产以来，以“两化、团队、无事故”为理念，对漏钢原因进行了分析研究；依靠结晶器漏钢预报系统和坚持连铸结晶器保护渣测量制度及结晶器液面流场活跃状况的监控，并对生产操作、设备运行精度、原辅材料、工艺制度等方面制定了一系列控制标准，职工干部认真执行工艺操作制度和车间文化理念，实现两台板坯铸机从 2015 年 2 月至今连续浇注 6 年多，铸钢 100000 余炉，累计产钢超过 970 余万吨无漏钢的世界记录。

关键词：板坯连铸；漏钢率；原因分析；措施

Exploration and Practice of Controlling Breakout of Slab Caster

Cai Changqing[1], Wang Longfei[2]

(Steelmaking Plant, Fujian Steel Minguang Co., Ltd., Sanming 365000, China)

Abstract: Since the slab caster was put into operation in the steelmaking plant of Fujian Sangang Minguang Co., Ltd., Based on the concept of "two modernizations, team, no accidents"，The causes of steel leakage are analyzed and studied; Depending on the prediction system of mold breakout, the measurement system of mold flux and the monitoring of mold liquid level flow field，In addition, a series of control standards are established for production operation, equipment operation accuracy, raw and auxiliary materials, process system, etc，Staff and cadres earnestly implement process operation system and workshop culture concept，Two slab casters have been continuously casting for 6 years since February 2015，Nearly 100000 heats of cast steel，The world record of producing more than 9.7 million tons of steel without leakage.

Key words: slab continuous casting; breakout rate; cause analysis; measures

铸坯余热压下技术对改善铸坯内部质量的研究

刘宏强，卫广运，张瑞忠，田志强，刘 崇

（河钢集团钢研总院，河北石家庄 050023）

摘 要：为了提高大断面铸坯内部的质量，减少铸坯疏松、缩孔等缺陷，本文依据铸坯余热压下原理进行中试生产试验。通过低倍、密度检测、金属原位和超声检测实验分析铸坯余热压下对铸坯质量的影响。实验结果表明：随着压下量从 0mm 增加至 24mm，缺陷总数量从 1746 个减少到 622 个，缺陷数量减少 64.37%，总缺陷尺寸从 651.2mm 减少至 242.5mm，减少了 62.76%；当压下量为 12mm 时，铸坯致密度最高，致密度相对于 1 号样提高 3.69%，统计疏松度降低 73.27%，铸坯内部质量最好。

关键词： 铸坯余热压下；中心疏松；中心缩孔；致密度

Study on the Improvement of Internal Quality of Slab by the Technology of Residual Hot Reduction

Liu Hongqiang, Wei Guangyun, Zhang Ruizhong,
Tian Zhiqiang, Liu Chong

(HeSteel Group Technology Research Institute, Shijiazhuang 050023, China)

Abstract: In order to improve the internal quality of large section slab and reduce slab porosity and shrinkage defects, in this paper, a pilot test is carried out according to the theory of residual hot reduction technology. The effect of residual hot reduction on the quality of slab was analyzed by means of acid pickling , density measurement original position analysis and ultrasonic nondestructive testing. The test results show that with the amount of reduction is increased from 0 mm to 24mm, the total defect decrease from1746 to 622, the number of defects decreased by 64.37%. The total defect length decrease from651.2mmto 242.5mm, the length of defects decreased by 62.76%. When the reduction is 12 mm, the density of slab is the highest, the density increases by 3.69% compared with sample 1, and the statistical porosity decreases by 73.27%, the internal quality of slab is the best.

Key words: billet residual hot reduction; center porosity; central pipe; density

板坯连铸高拉速生产面临的几个问题探讨

何宇明

（重庆钢铁股份有限公司炼钢厂，重庆 401258）

摘 要： 高拉速下中间包、结晶器中夹杂物上浮难度增大，更易出现坯壳与铜板间润滑不良，铸坯表面纵裂纹率增加，同一冷却区域坯壳更薄、鼓肚严重、内部偏析加剧，热装时易出现轧后裂纹，下线铸坯精整难度增大。通过分析高拉速下结晶器、二次冷却与常规拉速时的差异，提出强化和均匀冷却一、二次冷却的重要性和方法，做好坯壳的冷却和支撑工作等，含铝、钛、铌等裂纹敏感钢元素的铸坯淬火后热装入炉轧制，避免轧后热裂纹出现，高拉速得有相应的对策措施才能把不良后果控制在允许范围内。

关键词： 连铸；高拉速；铸坯质量；强化冷却；支撑

Discussion on Several Problems in High Casting Speed Production of Slab Continuous Casting

He Yuming

(Chongqing Iron and Steel Co., Ltd., Steelmaking Plant, Chongqing 401258, China)

Abstract: At high casting speed, it is more difficult for inclusions in tundish and mold to float up, which makes it easier to have poor lubrication between shell and copper plate, increase longitudinal crack rate on slab surface, make slab shell thinner, bulge more serious and internal segregation more severe in the same cooling area, and cause cracks after rolling during hot charging, which makes it more difficult to finish lower strand. By analyzing the differences between mold, secondary cooling and conventional casting speed at high casting speed, the importance and methods of strengthening and uniform cooling primary and secondary cooling are put forward, and the cooling and supporting work of billet shell are well done. The billet containing aluminum, titanium, niobium and other crack sensitive steel elements is hot charged into furnace after quenching to avoid hot cracks after rolling, In order to control the adverse consequences within the allowable range, the corresponding countermeasures must be taken.

Key words: casting; high casting speed; slab quality; enhanced cooling; prop up

轴承钢大方坯凝固末端重压下工艺研究与实践

孙忠权，曾令宇，谭奇峰

（宝武集团广东韶关钢铁有限公司，广东韶关 512123）

摘 要： 连铸重压下技术是改善轴承钢铸坯中心疏松、缩孔和中心偏析的一种最有效的方法。韶钢与东北大学合作研发大方坯连铸机重压下工艺控制技术，在轻压下工艺装备和拉矫机控制系统基础上，将 8 号、9 号拉矫辊改造成渐变曲率凸型辊，形成轻压下和重压下两段式压下技术。本文依据 320mm×425mm 断面轴承钢连铸坯凝固末端重压下工艺实践，研究在不同拉速和重压下工艺参数对铸坯内部质量的影响。实践表明，轴承钢在拉速 0.58m/min 时实施重压下工艺后，铸坯中心缩孔改善明显，中心疏松宽度由 125mm 缩减至 110mm，碳偏析由 0.85~1.09 降至 0.95~1.04，圆钢探伤合格率提升了 8.34%。实践结果为韶钢开发高质量、高性能、高附加值的特钢产品提供了重要的技术支撑。

关键词： 大方坯连铸；轴承钢；重压下技术；中心疏松；碳偏析

Research and Practice of Heavy Reduction Process at Solidification End of Bearing Steel Bloom

Sun Zhongquan, Zeng Lingyu, Tan Qifeng

(Baowu Grounp Guangdong Shaoguan Iron and Steel Co., Ltd., Shaoguan 512123, China)

Abstract: Continuous casting heavy reduction technology is one of the most effective methods to improve the central porosity, shrinkage cavity and central segregation of bearing steel slab. Shaoguan Iron and Steel Co., Ltd. cooperated with Northeast University to develop the heavy reduction process control technology of bloom caster. Based on the light reduction process equipment and the control system of tension leveler, the 8#, 9# tension leveler rollers were transformed into convex rollers with gradual curvature to form the two-stage reduction technology of light reduction and heavy reduction. This paper is based on 320mm × the process practice of heavy reduction at the solidification end of 425mm section bearing steel continuous casting slab was carried out, and the effects of process parameters on the internal quality of slab under different drawing speed and heavy reduction were studied. The practice shows that after the heavy reduction process is implemented at the drawing speed of 0.58m/min, the central shrinkage cavity of the billet is obviously improved, the central loose width is reduced from 125mm to 110mm, the carbon segregation is reduced from 0.85 ~ 1.09 to 0.95 ~ 1.04, and the flaw detection qualified rate of round steel is increased by 8.34%. The practice results provide important technical support for the development of special steel products with high quality, high performance and high added value.

Key words: bloom continuous casting; bearing steel; heavy reduction technology; central porosity; carbon segregation

高拉速板坯连铸凝固传热行为及其均匀化控制

朱苗勇，蔡兆镇

（东北大学冶金学院，辽宁沈阳　110819）

摘　要：高速连铸是发展新一代高效连铸的主题，是实现直轧、铸轧的前提保障，是实现钢铁制造高效、绿色的具体体现。目前我国板坯的实际工作拉速基本在1.8m/min以下，包晶钢拉速大都1.2～1.4m/min。拉速提升，影响结晶器凝固传热的不利因素更加凸显，热通量增加，保护渣的消耗量降低，凝固坯壳与结晶器铜壁间的润滑变得越来越差，高温凝固坯壳承受各种应力应变的能力变得越来越弱，漏钢和裂纹成为最大挑战。如何确保高拉速条件下结晶器内凝固坯壳的均匀性与安全性以及铸坯质量，是实现高速连铸必须要面对和解决的技术难题。本文从分析包晶钢高速连铸结晶器凝固传热行为特征入手，阐述高效传热结晶器凝固均匀化控制技术，为高速连铸技术开发和生产提供借鉴。

关键词：高拉速连铸；均匀凝固；高效传热结晶器

Heat Transfer Behavior and Homogenous Solidification Control for High-speed Slab Continuous Casting of Steel

Zhu Miaoyong, Cai Zhaozhen

(School of Metallurgy, Northeastern University, Shenyang 110819, China)

Abstract: High-speed continuous casting is the theme for developing new generation of high-efficiency continuous casting technology and the premise to make direct rolling or continuous casting and rolling come true, that is the embody for realizing high-efficiency and green steelmaking production line. Presently, the actual casting speed for slab in China is no more than 1.8 m/min and it is in the range of 1.2-1.4m/min for continuous casting of peritectic steel. With increasing casting speed, the negative factors affecting the solidification in continuous casting mold show more obvious, and the lubrication between solidifying shell and mold copper plate becomes worse and worse due to the increase of heat flux and decrease of mold flux consumption, therefore thecapability of solidifying shell for resisting the all kinds of stress and strain during casting becomes weaker and weaker, andthe occurrence of breakout and cracks with high frequency has been the most challenge. How to ensure homogenous growth and safety of solidifying shell and slab quality, the technical issues, should be solved for high-speed casting. The behavior of heat transfer and solidification in mold with high-speed casting was analyzed and the control technology for mold with high-efficiency heat transfer and homogenous solidification was presented and discussed in this paper.

Key words: high-speed continuous casting; homogenous solidification; continuous casting mold with high-efficiency heat transfer

SWRCH22A冷镦钢铸坯质量改进研究

郭峻宇，刘志龙，黎　莉，王　冠

（宝武集团广东韶关钢铁有限公司炼钢厂，广东韶关　512123）

摘　要：SWRCH22A冷镦钢盘条在冷加工过程中存在笔尖状开裂的现象。经检测在开裂处发现了大尺寸的夹杂物，成分主要为O、Al、Ca等。其可能来源于钢水大量的夹杂物聚集、水口塞棒附着夹杂物脱落、浇注过程夹杂物的聚集。根据夹杂物来源，对冶炼与连铸工艺进行了优化。通过提高转炉高拉碳水平，改善转炉出钢渣洗用渣，优化精炼渣以及钙处理工艺，优化中间包耐材、浸入式水口及塞棒，优化结晶器保护渣理化性能、铸坯冷却工艺等，SWRCH22A冷镦钢铸坯质量得到显著提升。

关键词：SWRCH22A冷镦钢；笔尖状断裂；大尺寸夹杂物；钙处理工艺；中间包耐材；浸入式水口；铸坯质量

Study on Improvement of Billet Quality of SWRCH22A Cold Heading Steels

Guo Junyu, Liu Zhilong, Li Li, Wang Guan

(Baowu Grounp Guangdong Shaoguan Iron and Steel Co., Ltd., Shaoguan 512123, China)

Abstract: SWRCH22A Cold Heading Wire Rod had Pencil-Tip Shaped Fracture during the drawing. Large-sized inclusions were found in the cracks, and the main components of the inclusions were O, Al, Ca and so on. There are three possible sources of large inclusions: (1) inclusions gathered in molten steel; (2) inclusions attached to the nozzle stopper; (3) inclusions gathered during the pouring process. According to the source of inclusions, the smelting and continuous casting processes were optimized. In the converter stage, the high carbon catching level were improved, and the washing slag were improved; In the refining stage, the refining slag and calcium treatment process were improved; In the continuous casting stage, optimizing tundish refractory material, submerged entry nozzle and stopper; At the same time, the physical and chemical properties of mold flux and the cooling process of casting slab were optimized. In the end, the quality of SWRCH22A cold heading steel billet has been significantly improved.

Key words: SWRCH22A cold heading steel; pencil-tip shaped fracture; large-sized inclusion; calcium treatment; refractory material; submerged entry nozzle; billet quality

连铸坯内宏观夹杂物缺陷的刨层实验研究

范英同[1]，徐国栋[1]，刘中秋[2]，阮晓明[1]，沈　燕[1]，李宝宽[2]

（1. 宝钢股份炼钢厂，上海　201900；2. 东北大学冶金学院，辽宁沈阳　1110819）

摘　要：为了掌握夹杂物在整个连铸坯内的空间分布规律，采用刨层分析方法对连铸坯进行整体断面解剖，探究了宏观夹杂物的形貌、尺寸及空间分布规律。结果发现，夹杂物按形貌可分为近球形夹杂物、不规则形貌夹杂物和簇群状夹杂物。夹杂物在铸坯试样内部的分布是不均匀、不对称的，大尺寸夹杂物主要分布在铸坯皮下位置，小尺寸夹杂物的分布更加弥散。铸坯中80%以上的夹杂物缺陷在进入连铸机弯曲段前已经形成。在垂直段，由于受当时钢液流场的影响更多的夹杂物分布在外弧侧；在弯曲段，夹杂物在自身浮力的作用下更容易在铸坯的内弧侧被捕捉。粒径介于0~150μm夹杂物占总量的88.2%，150~250μm占总量的9.49%，大于250μm仅占总量的2.31%。

关键词：刨层实验；宏观夹杂物；夹杂物形貌；空间分布

Study on Macro Inclusion Defects in Continuous Casting Slabs by Planing Experiment

Fan Yingtong[1], Xu Guodong[1], Liu Zhongqiu[2], Ruan Xiaoming[1], Shen Yan[1], Li Baokuan[2]

(1. Baoshan Iron & Steel Co., Ltd., Shanghai 201900, China;
2. School of Metallurgy, Northeastern University, Shenyang 110819, China)

Abstract: In order to grasp the spatial distribution rule of inclusions in the continuous casting slabs, the whole section anatomy of continuous casting slab was carried out by using the method of planing analysis. The macroscopic morphology, size and spatial distribution of inclusions were studied. The results show that the inclusions can be classified into nearly spherical inclusions, irregular inclusions and cluster inclusions. The distribution of inclusions is unevenly and asymmetrically in the slab sample. The larger inclusions are mainly distributed under the skin of the slab, while the distribution of the smaller inclusions is more diffuse. More than 80% of the inclusion defects in the slab are formed before they enter the bending section of the caster. In the vertical section, more inclusions are distributed in the outer arc side due to the influence of the current molten steel flow field. In the bending section, the inclusion is more easily captured in the inner arc side of the caster under the action of buoyancy. The inclusions with size between 0μm and 150μm accounted for 88.2% of the total, 150μm and 250μm accounted for 9.49%, larger than 250μm just accounted for2.31%.

Key words: planing experiments; macroscopic inclusions; inclusion morphology; spatial distribution

连铸导向段支架辅助定位制造技术实践应用

刘益民，齐盛文，王 玲，周 勇，王玲芳

（鞍钢重型机械有限责任公司，辽宁鞍山 114031）

摘 要：介绍了一种导向段支架的加工装配方法，通过在导向段支座的适当位置增加工艺销孔，再结合装配工艺轴的方式，将空间尺寸转换为可直接测量的尺寸，在地平铁平面上进行装配和检查调整，可以实现替代数控机床检测，节约数控机床台时，经过现场安装，完全符合现场精度要求。

关键词：辅助定位；导向段支架；工艺销孔；加工工艺优化

Application Practice of Using Auxiliary Positioning Method to Manufacture Guide Segment Bracket for Continuous Casting

Liu Yimin, Qi Shengwen, Wang Ling, Zhou Yong, Wang Lingfang

(Ansteel Heavy Machinery Co., Ltd., Anshan 114031, China)

Abstract: This article introduces a guide segment bracket inspection method which could be carry out on the benchmark platform. By machining and assemblingof guide segment bracket, and adding the process pin hole in the proper position of the guide bracket, alsocombined with assembling the process shaft,so that convert the space distance to a directly measurable dimension, made it possible to carry out assembly and inspection on the benchmark platform.By this method, work pieces measuring though CNC machine has been replaced, it shorten the time spent on CNC machine .After on-site installation, this method completely reachsthe accuracy requirements.

Key words: auxiliary positioning; guide bracket; process pin hole; process optimization

镀锡板 T5 钢结晶器液面波动控制实践

肖同达，毛会营，王凌晨

（宝钢股份武钢有限炼钢厂，湖北 武汉 430080）

摘　要：镀锡板 T5 钢浇铸过程中出现结晶器液面波动，被迫降低铸机拉速，但效果不明显，导致铸坯裂纹及漏钢事故发生，结晶器液面波动机理分析及过程控制愈加迫切。分析发现钢水在结晶器内凝固时发生包晶反应，导致坯壳生长不均匀，铸坯在扇形段内发生鼓肚是导致结晶器液面波动的主要原因。通过采取结晶器水、二冷水调整，钢水成份优化、铸机精度控制等措施，结晶器液面波动得到控制，铸坯质量得到改善。

关键词：包晶反应；结晶器；二冷水；液面波动；镀锡板

Control Practice of Mould Liquid Level Fluctuation of T5 Tinplate

Xiao Tongda, Mao Huiying, Wang Lingchen

(Steelmaking Plant, WISCO Co., Ltd., Baoshan Iron and Steel, Wuhan 430080, China)

Abstract: Tinplate T5 appeared in the process of casting mould steel liquid level fluctuation, forced to reduce casting machine speed, but the effect is not obvious, Lead to slab cracks and steel leakage accidents, mold level fluctuation mechanism analysis and process control of more and more urgent.It was found that the main factor for mold level fluctuation was the unevenness of strand shell caused by peritectic reaction during solidifying of molten steel and the billet is bulged in the fan-shaped segment. By the adjustment of second cooling water and mold water, the optimization of molten steel composition, accuracy control of casting machine, the problem had been settled and slab quantity was improved.

Key words: peritectic reaction; mold; second cooling water; mold level fluctuation; tinplate

高钛钢连铸过程的数值模拟及工艺研究

吴国荣[1,2,3]，祭　程[1,2]，朱苗勇[1,2]，陈天赐[1,2]

（1. 东北大学冶金学院，辽宁沈阳　110819；2. 东北大学多金属共生矿生态化冶金教育部重点实验室，辽宁沈阳　110819；3. 攀枝花钢铁研究院有限公司，四川攀枝花　617099）

摘　要：针对高钛钢板坯在连铸过程中不同过热度、拉速条件下的热/力学行为。利用有限元仿真技术建立高钛钢板坯连铸过程三维热/力学耦合模型，系统研究了该钢种的凝固传热与热收缩行为规律。结果表明，在高钛钢高过热度浇铸条件下，拉速需要≤1.0m/min，结晶器入口窄面与宽面开口度为 207mm 和 1218mm，结晶器出口窄面与宽面开口度为 205.4mm 和 1202mm。

关键词：高钛钢；凝固；数值模拟；连铸工艺

Numerical Simulation and Process Study on Continuous Casting Process of High Titanium Steel

Wu Guorong[1,2,3], Ji Cheng[1,2], Zhu Miaoyong[1,2], Chen Tianci[1,2]

(1. School of Metallurgy, Northeastern University, Shenyang 110819, China;
2. Key Laboratory for Ecological Metallurgy of Multimetallic Mineral (Ministry of Education), Northeastern University, Shenyang 110819, China;
3. Panzhihua Iron and Steel Research Institute Co., Panzhihua 617099, China)

Abstract: Aiming at the thermal/mechanical behavior of high titanium steel continuous casting slab under different superheat and casting speed conditions. The three-dimensional thermodynamic coupling model was established by finite element simulation technology, and the solidification and heat transfer law of the steel was studied by numerical simulation.The results show that under the condition of high superheat casting of high titanium steel, the casting speed needs to be less than 1.0m/min, the opening degree of narrow and wide surface at the entrance of the mold is 207mm and 1218mm, and the opening degree of narrow and wide surface at the exit of the crystallizer is 205.4mm and 1202mm.

Key words: high titanium steel; solidification; numerical simulation; continuous casting process

基于深度学习的连铸坯角部裂纹实时检测模型

孟晓亮，宋翰凌，罗　森，王卫领，朱苗勇

（东北大学冶金学院，辽宁沈阳　110819）

摘　要：为了实时准确检测高温连铸坯角裂纹，本文集成了 ShufflenetV2 网络结构和 Focus 模块，提出了 YOLOv5-SFA 模型。在本模型中，ShufflenetV2 网络结构可以有效提升训练速度，在 ShufflenetV2 网络结构的基础上，Focus 模块将输入数据进行切分，增加通道数量，使得通道之间的特征信息得到更加充分地学习，提高模型抗噪声干扰的能力，从而提高了模型的准确性。使用同一数据集分别训练 YOLOv5 模型和 YOLOv5-SFA 模型并进行检测。结果显示，相较于 YOLOv5 模型，YOLOv5-SFA 模型的 mAP 为 92.48%，提升了 1.7%；收敛所需 epoch 为 88，减少了 43.6%；loss 为 0.006%，减小了 75%；训练时间为 0.369h，缩短了 61%。并有效改善了原模型检测结果中出现的过检现象。

关键词：深度学习；YOLOv5-SFA 模型；训练时间；过检现象

A Real-time Detection Model for Corner Cracks of Continuous Casting Slab based on Deep Learning

Meng Xiaoliang, Song Hanling, Luo Sen, Wang Weiling, Zhu Miaoyong

(School of Materials and Metallurgy, Northeastern University, Shengyang 110819, China)

Abstract: In order to accurately detect the corner cracks of high-temperature continuous casting billets in real time, this paper integrates the ShufflenetV2 network structure and the Focus module, and proposes the YOLOv5-SFA model. In the model, the ShufflenetV2 network structure can effectively improve the training speed. On the basis of the ShufflenetV2 network structure, the Focus module divides the input data and increases the number of channels, so that the feature information between channels can be more fully learned, and the ability to resist noise interference can be improved, thereby improving the accuracy of the model. Use the same data set to train the YOLOv5 model and the YOLOv5-SFA model respectively and perform detection. The results show that compared with the YOLOv5 model, the mAP of the YOLOv5-SFA model is 92.48%, which is an increase of 1.7%; the epoch required for convergence is 88, which is a reduction of 43.6%; the loss is 0.006%, which is a reduction of 75%; the training time is 0.369h, which is reduced by 61%. Besides, it effectively improves the over-detection phenomenon in the original model detection results.

Key words: deep learning; YOLOv5-SFA model; training time; solute concentration

连铸凝固末端压下过程裂纹萌生扩展机理及风险预测研究

祭　程[1,2]，朱苗勇[1,2]

（1. 东北大学冶金学院，辽宁沈阳　110819；

2. 东北大学多金属共生矿生态化冶金教育部重点实验室，辽宁沈阳　110819）

摘　要：随着连铸断面在增宽加厚和产品质量性能需求的不断提升，大幅增加压下量、延长压下区间已成为凝固末端压下工艺发展的普遍共识，生产者们也愈加担心可能导致的压下裂纹缺陷。本文针对凝固末端压下过程中间裂纹萌生与角部裂纹扩展的特点，设计了两种裂纹的临界应变测定方法；基于宽厚板坯连铸全程三维热/力耦合有限元仿真模型分析了凝固末端压下过程铸坯凝固与变形规律，在此基础上开发了裂纹萌生扩展风险的预测模型，分析了不同压下方案下的裂纹风险，为压下工艺设计提供了定量数据支撑。

关键词：连铸；凝固末端压下；中间裂纹；角部裂纹；临界应变

Risk Prediction of Crack Formation and Propagation in Solidification End Reduction Process of Continuous Casting Slab

Ji Cheng[1,2], Zhu Miaoyong[1,2]

(1. School of Metallurgy, Northeastern University, Shenyang 110819, China;

2. Key Laboratory for Ecological Metallurgy of Multimetallic Mineral (Ministry of Education), Northeastern University, Shenyang 110819, China)

Abstract: It was gradually recognized that the reduction amount increase and reduction zone extension is necessary for improving center segregation and porosity of large section continuous casting slab and bloom, and the risk of crack formation and propagation was increased inevitably. In this work, combing with the mechanism of internal crack formation and transverse corner crack propagation during reduction process, the critical strain determination method of two kinds of cracks was designed. The solidification and deformation behavior of wide-thick continuous casting slab was analyzed using finite element method. The risk prediction model of crack model was presented, and the crack risk with different solidification end reduction conditions was simulated for quantitativelydesigning of reduction amount and zone parameters.

Key words: continuous casting; solidification end reduction; internal crack; transverse corner crack; critical strain

厚板坯宽面偏离角连续纵向凹陷的预测与分析

牛振宇，蔡兆镇，朱苗勇，解明明

（东北大学冶金学院，辽宁沈阳　110819）

摘　要：宽面偏离角纵向凹陷是厚板坯连铸中频发的表面质量缺陷，凹陷带常伴随有表面及皮下裂纹，严重影响轧材的质量。研究厚板坯宽面偏离角凹陷形成机理，从而提高厚板坯质量，对保障宽厚板的高质与高效化生产具有重要意义。本文采用数值仿真方法，建立了包括结晶器、二冷区、压下段在内的厚板坯连铸过程三维热/力耦合有限元模型，综合考量了铸坯非均匀传热、鼓肚变形、辊道支撑与压下等行为，基此对铸坯凝固过程的偏离角凹陷形成进行全流程预测与分析。研究结果表明：结晶器内坯壳收缩导致铸坯角部与偏离角区域非均匀传热，进而引发偏离角形成热点。当铸坯出结晶器窄面足辊区后，铸坯窄面发生鼓肚变形，宽面偏离角热点处向内弯曲，从而形成纵凹陷。在压下区间，凹陷带两端被压平，宽度和深度同时下降，窄面在辊夹持作用下发生褶皱，最终形成冷态铸坯所示的凹陷缺陷。

关键词：偏离角凹陷；鼓肚；非均匀传热

Prediction and Analysis of the Longitudinal Off-corner Depression during Slab Continuous Casting

Niu Zhenyu, Cai Zhaozhen, Zhu Miaoyong, Xie Mingming

(School of Metallurgy, Northeastern University, Shenyang 110819, China)

Abstract: The longitudinal off-corner depression frequently occurring in slab continuous casting has a significant influence on the surface quality. Accompanied with the depression, the surface and subsurface cracks also undermine the quality of the plant product. The investigation on the formation of longitudinal depression is urgent for the improvement of slab quality as well as the rolled plant. In the present work, a numerical model is developed to fully predict and analyze the formation of longitudinal off-corner depression with consideration of uneven heat transfer in mold, shell bulging and roll support etc. The results show the uneven heat transfer in mold leads to hot spots in the wide-face off-corners. Below the mold, the lateral support rolls constrain the shell deformation and prevent the longitudinal depression. Out of the foot roll segment, the narrow face bulges, which bends the hot spot at the wide-face off-corner inward and cause the longitudinal depression. In the reduction segments, both side of the depression are flattened by rolls, meanwhile, the width and depth of the depression decrease simultaneously.

Key words: longitudinal off-corner depression; shell bulging; uneven heat transfer

电磁作用下 20CrMnTi 钢 160mm×160mm 方坯内部质量控制研究

康吉柏，王卫领，朱苗勇

（东北大学冶金学院，辽宁沈阳　110819）

摘　要：本研究开展 160mm×160mm 20CrMnTi 方坯结晶器电磁搅拌(M-EMS)与凝固末端电磁搅拌(F-EMS)参数协同优化试验，通过低倍检测、金相检测和碳偏析检测探究方坯内部质量优劣。此外利用开源软件 OpenFOAM 预测方坯凝固传热过程，获取方坯平均冷却速率(C_R)与局部凝固时间(θ)，用以定量描述一次枝臂间距(PDAS)。结果表明：最优电搅参数为 M-EMS 250A/4.0Hz、F-EMS 300A/7.0Hz；方坯内、外弧侧柱状晶/等轴晶转变(CET)起始位置分别距方坯表面约 42mm 和 28mm，受 M-EMS 电流强度的影响很小；M-EMS 240 条件下 PDAS(λ_1)与平均冷却速率的关系为 λ_1=354.8$C_R^{0.28\lg(\theta/250)}$；碳元素存在皮下负偏析、CET 前沿正偏析、凝固中心邻近区域负偏析现象。另外，阐释了中心正负偏析形成机理。

关键词：20CrMnTi；一次枝臂间距(PDAS)；柱状晶/等轴晶转变(CET)；宏观偏析；电磁搅拌(EMS)

Study on Internal Quality Control of 160mm×160mm 20CrMnTi Billet

Kang Jibai, Wang Weiling, Zhu Miaoyong

(School of Metallurgy, Northeastern University, Shenyang 110819, China)

Abstract: The present work carried out industrial tests to optimize synergistically processing parameters of M-EMS and F-EMS，and used etching technology, microscopic examination and carbon-segregation testing to estimate quality of 160mm×160mm 20CrMnTi billets. In addition, average cooling rate notated as C_R and the local solidification time notated as $\theta$of steel was obtained by modeling the solidifying process of billet using open source software OpenFOAM, which helps to determine the primary dendrite arm spacing (PDAS) quantitatively. The results show that the optimized EMS parameters are M-EMS 250A/4.0Hz and F-EMS 300A/7.0Hz. And the initial locations of columnar to equaxied transition zone at inner arc and outer arc sides are fixed around 42 and 28 mm, respectively, and they are little affected by EMS. Besides, the PDAS notated as λ_1 is described as λ_1=354.8$C_R^{0.28\lg(\theta/250)}$under the condition of M-EMS current as 240A. There exits negativesegregation of carbon below the billet surface, however, carbon enriches in front of CET zone, and appears to be negative segregation at the vicinity of freezing center. Moreover, center segregation mechanism was also explained in the present work.

Key words: 20CrMnTi; primary dendrite arm spacing (PDAS); columnar to equaxied transition(CET); macro-segregation; electromagnetic stirring (EMS)

GCr15 轴承钢连铸冷却速率下凝固原位观察研究

罗腾飞，王卫领，朱苗勇

（东北大学冶金学院，辽宁沈阳　110819）

摘　要：GCr15 轴承钢在连铸凝固过程中的组织生长是碳化物液析的重要诱因，成为产品质量提升的关键。因此，本研究针对国内某钢厂 240mm×240mm GCr15 轴承钢的连铸过程，选取方坯表面下方 40mm、80mm 和 120mm 位置处的坯样为研究对象，首先建立二维凝固传热模型，结合红外测温实验，求解它们在糊状区的平均冷却速率，然后借助高温激光共聚焦扫描显微镜(HT-CSLM)原位观察它们在连铸条件下的凝固过程，研究揭示不同冷却速率对晶粒生长动力学的影响。结果表明：距方坯表面 40mm、80mm 和 120mm 位置处的糊状区平均冷却速率分别为 24.70℃/min、17.02℃/min 和 18.95℃/min，固相等效的 γ-Fe 晶粒生长速率分别为 1.043μm/s、0.973μm/s 和 1.015μm/s。

关键词：GCr15 轴承钢；二维凝固传热模型；糊状区平均冷却速率；凝固动力学

In-situ Observation of Solidification of GCr15 Bearing Steel Atcoolingrates of Continuous Casting

Luo Tengfei, Wang Weiling, Zhu Miaoyong

(School of Metallurgy, Northeastern University, Shenyang 110819, China)

Abstract: The growthofthesolidificationmicrostructure is important factorsfortheprecipitation ofthecarbide fromtheliquid duringthecontinuous castingof the GCr15 bearing steel, which becomes the key to improve thequalityoftheproduct. Therefore, thepresentworkfocusedonthe continuous casting process of GCr15 bearing billetwiththetransversesection of 240mm × 240mm in a domestic steel plant, and tooksamples 40, 80 and 120 mm below the billet surface as the research objects. First, atwo-dimensional solidification heat transfer model was developed, and the average cooling rate in the mushy zone was obtained withthecombinationofthetemperature measurement viatheinfrared thermal imager. Then, the*in-situ* observation of their solidification process under continuous casting conditions by means of high temperature laser confocal scanning microscopy (HT-CSLM)wascarriedout,and the effects of thecooling rate on grain growth kinetics was investigated. The results showthatthe average cooling rates are 24.7, 17.02 and 18.95 ℃/min at the positions of 40, 80 and 120mm below the billet surface, respectively.And, the growth rates of γ-Fe grains equivalenttothesolid phaseare 1.043, 0.973 and 1.015μm/s.

Key words: GCr15 bearing steel; two-dimensional solidification heat transfer model; average cooling rate in mushy zone; solidification kinetics

4　轧制与热处理

大会特邀报告
分会场特邀报告
矿业工程
炼铁与原燃料
炼钢与连铸
★　轧制与热处理
表面与涂镀
金属材料深加工
先进钢铁材料及其应用
粉末冶金
能源、环保与资源利用
冶金设备与工程技术
冶金自动化与智能化
冶金物流运输
其他

4.1 板 材

低温压力容器用低合金钢 15MnNiNbDR 研制与开发

邢梦楠，胡昕明，王 储，欧阳鑫，贾春堂

（鞍钢集团钢铁研究院，辽宁鞍山 114009）

摘 要：为进一步获得具有优良低温性能和高强度级别的低合金钢，通过合理优化成分设计、加热制度、控轧控冷以及热处理工艺，开发出在低温条件下具有良好的低温韧性、焊接性能以及高的强度级别的 15MnNiNbDR 低合金钢。结果表明，低温压力容器用低合金钢 15MnNiNbDR 化学成分简单且合理，钢质纯净度高，强韧性匹配良好，焊接性能优异，完全能够满足石化领域相关低温设备设计要求，具有良好的使用前景与经济价值。

关键词：压力容器；低温；低合金钢；力学性能；组织

Research and Development of 15MnNiNbDR Low Alloy Steel for Low Temperature Pressure Vessel

Xing Mengnan, Hu Xinming, Wang Chu, Ouyang Xin, Jia Chuntang

(Iron & Steel Research Institute of Angang Group, Anshan 114009, China)

Abstract: In order to further obtain low alloy steel with excellent low temperature performance and high strength grade, 15MnNiNbDR low alloy steel with good low temperature toughness, welding performance and high strength grade under low temperature condition was developed through reasonable optimization of composition design, heating system, rolling control and cooling control and heat treatment process. The results show that the low alloy steel 15MnNiNbDR used in the low temperature pressure vessel has simple and reasonable chemical composition, high purity of steel, good matching of strength and toughness, and excellent welding performance, which can completely meet the design requirements of relevant low temperature equipment in the petrochemical field, and has a good application prospect and economic value.

Key words: pressure vessel; low temperature; low alloy steel; mechanical properties; structure

精轧前机架 F12 自动预调平工艺研究

赵金凯，刘人溥，王 东，沈益钊，王 波

（宝钢湛江钢铁有限公司，广东湛江 524072）

摘　要：精轧轧制过程分为三段，分别是头部穿带、中部轧制及尾部抛钢，任何一段是否稳定直接影响着产线的顺行及质量的稳定发挥，其中头部穿带时的稳定性最为重要，且人工干预最多。本文对湛钢 2250 热轧精轧穿带时人工控制的现状进行了充分调研，结合现有的机械设备及检测设备，在 L1、L2 系统上开发了一种根据粗轧 R2 中心线偏差及油柱偏差，进行精轧前机架 F1、F2 自动预调平控制的功能，该功能目前在 2250 轧线上稳定投用。在不增加额外辅助设备的基础上，显著提高了精轧穿带时的稳定性，同时也对减轻操作劳动负荷、稳定现场生产、推进全自动轧钢有着重要意义。

关键词：精轧；自动预调平；全自动轧钢；中心线偏差

Study on the Automatic Pre-leveling Process of F1-F2 before Finishing Rolling

Zhao Jinkai, Liu Renpu, Wang Dong, Shen Yizhao, Wang Bo

(Baosteel Zhanjiang Iron & Steel, Zhanjiang 524072, China)

Abstract: The finishing rolling process is divided into three stages, namely head threading, middle rolling and tail polishing. Whether any one of the stages is stable directly affects the forward movement of the production line and the stable performance of the quality. Stability is the most important, and manual intervention is the most. This article fully investigates the current situation of manual control during strip threading of Zhangang's 2250 hot rolling and finishing rolling. Combining with the existing mechanical equipment and testing equipment, a system based on rough rolling R2 center line deviation and oil is developed on the L1 and L2 system. Column deviation, the function of automatic pre-leveling control of the frames F1 and F2 before finishing rolling. This function is currently in stable use on the 2250 rolling line. On the basis of not adding additional auxiliary equipment, the stability of the finishing rolling and threading is significantly improved, and it is also of great significance for reducing the operating labor load, stabilizing on-site production, and promoting fully automatic steel rolling.

Key words: finishing rolling; automatic pre-leveling; automatic steel rolling; center line deviation

降低支撑辊成本，提高使用寿命

王　磊，张彩霞，崔二宝，辛艳辉，何立东，尹玉京

（北京首钢股份有限公司，河北迁安　064400）

摘　要：支撑辊作为热轧生产的重要部件，直接影响生产的稳定和成本。本文针对支撑辊使用进行分析研究，对支撑辊材质改进及换辊制度、磨削制度进行优化，提高支撑辊使用寿命，有效降低成本。

关键词：热连轧；支撑辊；辊型；磨削

Reduce the Cost of Back-up Roll and Improve Its Service Life

Wang Lei, Zhang Caixia, Cui Erbao, Xin Yanhui, He Lidong, Yin Yujing

(Beijing Shougang Co., Ltd., Qian'an 064400, China)

Abstract: As an important part of hot rolling production, back-up roll directly affects the stability and cost of production. In this paper, the use of back-up roll is analyzed and studied, and the material improvement, roll changing system and grinding system ofback-up roll are optimized to improve theservice life of back-up roll and effectively reduce the cost.
Key words: hot rolling; back-up roll; roll forming; grinding

一种热连轧线卷取前侧导板的研发改造

王秋林

（首钢京唐公司热轧部，河北唐山　063000）

摘　要：为了缩短轧线计划停机时间提高作业率，本文就薄板热连轧生产线卷取前侧导板的研究改造进行了阐述。从简单地侧导板部件改善，完成了减少每次更换数量，最终实现了大幅缩短每日计划停机更换时间的目的，突破了多年来国内外类似轧线提高作业率的瓶颈问题。

此方案于 2018 年 5 月底在首钢京唐公司热轧 2250 线实施完成，在线投用一年多效果超出预期。通过对标，该项技术在国际同行业、同设备结构领域处于领先水平。
关键词：带钢；卷取前侧导板；斜形段；平行段

热轧加热炉装钢机装钢顺序控制的精度提升

王宇军

（武汉钢铁有限公司 热轧厂，湖北武汉　430083）

摘　要：热轧加热炉的生产是通过装钢机将钢坯从运输辊道装入加热炉内。装钢机的工作效率将直接影响整个加热炉生产的效率和稳定性。通过对装钢机的顺序控制进行深度的分析，并从控制的算法和逻辑进行优化，能够有效提高装钢机的运行效率，缩小加热炉内的钢坯间隙，提高加热炉的生产效率。
关键词：装钢机；定位精度；顺序控制；装钢间隙

Increased Accuracy of Steel Loading Sequence Control of Steel Loading Machine for Hot Rolling Furnace

Wang Yujun

(Wuhan Iron & Steel Co., Ltd., Wuhan 430083, China)

Abstract: The production of the hot rolling heating furnace is to load the billet from the transport roller table into the heating furnace through the steel loading machine. The working efficiency of the steel loading machine will directly affect the efficiency and stability of the entire heating furnace production. Through in-depth analysis of the sequence control of the steel loading machine, and optimization from the control algorithm and logic, the operating efficiency of the steel loading machine can be effectively improved, the billet gap in the heating furnace can be reduced, and the production efficiency of the heating furnace can be improved.

Key words: steel loading machine; positioning accuracy; sequence control; steel clearance

带热卷箱热连轧生产线功率加速度的应用

陈启发，徐朝辉，何　璋

（重庆钢铁股份有限公司轧钢厂，重庆　401220）

摘　要：本文介绍了某带热卷箱热连轧生产线通过功率加速度的应用，解决过程中出现的终轧温度命中问题，薄规格轧制过程中水量波动造成的厚度及终轧温度波动问题。达到了提高薄规格轧制稳定性以及提升轧制节奏，提高产量的目的。

关键词：热连轧；热卷箱；功率加速度

Application of Power Acceleration in the Hot Strip Mill with Hot Coil Box

Chen Qifa, Xu Zhaohui, He Zhang

(Rolling Mill Plant of Chongqing Iron and Steel Co., Ltd., Chongqing 401220, China)

Abstract: This paper introduces the application of power acceleration in a hot strip mill with hot coil box, which solves the problem of hitting the final rolling temperature in the process, and the problem of thickness and final rolling temperature fluctuation caused by water fluctuation in the thin gauge rolling process. Achieve the purpose of improving the rolling stability of thin gauge, improving the rolling rhythm and increasing the output.

Key words: hot strip mill; power acceleration; hot coil box; application

热轧产线卷取机智能消防环保功能的开发应用

王　君，张建华，王　超，高文刚，管宝伟

（首钢京唐钢铁联合有限责任公司热轧作业部，河北唐山　063200）

摘　要：常规热轧板带产线布置的地下卷取机，设备构造相对复杂，由机械、液压、电气等设备组成。如果卷取机内高温热卷与锂基脂接触发生火灾，导致设备损坏、烟气污染大气环境。不仅对产线造成一定损失，现场作业人员也同样存在安全隐患。经过摸索开发了一种高效智能的消防环保功能，降低了一系列火灾风险。

关键词：卷取机；锂基脂；火灾；钢卷；夹送辊；助卷辊；芯轴；环保

Development and Application of Intelligent Fire Protection and Environmental Protection Function for Coiler of Hot Rolling Line

Wang Jun, Zhang Jianhua, Wang Chao, Gao Wengang, Guan Baowei

(Shougang Jingtang United Iron & Steel Co., Ltd., Tangshan 063200, China)

Abstract: The down coiler, which is arranged in the conventional hot-rolled strip production line, is composed of mechanical, hydraulic and electrical equipment. If the high-temperature hot coil in the coiler is in contact with lithium grease, the equipment will be damaged and the smoke will pollute the atmosphere. Not only to the production line caused certain losses, the field workers also exist security risks. After exploring and developing an efficient and intelligent function of fire protection and environmental protection, a series of fire risks are reduced.

Key words: down coiler; lithium grease; fire; steel coil; pinch roll; wrapper roll; mandrel; environmental protection

精轧轧制力偏差与机架间带钢跑偏的研究

王　波，刘人溥，赵金凯，王　东

（宝钢湛江钢铁有限公司，广东湛江　524072）

摘　要：产线的稳定与各区域轧制的稳定性息息相关。精轧轧制分为三阶段，分别是穿带、轧制及抛钢，任何一阶段发生异常都会影响产线的稳定顺形及产能发挥。本文在分析热轧带钢轧制过程发生跑偏现象基础上，重点对各机架轧制力偏差进行研究，通过数据分析发现，轧制力偏差与机架间跑偏的相关性，通过进一步摸索正常轧制及发生跑偏时机架轧制力偏差的趋势性变化，为后续提前预判机架间跑偏进行自动加张功能及自动调平纠偏功能的开发奠定一定的数据基础，以此来提高带钢轧制过程稳定性，助力产线产能的发挥。

关键词：跑偏；轧制力偏差；预判；稳定性

Study on the Deviation of Finishing Rolling Force and the Deviation of Interstand Strip

Wang Bo, Liu Renpu, Zhao Jinkai, Wang Dong

(Baosteel Zhanjiang Iron & Steel, Zhanjiang 524072, China)

Abstract: The stability of production line is closely related to the rolling stability of each region. Finishing rolling is divided into three stages, namely, strip crossing, rolling and steel throwing. Any abnormality in any stage will affect the stable conformation of production line and production capacity. Based on the analysis of the rolling process of hot rolling steel strip running deviation occur, based on the key research on the rack to rolling force error, through data analysis found that the deviation between the rack and correlation of rolling force, by further grope for normal timing running deviation rolling and rolling force deviation trend changes, for subsequent anticipation ahead of rack automatic running deviation between a function and the development of the automatic leveling correction function of certain data basis, in order to improve the stability of strip rolling process, the power capacity of production line.

Key words: running deviation; rolling force deviation; pre-judge; stability

IF 钢铁素体轧制工艺在梅钢热轧 1422 产线的实践

廖松林，何义新

（上海梅山钢铁股份有限公司热轧板厂，江苏南京 210039）

摘 要： IF 钢具有优良的深冲性能，具有优异的塑性应变比、高的应变硬化指数、良好的伸长率及非时效特性。以其为基础开发的热镀锌、热镀铝锌等冷轧 IF 钢产品，几乎可满足各种形状复杂的冷冲压成型件的性能要求，广泛应用于家电、汽车制造行业。理论上 IF 钢热卷有两种热轧生产工艺路径，一种是传统的高温出炉奥氏体方式轧制，一种是低温出炉的铁素体方式轧制。随着节能降耗要求的提高、薄规格 IF 产品需求的出现，IF 钢铁素体轧制在热轧工业生产实践中逐渐得到应用，开发出了能替代传统奥氏体 IF 钢的铁素体 IF 钢和 2.0mm 的超薄铁素体 IF 钢。

关键词： IF 钢；深冲性能；奥氏体；铁素体

Practice of IF Steel Ferrite Rolling Process in 1422 Hot Rolling Plant of Meishan Iron & Steel

Liao Songlin, He Yixin

(Hot Rolling Plant of Meishan Iron & Steel Co., Nanjing 210039, China)

Abstract: If steel has excellent deep drawing performance, excellent plastic strain ratio, high strain hardening index, good elongation and non aging properties. The cold-rolled if steel products such as hot-dip galvanizing and hot-dip aluminum zinc can meet the performance requirements of various complex cold stamping parts, and are widely used in the household appliances and automobile manufacturing industries. Theoretically, if steel hot rolling has two hot rolling production paths, one is traditional high temperature out of furnace austenite rolling, the other is the ferrite rolling at low temperature. With the improvement of energy saving and consumption reduction and the emergence of thin if products, the ferrite rolling of IF steel has been gradually applied in the hot rolling industry. The ferrite if steel which can replace the traditional austenite if steel and the ultra-thin ferrite if steel of 2.0 mm have been developed.

Key words: IF steel; deep drawing performance; austenite; ferrite

表面特征对 DP590 钢腐蚀行为的综合作用

方百友

（宝钢日铁汽车板有限公司技术质量管理部，上海 201900）

摘 要： 本文运用金相显微镜、白光干涉显微镜、X 射线衍射、腐蚀电化学测试以及湿热试验等多种方法，就 DP590 双相组织与表面形貌和残余应力对其腐蚀行为的影响进行了研究。研究结果发现马氏体含量最高的试样没有表现出最快的腐蚀速度，这表明 DP590 钢的腐蚀行为并不仅仅受到马氏体含量的影响。事实上，粗糙的表面以及残余拉应力均会降低钢的耐蚀性能，而压应力能够有效的减缓腐蚀的萌生过程。因此具有较高含量的马氏体，却同时具有较为光滑的表面并受到残余压应力影响的试样表现出与马氏体含量较低的试样相似的耐蚀性能。也就是说，DP590钢的腐蚀速度并不仅仅依赖于某个单一因素，而是受到这些因素的综合作用。

关键词：腐蚀行为；动电位极化；残余应力；表面形貌；马氏体含量

Comprehensive Effects of Surface Features on Corrosion Behavior of Temper Rolling Processed DP590 Steel

Fang Baiyou

(Baosteel-Nippon Steel Automotive Steel Sheets Co., Ltd., Shanghai 201900, China)

Abstract: In most literatures, the corrosion rate of DP steels mainly depended on the volume fraction of martensite, in which the increased amount of martensite significantly lifted the corrosion rate. The effects of surface topography and residual stress on the corrosion rate of DP590, however, are seldom taken into consideration. In this paper, the effects of these two factors are discussed combining with that of the dual phase microstructure. The surface features of DP590 steel samples were characterized with optical interferometric microscopy and X-ray diffraction, respectively. And corrosion properties were evaluated using potentiodynamic polarization and damp heat tests. It was found that the sample with a highest volume fraction of martensite did not show the highest corrosion rate, vice versa. It was indicated that the volume fraction of martensite was not the only determinant factor in this case. In fact, rougher surface and tensile residual stress bear adverse effects on the corrosion resistance of the steels, while compressive residual stress is considered to effectively retard corrosion initiation. Therefore, the sample with a higher martensite proportion, a relatively smooth surface and a compressive residual stress may possess a similar corrosion behavior to the one with a lower martensite proportion. The experiment results in this study verified that he corrosion rate of DP590 was not determined by only one factor, but the comprehensive effect of these factors.

Key words: corrosion behavior;potentiodynamic polarization;residual stress;surface topography;martensite proportion

超宽管线钢板形控制研究

姚 震，于金洲，韩千鹏，王亮亮，韩 旭，李新玲，王若钢

（鞍钢股份中厚板事业部鲅鱼圈中厚板厂，辽宁营口 115007）

摘 要：鞍钢5500mm宽厚板生产线是目前国内最宽的宽厚板生产线之一，由于前期建设中未设置预矫直机及温矫直机，在超宽管线钢板形控制方面极具挑战，本文利用六西格玛质量工具，进行全流程关键工艺控制参数分析，最终找到超宽管线钢板形控制关键工艺技术参数，并使超宽管线钢板形控制得到明显提升。

关键词：管线钢；板形；6σ；超宽

Research on Steel Shape Control of Ultra-wide Pipeline

Yao Zhen, Yu Jinzhou, Han Qianpeng, Wang Liangliang, Han Xu, Li Xinling, Wang Ruogang

(Bayuquan Medium and Heavy Plate Plant, Medium and Heavy Plate Division of Anshan Iron and Steel Co., Ltd., Yingkou 115007, China)

Abstract: Anshan Iron and Steel's 5500mm wide and heavy plate production line is currently one of the widest wide and heavy plate production lines in China. Since there is no pre-straightening machine and warm straightening machine in the early construction, it is very challenging to control the shape of the ultra-wide pipeline. This article uses six sigma quality tools. , Carry out the analysis of the key process control parameters of the whole process, and finally find the key process technical parameters of the steel shape control of the ultra-wide pipeline, and significantly improve the control of the steel shape of the ultra-wide pipeline.
Key words: pipeline steel; plate shape; six sigma; optimization

厚规格 X80 级别管线钢低温 DWTT 性能控制研究

姚 震，张 坤，韩千鹏，王亮亮，韩 旭，李新玲，王若钢

（鞍钢股份中厚板事业部鲅鱼圈中厚板厂，辽宁营口 115007）

摘 要： 厚规格 X80 级别管线钢是目前管线钢生产的控制难点，本文主要从管线钢生产工艺方面切入，从加热、轧制及冷却工艺方面进行研究，开发出一套满足用户使用需求的厚规格 X80 级别管线钢低温性能的控制方法，同时钢板性能合格率及钢板制管后性能合格率均保证在98%以上。
关键词： 管线钢；DWTT；厚规格；工艺；性能合格率

Research on Low Temperature DWTT Performance Control of Heavy Gauge X80 Pipeline Steel

Yao Zhen, Zhang Kun, Han Qianpeng, Wang Liangliang, Han Xu, Li Xinling, Wang Ruogang

(Bayuquan Medium and Heavy Plate Plant, Medium and Heavy Plate Business Department of Anshan Iron and Steel Co., Ltd., Yingkou 115007, China)

Abstract: heavy gauge X80 pipeline steel is the control difficulty of pipeline steel production at present, this paper mainly cut in from the production process of pipeline steel, from the aspects of heating Research on rolling and cooling process, and develop a set of control method for low temperature performance of thick X80 pipeline steel to meet the needs of users. At the same time, the qualified rate of steel plate performance and the qualified rate of steel plate after pipe making are guaranteed to be more than 98%.
Key words: pipeline steel; DWTT; thickness specification; technology; qualified rate of performance

基于特厚板坯在加热炉生产现场实践

黄素军[1]，韩 旭[1]，于金洲[1]，张东明[1]，姜世伟[1]，刘常鹏[2]，孙守斌[2]

（1.鞍钢股份公司中厚板事业部鲅鱼圈中厚板厂，辽宁营口 115007；
2.鞍钢股份有限公司技术中心，辽宁鞍山 114009）

摘　要：本文对加热炉 500mm 复合坯生产问题进行分析，针对相关问题对鞍钢鲅鱼圈厚板部 5500 线加热炉的改造，提高加热炉自动化生产 500mm 复合坯的能力，增加 500mm 复合坯的产量，提高产品质量，并对加热炉改造后效果进行分析。

关键词：特厚板；加热炉；改造；工艺

The Practice of Extra Heavy Plate Reheating in Reheating Furnace

Huang Sujun[1], Han Xu[1], Yu Jinzhou[1], Zhang Dongming[1], Jiang Shiwei[1], Liu Changpeng[2], Sun Shoubin[2]

(1. Bayuquan Iron & Steel Subsidiary Company, Angang Steel Co., Ltd., Yingkou 115007, China
2. Technology Center of Ansteel Co., Ltd., Anshan 114009, China)

Abstract: In this paper, the production difficulty of 500mm composite heavy plate is analyzed, the reconstruction on reheating furnace in 5500mm heavy plate production line has been performed to overcome the technical difficulties, improve the reheating production in auto model and increase the 500mm composite heavy production and production quality, and the effect of reconstruction is also analyzed.

Key words: extra-thick plate; heating furnace; transformation; process

热轧钢板横向瓢曲形成机理研究

韩千鹏，姚　震，李靖年，李新玲，王嘉研，罗　军

（鞍钢股份公司中厚板事业部鲅鱼圈中厚板厂，辽宁营口　115007）

摘　要：为了研究热轧钢板的板形瓢曲机理，控制并避免产生钢板瓢曲缺陷。通过构建物理模型，对热轧钢板横向瓢曲进行系统性的理论分析，揭示了材料变形前后由于热不均造成的内应力释放而产生的宏观板形变化的规律。

关键词：热轧钢板；横向瓢曲；热应力；相变

Research on the Mechanism of Transverse Bending of Hot-rolled Steel Plate

Han Qianpeng, Yao Zhen, Li Jingnian, Li Xinling, Wang Jiayan, Luo Jun

(Medium & Heavy Plate Business Department of Angang Co., Ltd., Yingkou 115007, China)

Abstract: In order to explore the shape bending mechanism of hot-rolled steel plate and control the bending defects of steel plate. By building a physical model, the transverse bending of hot-rolled steel plate is systematically analyzed theoretically, and the macroscopic shape change law of temperature before and after material deformation due to the release of the internal stress caused by thermal unevenness is revealed.

Key words: hot-rolled steel plate;transverse bending; thermal stress;phase transition

温度-应力-相变一体化调控技术在800MPa厚规格水电钢板开发中的应用

王亮亮，李新玲，王若钢，张　磊，张　坤

（鞍钢股份公司中厚板事业部，辽宁营口　115007）

摘　要：对温度—应力—相变一体化调控装备即淬火/温控机的特性进行了研究，依此特性成功开发 800MPa 级厚规格水电钢板热处理工艺，并对此规格钢板相关力学性能的过程能力进行了分析。在800MPa级特厚规格水电钢板方面，进行了工艺的尝试，并对金相组织和力学性能进行了研究，取得了积极的效果。

关键词：温度；应力；相变；一体化调控；力学性能

The Characteristics of Temperature-stress-phase Transformation Integrated Control Equipment Used in 800MPa Super-thick Hydropower Steel Plate Processing

Wang Liangliang, Li Xinling, Wang Ruogang, Zhang Lei, Zhang Kun

(Medium and Heavy Plate Business Department of Angang Co., Ltd., Yingkou 115007, China)

Abstract: The characteristics of temperature-stress-phase transformation integrated control equipment, i.e., quenching/temperature controller, were studied. According to this characteristic, the heat treatment process of 800MPa thick hydropower steel plate was successfully developed, and the process capability of related mechanical properties of this steel plate was analyzed. In the aspect of 800MPa super-thick hydropower steel plate, the process was tried, and the metallographic structure and mechanical properties were studied, and positive results were obtained.

Key words: temperature; stress;phase trans formation; phase transformation; mechanical properties

中厚板轧制节奏与温控模型高度拟合的技术研究

王嘉研，韩　旭，罗　军，王若钢，刘宏博，韩千鹏，郭金平，姚　震

（鞍钢股份公司中厚板事业部鲅鱼圈中厚板厂，辽宁营口　115007）

摘　要：近几年中厚板的产品的规格和适用性在不断拓展，鲅鱼圈 3800 产线的品种结构同样也在不断调整，不同的工艺要求、工序路径对于现场实际生产中要求就比较高，产线的生产计划排产的难度是逐步增加的，排产的好坏直接影响现场生产节奏。目前在实际过程中岗位人员只能依靠温控模型计算的大概值和自身经验去大约估计和掌控，长期的误差累计下来对产量的影响巨大，另外也易造成工艺执行率偏低，影响产品质量。在疫情防疫全球经济都受影响的当前，中厚板行业降低成本很重要的措施就是提效增产。因此，研究中厚板温控模型与生产节奏高度拟合显得尤为重要，以此寻找一个快捷准确的粗精轧待温时间评估方法，为计划编排和节奏控制找到依托，实现现场

规范化、标准化生产，达到提高生产效率、降低生产成本的目的。

关键词：生产计划；温控模型；生产节奏；提效

Technical Research on Highly Fitting of Plate Rolling Rhythm and Temperature Control Model

Wang Jiayan, Han Xu, Luo Jun, Wang Ruogang, Liu Hongbo, Han Qianpeng, Guo Jinping, Yao Zhen

(Bayuquan Medium and Heavy Plate Plant, Medium and Heavy Plate Division of Anshan Iron and Steel Co., Ltd., Yingkou 115007, China)

Abstract: Plate of the specifications of the products in recent years and applicability in expanding, BaYuJuan production line of 5500 varieties also constantly adjust the structure, technological requirements of different process route for the scene in the actual production requirement is high, the difficulty of the production line production plan of production scheduling is gradually increased, the stand or fall of production scheduling directly affects the production rhythm is currently in the process of actual positions can only rely on temperature control model calculation about values and their own experience about estimation and control, the error of long-term cumulative down a huge impact on production, also easy to cause low process are enforced, affect the quality of the product As the global economy is affected by the epidemic prevention and epidemic prevention, the most important measure for the medium-thick plate industry to reduce costs is to improve efficiency and increase production. Therefore, it is particularly important to study the high fitting of the temperature control model of medium-thick plate with the production rhythm.

Key words: production planner; temperature cybernetic model; production rhythm; effect-raising

湛钢 2250 热轧粗轧打滑分析研究

李自强，曾龙华，豆伟晗，陈宇翔，谭文辉，王大云

（湛江钢铁热轧厂 2250 产线，广东湛江　524072）

摘　要：对湛江钢铁 2250 热轧粗轧的打滑现象进行研究，对打滑的原因进行分析；通过原因分析制定防打滑措施以及开发打滑后的控制功能。

关键词：热轧；粗轧；打滑原因；打滑控制

Study on Slip in Zhangang 2250 Hot-strip Rough Mill

Li Ziqiang, Zeng Longhua, Dou Weihan, Chen Yuxiang, Tan Wenhui, Wang Dayun

(Zhangang 2250 Hot-strip Rough Mill, Zhanjiang 524072, China)

Abstract: the skid phenomenon of 2250 hot rolling rough rolling of Zhanjiang iron and steel was studied, and the causes of skid were analyzed. Through the analysis of the causes, anti-skid Measures are formulated and the control function after skid is developed.

Key words: hot rolling; rough rolling; slip reasons; slip control

宽厚板边直裂的成因分析及控制方法

彭宁琦[1]，何 航[1]，黄远涛[2]，李培友[2]

（1. 湖南华菱湘潭钢铁有限公司 技术质量部，湖南湘潭 411101；
2. 湖南华菱湘潭钢铁有限公司 五米宽厚板厂，湖南湘潭 411101）

摘 要：通过不同连铸坯状态和轧制方式的工业性的对比试验，证实宽厚板边直裂的成因与连铸坯窄面上的足辊压痕线及其变形状况有关，由此提出了一种消除宽厚板边直裂的控制方法：在纵向的预成形轧制阶段，通过较多道次的平辊或立-平辊轧制，使连铸坯窄面上的足辊压痕线翻转至表面上，之后再进行后续的横向展宽轧制和纵向延伸轧制，即能有效避免“纵-横-纵”轧制模式下边直裂的产生。同时对足辊压痕线向表面翻转的数学模型及其工艺控制参数进行了研究，并获得实践应用，有效避免了边直裂的产生，提高钢板的成材率。

关键词：边直裂；边部线状缺陷；边部黑线；足辊压痕线；宽展

Cause Analysis and Control Method of Edge Straight Crack of Wide and Heavy Plate

Peng Ningqi[1], He Hang[1], Huang Yuantao[2], Li Peiyou[2]

(1. Technology and Quality Department, Hunan Valin Xiangtan Iron and Steel Co., Ltd., Xiangtan 411101, China; 2. Five Meters Wide and Heavy Plate Mill, Hunan Valin Xiangtan Iron and Steel Co., Ltd., Xiangtan 411101, China)

Abstract: Through industrial comparative tests of several different treated continuous casting billets and rolling modes, it is proved that the cause of edge straight crack of wide and heavy plate is related to the foot roll indentation line on the narrow surface of continuous casting billet and its deformation. Based on this, a control method to eliminate edge straight crack of wide and heavy plate is proposed. Using more passes of flat roll or vertical-flat roll rolling in longitudinal pre-forming rolling stage, first make the foot roll indentation line on the narrow surface of continuous casting billet turn over to the surface, and then carry out the subsequent transverse broadening rolling and longitudinal extended rolling, in this way, it is effective to avoid the emergence of edge straight crack in the "longitudinal - transverse - longitudinal" rolling mode. At the same time, the mathematical model of foot roll indentation line flipping to the surface and the specific control parameters of this process were studied, and the practical application was successfully implemented, which effectively avoided the occurrence of edge straight crack and improved the yield of steel plate.

Key words: edge straight crack; edge linear defects; edge black line; foot roll indentation line; spread

含磷钢拉伸断口表面起皮原因分析

孙 傲，刘志伟，张瑞琦，郭晓宏，王 鑫，吴成举，钟莉莉

（鞍钢集团钢铁研究院热轧产品研究所，辽宁鞍山 114009）

摘 要：针对含磷钢的拉伸试样在拉伸断裂后其断口表面处出现“起皮”开裂缺陷的问题，利用光学显微镜及扫描电镜进行了研究与分析。实验结果表明，“起皮”缺陷产生的主要原因是由于钢板浅表面层存在脱碳层以及大量的硅铝酸盐、硫化物夹杂以及少量含镁保护渣，这些内部缺陷在轧制过程中未被轧合，在拉伸外应力作用下，使得表层与基体分离，从而形成“起皮”开裂缺陷。

关键词：含磷钢；拉伸断裂；断口起皮；非金属夹杂物

Analysis on the Causes of Surface Skinning on the Tensile Fracture of P-containing Steel

Sun Ao, Liu Zhiwei, Zhang Ruiqi, Guo Xiaohong,
Wang Xin, Wu Chengju, Zhong Lili

(Ansteel Iron and Steel Research Institute, Anshan 114009, China)

Abstract: The skinning cracking defects on the fracture surface of p-containing steel after tensile fracture were studied and analyzed by optical microscope and scanning electron microscope. The experimental results show the main reason for the skinning defect is that ,the shallow surface layer of steels contain a decarburized layer, a large amount of aluminosilicate, sulfide inclusions, and a small amount of magnesium-containing mold powder. Because of these internal defects are not rolled during the rolling process and the surface layer is separated from the substrate, resulting in "skinning" cracking defects under the action of the external tensile stress.

Key words: P-containing weathering steesl; tensile fracture; fractur skinning; Non-metallic inclusions

鞍钢自主研发表检与进口表检在冷轧生产中的应用对比

张福义，王科峰，郭瀚明，李 赢

（鞍山钢铁集团有限公司冷轧四分厂生产作业区，辽宁鞍山 114001）

摘 要：冷轧有两道核心工序，一是联合机组的轧制过程，另一个是连退机组的退火过程。联合机组，过程的特点是：速度快，若发生断带事故，危害巨大。特点是：钢板生产速度相对慢，而对钢板表面质量要求极其高。鞍钢冷轧四分厂采用鞍钢自主研发表检和美国进口表检相互配合使用的方式，最大发挥各自优势，减少断带可能性，并提高产品表面质量。

关键词：表检；连退；轧机

Comparison of the Application of AISC Independent R & D and Imported Surface Inspection in Cold Rolling Production

Zhang Fuyi, Wang Kefeng, Guo Hanming, Li Ying

(Anshan Iron and Steel Co., Ltd. Cold Rolling No.4 Branch, Anshan 114001, China)

Abstract: There are two core processes in cold rolling, one is the rolling process of combined mill, the other is the annealing process of continuous annealing mill. The speed of the process of the combined unit is fast. If the belt break accident occurs, the harm is huge. The speed of continuous annealing mill is relatively slow, and the steel plate surface quality requirements are extremely high. Anshan Iron and Steel Co., Ltd. cold rolling No.4 branch adopts the way of cooperation between Anshan Iron and Steel Co., Ltd. self-developed surface inspection and American imported surface inspection, so as to give full play to their respective advantages, reduce the possibility of strip breakage, and improve the surface quality of products.

Key words: table inspection; continuous annealing; mill

微合金酸洗板表面麻点缺陷成因分析及控制技术浅析

孙　强

（上海梅山钢铁股份有限公司热轧厂，江苏南京　210039）

摘　要：本文对梅钢热轧厂 1780 产线的酸洗板的麻点缺陷进行了取样分析，并对其成因进行分析后，认为微合金酸洗板表面麻点缺陷主要是由于上游热轧原板表面的辊系铁皮遗传带来的，所以本文主要从热轧原板表面的辊系铁皮的成因机理着手，分析影响辊系铁皮的影响因素，如轧制润滑技术的投用，防剥落水的投用以及轧辊冷却水的水量和压力等，针对性的进行工艺参数的调整，从而有效消除了微合金酸洗板表面的麻点缺陷。

关键词：麻点；轧制润滑；辊系铁皮；防剥落水

梅钢 1422mm 产线卷取侧导板衬板磨损机理及优化

杨正鹏

（上海梅山钢铁股份有限公司热轧板厂，江苏南京　210039）

摘　要：本文针对 1422 卷取前侧导板使用周期及磨损无法满足现场需要的现状，梅钢热轧 1422 产线衬板磨损部位（工作面）采用普通合金焊条的电弧堆焊衬板，上线使用周期仅为 0.5 天，在使用周期内根据轧制的钢种和规格不同，磨损深度在 5~15mm（磨损深度标准≤8mm），不仅磨损达不到标准而且较深的磨损深度易造成带钢边部缺陷，同时每天更换 1 号、2 号机两套侧导板耗时 40min。本文重点阐述了采用合适的修复工艺及材质，延长卷取机前侧导板使用周期，降低劳动负荷，提高作业率及产品质量。通过新材质衬板的改进，卷取机前侧导板衬板工作面 5 天磨损量≤8mm，即更换周期由以往的 0.5 天延长到 5 天。

关键词：卷取机前侧导板；边部质量；磨损；衬板

Mechanism and Optimization of Coiler Side Guide Liner Wear in 1422mm Hot Rolling Plant in Meigang

Yang Zhengpeng

(Hot Rolling Plant of Meishan Iron & Steel Co., Nanjing 210039, China)

Abstract: Based on Coiler side guide liner using cycle and wear situation can't meet the needs of the present situation in 1422mm hot rolling plant. 1422mm hot rolling plant in Meigang whose liner wear parts (work side) adopt ordinary alloy electrode lining board. Using online cycle is only 0.5 day. In the use of cycle according to rolling steel grade and specification is different. The wear depth fluctuate from 5 to 15mm (wear depth standard is 8mm or less). Not only wear was not up to standard and deeper wear depth of the strip causing edge defects, but also lead to replace 1 # and 2 # machines every day. Two sets of side guide takes 40minutes. This paper focuses on the use of appropriate repair technology and materials, to extend the front side guide of coiler service cycle, reduce labor load, improve the rate of operation and product quality. Through the improvement of the new material lining board, the 5-day wear of the working side of the guide board lining board on the front side guide of the coiler is less than 8mm. In other words, the replacement cycle is extended from the previous 0.5 day to 5 days.

Key words: front side guide of coiler; side quality; wear; lining plate

低温取向硅钢边裂缺陷分析及改进

李庆贤[1]，张 华[2]，高 磊[1]，景 鹤[2]，李向科[3]，杨承宇[2]，王洪海[2]

（1.鞍钢集团钢铁研究院，辽宁鞍山 114009；2.鞍钢股份热轧带钢厂，辽宁鞍山 114021；
3.鞍钢股份制造管理部，辽宁鞍山 114021）

摘 要：利用金相显微镜、扫描电镜、能谱仪、热模拟、等分析手段对低温取向硅钢边裂的成因进行了探讨。结果表明，长时间或高温加热使得板坯边部晶粒异常长大，轧制过程中边部温度低，动态再结晶弱，边部塑形差，是造成边裂的主要原因。通过减少加热时间、校准加热数模、减少粗轧轧制道次，提高粗轧轧制速度，投入边部加热器对边部进行温度补偿等工艺措施予以改善。

关键词：低温取向硅钢；边裂；动态再结晶；边部加热器

Analysis and Improvement of Edge Cracking in Low-temperature Oriented Silicon Steel

Li Qingxian[1], Zhang Hua[2], Gao Lei[1], Jing He[2], Li Xiangke[3], Yang Chengyu[2], Wang Honghai[2]

(1.Ansteel Iron & Steel Research Institutes, Anshan 114009, China; 2.Hot Rolled Strip Steel Mill of Angang Steel Co., Ltd., Anshan 114021, China; 3.Department of Manufacture Management of Angang Steel Co., Ltd., Anshan 114021, China)

Abstract: By metallurgical microscope, scanning electron microscope, energy spectrometer, thermal simulation, and other analytical methods, the causes of edge cracking in low-temperature oriented silicon steel were discussed. The results show that long-term or high-temperature heating causes abnormal growth of grains at the edges of the slab, low edge temperature during rolling, weak dynamic recrystallization and poor edge shaping, which are the main causes of edge cracks. Improvements are made by reducing the heating time, calibrating the heating digital model, reducing the number of rough rolling passes, increasing the rough rolling speed, and putting in edge heaters to compensate the edge temperature.

Key words: low-temperature oriented silicon steel; edge cracking; dynamic recrystallization; edge heater

覆铝基板用钢的研制及开发

宋凤明，王　巍，胡晓萍

（宝山钢铁股份有限公司 研究院，上海　201900）

摘　要：覆铝板带既有钢的强度，又有铝耐蚀、质轻、美观和散热性好的特点，在散热器片、厨具、汽车零部件等领域具有广泛的市场需求。覆铝板带所用的基板要求具有良好的塑性和钢铝结合性能，以满足后续冲压加工及使用要求。介绍了覆铝基板的成分设计及覆铝钢生产流程，并对成品板材的微观组织、力学性能和钢铝结合性能做了分析。结果表明，覆铝基板的室温组织为均匀的等轴状铁素体，屈服强度在280MPa以下，延伸率超过40%，模拟钎焊后钢铝界面洁净，无铁铝化合物层形成，钢铝结合性能优良。

关键词：覆铝钢；铁铝化合物；钢铝结合；BAC300

Development of Steel for Al Cladding

Song Fengming, Wang Wei, Hu Xiaoping

(Research Institute, Baoshan Iron and Steel Co., Ltd., Shanghai 201900, China)

Abstract: Aluminum cladding strip has the strength of steel and the characteristics of aluminum, such as corrosion resistant, light weight, beautiful appearance and better thermal capacity. So it was widely applied in radiator, kitchen utensils, auto parts and other fields. Excellent plasticity and bonding performance of steel and aluminum is required for the steel used for aluminum cladding strip in order to meet the requirements of subsequent stamping process. The composition design of steel for aluminum cladding and the manufacture process of aluminum cladding steel was introduced, and the microstructure, mechanical properties and bonding performance of Fe-Al were analyzed. The results show that the microstructure of steel for aluminum cladding at room temperature is uniform equiaxed ferrite, the yield strength is less than 280mpa, and the elongation is more than 40%., the interface of steel and aluminum is clean after simulated brazing, and no Fe-Al compound layer is formed.

Key words: Al cladding steel; Al-Fe compounds; bonding of steel and aluminum; BAC300

高表面质量碳素钢柔性轧制生产实践

杨　玉[1]，张吉富[2]，王英海[1]，任俊威[2]，丛志宇[2]，马　锋[2]

（1. 鞍钢集团钢铁研究院，辽宁鞍山　114009；
2. 鞍钢股份有限公司鲅鱼圈钢铁分公司，辽宁营口　115007）

摘 要：本文主要介绍了优质碳素钢 45 的柔性热轧生产实践，通过在线控轧控冷柔性轧制，将 45 钢的抗拉强度从 600MPa 提高到 880MPa，可代替部分热处理高强钢直接使用。同时，该热轧高强钢板表面质量明显提升，取消喷砂工艺，电镀合格率 100%，避免了电镀霉斑的形成，节能环保。

关键词：电镀霉斑；热轧；显微组织；力学性能

Production Practice of Hot Rolled High Surface Quality Carbon Steel

Yang Yu[1], Zhang Jifu[2], Wang Yinghai[1], Ren Junwei[2], Cong Zhiyu[2], Ma Feng[2]

(1. Ansteel Iron & Steel Research Institutes, Anshan 114009, China;
2. Bayuquan Branch of Angang Steel Co., Ltd., Yingkou 115007, China)

Abstract: This paper mainly introduces the production practice of hot rolled high surface quality carbon 45 steel. The tensile strength of 45 steel is increased from 600MPa to 880mpa by online controlled rolling and cooling. They can be used directly instead of thermal treatment high-strength steel. The surface quality of the rolling high-strength steel is improved obviously. After canceling the sandblasting process, the qualitied rate of electroplating is 100%. There are no mildew spots on steel surface after electroplating. This process is energy saving and environmental protection.

Key words: electroplating mildew spots; hot rolling; microstructure; mechanical properties

中厚板轧制稳定性及板形控制研究

乔 馨[1]，闵成鑫[1]，李忠武[2]

（1. 鞍钢股份中厚板厂，辽宁鞍山 114021；2. 鞍钢股份有限公司，辽宁鞍山 114021）

摘 要：4300mm 厚板线生产过程中轧制稳定性直接影响产品质量，直接反映出来就是钢板镰刀弯情况。同时对于 4300mm 厚板线来说，厚规格钢板碎浪和控冷工艺钢板的横向瓢曲是目前的主要板形缺陷。本文通过对影响轧制镰刀弯、碎浪和横向瓢曲各因素进行分析，采取缩小轧机各部位尺寸间隙、AGC 油注偏差和改进加热、矫直和控冷工艺措施，可以明显减少钢板的镰刀弯、碎浪和横向瓢曲缺陷。

关键词：中厚板；镰刀弯；碎浪；横向瓢曲

Study on Rolling Stability and Strip Shape of Medium and Heavy Plates

Qiao Xin[1], Min Chengxin[1], Li Zhongwu[2]

(1. Medium and Heavy Plate Plant of Ansteel, Anshan 114021, China;
2. Angang Steel Co., Ltd., Anshan 114021, China)

Abstract: Rolling stability directly influence the quality and the camber of product in production line process of the heavy plates of 4300mm.For the heavy plate of 4300mm,the micro-waves and transverse waves are the main strip shape defects. By analysed the cause of camber, micro-waves and transverse waves, take measures of decreasing the gap of rolling machine and AGC machine, optimizing the technics of heating process, leveling process and cooling process, the camber, micro-waves and transverse waves defects can be reduced obviously.

Key words: medium and heavy plate; camber; micro-wave; transverse wave

冷却工艺对 Q450NQR1 耐候钢性能及组织的影响

王俊霖，梁　文，刘志勇，陶文哲，邓照军

（宝钢股份中央研究院武钢有限技术中心冷轧产品研究所，湖北武汉　430080）

摘　要：采用层流冷却的方式对厚度规格为 6mm 的 Q450NQR1 耐候高强钢进行轧制试验，研究了相同终冷温度下不同冷却速率和相同冷速下不同终冷温度对钢组织性能的影响；通过控制终冷温度为 650℃及冷速为 30℃/s，获得铁素体+珠光体组织，避免贝氏体相变破坏其延展性，使试验钢具有良好的强度及延展性，获得最佳工艺参数和为后续工艺调整提供理论依据。

关键词： Q450NQR1；冷速；终轧温度；组织；性能；析出

Influence of Cooling Processes on Mechanical Properties and Microstructure of Q450NQR1 Weathering Steel

Wang Junlin, Liang Wen, Liu Zhiyong, Tao Wenzhe, Deng Zhaojun

(R & D Centre of Wuhan Iron& Steel Co., Ltd., Baosteel
Central Research institute, Wuhan 430080, China)

Abstract: Laminar cooling techniques were used to process Q450NQR1 high strength weathering steel strip with a thickness of 6 mm. The evolutional disciplines of microstructure and mechanical properties were analyzed and discussed under different cooling rates and different final cooling temperatures. By controlling the final cooling temperature and cooling rate, ferrite and pearlite were attained, the experimental steel had good mechanical properties and optimal processing parameters of cooling were determined.

Key words: Q450NQR1; cooling rate; finish rolling temperature; microstructure; mechanical properties; precipitation

一种热镀锌光整机工作辊轴承座装配模式改进研究

曹七华，卢劲松，刘文杰，辛卫升，龚承志

（武钢有限冷轧厂，湖北武汉　430080）

摘　要：本文通过对某冷轧厂热镀锌光整机工作辊轴承座的装配模式进行研究，发现在此种设计模式下，操作人员费时费力，劳动强度大，作业效率低，轧辊上机后易出现生产事故。基于此现状，对原有设计模式进行改进优化，以达到降低劳动强度，提高生产效率，保障生产顺行的目的。

关键词：热镀锌；光整机；工作辊；轴承座；改进

Study on Improvement of Assembly Mode of Work Roll Bearing Block for Hot Dip Galvanizing Mill

Cao Qihua, Lu Jinsong, Liu Wenjie, Xin Weisheng, Gong Chengzhi

(Cold Mill of Wuhan Iron and Steel Ltd., Wuhan 430080, China)

Abstract: This article through to a hot dip galvanized light overall assembly model of work roll bearing abrasion, found that in this kind of design mode, the operator laborious, the intensity of labor is big, the efficiency is low, roll production accidents frequently occur after the computer, based on the status quo, to improve the original design pattern optimization, to achieve the reduction of labor intensity, improve production efficiency, Guarantee the production to run smoothly.

Key words: hot galvanizing; light overall; work roll; bearing block; improvement

基于多模型融合的焊接质量检测系统的设计与实现

孙悦庆[1]，苏统华[2]，李松泽[2]

（1. 宝山钢铁股份有限公司冷轧厂，上海 201900；
2. 哈尔滨工业大学软件学院，黑龙江哈尔滨 150001）

摘 要：冷轧带钢窄搭接焊缝质量检测是带钢拼接过程中的一个重要环节，正确地对焊缝质量进行检测能有效降低带钢的断带率，提高生产效率、生产质量，并降低生产成本。为了解决 BG 公司冷轧带钢窄搭接焊接设备在焊接过程中出现撕裂等通板故障，降低人工复核的次数，实现全自动化的焊接技术，提高焊接的效率与质量，本文提出了一种基于多模型融合的焊缝质量检测系统，利用传统的决策树、动态时间归整（DTW）与 K 最近邻法（KNN）的机器学习，并结合深度神经网络有监督学习的方法，通过多模型加权融合的方式，对 BG 公司现有的德国 MIEBACH、日本 TMEIC 以及宝悍搭接焊机焊接得到的带钢焊缝进行质量检测。结果表明，焊缝质量检测系统能准确、高效地检测出带钢缺陷焊缝，该系统能应用到实际生产工况。

关键词：决策树；K 最近邻法；动态时间规整；深度学习；神经网络；焊缝质量检测

Design and Implementation of Welding Quality Inspection System based on Multi Model Fusion

Sun Yueqing[1], Su Tonghua[2], Li Songze[2]

(1. Cold Rolling Plant of Baoshan Iron and Steel Co., Ltd., Shanghai 201900, China;
2. School of Software of Harbin Institute of Technology, Harbin 150001, China)

Abstract: The quality inspection of narrow lap welds of cold-rolled strip steel is an important part in the process of strip splicing. To reduce the defective rate of strip steel, improve production efficiency, improve production quality, and reduce production cost, an accurate weld quality inspection is needed. In order to solve the through-plate failure of BaoGang's cold-rolled strip steel narrow lap welding equipment during the welding process, reduce the number of manual rechecks, realize fully automated welding technology, and improve the efficiency and quality of welding, this paper proposes a weld quality inspection system based on multi-model fusion, using Decision Tree, Dynamic Time Warping (DTW) and K nearest neighbor (KNN) machine learning methods, combined with deep neural network supervised learning methods, to check the quality of the weld seam which obtained by Baosteel's existing German MIEBACH, Japan's TMEIC and Baohan lap welder machines. The results show that the weld seam quality inspection system can detect the bad product data of the strip steel welding seam accurately and efficiently, and the system can be applied to the actual production conditions.

Key words: decision tree; KNN; DTW; deep learning; neural network; weld seam quality inspection

精益即时化系统解决钢铁业“最后一公里”管理应用技术研究

刘德成[1]，孙悦庆[1]，李　森[2]，张新伟[2]

（1. 宝山钢铁股份有限公司冷轧厂，上海　201900；
2. 施耐德电气（中国）有限公司 绿色智能制造业务部，北京　100016）

摘　要：精益管理是在生产的基础上，将思想植入企业之中精益管理中，以消除企业管理中的不准确行为 ,杜绝任何形式的浪费在企业运营中发生，推进准确、最大化地创造社会价值，是向管理要效益、方法最大化地创造社会价值，是向管理要效益的一种新管理念和模式。精益管理对于国有企业尤其是钢铁行业提升企业管理水平、提升核心竞争能力具有十分重要的推动作用。

当前，钢铁行业竞争挑战加剧，行业发展转型升级需求更加紧迫，实施精益管理则成为了钢铁行业内加速发展的必由之路。BG1730 冷轧是 BG 股份“十一五”规划项目的主体项目，通过导入和应用精益管理，向管理要效益，不断挖掘企业内在潜能，提高了企业发展竞争力。本文通过运用精益即时化手段，探索了解决了企业管理“最后一公里”的应用方法。

关键词：精益；企业发展；企业运营绩效；即时化系统

The Applied Research of Lean Just-in-time System to Solve the Management Problem in Steel Industry

Liu Decheng[1], Sun Yueqing[1], Li Sen[2], Zhang Xinwei[2]

(1. Cold Rolling Plant of Baoshan Iron and Steel Co., Ltd., Shanghai 201900, China;
2. Schneider SSMF, Beijing 100016, China)

Abstract: Lean management is to eliminate inaccurate behaviors in business management and to prevent any form of wasteful practices in business operations by embedding the idea in the enterprise based on production. As a new management concept and model, lean management further creates social value accurately and maximally, improving management efficiency and maximizing management requirements. Lean management has a very significant role in promoting state-owned enterprises, especially in the steel industry, to improve their enterprise management level and enhance their core competencies.

At present, the competition in the steel industry has intensified. The needs for industry transformation and upgrading have become more urgent, so the implementation of lean management is the necessary path to accelerate the growth of the steel industry. Baosteel's 1730 Cold Rolling is Baosteel Group's main project of “11th Five-Year Plan”. The introduction and practice of lean management have become an important strategic task to accelerate the development of the company. This

article explores the application of Lean JIT to solve the "Last Mile" of enterprise management.
Key words: lean; enterprise development; performance; short interval management

中厚板厚度精度的优化

陈军平，渠秀娟，石锋涛
（鞍钢股份有限公司中厚板事业部，辽宁鞍山 114042）

摘 要：厚度精度项目进行了长度方向厚度优化（AGC 控制）、宽度方向厚度优化（凸度控制）、绝对厚度精度优化（模型精度）、测厚仪精度优化（补偿精度）等主要内容，最终的成熟技术已纳入了自动控制系统。
关键词：精度；补偿；控制；凸度

Thickness Accuracy Optimization of Mdium and Heavy Plate Plant of Ansteel Co., Ltd.

Chen Junping, Qu Xiujuan, Shi Fengtao
(Anshan Steel Co., Ltd. Medium and Thick Plate Business Department, Anshan 114042, China)

Abstract: The thickness accuracy project has carried out length direction thickness optimization (AGC controls) , width direction thickness optimization (crown controls) , absolute thickness accuracy optimization (model accuracy) , gama gauge accuracy optimization (compensate accuracy) , the ultimate mature technology has been brought into autocontrol system.
Key words: accuracy; compensates; controls; convexity

Q345B+S32205 双相不锈钢热轧复合板工业生产的工艺研究

石锋涛，隋松言，渠秀娟
（鞍钢股份公司中厚板事业部，辽宁鞍山 114042）

摘 要：采用真空轧制复合技术，成功生产出厚度（12+2）mm 的 Q35B+S32205 双相不锈钢异质复合钢板。结果表明，利用该工艺生产的钢板，结合度良好，超声波探伤没有明显缺陷存在，力学性能优异，复合钢板剪切强度超过 400MPa。在复合界面处存在不连续的小孔洞，经研究分析与复合坯组坯前表面粗糙度有关。研究表明在复合板轧制过程中，不锈钢在厚度方向的压延稍大于碳钢，可为工业批量生产时的坯料设计提供理论支撑。
关键词：真空轧制复合；异质复合钢板；复合界面；组织性能

Q345B+S32205 Duplex Stainless Steel Study on Industrial Production Process of Hot Rolled Clad Plate

Shi Fengtao, Sui Songyan, Qu Xiujuan

(Anshan Steel Co., Ltd. Medium and Thick Plate Business Department, Anshan 114042, China)

Abstract: Q345B+S32205 duplex stainless steel heterogeneous composite plate with thickness of (12+2) mm was successfully produced by vacuum rolling technology. The results show that the steel plate produced by this process has good bonding degree, no obvious defects in ultrasonic testing, excellent mechanical properties, and shear strength of composite steel plate exceeds 400 MPa. There are discontinuous small holes in the composite interface, which are related to the roughness before billet formation. The research shows that the rolling process of stainless steel in thickness direction is slightly larger than that of carbon steel in the process of composite plate rolling, which can provide theoretical support for blank design in industrial batch production.

Key words: vacuum rolling compounding; heterogeneous composite steel plate; composite interface; microstructure and properties

热连曲柄飞剪剪刃间隙控制分析及在线修复技术应用

吴长杰[1]，张会明[1]，醴亚辉[1]，郭维进[1]，东占萃[1]，王香梅[2]

（1. 北京首钢股份有限公司，河北迁安　064406；
2. 北京首钢机电有限公司迁安电气分公司，河北迁安　064406）

摘　要：热轧曲柄飞剪是热轧产线重点设备，其运行稳定性及设备精度直接关系到产线运行的稳定。现场曲柄飞剪受设备磨损、变形等影响，出现剪刃间隙偏差较大、剪刃磨损过快等问题，频繁出现剪切故障，对生产稳定产生较大影响。通过对现场设备运行条件分析及设备精度测量，确定设备故障的原因，并提出针对性解决措施，利用在线修复技术对飞剪设备进行在线修复，恢复设备精度，为热轧产线生产稳定提供设备保障。

关键词：热轧；飞剪；在线修复；精度

Analysis of Knife Clearance Control of Crank Crop Shear on HSM Line and Application of On-line Repair Technology

Wu Changjie[1], Zhang Huiming[1], Li Yahui[1], Guo Weijin[1], Dong Zhancui[1], Wang Xiangmei[2]

(1. Beijing Shougang Co., Ltd., Qian'an 064406, China;
2. Beijing Shougang Electromechanical Co., Ltd., Qian'an 064406 China)

Abstract: Hot-rolled crank fly shear is the key equipment of hot-rolled production line, and its operation stability and equipment accuracy are directly related to the stability of production line operation. On-site crank fly shear by equipment wear, deformation and other effects, the occurrence of cutting edge gap deviation is large, cutting edge wear too fast and other issues, frequent shear failure, production stability has a greater impact. Through the analysis of the operating conditions of the field equipment and the measurement of the accuracy of the equipment, determine the cause of the equipment failure, and put forward targeted measures to solve the problem, use the online repair technology to repair the fly-shear equipment online, restore the accuracy of the equipment, and provide equipment guarantee for the production stability of the hot-rolled production line.

Key words: Hot strip mill; crop shear; on-line repair; precision

热轧中高碳带钢加工开裂原因分析与改进

任俊威[1]，丛志宇[1]，杨 玉[2]，张吉富[1]

（1. 鞍钢股份有限公司鲅鱼圈钢铁分公司热轧部，辽宁营口 115007；
2. 鞍钢股份有限公司技术中心，辽宁鞍山 114000）

摘 要：结合用户在使用中高碳带钢过程中发生的开裂问题，分析了产生开裂的原因，评价了原料存在的夹杂、带状、脱碳等对裂纹产生的影响，提出带钢生产工艺过程的改进方法，以降低了开裂产生几率。

关键词：中高碳钢；开裂；热轧

Cause Analysis and Improvement of Processing Cracking of Hot Rolled Medium and High Carbon Steel Strip

Ren Junwei[1], Cong Zhiyu[1], Yang Yu[2], Zhang Jifu[1]

(1. Hot Rolling Department of Bayuquan Subsidiary of Angang Steel Company Limited, Yingkou 115007, China; 2. Technology Center of Angang Steel Company Limited, Anshan 114000, China)

Abstract: Combined with the cracking problems occurred in the processing of medium and high carbon steel，the causes of cracking were analyzed，the influence of the existing inclusions，segregation and decarbonization on the cracking was evaluated，and the provement method of strip production process was put forward，so as to reduce the cracking probability.

Key words: medium and high carbon steel; cracking; hot-rolling

冷轧带钢清洗停车斑控制技术

英钲艳，张福义，李忠华，孙铁旺

（鞍钢股份有限公司冷轧厂四分厂，辽宁鞍山 114021）

摘 要：为了进一步提升成材率，解决机组在焊接、剪切头尾停机时，带钢在工艺段中产生严重停车斑缺陷的问题。

通过现场跟踪，调研分析停车斑产生的位置和原因，在电解工作箱中增设喷梁，并优化配制碱液工艺，制定清透喷梁、喷嘴周期。通过上述措施，成功消除了机组短暂停机产生的停车斑缺陷。

关键词：冷轧；带钢；清洗停车斑；清洗液残留

The Technology of Control Cold Rolled Strip Cleaning Solution Residue

Ying Zhengyan, Zhang Fuyi, Li Zhonghua, Sun Tiewang

(Cold Strip Work Angang Steel Company Limited, Anshan 114021, China)

Abstract: To further improve the yield of strip steel, the cleaning solution residue should be removed when the unit is temporarily shut down. The location and cause of analysis are confirmed through adding spray beam, optimizing the preparation process of lye, setting up the cleaning cycle of spray beam and nozzle. Cleaning solution residue is removed.

Key words: cold rolled; strip steel; parking cicatrices; cleaning solution residue

集装箱薄材在热轧生产线稳定生产控制实践

范细忠，何士国，丛志宇，王　刚，张吉富，付青才

（鞍钢股份有限公司鲅鱼圈钢铁分公司，辽宁营口　115007）

摘　要：集装箱薄材（≤2.0mm）由于强度高、变形抗力大，在热轧产线生产过程中时常会出现头部轧破、尾部甩尾、头尾部顺折卡钢等情况，大幅降低生产效率，制约合同正常执行。结合鞍钢热轧产线集装箱薄材生产实践，通过采取工艺参数优化、设备精度严控管理等措施，大幅降低了集装箱薄材头尾部甩尾、顺折卡钢等事故几率，集装箱薄材一次轧制成功率达 95%以上，实现该钢在热轧产线稳定生产。

关键词：集装箱薄材；稳定生产；轧制成功率

Stable Production Control Practice of Container Thin Material in Hot Rolling Line

Fan Xizhong, He Shiguo, Cong Zhiyu, Wang Gang, Zhang Jifu, Fu Qingcai

(Bayuquan Iron & Steel Subsidiary Company of Angang Steel Co., Ltd., Yingkou 115007, China)

Abstract: Due to the high strength and high deformation resistance of container sheet (≤ 2.0 mm), the head is often broken, the tail is thrown off, and the head and tail are folded and jammed in the process of hot rolling production line, which greatly reduces the production efficiency and restricts the normal implementation of the contract. Combined with the production of container sheet in Angang hot rolling production line, through a lot of production practice, by adopting the measures of process parameter optimization, strict control and management of equipment precision, the accident probability of container sheet head and tail swing, folding and steel sticking was greatly reduced, and the power of container sheet rolling at one time was more than 95%, which realized the stable production of the steel in hot rolling production line.

Key words: container sheet; stable production; rolling success rate

热轧产线轧制节奏功能的研究与应用

刘　冬

（北京首钢自动化信息技术有限公司首迁运行事业部，河北迁安　064400）

摘　要：在热连轧机生产带钢过程中，实现热轧机自动出钢节奏控制可以提高机时产量，还可以降低操作工的劳动强度。本文通过对热轧 1580 生产线轧制节奏的研究，介绍了轧制节奏功能及实现方法及应用情况。轧制节奏功能不仅提高了热轧轧线的自动化水平，而且提高了轧制节奏的稳定性，使得 1580 热轧生产线的产能与产品质量都有一定程度的提高。

关键词：轧制节奏；自动模式；自动出钢

Research and Application of Mill Pacing Function in Hot Rolling Production Line

Liu Dong

(Shougang Automatic Information Technology Co.,Ltd., Qian'an 064400,China)

Abstract: In the strip production process of the hot rolling mill, the realization of the automatic mill pacing control of the hot rolling mill can increase the machine-hour output and reduce the labor intensity of the operators. This paper introduces the function of the mill pacing function and its implementation methods and applications through the research on themill pacing of Shougang Hot Rolling 1580 production line. The rolling mill pacing function not only improves the automation level of the hot rolling line, but also improves the stability of the mill pacing, so that the production capacity and product quality of the 1580 hot rolling line are improved to a certain extent.

Key words: mill pacing; automatic mode; automatic discharge

薄带钢冷连轧非焊缝断带机理及控制技术研究

王　畅[1,3]，于　洋[1,3]，王　林[1,3]，李　振[2]，宋浩源[2]，焦会立[2]

（1. 首钢集团有限公司技术研究院，北京　100043；2. 北京首钢股份有限公司，北京　100043；3. 绿色可循环钢铁流程北京市重点实验室，北京　100043）

摘　要：本文基于对非焊缝断带数据进行详细统计分析，发现热卷来料表面类缺陷包括热轧过程中折叠、氧化铁皮、夹杂等缺陷均可能造成轧制过程中高速断带；热轧原料头尾板形异常易于造成撕裂型断带；针对低碳铝镇静钢断带问题与其热卷边部组织落入单相区轧制密切相关。针对产生原因不同的各类型非焊缝断带，提出了相应的预防措施，有效地降低了非焊缝断带的发生率。

关键词：非焊缝断带；热轧；边部开裂；板形；组织

Study on the Breaking Mechanism and Control Technology of Non-weld Broken Cold Thin Strip

Wang Chang[1,3], Yu Yang[1,3], Wang Lin[1,3], Li Zhen[2], Song Haoyuan[2], Jiao Huili[2]

(1. Research Institute of Technology of Shougang Group Co., Ltd., Beijing 100043, China; 2. Beijing Shougang Co., Ltd., Beijing 100043, China; 3. Beijing Key Laboratory of Green Recyclable Process for Iron & Steel Production Technology, Beijing 100043, China)

Abstract: Based on the detailed statistical analysis of the non-weld strip breaking data, it was found that the surface defects including folding, scale, inclusion and other defects in the hot rolling process may cause high-speed strip breaking in the rolling process. The abnormal plate shape of the head and tail of hot rolled material was easy to cause tearing fracture. The strip breaking problem of low carbon aluminum killed steel is closely related to the structure of hot coiling edge falling into single-phase rolling zone. According to the different causes of non-weld broken belt, the corresponding preventive measures are put forward to reduce the incidence of non-weld broken belt effectively.

Key words: non-weld broken strip; hot rolling; edge cracking; strip shape; organization

热轧 1700ASP 加热炉双蓄热改造探索

魏秀东，苗　龙，徐小科

（鞍钢集团朝阳钢铁有限公司，辽宁朝阳　122000）

摘　要： 介绍了蓄热式燃烧技术在轧钢加热炉应用的特点及其工作原理，探索分析了该技术在 1700 ASP 热轧加热炉改造的可适性。加热炉采用蓄热式燃烧技术具有节约能源、降低 NOx 排放等优点，同时可获得较好的经济效益。

关键词： 蓄热式；燃烧技术；换向阀；加热炉

Exploration on Double Regenerative Transformation of 1700ASP Hot Rolling Furnace

Wei Xiudong, Miao Long, Xu Xiaoke

(Angang Chaoyang Iron and Steel Co., Ltd., Chaoyang 122000, China)

Abstract: The characteristics and working principle of regenerative combustion technology applied in reheating furnace for steel rolling are introduced, and the applicability of this technology in 1700 ASP hot rolling reheating furnace is explored and analyzed. The regenerative combustion technology used in heating furnace has the advantages of saving energy and reducing NOx emission, and can obtain better economic benefits.

Key words: regenerative type; combustion technology; directional valve; heating furnace

1580线SPA-H薄材单重提高后的稳定性研究

赵毓伟，丛志宇，王　刚，王　存，王　杰

（鞍钢股份鲅鱼圈钢铁分公司热轧部，辽宁营口　115007）

摘　要：对鞍钢股份有限公司鲅鱼圈钢铁分公司1580产线生产的SPA-H薄材单重提高后的稳定性进行研究，分析了单重提高后影响轧制稳定性的因素，结合产线设备和工艺特点，制定了优化计划编排,设备精度提升和升级改造，出钢温度、轧辊冷却水、精轧各机架负荷分配等工艺制度优化等解决措施，上述措施应用后产线SPA-H薄材产能显著提升，轧制稳定。

关键词：SPA-H薄材；单重提高；工艺优化；稳定性

Study on the Stability of SPA-H Thin Material after Weight Increase in 1580 Production Line

Zhao Yuwei, Cong Zhiyu, Wang Gang, Wang Cun, Wang Jie

(Angang Bayuquan Iron and Steel Co., Ltd., Yingkou 115007, China)

Abstract: The stability of SPA-H thin produced by Bayuquan 1580production line of Anshan Iron and Steel Co,Ltd. After single weight increase was studied,the factors influencing rolling stability were analyzed. Combined with the characteristics of production line equipment and process, the plan arrangement,equipment upgrade, technology optimization and other solution measures were developed. After the implementation of SPA-H sheet production capacity increased, rolling stability.

Key words: SPA-H thin material ;weight increased;process optimization;rolling stability

CPC高速钢轧辊在2250热轧的应用技术研究

吴真权，曾龙华，王　波，豆伟晗，喻圣男

（湛江钢铁有限公司热轧厂，广东湛江　524072）

摘　要：宝钢湛江2250热轧为成为全国最具竞争力的热连轧产线，进一步提升轧辊应用技术、降低成本、改善带钢轧制质量，2250热轧在精轧F1~4机架拓展应用CPC高速钢轧辊，主要从轧辊性能、稳定性轧制技术深入分析，输出CPC高速钢轧辊使用及轧制工艺技术。

关键词：2250热轧；CPC高速钢轧辊；轧辊使用

Expansion and Application of CPC High Speed Steel Roll in 2250 Hot Rolling

Wu Zhenquan, Zeng Longhua, Wang Bo, Dou Weihan, Yu Shengnan

(Hot rolling Mill of Zhanjiang Iron & Steel Co., Ltd., Zhanjiang 524072, China)

Abstract: Baosteel 2250 hot rolling in zhanjiang to become the most competitive strip production line, further enhance roll application technology, reduce cost and improve the quality of strip steel rolling, 2250 hot rolling in the finishing of F1-4 rack to expand application of CPC high speed steel roll, mainly from the roll rolling technology in-depth analysis, performance, stability output, CPC used high speed steel roll and rolling technology.

Key words: 2250 hot rolling; CPC high speed steel roll; roll rolling process

智能纠偏控制在安钢 1780mm 热连轧的研究与应用

饶　静，杨立庆，邓杭州

（安阳钢铁股份有限公司第二炼轧厂，河南安阳　455004）

摘　要： 辊缝水平调整是保证热连轧板形、防止轧件跑偏主要调整手段。针对纠偏控制依赖手动调整且存在滞后、方向预估错误的缺点，对轧机不对称跑偏的状况进行分析，得出轧机轧制力差的变化可作为轧机纠偏方向和程度的调整依据，设计一种智能纠偏控制系统，可以实现对轧制跑偏问题进行有效纠正，通过在安钢 1780mm 热连轧的工业试验，取得了良好的应用效果。

关键词： 热连轧；甩尾；智能；纠偏

Research and Application of Intelligent Deviation Correction Control in Angang Hot Strip Mill

Rao Jing, Yang Liqing, Deng Hangzhou

(Anyang Iron and Steel Co., Ltd, Anyang 455004, China)

Abstract: The horizontal adjustment of roll gap is the main means to ensure the shape of hot strip and prevent the deviation of rolled piece. Aiming at the shortcomings of manual adjustment, lag and direction estimation error in deviation correction control, this paper analyzes the asymmetric deviation of rolling mill, and concludes that the change of rolling force difference of rolling mill can be used as the basis for adjusting the direction and degree of deviation correction of rolling mill. An intelligent deviation correction control system is designed, which can effectively correct the problem of rolling deviation, Through the industrial test of 1780 mm hot strip mill in Anyang Iron and Steel Co., good application effect has been obtained.

Key words: hot rolling; tail flick; intelligence; correction

950 轧线精轧终轧温度二级模型的优化与控制

张　辉，张永亮，李　庆

（山东泰山钢铁集团有限公司，山东济南　271100）

摘　要：热轧带钢终轧温度的控制策略是根据生产和设备的实际情况，在精确预测机架内带钢温度的数学模型的基础上，通过调节轧制速度和机架间冷却水流量，有效地控制精轧机出口带钢温度。根据 950 轧钢生产现场工艺和设备实际情况,确定了热轧带钢终轧温度的控制策略。基于带钢在机架间温度变化的精确预报模型,通过调节轧制速度或机架间冷却水流量，有效地将带钢温度控制到目标范围内。

关键词：终轧温度；轧制速度；冷却水流量；数学模型

Optimization and Control of Secondary Model for Finishing Temperature of 950 Rolling Line

Zhang Hui, Zhang Yongliang, Li Qing

(Shandong Taishan Iron and Steel Group Co., Ltd., Ji'nan 271100, China)

Abstract: According to the actual situation of production and equipment, the final rolling temperature control strategy of hot strip mill is to effectively control the exit strip temperature of finishing mill by adjusting the rolling speed and cooling water flow between stands on the basis of the mathematical model of accurately predicting the strip temperature in stands. According to the actual situation of 950 steel rolling process and equipment, the control strategy of finishing temperature of hot strip is determined. Based on the accurate prediction model of strip temperature change between stands, the strip temperature can be effectively controlled to the target range by adjusting the rolling speed or cooling water flow between stands.

Key words: finishing rolling temperature; rolling speed; cooling water flow; mathematical model

炉卷轧机支撑辊剥落分析与辊型改进

杨立庆，饶　静，邓杭州

（安阳钢铁股份有限公司第二炼轧厂，河南安阳　455004）

摘　要：针对安钢 3500mm 炉卷轧机支撑辊剥落状况，通过分析四辊轧机的受力特点和原有倒角支撑辊辊型的特征，得出支撑辊辊型不良是辊间接触压力不均并造成边部剥落掉块的主要原因，同时弯辊力的施加更加剧了这一现象。通过设计一种新型 8 次方辊型曲线，在兼顾板形控制时减小了辊间接触压力，提高了轧机横向辊缝刚度，有效降低了轧辊剥落事故。

关键词：支撑辊；8 次方；弯辊力；辊型

Spalling Analysis and Roll Profile Improvement of Backup Roll of Steckel Mill

Yang Liqing, Rao Jing, Deng Hangzhou

(Anyang Iron and Steel Co., Ltd, Anyang 455004, China)

Abstract: According to the spalling condition of backup roll of 3500mm Steckel mill in Anyang Iron and Steel Co., Ltd., by analyzing the stress characteristics of four high mill and the characteristics of original chamfered backup roll profile, it is concluded that the poor backup roll profile is the main reason for uneven contact pressure between rolls and edge spalling, and the application of bending force aggravates this phenomenon. By designing a new type of octave roll profile curve, the contact pressure between rolls is reduced, the stiffness of transverse roll gap is improved, and the Roll Spalling accident is effectively reduced.

Key words: backup roll; eighth power; bending force; roll profile

高建钢厚板性能合格率的提升

冯　赞，廖宏义，张　娉

（湖南华菱湘潭钢铁有限公司技术质量部，湖南湘潭　411100）

摘　要：为提高湘钢高建钢厚板的一次性能合格率，通过调整成分以及轧制工艺，可以将屈服强度以及抗拉强度的波动范围缩小到 80MPa 以内，并提高冲击功平均值 100J 以上，为湘钢批量交付厚板 Q420GJD 提供了支撑。

关键词：高建钢；厚板；TMCP；性能合格率

Improvement of Qualified Rate of Heavy Plates for Building Structure Steel

Feng Zan, Liao Hongyi, Zhang Ping

(HunanValin Xiangtan Iron and Steel Company Technology and Quanlity Department, Xiangtan 411100, China)

Abstract: In order to improve the pass rate of the first performance of the heavy plate of Xiangtan Steel, the fluctuation range of yield strength and tensile strength can be reduced to within 80MPa by adjusting the composition and rolling process, and the average value of impact energy can be increased to more than 100J, which provides support for the mass delivery of heavy plate Q420GJD by Xiangtan Steel.

Key words: high construction steel; thick plate; TMCP; Qualified Rate

CSP 热轧汽车结构钢 QStE420 生产实践

郑海涛[1]，高 智[1]，田军利[1]，王 成[1]，程 曦[1]，魏 斌[2]

（1. 武钢有限条材厂 CSP 分厂，湖北武汉 430083；2. 武钢有限技术中心，湖北武汉 430083）

摘 要：介绍了武钢有限条材厂在 CSP 产线生产汽车结构钢 QStE420 的工艺流程及特点。针对过程控制难点制定现场生产指导措施；对 QStE420 生产过程中常见质量缺陷进行分析，提出控制手段，确保产品质量稳定。

关键词：CSP；QStE420；过程控制；质量

Production Practice of Hot-rolled Automobile Structural Steel QStE420 by CSP

Zheng Haitao[1], Gao Zhi[1], Tian Junli[1], Wang Cheng[1], Cheng Xi[1], Wei Bin[2]

(1. Plant of Long Product , Wuhan Iron & Steel Co., Ltd., Wuhan 430083, China;
2. Technology Center, Wuhan Iron & Steel Co., Ltd., Wuhan 430083, China)

Abstract: Introduce the process flow and characteristics of automobile structural steel QStE420 produced by CSP production line of WISCO.Work out on-site production guidance measures for process control difficulties. The common quality defects in the production of Qste420 are analyzed and the control measures are put forward to ensure the stability of product quality.

Key words: CSP; QStE420; process control; quality

电渣重熔+辊式淬火工艺提高特厚板全厚度组织性能均匀性

王庆海，刘 庚，叶其斌，田 勇，王昭东

（东北大学轧制技术及连轧自动化国家重点实验室，辽宁沈阳 110819）

摘 要：保证厚度方向的组织性能均匀性是特厚板生产领域一个巨大挑战。本文通过先进的电渣重熔（ESR）和辊式淬火（RQ）技术生产了 210mm 厚海洋工程用钢，该钢板展现了优异的组织和性能均匀性，实现了强度和韧性的良好匹配。与通过传统模铸（MC）和浸入式淬火（IQ）传统工艺生产的 178mm 钢板进行比较，淬火后的 ESR-RQ 钢马氏体在整个厚度方向上分布均匀，1/2 厚度马氏体比例为 39%，显著高于 MC-IQ 钢的 34%。ESR-RQ 钢在淬火态和回火态都显示出优异的综合机械性能，在 650℃回火的 ESR-RQ 钢在−60℃时的冲击功比 MC-IQ 钢高 150J（278%），在 600℃回火的 ESR-RQ 钢屈服强度比 MC-IQ 钢高 137MPa（18%）。

关键词：特厚板；淬火和回火；全厚度均匀性；电渣重熔；辊式淬火

Superior Through-thickness Homogeneity of Microstructure and Mechanical Properties of Ultra-heavy Steel Plate by Advanced Casting and Quenching Technologies

Wang Qinghai, Liu Geng, Ye Qibin, Tian Yong, Wang Zhaodong
(State Key Laboratory of Rolling and Automation, Northeastern University, Shenyang 110819, China)

Abstract: Keeping through-thickness homogeneity of mechanical properties has been a great challenge for producing heavy gauge steel plates. Here, we report a superior homogeneity of microstructure and hence strength and toughness achieved in a quenched and tempered (QT) 210mm thickness steel plate produced by advanced electro-slag remelting casting and roller quenching technologies (ESR-RQ). Some comparisons are made with a QT 178mm steel plate produced by conventional mold casting and immersion quenching in the water tank (MC-IQ). The martensite of the as-quenched ESR-RQ steel is distributed homogeneously across the thickness with a fraction of 39% at the 1/2 thickness, remarkable higher 34% than that of MC-IQ steel. ESR-RQ steel appears excellent comprehensive mechanical properties even for the as-quenched samples, as well as QT samples. Comparing with the MC-IQ steel, the Charpy impact energy at −60℃ is 150J (278%) higher for the samples tempered at 650℃, and the yield strength is 137MPa (18%) higher for the samples tempered at 600℃ than those of MC-IQ steel, respectively.

Key words: ultra-heavy steel; quenched and tempered; through-thickness homogeneity; electro-slag remelting; roller quenching

表面处理和热处理温度对高强钢氧化铁皮微观结构的影响

张亮亮[1,2]，于　洋[1,2]，李晓军[3]，徐德超[1,2]，齐　达[3]，王泽鹏[1,2]
（1. 首钢集团有限公司技术研究院，北京　100041；
2. 绿色可循环钢铁流程北京市重点实验室，北京　100041；
3. 首钢京唐钢铁联合有限责任公司，河北唐山　063200）

摘　要：实验室条件下，研究了热冲压高强钢在不同热处理温度下的氧化铁皮结构，同时探索原板不同表面形貌对氧化铁皮形成的影响，同时分析了不同的热处理温度下氧化铁皮界面平直度和元素富集情况。结果表明，随原板粗糙度变小，热处理后氧化铁皮厚度逐渐增加，界面不平度逐渐下降。但铁皮结合力存在先上升在下降的趋势，铁皮结合力和 Fe_3O_4 含量存在一定的正相关；860-950℃不同热处理温度下，Si、Mn 和 Cr 元素表面富集不同，920℃热处理温度的的铁皮中 Si、Cr 富集相对均匀，较为均匀的改变了界面平直度，且 Mn 未造成氧化铁皮的明显疏松，所以该温度下形成氧化铁皮结合力最佳。

关键词：表面处理；热处理温度；氧化铁皮；微观结构

Effect of Cooling Method after Coiling on Surface Quality and Performance of Shougang Jingtang P-containing Steel

Zhang Liangliang[1,2], Yu Yang[1,2], Li Xiaojun[3],
Xu Dechao[1,2], Qi Da[3], Wang Zepeng[1,2]

(1. Shougang Group Co., Ltd., Research Institute of Technology, Beijing 100041, China;
2. Beijing Key Laboratory of Green Recyclable Process for Iron & Steel Production Technology, Beijing 100041, China; 3. Department of Cold Rolling Operation of Shougang Jingtang Iron&Steel Company, Tangshan 063200, China)

Abstract: Under laboratory conditions, the scale structure of hot-stamped high-strength steel at different heat treatment temperatures was studied, and the influence of different surface morphologies of the substrate on the formation of scale was also explored. At the same time, the flatness and flatness of the scale interface at different heat treatment temperatures were analyzed. Element enrichment.The results show that as the roughness of the original plate becomes smaller, the thickness of the oxide scale gradually increases after heat treatment, and the unevenness of the interface gradually decreases. However, the iron sheet binding force has a tendency to first rise and then decrease, and there is a certain positive correlation between the iron sheet binding force and the Fe_3O_4 content;at different heat treatment temperatures of 860-950℃, the surface enrichment of Si, Mn and Cr elements is different. The concentration of Si and Cr in the iron sheet at the heat treatment temperature of 920℃ is relatively uniform, which changes the flatness of the interface more uniformly, and Mn does not cause The iron oxide scale is obviously loose, so the binding force of the iron oxide scale formed at this temperature is the best.

Key words: surface treatment; heat treatment temperature; scale; microstructure

退火温度对软质镀锡基板组织性能的影响

孙超凡[1,2]，方　圆[1,2]，王雅晴[1,2]，孙　晴[3]

（1. 首钢集团有限公司技术研究院，北京　100043；
2. 绿色可循环钢铁流程北京市重点实验室，北京　100043；
3. 首钢京唐钢铁联合有限责任公司镀锡板事业部，河北唐山　063200）

摘　要：为了研究微量 Ti、Nb 对软质镀锡基板组织性能的影响，为新品种开发制定合理的工艺参数提供参考，采用激光共聚焦显微镜观察了样品显微组织，采用 TEM 观察了样品析出相及其分布特点，采用拉伸试验机检测了样品力学性能。结果表明，微 Ti、微 Nb 处理超低碳软质镀锡基板分别在 705~720℃和 735~750℃完成再结晶，相比 Ti 微合金化，Nb 微合金化超低碳镀锡基板中第二相粒子尺寸更小，分布密度更大，再结晶形成的晶粒尺寸更细小。完全再结晶退火条件下，Ti、Nb 微合金化软质镀锡基板 α 织构最强组元均出现在{111}<110>组元附近，但 Ti 微合金化超低碳质镀锡基板 γ 织构整体强度更高，基板平面塑性应变比 r 值更大。

关键词：镀锡基板；退火温度；组织；性能

Influence of Annealing Temperature on the Microstructure and Mechanical Properties of TMBP

Sun Chaofan[1,2], Fang Yuan[1,2], Wang Yaqing[1,2], Sun Qing[3]

(1. Shougang Group Co., Ltd., Research Institute of Technology, Beijing, 100043, China; 2. Beijing Key Laboratory of Green Recyclable Process for Iron & Steel Production Technology, Beijing, 100043, China; 3. The Tinplate Department of Shougang Jingtang Iron and Steel Co., Ltd., Tangshan 063200, China)

Abstract: In order to study the effect of Ti、Nb microalloying on the microstructure and mechanical properties of TMBP , thus provide rational reference on the process parameters of the new breed steel, the microstructure of the steel was detected by OM, the morphology and distribution of the precipitated particle in the steel was observed by TEM, the mechanical properties of the steel was measured by tensile test. The result showed that TMBP added with Ti、Nb respectively completed its recrystallization between 705~720℃、735~750℃. Compared with the Ti microalloying TMBP, the Nb microalloying TMBP had a smaller average size and higher distribution density of precipitated particle, thus a smaller average recrystallized grain size.Under the condition of full annealing process ,TMBP with Ti、Nb microalloying both had the highest texture intensity near the {111}<112>component in the α fiber texture, but the Ti microalloying TMBP had a higher intensity of γ fiber texture，which led to a higher r during the tensile test.

Key words: TMBP; annealing temperature; microstructure; mechanical properties

高温回火对 590MPa 级高强钢低温冲击韧性的影响

孟繁霞，田　勇，叶其斌，王昭东，吴　迪

（东北大学轧制与连轧自动化国家重点实验室，辽宁沈阳　110819）

摘　要：调质工艺对实验钢的性能有着重要的影响。本文研究了高温回火对 590MPa 级高强度钢的低温冲击韧性和显微组织的影响。结果表明，550℃回火时，显微组织主要为回火屈氏体。随着回火温度的升高，回火屈氏体含量逐渐降低，回火索氏体含量增加。对试验钢的冲击断口形貌、宏观硬度和冲击韧性进行了对比分析，发现当回火温度为 630℃时，材料的冲击韧性最好。

关键词：回火温度；微观组织；断口形貌；冲击韧性；回火硬度

The Influence of High Temperature Tempering to the Low Temperature Impact Toughness of a 590MPa High Strength Steel

Meng Fanxia, Tian Yong, Ye Qibin, Wang Zhaodong, Wu Di

(State Key Laboratory of Rolling and Automation, Northeastern University, Shenyang 110819, China)

Abstract: The quenching and tempering process has an important effect on the properties of experimental steels.The effects of tempering temperature on the low temperature impact toughness and microstructure of a 590 MPa high-strength steel was studied. The results show that, when tempering at 550℃, the microstructure was mainly tempered troostite. With the increase of tempering temperature, the content of the tempered troostite body gradually decreased, the content of the tempered sorbite increased. The products were tempered sorbite when tempering above 650℃. The impact fracture morphology, macrohardness and impact toughness of the experimental steel were compared and analyzed, it is found that the best impacting toughness of the material could be obtained when tempering temperature was 630℃.

Key words: tempering temperature; microstructure; impact fracture morphology; low temperature impact toughness; macrohardness

两相区轧制工艺对低碳低合金钢组织与性能的影响

王益民，苏元飞，李慧杰，徐晓宁，叶其斌

（东北大学轧制技术及连轧自动化国家重点实验室，辽宁沈阳 110819）

摘 要：采用 THRP 工艺和 TMCP 工艺两种轧制工艺，通过金相和 EBSD 分析钢板不同厚度处的组织特征、晶粒尺寸，讨论了组织形成的相变过程；利用冲击实验对比研究了两块不同工艺钢板的冲击性能，重点分析了断口分层和晶粒细化对冲击性能的影响。结果表明：THRP 工艺有效细化了晶粒尺寸，尽管上平台冲击功降低，但是下平台冲击功显著提高，-140℃冲击功为 77 J 达到了 7Ni 钢标准，分层断裂是低温韧性提高的关键。

关键词：两相区轧制；晶粒细化；低温韧性；分层断裂

Effect of Intercritical Rolling on the Microstructure and Properties of Low-carbon Low-alloy Steel

Wang Yimin, Su Yuanfei, Li Huijie, Xu Xiaoning, Ye Qibin

(The State Key Laboratory of Rolling Automation, Northeastern University, Shenyang 110819, China)

Abstract: Two rolling processes, THRP process and TMCP process, were used to analyze the microstructure characteristics and grain size at different thicknesses of the steel plates by metallography and EBSD, and the phase transformation process of microstructure formation was discussed; the impact properties of the two steel plates with different processes were comparatively studied by impact experiments, focusing on the effects of fracture delamination and grain refinement on the impact properties. The results show that the THRP process effectively refines the grain size, and although the impact energy of the upper platform decreases, the impact energy of the lower platform increases significantly, and the impact energy of 77J at -140℃ reaches the standard for 7Ni steel, and delamination fracture is the key to low-temperature toughness improvement.

Key words: intercritical rolling; grain refinement; low temperature toughness; delamination fracture

SPFH590 不同热轧工艺对板形影响

卞 皓，殷 胜

（梅山钢铁股份有限公司，江苏南京 210039）

摘　要： SPFH590为低碳合金钢，轧后常出现浪形现象。针对SPFH590进行不同的热轧工艺试验，不同的工艺对轧后板形影响较大。研究发现SPFH590在不同热轧工艺下的材料金相组织、显微硬度都存在显著差异，通过分析制定了合理的热轧工艺，在保证性能的前提下，提高了轧后板形质量。

关键词： SPFH590；热轧工艺；金相组织；硬度

The Influence of Different Hot Rolling Process of SPFH590 on Strip Shape

Bian Hao, Yin Sheng

(Meishan Iron and Steel Limited by Share Lt, Nanjing 210039, China)

Abstract: SPFH590 is a low-carbon alloy steel, which often appears wave-shaped after rolling. Different hot rolling process tests are carried out for SPFH590, and different processes have a greater impact on the shape of the rolled strip. The study found that there are significant differences in the metallographic structure and microhardness of SPFH590 materials under different hot rolling processes. A reasonable hot rolling process has been established through analysis and the quality of the strip shape after rolling is improved under the premise of ensuring the performance.

Key words: SPFH590; hot rolling process; metallographic organization; hardness

Cr5材质中间辊剥落失效分析

崔海峰，胡现龙，乔建平

（宝钢轧辊科技有限责任公司，江苏常州　213012）

摘　要： 客户反馈我司生产的Cr5材质的中间辊在使用过程中发生了大面积剥落失效，剥落位置位于辊身中部，该产品设计硬度为79~81HSD，使用至报废直径时硬度要求为≥76HSD。本文通过现场剥落区域形貌观察，硬度检测，以及剥落位置取样组织检测，此次剥落是典型的浅表层多点疲劳源引起的剥落，由于此辊使用至接近报废直径时（剩余6mm）辊身组织中碳化物的分布局部区域不够均匀，且此部位正好位于此辊疲劳累积影响最严重的深度处（2～3mm），使得此部位应力集中，形成微裂纹，继续使用过程中裂纹不断扩展，最终导致剥落。

关键词： Cr5中间辊；碳化物；失效分析；辊身剥落；疲劳剥落

Spalling Failure Analysis of Cr5 Intermediate Roll

Cui Haifeng, Hu Xianlong, Qiao Jianping

(Baosteel Roll Technology Co., Ltd., Changzhou 213012, China)

Abstract: The customer feedback that the middle roll of Cr5 material produced by our company has large area peeling failure during the use process. The peeling position is located in the middle of the roller body. The design hardness of the product is 79-81HSD, and the hardness requirement when using to the scrap diameter is ≥ 76HSD. This paper, through the observation of the appearance of the peeling area, hardness test and sampling organization inspection of the peeling position, shows that the peeling is caused by the typical multi-point fatigue source on the superficial surface. The carbide distribution in the roller body is not uniform when the roller is used to the near scrap diameter (the remaining 6mm), And this part is located at the depth (about 2-3mm) which is the most serious influence of fatigue accumulation of this roller, which makes the stress concentration in this part, forming micro cracks, and the cracks continue to expand during the continuous use, which eventually leads to spalling.

Key words: 5% Cr content intermediate roll; carbide; failure analysis; roll body spalling; fatigue spalling

热基无花超厚锌层镀锌板生产工艺和技术研究

马占军，陈国涛，贾海超，李文超，唐嘉悦，陈相东

（承德钢铁集团有限公司，河北承德 067000）

摘 要：随着国民经济的快速提高和产业结构的优化调整，新能源、海绵城市等新兴领域的大力发展对镀锌产品的耐腐蚀性提出了更高的要求，而耐腐蚀性的提升方法主要是靠提高锌层厚度。河钢承钢冷轧公司已经将锌层厚度开发到 900g/m^2，本文主要对承钢热基无花超厚锌层的生产工艺和技术进行研究，同时介绍产线主要装备水平、产品定位、产品的耐腐蚀性评价等。

关键词：热基板镀锌；无锌花；超厚锌层；900g/m^2；生产工艺

Study on Process and Technology of Thermal-based Flowers-free Ultra-thick Galvanized Sheet

Ma Zhanjun, Chen Guotao, Jia Haichao, Li Wenchao, Tang Jiayue, Chen Xiangdong

(Chengde Iron and Steel Group Co., Ltd., Chengde 067000, China)

Abstract: With the rapid improvement of the national economy and the optimization and adjustment of the industrial structure, the vigorous development of new energy, sponge cities and other emerging fields has put forward higher requirements for the corrosion resistance of galvanized products.The method of improving the corrosion resistance is mainly to increase the thickness of the zinc layer.HBIS Chengsteel Cold Rolling Company has developed the thickness of the zinc layer to 900g/m^2. This article mainly studies the process and technology of Chengsteel thermal-based ultra-thick zinc layer Galvanized sheet, and introduces the main equipment level of the production line, products positioning, products corrosion resistance evaluation and application, etc.

Key words: thermal-basedgalvanizing; flowers-free zinc plating; ultra-thick zinc layer; 900g/m^2; process

短流程 CSP 热轧极薄材板形不良原因分析及控制方法

张亦辰，高 智，田军利，李 波，赵 强

（宝钢股份武钢有限公司条材厂，湖北武汉 430080）

摘　要： 通过 CSP 热轧带钢轧制数据和曲线的分析结合现场生产实际，对短流程热轧极薄材轧制过中带钢跑偏的原因进行了具体分析，并从（1）导板中心线，（2）速度控制，（3）板形控制、弯辊力、轧制力控制，（4）水系统，（5）轧辊这几个方面对轧制数据和曲线进行了着重分析给出相应的结论。并从热轧生产自动化的三级、二级、一级、零级优化及操作、维护等方面提出了相应的预防和减少极薄带钢板形不良的控制方法。生产实践结果表明，上述控制方法可有效减少带钢轧制过程中跑偏情况的发生，减少了热轧薄板轧破情况的发生，保证了精轧生产的连续性和稳定性。

关键词： 热连轧精轧；板形；短流程

Analysis and Control of CSP Hot Thin Strip Deformed during Rolling Process

Zhang Yichen, Gao Zhi, Tian Junli, Li Bo, Zhao Qiang

(Plant of Long Product, Wuhan Iron ＆ Steel Co., Ltd., Wuhan 430083, China)

Abstract: By analysis on data of hot strip rolling and curves, combining with production practice, the causes of hot thin strip deformed during rolling process were concretely analyzed. Analysis through 5 different sides, include (1) Centreline of guide plate, (2) Rolling speed, (3) Control of strip plate, WRB, roll force, (4) Water system, (5) Work rolls. Control methods of tertiary, secondary, primary, zero-level optimization of hot rolling production automation, operation, maintenance were put forward to prevent and reduce hot thin strip deformed during rolling process. Production practice result show that the above methods could effectively reduce hot thin strip deformed during rolling process, reduce hot-rolled thin plat tail cracks and reduce sudden change of wedge shape in steel rolling process, Which could ensure rolling continuity and stability of finish process.

Key words: hot strip mill; shape of strip; short process

提高 CSP 短流程带钢尾部宽度精度的控制方法

李　彪，姜　南，李　波

（宝武集团武汉钢铁有限公司条材厂，湖北武汉　430083）

摘　要： 通过 CSP 厂带钢轧制数据和曲线的分析以及结合现场生产实际，比较突出的问题是带钢的尾部超宽。本文从 CSP 厂实情出发重点针对原卷及出口材尾部宽度超差的原因进行了具体分析，根据热轧带钢尾部超宽的情况，从立辊侧压及短行程控制等方面进行分析验证，并提出了一些对策，有效地提高了宽度精度。

关键词： 尾部超宽；立辊；短行程；侧压力；宽度精度

Control Method of Improving the Precision of the Width of the End of Short Hot Strip

Li Biao, Jiang Nan, Li Bo

(Plant of Long Product, Wuhan Iron & Steel Co., Ltd., Wuhan 430083, China)

Abstract: Based on the analysis of strip rolling data and curve in CSP Plant and combined with the field production practice, the prominent problem is the over width of strip tail. In this paper, based on the actual situation of CSP plant, the cause of the excess width of the end of the original coil and the export material is analyzed in detail. According to the situation of the excess width of the end of the hot rolled strip, the side pressure of the vertical roll and the short stroke control are analyzed and verified, and some countermeasures are put forward to effectively improve the width accuracy and ensure the export of the original coil and the export material.

Key words: tail ultra wide; vertical roll; short stroke; side pressure; width accuracy

精轧 FDT 控制稳定性研究

于任飞，白 威，孙 伟，宁 腾，张 帅，刘德辉

（鞍钢股份有限公司鲅鱼圈钢铁分公司，辽宁营口 115007）

摘 要：鞍钢 1580 热轧线经过内部深层挖潜，轧制节奏步步提升，但轧制节奏提升后，前后钢板在 F7 轧机抛钢时间间隔缩短。为保证 F7 出口辊道冷却，将冷却水喷射时序提前，造成精轧出口高温计（FDT）进钢时水雾加大，高温计检测失真，影响层流区域温度控制（CTC）；同时精轧出口高温计（FDT）安装在精轧出口走桥，该位置距离层流第一组水幕距离仅为 4.2m，层流第一组水幕上集管投入时 FDT 区域水雾大，同样造成高温计温度失真，无法投入，影响部分钢种层流区域冷区能力和轧制质量。

关键词：FDT；高温计；失真；干扰

Study on FDT Control Stability of Finishing Rolling Mill

Yu Renfei, Bai Wei, Sun Wei, Ning Teng,
Zhang Shuai, Liu Dehui

(Angang Steel Company Limited Bayuquan Steel Branch,
Yingkou 115007, China)

Abstract: The rolling rhythm of Anshan Iron and Steel 1580 hot rolling line increases step by step after the deep tapping of the internal potential, but the time interval between the front and rear steel plates in the F7 rolling mill is shortened after the rolling rhythm is improved. In order to ensure the cooling of the roller table at the F7 exit, the timing sequence of cooling water injection is advanced, which causes the increase of water mist when the pyrometer (FDT) enters the steel at the finish rolling exit, resulting in the distortion of the pyrometer detection and affecting the temperature control in the laminar flow area (CTC). At the same time, the finishing rolling outlet pyrometer (FDT) is installed on the finishing rolling outlet bridge, which is only 4.2 meters away from the first group of laminar water curtain. When the upper collecting pipe of the first group of laminar water curtain is put into the FDT area, the water mist is heavy, which also causes the temperature distortion of the pyrometer and can not be put into the area, affecting the cold zone ability and rolling quality of some steel laminar flow areas.

Key words: FDT; pyrometer; distortion; interference

超低碳钢回火过程“析出塑性”行为形成的微观机制

丁文红[1]，王新东[2]，吴　挺[1]，孙　力[2]，刘天武[2]，潘　进[2]

（1. 武汉科技大学耐火材料与冶金国家重点实验室，湖北武汉　430081；
2. 河钢集团钢研总院，河北石家庄　050023）

摘　要：具有初始残余应力的材料在回火过程中伴随碳化物的析出会产生不可逆的塑性应变，使材料内的初始应力得到松弛。前期研究将这种与碳化物析出所伴生的塑性行为命名为“析出塑性”。尽管“析出塑性”对残余应力调控有着重要影响，但“析出塑性”形成的微观机制尚未揭示。本文以超低碳钢为对象，对“析出塑性”的形成机制展开研究。结果表明碳化物析出时在基体组织界面形成的空位会由界面向基体组织中扩散，并在应力作用方向上累积、崩塌，形成位错环，成为与碳化物析出相伴生不可逆塑性应变，即“析出塑性”。而且，碳化物尺寸越细小，“析出塑性”行为越显著。当碳化物尺寸由 14nm 减小至 11nm 时，析出塑性系数 K 增加 21%。减小析出相的尺寸，有利于回火残余应力的调控。

关键词：回火；相变塑性；碳化物析出；残余应力

Mechanism of “Precipitation Plasticity” for Ultra Low Carbon Steel during Tempering

Ding Wenhong[1], Wang Xindong[2], Wu Ting[1], Sun Li[2], Liu Tianwu[2], Pan Jin[2]

(1. The State Key Laboratory of Refractories and Metallurgy, Wuhan University of Science and Technology, Wuhan 430081, China; 2. HBIS Group Technology Research Institute, Shijiazhuang 050023, China)

Abstract: During the tempering process, the material with initial residual stress will produce an irreversible plastic strain accompanied by carbide precipitation, which makes the initial residual stress relax. The plastic behavior associated with carbide precipitation is named “Precipitation Plasticity”. Although “Precipitation Plasticity” plays an important role in the control of tempering residual stress, the mechanism of it has not been revealed. This paper focuses on the plastic behavior and microstructure of ultra-low carbon steel with 0.07wt% C during tempering. It is found that the interface vacancy which is formed at the interface between carbide and matrix when carbide precipitates during tempering, diffuses from the interface to the matrix and accumulates in the direction of the stress, resulting in irreversible plastic strain, which is the "Precipitation Plasticity". Moreover, the smaller the carbide size is, the more significant the “Precipitation Plasticit” is. When the size of carbide precipitated during tempering decreases from 14 nm to 11 nm, the precipitation plasticity coefficient K increases from 5.71×10^{-5} to 6.92×10^{-5}, increasing by 21%. Reducing the size of precipitates is beneficial to the control of tempering residual stress.

Key words: tempering; transformation plasticity; carbide precipitation; residual stress

中厚板风电钢轧制工艺对产品组织的试验性分析

马占福

（宝武集团新疆八一钢铁公司，新疆乌鲁木齐 830000）

摘 要：本文通过对风电钢轧制工艺参数加热温度、终轧温度、轧机压下量及加热温度与其产生相应的产品组织对比分析，过高的加热温度造成板坯奥氏体长大，导致轧后钢板晶粒粗大；降低终轧温度，有利于细化晶粒,且有利于提高组织均匀性；增大粗轧阶段道次压下率，同时降低终轧温度，得到产品组织的晶粒度均在9～10级，晶粒度平均提高1级左右。

关键词：风电钢；轧制工艺；组织；晶粒度

Experimental Analysis of Rolling Process on Product Structure of Plate Wind Power Steel

Ma Zhanfu

(Baowu Group, Xinjiang Bayi Iron & Steel Company, Urumqi 830000, China)

Abstract: In this paper, through the comparative analysis of the rolling process parameters of wind power steel, such as heating temperature, finishing temperature, rolling mill reduction and heating temperature, and the corresponding product structure, it is found that too high heating temperature results in austenite growth of slab and coarse grain of rolled steel plate; Decreasing finishing temperature is beneficial to grain refinement and microstructure uniformity; The results show that the grain size of the product is 9-10 grade by increasing the pass reduction rate in rough rolling stage and decreasing the finishing rolling temperature, and the grain size increases about one grade on average.

Key words: wind power steel; rolling process; microstructure; grain size

EDT 毛化效率和质量提升的研究

宫聚文，张立明

（本钢浦项冷轧薄板有限责任公司，辽宁本溪 117000）

摘 要：本文揭示了EDT电火花设备的工作原理。阐述了轧辊磨削后表面质量缺陷对毛化后表面质量的影响，说明了EDT设备自身影响毛化质量的原因及解决方法。用大量实验证明了各参数对轧辊表面的影响，通过优化各参数和改进工艺流程提高EDT打毛效率及提高轧辊表面毛化质量。从而保证各机组充足的供辊数量及为各机组生产高等级面板打下基础。

关键词：EDT电火花；毛化效率；粗糙度；PC值

Research on Improving EDT Texturing Efficiency and Improving Roll Texturing Quality

Gong Juwen, Zhang Liming

(Bx Steel Posco Cold-rolled Sheet Co., Ltd., Benxi 117000, China)

Abstract: This article reveals the working principle of EDT electric spark equipment.This paper expounds the influence of the surface quality defects after grinding on the surface quality after texturing, and explains the reasons and solutions of the influence of the EDT equipment itself on the texturing quality. A large number of experiments have proved the influence of various parameters on the roll surface. By optimizing each parameter and improving the process flow, the EDT roughening efficiency and the roll surface roughening quality are improved. In order to ensure the sufficient number of rollers for each unit and lay the foundation for the production of high-grade panels for each unit.

Key words: EDT electric spark; processing efficiency; roughness; PC value

减轻轧辊磨削划伤的方法

宫聚文，张立明

（本钢浦项冷轧薄板有限责任公司，辽宁本溪　117000）

摘　要：本文主要分析了轧辊磨削划伤的影响，轧辊磨削划伤的形成过程，经过长期跟踪观察寻找磨削划伤产生的因素，对磨床程序的各项参数进行试验分析，对各影响磨削划伤问题提出有效的解决措施，最终通过优化磨削程序、砂轮优选、磨削液洁净等确保轧辊磨削划伤减轻。通过后期大量磨削跟踪发现，轧辊磨削划伤较原来有较大改善，为稳定机组产品表面质量提供有效保障。

关键词：轧辊；划伤；砂轮；磨削；磨削液

Method to Reduce Grinding Scratches on Rolls

Gong Juwen, Zhang Liming

(Bx Steel Posco Cold-rolled Sheet Co., Ltd., Benxi 117000, China)

Abstract: This article mainly analyzes the influence of roll grinding scratches, the formation process of roll grinding scratches, and after long-term follow-up observation to find the factors that cause grinding scratches, test analysis of various parameters of the grinder program, and influence on each Effective measures are proposed to solve the problem of grinding scratches. Finally, the grinding process is optimized, the grinding wheel is optimized, and the grinding fluid is clean to ensure that the roll grinding scratches are reduced. Through the subsequent large-scale grinding tracking, it is found that the grinding scratches of the rolls have been greatly improved compared with the original ones, which provides an effective guarantee for the stable surface quality of the unit products.

Key words: roll; scratch; grinding wheel; grinding; grinding fluid

关于中间辊过渡弧磨削工艺改进研究

宫聚文，张立明

（本钢浦项冷轧薄板有限责任公司，辽宁本溪 117000）

摘 要：本文主要阐述了轧机中间辊的作用，中间辊过渡弧的作用，分析现有中间辊过渡弧磨削的方式和缺点。经过学习研究改进中间辊过渡弧磨削工艺，提升磨削效率和安全性。通过实验跟踪分析彻底改变现有工艺的缺点。为机组提供优质的轧辊，为现场工作人员提供舒适的工作方式。

关键词：轧辊；过渡弧；自动上刀；磨削

Research on Improvement of Over-arc Grinding Process of Intermediate Roll Research on Improvement of Over-arc Grinding Process of Intermediate Roll

Gong Juwen, Zhang Liming

(Bx Steel Posco Cold-rolled Sheet Co., Ltd., Benxi 117000, China)

Abstract: This article mainly describes the role of the middle roll of the rolling mill, the role of the excessive arc of the middle roll, and analyzes the existing methods and shortcomings of the excessive arc grinding of the middle roll. After studying and researching, improve the over-arc grinding process of the intermediate roll, and improve the grinding efficiency and safety. Through experimental tracking and analysis, the shortcomings of the existing technology are completely changed. Provide high-quality rolls for the unit, and provide comfortable working methods for on-site staff.

Key words: roll; excessive arc; automatic tool loading; grinding

有限元仿真技术在 DR 材性能改善及生产顺行的应用

康永华，莫志英，刘美丽，童建佳，刘 伟，张宝来

（首钢京唐钢铁联合有限责任公司镀锡板事业部，河北唐山 063200）

摘 要：利用有限元分析技术分析 DR 材用户应用及加工过程、规避连退工序瓢曲风险，在优化材料性能、低成本开发成品、降低试验风险和稳顺生产发挥重要作用。

关键词：有限元仿真；应用分析；瓢曲分析

Application of Finite Element Simulation Technology in DR Product Performance Improvement and Smooth Production

Kang Yonghua, Mo Zhiying, Liu Meili, Tong Jianjia,
Liu Wei, Zhang Baolai

(Tin-plate Business Division, Shougang Jingtang United Iron & Steel Co., Ltd., Tangshan 063200, China)

Abstract: Using finite element analysis technology to analyze user application, processing for DR material and avoiding buckle risk in continuous annealing process, it plays an important role in optimizing material performance, developing products with low cost, reducing test risk and stable production.

Key words: finite element simulation;application analysis; buckle analysis

铜/铝冷轧复合结合界面速度场分析

李晓青，魏立群，付　斌，冯　鑫，徐怀君，徐星星

（上海应用技术大学材料科学与工程学院，上海　201418）

摘　要：基于 MSC.Marc 有限元仿真软件，本文对 1060 纯铝和 TP2 纯铜双金属带冷轧复合界面的速度场进行模拟计算，确定了铜铝复合带在不同厚度配比和不同道次压下率时，其结合界面上的速度场变化与界面复合情况。有限元模拟计算分析表明：随着铜铝复合带厚度配比的增加，界面复合所需的搓动速度差范围变大，当铜带与铝带的厚度比为 1:4 时，在临界道次压下率下界面复合所需的搓动速度差介于 3.1×10^{-2}~1.2×10^{-1}mm/s 之间；当厚度配比为 1:5 时，铜铝复合带界面复合所需的搓动速度差介于 8.4×10^{-2}~4.5×10^{-1}mm/s 之间；当厚度配比为 1:6 时，界面复合所需的搓动速度差介于 9.1×10^{-2}~6.3×10^{-1}mm/s 之间。剥离试验分析表明：在铜带与铝带厚度比分别为 1:4，1:5，1:6 这三种情况下，对应的在临界道次压下率下界面均实现了良好的复合，而且铜铝复合带界面平均剥离强度大小随着道次压下率的增加而随之增加，其复合效果明显提高。

关键词：Marc 有限元；铜铝冷轧复合；界面速度场；速度差

Analysis of Velocity Field in the Interface of Copper/Aluminum Cold-rolled Composite Process

Li Xiaoqing, Wei Liqun, Fu Bin, Feng Xin, Xu Huaijun, Xu Xingxing

(School of Materials Science and Engineering, Shanghai Institute of Technology, Shanghai 201418, China)

Abstract: Based on the MSC.Marc finite element simulation software, this paper simulates the velocity field of the cold-rolled composite bonding interface of 1060 pure aluminum and TP2 pure copper bimetallic strips. The change of the velocity field on the bonding interface and the interface recombination of the copper-aluminum composite strips at different thickness ratios and different degree of reduction in pass are determined. The finite element simulation analysis shows that: with the increase of the thickness ratio of the copper-aluminum composite strip, the range of the rolling speed difference required for the interface composite becomes larger. When ratio of copper strip thickness to aluminum strip thickness was 1to4, the difference in rubbing speed required for interface recombination at the threshold degree of reduction in pass is between 3.1×10^{-2}~1.2×10^{-1}mm/s; When the thickness ratio was 1to5, the difference in rubbing speed required for the composite interface of the copper-aluminum composite strip is between 8.4×10^{-2}~4.5×10^{-1}mm/s; When the thickness ratio was 1to6, the difference in rubbing speed required for interface is between 9.1×10^{-2}~6.3×10^{-1}mm/s. The analysis of the peeling test showed that: in the three cases when ratio of copper strip thickness to aluminum strip thickness was 1to4, 1to5, and 1to6, the corresponding interface achieve a good composite strip at the threshold degree of reduction in pass.

Key words: Marc finite element; copper/aluminum cold-rolled composite; interface velocity field; velocity difference

Cr-Mo-V 系结构钢 CCT 曲线的测定与分析

冯丹竹，范刘群，黄　健，田　斌

（鞍钢股份有限公司，辽宁鞍山　114009）

摘　要：测定了一种 Cr-Mo-V 系合金结构钢的连续冷却转变曲线（CCT 曲线）。结果表明：当冷却速率为 0.2~0.5℃/s 时，该钢冷却至室温仅发生铁素体/珠光体相变；冷却速率增大到 1℃/s 时，先发生铁素体/珠光体相变，随后发生贝氏体转变；冷却速率达到 2℃/s 时，铁素体、贝氏体共存；冷速为 10~30℃/s 时，只存在贝氏体组织；冷却速率达到 50℃/s 时，获得少量马氏体组织。动态 CCT 曲线的测定为该钢种生产中控制轧制工艺和控制冷却工艺的制定提供理论依据。

关键词：Cr-Mo-V 系结构钢；CCT 曲线；控制冷却

Measurement and Analysis of CCT Curve of Cr-Mo-V Structural Steel

Feng Danzhu, Fan Liuqun, Huang Jian, Tian Bin

(Angang Steel Company Limited, Anshan 114009, China)

Abstract: The CCT curve of Cr-Mo-V alloy structural steel was measured. The results show that when the cooling rate is 0.2-0.5℃/s, only ferrite / pearlite phase transformation occurs when the steel is cooled to room temperature; When the cooling rate increases to 1℃/s, ferrite / pearlite transformation occurs first, and then bainite transformation occurs; When the cooling rate reaches 2℃/s, ferrite and bainite coexist; When the cooling rate is 10-30℃/s, only bainite exists; When the cooling rate reaches 50℃/s, a small amount of martensite is obtained. The measurement of dynamic CCT curve provides a theoretical basis for the formulation of controlled rolling process and controlled cooling process.

Key words: Cr-Mo-V structural steel; CCT curve; controlled cooling

关于单机架调试生产问题分析与质量改进

赵　刚，曹　垒，游　涌，张　杰

（张家港扬子江冷轧板有限公司，江苏张家港　215625）

摘　要：单机架机组作为沙钢冷轧中高牌号硅钢生产的核心机组之一，机组的连续稳定生产，对沙钢冷轧硅钢产品提升市场竞争力和占有率，具有重大意义。随着冷轧市场对中高牌号无取向硅钢需求的快速增长，硅钢正在往高质量、高磁感、低铁损要求发展，然而硅钢牌号越高，由于对性能要求越高，高牌号硅钢通常是高硅、高铝，且常化温度较高，轧制难度增加。另外，单机架机组是沙钢冷轧首次建设投产，很多技术和管理问题需要攻关解决，从调试情况看，遇到的技术问题，都是逐一制定措施攻关解决，单机架机组的产量也在稳步提升。本文重点分析单机架机组从调试生产，到质量稳定，再到产量提升的过程中，对每个出现的较大质量问题，如何发现攻关并解决。

关键词：单机架机组；中高牌号硅钢；调试生产；质量问题；解决

Analysis of Production Problems and Quality Improvement of Single Stand Commissioning

Zhao Gang, Cao Lei, You Yong, Zhang Jie

(Zhangjiagang Yangtze River Cold Rolled Plate Co., Ltd., Zhangjiagang 215625, China)

Abstract: As one of the core units in the production of medium and high grade silicon steel in cold rolling of Shagang, the continuous and stable production of stand-alone unit is of great significance to improve the market competitiveness and share of cold rolling silicon steel products of Shagang. With the rapid growth of demand for medium and high grade non oriented silicon steel in cold rolling market, silicon steel is developing towards high quality, high magnetic induction and low iron loss requirements. However, the higher the grade of silicon steel, the higher the performance requirements, the higher the grade of silicon steel is usually high silicon and high aluminum, and the higher the normalizing temperature, the more difficult the rolling. In addition, the stand-alone unit is the first time that Shagang cold rolling mill has been built and put into operation. Many technical and management problems need to be solved. From the commissioning situation, the technical problems encountered are all solved by formulating measures one by one, and the output of stand-alone unit is steadily increasing. This paper focuses on the analysis of the single stand unit from commissioning production, to quality stability, and then to the process of output improvement, how to find and solve the major quality problems of each unit.

Key words: single stand unit; medium and high grade silicon steel; commissioning production; quality problem; solve

退火方式对热镀锌微碳铝镇静钢组织性能影响

陈泓业，杨　平，王　滕，李伟钢，张百勇，
李　超，孙　霖，李志庆

（马鞍山钢铁股份有限公司，安徽马鞍山　243000）

摘　要：采用 CAG-III 热浸镀锌模拟试验机对碳含量百分数低于 0.04%的微碳铝镇静钢在不同退火方式模拟退火后的组织、性能以及抗时效性能进行对比研究。研究结果表明：采用美钢联法立式退火炉比改良森吉米尔法卧式退火炉生产热镀锌微碳铝镇静钢晶粒尺寸更粗大，具有更低的屈服强度，更优异的成型性能；而采用改良森吉米尔法卧式退火炉生产热镀锌微碳铝镇静钢的抗时效性能更加优异，更宜钢卷的长时间存放。

关键词：退火方式；微碳铝镇静钢；性能

Effect of Annealing Method on Microstructure and Properties of Micro Carbon Al Killed Hot Dip Galvanized Steel

Chen Hongye, Yang Ping, Wang Teng, Li Weigang, Zhang Baiyong, Li Chao, Sun Lin, Li Zhiqing

(Ma'anshan Iron and Steel Co., Ltd., Ma'anshan 243000, China)

Abstract: the microstructure, properties and aging resistance of micro carbon Al killed steel with carbon content less than 0.04% after simulated annealing with different annealing methods were studied by cag-iii hot dip galvanizing simulator. The results show that the grain size, yield strength and formability of micro carbon Al killed hot dip galvanized steel produced by vertical annealing furnace of American Steel United process are larger than those produced by horizontal annealing furnace of improved Sendzimir process; However, the aging resistance of hot-dip galvanized micro carbon Al killed steel produced by improved Sendzimir horizontal annealing furnace is better, and it is more suitable for long-term storage of steel coil.

Key words: Annealing method; micro carbon Al killed steel; mechanical properties

新钢连退炉低温区钢带跑偏分析

黄海生

（新余钢铁集团有限责任卷板厂，江西新余 338001）

摘 要：分析生产厚度小于 0.5mm 薄规格钢带在炉内均热段之后低温区跑偏原因，并提出解决方案。缓冷段钢带横断面冷却不均，并导致炉辊辊面温度不均，引发钢带在缓冷段跑偏；快冷段炉顶辊电辐射管加热器失控，炉辊辊面与钢带接触区域温度较非接触区温度低，炉辊凸度下降，炉辊自纠偏能力下降，导致钢带在快冷段跑偏；过时效钢带温度低，炉辊凸度下降，自纠偏能力下降，导致钢带在过时效 2 段及终冷段跑偏。

关键词：连退炉；钢带跑偏；分析

冷却工艺对级热镀锌 DP780 性能及表面的影响

富丰晶，刘宏亮，付东贺，周 航，闵 铜

（本钢集团有限公司研发院，辽宁本溪 117000）

摘 要：主要针对热镀锌生产过程中不同冷却工艺对热镀锌双相钢的性能和表面质量的影响进行了研究，分别采用了镀前形成马氏体及镀后形成马氏体两种工艺，设定了不同的快冷模式，得到不同冷却工艺条件下的组织、性能及表面情况。研究表明，当冷却工艺选择镀前形成马氏体时，采用快冷至马氏体形成温度以下，后感应加热至锌锅温度后进行镀锌处理，材料力学性能降低，但延伸率较好，镀后形成马氏体时采用中温转变工艺，在中温转变区，残余奥氏体中会富集碳、锰等合金元素，提高残余奥氏体淬透性，使材料在镀后冷却过程中获得更多的马氏体，同时铁素体的纯净性提高，得到更加良好的力学性能；但快冷温度超过490℃时，会造成锌锅温度增高，造成表面锌灰、锌渣等缺陷，所以快冷温度在470±5℃范围内，更有利于材料满足标准要求，同时具有较好的表面质量。

关键词： 冷却工艺；热镀锌双相钢；马氏体；淬透性

Effect of Rapid Cooling Temperature on Properties and Surface Quality of Hot Dip Galvanized DP780

Fu Yujing, Liu Hongliang, Fu Donghe, Zhou Hang, Min Tong

(Technology Research Institute of Benxi Steel Plate Co., Ltd., Benxi 117000, China)

Abstract: Effect of different cooling process on properties and surface quality of hot dip galvanized dual phase steel was researched. Mechanical properties, microstructure and surface analysis of the samples with different process parameters were analyzed by setting different rapid cooling outlet temperature. The results show that with the increase of the outlet temperature, elements such as carbon and manganese in ferrite diffuse to the grain of austenite to improve the harden ability of austenite and get more martensite, so the good mechanical properties can be obtained; but when the rapid cooling outlet temperature exceeds 490℃, it can lead to higher temperature of zinc pot causing surface defects such as zinc vapour ash, zinc dross. So the rapid cooling outlet temperature should be controlled at 470±5℃, the preferable surface quality can be obtained as well as the good mechanical properties.

Key words: cooling process; hot dip galvanized dual phase steel; martensite; harden ability

厚板轧制不锈钢头部形状改善研究

张敏文

（宝钢股份厚板部，上海　201900）

摘　要： 本文研究了厚板轧制过程中不锈钢头部形状规律，并分析了其与普碳钢的差异，同时给出适合不锈钢的展宽MAS设定值，为成材率提升做基础性工作。

关键词： 厚板轧制；不锈钢；头部形状

Research on Head Shape Control during Heavy Plate Rolling of Stainless Steel

Zhang Minwen

(Heavy Plate Mill Baosteel, Shanghai 201900, China)

Abstract: Research on head deformation of stainless steel, which is different from that of carbon steel, has been introduced in the paper, broadside rolling MAS parameters have been given as basic work for the yield.

Key words: heavy plate rolling; stainless steel; head shape

热连轧带钢卷取后冷却过程中残余应力分析

杨晓臻[1]，杨宴宾[1]，赵志毅[2]，白东亮[2]

（1. 宝钢股份有限公司热轧厂，上海　200000；
2. 北京科技大学材料科学与工程学院，北京　100083）

摘　要：热连轧带钢在精轧出口处检测到的平坦度不代表最终产品的平坦度，热轧高强钢冷却至室温后，易产生板形缺陷。热轧带钢轧后冷却过程存在温度、相变、应力耦合的变化，三者共同影响带钢残余应力的分布。本文利用有限元建立了热轧带钢轧后冷却模型，对带钢冷却层流冷却过程以及卷后冷却过程的应力演变与分布进行研究。研究表明，带钢横向的温度梯度和相变差异是影响应力变化的主要因素，对于厚度为5mm的带钢，在不考虑初始应力的状态下，冷却至室温后，带钢中部为16MPa的轧向拉应力，边部为-196MPa的轧向压应力，这种应力状态会导致带钢产生边浪。采用70I的微中浪控制可以使边部的压应力减小96MPa，减小比例为48.98%。

关键词：热轧带钢；钢卷冷却；有限元；残余应力

Analysis of Residual Stress in Hot-rolled Steel Strip during Cooling after Coiling

Yang Xiaozhen[1], Yang Yanbin[1], Zhao Zhiyi[2], Bai Dongliang[2]

(1. Baoshan Iron & Steel Co., Ltd., Shanghai 200000, China; 2. School of Material Science and Engineering, University of Science and Technology Beijing, Beijing 100083, China)

Abstract: Flatness detected at the exit of finishing rolling of hot rolled steel strip does not represent the flatness of the final product. when the hot-rolled high-strength steel is cooled to room temperature. There are coupling changes of temperature, phase transformation and stress in the cooling process of hot rolled strip, which together affect the distribution of residual stress of hot rolled strip.the cooling model of hot rolled strip after rolling is established by using finite element method, and the stress change of strip cooling layer flow cooling process and post-roll cooling process is studied in this paper. The research shows that the temperature gradient and phase change of the strip are the main factors affecting the stress change. For the strip with a thickness of 5mm, the middle part of the strip is 16MPa after cooling to room temperature without considering the initial stress. Rolling tensile stress, the edge is -196MPa rolling compressive stress, this stress state will cause the strip to generate side waves. The 70I micro-wave control can reduce the compressive stress of the edge by 96.52MPa, and the reduction ratio is 48.98%.

Key words: hot-rolled steel strip; coiling cooling; the FEA; residual stress

双蓄热式加热炉燃烧控制技术研究

包薪群

（攀钢集团西昌钢钒有限公司板材厂，四川西昌　615000）

摘　要： 攀钢集团西昌钢钒公司公司定位于高档汽车用钢生产基地，公司在进行建设之初，通过考察研究建设国内首批双蓄热式加热炉群。相比于常规燃烧和单蓄热燃烧，双蓄热体对高温烟气进行热量回收，并对冷空气和冷煤气同时进行预热，蓄热式加热炉的热效率可以达到70%以上，比传统加热炉高20%～30%。在使用初期，双加热炉热效率仅达到50%，通过与国内高校联合研究，基于蓄热式燃烧技术特点，研究双蓄热式加热炉燃烧控制技术，实现了蓄热式加热炉的高效节能效果。

关键词： 加热炉；双蓄热；燃烧控制；节能

Research on Combustion Control Technology of Double Regenerative Heating Furnace

Bao Xinqun

(Panzhihua Iron and Steel Group Xichang Steel and Vanadium Co., Ltd. Plate Plant, Xichang 615000, China)

Abstract: Panzhihua Iron and Steel Group Xichang Steel and Vanadium Company is positioned as a high-end automotive steel production base. Automotive panels require high surface quality. At the beginning of the construction, the company built the first batch of dual regenerative heating furnace groups in China through investigation and research. Compared with conventional combustion and single regenerative combustion, dual regenerators recover heat from high-temperature flue gas and preheat cold air and cold gas at the same time. The thermal efficiency of the regenerative heating furnace can reach more than 70%, which is higher than the traditional The heating furnace is 20%~30% high. In the initial stage of use, the thermal efficiency of the dual heating furnace only reached 50%, and the unit gas consumption was relatively high. Through joint research with domestic universities and based on the characteristics of regenerative combustion technology, the dual regenerative heating furnace combustion control technology was studied to achieve heat storage. The high-efficiency and energy-saving effect of the type heating furnace.

Key words: heating furnace; double heat storage; energy saving; thermal efficiency

屈服平台对冰箱门圆弧起棱缺陷的影响

杨士弘[1]，张　鹏[1]，杨　婷[1]，赵秀娟[2]

（1. 河钢集团钢研总院，河北石家庄　050023；2. 青岛河钢新材有限公司，山东青岛　266000）

摘　要： 低碳铝镇静钢是在家电领域应用最为广泛的一类钢种。但材料的拉伸曲线存在屈服平台现象，在材料成形过程中发生不均匀变形现象，导致缺陷发生。通过对材料的预变形，可以消除屈服平台，但由于长期存放或高温处理，均将导致屈服平台再度出现。采用彩涂板加工冰箱门板圆弧时，易于发生圆弧外侧出现起棱的缺陷。

关键词： 家电用钢；彩钢板；圆弧起棱；屈服平台

Effect of Yield Platform on Circular Arc Edge Forming Defect of Refrigerator Door

Yang Shihong[1], Zhang Peng[1], Yang Ting[1], Zhao Xiujuan[2]

(1. HBIS Techology Research Institute, Shijiazhuang 050023, China;
2. HBIS New Material, Qingdao 266000, China)

Abstract: Low carbon aluminum killed steel is the most widely used steel in the field of household appliances. However, the tensile curve of the material has the phenomenon of yield platform, which leads to uneven deformation defects in the process of material forming. The yield platform can be eliminated through the pre deformation of the material,, but due to long-term storage or high-temperature treatment, the yield platform will appear again. When using color coated plate to process the circular arc of refrigerator door plate, it is easy to have the defect of edge on the outside of the circular arc.

Key words: steel for household appliances; color steel plate; arc defect; yield platform

家电用钢的现状和发展趋势

杨士弘[1]，张　鹏[1]，杨　婷[1]，邹炎斌[2]

（1. 河钢集团钢研总院，河北石家庄　050023；2. 青岛河钢新材有限公司，山东青岛　266000）

摘　要：家电行业是钢材的主要高端下游行业之一。家电用钢相对于汽车用钢，钢种较少，以低碳、超低碳为主。通过不同的表面处理工艺，生产的镀层钢板、彩涂钢板等为家电的外观设计、环保等发展创造了条件。成熟应用于汽车行业的钢种，高强 IF 钢、高强低合金钢和锌铝镁镀层产品已经和正在引入家电制造。

关键词：家电用钢；彩钢板；涂镀产品；高强低合金钢

Present Situation and Development Trend of Steel for Household Appliances

Yang Shihong[1], Zhang Peng[1], Yang Ting[1], Zou Yanbin[2]

(1. HBIS Techology Research Institute, Shijiazhuang, 050023, China;
2. HBIS New Material, Qingdao 266000, China)

Abstract: Household appliance industry is one of the main high-end downstream industries of steel. Compared with automotive steel, steel grades for household appliances is less, mainly low-carbon and ultra-low-carbon. Through different surface treatment processes, the coated steel and color coated steel have created conditions for the development of appearance design and environmental protection of household appliances. Steel grades used in the automotive industry, high strength IF steel, HSLA steel and ZM coating products have been and are being introduced into household appliance manufacturing

Key words: steel for household appliances; color steel plate; coated products; high strength low alloy steel

热带无头轧制技术的进展与优势

田　鹏，康永林，梁晓慧，朱国明

（北京科技大学材料科学与工程学院，北京　100083）

摘　要：根据目前碳达峰碳中和的形势要求，对热带无头轧制技术的发展历程和现状进行了分析研究，通过对不同无头轧制产线的对比，总结分析了无头轧制在极限规格生产、高尺寸形状精度、低成本、节能减排等方面的优势，提出发展热带无头轧制技术的建议。

关键词：热带；无头轧制；低成本；节能减排

Progress and Advantages of Endless Hot Strip Rolling Technology

Tian Peng, Kang Yonglin, Liang Xiaohui, Zhu Guoming

(School of Materials Science and Engineering, University of Science and Technology Beijing, Beijing 100083, China)

Abstract: according to the current situation of carbon reaching peak and carbon neutralization, the development history and current situation of endless hot strip rolling technology are analyzed. Through the comparison of different endless rolling production lines, the advantages of endless rolling technology such as limited specification production, high precision of dimensional and shape, low cost, energy saving and emission reduction are summarized and analyzed, and the suggestions for developing endless hot strip rolling technology are put forward.

Key words: hot strip; endless rolling; low cost; energy saving and emission reduction

基于连续退火的 T3-IF 镀锡基板产品开发

黄　勇，周　密，刘晓锋，李学磊，李思全

（宝武集团武钢日铁镀锡板（武汉）有限公司，湖北武汉　430080）

摘　要：本文以基于我司连续退火工艺开发 T3-IF 镀锡基板新产品为研究对象。对钢种成分、热轧工艺、冷轧工艺、连退工艺进行了合理设计，成功开发出 T3CA-IF 镀锡基板新产品。

关键词：IF 钢；超低碳钢；连续退火；镀锡板

Development of T3-IF TMBP based on Continuous Annealin

Huang Yong, Zhou Mi, Liu Xiaofeng, Li Xuelei, Li Siquan

(Baowu Steel Grupo WISCO-NIPPON Steel Tinplate Co., Ltd., Wuhan 430080, China)

Abstract: The research object of this paper is to develop T3-IF tinplate based on our continuous annealing process. The composition of steel grade, hot rolling process, cold rolling process and continuous annealing process were reasonably designed. A new product of T3CA-IF tinplate was successfully developed.

Key words: IF steel; ultra-low carbon steel; continuous annealing; TMBP

首钢京唐公司镀锌高强度汽车板专用线关键工艺技术特点分析与研究

何云飞[1,3]，周玉林[1,3]，薛红金[2]，侯俊达[1,3]，张 征[1,3]，张乐峰[1,3]

（1. 北京首钢国际工程技术有限公司，北京 100043；2. 首钢京唐钢铁联合有限责任公司，河北唐山 063200；3. 北京市冶金三维仿真设计工程技术研究中心，北京 100043）

摘 要：本文探讨了首钢京唐镀锌高强度汽车板专用生产线的产品特点，关键工艺技术及设备，总体布置，产线生产实践及其发展前景等，有利于推广高强镀锌汽车板生产技术，对新建高强汽车板镀锌线的设备选型、先进技术采用与产品质量的提高具有借鉴意义。

关键词：工艺技术；纳米涂层；预氧化及还原；感应加热；超高强汽车板

Analysis and Research on the Key Processing Technology Characteristics of SGJT UHSS CGL

He Yunfei[1,3], Zhou Yulin[1,3], Xue Hongjin[2], Hou Junda[1,3], Zhang Zheng[1,3], Zhang Lefeng[1,3]

(1. Beijing Shougang International Engineering Technology Limited Corporation，Beijing 100043, China; 2. Shougang Jingtang Iron and Steel Co., Tangshan 063200, China; 3. Metallurgical Engineering 3-D Simulation Design Engineering Technology Research Center of Beijing, Beijing 100043, China)

Abstract: Product characteristics were discussed in this article, and it also introduced the key technologies and equipment, together with the general layout and actual producing practice, as well as the development prospects of the ultra-high-strength steel dedicated automotive sheet continuous hot-dip galvanizing line of Shougang Jingtang Iron and Steel Co., namely the SGJT UHSS CGL project.

This article has contributed to the popularization of the producing technology of high-strength automotive sheet, and has a research significance for the facility designation, advanced technology adoption, and product quality enhancement of newly built high-strength automotive sheet CGL project.

Key words: processing technology; NANO-coating; pre-oxidation and reduction; inductive heating; ultra-high-strength steel dedicated automotive sheet

新一代高技术轧机电工钢矩形断面板形控制创新研究

曹建国[1,2,3]，宋纯宁[1,2,3]，王雷雷[1,2,3]，赵秋芳[1,2,3]，王彦文[1,2,3]，肖 静[1,2,3]

（1. 北京科技大学机械工程学院，北京 100083；2. 北京科技大学人工智能研究院，北京 100083；3. 北京科技大学国家板带生产先进装备工程技术研究中心，北京 100083）

摘　要：电工钢是支撑国家机电、能源、国防和航空航天发展战略需求的重要软磁合金材料，“Dead flat”矩形断面超平材超高板形质量要求是电工钢等高端板带材板形控制前沿领域和瓶颈难题。本文分析了国际上不断探索的新一代高技术热、冷连轧机机型与板形控制技术特征及其日趋复杂化的难题，结合工业生产流程系统建立了无取向电工钢高温变形本构关系，建立了热、冷连轧机辊件三维有限元仿真模型，提出了适应所有工作辊长/短行程液压窜辊系统的新一代高技术热连轧全板形融合 π（PCFC All-in-one Integrated, PAI）机型、适应无工作辊液压窜辊系统的6辊冷连轧机边降与凸度紧凑 ECC-6（6-high Edge drop and Crown Compact）机型与热-冷连轧全过程一体化板形控制创新技术。和日本 k-WRS 热连轧机与 6 辊 UCM 冷连轧机等国际先进技术相比，结合生产实际投入到我国大型热-冷连轧骨干工业轧机生产应用，取得轧制单位显著扩大条件下边降、凸度和同板差等矩形断面超平材重要板形指标显著提高的生产实绩，为从根本上突破解决新一代高技术热-冷连轧机电工钢矩形断面板形控制瓶颈难题提供创新路径。

关键词：热连轧机；冷连轧机；矩形断面；板形控制

Innovation Research on Rectangular Section for Profile and Flatness Control of Electrical Steel in New-generation High-tech Rolling Mills

Cao Jianguo[1,2,3], Song Chunning[1,2,3],Wang Leilei[1,2,3], Zhao Qiufang[1,2,3], Wang Yanwen[1,2,3], Xiao Jing[1,2,3]

(1.School of Mechanical Engineering, University of Science and Technology Beijing, Beijing 100083, China; 2.Institute of Artificial Intelligence, University of Science and Technology Beijing, Beijing 100083, China; 3.National Engineering Research Center of Flat Rolling Equipment, University of Science and Technology Beijing, Beijing 100083, China)

Abstract: Electrical steel is an important soft magnetic alloy material to support the strategic needs of the development of national electromechanical, energy, national defense and aerospace. The quality requirement of "Dead Flat" rectangular section super-flat material with super-high shape is the frontier field and bottleneck problem of the shape control of high sheet and strip materials such as electrical steel. Based on the analysis of the new generation of high-tech hot and cold tandem rolling mills and the increasingly complicated technical characteristics of the shape control, the high temperature deformation constitutive relation of unoriented electrical steel was established in combination with the industrial production process system, and the three-dimensional finite element simulation model of hot and cold tandem rolling mill was established. A new generation of high-tech hot strip rolling full shape fusion π (PCFC All-in-One Integrated, PAI), 6-high Edge drop and Crown Compact ECC-6 model for work rolls for non-shifting of work rolls system and innovative integrated shape control technology for the whole process of hot and cold tandem rolling. And Japanese k-WRS strip machine with 6-high compared UCM cold tandem mill and other international advanced technology, combined with the actual production into large hot and cold tandem rolling backbone industry in China mill production application, obtains the rolling unit significantly expand the conditions below drop, crown and transverse thickness deviation rectangular super flat products have important shape index significantly improve production performance, it provides an innovative way to fundamentally break through the bottleneck problem of rectangular broken panel shape control of new generation of high technology hot and cold tandem mill electrical steel.

Key words: hot tandem rolling mill; cold tandem rolling mill; rectangular section; shape control

1420UCM 五机架连轧机振动抑制措施研究

黄 勇，龚 艺
（宝武集团武钢日铁镀锡板有限公司，湖北武汉 430083）

摘 要：通过对 FAT 期间轧机振动工艺调整过程的总结，发现 1420UCM 连轧机振动的特点，其振动主要发生在第四机架的辊缝处；并从通过提高辊缝润滑，降低轧制力可以抑制轧机振动，使得轧制镀锡板 T5 牌号速度最高可达 2000m/min。

关键词：轧机振动；辊缝润滑；UCM 轧机

Analyzing Reason and Controlling Measure of the Vibration in 1420UCM Cold Continuous Rolling Mill

Huang Yong, Gong Yi
(Baowu Steel Group WISCO-NIPPON Steel Tinplate Co., Ltd., Wuhan 430083, China)

Abstract: Summarizing the process of rolling mill vibration process adjustment during fat, it is found that the vibration characteristics of 1420ucm continuous rolling mill mainly occur at the roll gap of the fourth stand; The rolling mill vibration can be restrained by improving the roll gap lubrication and reducing the rolling force, so that the maximum speed of rolling T5 tinplate can reach 2000m/min.

Key words: rolling mill vibration; roll gap lubrication; UCM rolling mill

Mn13 高锰钢 TTT 曲线的测定与相变研究

黄冠勇[1]，刘建华[1]，何 杨[1]，肖爱达[2]，刘 宁[2]
（1. 北京科技大学工程技术研究院，北京 100083；
2. 涟源集团钢铁有限公司，湖南娄底 417009）

摘 要：采用热模拟试验、金相法以及 X 射线分析方法，测定了 Mn13 高锰钢静态 TTT 曲线以及研究 Mn13 高锰钢的相变规律。结果表明：在 400℃保温 7h 后，奥氏体基体开始析出针状碳化物；而在 450℃保温 0.8h 后，除了有针状碳化物之外，还有珠光体组织出现。在 TTT 曲线“鼻尖温度”575℃时效，碳化物和珠光体析出速度最快，仅需 0.2h。随着时效温度的升高，碳化物和珠光体逐渐回融到奥氏体中，当时效温度为 800℃时，奥氏体基体上的碳化物和珠光体完全消失。

关键词：高锰钢；热模拟；TTT 曲线；相变规律

Determination of TTT Curve and Study on Phase Transformation of Mn13 High Manganese Steel

Huang Guanyong[1], Liu Jianhua[1], He Yang[1], Xiao Aida[2], Liu Ning[2]

(1. Institute of Engineering Technology, University of Science and Technology Beijing, Beijing 100083, China; 2. Lianyuan Group Iron and Steel Co., Ltd., Loudi 417009, China)

Abstract: Using thermal simulation test, metallographic method and X-ray analysis method, the static TTT curve of Mn13 high manganese steel was measured and the phase transformation law of Mn13 high manganese steel was studied. The results show that after holding at 400℃ for 7 hours, the austenite matrix begins to precipitate needle-like carbides. After being kept at 450℃ for 0.8h, in addition to needle-like carbides, pearlite structures appeared. When the TTT curve "nose tip temperature" is effective at 575℃, the precipitation rate of carbide and pearlite is the fastest, only 0.2h. As the aging temperature increases, carbides and pearlite gradually melt into austenite. When the aging temperature is 800℃, the carbides and pearlite on the austenite matrix disappear completely.

Key words: high manganese steel; thermal simulation test; TTT curve; phase transition law

高强钢宽度异常分析及对策

肖 鹏

（河钢集团承钢公司 技术质量部，河北承德 067101）

摘 要：随着板带材市场竞争日益激烈，产品质量已经成为产线立足市场的基础。板带材的尺寸指标直接关系到用户产品质量和效益。本研究主要针对板带材宽度指标进行优化，重点解决粗轧宽度检测正常但经过精轧机轧制后成品宽度超宽的问题，最终给热轧用户及后续的工序创造更好的生产条件。

关键词：粗轧；宽度；精轧；立辊；减宽量

Analysis and Countermeasure of Abnormal Width of High Strength Steel

Xiao Peng

(Technical Quality Department, Chenggang Steel Company, Hegang Group, Chengde 067101, China)

Abstract: With the increasingly fierce competition in the plate and strip market, product quality has become the basis of the production line based on the market. The size index of plate and strip is directly related to the product quality and benefit of users. In this study, the width index of plate and strip was optimized, and the problem of normal rough rolling width detection but super wide finished product width after finishing mill rolling was mainly solved, so as to finally create better production conditions for hot rolling users and subsequent processes.

Key words: rough rolling; width; finish rolling; edging roll; width reduction of vertical roller

混凝土输送管用轧制复合板的组织与性能研究

吴真权[1]，温东辉[2]，闫 博[2]

（1. 宝钢股份热轧厂，上海 201900，2. 宝钢股份研究院，上海 201900）

摘 要：耐磨钢板市场应用广泛，其中在建筑行业，混凝土输送管作为重要部件需要频繁更换。目前常规使用的混凝土输送管采用机械套管方式，内管采用耐磨钢满足耐磨的功能性要求，外部采用普碳钢满足结构性要求。但此种方式下内管与外管之间的间隙会加剧内管的振动，导致磨损脱落，耐用性差，使用寿命低。采用轧制复合技术，开发了混凝土输送管用复合板，既能够满足耐磨性的要求，又能够具有良好的力学性能，满足制管、焊接等加工要求。极大地提升了输送管的稳定性与使用寿命。论文通过金相观测、拉伸、冷弯等力学性能测试，以及热处理实验，对混凝土输送管用轧制复合板的组织、拉伸性能、冷弯性能进行了分析讨论，复合界面结合良好，耐磨层硬度满足要求，加工性能优异。

关键词：复合轧制；耐磨；组织；力学性能

Study on Microstructure and Properties of Rolled Composite Plate for Concrete Pipeline

Wu Zhenquan[1], Wen Donghui[2], Yan Bo[2]

(1. Baosteel Hot Rolling Plant, Shanghai 201900, China;
2. Research Institute of Baosteel, Shanghai 201900, China)

Abstract: Wear-resistant steel sheet is widely used in the market. In the construction industry, concrete pipe as an important component needs to be replaced frequently. At present, the conventional concrete conveyor pipe adopts mechanical casing, the inner pipe uses wear-resistant steel to meet the functional requirements of wear resistance, and the outer pipe uses plain carbon steel to meet the structural requirements. But in this way, the gap between the inner tube and the outer tube will aggravate the vibration of the inner tube, resulting in wear and tear, poor durability and low service life. Using rolling composite technology, a composite plate for concrete conveyor pipes was developed, which not only meets the requirements of wear resistance, but also has good mechanical properties and meets the processing requirements of pipe making and welding. It greatly improves the stability and service life of the conveyor pipe. Through metallographic observation, tensile and cold bending tests, and heat treatment experiments, the structure, tensile properties and cold bending properties of the rolled composite plate for concrete pipe are analyzed and discussed. The composite interface is well bonded, the hardness of the wear resistant layer meets the requirements and the processing performance is excellent.

Key words: composite rolling; wear resistance; structure; mechanical properties

提高轧辊粗糙度精度的方法

宫聚文，张立明

（本钢浦项冷轧薄板有限责任公司，辽宁本溪 117000）

摘　要：本文主要分析了砂轮直径变化对轧辊粗糙度精度的影响，并针对其它关键因素进行跟踪记录，对磨床程序的各项参数进行试验分析，对各影响粗糙度精度的问题提出有效的解决措施，最终通过优化磨削程序确保轧辊磨削的粗糙度精度。通过后期大量磨削跟踪发现，轧辊粗糙度精度较原来有较大提高，为稳定机组产品表面精度提供有效保障。
关键词：轧辊；粗糙度；砂轮；直径；磨削

Methods to Improve the Accuracy of Roll Roughness

Gong Juwen, Zhang Liming

(Bengang Pohang Heading Steel Co., Ltd., Benxi 117000, China)

Abstract: This article mainly analyzes the influence of the change of the grinding wheel diameter on the accuracy of the roll roughness, and tracks and records other key factors, conducts an experimental analysis on the parameters of the grinder program, and proposes effective solutions to the problems that affect the roughness accuracy. , And finally ensure the roughness accuracy of roll grinding by optimizing the grinding program. Through a large number of grinding tracking in the later period, it is found that the roughness accuracy of the roll has been greatly improved compared with the original, which provides an effective guarantee for stabilizing the surface accuracy of the product of the unit.
Key words: roll; roughness; grinding wheel; diameter; grinding

酸洗轧机联合机组色差缺陷研究与分析

张彦雨，牟信博

（本钢浦项冷轧薄板有限责任公司，辽宁本溪　117000）

摘　要：随着高端汽车板市场需求量不断提升，其表面质量要求逐渐提高，为满足高端汽车板的高质量要求，我们通过调整酸洗速度、pH 值、喷嘴喷梁清洗和添加钝化剂，加强酸洗漂洗效果，同时调整乳化液浓度、喷嘴管理、流量控制、机架清洗，最终达到保证高端汽车面板表面质量的目的。
关键词：酸洗速度；pH 值；喷嘴喷梁；钝化剂；乳化液浓度；高端汽车板质量

Research and Analysis of Color Difference Defects in Pickling Mill Combined Unit

Zhang Yanyu, Mu Xinbo

(Bx Steel Posco Cold Rolled Sheet Co., Ltd., Benxi 117000, China)

Abstract: With the increasing market demand of high-end automobile panels, their surface quality requirements are gradually improved. In order to meet the high quality requirements of high-end automobile panels, we strengthen the pickling and rinsing effect by adjusting pickling speed, PH value, nozzle spray beam cleaning and adding passivator, and at the same time adjust emulsion concentration, nozzle management, flow control and frame cleaning, finally achieving the purpose of ensuring the surface quality of high-end automobile panels.
Key words: pickling speed；pH value；nozzle spray beam；passivator；emulsion concentration；high-end automobile plate quality

热轧酸洗结构钢黄斑缺陷的分析与控制

隋鹏飞，张 倩
（本钢板材股份有限公司冷轧总厂，辽宁本溪 117000）

摘 要：热轧酸洗板黄斑缺陷在国内酸洗机组结构钢生产中属于特性缺陷，通过自主研究，从钢种成份及前部工艺特点对应黄斑产生理论机理分析，反向研究钢种不同族群的工艺特性，结合工艺优化方向建立酸洗模型，可规避发生黄斑缺陷的临界条件，实现预先工艺控制模型。

关键词：酸洗板；黄斑；酸洗模型；预先控制

Analysis and Control of Corrosion Defect in Hot Rolled Pickling Structural Steel

Sui Pengfei, Zhang Qian
(BenGang Steel Plates Co.,Ltd.The Cold Rolling Mill, Benxi 117000, China)

Abstract: The corrosion defect belongs to characteristic defect in domestic pickling line. Through independent research and analysis in cold Rolling Mill, from the analysis of the theoretical mechanism of corrosion formation corresponding to the composition and front process characteristics of steel grades, the process characteristics of different groups of steel grades are inversely studied, and the pickling model is established in combination with the process optimization direction, so as to avoid the critical conditions for the occurrence of corrosion defects and realize the quality control Pre process control model.

Key words: pickling plate; corrosion; pickling model; pre control

基于不同连续退火机组 IF 钢性能同质化的工艺研究

张 倩，刘晓峰，崔 勇
（本钢板材股份有限公司冷轧总厂，辽宁本溪 117000）

摘 要：根据本钢两条连退线生产的 DC06 钢种的存在的性能差异问题，对 IF 钢性能的影响因素进行了分析并对 2 号连退进行降温及提高平整延伸率的试验以提高其强度。在保证产能的前提下，通过对 2 号连退退火温度降低 25 ℃及平整延伸率提高 0.1%的调整后，在实现了两条连退线 DC06 性能同质的同时又节约了能耗，降低了成本，提高了市场竞争力。

关键词：IF 钢；退火温度；平整延伸率；性能同质化

Study on Homogenization Process of IF Steel Echanical Properties based on Different Continuous Annealing Lines

Zhang Qian, Liu Xiaofeng, Cui Yong

(BenGang Steel Plates Co., Ltd. The Cold Rolling Mill, Benxi 117000, China)

Abstract: According to the echanical Properties difference of DC06 produced by two CAL of BenGang, the influencing factors of IF steel performance are analyzed and the tests of cooling down and increasing flat elongation of 2#CAL are carried out to enhance the strength of DC06. On the premise of ensuring the capacity, after temperature reducing by 25℃ and 0.1% increase of flat elongation, the DC06 of two CAL come to Mechanical Properties homogeneity, while saving energy consumption, reducing cost and improving market competitiveness.

Key words: IF steel; annealing temperature; elongation; mechanical properties homogenization

热轧高强钢板形质量控制研究

孙丽荣[1,2]，班晓阳[2]，王国栋[3]，万佳峰[2]，张吉庆[2]，王　峰[2]

（1. 东北大学轧制技术及连轧自动化国家重点实验室，辽宁沈阳　110000；2. 山东钢铁集团日照有限公司，山东日照　276800；3. 东北大学，辽宁沈阳　110000）

摘　要：板形质量是板带材的质量的重要方面，也是轧制领域关注的核心问题。板形质量控制水平不仅直接关系到板带产品质量，同时也是钢铁企业的轧制技术、装备及生产管理水平的重要标志。板带产品板形问题直接影响后工序及用户生产，是轧制领域关注的重点问题。高端板带材质量是钢铁企业生产能力的重要标志，关系到产品市场拓展和用户口碑。

高强钢由于强度较高，轧制过程容易产生不均匀变形，切割后内应力释放，隐形板形缺陷显现，对轧线板形控制带来了极大难度。由于传统的板形控制理论以机型和辊系为核心，较少考虑具体材料热轧过程塑性变形对板形的影响，而各类材料高温变形过程有其自身的特点，相关材料本构模型、高温相变规律没有完整可靠的文献资料。

为解决长期高强钢板形质量不高的问题，通过深入的现场调查，在统计测试、理论和试验研究的基础上，进行仿真模拟，揭示板形缺陷的原因，针对典型产品及缺陷，制定板形控制标准，修改模型参数设置，提出板形控制方案。

关键词：热轧高强钢；板形缺陷；模拟仿真

Study on Shape Quality Control of Hot Rolled High Strength Steel Plate

Sun Lirong[1,2], Ban Xiaoyang[2], Wang Guodong[3], Wan Jiafeng[2], Zhang Jiqing[2], Wang Feng[2]

(1. Northeastern University The State Key Laboratory of Rolling and Automation, Shenyang 110000, China; 2. Shandong Iron and Steel Group Rizhao Co., Ltd., Rizhao 276800, China; 3. Northeastern University, Shenyang 110000, China)

Abstract: The shape quality is an important aspect of the quality of sheet and strip, and is also the core issue in the field of rolling. The quality control level of strip shape is not only directly related to the quality of strip products, but also an important symbol of rolling technology, equipment and production management level in iron and steel enterprises. The shape problem of strip and plate products has a direct impact on the later process and customer production, which is the key problem in rolling field. High-end plate and strip quality is an important symbol of production capacity of iron and steel enterprises, which is related to product market expansion and user reputation.

Due to the high strength of high strength steel, it is easy to produce uneven deformation in the rolling process, and the internal stress is released after cutting, and the invisible plate defects appear. Therefore, it is very difficult to control the shape of rolling line. Moreover, because the traditional shape control theory is centered on the machine type and roll system, the influence of plastic deformation on the shape of specific materials during hot rolling process is seldom considered. However, the high temperature deformation process of various materials has its own characteristics, and there is no complete and reliable literature on relevant material constitutive model and high temperature phase transition law.

In order to solve the long-term problem of high strength steel plate shape quality is not high, to ease the difficulty of production in the next process, to respond to user concerns, to give full play to the production capacity of on-site equipment and process, on the basis of in-depth field investigation and theoretical analysis, By thermal simulation experiment and metallographic observation experiment methods, such as varieties of high-end plate strip for typical high temperature plastic deformation behavior, organizational structure, and the constitutive model of research, and then through the finite element numerical simulation reveals the hot rolling process of strip shape change law of flatness defect revealed a typical product of reason, and targeted plate shape control method is put forward, It is of great significance to develop high precision shape control technology which is suitable for the hot rolling of high-end sheet and strip products, to steadily improve the shape quality and potential shape quality of hot rolling HSS products, and to carry out the corresponding industrial test and production application, which is of great significance to improve the shape quality of hot rolling HSS products.

High strength steel due to high strength hot rolled, rolling process and uneven cooling uneven deformation of the internal stress is not easy to trigger a dominant flatness defects, and the internal stress release after cutting, then contact flatness defect, through in-depth field investigation, the statistical test, theoretical and experimental research, on the basis of simulation, reveals the cause of the flatness defect, in view of the typical product and defect, Make the shape control standard, modify the model parameter setting, and put forward the shape control scheme.

Key words: hot rolled high strength steel; plate type defect; analog simulation

4.2 长材和钢管

淬火油温对 40CrMnMo 钻杆的微观组织和性能的影响

钟 彬[1,2]，陈义庆[1,2]，李 琳[1,2]，高 鹏[1,2]，伞宏宇[1,2]，艾芳芳[1,2]，田秀梅[1,2]

（1. 海洋装备用金属材料及其应用国家重点实验室，辽宁鞍山 114009；
2. 鞍钢集团钢铁研究院，辽宁鞍山 114009）

摘 要：利用硬度测试、显微组织观察、力学性能及冲击功测试等研究淬火油温度对钻杆淬火态的金相组织和截面硬度，回火后的力学性能和冲击功的影响。结果表明：淬火油温度对钻杆的微观组织和硬度影响显著，进一步影响回火后的力学性能、屈强比和冲击功。随着淬火油温的升高，冷却能力先升高，后降低，油温 30~50℃为宜，淬透性好，微观组织均匀，晶粒细小，回火后力学性能好，屈强比高，具有较高的抗冲击性能和较好的韧性，综合性能良好。

关键词：机油温度；热处理；40CrMnMo；钻杆

The Effect of Quenching Oil Temperature on the Microstructure and Properties of 40CrMnMo Drill Pipe

Zhong Bin[1,2], Chen Yiqing[1,2], Li Lin[1,2], Gao Peng[1,2], San Hongyu[1,2],
Ai Fangfang[1,2], Tian Xiumei[1,2]

(1. State Key Laboratory of Metal Material for Marine Equipment and Application, Anshan 114009, China;
2. An-shan Iron & Steel Institute, Anshan 114009, China)

Abstract: The study of the effect of quenching oil temperature on the metallographic structure and section hardness of the quenched drill pipe, as well as the mechanical properties and impact energy after tempering was carried out by means of hardness test, microstructure observation, mechanical properties and impact energy test. The results show that the quenching oil temperature has a significant effect on the microstructure and hardness of the drill pipe, and also has effect on the mechanical properties, yield ratio and impact energy of the drill pipe after tempering. With the increase of quenching oil temperature, the cooling capacity increases firstly and then decreases, the appropriate oil temperature is 30~50℃, the drill pipe has excellent hardenability, uniform microstructure, fine grains. The drill pipe has excellent mechanical properties, high yield ratio, high impact resistance and excellent toughness after tempering, comprehensive performance of the drill pipe is excellent.

Key words: engine oil temperature; heat treatment; 40CrMnMo steel; drill pipe

带肋钢筋成品弯头冲出口影响因素分析及控制

李存林，陈祖政，周汉全，巫献华，黄育坚

（宝武集团广东韶钢松山股份有限公司，广东韶关 512123）

摘　要：针对韶钢棒三生产线频繁出现的成品弯头冲出口问题，通过大量的现场跟踪发现造成成品弯头冲出口的主要原因：（1）轧件温度不均匀；（2）切分线差大；（3）导卫装配不良；（4）成品孔型设计不合理。为此，对工艺进行了改进：（1）钢头温度要确保均匀，且钢头温度要比其它部位烧高 20～30℃；（2）优化切分线差操作技术，将切分线差控制在 10cm 以内；（3）提高导卫装配精度，确保导卫与孔型的对中性；（4）优化成品孔型设计，保证轧件头部更好地脱槽。采用上述措施后，有效解决了热轧带肋钢筋成品弯头冲出口问题。

关键词：带肋钢筋；头部弯曲；温度；切分线差；导卫；孔型设计

Influence Factors and Improving Measures of the Shocking the Outlet by Finished Product Head Warping of Rolled Ribbed Bars

Li Cunlin, Chen Zuzheng, Zhou Hanquan, Wu Xianhua, Huang Yujian

(Baowu Group SGIS Songshan Co., Ltd., Shaoguan 512123, China)

Abstract: The main causes of the shocking the outlet by finished product head warping by the third bar production workshop of Shaoguan Iron and Steel Co. were found through a lot of field tracking. There were four main reasons. The first was the temperature of rolled piece is not uniform. The second was slitting line much difference. The third was bad assembling of guide and guard. The fourth was The pass design of finished product is not reasonable. According to these four main reasons, some process measures were proposed.The first ,Ensure uniform temperature of steel head, And the temperature of the steel head is 20-30℃ higher than that of other parts. The second, optimized theslitting line differenceoperation technique, controlled the slitting line differencewithin 10cm. The third, improve the accuracy of guide and guard assembly, make sure the guide and the pass are alignment of center line. The fourth , optimize pass design of finished products, ensure that the head of rolled piece is better off the roll groove. Through the above process measures, effectively solve the problem of affect the shocking the outlet by finished product head warping of hot rolled ribbed bars.

Key words: rolled ribbed bars; head bending; temperature; slitting line difference; guide; pass design

不锈钢复合螺纹钢筋的生产实践

陈祖政，郑家贤，张 鑫

（韶钢松山股份有限公司特轧厂，广东韶关 512123）

摘 要：工程上采用一般螺纹钢因腐蚀而需要高昂的维护费用，采用双相不锈钢成本又过高，不锈钢复合螺纹钢筋可兼有成本低廉又能满足使用要求的特点。本文介绍了韶钢采用覆层为316不锈钢，基体为HRB400E低合金钢的圆坯料轧制复合螺纹钢的生产实践。首次采用方坯、圆坯交替入炉的方式，克服了圆坯料在传统方坯步进加热炉内容易滚动的问题；通过计算采用了合适的圆坯料尺寸，将1#、2#轧机孔型改成椭圆—圆孔型系统，实现了方坯与圆坯钢采用相同料型连续轧制，生产出复合钢筋成品具有良好的冶金结合特性，物理性能符合相关标准。

关键词：不锈钢；圆坯轧制；复合钢筋；孔型设计

Production Practice of Stainless Steel Composite Rebar

Chen Zuzheng, Zheng Jiaxian, Zhang Xin

(Baowu Group Guangdong Shaoguan Iron and Steel Co., Ltd., Shaoguan 512123, China)

Abstract: The use of general rebar in the project requires high maintenance costs due to corrosion, and the cost of using duplex stainless steel is too high. Stainless steel composite rebar can have the characteristics of low cost and meeting the requirements of use. This article introduces the production practice of Shaoguan Steel's rolling composite rebar with round billets with 316 stainless steel cladding and HRB400E low-alloy steel. It is the first time that the billet and round billet are alternately fed into the furnace, which overcomes the problem that the round billet is easy to roll in the traditional billet stepping heating furnace; through calculation, the appropriate round billet size is adopted, and the pass of the 1# and 2# rolling mills It is changed to an ellipse-round pass system to realize the continuous rolling of billet and round billet steel with the same material type, producing composite steel bars with good metallurgical bonding characteristics and physical properties in compliance with relevant standards.

Key words: stainless steel; round billet rolling; composite steel; pass design

冲击器用 45CrNiMoV 钢制管实验研究

解德刚，吴　红，赵　波，王善宝，袁　琴

（鞍钢集团钢铁研究院，辽宁鞍山　114009）

摘　要：本文对 45CrNiMoV 进行了制管生产的尝试。工艺参数对力学性能影响的研究结果表明，该钢在 1100℃以下的热强度较高，达到 200kgf 以上，并且热强度的升高速率随温度的降低显著增大，制管时应避免低温轧制，控制轧管温度不低于 1100℃为宜；该钢的热轧态硬度较高，达到 40HRC，不利于机械加工，经过 860℃退火处理后，硬度可降低到 25~28HRC；该钢在 860℃正火+880℃淬火+不低于 610℃回火时的冲击功可达到 90J 以上， 回火温度 610℃为冲击功升高速率由小变大的拐点。

关键词：45CrNiMoV；钢管；力学性能；工艺参数

Experimental Study on 45CrNiMoV Steel Pipe for Impactor

Xie Degang, Wu Hong, Zhao Bo, Wang Shanbao, Yuan Qin

(Ansteel Iron & Steel Research Institutes, Anshan 114009, China)

Abstract: In this paper, the production of 45CrNiMoV pipe is attempted. The effect of process parameters on mechanical properties is studied. The thermal strength of the steel is high below 1100℃, reaching over 200 KGF. The increasing rate of thermal strength increases with the decrease of temperature. During production, it is advisable to avoid too low rolling temperature and control the temperature no less than 1100℃. The hot rolling hardness of the steel is high, reaching 40HRC, which is not conducive to machining. After annealing at 860℃, the hardness can be reduced to 25-28HRC. The impact energy of the steel after normalizing at 860℃ and quenching at 880℃ and tempering at 610℃ or above can reach above 90J. When the tempering temperature is 610℃, the inflexion point of high rate of impact energy rising from small to large can reach above 90J.

Key words: 45CrNiMoV; steel pipe; mechanical properties; process parameters

40Cr 棒材奥氏体混晶问题冶金实践

杜东福，张　磊

（凌源钢铁股份有限公司，辽宁凌源　122500）

摘　要：结合凌钢工艺装备，对合金结构钢 40Cr 圆钢不同机组混晶问题进行了调查分析，找出了混晶的主要原因 Als 含量低、轧钢加热操作不稳定等。通过优化 Als 设计，成分下限由 0.005%提高至 0.018%，控制与 N 结合的铝含量达 0.008%以上，同时加严轧钢“闷炉”时降温制度，最终 40Cr 混晶问题得到了解决。

关键词：混晶；酸溶铝；氮含量；晶粒度

Metallurgical Practice of Mixed Austenite Grain Problem in 40Cr Bar

Du Dongfu, Zhang Lei

(Lingyuan Iron & Steel Co., Ltd., Lingyuan 122500, China)

Abstract: Combined with Linggang's process equipment, The mixed grain problem of alloy structural steel 40Cr round steel was investigated and analyzed, and the main reasons for the mixed grain were found to be low Als content and unstable heating operation of rolling steel. By optimizing the Als design, the lower limit of composition was determined by Increased from 0.005% to 0.018%, controlled the aluminum content combined with N to over 0.008%, and at the same time tightened the cooling system during the "Stuffy Furnace" of steel rolling, and finally the 40Cr mixed crystal problem was solved.

Key words: mixed grain; acid soluble aluminum; nitrogen content; grain size

滚动导卫失效原因分析

王扬发，张 鑫，李 班，何海峰

（宝武集团广东韶关钢铁有限公司特轧厂，广东韶关 512123）

摘 要：针对韶钢高线粗轧机组 4#轧机入口滚动导卫频繁失效问题，对导辊和销轴进行断口分析，判断韧性断裂非主要原因；长期承受冲击载荷和交变载荷的共同作用为导辊失效的主要原因；销轴材料中存在魏氏体组织使其脆性提高为销轴失效的主要原因。结合产线特点提出改善导辊冷却和对前机架进行孔型优化等预防措施。

关键词：导辊；销轴；失效；断裂；魏氏体组织

Failure Analysis of Rolling Guide and Guard

Wang Yangfa, Zhang Xin, Li Ban, He Haifeng

(Baowu Group Guangdong Shaoguan Iron and Steel Co., Ltd., Shaoguan 512123, China)

Abstract: In view of the frequent failure of rolling guide at the inlet of 4# rolling mill of Shaoguan Iron and Steel Co. LTD. the fracture of guide roller and pin shaft was analyzed to determine that the ductile fracture was not the main reason.The combined action of long-term impact load and alternating load is the main reason of guide roll failure.The main reason of pin failure is the increase of brittleness due to the existence of weissenite structure in pin material.According to the characteristics of the production line, preventive measures such as improving the cooling of guide roll and optimizing the pass profile of the front frame are put forward.

Key words: guide roller; pin shaft; failure; fracture; widmanstatten structure

耐酸性腐蚀性能良好的低温无缝钢管的研制

赵　波，吴　红，解德刚，袁　琴，王善宝

（鞍钢集团钢铁研究院，辽宁鞍山　114009）

摘　要：依据美国 ASTM A333 标准要求，通过合金成分设计、冶炼轧制工艺优化、热处理制度筛选等手段，试制了低温冲击性能及抗 H_2S 腐蚀性能优良的低温无缝钢管。结果表明：试制的低温无缝钢管热轧态及正火态低温力学性能及抗腐蚀性能满足标准要求，热轧态组织为珠光体+铁素体，晶粒均匀细小，析出相以 TiN 颗粒为主，–60℃发生韧脆转变，经过 900℃正火热处理后，晶粒更为细小，析出相主要为 TiN+少量的 VN 颗粒，–80℃仍未发生韧脆转变。

关键词：耐低温；耐酸性腐蚀；韧脆转变；无缝管

Research of Low-temperature Service Seamless Steel Pipe with Fine Sour Corrosion Resistance

Zhao Bo, Wu Hong, Xie Degang, Yuan Qin, Wang Shanbao

(Iron and Steel Research Institute of Angang Group, Anshan 114009, China)

Abstract: According to ASTM A333 standard, the seamless line pipe which has fine low temperature impact properties for Low-temperature Service and H2 S corrosion resistance was researched through the means of chemical composition designing, optimizing smelting and rolling processes, and screening heat treatment processes. The results show that the low temperature mechanical properties and anti-H_2S corrosion properties of hot-rolled trial-pipe and normalized trial-pipe can meet ASTM A333 standard. The structure of hot-rolled trial-pipe is Pearlite + ferrite, and the grain is uniform and fine. The main precipitates phase are TiN particles in hot-rolled trial-pipe. Ductile-brittle transition occurs at –60℃. The grain is finer after normalized at 900℃. The main precipitates are TiN particles and a few VN particles in normalized trial-pipe, Ductile-brittle transition occurs at lower –80℃.

Key words: low temperature resistance; sour corrosion resistance; ductile-brittle transition; seamless steel pipe

帕德玛大桥用 16#特种槽钢的研制与开发

丁　宁[1]，刘思洋，李忠武[2]

（1. 鞍钢股份有限公司大型总厂，辽宁鞍山　114021；2. 鞍钢股份有限公司，辽宁鞍山　114021）

摘　要：鞍钢股份有限公司大型总厂成功开发了孟加拉帕德玛大桥压浆槽用 16#特种槽钢，介绍了该槽钢的化学成分、力学性能及工艺设计和试生产情况。

关键词：特种槽钢；孔型设计；开发

Research and Development of 16# Special Channel Steel for Padma Bridge

Ding Ning[1], Liu Siyang[1], Li Zhongwu[2]

(1. Heavy Section Mill Plant of ANSTEEL, Anshan 114021, China;
2. Angang Steel Co., Ltd., Anshan 114021, China)

Abstract: Angang Steel Company limtied has successfully developed the 16# special tank for the grouting tank of Bangladesh Padma Bridage. The chemical composition, mechanical properties, process design and production of the channel steel are introduced.

Key words: special channel steel; groove design; develop

韶钢高一线 CH1T 钢成材率影响因素及提高措施研究

张广化[1]，戴杰涛[2]

（1. 宝武集团广东韶关钢铁有限公司，广东韶关 512123；
2. 广州大学机械与电气工程学院，广东广州 510006）

摘　要：在韶钢高一线重点开发冷镦钢等高等级工业线材的背景下，本文针对超低碳冷镦钢 CH1T 成材率低的问题进行了研究，采用人机料法环的分析方法确定了导致 CH1T 冷镦钢成材率低的原因轧制工艺控制基础差、料型设计不合理和控轧控冷温度设定不合理。针对这些存在的问题，通过理论研究、模拟试验和现场测试相结合的手段对其产生的原因进行了研究，在此基础上制定了工艺改进措施。通过现场实践表明提出的工艺改进措施大大改善了韶钢高一线超低碳冷镦钢 CH1T 成材率低的问题，取得了显著的经济效益。

关键词：CH1T；料型；成材率；控轧控冷

Research on Influencing Factors and Improving Measures of CH1T Steel Yield Rate of High Speed Wire 1# Shaoguan Steel

Zhang Guanghua[1], Dai Jietao[2]

(1. Shaoguan Iron & Steel Co., Baowu Steel Group, Shaoguan 512123, China;
2. School of Mechanical and Electrical Engineering, Guangzhou University, Guangzhou 510006, China)

Abstract: In the context of focusing on the development of high-grade industrial wire such as cold heading steel, In this paper, the problem of low yield rate of ultra-low carbon cold heading steel CH1T is studied. Through the use of man-machine material method analysis method, it is determined that the reason for the low yield rate of CH1T cold heading steel is the poor rolling process control foundation, the unreasonable material type design and the unreasonable setting of the controlled rolling and controlled cooling temperature. In response to these existing problems, the reasons for their occurrence were studied through a combination of theoretical research, simulation test and field test, and on this basis, process improvement measures were formulated. Field practice shows that the proposed process improvement measures have greatly improved the low yield rate of Shaogang's first-line ultra-low carbon cold heading steel CH1T, and achieved significant economic benefits.

Key words: CH1T; material type; yield rate; controlled rolling and controlled cooling

改善地铁车轮多边形问题的生产实践

国新春，陈　刚，邓荣杰，王　健，鲁　松，宁　珅

（宝武集团马钢轨交材科技有限公司，安徽马鞍山　243000）

摘　要： 车轮多边形磨问题在地铁车轮运用过程中表现的越发突出，也是待予解决难题之一。地铁公司一般采用镟修的方式来缓解车轮多边形磨损对车辆运营带来的影响，这种方式增加了运营成本。本文从车轮实现出发，通过优化车轮材质、热处理工艺，提高车轮硬度和硬度均匀性，进而改善地铁车轮多边形问题。

关键词： 地铁；车轮；多边形；硬度均匀性

Development of Damping Ring Wheel for Urban Rail Transit

Guo Xinchun, Chen Gang, Deng Rongjie, Wang Jian, Lu Song, Ning Shen

(Baowu Group Ma Rail Material Technology Co., Ltd., Ma'anshan 243000, China)

Abstract: The polygonal wheel grinding problem is more and more prominent in the process of Metro wheel application, which is also one of the problems to be solved. Metro companies generally adopt the method of turning to alleviate the impact of wheel polygonal wear on vehicle operation, which increases the operation cost. Starting from the realization of the wheel, this paper improves the hardness and hardness uniformity of the wheel by optimizing the wheel material and heat treatment process, so as to improve the polygon problem of the metro wheel.

Key words: metro; wheel; polygon; hardness uniformity

铁路桥梁用高强钢绞线盘条轧制控冷工艺研究

赵晓敏[1]，石　龙[2]，涛　雅[3]

（1. 内蒙古包钢钢联股份有限公司技术中心，内蒙古 包头　014010；2. 中国铁道科学研究院，北京　100081；3. 内蒙古包钢钢联股份有限公司长材厂，内蒙古 包头　014010）

摘　要： 包钢自主设计了铁路桥梁用 BG87 高强钢绞线盘条，通过轧制控冷工艺、静态 CCT 曲线的测定、拉伸试验以及金相等手段研究了其力学性能、显微组织以及拉拔性能。结果表明 ：开发的高碳、高硅钢盘条强度与拉拔性能可以满足 2300 MPa 级预应力钢绞线的使用要求，产品得到应用，实现了国内 2300MPa 高强钢绞线产品的突破。

关键词： 2300MPa 级；高强钢绞线；铁路桥梁；盘条

Research on Controlled Cooling Process of High-strength Steel Stranded Wire Rod Rolling for Railway Bridges

Zhao Xiaomin[1], Shi Long[2], Tao Ya[3]

(1. Technical Center of Steel Union Co., Ltd. of Baotou Steel (Group) Corp., Baotou 014010, China; 2. China Academy of Railway Science Corporation Limited, Beijing 100081, China; 3. Long Products Plant of Steel Union Co., Ltd. of Baotou Steel (Group) Corp., Baotou 014010, China)

Abstract: Baotou Iron & Steel independently designed BG87 high-strength steel stranded wire rod for railway bridges, and studied its mechanical properties, microstructure, tensile properties and pull performance by means of rolling controlled cooling process, static CCT curve measurement, tensile test and gold equivalent methods. The results show that the strength and drawing performance of the developed high-carbon, high-silicon steel wire rod can meet the requirements of the use of 2300 MPa pre-stressed steel strands, and the product has been applied, achieving a breakthrough in domestic 2 300 MPa high-strength steel strand products.

Key words: 2300MPa class; high-strength steel strand; railway bridge; wire rod

SWRH82B 盘条冬季脆断问题研究

李建龙，刘磊刚，车　安，尹　一

（鞍钢股份有限公司线材厂，辽宁鞍山　114042）

摘　要：本文介绍了预应力钢绞线 SWRH82B 盘条，生产时存在线架脆断情况，对线架脆断盘条影响因素进行分析总结，并对改善脆断问题，制定相关措施。

关键词：化学成分；SWRH82B；控冷；应力

Study on Brittle Fracture of SWRH82B Wire rod in Winter

Li Jianlong, Liu leigang, Che An, Yin Yi

(Ansteel Wire Rod Mill, Anshan 114042, China)

Abstract: This paper introduces the brittle fracture of wire frame in the production of SWRH82B wire rod of prestressed steel wire.This paper analyzes and summarizes the influencing factors of brittle wire in use, and formulates relevant measures to improve the brittle fracture problem.

Key words: chemical composition; SWRH82B; controlled cooling; residual stress

高速线材轧制事故分析及控制

尚俊男，张　欢，刘磊刚，郭思聪

（鞍钢股份有限公司线材厂，辽宁鞍山　114009）

摘　要：根据近几年鞍钢高速线材厂 2#线生产过程中，发生的一些常见轧制事故，针对坯料事故、粗中轧区域、预精轧区域、高速段区域，分析事故产生原因，提出减少事故的控制措施，从而达到提效增产的目的。

关键词：高速线材；轧制；堆钢；事故控制

Rolling Accidents Analysis and Control in Rolling High Speed Wire Rod

Shang Junnan, Zhang Huan, Liu Leigang, Gao Sicong

(Anshan Iron and Steel Co., Ltd., Anshan 114009, China)

Abstract: According to some common rolling accidents occurred in the production process in Ansteel high speed wire rod in recent years，aiming at billet accidents，the rough and medium rolling area, pre finishing rolling area, high-speed rolling area，the causes of the accidents are analyzed, and the control measures to reduce the accidents are put forward, so as to achieve the purpose of increasing production.

Key words: high-speed wire rod; rolling; steel-heaping; accident analysis and control

复杂工况下抗腐蚀油井管设计与生产实践

陈志刚，白喜峰，洪　汛，韩久富，刘海齐，张　尧，邸　军

（鞍钢集团 无缝钢管厂，辽宁鞍山　114021）

摘　要：鞍钢无缝厂依据美国石油协会 API Spec 5CT 标准，通过优化成分设计、冶炼、轧制和热处理工艺，试制并生产了力学性能及抗腐蚀性能优良的无缝石油管。检验结果表明：产品的力学性能及抗酸性腐蚀性能均满足 API Spec 5CT 及抗腐蚀标准要求。

关键词：无缝管；石油管；抗腐蚀；力学性能

Design and Production Practice of Anti-corrosion Oil Well Pipe under Complicated Working Conditions

Chen Zhigang, Bai Xifeng, Hong Xun, Han Jiufu, Liu Haiqi, Zhang Yao, Di Jun

(Seamless Pipe Factory of Anshan Iron and Steel(Group)Co., Anshan 114021, China)

Abstract: In accordance with the American Petroleum Institute API Spec 5CT standard, Angang Seamless Plant trial-produced and produced seamless petroleum pipes with excellent mechanical properties and corrosion resistance through optimized composition design, smelting, rolling and heat treatment processes. The inspection results show that the mechanical properties and acid corrosion resistance of the product meet the requirements of API Spec 5CT and corrosion resistance standards.

Key words: seamless pipe; petroleum pipe; corrosion resistance; mechanical properties

长周期排管锯片的生产工艺研究

洪 汛，白喜峰，韩久富，陈志刚，刘海齐，傅 强，邸 军，陈 英

（鞍钢集团公司大型总厂无缝厂，辽宁 鞍山 114021）

摘 要：分析了排管锯片钎焊中产生表面气孔、焊接裂纹、接头性能降低等现象的原因，提出相应的解决办法，有效地解决了鞍钢无缝钢管厂排管锯片在生产中的掉刀头现象。排管锯片由以前的平均1400片/月降将低到现在的900片/月；而且刀头焊接质量提高后，减少了生产中的换刀时间，提高了工作效率，有效地缓解了排管锯能力不足的问题。新的钎焊生产工艺通过技术创新节约了成本，响应了国家建设绿色工厂的号召。

关键词：硬质合金；刀头；钎焊；无缝钢管

Research on Production Technology of Long-period Steel Pipe Saw Blade

Hong Xun, Bai Xifeng, Han Jiufu, Chen Zhigang, Liu Haiqi, Fu Qiang, Di Jun, Chen Ying

(Seamless Pipe Factory of Anshan Iron and Steel (Group) Co., Anshan 114021, China)

Abstract: after analyzing the reason of gas crack on the face、welding crack and the reduce of performance at the stick point of the pipe cutting knives during braze welding , An effective five-step method to manufacturing cutting knives applied for solving the break-off problem of knives heads on the cutting machine in the seamless tube plant of AnShan Iron &Steel Group Corporation during production. The utility rate of the knives was 1200 pieces each week at present, which decreased 200 pieces each week than before. Moreover, with the increasing of welding quality , the time for the cutting machine operator to change the knives was reduced during manufacture, and the work efficiency was increased, which reduced the ability problem of cutting machine effectively. The five-step method to manufacturing cutting knives decreased the manufacture cost, respond the call by the government to build the green workshop through technology innovation .

Key words: cemented carbides; knife head; brazing process; seamless pipe

高强抗震螺纹钢直接轧制孔型研究

戴江波[1]，王保元[1]，卓 见[1]，金培革[2]，庄继超[2]，刘 宏[2]

（1. 中冶南方武汉钢铁设计研究院有限公司轧钢事业部，湖北武汉 430083；

2. 宝武集团襄樊钢铁重材有限公司，湖北襄阳 441100）

摘　要：宝武集团襄樊钢铁长材有限公司采用废钢电炉冶炼直接轧制生产工艺，本文结合襄钢生产实际，利用大型有限元 MARC 软件，基于旋转轧辊刚性面接触摩擦引领钢坯运动、钢坯热轧塑性变形的热力耦合有限元法,建立了在 825℃至 1050℃的钢坯中轧有限元仿真模型，为直接轧制孔型设计提供了理论计算的基础，为改善直接轧制产品质量提出了粗轧生产有利孔型的配置。

关键词：直接轧制；高强螺纹钢；孔型；温度；有限元

The Research of the Hole Type's Design about the Direct Rolling Mill for the Thread Steel

Dai Jiangbo[1], Wang Baoyuan[1], Zhuo Jian[1], Jin Peige[2], Zhuang Jichao[2], Liu Hong[2]

(1. WISDRI Wuhan Iron and Steel Design Institute Co., Ltd. WuHan 430083, China; 2. Baowu Group Xiangyang Steel Co., Ltd., Xiangyang 441100, China)

Abstract: Xiangfan Iron & Steel excellent talent Co., Ltd. of Baowu Group adopts the direct rolling process of scrap steel smelting in electric furnace. Combining with the production practice of Xianggang, this paper uses large-scale finite element MARC software, and based on the thermo-mechanical coupling finite element method that the rigid surface contact friction of rotating roller leads the billet movement and the plastic deformation of billet hot rolling, establishes the finite element simulation model of billet intermediate rolling at 825℃ to 1050℃, which provides the theoretical calculation basis for the design of direct rolling pass and puts forward the configuration of favorable pass for rough rolling production to improve the quality of direct rolling products.

Key words: the direct rolling; high strength thread steel; the hole type; temperature; finite rlement

稀土微合金化 Q345NQR2 热轧 H 型钢试制

卜向东，涛　雅，唐建平，宋振东，张达先

（内蒙古包钢钢联股份有限公司 技术中心，内蒙古包头　014010）

摘　要：本文通过在包钢现有的 Q345NQR2 热轧 H 型钢基础上，加入 0.0050%铈铁合金，对比分析稀土合金加入对铸坯高温性能、常规力学性能及耐腐蚀性能的影响，对比结果表明，铈铁合金的加入有效改善了 Q345NQR2 异型坯高温脆性区，–40℃低温冲击韧性及耐大气腐蚀性能略有提高，改善效果不明显。

关键词：稀土；微合金化；Q345NQR2；H 型钢

低淬火开裂个性化的 40Cr 棒材开发

陈定乾，蒋　骏

（湘潭钢铁有限公司，湖南湘潭　411101）

摘　要：从淬火开裂试样分析，不能准确判断开裂原因，但是从裂纹宏观和微观形态来看，明显的淬火裂纹，淬火开裂均发生在应力比较集中的位置。由于不能确定裂纹源，无法分析其具体产生的原因。从生产企业的反馈，开发之前的 40Cr 开裂比例较大，严重影响生产效率，需要开发个性化低淬火开裂的钢材。本文通过从冶炼、轧制方面控制入手，细化晶粒、消除应力。这种低淬火开裂个性化 40Cr 棒材加工后，淬火开裂的比例大幅度降低。

关键词： 低淬火开裂；棒材；个性化；40Cr

The Development of 40Cr Bar with Low Quenching Cracking and Personalized

Chen Dingqian, Jiang Jun

(Xiangtan Iron and Steel (XISC) Co., Ltd., Xiangtan Hunan 411101, China)

Abstract: Through the analysis of quenchingcracking of specimens, the cause of cracking cannot be accurately judged.From the perspective of the macro and micro morphology of cracks, obvious quenching cracks andcracking occurat stress concentration locations. Since the source ofcracking cannot be determined, it is impossible to analyze the specific cause. The feedbacksfrom the manufacturers showthat the carcking ratio of 40Cr before development is bigger. The large proportions of carcking ratio seriously affectthe production efficiency. Therefore, it is necessary to develop bar with low quenchingcracking. This paper proceeds with control from smelting and rolling, pass schedule, refines grain finenessand relieves stress. When this steel bar with lowquenching cracking and personalized is processed, the proportion of quenching cracking fall below.

Key words: low quenching cracking; bar; personalized; 40Cr

钢轨在线超声探伤检测系统研发

马 东[1]，刘 鹤[1]，贾宏斌[2]，刘思洋[1]

（1. 鞍钢股份有限公司大型总厂，辽宁鞍山 114021；
2. 鞍钢集团钢铁研究院长材产品研究所，辽宁鞍山 114009）

摘 要： 钢轨在检测过程中在水平或垂直方向瞬间产生位移时，进口钢轨在线超声探伤系统的探头起落装置无法实时驱动探头产生相应的跟随位移，同时探头倾斜角度调试完成后采用机械装置锁死固定，造成检测过程中探头工作面与钢轨被检测面不能实时贴合，形成误漏报警。其机械装置无法满足鞍钢股份有限公司大型总厂现场的具体生产工艺，所以作者主持开发了一套钢轨在线超声探伤系统。该系统已在线应用近 10 年，检测准确率高，系统运行稳定，至今无漏检发生。本文全面地介绍了高速铁路用钢轨超声探伤系统的各个组成部分和设计原理。

关键词： 钢轨；超声探伤；探头随动装置；检测系统；检测准确率

Development of Rail on Line Ultrasonic Testing System

Ma Dong[1], Liu He[1], Jia Hongbin[2], Liu Siyang[1]

(1. The Heavy Steel Rolling Plant of Angang Steel Co., Ltd, Anshan 114021, China;
2. Ansteel Group Iron and Steel Research Institue, Anshan 114009, China)

Abstract: When the rail moves instantaneously in the horizontal or vertical direction in the detection process, the probe landing device of the imported rail on-line ultrasonic flaw detection system can not drive the probe to generate the corresponding following displacement in real time. At the same time, after the probe tilt angle debugging is completed, the mechanical device is used to lock and fix it, resulting in the probe working surface and the rail detected surface can not fit together in real time in the detection process, Form a false alarm. The mechanical device can not meet the specific production process of the large-scale general plant of Angang, so the author presided over the development of a set of on-line ultrasonic flaw detection system for rail. The system has been applied online for nearly 10 years, with high detection accuracy, stable operation and no missed detection. This paper introduces the components and design principle of the ultrasonic flaw detection system for high speed railway rails.

Key words: rail; ultrasonic testing; probe follower; detection system; detection accuracy

武汉地铁 2 号线用钢轨的踏面剥离掉块成因分析与改进建议

叶途明，王金平，易卫东

（宝钢股份武汉钢铁有限公司条材厂，湖北 武汉　430083）

摘　要： 武汉地铁 2 号线在某曲线段运行 3 年后出现了踏面剥离掉块伤损。本文通过对缺陷轨的宏观形貌、显微组织、化学成分、力学性能等进行了检验。结果表明：钢轨的踏面剥离掉块主要是由于车轮和钢轨表面的滚动接触疲劳造成的。通过提高钢轨性能、选用热处理钢轨、进行钢轨打磨、选用新轨头廓形 60N 钢轨等措施，可有效改善该缺陷的形成和扩展。

关键词： 地铁钢轨；踏面；剥离掉块；分析与改进

Analysis and Improvement Technology about Shelling Defects on Subway-Rails Of Wuhan Metro Line 2

Ye Tuming, Wang Jinping,Yi Weidong

(Plant of Long Product, Wuhan Iron & Steel Co., Ltd., Wuhan 430083, China)

Abstract: The shelling defects occurred on curve segment at Subway-rails Of Wuhan Metro Line 2 after 3 years. The macro-morphology, the microstructure, the chemical composition and the mechanical properties of the rail were inspected. The results showed that the shelling defects on curve segment of subway-rails were due to the Contact Fatigue of wheel-rail. Through increasing performances of rails, choosing head-hardened rails, grinding rails, choosing 60N new-profile rails etc, the formation and spreading of shelling defects can be improved effectively.

Key words: subway-rail; rail surface; shelling defect; analysis and improvement

控制重轨踏面周期性轧痕的攻关实践

何　彬，段　文，董茂松，周意平，乘　龙，叶佳林，
陈　楠，余选才，黄海滨，刘　义

（武钢有限条材厂，湖北武汉　430083）

摘　要：通过对重轨踏面周期性轧痕迹的系统分析，对重轨精轧工艺进行优化，并试验改进轧辊材质，最终有效控制了踏面周期粘肉难题。

关键词：重轨；周期轧痕；精轧

Practice of Controlling the Periodic Rolling Mark of Heavy Rail Tread

He Bin, Duan Wen, Dong Maosong, Zhou Yiping, Cheng Long, Ye Jialin,
Chen Nan, Yu Xuancai, Huang Haibin, Liu Yi

(Plant of Long Product , Wuhan Iron & Steel Co., Ltd., Wuhan 430083, China)

Abstract: Based on the systematic analysis of the periodic rolling marks of heavy rail tread, the fine rolling process of heavy rail is optimized, and the roll material is improved by experiment. Finally, the problem of periodic meat sticking on the tread is effectively controlled.

Key words: heavy rail; periodic rolling mark; finishing

帘线钢 T5 温度达标率提升方法剖析

郑传昱，高江明，盛光兴，郭　磊，罗　闯，熊志辉，胡　伟，陈建飞

（武钢有限条材厂，湖北武汉　430083）

摘　要：吐丝温度 T5 是高速线材性能控制的关键节点温度，通过对如何提高 T5 温度达标率的目标进行分析、分解、实验、总结,找出了影响 T5 达标率的核心要素，运用技术和管理等手段，在人、机、料、法、环方面多管齐下，实现吐丝温度 T5 连续两年稳步提升的目标。

关键词：吐丝温度 T5；达标率；技术和管理；稳步提升

Analysis of Raising Method of T5 up to Standard in Cord Steel

Zheng Chuanyu, Gao Jiangming, Sheng Guangxing, Guo Lei,
Luo Chuang, Xiong Zhihui, Hu Wei, Chen Jianfei

(Plant of Long Product , Wuhan Iron & Steel Co., Ltd., Wuhan 430083, China)

Abstract: Spinning temperature T5 is the key to the high speed wire rod performance control node. The thesis analyzes on how to improve the rate of reaching the standard of T5 to find out the core factors affecting it, by means of technology and management, in man, machine, material, method, ring multiprocessing, thus, spinning temperature T5 is to realize the goal of ascension steadily for two consecutive years.

Key words: spinning temperature T5; rate of reaching the standard; technology and management; ascension steadily

不锈钢-碳钢复合方坯轧制工艺研究

张　瑜[1]，任玉辉[2]，廖德勇[1]，王　杨[1]，金纪勇[1]，徐　曦[1]

（1.鞍钢集团钢铁研究院，辽宁鞍山　114009；
2.鞍钢股份有限公司制造管理部，辽宁鞍山　114001）

摘　要： 每年我国都因材料腐蚀承受较大的经济损失，在研究添加不同元素提高原材料的耐蚀性能的同时，更多的研究者和使用者将注意集中在复合材料的研发，复合材料凭借其优异的力学性能、较高的经济效益和优秀的耐蚀性能备受青睐。然而，复合材料中型材的发展较板材发展缓慢，制备工艺的复杂是关键原因之一。本文提出一种新的复合型材制备思路—复合方坯轧制，接轨实际生产，对复合方坯的轧制进行试验及检测评价，对今后复合型材的生产奠定基础。

关键词： 复合材料；方坯轧制；复合型材

Research on Rolling Technology of Stainless Steel and Carbon Steel Composite Square Billet

Zhang Yu[1], Ren Yuhui[2], Liao Deyong[1], Wang Yang[1], Jin Jiyong[1], Xu Xi[1]

(1. Ansteel Group Iron and Steel Research Institue, Anshan 114009, China;
2; Manufacturing and Management Department of Angang Steel Co., Ltd., Anshan 114001, China)

Abstract: We suffer enormous economic losses because of material corrosion every year in China，which leads more researchers and users to focus on developing composite materials while studying how to improve the corrosion resisting property through adding different elements. Composite materials gained lots of popularity for their excellent mechanical properties, high economic benefits and good corrosion resisting properties, however, development of composite materials used in section steel remains inactive, and the complex manufacturing process is one of the main factors. This paper presents a new method of manufacturing composite section, which is composite square billet rolling. The method is well in line with practical manufacturing, which will provide a solid foundation in section manufacturing through testing and evaluating after rolling.

Key words: composite material; square billet rolling; composite section

轴承钢球化退火过程不同保温时间对脱碳层厚度的影响规律

蒋国强[1]，莫杰辉[1]，孙应军[1]，苏福永[2]

（1. 宝武杰富意特殊钢有限公司，广东韶关 512123；
2. 北京科技大学能源与环境工程学院，北京 100083）

摘 要：轴承钢在球化退火过程中的脱碳现象会使轴承钢表面形成一层脱碳层，降低轴承钢产品的硬度和耐磨性，严重影响产品的质量。本文针对轴承钢球化退火过程不同保温时间下的脱碳现象开展研究，设计了脱碳过程实验平台和实验方案，对轴承钢温度为 810℃炉内气氛为 0.3%含水率时保温 20～70min 后的脱碳状况进行了实验研究，结果表明脱碳层厚度随保温时间增加而增大，脱碳层厚度与保温时间的平方根成正比。

关键词：轴承钢；表面脱碳；球化退火

Influence of Different Holding Time on Decarburization Layer Thickness of Bearing Steel during Spheroidizing Annealing

Jiang Guoqiang[1], Mo Jiehui[1], Sun Yingjun[1], Su Fuyong[2]

(1. Baowujiefuyi Special Steel Co., Ltd., Shaoguan 512123, China; 2. School of Energy and Environmental Engineering, University of Science and Technology Beijing, Beijing 100083, China)

Abstract: the decarburization phenomenon of bearing steel during spheroidizing annealing will form a decarburization layer on the surface of bearing steel, reduce the hardness and wear resistance of bearing steel products, and seriously affect the quality of products. In this paper, the decarburization phenomenon of bearing steel during spheroidizing annealing at different holding time was studied. The experimental platform and scheme of decarburization process were designed. The decarburization status of bearing steel at 810℃ with 0.3% moisture content in furnace atmosphere after holding for 20-70min was studied. The results show that the thickness of decarburized layer increases with the increase of holding time, and the thickness of decarburized layer is proportional to the square root of holding time.

Key words: bearing steel; surface decarburization; spheroidizing annealing

棒线材连轧生产线钢坯无头焊接轧制改造模式探讨

徐言东[1]，程知松[1]，洪荣勇[2]，徐正斌[1]，王兆辉[1]

（1. 北京科技大学工程技术研究院，北京 100083；
2. 福建三钢闽光股份有限公司，福建 三明 365000）

摘 要：棒线材连轧生产线在钢铁生产中在占有很大比重，其生产成本高低直接影响着行业的利润。国内外相关研究机构和企业一直在进行如何降低成本的研究，通过增加或改造一些设备和控制系统的关键技术，融入到现有生产线工艺中进行探索。经过近年来的实践，基于发现钢坯无头焊接轧制技术的应用已经取得了较好的效果，对棒线材的连轧生产线进行改造，可以不改变现有生产线主体设备布局，仅在现有的加热炉和粗轧机组之间增加钢坯无头焊接设备，本模式停工时间短，新增设备可部分国产，投资少，见效快。

关键词：棒线材连轧生产线；钢坯无头焊接轧制；成本

Discussion on Transformation Mode of Endless Welding Rolling of Billet in Continuous Bar and Wire Rolling Line

Xu Yandong[1], Cheng Zhisong[1], Hong Rongyong[2], Xu Zhengbin[1], Wang Zhaohui[1]

(1. Engineering and Technology Research Institute, University of Science and Technology Beijing, Beijing 100083, China; 2. Fujian Min Guang Steel Co., Ltd., Sanming 365000, China)

Abstract: Bar and wire continuous rolling line occupies a large proportion in iron and steel production, and its production cost directly affects the profits of the industry. Relevant research institutions and enterprises at home and overseas have been research how to reduce the cost, by adding or transforming some crucial technologies of equipment and control system, and integrating them into the existing production line process. Based on the research in recent years, it is found that the application of endless welding rolling technology has got satisfactory effects. The continuous rolling production line of bar and wire can be reformed without changing the main equipment layout of the existing production line, and only adding endless welding equipment between the existing heating furnace and roughing mill. The downtime of this mode is short, and the newly added equipment can be partially made in China with less investment and quick effect.

Key words: continuous bar and wire rolling line; billet endless welding rolling; cost

鞍钢线材ϕ6.5mm 规格柔性轧制生产实践研究

张 欢[1]，刘磊刚[1]，尚俊男[1]，安绘竹[2]，李建龙[1]，李 凯[3]

（1. 鞍钢股份有限公司线材厂，辽宁 鞍山 114042；2. 鞍钢集团钢铁研究院，辽宁 鞍山 114009；3. 鞍钢股份产品发展部，辽宁 鞍山 114021）

摘 要：利用现有的孔型研究 ϕ6.5mm 规格不同轧制路径的可能性，为生产组织提供更多选择；对不同工艺路径优劣性进行对比，选择最佳的工艺路径和最经济的生产组织方式。

关键词：高速线材；功率；工艺路径；尺寸精度；传动比

Research on ϕ6.5mm Production Practice of Flexible Rolling of Aneteel

Zhang Huan[1], Liu Leigang[1], Shang Junnan[1], An Huizhu[2], Li Jianlong[1], Li Kai[3]

(1. Wire Rod Mill of Angang Steel Co., Ltd., Anshan 114042, China; 2. Ansteel Iron & Steel Research Institues, Anshan 114009, China; 3. Product Development Department of Angang Steel Co., Ltd., Anshan 114021, China)

Abstract: Withing the available grooves to research the possibility of different rolling Φ6.5mm process paths, offer more options for the production organization. Comparise advantages and disadvantages of different process paths, choose the best process path and the most economical production organization.
Key words: high-speed wire rod; power; process path; accurate size; gear ratio

高线轧制 CH1T 钢冷墩开裂的原因及对策探讨

张焕远，黄锦标，张建华

（宝武集团广东韶关钢铁有限）

摘　要：高线轧制的 CH1T 钢成品线卷，取样经做冷墩试验后，出现的冷墩开裂问题，不仅影响工业材的成材率，还会影响客户二次加工问题，发生质量异议。本文首先分析了轧制 CH1T 钢成品线卷出现冷墩开裂的原因，然后指出解决的对策，最后总结轧制过程工艺控制措施，以供参考。

关键词：CH1T 钢；冷墩开裂；解决对策；工艺控制

Discussion on the Reason and Countermeasure of Cracking of Cold Pier in High Wire Rolling CHIT Steel

Zhang Huanyuan, Huang Jinbiao, Zhang Jianhua

(Baowu Group Guangdong Shaoguan Iron & Steel Co., Ltd.)

Abstract: After the cold pier test, the crack problem of cold pier appeared in the high wire rolled CH1T steel wire coil samples, which not only affected the yield of industrial materials, but also affected the secondary processing problems of customers, resulting in quality objections. In this paper, the causes of cold pier cracking in rolled CH1T steel wire coil are analyzed, and the solutions are pointed out. Finally, the rolling process control measures are summarized.
Keywords: CH1T steel ; cracking of cold pier

铸坯缺陷对易切削钢冷拔裂纹影响及改进措施

李富强，罗新中，朱祥睿，章玉成

（宝武集团广东韶关钢铁有限公司检测中心，广东韶关　512123）

摘　要：针对 1215MS 易切削钢线材冷拔过程中出现开裂的质量问题，采用光学显微镜和扫描电镜等分析手段，对 1215MS 易切削钢开裂原因进行分析。分析结果表明：1215MS 线材冷拔开裂的的原因有铸坯近表面夹渣、水口侵蚀、皮下气泡等铸坯缺陷。通过使用易切削钢专用保护渣，结晶器采用正弦振动、降低连浇炉数、控制精炼自由氧含量等工艺措施，可有效避免冷拔纵裂纹的发生。

关键词：1215MS；纵裂；近表面夹渣；水口侵蚀；皮下气泡

Effect of Slab Defects on Cold-drawn Cracks of Free-cutting Steel and Improvement Measures

Li Fuqiang, Luo Xinzhong, Zhu Xiangrui, Zhang Yucheng

(Baowu Steel Group Guangdong Shaoguan Iron and Steel Co., Ltd., Shaoguan 512123, China)

Abstract: Aiming at the quality problems of cracking during cold drawing of 1215MS free-cutting steel wire, optical microscope and scanning electron microscope were used to analyze the cracking causes of 1215MS free-cutting steel. The analysis results show that the causes of cold-drawn cracking of the 1215MS wire include slag inclusion near the slab, erosion of the nozzle, and subcutaneous bubbles. The occurrence of cold-drawn longitudinal cracks can be effectively avoided by using special protective slag for free-cutting steel, the mold with sinusoidal vibration, reducing the number of continuous casting furnaces, and controlling the free oxygen content of the refining process.

Key words: 1215MS; longitudinal crack; slag inclusion near the surface; erosion of the nozzle; subcutaneous bubbles

国内中高碳硬线环保免酸洗工艺技术发展现状

王海宾[1]，贾建平[1]，王建忠[1]，曹光明[2]，张朝磊[3]

（1. 河钢集团宣钢公司，河北 宣化 075100；2. 东北大学 轧制技术及连轧自动化国家重点实验室，辽宁 沈阳 110819；3. 北京科技大学 材料科学与工程学院，北京 100083）

摘　要：总结国内外中高碳硬线环保免酸洗工艺技术取得的研究性成果，介绍近年来环保免酸洗中高碳硬线氧化铁皮去除设备和氧化铁皮控制技术取得的一些进步，以 82B 为例结合实验和工业化生产得出，适合于机械剥壳的 82B 最优氧化铁皮结构是以 FeO 为主，Fe_3O_4所占比例应控制在 30%～40%，并指出了环保型免酸洗中高碳硬线的发展前景和重要意义。

关键词：免酸洗；中高碳硬线；氧化铁皮；机械剥壳；抛丸

Development Status of Domestic Medium and High-Carbon Hard Wire Environmentally Friendly Pickling-Free Process Technology

Wang Haibin[1], Jia Jianping[1], Wang Jianzhong[1], Cao Guangming[2], Zhang Chaolei[3]

(1. HBIS Group Xuansteel Compan, Xuanhua 075100, China; 2. The State Key Laboratory of Rolling Automation, Northeastern University, Shenyang 110819, China; 3. School of Materials Science and Engineering, University of Science and Technology Beijing, Beijing 100083, China)

Abstract: Summarize the research results of environmentally friendly pickling-free process technology for medium and high-carbon hard wire at home and abroad. Take 82B as an example to introduce some progress made in recent years in environmentally friendly pickling-free wire 82B scale removal equipment and scale control technology, combined with experiments According to industrial production, the optimal scale structure of 82B suitable for mechanical peeling is mainly FeO, and the proportion of Fe_3O_4 should be controlled at about 36%. A large number of laboratory research and industrial production data show that the optimal scale and proportion of medium and high-carbon hard wire suitable for mechanical peeling basically follow this rule. It also pointed out the development prospects and significance of environmentally friendly pickling-free medium and high-carbon hard wires.

Key words: no pickling; medium and high carbon hard wire; oxide scale; mechanical peeling; shot blasting

微张力在轧制过程中的应用

马 越

（宣钢二钢轧厂，河北张家口 075100）

摘 要：宣钢二钢轧一棒生产线轧机为全连续式小型轧机，共 18 架，呈平立交替式布置。主轧线的主要设备有：冷热坯上料设备、步进梁式加热炉、高压水除鳞系统、主轧机、切头碎断飞剪、穿水冷却系统、倍尺剪、步进齿条式冷床、定尺剪、计数、打捆、称重、收集系统及液压润滑系统。上述主要设备引进世界上著名的冶金设备设计公司 POMINI 公司，其余部分为国外设备国内制造。

在棒材热轧生产线中，为保证产品尺寸精度，提高产品质量，避免由于各种原因产生的拉钢和堆钢现象，在粗轧区 1～10 架的轧机间引入了微张力控制思想，这是保证产品的关键自动控制环节。由于轧制过程中的工艺参数很多，如轧制速度、轧制力矩、轧制温度等，控制较复杂。

关键词：微张力控制；转矩；活套控制

Application of Micro Tension in Roling Process

Ma Yue

(Xuanhua Steel, Zhangjiakou 075100, China)

Abstract: Xuanhua Iron & Steel Co. , Ltd. . The rolling mill of No. 2 rolling and No. 1 bar production line is 18 continuous mini-mills, which are arranged alternately in horizontal and vertical position. The main rolling line of the main equipment are: hot and cold billet feeding equipment, walking beam heating furnace, high-pressure water phosphorus removal system, the main rolling mill, cut-and-break flying shear, through water cooling system, double-length Shear, step-rack cooling bed, fixed-length shear, counting, Baling, weighing, collection system and hydraulic lubrication system. The main equipment introduced to the world famous Metallurgical Equipment Design Company Pomini , the rest of the equipment for foreign domestic manufacturing.

In order to ensure the dimensional accuracy of products, improve the quality of products and avoid the phenomena of pulling and piling up steel caused by various reasons, the idea of micro-tension control is introduced into the rolling mills of 1-10 stands in rough rolling area, this is the key to ensure the automatic control of the product. Because there are many technological parameters in the rolling process, such as rolling speed, rolling Torque, rolling temperature, etc. , the control is more complex.

Key words: micro-tension control; torque; loop control

矿用 U 型钢生产工艺在万能轧机系统中的实践与应用

余选才

（武钢有限条材厂，湖北武汉 430081）

摘 要：本文从矿用 U 型钢生产工艺流程出发，以 36U 为例阐述了其生产工艺在万能轧机系统中的应用与实践，分析了生产过程中常见问题及解决方法。
关键词：矿用钢；36U；万能轧机；扭转

Practice and Application of Mining U-shape Steel Production Process in Universal Rolling Mill System

Yu Xuancai
(Wuhan Iron and Steel Co., Ltd., Wuhan 430080, China)

Abstract: Based on the production process of mine U-shape steel, the application and practice of 36U production process in universal rolling mill system are expounded, and the common problems and solutions in production process are analyzed.
Key words: mine steel; 36U; universal mill; twist

HRB400 热轧带肋钢筋弯曲脆断原因分析

张 阳，王 潇，刘明辉，向思宇
（武汉钢铁有限公司质量检验中心，湖北武汉 430080）

摘 要：热轧带肋钢筋 HRB400 弯曲试验时出现脆断现象，本文通过拉伸试验、化学成分分析、金相分析的方法，对造成脆断的原因进行分析。结果表明：造成弯曲脆断是由于试样内部组织存在大量魏氏体，因而脆性增加，韧性变差。
关键词：热轧带肋钢筋；脆断；魏氏体

Causes Analysis on Brittle Fracture in Bend Test of HRB400 Hot Rolled Ribbed Steel Bars

Zhang Yang, Wang Xiao, Liu Minghui, Xiang Siyu
(Quality Test Centre of Wuhan Iron and Steel Co., Ltd., Wuhan 430080, China)

Abstract: The phenomenon of brittleness fracture in HRB400 bending test of hot rolled ribbed steel bar is analyzed in this paper by tensile test, chemical composition analysis and metallographic analysis.The results show that there is a large amount of upper Widmannstatten structure in the microstructure of the specimen, which leads to the increase of brittleness and the deterioration of toughness.In order to avoid this situation, cooling temperature and rolling speed should be strictly controlled to avoid a large amount of Widmannstatten structure structure, so as to ensure good mechanical properties.
Key words: hot rolled ribbed steel bars; brittle fracture; Widmannstatten structure

车轮压轧过程数值模拟研究

国新春，陈 刚，王 健，宁 珅

（宝武集团马钢轨交材科技有限公司，安徽马鞍山 243000）

摘 要：本文对轮坯的压制过程进行了模拟分析，从预锻、终锻过程反推的方法，揭示了几种典型缺陷的成因。从而得出车轮轧制本身是一个开放性的塑性变形过程，只要轮坯几何形状合理、进给率适当，车轮轧制过程本身并不会使发生折叠。

关键词：车轮；模拟；轧制；折叠

Numerical Simulation of Wheel Rolling Process

Guo Xinchun, Chen Gang, Wang Jian, Ning Shen

(Baowu Group Ma Rail Material Technology Co., Ltd., Ma’anshan 243000, China)

Abstract: In this paper, the pressing process of wheel blank is simulated and analyzed, and the causes of several typical defects are revealed by the method of reverse deduction from the process of pre forging and final forging. It is concluded that the wheel rolling itself is an open plastic deformation process. As long as the wheel blank geometry is reasonable and the feed rate is appropriate, the wheel rolling process itself will not cause folding.

Key words: wheel; simulation; rolling; folding

微合金化生产 HRB400E 热轧带肋钢筋时效的研究及应用

高红星，陈 煜，韩北方，牛晓翠

（首钢长治钢铁有限公司，山西长治 046031）

摘 要：通过对首钢长钢微合金化生产 HRB400E 热轧带肋钢筋自然时效 3 天、20 天后的屈服强度进行检测，发现采用该工艺生产的钢筋屈服强度呈下降规律，为确保用户在进场检验时获得合格钢筋，首钢长钢的内控标准要求出厂的热轧带肋钢筋屈服强度下限高于 430MPa。

关键词：HRB400E；热轧带肋钢筋；时效；屈服强度；预警值

Study and Application of Aging in Microalloying HRB400E Hot-rolled Ribbed Steel Bars

Gao Hongxing, Chen Yu, Han Beifang, Niu Xiaocui

(Shougang Changzhi Iron & Steel Co., Ltd., Changzhi 046031, China)

Abstract: The yield strength of the HRE400E hot rolled ribbed steel bar after aging treatment, which produced by the microalloying in Shougang Changgang, has been investigated. The results show that the yield strength decrease. In order to obtain qualified steel bar for consumer, the internal control standard of the yield strength for in shougang changgang is higher than 430 MPa.

Key words: HRB400E; hot-rolled ribbed bar; aging; yield strength; early warning value

国产高速上钢系统的应用与提高

刘　强，于　杨，胡　洪

（首钢长钢公司，山西长治　046031）

摘　要：高速上钢系统主要应用在高速棒材精轧机后将倍尺材以18～40m/s的成品速度输送到冷床矫直装置，使小型单线棒材12～16mm机时产量由60t提高至110t。国产高速上钢系统在制造水平、控制系统工艺、操作维护水平较进口设备(达涅利、西马克等)存在一定的优势与劣势。通过在应用中的不断改进与提高，必将全面达到或超过进口设备的水平。

关键词：高速棒材；上钢系统；应用与提高

Application and Improvement of Domestic High Speed Steel Feeding System

Liu Qiang, Yu Yang, Hu Hong

(Shougang Changzhi Iron & Steel Company, Changzhi 046031, China)

Abstract: The high-speed steel feeding system is mainly used in the high-speed bar finishing mill to transport the double length bar to the cold bed straightening device at the rate of 18-40m/s, so that the output of small single-wire bar 12-16mm is increased from 60 tons to 110 tons per hour. Domestic high speed steel feeding system at the manufacturing level, control system technology, operation and maintenance level has certain advantages and disadvantages compared with imported equipment (Danieli, Simak, etc.). Through continuous improvement in application,it will reach or overpass the level of imported equipment.

Key words: high speed rolling bar; steel feeding system; application and improvement

高速棒材与切分棒材生产线优势论述

詹卫金，王　伟

（首钢长治钢铁有限公司轧钢厂，山西长治　046031）

摘　要：本文主要讲述高速棒材与切分棒材生产线的区别，从工艺流程、力学性能、金相组织、尺寸精度、重量偏差、成材率指标方面进行对比，全面阐述了高速棒材生产线的生产优势。

关键词：高速棒材；切分棒材；对比；效益

Discuss the Advantages of High Speed Rolling Compared with Splitting Rolling Bar Production Line

Zhan Weijin, Wang Wei

(Rolling Mill of Shougang Changzhi Iron & Steel Co., Ltd., Changzhi 046031, China)

Abstract: This paper mainly describes the difference between high speed rolling and splitting rolling bar production line, from the technology process, mechanical properties, metallographic structure, dimension accuracy, weight deviation and yield, and it totally describes the production advantages for high speed rolling bar production line.

Key words: high speed rolling bar; splitting rolling bar; contrast; benefit

R260 出口钢轨的开发实践

王瑞敏[1,2]，周剑华[2]，朱　敏[2]，费俊杰[2]，赵国知[2]

（1. 武汉科技大学 材料与冶金学院，湖北武汉　430081；
2. 宝钢股份中央研究院长材所，湖北武汉　430080）

摘　要：R260 钢轨主要用于客运线路或货运线路的直线段，要求有较高的强度和良好的韧性，针对国际出口钢轨市场需求，通过设计合理的内控化学成分和冶炼轧制工艺，成功开发了欧标 R260 钢轨，经工业化批量生产，产品实物质量控制良好，各项性能均符合欧洲钢轨标准 EN13674.1 的要求。

关键词：R260 钢轨；出口；开发

Development of Exported R260 Rail

Wang Ruimin[1,2], Zhou Jianhua[2], Zhu Min[2], Fei Junjie[2], Zhao Guozhi[2]

(1. School of Materials and Metallurgy, Wuhan University of Science and Technology, Wuhan 430081, China; 2. Institute of Long Steel Product of Baosteel Central Research Institute, Wuhan 430080, China)

Abstract: R260 rail is mainly used in passenger lines and straight section of freight lines, which requires high strength and good toughness. According to the needs of the international export rail market, through the design of reasonable internal control chemical composition and smelting and rolling processes, the R260 rail has been successfully developed. Through industrial mass production, the product quality is well controlled, and all properties meet the requirements of European rail standard EN13674.1.

Key words: R260 rail; exported; development

BD轧机推床传动系统受力分析与改进

余富春

（安宁昆钢型材厂，云南昆明 650300）

摘 要：本文简要介绍了热轧中、小型型钢的生产工艺流程，详细介绍了BD开坯轧机的辅助系统——推床装置的传动系统结构及功能，以及推床装置的控制要求，对推床装置传动减速机频繁损坏的原因进行受力分析和计算，通过校核对比，找到导致减速机损坏的原因，提出了改进方案，实施效果良好。

关键词：推床；传动装置；矫直；缓冲装置

Force Analysis and Improvement of Transmission System

Yu Fuchun

(Anning Kungang Profile Factory, Kunming 650300, China)

Abstract: This article briefly introduces the production process of hot-rolled medium and small section steel. The auxiliary system of the BD blooming mill—the transmission system structure of the pusher device is introduced in detail. Structure and function, as well as the control requirements of the pusher device, the frequency of the reducer of the pusher device is .The cause of the complicated damage is analyzed and calculated, and through the check and comparison, the cause of the reduction is found. The reason for the damage of the speed machine, the improvement plan is proposed, and the implementation effect is good.

Key words: push bed; transmission device; straightening; buffer device

螺纹钢轧后冷却规律研究及工艺优化

庞博文[1]，吴光行[1]，戴江波[1]，卓 见[1]，孙 晶[2]，李 兴[2]

（1. 中冶南方武汉钢铁设计研究院有限公司，湖北武汉 430081；
2. 宝武集团鄂城钢铁有限公司轧材厂，湖北鄂州 436000）

摘 要：围绕冷床冷却能力不足这一问题，用仿真软件模拟研究某厂20mm规格螺纹钢轧后冷却温度场演变规律，基于此提出工艺优化，增加喷雾冷却系统，改善冷床工作环境，满足提高产量的同时又保证了钢材性能。

关键词：螺纹钢；喷雾冷却；冷床；CFX

Research on the Cooling Law of Rebar after Rolling and Process Optimization

Pang Bowen[1], Wu Guangxing[1], Dai Jiangbo[1], Zhuo Jian[1], Sun Jing[2], Li Xing[2]

(1. WISDRI South Wuhan Iron and Steel Design and Research Institute Co., Ltd., Wuhan 430081, China;
2. Rolling Mill of Baowu Group Echeng Iron and Steel Co., Ltd., Ezhou 436000, China)

Abstract: Focusing on the problem of insufficient cooling capacity of the cooling bed, the evolution law of the cooling temperature field after rolling of a 20mm rebar of a certain factory was studied through CFX simulation, and the cooling water volume was modified to increase the spray cooling system, improve the working environment of the cooling bed, and meet the requirements of increasing production. At the same time, the performance of steel is guaranteed.

Key words: rebar; spray cooling; cooling bed; CFX

SCM440 轧后钢材弯曲原因分析

李崇建，尹衍成

（青岛特殊钢铁有限公司棒材所，山东青岛 266409）

摘　要：某公司生产的 SCM440 轧后棒材产生弯曲。通过现场观察，化学成分分析、金相组织观察、SEM 分析等方法，分析了轧材的弯曲原因。结果表明其组织不对称引起的轧后相变而产生组织应力不对称是导致轧材产生弯曲的根本原因，并提出了相应的解决办法。

关键词：SCM440；轧材；弯曲；组织应力

Analysis of Bending Reason of SCM440 Rolled Steel

Li Chongjian, Yin Yancheng

(Qingdao Special Iron & Steel Co., Ltd. Bar Division, Qingdao 266409, China)

Abstract: The SCM440 bar produced by a company is bent after rolling. The bending reason of rolled material was analyzed by means of field observation, chemical composition analysis, metallographic observation and SEM analysis. The results show that the microstructure stress asymmetry caused by the phase transition after rolling is the fundamental cause of bending of the rolled material, and the corresponding solutions are put forward.

Key words: SCM440; rolled bars; bending defect; microstructure stress

600MPa 级高强锚杆用钢筋开发

刘效云[1,2]，靳刚强[1,2]，韩伟娜[1]，马海峰[1,2]，贾元海[1,2]，王晓飞[1]

（1. 河钢股份有限公司承德分公司，河北承德 067002；
2. 河北省钒钛工程技术研究中心，河北承德 067002）

摘　要：通过分析合金元素对锚杆钢筋组织性能的影响及 YB/T4364—2014《锚杆用热轧带肋钢筋》标准，结合我公司钒资源优势，成功开发了 MG600 高强度锚杆钢筋，在保证产品强度指标的同时，具有更高的冲击韧性，可提高支护强度，尤其提高承受巨大切应力的能力，降低井巷支护密度，可进一步保障安全，减少钢材消耗。

关键词：MG600 锚杆钢筋；冲击韧性；力学性能

Development of High Strength and Toughness 600 Anchor Bar

Liu Xiaoyun[1,2], Jin Gangqiang[1,2], Han Weina[1], Ma Haifeng[1,2],
Jia Yuanhai[1,2], Wang Xiaofei[1]

(1. Chengde Branch of Hebei Iron and Steel Group Co., Ltd, Chengde, 067002, China;
2. Hebei Vanadium Titanium Engineering Technology Research Center, Chengde, 067002, China)

Abstract: By analyzing the effect of alloying elements on the microstructure and properties of anchor bar and the YB / T4364—2014 "hot rolled ribbed anchor bar" standard, Combined with my company's advantages of vanadium resources, MG600 high-strength anchor bar was successfully developed. At the same time to ensure the strength of the product, MG600 high-strength anchor bar have higher impact toughness, Can improve the strength of support, Especially to improve the ability to bear enormous shear stress, Reducing the density of well lane，Can further protect the safety, reduce steel consumption.

Key words: MG600 high-strength anchor bar; impact toughness; mechanical properties

HRB500E 高强抗震盘螺的研制

韩伟娜[1]，何文心[1]，贾元海[1,2]，高　敏[1]，刘效云[1,2]

（1. 河钢股份有限公司承德分公司，河北承德 067002；
2. 河北省钒钛工程技术研究中心，河北承德 067002）

摘　要：通过采用钒微合金化方式，成功的开发了 HRB500E 高强抗震盘螺，产品性能均达到 GB/T1499.2—2018 标准规定的抗震要求，500MPa 盘螺研制并广泛应用市场，具有良好的社会效益

关键词：500MPa 级；高强盘螺；研制

Research and Development of 500MPa Grade High Strength Plate

Han Weina[1], He Wenxin[1], Jia Yuanhai[1,2], Gao Min[1], Liu Xiaoyun[1,2]

(1. Chengde Branch of Hebei Iron and Steel Group Co., Ltd, Chengde, 067002, China; 2. Hebei Vanadium Titanium Engineering Technology Research Center, Chengde, 067002, China)

Abstract: Through vanadium microalloy method, HRB500E high-strength seismic coil was successfully developed, product performance met the seismic requirements specified in GB/T1499.2—2018 standard, 500MPa coil was developed and widely used in the market, with good social benefits.

Key words: 500MPa grade; high and strong seismic resistance; valve snail

5 表面与涂镀

大会特邀报告
分会场特邀报告
矿业工程
炼铁与原燃料
炼钢与连铸
轧制与热处理
★ 表面与涂镀
金属材料深加工
先进钢铁材料及其应用
粉末冶金
能源、环保与资源利用
冶金设备与工程技术
冶金自动化与智能化
冶金物流运输
其他

热轧低碳钢带表面氧化铁皮分析及酸洗生产实践

张 星[1]，刘 欢[2]，高小尧[1]，王俊伟[2]，侯明山[1]

（1. 唐山钢铁集团有限责任公司 技术中心，河北唐山 063016；
2. 唐山钢铁集团有限责任公司 冷轧薄板厂，河北唐山 063016）

摘 要：基于不同热轧产线生产的低碳钢带表面氧化铁皮厚度和微观结构分析设计了模拟酸洗实验并组织了工业生产实践，结果表明：2050线相对1580线热轧温度更高、轧后冷却距离更长，导致钢带表面氧化铁皮更厚、共析转变更充分、转变组织更致密，因此钢带适宜酸洗温度更高、酸洗时间更长；根据相关理论和实践经验对2050线热轧钢带氧化铁皮难以酸洗的问题提出了生产工艺改进建议。

关键词：低碳钢；氧化铁皮；酸洗；微观结构

Analysis of Surface Scales and Industrial Pickling Practice on Low Carbon Steel Hot Rolled Strip

Zhang Xing[1], Liu Huan[2], Gao Xiaoyao[1], Wang Junwei[2], Hou Mingshan[1]

(1. Tangshan Iron and Steel Group Co., Ltd., Technology Center, Tangshan 063016, China;
2. Tangshan Iron and Steel Group Co., Ltd., Cold Rolling Plant, Tangshan 063016, China)

Abstract: Surface scales thickness and microstructure analysis were carried on low carbon steel hot rolled strips from different production lines; and then simulated experiment and industrial pickling practice were investigated based on the result. The result shows that: because of higher hot rolling temperature and longer cooling distance, the surface scale on the strip has thicker, more complete and denser microstructure from 2050 line than that from 1580 line, which resulted to higher temperature and more time when acid pickling. In view of related theory and practice, some producing progress recommendations are given aiming to solve the problem of scale removing.

Key words: low carbon steel; scale; pickling; microstructure

几种GA板用防锈油的性能对比和选择

黎 敏，赵晓非，刘永壮，刁鑫林

（首钢集团有限公司技术研究院，北京 100041）

摘 要：通过挥发性试验、中性盐雾试验、摩擦因数测试和脱脂性能测试对比了6种合金化热镀锌板用防锈油的挥发性、润滑性、耐蚀性和可清洗性。结果表明：防锈油的黏度对其挥发性有较大的影响，在自然环境下都会快速挥发，在储存时应做好防护，避免防锈油挥发而导致防锈、润滑等作用消失，油溶性缓蚀剂中酯类成分的添加有利于热镀锌板表面摩擦性能的改善，黏度对其摩擦性能也有一定的影响，黏度适当增大有利于油膜吸附在基材表面，并在外力作用时减小磨损。影响油品可清洗性主要是油溶性缓蚀剂与金属结合力的强弱，与基础油黏度关系不大。

关键词：GA 板；防锈油；黏度；耐腐蚀性

Comparison of Properties between Several Kinds of Antirust Oils and How to Select Them for Galvannealed

Li Min, Zhao Xiaofei, Liu Yongzhuang, Diao Xinlin

(Shougang Research Institute of Technology, Beijing 100041, China)

Abstract: The volatility, lubricity, corrosion resistance, and cleanability of six kinds of antirust oils for Galvannealed products were studied by volatility test, neutral salt spray test, friction coefficient test, and degreasing test, respectively. The results show that the viscosity of anti-rust oil has a great influence on its volatility, and it will volatilize rapidly in natural environment. During storage, it should be protected to avoid the disappearance of anti-rust and lubrication caused by volatilization of anti-rust oil. The addition of ester components in oil-soluble corrosion inhibitor is beneficial to improve the surface friction performance of hot-dip galvanized sheet, and viscosity also has a certain influence on its friction performance. Appropriate increase of viscosity is beneficial to the adsorption of oil film on the surface of substrate, and reduces wear when external force acts. The washability of oil products is mainly affected by the combination of oil-soluble corrosion inhibitor and metal.

Key words: galvannealed; antirust oil; volatility; corrosion resistance

船用钢板表面防锈蚀涂层的制备及性能

高　鹏[1,2]，陈义庆[1,2]，武裕民[1,2]，艾芳芳[1,2]，李　琳[1,2]，钟　彬[1,2]，
伞宏宇[1,2]，苏显栋[1,2]，沙楷智[1,2]，田秀梅[1,2]

（1.海洋装备用金属材料及其应用国家重点实验室，辽宁鞍山　114009；
2.鞍钢集团钢铁研究院 焊接与腐蚀研究所，辽宁鞍山　114009）

摘　要：采用水性无机自固化涂料在船用 AH32 钢板表面制备了防锈蚀涂层，对涂料涂覆性能及涂层防护性能进行了研究。研究结果表明，水性无机自固化防锈蚀涂层具有优异的耐热性能，适于船用 AH32 钢板的在线处理；在高温高湿及高氯离子环境下涂层对船用 AH32 钢有良好的防护性能并可有效防止腐蚀深坑的生成。

关键词：水性无机自固化涂料；船用钢；防锈蚀涂层

Preparation and Property of Anti-rust Coating on Ship Steel Plate

Gao Peng[1,2], Chen Yiqing[1,2], Wu Yumin[1,2], Ai Fangfang[1,2], Li Lin[1,2], Zhong Bin[1,2], San Hongyu[1,2], Su Xiandong[1,2], Sha Kaizhi[1,2], Tian Xiumei[1,2]

(1. State Key Laboratory of Metal Material for Marine Equipment and Application, Anshan 114009, China;
2. Iron & Steel Research Institute of Angang Group, Anshan 114009, China)

Abstract: Waterborne self-cure inorganic coatings were coated on AH32 ship steel plate. The coating property and properties of self-cure inorganic coatings were studied. The results show that the waterborne self-cure inorganic coatings have been excellent thermal resistance, it can be applied to online processing of AH32 ship steel plate. The protective property of coatings exhibits excellent performance in environment of high temperature, high humidity and high chlorine ion. The anti-rust coating can effectively prevent formation of corrosion pit.

Key words: waterborne self-cure inorganic coatings; ship steel plate; anti-rust coating

合金化热镀锌板表面白斑缺陷的形成机理

蔡顺达[1]，宋利伟[1]，孙荣生[1]，张 健[2]，李 岩[1]，钟莉莉[1]

（1. 鞍钢集团钢铁研究院，辽宁鞍山 114009；2. 鞍钢股份有限公司，辽宁鞍山 114021）

摘 要：通过扫描电镜对合金化热镀锌板表面白斑缺陷进行微观形貌分析，运用 EDS 能谱仪对镀锌板表面正常区和白斑缺陷的元素含量进行对比分析，同时结合辉光放电光谱对合金化镀层进行研究，最终对合金化热镀锌板表面白斑缺陷的形成机理进行阐述。研究结果表明冷轧板表面点锈缺陷的存在是产生合金化热镀锌板表面白斑缺陷的主要原因，点锈缺陷会在合金化过程中阻碍锌铁相互扩散并造成表面异常凸起，合金化热镀锌板在局部区域锌铁比提升 7%及微观形貌的凹凸不平的交互作用下，最终宏观形貌表现为白斑缺陷。通过控制轧机架油泥滴落，降低乳化液残留和存放成品库房温度和湿度等措施，可有效控制冷轧板点锈缺陷的产生进而消除合金化热镀锌板表面白斑缺陷的产生。

关键词：合金化热镀锌；表面缺陷；点锈；白斑缺陷

Mechanism of the White-spot Defects on the Surface of Alloyed Hot-dip Galvanized Sheets

Cai Shunda[1], Song Liwei[1], Sun Rongsheng[1], Zhang Jian[2], Li Yan[1], Zhong Lili[1]

(1. Ansteel Iron&Steel Research Institute, Anshan 114009, China;
2. Angang Steel Co., Ltd., Anshan 114021, China)

Abstract: The microscopic morphology analysis of the white spot defects on the surface of the galvanized hot-dip galvanized sheet was carried out by scanning electron microscope, and the element content of the normal area and the white spot defect on the surface of the galvanized sheet was compared and analyzed by the EDS energy spectrometer. At the same time, the alloying was combined with the glow discharge spectroscopy. The coating is studied, and finally the formation mechanism of the white spot defect on the surface of the alloyed hot-dip galvanized sheet is explained. The research results show that the existence of spot rust defects on the surface of cold-rolled sheets is the main cause of white spot defects on the surface of alloyed hot-dip galvanized sheets. The spot rust defects will hinder the mutual diffusion of zinc and iron during the alloying process and cause abnormal surface protrusions and alloying. In the hot-dip galvanized sheet, under the interaction of a 7% increase in the zinc-to-iron ratio in the local area and the unevenness of the microscopic morphology, the final macroscopic morphology appears as a white spot defect. By controlling the dripping of oil sludge in the rolling stand, reducing the emulsion residue and storing the temperature and humidity of the finished product warehouse, the occurrence of spot rust defects in cold-rolled sheets can be effectively controlled and the white spots on the surface of alloyed hot-dip galvanized sheets can be eliminated.

Key words: alloyed hot-dip galvanized sheets; surface defects; rust; white-spot defects

镀铝锌机柜钢产品锌层折弯开裂机理研究及控制手段优化

宋利伟[1]，蔡顺达[1]，姜丽丽[2]，孙荣生[1]，杨洪刚[1]，李　岩[1]
（1. 鞍钢集团钢铁研究院，辽宁鞍山　114009；2. 鞍钢股份冷轧厂，辽宁鞍山　114021）

摘　要：镀铝锌机柜钢产品（钢质 DX51D，厚度 1.2mm 以上，锌层重量 A120 以上）在用户加工使用过程中，频繁发生 180°折弯后锌层开裂问题，影响用户使用。该研究从理论上分析了鞍钢镀铝锌机柜钢产品折弯后锌层开裂问题的原因，并对锌层韧性的机理及其影响因素进行了深入研究，最后提出了提高机柜钢产品锌层韧性，防止折弯使用时出现锌层开裂缺陷的工艺控制方法。

关键词：镀铝锌；锌层开裂；锌层韧性；合金过渡层

Study on Benging Cracking Mechanism of Zinc Layer of Aluminum-zinc Cabinet Steel Products and Optimization of control Measures

Song Liwei[1], Cai Shunda[1], Jiang Lili[2], Sun Rongsheng[1], Yang Honggang[1], Li Yan[1]
(1. Iron and Steel Research Institute of Angang Group, Anshan 114009, China;
2. Cold Rolling Plant of Ansteel, Anshan 114021, China)

Abstract: Aluminum-zinc-plated cabinet steel products (steel DX51D, thickness above 1.2mm, zinc layer weight above A120) are frequently cracked after 180 bending, which affects users' use. In this study, the causes of zinc layer cracking after bending of aluminum-zinc cabinet steel products in Angang were analyzed theoretically, and the mechanism of zinc layer toughness and its influencing factors were deeply studied, Finally, the process control methods to improve the zinc layer toughness of cabinet steel products and prevent zinc layer cracking defects during bending were put forward.

Key words: aluminum and zinc plating; the zinc layer cracks; toughness of zinc layer; alloy transition layer

镀铬工作辊在冷连轧机组的应用

毛玉川[1]，孙荣生[2]，刘英明[1]，张一凡[1]，张福义[1]，郑麟飞[1]
（1. 鞍钢股份有限公司冷轧厂，辽宁 鞍山　114021；
2. 鞍钢集团钢铁研究院，辽宁 鞍山　114009）

摘　要：为了有效的利用轧辊表面镀铬特性，使其在冷连轧机组上得到更好的应用，本文首先对表面镀铬层的厚度进行了确定，并对镀铬辊在冷连轧机组上的使用情况进行了跟踪，发现使用镀铬工作辊能够有效提高轧辊轧制周期，减少换辊时间，同时能够提高带钢的实物质量，最后对在镀铬辊上完成自由轧制提出了建议。

关键词：冷连轧；镀铬辊；应用；轧制周期

The Appliance of the Work Roll by Chromate Treatment on the TCM

Mao Yuchuan[1], Sun Rongsheng[2], Liu Yingming[1], Zhang Yifan[1], Zhang Fuyi[1], Zheng Lingfei[1]

(1.Cold Rolled Strip Steel Mill of Angang Steel Co., Ltd., Anshan 114021, China;
2.Ansteel Iron and Steel Research Institutes, Anshan 114009, China)

Abstract: For using the characteristic of the chromate-roll validly and ensuring the appliance on the TCM, the thickness of the chromate was confirmed in this paper firstly, and the using of the chromate-roll on the TCM was tracked, the fact is that increasing the rolling time, reduce the changing roll time and heightening the quality of the rolling strip by using of the chromate-roll on the TCM was found, and the suggestion for liberally rolling on the TCM at last.

Key words: the tandem cold mill; chromate-roll; appliance; rolling cycle

热镀锌生产线的环保抑尘处理技术应用探索

徐言东[1]，王兆辉[1]，马　骏[2]，王占坡[3]，马树森[4]

（1. 北京科技大学工程技术研究院，北京　100083；2. 唐山市开平鑫德热镀锌技术有限公司，河北唐山 0630213；3. 冶金自动化研究设计院，北京　100071；
4. 大连中锌机械设备有限公司，辽宁大连　116000）

摘　要：热镀锌生产线在作业过程中，会出现大量锌烟雾，弥漫在车间的空气中，如果不加以捕集和处理，不仅污染生产环境，还会对生产人员造成职业健康危害，而且会污染外部环境。随着环保政策的标准提高，对热镀锌企业的环保治理实行一票否决制，国内外相关研究机构和企业加强了对锌烟雾环保治理研究投入，经过调研，发现各种捕集方式纷杂，需要根据生产线的工艺特性进行合理选择；目前脉冲布袋式除尘系统仍占据锌烟雾处理的主流，旋风布袋组合式、湿法静电式、多段分离式除尘器等其他方式偶有应用案例，仍然处于探索阶段，并未形成工业化、规模化应用，钢铁行业轧钢生产线采用的塑烧板除尘器、日本某公司 BEC 粉尘环保抑制处理工艺设备进行跨行业试验应用，取得了较好的效果，如果加以推广，锌烟雾环保抑尘技术必将形成技术突破得以更好的促进整个热镀锌行业锌烟雾处理的技术革命。

关键词：热镀锌；环保抑尘；锌烟雾

Exploration and Application of Environmental Protection and Dust Suppression Technology in Hot Dip Galvanizing Production Line

Xu Yandong[1], Wang Zhaohui[1], Ma Jun[2], Wang Zhanpo[3], Ma Shusen[4]

(1. Engineering and Technology Research Institute, University of Science and Technology Beijing, Beijing 100083, China; 2. Kaiping Co., Ltd., Tangshan 0630213, China; 3. Automation Research and Design Institute of Metallurgical Industry Beijing 100071, China;
4. Dalian Zhongzin Machinery Equipment Co., Ltd., Dalian 116000, China)

Abstract: During the operation of hot dip galvanizing line, a large amount of zinc smoke will appear in the air of workshop. If it is not captured and treated, it will not only pollute the production environment, but also cause occupation health hazards to the production personnel, and pollute the external environment. With the improvement of the standard of environmental protection policy, the one vote veto system is implemented for the environmental protection treatment of hot-dip galvanizing enterprises. Relevant research institutions and enterprises at home and abroad have strengthened the investment in the research of environmental protection treatment of zinc smoke. After investigation, it is found that various capture methods are complex and need to be reasonably selected according to the process characteristics of the production line; At present, the pulse bag type dust removal system still occupies the mainstream of zinc smoke treatment. There are occasional application cases of cyclone bag combination type, wet electrostatic type, multi-stage separation type dust remover and other methods, which are still in the exploration stage and have not formed industrialized and large-scale application. The cross industry test and application of plastic burning plate dust collector and Japan BEC dust environmental protection suppression treatment process and equipment used in steel rolling production line of iron and steel industry have achieved good results. If it is promoted, zinc smoke environmental protection and dust suppression technology will form a technological breakthrough, which can better promote the technological revolution of zinc smoke treatment in the whole hot dip galvanizing industry.

Key words: hot dip galvanizing; environmental protection and dust suppression; Zinc smog

连退板表面磷化膜 P 比测定方法的研究

蔡 宁，郝玉林，任 群，龙 袁，姚士聪，曹建平

（首钢集团有限公司技术研究院、绿色可循环钢铁流程北京市重点实验室，北京 100043）

摘 要： 钢铁材料经过磷化处理后可以具有防锈、减摩、增加涂漆附着力等作用，因此磷化处理是钢铁材料一种重要的表面处理技术，尤其在汽车板领域应用广泛。汽车板表面磷化处理通常采用锌系磷酸盐，形成的磷化膜中通常包含两相，分别为磷酸锌（Hopeite，简称为 H 相，化学式为 $Zn_3(PO_4)_2 \cdot 4H_2O$）、磷酸二锌铁（phosphophyllite，简称为 P 相，化学式为 $Zn_2Fe(PO_4)_2 \cdot 4H_2O$）。通常采用 P 比分析磷化膜中磷酸二锌铁（P 相）所占的比例。有关 P 比测试方法的研究报道相对较少，本文采用 X 射线衍射（XRD）法系统分析了 Cu、Cr 二种靶材、扫描步长、单步信号采集时间等对 P 比测试结果的影响，以及对比直接计算法与模拟分峰法对 P 比测试结果的影响，为 P 比测试方法的选择提供参考。结果发现 Cu 靶测试中 P 相与 H 相最强峰角度差 0.24°，重峰明显，采用 Cr 靶测试可将两相的角度差提高到 0.35°，重峰现象得到改善，进而提高了 P 比测量精度。

关键词： 磷化处理；H 相；P 相；P 比；X 射线衍射

Study on the Determination of P Ratio for Phosphating Film on Continuous Annealing Plates

Cai Ning, Hao Yulin, Ren Qun, Long Yuan, Yao Shicong, Cao Jianping

(Research Institute of Technology of Shougang Group Co., Ltd., Beijing Key Laboratory of Green Recyclable Process for Iron & Steel Production, Beijing 100043, China)

Abstract: Phosphating can increase corrosion resistance, decrease friction coefficient and increase paint adhesion for iron and steel materials. Therefore, phosphating is an important surface treatment technology of iron and steel materials, especially for automobile plates. Zinc phosphates are usually used for the surface phosphating treatment of automobile plates. The phosphating film usually contains two phases, namely, hopeite (H phase for short, the chemical formula is $Zn_3(PO_4)_2$•$4H_2O$) and phosphorhophyllite (P phase for short, the chemical formula is $Zn_2Fe(PO_4)_2$•$4H_2O$). P ratio is usually used to analyze the proportion of P phase in the phosphating film. There were few reports about p-ratio test method. In this paper, X-ray diffraction (XRD) method was used to analyze the influence of Cu and Cr targets, scanning step, step signal acquisition time, etc. on p-ratio test results, as well as the influence of direct calculation method and simulated peak splitting method on p-ratio test results, so as to provide a reference for the selection of p-ratio test methods. The results show that the angle difference between the strongest peak of P phase and H phase is 0.24°, and the overlap of the two peaks is obvious. The angle difference between the two phases can be increased to 0.35° by Cr target test, and the phenomenon of the peak overlap can be decreased, and the accuracy of P ratio measurement can be improved.

Key words: phosphating; hopeite; phosphorhophyllite; p-ratio; X-ray diffraction

双层焊管表面黑灰原因分析

李 雯[1]，王志登[1]，阎元媛[2]

（1. 宝钢研究院 梅钢技术中心，江苏 南京 210039；2. 宝钢研究院，上海 201900）

摘 要：针对双层卷焊管焊后表面发黑的现象，首先采用扫描电镜、热重分析对助焊黑漆的成分、热物性进行了研究，结果表明，助焊黑漆一方面可在钎焊过程中辅助吸热，另一方面碳化成灰膜后包裹焊管，可提高熔融铜膜层的表面张力，防止铜膜融化后在焊管表面流动影响表面质量。采用三维共聚焦显微镜观察发黑焊管表面与正常焊管表面，发现发黑焊管的铜膜层呈现明显的由于成分过冷而形成的胞状晶，而正常表面则为平面晶。利用 DIL805A/D 热膨胀仪研究了不同表面状态下、不同钎焊工艺下涂覆黑漆的焊管经高温模拟钎焊后的表面黑灰残留及铜膜层的微观组织情况，结果发现，焊管表面粗糙度及轧制油对黑灰残留有直接影响，没有轧制油及较高的表面粗糙度，均会造成明显的黑灰残留；同时，较高的加热温度及较慢的冷却速度也会造成一定的黑灰残留，实验结果与表面发黑焊管实际观察到的胞状晶的形成原因较为吻合。

关键词：双层卷焊管；表面发黑；残余黑灰；助焊漆

Analysis on Causes for Ashes Remained on the Double Wall Brazed Tube Surface

Li Wen[1], Wang Zhideng[1], Yan Yuanyuan[2]

(1. R&D Center for Meisteel of Baosteel Research Institute, Nanjing 210039, China;
2. Baosteel Research Institute, Shanghai, 201900, China)

Abstract: Aiming at the phenomenon of the blackening surface on the double wall brazed tube,SEM was used to analyse the ingredients of the auxiliary welding paint,thermogravimetric analysis was used to analyse paint thermophysical properties,the results show that auxiliary paint could help heat absorption and residual paint ashes could wrap the tube and prevent copper plating from flowing when the temperature reaches the melting point.3D measuring laser microscope was used to observe the surface microstructure of the bright and blackening tube, plane grain presented in the former and the typical cellular crystal originated from compenent supercooling presented in the the latter.DIL805A/D thermal dilatometer was used to simulate the weiding process under different tube surface and different heating processes.The research indicated the higher sruface roughness and the less rolling oil,the more residual ashes, at the same time,higher heating temperature and slower cooling rate ,the more ashes too,which surely coincided with the cellular crystal analysis of the blackening tube surface.

Key words: double wall brazed tube; blackening surfae;residual ashes ;auxiliary welding paint

热轧板带平整机矫直辊修复的新型表面复合强化技术

解明祥，倪振航，胡小红，赵慧颖

（安徽马钢表面技术股份有限公司，安徽马鞍山　243000）

摘　要：本文通过分析热轧板带平整机矫直辊的工况和表面失效形式，综合表面技术特点，创新堆焊硬化层加超音速喷涂碳化钨涂层的复合强化技术，开展了工艺路线设计、材料选择和关键工序参数的摸索，并制作了试块对复合强化层的金相组织、硬度、涂层结合强度、磨损性能等进行了检测，结果显示复合强化层的各项指标优良，最后通过复合强化技术在矫直辊上实际应用和验证，有效改善了矫直辊表面使用效果、延长了其使用寿命。

关键词：平整机；矫直辊；堆焊；超音速火焰喷涂

New Surface Compound Strengthening Technology for the Repairment of Straightening Roll Used in Hot Strip Skin-pass Mill

Xie Mingxiang, Ni Zhenhang, Hu Xiaohong, Zhao Huiying

(Anhui Ma Steel Surface Technology Co., Ltd., Ma'anshan 243000, China)

Abstract: The working conditions and surface failure modes of straightening roll used in hot strip temper mill were analyzed in this article. The compound strengthening technology which combine welding hardened layer and HVOF sprayed tungsten carbide coating is innovated after comprehensive considering the characteristics of each surface technology. This article comprehensively innovate a new technology, which adopts overlaying welding hardened layer and HVOF spraying tungsten carbide coating, to strengthen the surface during the repairing of the straightening roll. The process route design, material selection and key process parameters exploration were carried out. Then the compound strengthening layer was deposited on samples. The metallography, hardness, coating bonding strength and wear resistance were tested respectively. The result shows that each index of the compound strengthening layer is excellent. Finally, the compound strengthening technology was carried out on the surface of straightening roll. It exhibits excellent performance after on-line service. And the service life of the roll is prolonged effectively.

Key words: hot strip skin-pass mill; straightening roll; overlaying welding; HVOF spraying

环保型聚酯涂层建筑彩涂钢板工艺技术研究

施国兰，钱婷婷，谷　曦

（马鞍山钢铁股份有限公司，安徽马鞍山　243000）

摘 要：为顺应国家标准对卷材涂料有害物质的限量要求，开展了环保型聚酯涂层彩涂钢板工艺试制，并从涂层物理性能、耐中性盐雾腐蚀性能、耐酸碱性能等方面进行了评价；从结果看，环保型聚酯彩涂钢板的性能满足标准的要求，但部分性能仍与含铬体系产品有差距，需要继续优化。

关键词：环保；无铬；聚酯彩涂板

Study on Technology of Environmentally Friendly Polyester Color Coated Steel Sheet for Buildings

Shi Guolan, Qian Tingting, Gu Xi

(Ma'anshan Iron & Steel Co., Ltd., Ma'anshan 243000, China)

Abstract: In order to meet the requirements of national standards for the limit value of environmental harmful substances in coil coatings, the trial production of environmentally friendly polyester color coated steel sheet was carried out, and the physical properties, neutral salt spray corrosion resistance, acid and alkali resistance of the coating were evaluated; From the results, the performance of the environmental protection polyester color coated board meets the requirements of the standard, but some properties are still behind the products with chromium system, which need to be further optimized.

Key words: Environmentally friendly; Chromium free; Polyester color coated sheet

水性丙烯酸聚氨酯面漆在轨道交通车轮上的应用研究

周立强，陈 刚，张 磊，吴 争，魏 天

（宝武集团马钢轨交材料科技有限公司，安徽马鞍山 243000）

摘 要：以水性羟基丙烯酸二级分散体与疏水性多异氰酸酯和亲水性多异氰酸酯交联制得轨道交通车轮用水性丙烯酸聚氨酯面漆，研究了基料的选型、固化剂混拼不同质量比、n（-NCO）/n（-OH）、涂装作业环境对涂层外观以及性能的影响，简述了水性丙烯酸聚氨酯面漆在轨道交通车轮上的应用。

关键词：轨道交通车轮；水性羟基丙烯酸分散体；异氰酸酯；涂装工艺

Application of Waterborne Acrylic Polyurethane Topcoat in Rail Transit Wheels

Zhou Liqiang, Chen Gang, Zhang Lie, Wu Zheng, Wei Tian

(Baowu Group Masteel Rail Transit Materials Technology Co., Ltd., Ma'anshan 243000, China)

Abstract: Waterborne acrylic polyurethane topcoat for rail transit wheels was prepared by crosslinking waterborne hydroxyacrylic acid secondary dispersion with hydrophobic polyisocyanate and hydrophilic polyisocyanate. The effects of base material selection, curing agent mixing, n(-NCO)/n(-OH) and coating environment on the appearance and properties of the coating were studied, The application of waterborne acrylic polyurethane topcoat on rail transit wheels was introduced.

Key words: rail transit wheels; aqueous hydroxy acrylic acid dispersion; isocyanate; coating process

X 荧光的 K 谱线法测定锌铁合金镀层铁含量的方法研究

华 犇，范 纯，朱子平

（宝钢股份有限公司制造管理部，上海 201900）

摘 要：采用 K 谱线法测定锌铁合金镀层铁含量精度以及准确性相比于 L 谱线有很大的提高，而采用经验法测定受到锌层重量差异的影响较大不能得到准确结果。本方法建立基本参数法进行多层合金薄膜的成分和膜厚的同时分析，同时考虑表面涂油量对检测结果的影响。GB/T24514 给出了火焰原子吸收光谱法测定镀层厚度和化学成分的方法，通过与国标方法的比对验证验证方法的准确性。

关键词：锌铁合金；X 荧光法；K 谱线；铁含量

Determination of Iron Content in Zn-Fe Alloy Coating by X-ray Fluorescence Spectrometry

Hua ben, Fan Chun, Zhu Ziping

(Bao Iron & Steeel Co., Ltd., Shanghai 201900, China)

Abstract: The precision and accuracy of determination of iron content in Zn-Fe alloy coating by K-spectrum method are much higher than that by L-spectrum method. The basic parameter method is established for the simultaneous analysis of the composition and thickness of multi-layer alloy films, and the influence of the amount of oil applied on the results is considered. GB/T24514 gives the flame atomic absorption spectroscopy method to determine the thickness and chemical composition of the coating, and verifies the accuracy of the method by comparing with the national standard method.

Key words: Zn-Fe alloy coating; X ray method; K-spectrum; iron content

光伏支架用后浸镀锌和连续热镀锌铝镁钢带耐蚀性对比研究

于程福[1]，梅淑文[1]，高小尧[1]，肖 伟[2]，张 鹏[3]

（1. 唐山钢铁集团有限责任公司 技术中心，河北唐山 063016；2. 唐山钢铁集团有限责任公司冷轧薄板厂，河北唐山 063016；3. 河钢集团钢研总院，河北石家庄 052160）

摘 要：为研究钢带表面镀层对耐蚀性的影响规律，对不同镀层厚度的后浸镀锌钢带与镀层重量 275g/m^2 的连续热镀锌铝镁钢带加工的 C 型光伏支架进行了盐雾试验和电化学试验，并对镀层组织、成分进行了分析，结果表明：镀层重量为 275g/m^2 的锌铝镁钢带与镀层厚度 85μm 的后浸镀锌钢带耐蚀性与使用寿命相当；100μm 的后浸镀锌钢带因合金层组织异常导致其早期出现轻微锈点，但耐蚀性最强，使用寿命最长；而 65μ 的后浸镀锌钢带因镀层厚度过小导致耐蚀性最差，使用寿命最短。

关键词：锌铝镁；后浸镀锌；盐雾试验；镀层组织

Comparative Study on Corrosion Resistance of Post-dip Galvanizing and Continuous Hot-dip Galvanizing Zn-Al-Mg Steel Strip For Photovoltaic Support

Yu Chengfu[1], Mei Shuwen[1], Gao Xiaoyao[1], Xiao Wei[2], Zhang Peng[3]

(1. Tangshan Iron and Steel Group Co., Ltd. Technology Center, Tangshan 063016, China;
2. Tangshan Iron and Steel Group Co., Ltd. Cold Rolling Plant, Tangshan 063016, China;
3. HBIS Group Technology Research Institute, Shijiazhuang 052160, China)

Abstract: In order to study the steel strip surface coating on corrosion resistance of the influence law. Dipped galvanized steel strips of different coating thickness and coating weight of 275g/m^2 continuous hot dip galvanized Zn-Al-Mg strip processing photovoltaic salt-fog experiment and Electrochemical experiment was carried out with a C type support, And after galvanizing coating organization and composition are analyzed, The results show that the corrosion resistance and service life of Zn-Al-Mg steel strip with coating weight of 275g/m^2 is similar to that coating thickness of 85μm. Due to the abnormal microstructure of the alloy layer, the red rust time of the 100μm post-dip galvanized steel strip is shorter, but the corrosion resistance is the strongest, the longest service life ;while the corrosion resistance and service life of the 65μm post-dip galvanized steel strip is the worst due to the small coating thickness.

Key words: hot-dip galvanizing zn-al-mg steel; post-dip galvanized; salt spray test; coating organization

浅谈先进彩涂板发展现状与工程创新实践

王海东，严江生，朱嘉芳

（宝钢工程技术集团有限公司 冷轧事业部，上海 201900）

摘 要：本文介绍了先进彩涂板的发展现状与趋势，并简要论述了典型先进彩涂板产线自主集成与工程创新实践。中国宝武典型先进彩涂板产线的建设对于对于后续建设或改造同类产线，具有很好的借鉴价值和示范引领作用。

关键词：彩涂板；创新；实践

Present Development and Engineering Innovation Practice of Color Coated Steel Sheet

Wang Haidong, Yan Jiangsheng, Zhu Jiafang

(Cold Rolling Department of Baosteel Engineering & Technology Group Co., Ltd, Shanghai 201900, China)

Abstract: The present development of color coated steel sheet is introduced. The development of advanced color coated steel sheet and engineering innovation practice of typical advanced color coated steel sheet production line is described. The construction of Baowu typical advanced color coated steel sheet production line is of great strategic significance for improving the production technology level of color coated steel sheet in China and leading the production technology progress of color coated steel sheet in China.

Key words: color coated steel sheet; innovation; practice

带钢连续退火炉喷气快冷技术分析

田茂飞[1]，许秀飞[2]

（1. 重庆赛迪工程咨询有限公司，重庆　400013；中冶赛迪集团中央研究院，重庆　401122）

摘　要：本文对三种主流带钢连续退火炉喷气快冷技术进行了比较和分析。认为：喷缝式冷却风箱与传统的喷箱相比无论是热效率还是冷却速率都有了大幅度的提高；圆筒形的炉膛和喷梁开孔结构的高速冷却模块，其炉膛和冷却模块的刚度都很高，非常适合采用高氢介质实现带钢的快速冷却；高速"气线"分区冷却箱热气溢流速度很快、换热效率很高，而且带钢横向温度均匀性好，不会产生瓢曲问题。

关键词：带钢连续退火；高强钢；喷气快冷

Analysis of Jet Cooling Technology of Continuous Annealing Furnace for Strip Steel

Tian Maofei[1], Xu Xiufei[2]

(1. Chongqing CISDI Engineering Consulting Co., Ltd., Chongqing 400013, China;
2. Central Research Institute of MCC CISDI Group, Chongqing 401122, China)

Abstract: In this paper, three mainstream jet cooling technologies of continuous annealing furnace for strip steel are compared and analyzed. It is concluded that the thermal efficiency and cooling rate of the spray-slot cooling bellows are greatly improved compared with the traditional spray cooling bellows. The rigidity of the cylindrical furnace and the high-speed cooling module with nozzle beam opening structure is very high, which is very suitable for the rapid cooling of strip steel with high hydrogen medium; The high-speed "gas line" zoned cooling box has a fast hot gas overflow speed, high heat exchange efficiency, and good uniformity in the transverse temperature of the strip, which will not cause buckling problems.

Key words: continuous annealing of strip steel; high-strength steel; jet cooling

某镀铝锌线带钢加工脱锌缺陷分析

郭海涛，晋红革，王雅伟

（首钢京唐钢铁联合股份有限公司冷轧作业部，河北唐山　063200）

摘　要：镀铝锌产品是钢带进入到成分为 55%Al - 43.4%Zn - 1.6%Si 的镀液中制作的钢板，和镀锌钢板相比，具有靓丽的外观，并具备相同的加工性能和更高的耐蚀性、反射率和硬度。用户反映镀铝锌产品在加工过程中，产生了脱锌问题，本文对比脱锌样板和正常样板对镀层结构、结合层状态、钢基表面成分进行电镜分析，从清洗质量、退火炉气密性和工艺参数控制三方面阐述了脱锌的原因。

关键词：镀铝锌钢板；合金层；粘附性；脱锌

Aluminized Zinc Steel Dezincification Analyse

Guo Haitao, Jin Hongge, Wang Yawei

(Shougang Jingtang Steel Federation Limited Liability Company Cold Rolling Department, Tangshan 063200, China)

Abstract: Compared with galvanized steel plate, 55%Al - 43.4%Zn - 1.6%Si Aluminized zinc steel has Excellent corrosion resistance, beautiful appearance, and has the same processing performance and higher corrosion resistance, reflectivity and hardness. Users report that the problem of dezincification occurs during the processing of aluminized zinc products. This paper analyzes the structure of the coating, the state of the bonding layer and the surface composition of the steel base electron by microscopy, then expound the reasons of dezincification from three aspects of cleaning quality, air tightness of annealing furnace and process parameter control.

Key words: aluminized zinc steel; alloy layer; adhesion; dezincification

表面选择性氧化对热镀锌 DP980 镀层/基板界面抑制层的影响

朱 敏[1]，金鑫焱[2]，陈 光[3]

（1. 宝钢湛江钢铁有限公司制造管理部，湛江 524072；2. 宝山钢铁股份有限公司中央研究院，上海 201999；3. 宝山钢铁股份有限公司制造管理部，上海 201999）

摘　要：以热镀锌双相钢 DP980 为研究对象，采用连续退火及热浸镀模拟的方法，研究了退火钢板表面选择性氧化对热镀锌钢板镀层/基板界面抑制层的影响。采用 GD-OES 分析了退火板、镀锌板元素深度分布，使用 SEM 观察了溶锌后基板表面抑制层形貌，使用 TEM 观察了 FIB 制备的截面试样镀层/基板界面抑制层的微观结构。结果表明，当退火气氛露点为–10℃时，退火试样表面形成了不连续的 Mn 外氧化及 Mn、Si 内氧化。在热浸镀过程中，局部露出铁基体的位置 Al、Fe 反应较充分，形成了相对粗大的 Fe-Al 抑制层；局部 Mn 外氧化较厚的位置，影响了 Al、Fe 扩散，显著阻碍了抑制层形成。

关键词：DP980；选择性氧化；热镀锌；抑制层

Effect of Selective Oxidation on the Formation of Inhibition Layer During Hot Dip Galvanizing of DP980 Steel Sheet

Zhu Min[1], Jin Xinyan[2], Cheng Guang[3]

(1. Manufacturing Management Department, Baosteel Zhanjiang Iron & Steel Co., Ltd., Zhanjiang 524072, China; 2. Central Research Institute, Baoshan Iron and Steel Co., Ltd., Shanghai 201999, China; 3. Manufacturing Management Department, Baoshan Iron and Steel Co., Ltd., Shanghai 201999, China)

Abstract: Annealing and hot dip galvanizing simulation experiments were conducted to investigate the effect of selective oxidation on the formation of inhibition layer during hot dip galvanizing of dual phase DP980 steel sheet. Both the alloy elemental depth profiles of the annealed and galvanized samples were tested by GD-OES, the surface morphology of the inhibition layer on the substrate was observed by SEM, and the cross section was prepared by FIB and observed by TEM. When it is annealed in an atmosphere having a dew point of -10℃, a discontinuous external Mn oxidation layer and internal Mn, Si oxidation form on the surface and in the subsurface of the steel. The exposed substrate which is not covered by external oxidation promotes the nucleation and growth of Fe-Al intermetallic phase, forming coarse grain inhibition layer. Thick MnO external oxidation layer decrease the diffusion rate of Al and Fe, significantly suppressing the formation of inhibition layer.

Key words: DP980; selective oxidation; hot dip galvanizing; inhibition layer

连续退火炉麻点缺陷分类及产生机理分析

瞿作为，陈　平，杨　力，曲慧铃，陈　佳，关洪星

（宝钢股份武钢有限冷轧厂，湖北 武汉　430080）

摘　要：本文总结归纳了汽车厂用户反映的普冷汽车用钢板连续退火炉表面“麻点”缺陷，通过对不同用户反馈的“麻点”缺陷使用某型号显微分析仪进行微观形貌特征分析，进行微观分类辨识，并对各类型麻点形成机理进行研究分析，总结出四类不同形貌下的“麻点”真实形态及其对应的产生机理。

关键词：连续退火；麻点；缺陷分类；产生机理

Classification and Mechanism Analysis of Pitting Defects in Continuous Annealing Furnace

Qu Zuowei, Chen Ping, Yang Li, Qu Huilin, Chen Jia, Guan Hongxing

(Cold Mill of Baosteel Wuhan Iron and Steel Ltd., Wuhan 430080, China)

Abstract: In this paper, the “pitting” defects on the surface of steel plate continuous annealing furnace reported by users in automobile factories are summarized. Through the “pitting” defects reported by different users, a certain type of microanalyzer is used to analyze the micro-morphology characteristics, carry out microscopic classification and identification, and the formation mechanism of various types of pitting is studied and analyzed. The real morphology and corresponding formation mechanism of four kinds of "pitting" with different morphologies are summarized.

Key words: continuous annealing; pitting; defect classification; the mechanism

重量法测定镀锌钢板无铬钝化膜质量

吴镇君

（本钢板材股份有限公司检化验中心，辽宁 本溪 117000）

摘 要：钝化作为镀锌钢板的一种表面后处理方式，增强了镀锌板的耐腐蚀性能。钝化液的涂覆量是镀锌板生产过程中必须控制的工艺参数之一，所以需对镀锌板的钝化膜质量进行分析测定。该论文研究了用重量法测定镀锌钢板的无铬钝化膜质量。用配制好的剥离液将试样中含有钝化膜的镀锌层剥离溶解，利用试样前后的重量差和试样表面积求出锌层和钝化膜层总质量，再用电感耦合等离子体发射光谱法测定镀层中锌含量，进而求得钝化膜的质量。该方法检测结果准确可靠，而且排除了钝化液成分波动的影响。

关键词：重量法；镀锌板；钝化膜质量；电感耦合等离子体；锌层质量

Determination of the Mass of Chromium-free Passivation Film on Galvanized Steel Plate by Gravimetric Method

Wu Zhenjun

(Inspection Center in Bengang Steel Plates Co., Ltd, Benxi 117000, China)

Abstract: Passivation, as a surface post-treatment method of galvanized steel plate, enhances the corrosion resistance of galvanized steel plate. The coating amount of passivation solution is one of the process parameters that must be controlled in the production process of galvanized plate. Therefore, it is necessary to determine the mass of passivation film on galvanized steel plate. In this paper, the determination of the mass of chromium-free passivation film on galvanized steel plate by gravimetric method was studied. The prepared stripping solution was used to peel and dissolve the zinc coating with passivation film of the sample. The total mass of zinc coating and passivation film can be calculated by the lost weight and the surface area of the sample. The mass of zinc coating was determined by inductively coupled plasma emission spectrometry. Then the mass of passivation film was obtained. The test results by this method are accurate and reliable, and the influence of composition fluctuation of passivation solution is eliminated.

Key words: gravimetric method; galvanized steel plate; mass of passivation film; inductively coupled plasma; mass of zinc coating

热镀锌板表面细小点状缺陷分析与改善

丁 涛，张 云，任彦峰

（武钢有限冷轧厂，湖北武汉 430083）

摘 要：某热镀锌机组产品长期存在下表面密集细小点状缺陷，影响某重点汽车用户产品认证。本文主要介绍这类轻微点状缺陷的排查和改善方法：利用超景深三维显微镜和扫描电镜等对缺陷进行微观分析，通过缺陷形态、微观形貌分析，锁定缺陷发生区域。通过对锁定区域进一步分段排查和取样分析，锁定缺陷的具体发生位置，进而针对性地对辊系位置进行测量确认和位置调整，最终细小点状缺陷消除。

关键词：点状缺陷；微观分析；防皱辊；位置调整

Analysis and Improvement of Surface Tiny Spot Defects of Hot-dip Galvanized Strip

Ding Tao, Zhang Yun, Ren Yanfeng

(Cold Rolling Mill of WISCO, Wuhan 430083, China)

Abstract: The lower surface tiny spot defects of strip in a hot-dip galvanized production line existed in a long-term,which affects the product certification of a key automobile user.This paper mainly introduces the inspection and improvement methods of this kind of tiny spot defects: micro analysis of the defects by using ultra depth of field three-dimensional microscope and scanning electron microscope, through the analysis of defect morphology and micro morphology, lock the position where the defects take place.Through further segmented investigation and sampling analysis of the locking area, the specific location of the defects is locked, and then the roll position is confirmed and adjusted, finally the tiny spot defects are eliminated.

Key words: spot defects; micro analysis; anti-wrinkle roll; position adjustment

热镀锌带钢边部锌层增厚缺陷成因分析及控制措施

宋青松，郑艳坤，马幸江，候耿杰，段晓溪

（首钢京唐钢铁联合有限责任公司冷轧作业部，河北唐山　063200）

摘　要：结合某镀锌产线实际情况，通过对带钢边部增厚产生机理、影响因素等几方面进行了深入的研究，从工艺技术和设备方面提出了改善措施，使带钢边部增厚缺陷得到了有效的控制。

关键词：热镀锌；边部增厚；气流分散；边部挡板

Hot Dip Galvanized Strip Edge Thickening Defect Analysis and Control Measures

Song Qingsong, Zheng Yankun, Ma Xingjiang, Hou Gengjie, Duan Xiaoxi

(Jingtang Iron and Steel Union Company Limited, Capital Iron and Steel Company, Tangshan 063200, China)

Abstract: Based on the actual situation of a certain galvanizing production line, the mechanism and influencing factors of strip over-thickening were studied in depth, and the improvement measures were put forward from the aspects of technology and equipment, so that the defects of strip over-thickening were effectively controlled.

Key words: Hot dip galvanizing; Edge thickening; Airflow dispersion; Barrier

先进高强钢连续热浸镀锌铝镁技术的研究进展

周　欢[1,2]，赵爱民[1]，肖　俊[1]，田　耕[1]，郄镕鉴[1]，崔译夫[1]

（1. 北京科技大学钢铁共性技术协同创新中心，北京　100083；
2. 首钢京唐钢铁联合有限责任公司，河北唐山　063200）

摘　要：热浸镀锌铝镁高强钢具有十分优异的耐腐蚀性能和良好的综合力学性能，对先进钢铁材料的发展具有重要意义。本文介绍了热镀锌铝镁技术的研究及发展历程，系统地梳理了近年来热浸镀锌铝镁的商业化应用。详细汇总了近年来热镀锌在国内外的研究进展，包括镀层成分设计、组织演变、界面结构和腐蚀性能与机理等。重点阐述了锌铝镁镀层耐蚀机理，利用相图计算和热力学分析指导锌铝镁镀层合金的成分优化；探讨先进高强钢与锌铝镁液态合金的润湿规律和界面反应；热浸镀过程中薄层液膜的凝固机理以及镀层与 AHSS 钢板的组织演变规律。分析了国内热浸镀锌铝镁在发展过程中存在的主要问题，提出了相应的建议，从而为先进高强钢热浸镀锌铝镁技术的应用研究提供参考。

关键词：锌铝镁镀层；热浸镀；先进高强钢；组织与性能

Research Progress of Continuous Hot-dip Galvanizing of Aluminum and Magnesium on Advanced High-strength Steel

Zhou Huan[1,2], Zhao Aimin[1], Xiao Jun[1], Tian Geng[1], Qie Rongjian[1], Cui Yifu[1]

(1. University of Science and Technology Beijing, Beijing, 100083, China;
2. Shougang Jingtang Iron and Steel United Co., Ltd., Tangshan, 063200, China)

Abstract: Hot-dip galvanized aluminum-magnesium high-strength steel has excellent corrosion resistance and good comprehensive mechanical properties, which is of great significance to the development of advanced steel materials. The research and development history of hot-dip galvanizing aluminum-magnesium technology are introduced, and the commercial application of hot-dip galvanizing aluminum-magnesium in recent years is systematically reviewed. The research progress of hot-dip galvanizing at home and abroad in recent years is summarized in detail, including coating composition design, microstructure evolution, interface structure, corrosion performance and mechanism, etc. The focus is on the corrosion resistance mechanism of zinc-aluminum-magnesium coatings, the phase diagram calculation and thermodynamic analysis are used to guide the composition optimization of zinc-aluminum-magnesium coating alloys; the wetting law and interface reaction of advanced high-strength steel and zinc-aluminum-magnesium liquid alloys and the hot-dip coating process are discussed The solidification mechanism of the thin liquid film and the microstructure evolution law of the coating and AHSS steel sheet. The main problems in the development process of domestic hot-dip galvanizing aluminum-magnesium are analyzed, and corresponding suggestions are put forward, so as to provide reference for the application research of advanced high-strength steel hot-dip galvanizing aluminum-magnesium technology.

Key words: zinc-aluminum-magnesium coating; hot dip coating; advanced high-strength steel; organization and performance

基于改进粒子群算法的热镀锌带钢镀层质量智能预报

任新意[1]，尹显东[1]，徐海卫[1]，于　孟[1]，高慧敏[1]，黄华贵[2]

（1. 首钢京唐钢铁联合有限责任公司技术中心，河北唐山　063210；
2. 燕山大学国家冷轧板带装备及工艺工程技术研究中心，河北秦皇岛　066004）

摘　要：针对热镀锌带钢锌层厚度控制系统的非线性、多变量、大时滞的特性，通过对锌层测厚仪检测数据预处理实现了气刀工艺参数和带钢速度与锌层厚度的匹配。建立了关于带钢锌层厚度的 BP 神经网络预报模型，并提出了一种自适应混沌粒子群算法对锌层厚度神经网络预报模型的结构和参数进行优化。现场应用结果表明，采用自适应混沌粒子群算法优化的锌层厚度神经网络预报模型具有更好的精度，从而为提高产品镀层质量和降低生产成本提供了理论依据和应用参考。

关键词： 热镀锌；锌层厚度；数据延迟；粒子群；神经网络；预报模型

Intelligent Prediction of Coating Quality of Hot Dip Galvanized Strip based on Improved Particle Swarm Algorithm

Ren Xinyi[1], Yin Xiandong[1], Xu Haiwei[1], Yu Meng[1], Gao Huimin[1], Huang Huagui[2]

(1. Technology Center, Shougang Jingtang United Iron and Steel Co., Ltd., Tangshan 063210, China;
2. National Engineering Research Center for Equipment and Technology of Cold Strip Rolling, Yanshan University, Qinhuangdao 066004, China)

Abstract: In view of the nonlinear, multivariable and large time delay characteristics of the zinc layer thickness control system of hot-dip galvanized strip, the matching of air knife process parameters, strip speed and zinc layer thickness is realized by preprocessing the detection data of zinc layer thickness gauge. A BP neural network prediction model for zinc layer thickness of strip steel is established, and an adaptive chaotic particle swarm optimization algorithm is proposed to optimize the structure and parameters of the neural network prediction model for zinc layer thickness. The field application results show that the neural network prediction model of zinc coating thickness optimized by adaptive chaotic particle swarm optimization has better accuracy, which provides theoretical basis and application reference for improving product coating quality and reducing production cost.

Key words: hot dip galvanizing; zinc coating thickness; data delay; particle swarm optimization; neural network; forecasting model

GA 汽车板冲压脱锌缺陷的改善

杨 芃[1]，陈园林[1]，杜小峰[2]

（1. 宝钢股份武汉钢铁有限公司，湖北武汉 430083；
2. 宝钢股份中央研究院，湖北武汉 430083）

摘　要： 通过镀层结构分析、脱锌与不脱锌样品各项性能对比、模具状态确认，确定了合金化（GA）汽车板冲压脱锌主要原因为钢板与模具之间摩擦力过大，从而导致镀层在冲压时刮擦剥落。采取镀层铁含量控制在 10%~13%、产品落料后放置时间缩短至 1 周以内、保持冲压模具表面光洁等措施，成功解决了合金化热镀锌钢板冲压脱锌问题。

关键词： 合金化热镀锌；汽车板；冲压；脱锌

Improvement of Galling Defect during Stamping of GA Automotive Sheet

Yang Peng[1], Chen Yuanlin[1], Du Xiaofeng[2]

(1. Wuhan Iron and Steel Co., Ltd. of Baosteel, Wuhan 430083, China;
2. Center Research Institute of Baosteel, Wuhan 430083, China)

Abstract: By analyzing coating microstructure, comparing the properties of sample with galling defect and sample without the defect and checking the condition of the stamping die, the main reason of galling defect during stamping of GA automotive sheet is the big fiction force between die and sheet, leading to the peel-off of the coating during Stamping. Adopting several measures, such as 10-13% control range of coating iron content, storage time within 1 week after sheet blanked,and keeping die surface smooth, the galling defect during stamping of GA automotive sheet is eliminated successfully.

Key words: hot-dip galvannealing; automotive sheet; stamping; galling

极低锡量镀锡板上铬-磷复合膜层的制备与表征

尹显东[1]，彭大抗[2]，黎德育[2]，方 圆[3]

（1. 首钢京唐钢铁联合有限责任公司技术中心，河北唐山 063000；2. 哈尔滨工业大学化工与化学学院，黑龙江哈尔滨 150001；3. 首钢集团有限公司技术研究院，北京 100043）

摘 要：为提高极低锡量镀锡板的耐蚀性能，在铬酐钝化液中加入磷酸二氢铝，通过阴极电解的方式在极低锡量镀锡板上形成钝化膜，并采用扫描电镜、盐雾试验箱、电化学工作站对极低锡量镀锡板的表面形貌、耐蚀性能、成膜过程进行研究。试验结果表明，铬磷钝化液较佳的钝化工艺条件为pH=2.6、钝化电流1A/dm2、钝化时间8s、钝化温度50℃。铬磷钝化液在该工艺下的钝化膜耐蚀性能优于铬酐钝化液，在盐雾3h后，铬磷钝化后的极低锡量镀锡板盐雾等级仍在5级以上，铬酐钝化后的镀锡板盐雾等级在3级左右。这与加入磷酸二氢铝改善钝化液在钢板表面的成膜过程，增强极低锡镀锡板漏铁处的耐蚀性能有关。

关键词：镀锡板；极低锡量；铬酸盐；磷酸盐；耐蚀性；成膜过程

Preparation and Properties of Chromium-phosphorus Composite Film Extremely Low Tinplate Steel

Yin Xiandong[1], Peng Dakang[2], Li Deyu[2], Fang Yuan[3]

(1. Shougang Jingtang Iron and Steel United Co., Ltd., Tangshan 063000, China;
2. Harbin Institute of Technology, Harbin 150001, China;
3. Research Institute of Technology, Shougang Group Co., Ltd., Beijing 100043, China)

Abstract: In order to improve the corrosion resistance of the tinplate with extremely low tin content, aluminum dihydrogen phosphate is added to the chromic anhydride passivation solution, and a passivation film is formed on the tinplate with extremely low tin content by cathodic electrolysis. And use scanning electron microscope, salt spray test box, electrochemical workstation to study the surface morphology, corrosion resistance and film forming process of extremely low tinplate steel. The test results show that the best passivation process conditions for chromium-phosphorus passivation solution are pH=2.6, passivation current 1A/dm^2, passivation time 8s, and passivation temperature 50℃. The corrosion resistance of the passivation film of the chromium-phosphorus passivation solution under this process is better than that of the chromium-anhydride passivation solution. After 3 hours of salt spray, the salt spray level of the extremely low tinplate steel after chromium-phosphorus passivation is still at level 5. Above, the salt spray grade of tinplate after chromic anhydride passivation is about 3. This is related to the addition of aluminum dihydrogen phosphate to improve the film-forming process of the passivation solution on the surface of the steel plate and to enhance the corrosion resistance of the extremely low tinplate steel.

Key words: tinplate; very low tin content; chromate; phosphate; corrosion resistance; film formation process

钝化电量分配对 EOE 用镀锡板附着力的影响

黄 勇，张东方

（宝武集团武钢日铁镀锡板有限公司，湖北武汉 430080）

摘 要：EOE 产品对附着力要求严格，通常采用低钝化膜工艺，但附着力仍不能满足高要求的客户。本文通过采用重铬酸钠对镀锡量为 2.8 g/m^2 和 5.6g/m^2 镀锡板基板进行阴极钝化，钝化过程使用两个钝化槽，选取了 3∶5；1∶0；0∶1；三种电流密度分配进行研究，对钝化后的试样进行了模拟涂装烘烤，测试其附着力。结果表明，钝化电流分配为 1∶0 时，产品附着力为一级。

关键词：镀锡板；钝化；附着力；EOE

Effect of Passivation Charge Distribution on Adhesion of Tinplate for EOE

Huang Yong, Zhang Dongfang

(Baowu Steel Group WISCO-NIPPON Steel Tinplate Co., Ltd., Wuhan 430080, China)

Abstract: EOE products have strict requirements on adhesion, usually using low passive film process, but the adhesion still can not meet the high requirements of customers. In this paper, sodium dichromate was used for cathodic passivation of tinplate substrate with tin plating amount of 2.8 g/m^2 and 5.6 g/m^2. Two passivation tanks were used in the passivation process, and three current density distributions (3∶5; 1∶0; 0∶1) were selected to study. The passivated samples were simulated by coating and baking, and their adhesion was tested. The results show that when the passivation current distribution is 1:0, the adhesion of the product is first order.

Key words: tinplate; passivation; adhesion strength; EOE

热镀锌板锌灰的对表面质量的影响及控制措施

关洪星，陈 平，陈 佳，瞿作为，任彦峰

（宝钢股份武汉钢铁有限公司，湖北武汉 430083）

摘 要：结合冷轧热镀锌产线的生产工艺和设备，对热浸镀锌的锌灰形成机理进行分析研究，探讨在实际生产中锌灰对板面质量的影响和提出了优化工艺进行控制，提高热镀锌板表面质量满意度。

关键词：热镀锌；锌灰；工艺优化

The Effect of Zinc Dust to Product Surfacein Process of Hot Dip Galvanizedand Its Purification

Guan Hongxing, Chen Ping, Chen Jia, Qu Zuowei, Ren Yanfeng

(Baosteel Wuhan Iron and Steel Company Limited, Wuhan 430083, China)

Abstract: Combined with the production process and equipment of cold rolling and hot galvanizing line, The formation mechanism of zinc dust in hot dip galvanizing was studied, The influence of zinc dust on the quality of board surface in practical production was discussed, and put forward to optimize the process to control, improving the satisfaction of surface quality of hot dip galvanized sheet.

Key words: hot dip galvanized steel sheet; zinc dust; process optimization

本钢 800MPa 级热镀锌双相钢产品生产工艺优化

马 峰，李鸿友，崔 勇，刘晓峰，张晶晶

（本钢板材股份有限公司冷轧总厂，辽宁本溪 117000）

摘 要： 为了控制热镀锌双相钢 800MPa 级别产品表面质量，提高热镀锌表面锌层附着力，减少由于合金元素发生选择性氧化而导致的析出缺陷，利用本钢热镀锌机组预氧化段进行表面预氧化控制，从而提高产品的表面质量。针对合金元素产生的选择性氧化，该研究内容通过对炉内露点、气氛以及氧含量等关键参数进行调整，从而实现材料中合金元素的内部氧化，并且配合合理的机组其他工艺参数，在热镀锌双相钢 780MPa 产品生产过程中进行氧化物埋层处理，从而提高产品的表面质量。同时结合生产计划编排，实现 DP780 产品连续稳定生产。

关键词： 预氧化腔；外氧化；内氧化；露点

Optimization of Production Process of 800MPa Grade Hot-dip Galvanized Dual-phase Steel at Ben Steel Group

Ma Feng, Li Hongyou, Cui Yong, Liu Xiaofeng, Zhang Jingjing

(Bengang Steel Plates Co.,Ltd The Cold Rolling Mill, Benxi, 117000, China)

Abstract: In order to control the surface quality of hot-dip galvanizing DP780 products, improve the adhesion of zinc coating on surface, and reduce the precipitation defects caused by the selective oxidation of alloy elements, the surface preoxidation control is carried out in the preoxidation section of the hot dip galvanizing line of Ben Steel Group, so as to improve the surface quality of products.The selective oxidation of oxyphilic elements was controlled by adjusting the key parameters such as dew point and oxygen content in the furnace, and cooperates with other reasonable technological parameters of the unit, then the surface quality would be improved. At the same time, the continuous and stable production of DP780 products is realized by combining production planning.

Key words: oxidation chamber; internaloxidation;external oxidation; dew point

本钢 3 号镀锌机组 DP 钢表面析出缺陷的分析与研究

艾厚波[1]，刘晓峰[1]，崔　勇[1]，付东贺[2]

（1. 本钢板材股份有限公司冷轧总厂，辽宁本溪　117000；
2. 本钢板材股份有限公司技术研究院，辽宁本溪　117200）

摘　要：本文从原理上分析了本钢 3 号镀锌机组双相钢表面产生析出缺陷的原因，针对缺陷产生原因对酸轧工序的拉矫延伸率、换酸量、酸洗温度及漂洗 pH 值和热镀锌工序的工艺速度、出缓冷温度、锌液 Al 含量和气刀距离等相关工艺参数进行调整试验，使双相钢表面析出缺陷得到缓解。

关键词：热镀锌；双相钢；表面析出

Analysis and Research on Surface Precipitation Defects of DP Steel in No.3 Galvanizing Line of Benxi Steel

Ai Houbo[1], Liu Xiaofeng[1], Cui Yong[1], Fu Donghe[2]

(1. Bengang Steel Piates Co., Ltd. The Cold Rolling Mill, Benxi, 117000, China,;
2. R&D Institute of Benggang Steel Plates Co., Ltd, Benxi, 117000, China)

Abstract: This paper analyzes the causes of precipitation defects on the surface of dual phase steel in No.3 galvanizing line,According to the causes of the defects, the relevant process parameters of the acid rolling process, such as the elongation of tension straightening, the amount of acid exchange, the pickling temperature and the pH value of rinsing, the process speed of hot galvanizing process, the slow cooling temperature, the Al content of liquid zinc and the air knife distance, are adjusted and tested, The precipitation defects on the surface of the dual phase steel are relieved.

Key words: hot dip galvanizing; dual phase steel; surface precipitation

6　金属材料深加工

固定短芯棒冷拔钢管残余应力仿真分析与应用研究

王家聪[1]，覃海艺[2]，刘贤翠[1]，刘 洋[1]

（1. 徐州徐工液压件有限公司，江苏徐州 221004；
2. 上海交通大学塑性成形技术与装备研究院，上海 200030）

摘 要：基于有限元分析模拟钢管的冷拔过程，提出厚度方向至少采用32层低阶单元或16层高阶单元、通过修正法确定摩擦系数、采用轴对称模型的建模方法。并采用降温法给钢管表面施加残余应力，获得了残余应力分布状态对椭圆度的影响规律。计算结果显示：（1）模具入口角度越小，残余应力分布越均匀；（2）内外模定径带起始点的相对位置对位移及残余应力影响都很大；（3）长定径带更有利于残余应力的均匀分布；（4）内外模定径带终点的相对位置对残余应力影响很大，但对位移影响较小。

关键词：固定短芯棒冷拔；残余应力；椭圆度；降温法；有限元

Influence of Fixed Plug Drawing Die on Residual Stress of Steel Pipe

Wang Jiacong[1], Qin Haiyi[2], Liu Xiancui[1], Liu Yang[1]

(1. Xuzhou XCMG Hydraulics Co., Ltd., Xuzhou 221004, China;
2. Institute of Forming Technology & Equipment, Shanghai Jiao Tong University, Shanghai 200030, China)

Abstract: Based on the finite element analysis to simulate the drawing process of steel pipe, a modeling method is proposed, which uses at least 32 layer low-order elements or 16 layer high-order elements in the thickness direction, determines the friction coefficient through the correction method, and adopts the axisymmetric model. The residual stress is applied on the surface of steel pipe by cooling method, and the influence of residual stress distribution on roundness is obtained. The analysis results show that: (1) The smaller the die entrance angle is, the more uniform the residual stress distribution is; (2) The relative position of the starting point of the inner die sizing zone and outer die sizing zone has a great influence on the displacement and residual stress; (3) The long sizing zone is more conducive to the uniform distribution of residual stress; (4) The relative position of the end point of the inner die sizing zone and outer die sizing zone has a great influence on the residual stress, but it has little effect on displacement.

Key words: fixed plug drawing; residual stress; roundness; cooling method; finite element

微合金钢组织性能控制技术研究与应用

丁 茹，张鹏武，康新成，陈一峰

（武汉钢铁有限公司 热轧厂，湖北武汉 430083）

摘 要：本文分析了微合金元素的作用机理及组织、性能全流程控制技术，从板坯、加热、压下、终轧温度、冷却速率、卷取温度、热处理七个核心环节详细讲述微合金化技术的关键，并结合生产实践举例说明，对同行业工作者具有一定的借鉴意义。

关键词： 微合金钢；组织；性能；控制技术

Research and Application of Tuning Microstructure and Mechanical Properties of Microalloyed Steel

Ding Ru, Zhang Pengwu, Kang Xincheng, Chen Yifeng
(Wuhan Iron & Steel Co., Ltd., Wuhan 430083, China)

Abstract: This article analyzes the effect of micro-alloyed elements on microstructure and mechanical properties. The tuning procedure of microstructure and mechanical properties is further proposed considering the whole production process of microalloyed steel, including the slab, heating, reduction, final rolling temperature, cooling rate, coiling temperature, and heat treatment. The influencing mechanism of micro-alloyed elements is detailed described and some production cases are provided, which sheds light on the production in the similar company.

Key words: microalloyed steel; microstructure; mechanical properties; tuning procedure

热轧商品材切割翘曲原因分析及控制措施

张鹏武，刘　亮，袁　金，徐　浩，王　跃，卢家麟
（武汉钢铁有限公司　热轧厂，湖北武汉　430083）

摘　要： 本文针对影响热轧商品材切割翘曲的三大因素开展研究：轧线横断面温度差异、层流冷却不均、横切矫直工艺，通过系统攻关找出三大因素对切割翘曲的影响及控制方法，解决了热轧商品材切割翘曲缺陷，对同类产线具有指导意义。

关键词： 切割翘曲；温度差异；层流冷却；横切矫直

Analysis and Control Measures of Cutting Warping of Hot-rolled Commercial Steel Products

Zhang Pengwu, Liu Liang, Yuan Jin, Xu Hao, Wang Yue, Lu Jialin
(Wuhan Iron & Steel Co., Ltd., Wuhan 430083, China)

Abstract: This paper investigates the effect of three factors, namely the temperature difference of cross-section, the uneven laminar cooling, and the cross-cut straightening process, on the cutting warping of hot-rolled commercial steel products. Based on the systematic research, reasons and control methods are proposed for cutting warping. The quality of steel products can be promoted, which sheds light on the control of similar hot-rolled steel products.

Key words: cutting warpage; temperature difference; laminar cooling; cross-cut straightening

热连轧产线生产效率提升关键技术研究与应用

周一中，丁　茹，张鹏武，康新成，陈一峰

（武汉钢铁有限公司　热轧厂，湖北武汉　430083）

摘　要：本文针对影响热连轧产线生产效率提升的五大核心要素：单重、热装、倒垛、节奏、故障展开研究，找出五大要素之间的相互制约、协同关系，探索出对产能提升具有重大意义的核心参数配置，可推广应用至同类热连轧产线，提升生产效率。

关键词：单重；热装；倒垛；节奏；故障；生产效率

Research and Application of Key Technologies for Improving Production Efficiency of Hot Strip Production Line

Zhou Yizhong, Ding Ru, Zhang Pengwu, Kang Xincheng, Chen Yifeng

(Wuhan Iron & Steel Co., Ltd., Wuhan 430083, China)

Abstract: This paper investigates the effects of five vital factors on production efficiency of hot strip production line: unit weight, hot loading, restacking, rhythm and failure. Interaction and synergy relationship of the five factors are observed and the optimal parameters are obtained. The acquired parameters is of significance to productivity and can be applied to similar hot strip production lines for improving production efficiency.

Key words: single weight; hot loading; inverted stack; rhythm; failure; production efficiency

热轧钢卷尾部层间挫伤原因分析及改善

陈　刚

（宝钢股份武汉钢铁有限公司，湖北武汉　430083）

摘　要：某热轧生产线厚窄规格钢种在精整开卷尾部 100m 左右存在层间挫伤，经过多次跟踪试验，确认了原卷尾部卷紧度低是导致原卷尾部产生热态层间挫伤缺陷的原因，并且发现卷紧度与尾部带钢凸度有相关性。由于该规格精轧负荷低，通过调整精轧设备无法达到目标凸度，因此通过采取降低 RT4、增加中间坯厚度等措施提高尾部凸度，提高卷紧度，有效地解决了厚窄规格钢尾部层间挫伤缺陷问题。

关键词：层间错动；层间挫伤；松卷；凸度；热轧

Analysis and Improvement of Interlayer Scratch at Tail of Hot-rolled Strip Coil

Chen Gang

Abstract: There exists interlayer scratch on a hot-rolle steel production after uncoiling. The position of the scratch is about 100 meters away from tail of the steel production. After series of tracking tests, it is confirmed that the low degree tightness of the original coiling is the reason for the scratch defect, and the tightness of coiling correlates well with the convexity of the steel strip. The target convexity could not be achieved by adjusting the final rolling equipment due to the low loading requirement of the steel during final rolling process. Therefore, measures, such as reducing RT4 and increasing the thickness of the intermediate billet, are taken to improve the convexity of the tail of the steel and increase the tightness of the steel coiling. The measures effectively solve the interlayer scratch at tail of the thick and narrow steel. KEYWORDS interlayer motion interlayer scratch loose coiling convexity hot rolling.

Key words: interlayer dislocation; interlayer scratch; loose coiling; convexity; hot rolled

电渣锭型与变形量对 Cr12MoV 钢中共晶碳化物的影响

朱　斌[1]，胡　瑜[1]，唐远寿[2]，曹鹏军[2]，栗克建[2]，陈知伟[1]

（1. 重庆钢铁研究所有限公司，重庆　404100；
2. 重庆科技学院 冶金与材料工程学院，重庆　401331）

摘　要：本研究选用直径分别为 Φ150mm、Φ225mm、Φ275mm 的 Cr12MoV 3 只钢锭，对其铸态组织中莱氏体共晶网络最大厚度、直径及比表面积进行测定，在相同的工况下进行加工变形，然后评定 3 者的共晶碳化物不均匀度，检测 3 者的碳化物最大水平弦长，并通过扫描电镜能谱扫描、XRD 分析表征了变形前后共晶碳化物的形貌特征、分布、类型及合金元素偏聚状况。实验结果表明：共晶碳化物的不均匀度级别随锭型直径及变形量的增加而降低，其分布、均匀性得以改善；但碳化物最大水平弦长则表现出与之相反的趋势。

关键词：Cr12MoV；锭型；变形量；共晶碳化物；不均匀度；弦长

Influence of Electroslag Ingot and Deformation Amount on Eutectic Carbide of Cr12MoV Steel

Zhu Bin[1], Hu Yu[1], Tang Yuanshou[2], Cao Pengjun[2], Li Kejian[2], Chen Zhiwei[1]

(1. Chongqing Iron and Steel Research Institute Co., Ltd., Chongqing 404100, China;
2. School of Metallurgy and Material Engineering, Chongqing University of Science and Technology, Chongqing 401331, China)

Abstract: The maximum thickness, diameter and specific surface area of the eutectic carbide net of Cr12MoV as cast ingots with diameters of Φ150, Φ 225 and Φ 275 were measured. Then, after deformation under the same working condition, the eutectic carbide inhomogeneity of the three was evaluated and the maximum carbide chord length of the three was calculated. Finally, the morphology, distribution, type, content and segregation of alloying elements of eutectic carbides before and after deformation were characterized by SEM and XRD analysis. The experimental results show that the unevenness of eutectic carbide decreases with the increase of ingot diameter and deformation, and the distribution and uniformity of eutectic carbide are improved. However, the maximum horizontal chord length of carbides shows the opposite trend.

Key words: Cr12MoV; ingot shape; deformation; eutectic carbide; non-uniformity; chord length

初始残余应力对高强钢冲压回弹影响机理

崔金星，梁增帅，彭　艳，孙建亮

（燕山大学 国家冷轧板带装备及工艺工程技术研究中心，河北秦皇岛　066004）

摘　要：研究了高强钢板初始残余应力对冲压加工过程回弹变形影响机理。基于平面应变和幂指函数材料模型，建立了考虑初始残余应力影响的板带冲压回弹数学模型，定义了回弹角偏差值计算方法表征残余应力对板带冲压回弹的影响机制。通过模型分析了不同材质、不同加工参数和不同板带强度的高强板冲压回弹角和回弹角差值的大小，揭示了初始残余应力对高强钢板冲压回弹影响机制。结果表明：初始残余应力改变了板内应力的分布和大小，使回弹角值沿板宽方向产生波动；初始残余应力对回弹结果的影响大小随着板厚和材料强度的增大而减小。本研究对初始残余应力状态下高强钢板冲压回弹问题提供力理论依据。

关键词：初始残余应力；平面应变理论；冲压回弹；回弹角差值

Mechanism of Residual Stress Affecting Springback of High Strength Steel During Stamping Process

Liu Jinjing, Liang Zengshuai, Peng Yan, Sun Jianliang

(National Engineering Research Center for Equipment and Technology of Cold Rolled Strip, School of Mechanical Engineering, Yanshan University, Qinhuangdao 066004, China)

Abstract: The mechanism of initial residual stress affecting springback deformation of high strength steel sheet during stamping process was studied. Based on the plane strain and power exponent function material model, the springback mathematical model considering the effect of initial residual stress was established. The calculation method of springback angle deviation was defined to characterize the influence mechanism of residual stress on springback. The springback angle and the difference of springback angle of the high strength steel sheet with different materials, processing parameters and strip strength were analyzed using the model. The mechanism of initial residual stress affecting springback was then revealed. Results show that the initial residual stress is related to the distribution and size of the internal stress, and makes the springback angle fluctuate along the width direction of the plate. The influence degree of the initial residual stress on the springback decreases with the increase of the plate thickness and material strength. This study provides a theoretical basis for the springback of high strength steel sheet in the initial residual stress state.

Key words: initial residual stress; plane strain theory; springback; angle difference

含硼冷镦钢 SAE10B21A 螺帽开裂原因分析

袁桥军，李凌锋，杜成建

（湖南华菱湘潭钢铁公司，湖南湘潭　411101）

摘　要：材质为含硼冷镦钢 SAE10B21A 盘条在经过拉拔后冷镦加工成带法兰面螺帽的过程中，部分螺帽在法兰面处出现开裂。对开裂试样进行了宏观检测、金相检测、扫描电镜及能谱检测分析，结合对应炉批次盘条生产工艺质量控制情况，分析认为材料近表面大尺寸夹杂物是导致法兰面开裂的主要原因。

关键词：含硼冷镦钢 SAE10B21A；螺帽；开裂

Analysis on Cracking of SAE10B21A Nut of Boron Bearing Cold Heading Steel

Yuan Qiaojun, Li Lingfeng, Du Chengjian

(Hunan Valin Xiangtan Steel, Xiangtan 411101, China)

Abstract: During the cold upsetting process to make the drawn boron bearing cold heading steel SAE10B21A into nuts with flange surface, cracks occur at flange surface. The cracking specimens are examined by the real-time observation, optical microscope, scanning electron microscope and energy spectrum. The production process parameters of the corresponding batches are further taken into consideration. Results show that the large size inclusions near the surface of the material are responsible for the cracking of the flange surface.

Key words: boron bearing cold heading steel SAE10B21A; nut; cracking

不同级别高强度紧固件的疲劳性能及断口分析

董　庆[1,2]，张　鹏[1,2]，陈继林[1,2]

（1. 河北省线材工程技术创新中心，河北邢台　054027；
2. 邢台钢铁有限责任公司，河北邢台　054027）

摘　要：参考 GB/T 13682—92 试验方法，使用高频疲劳试验机测试了 8.8 级、10.9 级、12.9 级和 14.9 级共四种不同强度级别紧固件的疲劳性能。不同的交变载荷对疲劳性能的影响很大。在过高的交变载荷作用下，14.9 级紧固件的疲劳性能未满足国标，但满足客户技术要求。使用扫描电子显微镜分析了过高交变载荷作用下疲劳断口的形貌特征。其中，10.9、12.9、14.9 级紧固件的疲劳源均为应力集中的疲劳台阶区域，而瞬断区则各有特点，10.9 级紧固件的瞬断区为圆形横截面的心部区域，12.9 级、14.9 级紧固件的瞬断区为类似拉伸断口剪切唇的局部区域。

关键词：高强度；紧固件；疲劳性能；断口分析

Fatigue Performance and Fracture Analysis of Different Grades of High-strength Fasteners

Dong Qing[1,2], Zhang Peng[1,2], Chen Jilin[1,2]

(1. Hebei Engineering Innovation Center For Wire Rod, Xingtai 054027, China;
2. Xingtai Iron & Steel Co., Ltd., Xingtai 054027, China)

Abstract: With reference to the test method proposed in GB/T 13682—92, a high-frequency fatigue testing machine was used to acquire the fatigue performance of four different strength grade fasteners of 8.8, 10.9, 12.9 and 14.9 grades. Alternating loads have a great influence on fatigue performance. Under the condition of higher alternating loading, the fatigue performance of 14.9 fasteners does not meet the requirements of the national standard, but meets the customers' technical requirements. Scanning electron microscope was used to analyze the morphological characteristics of fatigue fracture under high alternating loading. Results show that the fatigue sources of 10.9, 12.9, and 14.9 fasteners are all stress-concentrated fatigue step regions, while the transient regions display different characteristics. The transient regions of 10.9 fasteners are the central regions of circular cross-sections. The transient regions of grade 12.9 and grade 14.9 fasteners is a local area similar to the shear lip of tensile fracture.

Key words: high strength; fasteners; fatigue performance; fracture analysis

无粘结镀锌钢绞线斜拉索的技术研究及应用

游胜意[1]，张海良[2]，汤 亮[2]，倪晓峰[1]，周生根[1]，张世昌[1]，金 芳[2]

（1. 奥盛(九江)新材料有限公司，江西九江 332000；
2. 上海浦江缆索股份有限公司，上海 200120）

摘 要： 介绍选用沙钢 S87B-T 盘条生产 ϕ15.20mm、1960MPa 高性能无粘结镀锌钢绞线斜拉索产品的技术要求和研制过程，针对产品的高强度、高防腐性能、高精度的要求及优化拉拔技术、内加热镀锌技术、防腐油脂 PE 套双控技术、光纤压力传感技术、充气式拉索等关键技术进行创新，经检验和应用产品各项性能指标满足相关标准的要求。

关键词： 镀锌钢绞线；光纤检测；防腐；充气式拉索；无粘结

Research and Application of Unbonded Galvanized Steel Strand Stay Cable

You Shengyi[1], Zhang Hailiang[2], Tang Liang[2], Ni Xiaofeng[1], Zhou Shenggen[1], Zhang Shichang[1], Jin Fang[2]

(1. Ossen (Jiujiang) Innovation materials Co., Ltd., Jiujiang 332000,China;
2. Shanghai Pujiang Cable Co., Ltd., Shanghai 200120, China)

Abstract: It is introduced that the technical requirements and research process of high performance unbonded galvanized steel strand stay cable with Φ15.20mm、1960MPa that produced by S87B-T Rod of Shagang.Innovation methods have been proposed based on the requirements of the product such as high strength, high corrosion resistant performance ,and high accuracy, Optimized drawing technology, interior heating galvanizing technology, anti-corrosion grease + PE sleeve double control technology, optical fiber pressure sensory technology, inflatable cable are also used to develop the innovation methods. All the technical indexes are in accordance with the correlative standards after sorts of tests and actual application.

Key words: galvanized steel strand; optical fiber detection; corrosion resistant protection; inflatable cable; unbonded

新材质重载货车轮服役性能研究

庞晋龙[1]，陈　刚[1]，邓荣杰[1]，高　伟[2]

（1. 宝武集团马钢轨交材料科技有限公司，安徽马鞍山　243000；
2. 马鞍山钢铁股份有限公司技术中心，安徽马鞍山　243000）

摘　要：重载运输是世界铁路货运的发展方向，但随着车辆轴重的增大，车轮的运行条件更为苛刻，需承受更高的轮轨接触应力和制动热负荷，对车轮提出更高要求。本文针对 C80E 型车辆用重载车轮在我国大秦铁路的服役情况，通过研究新老材质车轮的成分、力学性能、组织等综合性能特性，主要从磨耗量、热裂纹、抗剥离等服役特性入手，综合对比评价了两种材质车轮的运行服役情况，结果表明：新材质 CL70 车轮具有优良的抗接触损伤、抗磨损和抗热损伤能力，实物综合性能高，各方面服役性能都明显优于常规 CL60 材质车轮，更适合我国重载货运的需求，对于我国铁路货运发展战略的快速推进具有重要的支撑作用。

关键词：C80E 型车辆；CL70 新材质；重载货车轮；服役性能

Research on Service Performance of New Material Heavy-haul Freight Trains Wheels

Pang Jinlong[1], Chen Gang[1], Deng Rongjie[1], Gao Wei[2]

(1. Baowu Group Masteel Rail Transit Materials Technology Co., Ltd, Ma’anshan, 243000, China;
2. Technology Center, Ma’anshan Iron and Steel Co., Ltd, Ma’anshan, 243000, China)

Abstract: Heavy-haul is the development direction of railway freight transportation in the world. However, with the increase of axle weight of trains, wheels need to be subjected to high stress and braking heat load. It means that the service conditions of wheels are more stringent. This paper comprehensively investigates the service performance of two kinds of material heavy-haul wheels based on the service condition of C80E heavy-haul train wheel in Daqin Railway Line. Compositions, mechanical properties and microstructure of new and old material wheels are analyzed. Service performances, such as of wear, hot crack, anti-peeling and so on, are evaluated. Results show that the new material CL70 wheel has high resistance to contact damage, wear and thermal damage. High physical comprehensive performance and service performance of CL70 have been illustrated, which is significantly better than the conventional CL60 wheel. The aforementioned result indicates that CL70 is more suitable for Daqin railway heavy-haul freight demand, and plays an important supporting role in the rapid promotion of China's railway freight development strategy.

Key words: C80E type vehicle; CL70 new material; heavy-haul freight car wheel; service performance

挡泥墙体用钢管钢筋网 SP-CIP

罗 晔

（宝钢中央研究院武钢有限技术中心，湖北武汉 430080）

摘 要：为了解决普通异形钢筋重量偏重的问题，韩国 POSCO 公司联合下游客户共同开发了钢管钢筋网 SP-CIP（Smart Pipe-Cast In Placed pile）。主筋采用 STG800 钢管，而 STG800 钢管的原材料为热轧高强钢 PosH690。SP-CIP 钢管钢筋网同时具备价格竞争力和优异性能。首先，STG800 钢管的强度与普通钢筋相同，截面积却减少一半，每米单价比钢筋便宜 5%~10%。其二，重量比钢筋轻 50%以上，从而极大减轻了现场施工人员的负担。此外，SP-CIP 钢管钢筋网可直接进行焊接加工，具有较高的稳定性，在工厂完成制作后即可进行现场安装。预计今后 SP-CIP 钢管钢筋网将会是极具潜力的替代产品。

关键词：钢管钢筋网 SP-CIP；STG800 钢管；PosH690；优势

Smart Pipe-cast in Placed Pile for Mud Retaining Wall

Luo Ye

(R&D Center of Wuhan Iron & Steel Co., Ltd., Baosteel Central Research Institute, Wuhan 430080, China)

Abstract: In order to solve the problem of heavy weight of ordinary special-shaped steel bars, POSCO Company of South Korea, together with downstream customers, jointly developed Smart Pipe-Cast In Placed pile（SP-CIP）.The main steel bar is STG800 steel pipe, while the STG800 steel pipe is made of hot rolled high strength steel PosH690. SP-CIP has both price competitiveness and excellent performance. First of all, the strength of STG800 steel pipe is the same as that of ordinary steel bar, but the cross-sectional area is reduced by half, and the unit price per meter is 5%~10% cheaper than steel bar. Second, the weight is more than 50% lighter than steel bar, thus greatly reducing the burden of on-site construction personnel. In addition, SP-CIP can be welded directly, which has high stability and can be installed on site after the completion of production in the factory. It is expected that SP-CIP will be a potential substitute in the future.

Key words: smart pipe-cast in placed pile; STG800 steel pipe; PosH690; advantage

降低金属制品材镦打开裂的研究与实践

蔡常青

（福建三钢闽光股份有限责任公司炼钢厂，福建三明 365000）

摘 要：随着经济的快速发展，福建省内和周边工业金属制品用材的用量迅速增加，同时用户对工业用材质量的要求也越来越高，但工业用材在用户手中也会发生质量异议，而镦打开裂是其中主要的质量缺陷，本文对近3年来产生冷顶锻开裂的原因进行研究分析，通过进一步完化与完善生产工艺，降低了镦打的开裂率。

关键词： 金属制品材；镦打开裂；非稳态；质量异议降低

Research and Practice on Reducing Upsetting Crack of Metal Products

Cai Changqing

(Steelmaking Plant, Fujian Steel Minguang Co., Ltd., Sanming 365000, China)

Abstract: With the rapid development of economy. the types and quantities of metal products produced by Fujian Sangang are increasing year by year. users are demanding higher and higher quality. quality objections inevitably arise in the hands of users. the upsetting crack is one of the main quality objections. In this paper, the causes of cold upset forging cracking in the past three years are studied and analysis. through completion and improvement of production process. reducing the cracking rate of cold heading. external quality objections have diminished.

Key words: metal products; upsetting crack; unsteady state; decreased quality objection

镀锌钢板冲压开裂原因分析及改善措施

施刘健，张喜秋，李凯旋，杨少华，王卫远

（马鞍山钢铁股份有限公司制造管理部，安徽马鞍山　243003）

摘　要： 镀锌机组生产的钢板在冲压时出现开裂缺陷，直接影响了使用效果。取样检测分析后发现主要原因是基板芯部位置存在条状夹杂，在外力作用下出现开裂并向两侧撕裂。

关键词： 镀锌；冲压；开裂；夹杂

The Reason and Improvement Measures of Stamping Cracking of Galvanized Steel Sheet

Shi Liujian, Zhang Xiqiu, Li Kaixuan, Yang Shaohua, Wang Weiyuan

(Manufacturing Management Department, Ma'anshan Iron & Steel Co., Ltd., Ma'anshan 243003, China)

Abstract: The steel produced by hot dip galvanizing line displays cracking defects during stamping, which directly affects the application of the steel. Sampling analysis showed that the main reason for the defect is the strip inclusions in the center position. Inclusions result in the micro-crack and the crack propagates into the two side of inclusions under the external force.

Key words: galvanization; stamp; crack; inclusion

800MPa 级两种流程复相钢显微组织对力学性能的影响

方 幸[1]，杨晓宇[1]，米振莉[1]，张瀚龙[2]

（1. 北京科技大学 工程技术研究院，北京 100083；
2. 宝山钢铁股份有限公司 中央研究院，上海 201900）

摘 要：薄板坯连铸连轧（CSP）工艺和传统铸轧（CCR）工艺生产的 800MPa 级热轧复相钢在硬度及其强度等力学性能上的表现不尽相同。本文利用光学金相显微镜（OM）和扫描电子显微镜（SEM）等微观表征手段，系统研究了两种流程生产的复相钢的显微组织特征，并尝试从微观角度解释其宏观性能产生差异的原因。两种流程复相钢晶粒细小且均匀，其中 CSP 生产的板材平均晶粒尺寸 2.79μm，CCR 生产的板材平均晶粒尺寸 2.68μm；两种流程复相钢的显微组织都以贝氏体为基体，伴有多边形铁素体及少量 M/A 岛；其中 CSP 工艺获得的组织以粒状贝氏体为主，而 CCR 工艺获得的组织以粒状贝氏体混合板条贝氏体为主，因此在微观形貌的表征上产生了一定的差异。CSP 及 CCR 工艺生产的板材抗拉强度分别为 781.12MPa 和 833.22MPa，硬度分别为 219HV 和 223HV，屈强比分别为 0.89 和 0.84；相比于 CCR 工艺，CSP 工艺生产的板材虽然在强度和硬度值稍显降低，但其屈强比有明显提升。

关键词：复相钢；CSP 流程；CCR 流程；粒状贝氏体；力学性能；显微组织；屈强比

Investigation of Microstructure and Mechanical Properties of 800MPa Grade Multiphase Steel with Different Processes

Fang Xing[1], Yang Xiaoyu[1], Mi Zhenli[1], Zhang Hanlong[2]

(1. Institute of Engineering Technology, University of Science & Technology Beijing, Beijing 100083, China; 2. Academia Sinica, Baoshan Iron & Steel Co., Ltd., Shanghai 201900 China)

Abstract: The hardness and strength of 800MPa hot rolled multiphase steels produced by compact strip production (CSP) process and traditional casting and rolling (CCR) process are different. In this paper, by means of optical microscopy (OM) and scanning electron microscopy (SEM), the microstructure characteristics of the duplex steels produced by the two processes were systematically studied, and reasons for the difference of the macroscopic properties of the duplex steels were explained from the microscopic point of view. The average grain size of the two processes is 2.79μm for CSP and 2.68μm for CCR. The microstructure of the two processes is bainite matrix accompanied by polygonal ferrite and a small amount of M/A islands. The microstructure obtained by CSP process is dominated by granular bainite, while the microstructure obtained by CCR process is dominated by granular bainite mixed lath bainite. Therefore, differences in the microstructure have been observed. The tensile strength of CSP and CCR sheets are 781.12MPa and 833.22MPa, the hardness is 219HV and 223HV, and the yield ratio are 0.89 and 0.84, respectively. Compared with the CCR process, the strength and hardness values of the plates produced by CSP process are slightly decreased, but the yield ratio is significantly improved.

Key words: complex phase steel; CSP process; CCR process; granular bainite; mechanical properties; microstructure; yield ratio

双层 X 形管内高压成形加载路径的智能优化方法

冯莹莹，骆宗安，谢广明，吴庆林

（东北大学，轧制技术及连轧自动化国家重点实验室，辽宁沈阳　110819）

摘　要：本文针对双层 X 形管内高压成形过程加载路径的智能优化方法开展研究。将多目标动态优化方法用于双层 X 形管工艺质量参数的动态调整与优化。通过建立内层管最大减薄率、外层管最大减薄率、极限圆角半径、支管高度四个主要评价因素的柔性可调整区间，协调优化参数，使双层 X 形管的变形尽可能处于最佳状态。采用线性加权法建立双层 X 形管工艺质量参数的多因素综合目标函数，并运用遗传算法和 BP 神经网络控制方法对多目标动态优化函数进行推导求解，优化了学习效率，提高了计算精度，改进了与实际工艺的匹配程度。模拟结果与实验结果的误差在±7%以内，说明此双层 X 形管在内高压成形过程的加载路径优化控制方法具有较高的精度和可行性。

关键词：双层 X 形管；内高压成形；加载路径；多目标动态优化；神经网络

Intelligent Optimization Method for Loading Path of Double-layer X-tube Hydroforming Process

Feng Yingying, Luo Zongan, Xie Guangming, Wu Qinglin

(The State Key Laboratory of Rolling and Automation, Northeastern University, Shenyang 110819, China)

Abstract: In this paper, the intelligent optimization method for the loading path of double-layer X-tube hydroforming process is studied. The multi-objective dynamic optimization method was applied to the dynamic adjustment and optimization of process quality parameters of double-layer X-tube.By establishing the flexible adjustable range of the maximum thinning rate of inner tube, the maximum thinning rate of outer tube, the limit fillet radius and the height of the branch tube, the process quality parameters were coordinated and optimized to make the deformation of the double-layer X-shaped tube in the best state as possible. The linear weighting method was used to establish the multi factor comprehensive objective function of the process quality parameters of double-layer X-tube, and the genetic algorithm and BP neural network control method was used to deduce and solve the multi-objective dynamic optimization function, which optimized the learning efficiency, improved the matching degree and calculation accuracy with the actual process. The error between the simulation results and the experimental results is within ± 7%, which shows that the optimal control method of loading path for double-layer X-tube hydroforming process has high accuracy and feasibility.

Key words: double-layer X-tube; hydroforming; loading path; multi-objective dynamic optimization; neural network

型材三维多点拉弯成形工艺开发与制造装备研究

梁继才，李　义，梁　策

（吉林大学材料科学与工程学院，吉林长春　130022）

摘　要：型材三维弯曲成形零件能够提供轻量化的车身结构，良好的空气动力学性能，以及更加安全舒适的乘坐空间等，越来越受到人们的关注。然而，由于型材三维弯曲成形的复杂性和成形件形状难于控制，其加工工艺制约了其快速发展。本文提出了一种新型的柔性三维拉弯成形工艺。新工艺采用离散化多点模具柔性成形方法，并研制了成形制造装备。基于变形叠加理论，将型材三维变形分解为水平和垂直两个平面内的变形分量，分步实现了铝型材复杂三维空间构型的成形工艺。最后通过进行制件加工制造，实现了柔性三维拉弯成形零件的精确成形。

关键词：型材；多点成形；三维拉弯；柔性制造

The Study of Forming Process and Equipment for Three Dimensional Multi-point Stretch Bending Profiles

Liang Jicai, Li Yi, Liang Ce

(School of Materials Science and Engineering, Jilin University, Changchun 130022, China)

Abstract: The light-weight vehicle structure, good aerodynamic property and more safety & conformable riding space can be provided by three dimensional bending profiles. Therefore, this kind of forming parts is getting more and more attention. But, the forming technology has been a bottleneck restricting its development for the forming complexity and difficulty to control the shape of forming parts. In this paper, a novel flexible three dimensional stretch bending process was proposed. The forming equipment was developed by using the flexible forming method of discretized multi-point die. Based on deformation superposition theory, the three dimensional deformation could be further decomposed as the deformation component on the horizontal plane and the deformation component on the vertical plane. The aluminum profile of complex spatial shape was finally manufactured and the precise forming of flexible three dimensional parts was achieved.

Key words: profile; multi-point forming; three dimensional stretch bending; flexible manufacturing

高光亮表面优质冷轧导轨钢批量化生产研究

张　华，白晓东，郑永春，王欣龙

（内蒙古包钢钢联股份有限公司稀土钢板材公司，内蒙古包头　014010）

摘　要：高光亮表面冷轧优质导轨钢主要具有极低粗糙度、厚度控制精度高、板形及表面质量优良等特点，需涉及多个工序进行窄工艺窗口控制，批量化生产难度较大。为此本文针对生产过程中存在粗糙度偏高、厚度偏薄、印痕表面质量缺陷的问题，通过采取相应的改进措施，使本厂生产的导轨钢达到了极低粗糙度、厚度控制稳定的高光亮表面的技术指标要求，满足了用户的使用需求，并实现了工业化批量生产。

关键词：导轨钢；粗糙度；板形；表面质量

Research of Industrial Production of High-quality Cold-rolled Guide Rail Steel with High-bright Surface

Zhang Hua, Bai Xiaodong, Zheng Yongchun, Wang Xinlong

(Rare-earth Plate Company of Inner Mongolia Baotou Steel Union Co., Ltd., Baotou 014010, China)

Abstract: Cold-rolled high-quality guide rail steel with high-bright surface has the main characteristics of extremely low roughness, high thickness control accuracy, excellent plate shape, surface quality and so on. It requires multiple processes for narrow process window control, and mass production is difficult. This article would like to solve the problems of high roughness, thinner thickness, and surface quality of imprints in the mass production process. By taking improvement measures, the guide rail steel produced by our factory has reached the technical index requirements, included extremely low roughness and high brightness surface and stable thickness control. The needs of users are met and the industrialized mass production is achieved.

Key words: cold-rolled guide steel; roughness; flatness; surface quality

QP980 高强钢电阻点焊工艺对接头性能及组织的影响规律研究

卫志超[1]，李 免[1]，米振莉[1]，钟 勇[2]，王恩茂[1]

（1. 北京科技大学工程技术研究院，北京 100083；
2. 宝山钢铁股份有限公司中央研究院，上海 201900）

摘 要：随着汽车轻量化的需求，汽车结构件越来越广泛的采用先进高强钢，而高强钢构件中的连接技术愈发显得重要，其中白车身生产中广泛采用的电阻点焊成为高强钢深加工技术领域的研究重点之一。本文针对 QP980 高强钢的电阻点焊工艺对接头性能和组织的影响进行了系统研究。研究表明，通过拉断试验，QP980 高强钢点焊接头存在两种破坏模式：界面破坏和熔核剥离破坏。随着焊接电流和焊接时间的增加，焊点可承受的最大剪切力逐渐增大，断裂模式由界面破坏转变为熔核剥离断裂。在较优参数下 QP980 接头的熔核区、粗晶区和细晶区均为马氏体组织，且硬度均高于母材；亚临界热影响区的组织为马氏体、回火马氏体、铁素体和残余奥氏体的混合组织，导致硬度显著降低；该区域的软化促使了熔核剥离破坏的产生。

关键词：QP980 高强钢；电阻点焊；失效模式；显微组织；力学性能

Effects of Resistance Spot Welding Process on Microstructure and Mechanical Properties of QP980 High-strength Steel

Wei Zhichao[1], Li Mian[1], Mi Zhenli[1], Zhong Yong[2], Wang Enmao[1]

(1. Institute of Engineering Technology, University of Science & Technology Beijing, Beijing 100083, China; 2. Central Research Institue, Baoshan Iron & Steel Co., Ltd., Shanghai 201900, China)

Abstract: With the demand for lightweight automobiles, advanced high-strength steels are widely used in automobile structural parts, and the connection technology in high-strength steel components is becoming more and more important. Among them, resistance spot welding, which is widely used in the production of body-in-white, has become one of the research focuses in the field of high-strength steel deep processing technology. In this paper, the influence of resistance spot welding process of QP980 high-strength steel on microstructure and mechanical properties is systematically studied. Results shows that there are two failure modes in the tensile test of spot welded joints: interface failure and nugget peeling failure. With the increase of welding current and welding time, the maximum shear force that the solder joint can bear gradually increases, and the interface damage changes to the nugget peeling fracture. Under the better parameters, the nugget zone, coarse grain heat-affected zone and fine grain heat-affected zone of QP980 joints are all martensite, and the corresponding hardness is higher than that of the base metal. The microstructure of the subcritical heat-affected zone is martensite, tempered martensite , ferrite and retained austenite, which leads to the decrease of hardness. The softening of this area promotes the generation of nugget peeling damage.

Key words: QP980 high-strength steel; resistance spot welding; failure mode; microstructure; mechanical properties

模具冷却及多段拉拔对拉拔温升及性能的影响

张洪龙[1]，张峰山[1]，李振红[2]

（1. 中钢集团郑州金属制品研究院有限公司，河南郑州 450001；
2. 河南机电职业学院信息工程学院，河南郑州 451150）

摘 要：碳素冷拉弹簧钢丝在拉拔过程中，由于材料变形及钢丝表面与拉拔模具摩擦的存在，发热是不可避免的，特别是高速连续拉拔过程中，钢丝的热量经多次的累积，温度升高明显。钢丝在拉拔过程中的温升一方面可以改善润滑条件，降低模具与钢丝的摩擦力，但过高的温升会严重影响钢丝的力学性能。通过对模具进行冷却及采用多段式拉拔能明显降低钢丝拉拔温升，改善钢丝性能。

关键词：多段拉拔；温升；模具冷却；钢丝性能

Influence of Die Cooling and Multi-stage Drawing on Temperature Rise and Performance of Steel Wire

Zhang Honglong[1], Zhang Fengshan[1], Li Zhenghong[2]

(1. Sinosteel Zhengzhou Research Institute of Steel Wire Products Co., Ltd., Zhengzhou 450001, China;
2. School of Information Engineering, Henan Mechanical and Electrical Vocational College, Zhengzhou 451150, China)

Abstract: Due to material deformation and friction between steel wire surface and drawing die, temperature rise is inevitable on the drawing process of cold-drawn non-alloy spring steel wire. Especially in the process of high-speed continuous drawing, the temperature of steel wire increases obviously after many passes of heat accumulation. The temperature rise of steel wire on the drawing process can improve the lubrication conditions and reduce the friction between the die and the steel wire, but over high temperature rise will seriously reduce the mechanical properties of the steel wire. By cooling the die and adopting multi-stage drawing, the temperature rise of steel wire can be obviously reduced and the properties of steel wire can be improved.

Key words: multi-stage drawing; temperature rise; die cooling; properties of steel wire

双层板渐进成形工艺下的表面质量研究

孔建非，吴 琦，王会廷，沈晓辉

（安徽工业大学冶金工程学院，安徽马鞍山 243032）

摘　要：为了分析工艺参数对双板渐进成形工艺中1060铝合金内层板的表面质量影响，本文在数控铣床上进行了圆锥件单点渐进成形实验。以304SUS为辅助板，1060铝合金为目标板，检测了目标成形件的表面粗糙度和表面形貌，研究了工艺参数（工具头直径、增量步长、成形角、和润滑因素）对1060内层板目标成形件表面质量的影响。结果表明：目标成形件的表面粗糙度随着成形工具头直径的增大而降低；成形角、增量步长和润滑因素因素对目标成形件的表面粗糙度不敏感，采用较大的增量步长可以在不降低表面质量的同时显著提高加工效率。

关键词：双层板；目标成形件；工艺参数；表面质量

Study on Surface Quality by Double-layer Incremental Forming Process

Kong Jianfei, Wu Qi, Wang Huiting, Shen Xiaohui

(School of Metallurgical Engineering, Anhui University of Technology, Ma'anshan 243032, China)

Abstract: In order to analyze the effect of process parameters on the surface quality of 1060 aluminum alloy inner sheet in double-plate incremental forming process, the single point incremental forming experiments of conical parts were carried out on a NC milling machine. With 304SUS as auxiliary sheet and 1060 aluminum alloy as target sheet, the effects of process parameters (tool size, step size, wall angle, and lubrication) on the surface quality of 1060 inner sheet metal were investigated. The results show that the Surface roughness of the target part decreases with the increase of the tool size, and the wall angle, step size and lubrication factor are not sensitive to the Surface roughness of the target part, the machining efficiency can be improved obviously without reducing the surface quality by using larger step size.

Key words: double-layer sheet; target forming part; process parameter; surface quality

镁含量对DP780锌铝镁镀层组织和耐蚀性能的影响

朱泽升，米振莉，周大元，朱　蓉

（北京科技大学 工程技术研究院，北京　100083）

摘　要：本研究通过向锌液中添加Al、Mg元素，在热镀锌基础上制备了Zn-2wt%Al-xwt%Mg（x=2、3、4）合金镀层，利用扫描电镜、X射线衍射仪研究了锌铝镁镀层表面及截面微观结构、合金层形貌及物相组成，采用电化学工作站分析了不同镁含量镀层的耐蚀性能及耐蚀机理。结果表明：锌铝镁镀层主要由初生Zn相、Zn/$MgZn_2$二元共晶组织和Zn/$MgZn_2$/Al三元共晶组织组成。镁含量的提高会使钢板表面的各相组织含量及种类发生较大变化，截面处的枝晶状组织会随着镁含量的增加而逐渐粗大。与其他两种镀层相比，Zn-2wt%Al-2wt%Mg镀层镁富集区域较少，局部腐蚀不显著，因此耐蚀性能更好。

关键词：DP780钢；锌铝镁镀层；组织形貌；耐蚀性能

Effect of Magnesium Content on Microstructure and Corrosion Resistance of Zn–Al–Mg Alloy Coated DP780 Steel Sheet

Zhu Zesheng, Mi Zhenli, Zhou Dayuan, Zhu Rong

(Institute of Engineering Technology, University of Science and Technology Beijing, Beijing 100083, China)

Abstract: In this article, Zn-2wt%Al-xwt%Mg (x=2, 3, 4) alloy coatings were prepared by adding Al and Mg elements to zinc solution of hot dip galvanizing process. The surface and cross-sectional microstructure, phase composition of Zn-Al-Mg coating were analyzed by scanning electron microscope and X-ray diffractometer. The corrosion resistance of the coatings with different magnesium contents was analyzed using electrochemical workstation machine. Results show that the Zn-Al-Mg coating is mainly composed of primary Zn phase, Zn/$MgZn_2$ binary eutectic structure and Zn/$MgZn_2$/Al ternary eutectic structure. With the increase of Mg content, the microstructure on the surface of the coatings are significantly changed, and the dendritic microstructure at the cross section are gradually expand. Compared with the other two coatings, Zn-2wt%Al-2wt%Mg coating has less magnesium enrichment area and less local corrosion. Therefore, the Zn-2wt%Al-2wt%Mg coating acquires the best corrosion resistance performance.

Key words: DP780 steel; Zn-Al-Mg alloy coating; coating microstructure; corrosion resistance

考虑接触压强分布影响的热冲压仿真与工艺优化

袁俞哲[1]，杨海波[1,3,4]，薛 飞[2]，罗 松[2]，丁仁根[3]，李书志[3]

（1. 北京科技大学机械工程学院，北京 100083；2. 东莞市中泰模具股份有限公司，广东东莞 523475；3. 东莞材料基因高等理工研究院，广东东莞 523808；4. 流体与材料相互作用教育部重点实验室，北京 100083）

摘 要：针对复杂型面热冲压模具提出了一种考虑实际接触压强分布影响的热冲压仿真方法：利用 Python 二次开发对 ABAQUS 传热分析中板料与模具表面的换热系数进行定义以考虑接触压强分布的影响。利用该仿真方法对汽车 A 柱下加强板热冲压模具进行了仿真，分析了保压力和冷却水温度的影响：在一定范围内，保压力越大，板料淬火时冷却速率越大；冷却水温度对出模阶段模具冷却影响较大，连续工作时模面温度与冷却水温度呈正相关。优选的保压力为 5500kN，冷却水温度为 8℃。结果表明：优化条件下模具在第 10 个工作周期进入温度稳定状态，板料冷却速率满足马氏体完全转变条件，该设计符合连续生产要求。

关键词：热冲压；热冲压模具；数值仿真；汽车高强钢

Hot Stamping Simulation Considering the Influence of Contact Pressure Distribution and Process Optimization

Yuan Yuzhe[1], Yang Haibo[1,3,4], Xue Fei[2], Luo Song[2], Ding Rengen[3], Li Shuzhi[3]

(1. School of Mechanical Engineering University of Science and Technology Beijing, Beijing 100083, China; 2. Vision Tool & Mould Co., Ltd., Dongguan 523475, China; 3. Dongguan Institute of Materials and Genes Advanced Institute of Technology, Dongguan 523808, China; 4. Key Laboratory of the Interaction of Fluids and Materials, Ministry of Education, Beijing 100083, China)

Abstract: A simulation method of hot stamping considering the influence of the actual contact pressure distribution is proposed for the complex hot stamping die. The heat transfer coefficient between the sheet metal and the die surface is defined by Python secondary development. The proposed method is used to simulate the hot stamping die of the reinforced plate under A-pillar of automobile. Effects of pressure maintaining and cooling water temperature are analyzed. In a certain range, the greater the pressure is, the higher the cooling rate is when the plate is quenched. The cooling water temperature has a great influence on the cooling of the dies in the interval stage. The temperature of the mold surface is positively correlated with the cooling water temperature when the die works continuously. The optimal pressure is 5500kN and the cooling water temperature is 8℃. Results show that the dies enter the stable temperature in the 10th working cycle under the optimized condition. The cooling rate of the sheet meets the condition of martensite complete transformation, and the design meets the requirements of continuous production.

Key words: hot stamping; hot stamping die; numerical simulation; automotive high-strength steel

Deformation of Non-metallic Inclusions during Hot-rolling and Inclusion-microstructure Correlation in a Medium Mn Steel: A Case Study Contributes to Inclusion Engineering

Wang Yong[1], Yang Yonggang[1,2], Dong Zhihua[1,3], Park Joo Hyun[1,4], Mi Zhenli[2], Mu Wangzhong[1,5]

(1. Department of Materials Science and Engineering, KTH Royal Institute of Technology, SE-10044, Stockholm, Sweden; 2. Beijing Advanced Innovation Center for Materials Genome Engineering, National Engineering Research Center for Advanced Rolling Technology, University of Science and Technology Beijing, Beijing 100083, China; 3. State Key Laboratory of Mechanical Transmissions, College of Materials Science and Engineering, Chongqing University, Chongqing 400044, China; 4. Department of Materials Science and Chemical Engineering, Hanyang University, 15588, Ansan, Korea;.Institute of 5.Multidisciplinary Research for Advanced Materials (IMRAM), Tohoku University, Sendai, Japan)

Abstract: Inclusion engineering is a comprehensive concept dealing with the control of amount, size distribution, and chemical composition of non-metallic inclusions in the liquid steel and during solidification. Furthermore, it also concerns the correlation between inclusion, microstructure and property on the quality control of the final product, e.g. advanced high strength steel (AHSS). Medium-Mn steel (MMS)which usually contains 3-11% Mn is a typical and new category of 3rd generation AHSS developed in the recent 1-2 decades due to a unique trade-off of strength and ductility. This work provides a fundamental study of inclusion and microstructure correlation in a new designed MMS. Specifically, the effect of hot-rolling conditions on the deformation behavior of different types of inclusions (MnS, $MnSiO_3$, and complex oxy-sulfide) is investigated in this work. Furthermore, the evolution of grain size as well as the inclusion aspect ratio during the rolling is quantitative discussed. In addition, the microstructure characteristics are characterized in different scales. This work aims to contribute in 'inclusion engineering' of new generation AHSS applied in different industries, e.g. automotive.

Inclusions in the as-cast and hot-rolled MMS samples were collected by the electrolytic extraction (EE) method and was characterized by scanning electron microscopy (SEM) equipped with energy dispersive spectroscopy (EDS) for the composition and size analysis. The microstructure characteristic is identified by electron backscatter diffraction (EBSD). Coefficient of thermal expansion (CTE) is characterized by dilatometer, micro-hardness is also detected by the hardness indenter with a load of 100g. Besides, the inclusion formation is also predicted by thermodynamic calculation using Thermo-Calc, and other related physical parameters, e.gYoung's modulus is calculated by density function theory (DFT) method.

Both of $MnSiO_3$ and MnS are soft inclusions which are able to be deformed during the hot-rolling, the aspect ratio increases significantly from as-cast to hot-rolled condition. When the maximum size of different inclusions is similar, MnS deforms more than $MnSiO_3$ does, due to a joint influence of different physical parameters (hardness, Yong's modulus and difference of CTE between matrix and inclusions). However, when the maximum size of one type inclusion (e.g. $MnSiO_3$) is much larger than another one (e.g. MnS), the maximum size of larger soft inclusions plays a dominant role than other reasons. In addition, the deformation behavior of mixed phase (dual phase?) inclusion is dependent on the major phase, i.e., either oxide or sulfide. Last but not least, the reduction ratio of (?) grain size and aspect ratio of different types of inclusion are provided quantitatively.

不同盘条对加工硬化率的影响

周志嵩，姚海东，吕 辉，张喜泽，寇首鹏

（江苏兴达钢帘线股份有限公司，江苏泰州 225721）

摘 要：本文对比了 6 种盘条加工的不同强度的镀铜丝对湿拉钢丝强度的影响。随着片层间距逐渐减小，镀铜丝抗拉强度升高。在相同的盘条下，镀铜丝强度越高，湿拉拉拔的加工硬化率也越大。不同的盘条，如果镀铜丝强度控制在同一水平，那么湿拉拉拔的加工硬化率基本相近。不同化学成分的盘条均有比较容易获得的强度和组织范围，随着碳含量或铬含量的增加，对应镀铜丝的强度较高。6 种盘条的碳含量在共析点 0.77%附近，在 3.41-3.73 的总应变范围时，钢丝的初始强度越高，拉拔加工硬化率也就越大。

关键词：抗拉强度；片层间距；加工硬化率

The Effect of Different Wire Rods on Work Hardening Rate

Zhou Zhisong, Yao Haidong, Lü Hui, Zhang Xize, Kou Shoupeng

(Jiangsu Xingda Steel Tyre Cord Co., LTD., Taizhou 225721, China)

Abstract: The plated wires with different strengths processed by 6 kinds of wire rods on the strength of wet drawn steel wires are compared. As the lamella spacing gradually decreases, the tensile strength of the plated wire increases. Under the same wire rod, the higher the strength of the plated wire, the greater the work hardening rate of wet drawing. Under different wire rods, when the strength of the plated wire is controlled at the same level, the work hardening rate of wet drawing is basically similar. Wire rods with different chemical compositions have relatively easy-to-obtain strength and microstructure. The addition of carbon content or chromium content results in higher strength of the plated wire. The carbon content of the 6 kinds of wire rods are near the eutectoid point 0.77%, and when the total strain range is from 3.41 to 3.73, the higher the initial strength of the steel wire, the greater the drawing work hardening rate.

Key words: tensile strength; lamella spacing; work hardening rate

拉弯工艺参数对回弹的影响研究

户全超[1]，韩 飞[1]，徐 根[2]

（1. 北方工业大学机械与材料工程学院，北京 100144；
2. 张家港市同力冷弯金属有限公司，江苏 张家港 215600）

摘 要：拉伸量是重要的拉弯成形工艺参数，本文从理论解析方面分析了随着拉伸量变化，弯曲截面高度方向上应力应变的分布及变化情况，对拉伸量关于回弹的影响规律和机理进行了分析。并通过 ABAQUS 仿真软件进行了仿真实验，从云图上比较直观的观测到应力应变中性层的偏移及截面上应力的分布情况，通过拉弯实验验证了有限元模型的准确性。研究发现：随着拉伸量的增加，截面上应力分布由内外侧反向变为内外侧同向，最终都进入塑性区域；型材截面内外侧应力差值逐渐减小，从而使回弹量减小。

关键词：拉弯成形；数值模拟；回弹；工艺参数

Influence of Stretch Bending Process Parameters on Springback

Hu Quanchao[1], Han Fei[1], Xun Gen[2]

(1. College of Mechanical and Materials Engineering, North China University of Technology, Beijing 100144, China; 2. Zhangjiagang Tongli Cold Formed Metal Co., Ltd., Zhangjiagang 215600, China)

Abstract: Drawing amount is an important process parameter of stretch bending. In this paper, with the change of drawing amount, the distribution and change of stress and strain along the height direction of bending section are analyzed theoretically, and the influence law and mechanism of drawing amount on springback are analyzed. Through the ABAQUS simulation software, the displacement of the stress-strain neutral layer and the stress distribution on the section can be observed intuitively from the cloud image, and the accuracy of the finite element model is verified by the stretch bending experiment. The results show that: with the increase of tensile strength, the stress distribution on the cross section changes from inside and outside in the opposite direction to inside and outside in the same direction, and finally enters the plastic region; The stress difference between the inner and outer sides of the profile section decreases gradually, so that the springback decreases.

Key words: stretch bending; numerical simulation; springback; process parameters

DP980电阻点焊工艺仿真及焊点材料卡片开发

王铭泽，郭　晶，李科龙，孙　洋

（本钢板材股份有限公司研发院，辽宁本溪　117000）

摘　要：采用有限元仿真的方式对上下板厚均为1.2mm的DP980钢板点焊过程进行数值模拟，分析其焊接温度、焊核直径等参数；开发出基于DP980-DP980焊点力学性能的LS-Dyna *Mat#100材料卡片，并结合仿真所得的焊核直径制成剪切拉伸与十字拉伸断裂试验仿真数模，以对标与优化材料卡，确保材料卡在实际应用中的准确性；通过与试验结果的对比发现，仿真结果与试验得出焊核直径相差0.61%，优化后的剪切试验仿真与试验结果最大载荷相差1.2%，十字拉伸试验最大载荷相差2.5%，误差较小；证明利用仿真手段可较为准确预测焊点形貌与尺寸，同时可制成供整车碰撞分析使用的材料卡片。

关键词：点焊；双相钢；焊点；材料卡；LS-Dyna

The Simulation of Resistance Spot Welding Process and Development of Spot Welding Material Card for DP980

Wang Mingze, Guo Jing, Li Kelong, Sun Yang

(Technology Research Institute of Benxi Steel Plate Co., Ltd., Benxi 117000, China)

Abstract: Based on the finite element method, the simulation of the spot welding process was carried out using two 1.2-mm-thickness DP980 steel plates as the upper and lower plate. The welding temperature and the welding nugget diameter were analyzed. The material card LS-Dyna Mat#100 based on the mechanical properties of DP980-DP980 welding spots was developed. Combined with the nugget diameter, the simulation models of shear tensile and cross tensile fracture test were made for benchmarking and optimizing material card, which ensure the accuracy of material card in application. The test results agree well with simulation analysis, with the nugget diameter difference of 0.61%, the maximum load difference of shear test of 1.2%, and the maximum load difference of cross tensile test of 2.5%. It is proved that the size and morphology of welding spots can be predicted accurately by the simulation method that is used in current research. Material card developed in this article can be further used for vehicle crash analysis.

Key words: spot welding; DP steel; welding spots; material card; LS-Dyna

棒线材深加工技术应用及发展趋势

王卫卫，赵　舸，侯中晓

（钢铁研究总院冶金工艺研究所，北京　100081）

摘　要：本文通过针对深加工原材料、设备、工艺最新研究进展的技术调研以及本研究团队所做的研发工作总结，较为全面的阐述了棒线材深加工技术应用及发展趋势，也为今后的高品质棒线材的深加工技术的进一步发展提供技术参考。

关键词：深加工；新工艺；新设备；高品质棒线材；发展趋势

Application and Development Trend of Hot Rolled Rebar after Implementation of New Standard

Wang Weiwei, Zhao Ge, Hou Zhongxiao

(Metallurgical Technology Institute of Central Iron & Steel ResearchInstitute, Beijing 100081, China)

Abstract: This paper is based on the latest research progress of raw materials, equipment and technology for deep processing. Research results done by our research team are also taken into consideration. This paper comprehensively clarifies the application and development trend of rod and wire deep processing technology. This provides technical reference for the further development of high-quality rod and wire deep processing technology in the future.

Key words: deep-processing; new technology; new equipment; high-quality wire rod and bar;development trend

304H 不锈钢丝连续拉拔下的微观组织和织构变化规律

彭　科，刘　静，程朝阳，彭志贤

（武汉科技大学耐火材料与冶金国家重点实验室，湖北武汉　430081）

摘　要：本文以 304H 奥氏体不锈钢为研究对象，采用连续拉拔工艺，对各道次出模钢丝进行微观组织结构、织构分析，讨论了拉拔变形对奥氏体不锈钢丝不同区域的形变和微区结构的影响规律。结果表明，在连续冷拉拔过程中，钢丝晶粒尺寸发生了明显的降低，同时钢丝内部将持续形成{111}//ND 与{110}//ND 织构；对比钢丝心部和边部的织构形成情况，钢丝心部表现出较边部更强的织构。

关键词： 304H 奥氏体不锈钢；拉拔钢丝织构；冷拉拔

Microstructure and Texture evolution of 304H Stainless Steel Wire During Continuous Drawing

Peng Ke, Liu Jing, Cheng Zhaoyang, Peng Zhixian

(State Key Laboratory of Refractories and Metallurgy, Wuhan University of Science and Technology, Wuhan 430081, China)

Abstract: The grain size, microstructure and texture of 304H austenitic stainless steel wire were analyzed by optical microscope and scanning electron microscope. The influence of drawing process on deformation and microstructure in different regions of austenitic stainless steel wire was discussed. Results show that the grain size of 304H austenitic stainless steel decreases obviously during continuous cold drawing, and the {111}//ND and {110}//ND textures are formed in the steel wire. Compared with the texture in the center and edge of steel wire, the distribution of {111}//ND texture in the center of steel wire is stronger than that in the edge. The change of {110}//ND texture is not obvious with the increase of drawing deformation.

Key words: 304H austenitic stainless steel; drawing steel wire texture; cold drawing

辊弯成形工艺对 Q&P980 高强钢宏微观性能的影响研究

孟伊帆，韩 飞

（北方工业大学机械与材料工程学院，北京 100144）

摘 要： 本文主要研究了辊弯成形工艺对 Q&P980 高强钢宏微观性能的影响。对 Q&P980 高强钢采用辊弯成形工艺，设计 V 型件成形角度分别为 10°、20°、30°，对其回弹角及弯曲部位硬度分别进行测试，采用 OM、SEM、XRD 对弯曲部位进行了微观组织分析，并且对工件进行了单轴拉伸实验，分析了 Q&P980 高强钢的力学性能。研究发现，在单道次辊弯成形中，随着弯曲角度地增加，Q&P980 高强钢的回弹角度呈先升后降的趋势；奥氏体发生相变，马氏体含量的增多，进而提高了型材的硬度。

关键词： 辊弯成形工艺；Q&P980 高强钢；回弹；相变

Study on the Influence of Roll Bending Process on Macro and Micro Properties of Q&P980 High Strength Steel

Meng Yifan, Han Fei

(College of Mechanical and Materials Engineering, North China University of Technology, Beijing 100144, China)

Abstract: This paper mainly studies the influence of roll forming process on the macro and micro properties of Q&P980 high strength steel. The roll forming method was used for the forming of Q&P980 high strength steel. The V-shaped parts were designed to have a forming angle of 10°, 20° and 30° respectively. The springback angle and the hardness of the forming parts were tested respectively. The mechanical properties of Q&P980 high strength steel were analyzed. It was found that the springback angle of Q&P980 HSS increased first and then decreased with the increase of forming angle in the single-pass roll forming process. Austenite transformed into martensite, resulting in the increasing martensite volume fraction, which improves the hardness of the profiles.

Key words: roll forming process; Q&P980 high strength steel; springback; phase change

特厚规格热冲压材料与热冲压工艺的设计开发

刘宏亮[1]，李龙泽[2]，陈　宇[1]，杨　波[1]，李春诚[1]，关　琳[1]，焦　坤[1]

（1. 本钢板材股份有限公司研发院，辽宁本溪　117000；

2. 吉林省正轩车架有限公司，吉林辽源　136699）

摘　要：为了解决特厚规格热冲压钢采用常规工艺生产出的车辆底盘零件无法满足使用要求的问题，本研究设计了适合车辆底盘零件使用的特厚规格热冲压材料和热冲压工艺，并分别对由传统6.0mm厚度PHS1500和热冲压工艺制备的零件与本工作设计的相同厚度的PHS1500T和热冲压工艺制备的零件进行了微观组织、显微硬度以及综合力学性能测试。结果表明：采用传统厚规格PHS1500和热冲压工艺制备的零件芯部存在铁素体，芯部与边部硬度差显著增加，影响材料综合力学性能；采用本工作设计的厚规格PHS1500T和热冲压工艺生产的零件芯部和边部均为马氏体，芯部和边部硬度差明显降低，材料综合力学性能满足零件使用要求。在此基础上，结合表面处理及夹杂物变质技术，最终开发出了适合扭力梁生产的热冲压材料和热冲压工艺方案。

关键词：热冲压；工艺；组织；厚规格

Design and Development of Extra Thick Hot Stamping Material and Process

Liu Hongliang[1], Li Longzhe[2], Chen Yu[1], Yang Bo[1], Li Chuncheng[1], Guan Lin[1], Jiao Kun[1]

(1. Products Research Institute of Benxi Steel Plates Co., Ltd., Benxi 117000, China;

2. Jilin Zhengxuan Car Frame Co., Ltd., Liaoyuan 136699, China)

Abstract: In this work, to solve the problem that the automobile chassis parts produced by the conventional process of extra thick hot stamping steel can not meet the using requirements, the extra thick hot stamping materials and hot stamping process suitable for automobile chassis parts are designed. Meanwhile, the microstructure, microhardness and comprehensive mechanical properties of the parts prepared by the traditional 6.0mm thickness PHS1500 and hot stamping process and the parts prepared by the same thickness PHS1500 and hot stamping process designed in this work are investigated, respectively. The results show that there is ferrite in the core of the part prepared by traditional thickness specification PHS1500 and hot stamping process, and the hardness difference between the core and the edge increases significantly, which affect the comprehensive mechanical properties of the materials. The core and edge of the part produced by the thickness specification PHS1500T and the hot stamping process of this work is martensite, the hardness difference between the core part and edge part is reduced obviously, and the comprehensive mechanical properties of the material meet the requirements of the parts. On this basis, combined with the removal of surface treatment and inclusion modification technology, the hot stamping material and hot stamping process suitable for torsion beam production are finally developed.

Key words: hot stamping; process; microstructure; thick specification

一种耐大气腐蚀深冲钢的开发

刘　恒，吴彦欣，米振莉

（北京科技大学工程技术研究院，北京　102200）

摘　要：深冲钢因其优异的深冲性能被广泛应用于复杂形状零件当中，但其耐大气腐蚀性能不佳，尤其长时间在海洋大气中服役会破坏产品的性能，因此普通深冲钢已不能满足工业的需求。本文基于传统DC01钢，通过添加适量的合金元素，开发了一种兼具良好耐海洋大气腐蚀性能和深冲性能的新型耐候深冲钢，并通过模拟连续退火的方式对该新型耐候深冲钢在不同退火温度下的力学性能以及成形性能展开研究，同时利用电化学检验试退火后的新型耐候深冲钢在 Cl^- 离子环境下的耐蚀能力，分析退火温度对新型耐候深冲钢耐海洋大气腐蚀能力的影响。结果表明：在一定的退火温度范围内，随着退火温度的升高，该耐候深冲钢的屈服强度和抗拉强度降低，而n值和延伸率呈上升趋势。在750℃的退火温度下，r值与自腐蚀电位最大，此时成形性能最好，受大气腐蚀的倾向性最小，为新型耐候深冲钢的最佳退火温度。

关键词：深冲钢；耐大气腐蚀性能；力学性能；成形性能；退火温度；织构；电化学

Development of a Corrosion Resistant Deep Drawing Steel

Liu Heng, Wu Yanxin, Mi Zhenli

(University of Science and Technology Beijing, Engineering Technology Research Institute, Beijing 102200, China)

Abstract: Deep drawing steel is widely used in complex shape parts due to its excellent deep drawing performance, but its atmospheric corrosion resistance is not good, especially in the ocean atmosphere for a long time service will damage the performance of the product, so the ordinary deep drawing steel has been unable to meet the needs of industry. In this paper, based on the traditional DC01 steel, a new type of weathering deep drawing steel with good oceanic and atmospheric corrosion resistance and deep drawing performance was developed by adding an appropriate amount of alloying elements. The mechanical properties and forming properties of the new type of weathering deep drawing steel under different annealing temperatures were studied by simulating continuous annealing. At the same time, the corrosion resistance of the new weathering deep drawing steel after annealing in Cl^- ion environment was tested by electrochemical test, and the influence of annealing temperature on the corrosion resistance of the new weathering deep drawing steel in ocean atmosphere was analyzed. The results show that in a certain range of annealing temperature, with the increase of annealing temperature, the yield strength and tensile strength of the weather-resistant deep drawing steel decrease, while the N value and elongation increase. Under the annealing temperature of 750℃, the R value and self-corrosion potential are the maximum, the forming property is the best, and the tendency to be corroded by atmosphere is the least, which is the best annealing temperature for the new type of deep drawing steel.

Key words: deep drawing steel; weatherability; mechanical properties; forming property; annealing temperature; texture; electrochemistry

终轧温度对压缩机用热轧酸洗板成形性能的影响

崔凯禹[1,2]，李正荣[1]，刘序江[1]，叶晓瑜[1]，汪创伟[1]，胡云凤[1]

（1. 攀钢集团攀枝花钢铁研究院有限公司，四川攀枝花　617000；
2. 哈尔滨工业大学材料科学与工程学院，黑龙江哈尔滨　150000）

摘　要：本研究对压缩机用热轧酸洗板采用 850℃和 880℃的终轧温度进行生产，结合热力学平衡相图和终轧温度分布模拟仿真以及金相组织和综合性能的检测结果，分析了终轧温度对压缩机用热轧酸洗板成形性能的影响。结果表明，采用 880℃进行终轧时材料组织性能优良，但采用 850℃进行终轧时，板宽距边部 40mm 处温度降低到约 820℃，低于相变点温度 A_{r3}，使终轧处于奥氏体-铁素体两相区轧制，产生混晶组织，最终导致材料的断后伸长率 A_{50mm} 和均匀伸长率 A_{gt} 出现各向差异且数值降低，加工硬化指数 n 减小，塑性应变比 r 值在各个方向上的差异增大且最小 r 值减小，塑性应变比加权平均值 $\bar{r}$ 值降低，以及塑性应变比各向异性度 $|\Delta r|$ 值增大，从而显著恶化了材料的成形性能。

关键词：热轧酸洗板；终轧温度；成形性能；混晶

Effect of Finishing Temperature on Forming Properties of Hot Rolled Pickling Steel Plate for Compressor

Cui Kaiyu[1,2], Li Zhengrong[1], Liu Xujiang[1], Ye Xiaoyu[1], Wang Chuangwei[1], Hu Yunfeng[1]

(1. Pangang Group Research Institute Co., Ltd., Panzhihua 617000, China;
2. Harbin Institute of Technology, School of Materials Science and Engineering, Harbin 150000, China)

Abstract: This paper investigates the effect of finishing temperature on forming properties of hot rolled pickling steel plate for compressor. The used steel plate were produced using 850℃ or 880℃ as the finishing temperature. The thermodynamic equilibrium phase diagram is calculated and the simulation showing finishing temperature is established. Microstructure is observed and mechanical properties are revealed. Results show that the tested steel produced with the finishing temperature of 880℃ acquires homogeneous microstructure and outstanding properties. However, when the tested steel produced with the finishing temperature of 850℃, the temperature at the position that is 40mm apart from plate edge decreases to around 820℃. The temperature is lower than austenite-ferrite phase transformation temperature, namely the finishing rolling temperature is in two phase region, which resulting in the heterogeneous grain size in total length direction. The aforementioned phenomenon causes that elongation A_{50mm} and uniform elongation A_{gt} show difference in the length directions. In details, the A_{50mm} and A_{gt} decrease, work hardening exponent decreases, the difference in plastic strain ratio r increases with the values decrease, plastic strain ratio weighted average decreases, and plastic strain ration anisotropy degree $|\Delta r|$ increases. All parameters indicate the decreasing forming properties.

Key words: hot rolled pickling steel plate; finishing temperature; forming properties; heterogeneous grain size

高线大规格线卷尾部吐丝及圈形控制

经勇明，郑团星，王扬发，何海峰

（宝武集团中南钢铁韶钢松山股份有限公司特轧厂，广东韶关 512123）

摘 要：本文主要针对大规格线卷尾部大圈、滞留通道等故障进行分析，重新优化夹送辊、吐丝机控制逻辑及参数，有效解决尾部圈形问题，提高线卷外观质量及产品形象。

关键词：大规格；尾部；圈形；控制

Spinning and Coil Shape Control at the Tail of Large Size Steel Wire Coils

Jing Yongming, Zheng Tuanxing, Wang Yangfa, He Haifeng

(Baowu Group Shaogang Songshan Co., Ltd. Special Rolling Mill., Shaoguan 512123, China)

Abstract: This paper mainly analyzes the overlarge size of coil at the tail of the steel wire and its retain on the channel. Adjusting the procedure and parameters of the pinching roller and spinneret effectively solve the problem of the coil shape and improve the quality of the coil and product.

Key words: large size; tail; coil shape; control

薄带钢轧后冷却温度场及相变的数值模拟

张茂才，米振莉，苏 岚，王 迈，常 江

（北京科技大学工程技术研究院，北京 100083）

摘 要：带钢卷取温度对带钢最终的组织性能有重要影响，为了保证轧后冷却能达到设计要求的卷取温度，本文结合某钢厂 CSP 产线的 SPHC 钢种的现场实测数据，对薄带钢建立二维对称的 1/4 横截面有限元模型，模拟分析薄带钢轧后冷却过程中的温度场变化和冷却结束后的组织分布特征。计算得到 3.52mm 厚度的带钢上表面温度为 681.1℃，实测温度为 682℃，计算值与实测值相差 0.9℃，相对误差为 0.13%，3.02mm 和 3.82mm 厚的计算相对误差分别为 1.62%和 0.73%，本文所建立模型具有一定的准确性，且带钢在厚度方向的温度分布较为均匀。通过计算发现，带钢在粗调第一段的冷却效率最高，且在冷却过程中发生奥氏体向铁素体的转变，相变潜热使带钢在空冷阶段出现回温，冷却速度降低，当整个冷却过程结束时带钢相变也已进行完全，最终组织几乎为全铁素体且分布均匀。

关键词：CSP；轧后冷却；温度场；相变；有限元

Numerical Simulation of Temperature Field and Phase Transformation in the Cooling Process of Thin Strip Steel

Zhang Maocai, Mi Zhenli, Su Lan, Wang Mai, Chang Jiang

(Institute of Engineering Technology, University of Science and Technology, Beijing 100083, China)

Abstract: The coiling temperature of the strip has an important influence on the final microstructure and properties of the strip. In order to ensure that temperature of the strip after cooling can reach the required coiling temperature, this paper establishes the finite element analysis model based on the measured data of SPHC steel that is produced in one of CSP production lines. The established model is a two-dimensional symmetrical 1/4 cross-section finite element model. It can be used to analyze the temperature field changes during the cooling process of the thin strip steel after rolling. The microstructure distribution after cooling can also be revealed using the model. The temperature of the upper surface of the strip with a thickness of 3.52mm is calculated to be 681.1℃, and the measured temperature is 682℃. The difference between the calculated value and the measured value is 0.9℃, and the relative error is 0.13%. For the strip of 3.02mm and 3.82mm thickness, the calculated relative errors are 1.62% and 0.73% respectively. The established model in this paper has relatively high accuracy, and the revealed temperature distribution of the strip in the thickness direction is relatively homogeneous. Moreover, it is found that the cooling efficiency of the strip in the first stage of coarse adjustment is the highest, and the phase transformation from austenite to ferrite occurs during the cooling process. The phase change causes that the temperature the strip increases in the air cooling stage, and thus the cooling rate is reduced. When the entire cooling process completes and the strip phase transformation finishes, the almost all structure is ferrite and homogeneously distributed.

Key words: CSP; cooling after rolling; strip temperature; phase transformation; finite element analysis

稀土耐腐蚀热镀锌板开发

薛　越，何建中，辛广胜，高　军，郭　勇

（包钢股份薄板坯连铸连轧厂，内蒙古包头　014000）

摘　要：本文通过稀土对热镀锌板腐蚀形貌、腐蚀行为及电化学测试等机理分析与讨论，进行了生产工艺转化，从稀土锌锭制备、稀土合金加入及溶解、稀土收得率提高和腐蚀效果等方面进行了系统地试验，成功开发了稀土耐腐蚀热镀锌钢板，进一步验证了稀土提高热镀锌钢板的耐腐蚀性能，稀土热镀层腐蚀失重速率降低 26.32%。

关键词：稀土元素；热镀锌板；耐腐蚀性能

Development of Rare Earth Corrosion Resistant Hot Dip Galvanized Plate

Xue Yue, He Jianzhong, Xin Guangsheng, Gao Jun, Guo Yong

(Baotou Steel Compact Strip Production Plant, Baotou 014000, China)

Abstract: In this paper, the corrosion morphology, corrosion behavior and electrochemical test mechanism of rare earth on hot-dip galvanized sheet were analyzed and discussed, and the production process was transformed. Systematic tests were carried out from the aspects of preparation of rare earth zinc alloy, addition and dissolution of rare earth zinc ingot, improvement of rare earth yield and corrosion effect, etc. The rare earth corrosion-resistant hot-dip galvanized sheet was successfully developed, which further verified that rare earth improved the corrosion resistance of hot-dip galvanized sheet and reduced the corrosion weight loss rate of rare earth hot-dip galvanized sheet by 26.32%.

Key words: rare earth; hot dip galvanized sheet; corrosion resistance

1300MPa 级超高强钢焊接性能研究

程浩轩

（华菱湘钢技术质量部，湖南湘潭　411100）

摘　要：本文以某钢厂生产的 Q1300E 超高强钢板为试验材料，按照“低强匹配”原则选用 120kg 级焊丝，制定适合的焊接工艺进行焊接评定试验。结果表明，相同热输入条件下，焊后热处理能明显改善焊接接头的性能。试验过程中需严格控制板材下料、坡口加工、组对成型、焊前预热、焊接工艺及焊后热处理等工序，采用同一型号焊丝进行打底、填充及盖面焊接，经对试板进行力学性能试验发现，按此焊接工艺可实现 Q1300 超高强钢的高效、高质量、高稳定性焊接，最大限度避免焊接裂纹，消除焊接应力，且为机器人焊接提供了技术支撑。

关键词：超高强钢；冷裂纹；热输入；热处理

Study on Weldability of 1300MPa Ultra High Strength Steel

Cheng Haoxuan

(Technology and Quality Department of Valin Xiangtan Steel, Xiangtan 411100, China)

Abstract: In this paper, the Q1300E ultra-high strength steel plate produced by a steel company is used as the test material. According to the principle of “low strength matching”, 120kg welding wire is selected, and the suitable welding procedure is set for the welding qualification test. Results show that the properties of the welded joint can be improved by post welding heat treatment under the same heat input. During the test process, it is necessary to strictly control the plate blanking, groove processing, assembly forming, preheating before welding, welding process and post welding heat treatment, etc. The same type of welding wire is used for backing, filling and cover welding. Through the mechanical property test of the plate, it is found that the high efficiency, high quality and high stability welding of Q1300 ultra-high strength steel can be obtained. Almost all welding cracks can be avoided and the welding stress can be eliminated. In addition, the technical support for robot welding can be further developed based on current research.

Key words: ultra high strength steel; cold crack; heat input; heat treatment

SCM435 钢丝拉拔笔尖状断裂原因分析及应对措施

朱祥睿，罗新中，李富强

（广东韶钢松山股份有限公司检测中心，广东韶关　512123）

摘 要：材质为SCM435合金冷镦钢盘条在拉拔加工过程中出现笔尖状断裂。对拉拔断裂钢丝取样进行宏观检测、金相检测、扫描电镜及能谱分析等理化检测分析，检测结果表明，钢丝心部存在大尺寸硬脆的铝酸钙盐夹杂物是导致钢丝拉拔笔尖状断裂的主要原因。由于钢丝心部存在大尺寸球状铝酸钙盐夹杂物，该类型夹杂物硬而脆，变形性能极差，导致钢丝在拉拔变形过程中容易在球状铝酸钙盐夹杂物部位产生孔隙，随着钢丝拉拔变形的继续，夹杂物部位的孔隙逐渐扩展为裂纹，裂纹从心部往周边延伸，最终导致钢丝拉拔呈笔尖状断裂。

关键词：冷镦钢；笔尖状断裂；夹杂物

Analysis of Nib-like Fracture of Drawing for SCM435 Steel Wire and Its Solutions

Zhu Xiangrui, Luo Xinzhong, Li Fuqiang

(Testing Center of Guangdong Shaoguan Steel Songshan Co., Ltd., Shaoguan 512123, China)

Abstract: The cold heading steel wire rod made of SCM435 alloy has nib-like fracture during the drawing process. Samples from drawing broken steel wire were analyzed by macroscopic observation, metallographic examination, scanning electron microscope and energy spectrum analysis. Results show that the existence of large-size hard and brittle calcium aluminate inclusions in the center position of the steel wire is the main reason for the nib-like fracture. The inclusions are hard and brittle, and their elongation is very limited, which leads to the occurrence of holes. With further deformation, holes gradually expand into cracks, and the cracks grow from the center to the surface, resulting in the nib-like fracture of the steel wire.

Key words: cold heading steel; nib-like fracture; inclusion

难变形金属温轧试验机的研发

孙 涛，牛文勇，王贵桥，矫志杰，杨 红，李建平

（东北大学轧制技术及连轧自动化国家重点实验室，辽宁沈阳 110819）

摘 要：温轧是提高难变形金属薄带成形性能与质量的重要手段，科研人员针对温轧工艺开展了大量研究。传统的温轧实验是在轧机旁设置加热炉，将金属片加热至所需温度后，人工夹持送入轧辊进行轧制。这种方法效率低、降温快，而且不能获得精准的工艺参数，无法模拟生产过程中带张力在线温轧。为了解决这一问题，东北大学在液压张力冷轧试验机的基础上，增加接触式测温、轧件在线电阻加热、轧辊在线加热、液压微张力控制、变形区温度预测、厚度软测量等功能，研发出液压张力温轧试验机，针对单片金属薄带可实现带张力温轧工艺研究。该试验机已推广应用至10余家国内外企业研究院所及高校，可以满足难变形金属温轧工艺研究和新材料研发的需求。

关键词：温轧；宽展；薄带；在线加热

Development of Warm Rolling Mill for Difficult-to-Deform Metal

Sun Tao, Niu Wenyong, Wang Guiqiao, Jiao Zhijie, Yang Hong, Li Jianping

(The State Key Laboratory of Rolling and Automation, Northeastern University, Shenyang 110819, China)

Abstract: Warm rolling is an important method to improve the formability and quality of difficult-to-deform metal strip. By the traditional warm rolling method,the sheet was heated to the required temperature in a furnace beside the rolling mill, and then was manually clamped and sent to the gap of the rolling mill. This method has the disadvantage of low efficiency, fast cooling, inaccurate process parameters, and is impossible to simulate the production process of warm rolling with tension. In order to solve this problem, on the basis of the hydraulic tension cold rolling mill, Northeastern University has developed a hydraulic tension warm rolling mill by adding the functions of contact temperature measurement, on-line resistance heating of rolled piece, on-line heating of rolls, hydraulic micro tension control, temperature prediction of deformation zone, soft measurement of thickness, etc. The pilot mill has been applied to more than 10 enterprises, research institutes and universities at home and abroad.It can meet the needs of research on warm rolling process of difficult-to-deform metals, and development of new metal materials.

Key words: warm rolling mill; difficult to deform; metal strip; online heating

浅析单机架机组轧制硅钢的同板差情况

赵　刚，曹　垒，游　涌，张　杰

（张家港扬子江冷轧板有限公司，江苏张家港　215625）

摘　要：硅钢同板差指标包含纵向厚度波动和横向同板差，纵向厚度波动根据在线测厚仪检测厚度曲线（单机架和硅钢工序两个曲线），而横向同板差主要靠硅钢多点测厚仪取样检测。为进一步了解目前沙钢冷轧单机架硅钢同板差实际水平，随机抽取 2021 年 2-3 月份单机架轧制典型的硅钢卷，根据单机架测厚仪和硅钢测厚仪检测厚度曲线对比，同时利用多点测厚仪检测横向同板差，与酸轧厚度曲线同口径进行对比，分析沙钢冷轧单机架硅钢同板差水平及改进方向。

关键词：单机架机组；硅钢；厚度波动

Analysis on the Same Plate Difference of Silicon Steel Rolled by Single Stand Mill

Zhao Gang, Cao Lei, You Yong, Zhang Jie

(Zhangjiagang Yangtze River Cold Rolled Plate Co., Ltd., Jiangsu Zhangjiagang 215625, China)

Abstract: The thickness deviation, one of the indices of silicon steel, includes the longitudinal thickness deviation and the transverse thickness deviation. The longitudinal thickness deviation is revealed based on two thickness curves that are related to the single rolling stand and the silicon steel process, respectively. The thickness curves are measured by online thickness gauge. The transverse thickness deviation is mainly measured by multi-point measurement of the thickness gauge. In order to grasp the actual thickness deviation of silicon steel fabricated using the single cold rolling stand of Sha steel company, the typical silicon steel coil, rolled by single rolling stand during February to march in 2021, was randomly selected. The thickness curves that are related to the single rolling stand and the silicon steel process are measured. The transverse thickness deviation is obtained using the multi-point measurement method of the thickness gauge. All the measured results are compared with the results of the same-size steel produced using the picking and rolling method. The degree of thickness deviation and the improvement method are analyzed based on the measured results.

Key words: single rolling stand; silicon steel; thickness deviation

双碳背景下钢材深加工产业的现状与发展趋势

董馨浍，张光明，杨梅梅，于治民

（冶金工业信息标准研究院冶金信息研究所，北京 100006）

摘 要：“碳达峰”和“碳中和”决策部署是我国统筹国际、国内两个大局的重要战略决策，是对生产、消费、技术、经济、能源体系的历史性革命。钢铁行业作为我国落实碳减排目标的重要责任主体，要实现双碳目标任重道远。当前，我国钢铁工业已处在减量发展、联合重组和强化环保三期叠加阶段，正在向高质量低碳阶段演进。“十四五”期间，无论是国家层面还是地方层面均强调把实现高质量发展摆在更加突出的位置，对我国钢铁工业提出了新的更高的发展要求。面对新形势，钢铁企业普遍开展的多元发展战略为钢材深加工提供了机遇，装备制造业集群化发展也为发展钢材深加工提供了空间。本文论述了国内外钢材深加工产业发展的现状，分析了双碳背景下钢材深加工产业面临的机遇及挑战，并提出“十四五”期间我国钢材深加工产业的重点发展方向。

关键词：双碳；钢材深加工；钢材深加工产业趋势

Present Situation and Development Trend of Steel Deep Processing Industry Under the Background of Carbon Emission Peak and Carbon Neutrality

Dong Xinhui, Zhang Guangming, Yang Meimei, Yu Zhimin

(China Metallurgical Information and Standardization Institute, Metallurgical Information Research Department, Beijing 100006, China)

Abstract: The decision and deployment of "carbon peak" and "carbon neutral" is an important strategic decision for China to coordinate the international and domestic situation, and a historic revolution in production, consumption, technology, economy and energy system. Steel industry as an important responsibility to implement the carbon emission reduction target in China, to achieve the double carbon target has a long way to go. At present, China's steel industry has been in the reduction of development, joint restructuring and strengthening environmental protection three superposition stage, is to the high quality and low carbon stage evolution. During the 14th Five-year Plan period, both the national level and the local level emphasize the realization of high-quality development in a more prominent position, and put forward new and higher development requirements for China's iron and steel industry. Facing the new situation, the diversified development strategy of steel enterprises has provided opportunities for the deep processing of steel, and the cluster development of equipment manufacturing industry has also provided space for the development of deep processing of steel. This paper discusses the development status of steel deep processing industry at home and abroad, analyzes the opportunities and challenges faced by steel deep processing industry under the background of "carbon peak" and "carbon neutral", and puts forward the key development direction of China's steel deep processing industry during the "14th Five-year Plan".

Key words: carbon peak and neutrality; steel deep processing; development trend of steel deep processing industry

电镀锌汽车油箱板生产工艺的研究

孙晨航，曹　洋，车晓宇

（本钢浦项冷轧薄板有限责任公司，辽宁本溪　117000）

摘　要：单面钝化工艺为新兴工艺，其生产要求超出本钢电镀锌原有设计大纲，为实现其稳定供货就需要对电镀锌汽车油箱板生产工艺进行研究，本项目从如何实现单面钝化工艺、如何优化提速后的单面镀工艺参数、如何提升未镀面质量等问题出发研究，形成一整套适用于本钢特点的电镀锌汽车油箱板的生产技术、组织模式、产品规范和技术诀窍。

关键词：单面钝化；钝化膜厚；控制要点；未镀面质量

Study on the Production Technology of Electric Galvanized Steel for Automobile Oil Tanks

Sun Chenhang, Cao Yang, Che Xiaoyu

(Bx Steel Posco Cold Rolled Sheet Co.,ltd., Benxi 117000, China)

Abstract: The single passivation technology is a novel technology and its production requirements exceed the original design outline of electric galvanized steels produced in Bx Iron and Steel. In order to acquire stable production, the processing parameter of electric galvanized steel should be investigated, which is related to the implementation of the single passivation technology, the optimization of parameters of the electric galvanized process, and the improvement of the surface quality of uncoated steel plates. The production technology, organizational model, product specifications and technical know-how of electro-galvanized automobile fuel tank steel plates for Benxi Iron and Steel are obtained in the current research.

Key words: single side passivation；passivation film thickness；control points；unplated surface quality

7 先进钢铁材料及其应用

7.1 汽车用钢

汽车底盘用 500MPa 级热轧酸洗钢板的研制开发

孙成钱，时晓光，董 毅，刘仁东，韩楚菲，王俊雄

（鞍钢集团钢铁研究院，辽宁鞍山 114009）

摘 要： 通过热轧厂热轧和冷轧厂盐酸酸洗，试制了含钛和不含钛的两种热轧酸洗钢板。采用金相显微镜、透射电镜和拉伸试验机研究了两种试制钢板的显微组织和性能。结果表明：两种钢板的显微组织均由铁素体和珠光体组成，其屈服强度达到 400MPa 以上，抗拉强度高于 500MPa，断后伸长率大于 30%。与不含钛钢相比，含钛钢的晶粒细小，屈服强度和抗拉强度高，断后伸长率低，屈强比高。

关键词： 热轧酸洗板；钛；显微组织；性能

Development of 500MPa Grade Hot-rolled and Pickled Steel Plate for Automobile Chassis

Sun Chengqian, Shi Xiaoguang, Dong Yi, Liu Rendong, Han Chufei, Wang Junxiong

(Iron and Steel Research Institute of Ansteel Group Corporation, Anshan 114009, China)

Abstract: Two kinds of hot-rolled pickled steel plates with and without titanium were trial-produced by hot-rolling in a hot rolling plant followed by pickling with hydrochloric acid in a cold rolling plant. The microstructure and properties of two trial-produced steel plates were studied by means of optical microscope, transmission electron microscope and tensile testing machine. The results showed that the microstructure of two steel plates was composed of ferrite and pearlite, and the yield strength was above 400MPa, the tensile strength was above 500MPa, and the elongation was over 30%. Compared with the steel without Ti addition, the Ti-containing steel exhibited finer grain, and thus higher yield strength, tensile strength and yield ratio, but lower elongation.

Key words: hot-rolled and pickled plate; titanium; microstructure; property

热处理温度对 Fe-Mn-Al-C 系轻质钢中铝系碳化物的析出影响规律探究

孟静竹，刘仁东，郭金宇，徐荣杰，王科强，金晓龙

（鞍钢集团钢铁研究院，辽宁鞍山 114001）

摘 要： Fe-Mn-Al-C 系轻质钢中由于添加了较多的 Al 元素，在某些条件下会生成铝系碳化物，导致性能急剧下降。

本文通过使用 OM、SEM、XRD、TEM、EPMA、拉伸试验，维氏硬度测试等手段对不同处理温度下的 Fe-18Mn-0.8C-0.3Si-(3,6,9)Al 实验钢中含铝化合物析出的规律以及对性能的影响进行研究。发现在 900℃以下的热处理温度下 K-相是存在，而当热处理温度高于 900℃时，K-相就不再析出。在生产中应避免粗大的 K-相在钢中析出，应采用较高的轧制温度以及热处理温度就能避免 K-相的析出。

关键词：Fe-Mn-Al-C 钢；K-相；热处理温度；力学性能

The Research of the Aluminized Carbide's Precipitated Law in the Fe-Mn-Al-C Light Weight Steel under Different Heat Treatment Temperature

Meng Jingzhu, Liu Rendong, Guo Jinyu, Xu Rongjie, Wang Keqiang, Jin Xiaolong

(Ansteel Group Iron and Stell Research Institute, Anshan 114001, China)

Abstract: Aluminum series carbides will be formed in Fe-Mn-Al-C series light weight steel with more Al element under certain conditions,which can cause a sharp drop in performance. Mechanical properties and microstructure evolution of the law of Aluminized carbide-KAPPA phase precipitation in Fe-18Mn-0.8C-0.3Si-(3,6,9)Al experimental steel undergone different heat treatment temperature was investigated by means of OM, SEM, XRD, TEM, EPMA, tensile test and Vickers hardness tester.The results shows that the KAPPA phase existed at heat treatment temperature below 900℃.It will dissolved when the temperature comes to above 900℃. The thick K-phase should be avoided from precipitation in steel during production,which can be avoided by using higher hot rolling temperatures and heat treatment temperatures.

Key words: Fe-Mn-Al-C steel; K phase; heat treatment temperature; mechanical properties

红外碳硫仪测定汽车板、管线钢中硫元素的探讨

李青青，刘步婷

（武钢有限质检中心，湖北武汉 430080）

摘　要：以硫含量较低的普通汽车板和管线钢作为分析目标，CS844 型碳硫仪作为分析仪器，采用高频感应加热炉燃烧样品，利用红外线吸收法测试样品中硫元素的质量分数。通过对分析条件、空白实验、精度实验和生产样进行试验分析研究。结果表明：在设定分析条件下，碳硫仪分析屑状样和柱状样得到的硫元素含量准确度好，RSD<0.5%，精密度高，因此可广泛用于钢厂汽车板和管线钢的分析测定。

关键词：碳硫仪；硫；汽车板；管线钢

Determination of Sulfur in Ordinary Automobile Plate and Pipeline Steel by Infrared Absorption Spectrum

Li Qingqing, Liu Buting

(Quality Inspection Center of Wuhan Iron and Steel Group Company, Wuhan 430083, China)

Abstract: The mass fraction of sulfur in the sample of ordinary automobile plate and pipeline steel was measured by high-frequency infrared absorption method using the CS844 type carbon-sulfur meter as the analysis instrument. The analysis conditions,blank and precision experiment and production sample were studied.The results indicate that the good accuracy and high precision of sulfur content of chip and columnar sample under the given analytical conditions, and the RSD<0.5%. Therefore, the CS844 type carbon-sulfur meter can be widely used in the determination of ordinary automobile plate and pipeline steel.

Key words: carbon sulfur meter; sulfur; automobile plate; pipeline steel

应变速率对δ-TRIP980 钢动态拉伸变形行为的影响

徐 鑫[1,2]，梁 笑[1]，李春林[1]，陆晓锋[1]，林 利[1]，刘仁东[1]

（1. 鞍钢集团钢铁研究院，辽宁鞍山 114009；
2. 东北大学轧制技术及连轧自动化国家重点实验室，辽宁沈阳 110819）

摘 要：利用液压伺服高速拉伸试验机对 δ-TRIP980 钢进行不同应变速率下的拉伸变形实验，结合 SEM 和 XRD 等手段，研究了应变速率对 δ-TRIP980 钢动态拉伸性能及变形行为的影响规律及机制。结果表明，δ-TRIP980 在不同拉伸速率下其力学性能存在着明显差异，随着应变速率的增加，屈服强度、均匀延伸率和断裂延伸率升高，但由于在高应变速率下未完成相变的残余奥氏体含量逐渐增加，材料的加工硬化率降低，导致抗拉强度降低。

关键词：δ-TRIP980 钢；动态拉伸；应变速率；残余奥氏体

Effect of Strain Rate on the Dynamic Tensile Deformation Behavior of δ-TRIP980 Steel

Xu Xin[1,2], Liang Xiao[1], Li Chunlin[1], Lu Xiaofeng[1], Lin Li[1], Liu Rendong[1]

(1. An steel Group Iron and Steel Research Institute, Anshan 114009, China; 2. State Key Laboratory of Rolling Technology and Automation, Northeastern University, Shenyang 110819, China)

Abstract: The tensile deformation tests of δ-TRIP 980 steel under different strain rates were carried out by using a servo hydraulic high speed tensile testing machine, the effect of strain rate on dynamic tensile properties and deformation behavior of δ-TRIP980 steel was investigated by means of SEM and XRD. The results show that there are significant differences in the mechanical properties of δ-TRIP980 at different tensile rates. With the increase of strain rate, the yield strength, uniform elongation and total elongation increase, however, as the content of retained austenite that has not undergone transformation at high strain rates gradually increases, the work hardening rate of the material decreases, resulting in the decrease of tensile strength.

Key words: δ-TRIP980 steel; dynamic tensile; strain rate; retained austenite

配分时间对 1000MPa 级 Q&P 钢组织性能的影响

王亚东，孟庆刚，崔宏涛，左海霞

（本溪钢铁集团有限责任公司技术研发院，辽宁本溪 117000）

摘　要：通过两相区退火结合一步淬火配分热处理工艺，在热模拟条件下对1000MPa级Q&P钢进行不同配分时间下的Q&P工艺处理，利用扫描电镜和X射线衍射仪分析配分时间对相组成、残余奥氏体体积分数及其碳含量的影响，通过拉伸试验测定Q&P钢的力学性能，研究组织演变对性能的影响。结果表明：实验材料的显微组织均主要由铁素体、马氏体和残余奥氏体构成，随着配分时间的延长，抗拉强度呈下降趋势，马氏体自身属性在配分过程中的变化，残余奥氏体体积分数及其碳含量共同决定断后伸长率呈先升高后降低最后又升高的趋势，在配分时间为600s时实验材料可获得最佳的强塑性匹配。

关键词：Q&P钢；配分时间；组织；性能

Effect of Partitioning Time on Microstructure and Properties of 1000MPa Q&P Steel

Wang Yadong, Meng Qinggang, Cui Hongtao, Zuo Haixia

(Technical Research Institute, Benxi Iron and Steel (Group) Co., Ltd., Benxi 117000, China)

Abstract: The 1000 MPa grade Q&P steel was treated by intercritical annealing combined with one-step quenching and partitioning heat treatment under thermal simulation conditions. The effects of partitioning time on phase composition, volume fraction of retained austenite and its carbon content were analyzed by scanning electron microscope and X-ray diffraction. The mechanical properties of were tested, and the effect of microstructure evolution on mechanical properties was studied. The results show that the microstructure of the experimental materials are mainly composed of ferrite, martensite and retained austenite, with the extension of the partitioning time, the tensile strength shows a downward trend, the change of martensite properties in the partitioning process, the volume fraction of retained austenite and its carbon content determine the trend of elongation, the best strength and elongation matching can be obtained when the partitioning time is 600s.

Key words: Q&P steel; partitioning time; microstructure; properties

热镀锌双相钢DP780电阻点焊工艺研究

王亚东，孟庆刚，王亚芬，杨天一

（本钢技术研究院，辽宁本溪　117000）

摘　要：为了改善热锌双相钢DP780电阻点焊接头微观组织与力学性能，在电阻点焊之后通过增加回火脉冲，对比不同点焊工艺对点焊接头组织、硬度、拉伸载荷和吸收能的影响。结果表明单脉冲工艺下，熔核区组织为粗大的板条马氏体，拉伸载荷较低，增加回火脉冲使熔核区组织细化，熔核区和热影响区的马氏体软化，塑韧性提高，脆硬倾向减小，拉伸试验过程中提高了对断裂发展的抵抗力，拉伸载荷提高，当回火脉冲电流为5kA，回火脉冲时间为200ms参数下，可获得满意的焊接力学性能。

关键词：热镀锌双相钢；电阻点焊；力学性能

Study on Resistance Spot Welding Process of Hot Dip Galvanized Dual Phase Steel DP780

Wang Yadong, Meng Qinggang, Wang Yafen, Yang Tianyi

(Bengang Technology Research Institute, Benxi 117000, China)

Abstract: In order to improve the microstructure and mechanical properties of hot galvanized steel DP780 resistance spot welded joints, tempering pulses were added to compare the effects of different spot welding processes on the microstructure, hardness, tensile load and energy absorption of spot welded joints after resistance spot welding. The results show that the microstructure of the nugget is coarse lath martensite under the single-pulse process, and the tensile load is lower. Add tempering pulse makes the nugget microstructure refined, martensite softed in the nugget and the heat affected zone. The toughness is improved, brittle tendency is reduced, the resistance to fracture development is improved during the tensile test, and the tensile load is increased. When the tempering pulse current is 5kA and the tempering pulse time is 200ms, satisfactory welding mechanical properties can be obtained.

Key words: hot galvanized DP steel; resistance spot welding; mechanical properties

1300MPa 级马氏体钢连续冷却组织转变研究

金晓龙[1]，张福义[2]，孙荣生[1]，郑飞龙[2]，蔡顺达[1]，宋利伟[1]

（1. 鞍钢集团钢铁研究院，辽宁鞍山 114009；2. 鞍钢股份有限公司冷轧厂，辽宁鞍山 114021）

摘 要： 利用 Gleeble-3800 型热模拟试验机测定了 Ms1300 马氏体压缩变形后动态连续冷却转变（CCT）曲线，并模拟其热轧生产工艺过程来研究轧制工艺参数对显微组织的影响，结果表明，冷速为 0.5℃/s 时，组织为铁素体和贝氏体，硬度值为 287HV；冷速为 1、2℃/s 时，组织为单一贝氏体，硬度值为 308HV 和 386HV；冷速为 5℃/s、10℃/s、20℃/s、30℃/s 时，组织为贝氏体和马氏体，硬度值在 428～462HV 之间；冷速为 50℃/s 时，组织全为马氏体，硬度达到 470HV。

关键词： 马氏体钢；热模拟；动态 CCT 曲线；组织

Study on Transformation of Martensite Steel 1300MPa after Continuous Cooling

Jin Xiaolong[1], Zhang Fuyi[2], Sun Rongsheng[1], Zheng Feilong[2], Cai Shunda[1], Song Liwei[1]

(1. Iron & Steel Research Institute, Ansteel Group, Anshan 114009, China;
2. Cold Rolling Plant, Ansteel Co., Ltd., Anshan 114021, China)

Abstract: Dynamic continuous cooling transformation (CCT) curves of Ms1300 steel were determined on Gleeble-3800 thermal simulator. The hot rolled process was simulated to study the effect of rolling parameters on microstructure. The results show that the microstructure of the steel are ferrite and bainite with cooling rate 0.5℃/s ,the value of the hardness is 287HV; the microstructure is bainite when The cooling rate are 1 and 2℃/s, the value of the hardness are 308 and 386HV; the microstructure is bainite and martensite when The cooling rate are 5, 10, 20 and 30℃/s, the value of the hardness is from 428 to 462HV; the microstructure is martensite when The cooling rate is 50℃/s, the value of the hardness is 470HV.

Key words: martensite steel; thermal simulation; dynamic CCT curve; microstructure

Nb-V 微合金化对 Fe-Mn-Al-C 奥氏体钢热加工行为的影响

赵 婷[1]，荣盛伟[1]，郝晓宏[1]，王天生[1,2]

（1. 燕山大学亚稳材料制备技术与科学国家重点实验室，河北秦皇岛 066004；
2. 燕山大学国家冷轧板带装备及工艺工程技术研究中心，河北秦皇岛 066004）

摘　要：本文对比研究了 Fe-Mn-Al-C 奥氏体钢和 Nb-V 微合金化 Fe-Mn-Al-C 奥氏体钢在单轴热压缩变形过程中的流变行为，分析了两种实验钢的热加工参数并建立了本构方程与 3D 热加工图。结果表明：在低温低应变速率下，Nb-V 微合金化钢的流变软化现象更加明显，Nb-V 微合金化使得 Fe-Mn-Al-C 钢更容易发生动态再结晶。Nb-V 微合金化后 Fe-Mn-Al-C 钢的热变形激活能略有降低，高功耗区略小但稳定区增大，即添加 Nb-V 后 Fe-Mn-Al-C 钢的热加工稳定性提高。Nb-V 微合金化 Fe-Mn-Al-C 奥氏体钢在应变 0.7 的最优热加工区域为 894～1025℃, 0.01～0.14 s^{-1} 和 1050～1200℃, 0.03～0.95 s^{-1}。

关键词：Fe-Mn-Al-C 奥氏体钢；Nb-V 微合金化；热压缩；流变曲线；加工图

Effect of Nb-Vmicro-alloying on Hot Deformation Behavior of Fe-Mn-Al-C Austenitic Steel

Zhao Ting[1], Rong Shengwei[1], Hao Xiaohong[1], Wang Tiansheng[1,2]

(1. State Key Laboratory of Metastable Materials Science and Technology, Yanshan University, Qinhuangdao 066004, China; 2. National Engineering Research Center for Equipment and Technology of Cold Strip Rolling,Yanshan University, Qinhuangdao 066004, China)

Abstract: In the present paper, flow behaviorsof Fe-Mn-Al-C austenitic steel and Nb-V micro-alloyed Fe-Mn-Al-C austenitic steelduring uniaxial hot compression deformation werestudied. The hot working parameters of two tested steels were analyzed, and constitutive equation and 3D processing map were established.Results showed that, obvious flow softening occurred in the Nb-V micro-alloyed Fe-Mn-Al-C steel at low temperature and low strain rate, indicating a promoting effect ofNb-V micro-alloyed on the dynamic recrystallization.The hot deformation activation energy of Fe-Mn-Al-C steelwas decreasedslightly because of the addition ofNb-V. Meanwhile, the high power efficiency region was reduced but the stable region was broadened, which indicateda positive effect of Nb-Vmicro-alloying on the hot deformation stability. The optimum processing domain for hot deformation of the Nb-V micro-alloyed Fe-Mn-Al-C steel at strain of 0.7 are 894-1025℃, 0.01-0.14 s^{-1} and 1050-1200℃, 0.03-0.95 s^{-1}.

Key words: Fe-Mn-Al-C austenitic steel; Nb-V micro-alloying;hot compression; flow curve; processing map

IF 钢汽车外板“麻面”缺陷原因分析与工艺优化研究

李建新，王　波

（宝钢湛江钢铁有限公司热轧厂，广东湛江　524000）

摘　要：根据冷轧反馈 IF 钢汽车外板存在“麻面”缺陷，通过对缺陷宏观形貌、分布规律及扫描电镜微观分析，确定该缺陷为热轧原板晶界氧化导致镀锌基板形成微观翘皮，宏观表现为“麻面”缺陷。结合晶界氧化改善机制并对比不同出钢记号 IF 钢外板缺陷发生情况，通过快速工序验证，明确添加 B 元素及降低热轧出炉温度可减少或避免“麻面”缺陷发生。

关键词：IF 钢；麻面；晶界氧化；B 元素；出炉温度

Improvement and Reasons on “the Pitting Surface” of the Outer Plate of IF Steel

Li Jianxin, Wang Bo

(Hot rolling plant of Baosteel Zhanjiang iron and Steel Co., Ltd., Zhanjiang 524000, China)

Abstract: According to the defect of “pitting surface” in cold-rolled feedback IF steel automobile exterior sheet, the macro-morphology, distribution rule and scanning electron microscopy analysis of the defect indicate that the defect is caused by grain boundary oxidation of hot-rolled original sheet, which results in micro-warping of galvanized substrate, and the macro-appearance is “flax surface” defect. Combining with the mechanism of grain boundary oxidation improvement and comparing the occurrence of defects in IF steel with different tapping marks, it is clear that adding B element and lowering tapping temperature during hot rolling can reduce or avoid the occurrence of flake surface defects through rapid process verification.

Key words: IF steel; the pitting surface; oxidation of the grain boundary; B element; tapping temperature

先进耐磨钢批量稳定生产技术研究

李建新，王　波，刘亚会

（宝钢湛江钢铁有限公司热轧厂，广东湛江　524000）

摘　要：先进耐磨钢拓展初期面临 F1 尾部抛钢轧制力大、弯辊设定到极限无调节余量废钢、卷取温度波动大等一系列制约该钢种批量稳定生产的关键核心问题，通过优化出炉温度、粗轧速度、精轧速度、精轧用水、终轧温度、层冷用水进而提高了先进耐磨钢的生产、质量稳定性。

关键词：先进耐磨钢；批量稳定；温度；速度

Research on Batch Stable Production Technology of Advanced Wear-resistant Steel

Li Jianxin, Wang Bo, Liu Yahui

(Hot Rolling Plant of Baosteel Zhanjiang Iron and Steel Co., Ltd., Zhanjiang, 524000, China)

Abstract: In the early stage of development of advanced wear-resistant steel, a series of key problems that restrict the stable production of the steel are faced, such as the large rolling force of F1 tail steel, the large fluctuation of coiling temperature and so on. The production and quality stability of advanced wear-resistant steel are improved by optimizing the tapping temperature, rough rolling speed, finishing rolling speed, finishing rolling temperature and water for layer cooling.

Key words: advanced wear resistant steel; batch stable; temperature; speed

横向稳定杆用 34MnB5 淬火回火组织及性能

董现春[1,2]，郭占山[1,2]，蔡 宁[1,2]，张 衍[1,2]，王凤会[1,2]，赵英建[1,2]

（1. 首钢集团有限公司技术研究院，北京 100043；
2. 绿色可循环钢铁流程北京市重点实验室，北京 100043）

摘 要：对热轧态横向稳定杆用 34MnB5 钢板进行 950℃水冷淬火，在 250℃、350℃、450℃、500℃、600℃温度下保温回火，观察金相组织，进行拉伸试验、冲击试验。结果表明，随着回火温度的升高，组织由淬火马氏体，转变为回火马实体、回火屈氏体、回火索氏体，碳化物由针状逐渐变为短杆状。抗拉强度由 1524MPa 降低至 743MPa，$R_{p0.2}$ 由 1283MPa 降低至 617MPa，断后伸长率由 6.5%提高至 16%，−20℃夏比 V 型冲击吸收功由 5J 提升至 56J（55mm×10mm×5mm）。

关键词：横向稳定杆；34MnB5；淬火回火；组织及性能

Microstructure and Properties of 34MnB5 Quenched and Tempered for Stabilizer Bar

Dong Xianchun[1,2], Guo Zhanshan[1,2], Cai Ning[1,2], Zhang Yan[1,2], Wang Fenghui[1,2], Zhao Yingjian[1,2]

(1. Shougang Research Institute of Technology, Beijing 100043, China; 2. Beijing Key Laboratory of Green Recyclable Process for Iron & Steel Production Technology, Beijing 100043, China)

Abstract: The hot rolled 34MnB5 steel plate for transverse stabilizer bar was quenched at 950℃, and tempered at 250℃, 350℃, 450℃, 500℃ and 600℃. The microstructure, tensile test and impact test were observed. The results show that with the increase of tempering temperature, the microstructure changes from quenched martensite to tempered martensite, tempered troostite and tempered sorbite, and carbide gradually changes from needle to short rod. The tensile strength

decreased from 1524MPa to 743MPa, $R_{p0.2}$ decreased from 1283MPa to 617MPa, and the elongation increased from 6.5% to 16%, the Charpy V-type impact energy at −20℃ is increased from 5J to 56J(55mm×10mm×5mm).
Key words: stabilizer bar; 34MnB5; quench and tempering; microstructure and properties

大规格 48MnV 曲轴磁痕产生原因浅析

周成宏，刘年富，廖子东，董凤奎，李华强

（宝武杰富意特殊钢有限公司，广东韶关　512123）

摘　要：非调质钢 48MnV 棒材的生产流程为：BOF-LF-RH-连铸-加热-连轧-精整-棒材（Φ130~160mm）。棒材经锻造加工成曲轴成品后，在磁粉探伤时发现存在批量磁痕。通过采用直读光谱仪，金相显微镜、布氏硬度计、扫描电子显微镜及能谱（EDS）等方法，对 48MnV 曲轴磁痕产生的原因进行了检验和分析。结果表明：棒材经锻造和机加工后，其心部元素偏析带移至连杆颈表面，在后续轴颈淬火过程中产生较多的残余奥氏体，曲轴磁粉探伤时，无磁性的残余奥氏体形成漏磁场，最终导致磁痕的产生。针对此类磁痕，提出了相应的技术改进措施。
关键词：非调质钢；曲轴；磁痕；中心偏析

Analysis on the Cause of Magnetic Mark of Large Size 48MnV Crankshaft

Zhou Chenghong, Liu Nianfu, Liao Zidong, Dong Fengkui, Li Huaqiang

(BaoWu JFE Special Steel Co., Ltd., Shaoguan 512123, China)

Abstract: Non-quenching and tempering steel 48MnV bar production process is:BOF-LF-RH-CC-Heating-Continuous rolling-Finishing-Bar(Φ130~160mm). After the bar was forged into the finished crankshaft, batch magnetic marks were found in the magnetic particle inspection. By means of direct reading spectrometer, metallographic microscope, Brinell hardness tester, scanning electron microscope and energy dispersive spectroscopy (EDS), The causes of magnetic marks of 48MnV crankshaft are examined and analyzed. The results show that the core element segregation belt moves to the surface of the connecting rod neck after forging and machining, In the subsequent quenching process of journal, more residual austenite is produced, In crankshaft magnetic particle inspection, the magnetic field leakage is formed by the non-magnetic residual austenite, which eventually leads to the magnetic mark. In view of this kind of magnetic trace, the corresponding technical improvement measures are put forward.
Key words: non-quenched and tempered steel; crankshaft; magnetic mark; center segregation

施焊位置对镀锌板焊点表面堆锌缺陷的影响

张永强，付　参，伊日贵，王鹏博，鞠建斌，陈炜煊

（首钢集团有限公司技术研究院，北京　100043）

摘　要：镀锌板焊点堆锌是汽车可视件焊点外观质量提升的瓶颈问题。针对镀锌汽车板，分析了焊点表面堆锌的形态和成分，研究了施焊位置对焊点堆锌缺陷的影响规律，结果表明：堆锌缺陷发生在焊点周围靠近板材边缘一侧。建立了电阻点焊模型，模拟了点焊过程中的温度场，发现焊点形核的同时表面锌层熔化，且焊接电流越大表面锌层熔化区域越大。最后，通过铁粉追踪了不同位置施焊过程中焊点周围磁场的分布，揭示了堆锌缺陷产生机理：焊接电流在施焊位置周围的磁场分布决定了表面堆锌形态。

关键词：镀锌板；电阻点焊；堆锌；模拟；温度场；磁场

Effect of Welding Position on Zinc Bulge Defect of RSW Weld for Galvanized Sheet

Zhang Yongqiang, Fu Can, Yi Rigui, Wang Pengbo, Ju Jianbin, Chen Weixuan

(Shougang Group Co., Ltd. Research Institute of Technology, Beijing 100043, China)

Abstract: The automobile industry has high requirements for the weld surface quality, especially for visual parts. For galvanized sheets, the defects of zinc bulge always occurred around the resistance spot weld. In this article, the zinc bulges were analyzed by SEM and EDS. The effect of welding position on zinc bulge was studied. The weld close to the edge was prone to zinc bulge. And the position of zinc bulge was always around the weld and close to the edge side. A model was established to simulate the temperature field in the process of spot welding. It was found that the surface zinc layer melted at the same time with the nugget formation. When the welding current increased, the melting area of the surface zinc layer became larger. The distribution of the magnetic field around the welds during RSW process at different positions was tracked by iron powder. The mechanism of zinc bulge defect was revealed, the distribution of magnetic field around welding current determined the shape of zinc deposit on the surface.

Key words: galvanized sheet; resistance spot welding; zinc bulge; numerical simulation; temperature field; magnetic field

显微分析在 Al-Mg-Si 铝合金低倍异常组织分析中的应用

薛　菲，单长智，冯伟骏

（中国宝武集团宝山钢铁股份有限公司 中央研究院 新材料产业创新中心，上海　201900）

摘　要：针对 AlMgSi 系铝合金大规格扁铸锭常见低倍粗晶及组织不均匀典型缺陷，采用传统光学显微镜（OM）与电子显微分析术扫描电镜（SEM）、能谱仪（EDS）、波谱仪（EPMA）以及痕量分析如飞行时间二次离子质谱（TOP-SIMS）和无损 X 射线显微术（XDCT）等多类型原理和具有不同检测范围的显微定量分析技术相结合对铝合金大生产扁铸锭低倍异常组织、成分、痕量元素分布等因素的统计性规律进行表征。结果表明：显微分析技术的综合应用在解析宏观组织异常缺陷以及溯源分析有较好作用。比对结果显示，粗晶与细晶区域内痕量元素及主合金元素分布不均匀，孔隙率有较大差异。缺陷与细化剂质量有遗传关系。

关键词：低倍组织异常；显微分析；Al-Mg-Si 铸锭

Application of Microanalysis on the Failure Analysis of Macro Texture Abnormal in Al-Mg-Si Aluminum Alloy Ingot

Xue Fei, Shan Changzhi, Feng Weijun

(New Material Industry Innovation Center, Baoshan Steel Corporation Limited, Baowu Group, Shanghai 201900, China)

Abstract: Aiming at the typical defects of coarse grain and inhomogeneous microstructure in large size flat ingot of AlMgSi aluminum alloy, traditional optical microscope (OM), scanning electron microscope (SEM) and energy dispersive spectrometer (EDS) , electron probe micro analysis (EPMA), time-of-flight secondary ion mass spectrometry (TOF-SIMS) and Xradia Context microCT (XDCT)were used to characterize qualitatively and quantitatively about the difference of structure, composition and element distribution. The results show that the comprehensive application of microanalysis technologies have a good effect in the analysis of macro structure such as abnormal defects and trace element analysis. The results show that the distribution of trace elements and main alloy elements in coarse-grained and fine-grained regions is inhomogeneous, and the porosity is quite different. There is a genetic relationship between the defects and the quality of refiner.

Key words: abnormal structure; micro analysis; Al-Mg-Si ingot

汽车用钢 QStE500TM 高低周疲劳性能研究

韩 丹

（本钢板材研发院，辽宁本溪 117000）

摘 要：对汽车用钢 QStE500TM 进行高周疲劳试验、低周疲劳试验，获得了 QStE500TM 相应疲劳性能参数。试验结果表明：应力一寿命曲线为：$\log\sigma_{max}$=3.3849-0.12201logN，疲劳极限为 $\sigma_{0.1}$=483.75MPa；应变一寿命曲线为：$\Delta\varepsilon_t/2=(1037.26/E)\times(2N_f)^{-0.08782}+0.81132\times(2N_f)^{-0.69459}$，循环应力一应变曲线：$\Delta\sigma/2=847.93\times(\Delta\varepsilon_p/2)^{0.10187}$。通过对疲劳断口的观察分析，初步给出了疲劳断裂破坏特征。QStE500TM 疲劳性能的研究为高强钢疲劳行为分析和寿命预测提供了数据支持。

关键词：低周疲劳；高周疲劳；QStE500TM

Study on High and Low Cycle Fatigue Properties of Automobile Steel QStE500TM

Han Dan

(Benxi Iron & Steel Co., Ltd., Benxi 117000, China)

Abstract: The high cycle fatigue test and low cycle fatigue test of cold formed hot-rolled pickling QStE500TM steel were carried out, and the corresponding fatigue performance parameters of QStE500TM steel were obtained. The test results show that the stress-life curve is: $\log\sigma_{max}$=3.3849-0.12201logN, the fatigue limit is $\sigma_{0.1}$=483.75MPa; the strain-life curve is

$\Delta\varepsilon_t/2=(1037.26/E)\times(2N_f)^{-0.08782}+0.81132\times(2N_f)^{-0.69459}$, the cyclic stress-strain curve: $\Delta\sigma/2=847.93\times(\Delta\varepsilon_p/2)^{0.10187}$. Through the observation and analysis of the fatigue fracture, the failure characteristics of the fatigue fracture are initially given. The research on fatigue performance of QStE500TM provides data support for fatigue behavior analysis and life prediction of high-strength steel.

Key words: low cycle fatigue; high cycle fatigue; QStE500TM

采用图像法测定车轮用钢金相组织过程中不同放大倍数和不同视场的对比分析

王艳阳，后宗保，付声丽，王仲琨，汪良俊，程亚南

（马鞍山钢铁股份有限公司检测中心，安徽马鞍山 243000）

摘　要：本文利用 ZEISS 光学显微镜，通过 GB/T 15749—2008《定量金相测定方法》标准，探讨不同放大倍数下，车轮钢中铁素体组织含量的变化情况。实验结果表明：放大倍数越大，测定的组织含量越高；同倍数下，减小测定视场面积，测定的组织含量增加；增加图片衬度，测定的组织含量无明显变化。

关键词：组织定量；放大倍数；测定面积；衬度

Comparative Analysis of Different Magnification and Different Field of View in the Determination of Metallographic Structure of Wheel Steel by Image Method

Wang Yanyang, Hou Zongbao, Fu Shengli, Wang Zhongkun, Wang Liangjun, Cheng Ya'nan

(Testing Center of Ma'anshan Iron and Steel Co., Ltd., Ma'anshan 243000, China)

Abstract: In this paper, by using ZEISS optical microscope and using GB/T 15749—2008 Quantitative Metallographic Determination Method, the changes of ferrite tissue content in wheel steel at different magnification ratios were discussed. The results show that the higher the magnification, the higher the tissue content. At the same multiple, the measured field area decreased and the measured tissue content increased. There was no significant change in tissue content when contrast was increased.

Key words: organizational quantification; amplification factor; measuring area; contrast

双层焊管开裂缺陷分析

张志敏[1]，刘顺明[2]，杨利斌[2]，龙佳明[2]，余　璐[3]

（1. 首钢集团有限公司技术研究院，北京　100043；2. 首钢京唐钢铁联合有限责任公司，河北唐山　063200；3. 北京首钢股份有限公司，北京　100043）

摘 要：双层焊管开裂处，外层管铁素体晶粒尺寸较大，局部晶粒尺寸超过 100μm，内层管铁素体晶粒尺寸 20~30μm。外层硬度偏低，外层硬度比内层低 27.4HV。双层焊管开裂的原因，一是钎焊时液相铜扩散进入奥氏体晶界引起晶界脆化，成为裂纹萌生源头；二是钎焊后铁素体晶粒异常粗大，强度降低，导致弯曲时在变形较小的情况下发生开裂。
关键词：双层焊管；裂纹；钎焊；应力腐蚀

Crack Defect Analysis of Double Wall Brazed Tube

Zhang Zhimin[1], Liu Shunming[2], Yang Libin[2], Long Jiaming[2], Yu Lu[3]

(1. Shougang Group Co., Ltd. Research Institute of Technology, Beijing 100043, China; 2. Shougang Jingtang United Iron & Steel Co., Ltd, Tangshan 063200, China; 3. Beijing Shougang Co., Ltd, Beijing 100043, China)

Abstract: Ferrite grains of outer layer of double wall brazed tube are larger than inner layer in crack region. Size of some grains of outer layer is larger than 100μm. Grain size of inner layer is 20~30μm. The hardness of outer layer is 27.4HV lower than inner layer. There are two reasons for crack defects of double wall brazed tube. One reason for cracking is liquid copper diffused into the austenite grain boundaries and grain boundary embrittlement was caused during brazing process. These grain boundaries are sources of crack initiation. The other is that strength of tube is weakened by the coarse ferrite grains after brazing. So the tube cracked under low stress.
Key words: double wall brazed tube; crack; braze; stress corrosion

热成形钢高温力学性能和断口形貌研究

马闻宇，李学涛，郑学斌，徐德超，张博明，张永强

（首钢集团有限公司技术研究院，绿色可循环钢铁流程北京市重点实验室，
北京能源用钢工程研究中心，北京 100043）

摘 要：本文采用 Gleeble-2000 型热模拟试验机进行了铝硅镀层热成形钢的不同变形温度和应变速率下的拉伸试验。变形温度分别为 600℃、700℃和 800℃，应变速率分别为 0.1 s^{-1}、1 s^{-1} 和 10 s^{-1}。试验后可以获得材料的高温应力应变曲线。由结果可知，热成形钢的应力随应变速率的提高而增加。同一变形温度下，应力随着应变速率的增加而增加。同一应变速率下，应力随着变形温度的下降而增加。为深入理解材料高温变形和断裂机理，对所有试样的断口形貌进行了观察和分析。断口形貌主要由不同大小和深度的韧窝组成，低温时存在一定的解理面。材料的高温损伤断裂机制主要为微孔洞的形核，长大和聚合最终导致断裂，断裂方式为韧性断裂。
关键词：热成形钢；高温力学性能；断口形貌；断裂机理

Research on the Thermal Mechanical Property and Fracture Morphology of Hot Stamping Steel

Ma Wenyu, Li Xuetao, Zheng Xuebin, Xu Dechao,
Zhang Boming, Zhang Yongqiang

(Research Institute of Technology of Shougang Group Co., Ltd, Beijing Key Laboratory of

Green Recycling Process for Iron & Steel Production Technology, Beijing Engineering Research Center of Energy Steel, Beijing 100043, China)

Abstract: In this paper, the thermal-mechanical simulator Gleeble-2000 was employed to conduct the tension tests AlSi-coated hot stamping steel under different forming temperatures and strain rates. The forming temperatures were respectively 600℃, 700℃ and 800℃, and the strain rates were $0.1s^{-1}$, $1s^{-1}$ and $10s^{-1}$. The stress-strain curves were obtained after tests. The results show that the stress of hot stamping steel increases with the strain. At a constant forming temperature, the stress increases with the increase of the strain rate. At a constant strain rate, the stress increases with the decrease of the forming temperature. To further understand the thermal forming and crack mechanism, the fracture morphology of all the specimens was observed and analyzed. The fracture morphology consists of dimples with different sizes and depths. When the forming temperature is low, there is cleavage surface to a certain extent. The thermal damage and crack mechanism is mainly the nucleation, growth and polymerization of micro-voids. The fracture mode is ductile fracture.

Key words: hot forming steel; thermal mechanical property; fracture morphology; crack mechanism

过时效温度对 1000MPa 级冷轧双相钢组织和力学性能的影响

王亚东，孟庆刚，崔宏涛，左海霞

（本钢板材股份有限公司技术研究院，辽宁本溪 117000）

摘 要： 针对 1000MPa 级冷轧双相钢，进行不同温度下的过时效试验，使用拉伸试验机测定力学性能，利用扫描电镜分析组织结构、断口形貌及变形和断裂特征，研究过时效温度对组织性能的影响。结果表明：随过时效温度的升高，屈服强度和伸长率呈上升趋势，抗拉强度呈下降趋势，试验材料呈韧性断裂特征，过时效温度为300℃时综合性能良好，为 1000MPa 级冷轧双相钢产品优化提供实际指导。

关键词： 过时效；双相钢；组织；性能

Effects of Overaging Temperatures on Microstructure and Mechanical Properties of 1000MPa Cold Rolled Dual Phase Steel

Wang Yadong, Meng Qinggang, Cui Hongtao, Zuo Haixia

(Technical Research Institute, Benxi Iron and Steel (Group) Co., Ltd., Benxi 117000, China)

Abstract: For 1000MPa cold rolled dual phase steel overaging test at different temperature were conducted. Mechanical properties were measured by tensile testing machine. The microstructure, fracture morphology, deformation and fracture characteristics were analyzed by scanning electron microscope. The results showed that with the increase of overaging temperature the yield strength and elongation show upward trend, the tensile strength shows downward trend, the refinement of microstructure is beneficial to the improvement of toughness, the test material shows the characteristics of ductile fracture. Test material has good comprehensive properties when overaging temperature is 300℃, providing practical guidance for the optimization of 1000MPa cold rolled dual phase steel.

Key words: overaging; dual phase steel; microstructure; properties

第三代汽车钢组织和性能研究

亢 泽[1]，张志朋[1]，贾丽英[2]，徐惠敏[1]，孙 乐[1]，樊立峰[1]

（1. 内蒙古工业大学 材料科学与工程学院，呼和浩特 010051；
2. 河钢集团邯钢公司技术中心，邯郸 056015）

摘 要： 课题组前期已经研究了 5%Mn 钢热轧板、冷轧板以及逆相变退火后实验钢的力学性能。本文基于前期实验结果，借助金相显微镜、扫描电镜，X 射线衍射仪、以及透射电镜，系统研究了 5%Mn 钢热轧板、常化板、冷轧板及逆相变退火后材料的组织演变规律，研究结果如下：热轧板以及冷轧板组织均为板条马氏体和变形铁素体，但冷轧板组织相对热轧板组织晶粒更细小，同时在冷轧板组织中马氏体晶界处有细小的碳化物析出；经 930℃保温淬火+675℃逆相变退火后，得到超细晶铁素体、奥氏体以及板条马氏体混合组织，奥氏体含量为 22.34%。

关键词： 逆相变退火；马氏体；奥氏体；中锰钢；TRIP 效应

Study on Microstructure and Properties of the Third-generation Automotive Steel

Kang Ze[1], Zhang Zhipeng[1], Jia Liying[2], Xu Huimin[1], Sun Le[1], Fan Lifeng[1]

(1.School of Materials Science and Engineering, Inner Mongolia University of Technology, Hohhot 010051, China; 2.Technology Center, HBIS Group Hansteel Company, Handan 056015, China)

Abstract: Previously, our research group has already studyed the mechanical properties of the hot rolling production, the cold rolling production and the experimental steel after reverse transformation annealing. Themicrostructureof the hot rolling production, the normalizedsample, the cold rolling production and the experimental steel after reverse transformation annealing wereinvestigatedwith opticalmicroscopy based on the results of pre-test interview.The results are demonstrated as follows: The microstructure of the hot sheet and cold sheet comprises ferrite and martensite, and the grain size of cold rolled sheet is smaller than that of hot rolled sheet. There are fine carbides precipitated at the grain boundary of martensite in cold rolled sheet. The microstructure after reverse transformation annealing at 675℃ mainly consists of ultra-fine-grained ferrite and austenite, with the volume fraction of austenite of 22.34%.

Key words: reverse transformation annealing process; martensite; austenite; medium manganese steel; TRIP effect

热处理对 DP 钢氢扩散及氢脆敏感性的影响

王 贞[1,2]，刘 静[1,2]，黄 峰[1,2]，张施琦[1,2]

（1. 武汉科技大学省部共建耐火材料与冶金国家重点实验室，湖北武汉 430081；
2. 武汉科技大学湖北省海洋工程材料及服役安全工程技术研究中心，湖北武汉 430081）

摘　要：利用慢应变速率拉伸及氢渗透实验，结合 OM、SEM 等手段，研究了淬火温度及回火温度对 DP 钢氢扩散及氢脆敏感性的影响。结果表明，随着淬火温度的升高，马氏体含量增加，使 DP 钢中有效氢扩散系数降低，氢脆敏感性增强，但其变化幅度会受马氏体分布的影响。随着回火温度的升高，马氏体逐渐发生分解并析出碳化物，弥散分布的碳化物作为不可逆氢陷阱可有效捕获氢原子，且使捕获的氢原子分布均匀，导致材料的抗氢脆能力得到明显提升。综合考虑回火温度对氢脆敏感性及强度的影响，最佳回火温度为 330℃。

关键词：DP 钢；淬火温度；回火温度；氢扩散；氢脆敏感性

Effect of Heat Treatment on Hydrogen Diffusion and Hydrogen Embrittlement Susceptibility of DP Steel

Wang Zhen[1,2], Liu Jing[1,2], Huang Feng[1,2], Zhang Shiqi[1,2]

(1. The State Key Laboratory of Refractories and Metallurgy, Wuhan University of Science and Technology, Wuhan 430081, China; 2. Hubei Engineering Technology Research Center of Marine Materials and Service Safety, Wuhan University of Science and Technology, Wuhan 430081, China)

Abstract: Influence of heat treatment on hydrogen diffusion and hydrogen embrittlement susceptibility of DP steel was studied by using slow strain rate tensile test and hydrogen permeation test combined with OM and SEM. The results showed that with increasing of quenching temperature and martensite content, effective hydrogen diffusion coefficient decreased and hydrogen embrittlement susceptibility increased, but the variation range was affected by martensite distribution. With increasing of tempering temperature, martensite gradully decomposed and carbides perciptitated. The diffused carbides whichacted as irreversible hydrogen traps could effectively capture hydrogen and make hydrogen atoms evenly distributed, thus the anti-hydrogen embrittlement ability improved. Considering influence of tempering temperature on strength and hydrogen embrittlement susceptibility, the optimum tempering temperature was 330℃。

Key words: DP steel; quenching temperature; tempering temperature; hydrogen diffusion; hydrogen embrittlement susceptibility

低合金高强钢中铌析出行为研究

崔　磊[1]，王文军[2]，陈德顺[1]

（1. 马鞍山钢铁股份有限公司技术中心，安徽马鞍山　243000；
2. 中信金属微合金化技术中心，北京　100000）

摘　要：对 Nb-Ti 复合、单 Nb 两种微合金化的低合金高强钢在不同卷取温度、退火温度下的钢板，采用透射电镜（TEM）对第二相析出行为进行研究。结果表明，随着热轧卷取温度的降低，Nb 的析出受到抑制，尺寸小于 18nm 的析出物减少；冷轧退火过程中，固溶铌重新析出，不同退火温度下铌的析出比例和尺寸小于 18nm 的析出物变化不大。与 Nb-Ti 复合微合金化相比，单 Nb 微合金化更有利于 18nm 以下 Nb 析出。

关键词：Nb 微合金化；析出；卷取温度；退火温度

Study on Niobium Precipitation Behavior in HSLA Steel

Cui Lei[1], Wang Wenjun[2], Chen Deshun[1]

(1. Technology Center, Ma′anshan Iron & Steel Co., Ltd., Ma′anshan 243000, China;
2. CITIC Metal Microalloying Technology Center, Beijing 100000, China)

Abstract: A transmission electron microscope (TEM) was used to study the precipitation behavior of the second Nb-Ti HSLA and Nb HSLA steel under different coiling temperature and annealing temperature. As the coiling temperature decreases, the precipitation of Nb is suppressed, and the precipitates below 18nm decrease; during the cold rolling annealing process, solid solution niobium re-precipitates, the precipitation ratio of niobium at different annealing temperatures and the precipitates below 18nm not changed. Compared with Nb-Ti composite microalloying, Nb microalloying is more conducive to the Nb precipitation below 18nm.

Key words: Nb microalloying; precipitation; coiling temperature; annealing temperature

冷轧压下率对低成本第三代汽车钢微观组织和成型性能的影响

董瑞峰，赵庆波，毕晓宏，芦永发，陈子帅

（内蒙古工业大学材料科学与工程学院，内蒙古呼和浩特 010051）

摘　要： 采用拉伸试验、扫描电镜、XRD 等手段，研究了低成本第三代汽车钢在不同冷轧压下率条件下微观组织特点和成型性能。压下率分别为 66%、73%和 79%的冷轧实验钢经 630℃-10min 退火后微观组织以奥氏体+铁素体为主。不同冷轧压下率下的组织相似，常规力学性能相近。但随着压下率的增大，有利于冲压成型性能的{111}和{110}的织构组分增加，而{100}不利织构组分减少。且塑性应变比 r 值随压下率的增加而增大。当压下率为 79%时，{111}和{110}有利织构含量最多，分别为 17.7%和 25.5%，此时 r 值最大为 0.961，实验钢具有较好的成型性能。

关键词： 第三代汽车钢；压下率；织构；塑性应变比

The Effect of Cold Rolling Reduction Rate on the Microstructure and Forming Properties of Low-cost Third-generation Automotive Steel

Dong Ruifeng, Zhao Qingbo, Bi Xiaohong, Lu Yongfa, Chen Zishuai

(School of Materials Science and Engineering,
Inner Mongolia University of Technology, Hohhot 010051, China)

Abstract: Low-cost third-generation automotive steel was rolled by different reduction rate. Its microstructure and forming properties were investigated by means of tensile testing, scanning electron microscopy, XRD. The microstructures of the cold-rolled experimental steels with 66%, 73%, and 79% reduction rates were annealed at 630℃-10min, and the microstructure was mainly austenite and ferrite. The microstructures under different cold rolling reduction ratios are similar, and the conventional mechanical properties are similar. However, with the increasing downpressure rate, the texture components of {111} and {110} increase, while {100} unfavorable texture components decrease. And the plastic strain ratio *r* value increases with the increase of the reduction rate. When the reduction rate is 79%, {111} and {110} have the most favorable texture content, 17.7% and 25.5%, respectively. At this time, the maximum *r* value is 0.961, and the experimental steel has better formability.

Key words: Third-generation automotive steel; reduction rate; texture; plastic strain ratio

汽车用 TRIP 钢奥氏体逆相变基础研究

张献光[1]，刘　欢[1]，任英杰[1]，杨文超[1]，Miyamoto Goro[2]，Furuhara Tadashi[2]

（1. 北京科技大学冶金与生态工程学院，北京　100083；
2. 日本东北大学金属材料研究所，仙台　980-8577）

摘　要：马氏体基相变诱发塑性（TRIP）钢由于其优异的强塑性匹配及拉伸凸缘性能而成为汽车钢的理想材料。TRIP 钢的力学性能很大程度上取决于等温淬火过程中形成的残余奥氏体的体积分数及其稳定性，逆转变奥氏体的化学成分、尺寸和组织形貌对马氏体基 TRIP 中残余奥氏体的形貌与稳定性具有重要影响。

从板条马氏体或贝氏体到奥氏体的逆相变过程中，有两种形貌的奥氏体形成，即针状和块状。过去的研究表明[1-2]，逆相变前的初始组织，如淬火马氏体、回火马氏体或贝氏体，直接影响逆转变奥氏体的相变动力学和组织形貌。然而，目前有关初始组织对 Fe-Mn-Si-C 系合金中奥氏体逆相变的影响尚未系统地研究。本文通过系统研究不同初始组织对奥氏体逆相变行为的影响，以揭示渗碳体尺寸、分布、化学成分及残余奥氏体对逆相变行为的影响机制。

本研究所使用的合金为 Fe-2Mn-1.5Si-0.3C（质量百分数），样品经奥氏体化后，将样品投至冰盐水浴中进行淬火处理，以获得全马氏体组织。此后，将一部分淬火马氏体样品分别在 623K 和 923K 下回火 1h 以获得预回火马氏体。另一方面，在奥氏体化之后在 673K 下进行等温淬火处理 30 分钟或 10 小时，以获得分别约 20%残余奥氏体或贝氏体铁素体（BF）+碳化物的贝氏体。将上述初始组织在 1023K 或 1048K 下进行不同时间的临界退火，然后进行淬火。

研究结果表明，渗碳体粒径在块状奥氏体的形核中起着重要作用，通过预回火或较长时间等温淬火形成的较大尺寸渗碳体颗粒由于提高了块状奥氏体的形核速率从而促进了块状奥氏体的形成。此外，先存残余奥氏体由于抑制了渗碳体颗粒的析出，进而抑制了块状奥氏体的形成；20%的先存残余奥氏体对奥氏体逆相变动力学的影响基本没有影响，针状奥氏体中合金元素的分布特征表明针状奥氏体可能直接从残余奥氏体中长大形成的。高温回火导致的渗碳体/铁素体之间的严重合金元素配分或较长时间的等温淬火形成的粗大渗碳体颗粒对逆相变动力学有抑制作用。

高品质酸洗板 A510L 表面色差缺陷机理分析

李　岩[1,2]，夏　垒[3]，孙荣生[2]，蔡顺达[2]，宋利伟[2]，金晓龙[2]

（1. 海洋装备用金属材料及其应用国家重点实验室，辽宁鞍山　114009；2. 鞍钢集团钢铁研究院，
辽宁鞍山　114009；3. 辽宁科技大学材料与冶金学院，辽宁鞍山　114051）

摘 要：针对热轧酸洗板A510L酸洗后表面条纹色差缺陷，采用激光共聚焦显微镜（CLSM）对酸洗后带钢表面不同位置的微观形貌特征进行对比分析，运用高分辨扫描电子显微镜和EDS能谱仪对酸洗板A510L酸洗后色差缺陷的产生机理进行研究。结果表明，酸洗板A510L表面色差缺陷形成的主要原因与热轧原料基板和氧化铁皮界面状态密切相关，酸洗后部分氧化铁皮残留导致色差缺陷的产生，通过优化热轧卷曲温度并调整酸洗工艺段参数，最终成功消除该种表面缺陷。

关键词：热轧酸洗板；表面色差缺陷；盐酸洗；氧化铁皮

Analysis of the Mechanism of Chromatic Aberration Defects on A510L Surface of High Quality Hot Rolled Pickled Strip

Li Yan[1,2], Xia Lei[3], Sun Rongsheng[2], Cai Shunda[2], Song Liwei[2], Jin Xiaolong[2]

(1. State Key Laboratory of Metal Material for Marine Equipment and Application, Anshan 114009, China;
2. Iron & Steel Research Institutes of Ansteel Group Corporation, Anshan 114009, China;
3. School of Material and Metallurgy, University of Science and Technology Liaoning, Anshan 114051, China)

Abstract: The surface chromatic aberration defects of A510L hot-rolled pickling plate after pickling were analyzed by laser confocal microscope (CLSM). The generation mechanism of chromatic aberration defects on the surface of A510L pickling plate was studied by SEM and EDS. The results show that the main reason for the formation of chromatic aberration defects on the surface of the A510L pickled plate is closely related to the interface state of the hot-rolled raw material substrate and the oxide scale. The surface chromatic aberration defects are caused by some oxide scale residue after pickling. Finally, the surface defects were successfully eliminated by optimizing the crimping temperature of hot rolling and adjusting the parameters of pickling process.

Key words: hot rolled pickled strip; surface chromatic aberration defects; hydrochloric-acid pickling; oxidized scale

汽车雨刷臂用钢冷加工开裂原因分析与工艺改进

隋晓亮，赵东记，郭洛方，阎超楠，高永彬

（青岛特殊钢铁有限公司，山东青岛 266043）

摘 要：对用于生产汽车雨刷臂的ϕ9/10mm SWRH45B盘条在冷加工拍扁工序出现加工开裂进行检测分析，并通过改进措施，有效解决冷加工拍扁开裂问题。结果表明：造成冷加工拍扁开裂的主要原因为钢材内部出现的白亮带和严重的中心碳偏析；通过优化连铸工艺参数，改善了铸坯内部质量，消除了钢材内部的白亮带和中心偏析缺陷，客户反馈优化后的盘条在冷加工拍扁工序再未出现开裂，质量得到明显改善。

关键词：汽车雨刷臂；中心碳偏析；白亮带；末端电磁搅拌

Cause Analysis on Cold Deformation Cracking of Steel for Car Wiper Arms and Process Improvement

Sui Xiaoliang, Zhao Dongji, Guo Luofang, Yan Chaonan, Gao Yongbin

(Qingdao Special Iron & Steel Co., Ltd., Qingdao 266043, China)

Abstract: The ϕ9/10mm SWRH45B wire rod used in the production of automobile wiper arms was detected and analyzed for processing cracks in the cold processing flattening process, and improved measures were taken to effectively solve the cold processing flattening cracking problem. The results show that the main cause of cold-working flattening cracking is the white bright band and severe central carbon segregation in the steel; by optimizing the continuous casting process parameters, the internal quality of the cast slab is improved, and the white bright band and central segregation defects in the steel are eliminated. According to customer feedback, the optimized wire rod did not crack in the cold processing and flattening process, and the quality was significantly improved.

Key words: car wiper arms; central carbon segregation; white band; final electromagnetic stirring

热轧氧化铁皮易酸洗研究进展

张理扬[1,2]

（1. 宝山钢铁股份有限公司冷轧厂，上海　200941；
2. 湛江钢铁有限公司冷轧厂，广东湛江　524072）

摘　要：对于许多酸连轧机组生产的瓶颈环节往往是酸洗段，为了提高机组产能，通常从拉矫工艺改进、酸洗工艺改进等方面开展产能提升工作。其实还可以从提高热轧来料表面氧化铁皮的易酸洗性方面开展产能提升工作。本文对目前热轧氧化铁皮易酸洗性的研究进展进行了介绍，指出为了提高热轧来料的易酸洗性，可以从控制 Fe_3O_4 的生成量和减少氧化铁皮的厚度进行改进，具体生产控制上可以从基板化学成分优化、轧制温度和轧制速度控制、层流冷却模式改进、冷却均匀性优化、卷取温度优化、热轧卷取后的冷却速度控制等方面进行探索和改进。

关键词：易酸洗；氧化铁皮；热轧；层流冷却；卷取温度

Research Process on Easy Pickling of Hot-Rolled Oxide Scale

Zhang Liyang[1, 2]

(1. Baoshan Iron & Steel Co., Ltd., Shanghai 200941, China;
2. Cold Rolling Plant of Zhanjiang Iron & Steel Co., Ltd., Zhanjiang 524072, China)

Abstract: For many tandem acid rolling mills, pickling is the bottleneck of production. In order to improve the production capacity of the lines, the improvement of tension and straightening process and pickling process is usually carried out to improve the production capacity. In fact, it is also possible to improve the easy pickling of hot-rolled surface oxide scale. In this paper, the current progress of easy pickling of hot-rolled oxide scale was introduced. It was pointed out that in order to improve the easy pickling of hot rolled strips, the generation of Fe_3O_4 can be controlled and the thickness of the oxide scale can be reduced. The specific production measures include substrate chemical composition optimization, the rolling temperature and rolling speed control, laminar cooling model improvement, cooling uniformity optimization, coiling temperature optimization, the cooling speed control after coiling et al.

Key words: easy pickling; oxide scale; hot-rolling; laminar cooling; coiling temperature

Fe-22Mn-0.6C 钢拉伸变形过程中的动态应变时效行为

刘　帅[1]，刘焕优[1]，王明明[1]，冯运莉[1]，张福成[1,2]

（1. 华北理工大学冶金与能源学院，河北唐山　063210；
2. 燕山大学亚稳材料国家重点实验室，河北秦皇岛　066004）

摘　要：利用室温单向拉伸实验，结合 OM、XRD 等检测手段，研究了 Fe-22Mn-0.6C 钢拉伸变形过程中的锯齿产生的临界应变值以及锯齿幅度、局部应变集中和位错密度的变化情况。结果表明：Fe-22Mn-0.6C 钢产生锯齿的临界应变值为 0.03；随应变增加，锯齿幅度以及局部应变集中程度显著增大，产生更加强烈的动态应变时效（DSA）现象；通过 XRD 分析发现，随应变增加，实验钢中逐渐产生的大量位错以及产生的更加密集的 C-Mn 短程序结构是其动态应变时效现象增强的主要原因。

关键词：高锰钢；孪晶诱发塑性钢；动态应变时效；PLC 斑；位错密度

The Dynamic Strain Aging Behavior of Fe-22Mn-0.6C Steel in the Tensile Deformation Process

Liu Shuai[1], Liu Huanyou[1], Wang Mingming[1], Feng Yunli[1], Zhang Fucheng[1,2]

(1. College of Metallurgy and Energy, North China University of Science and technology, Tangshan 063210, China; 2. State Key Laboratory of Metastable Materials Science and Technology, Yanshan University, Qinhuangdao 066004, China)

Abstract: High manganese twinning-induced plasticity (TWIP) steels, which possess high strength and excellent ductility, have been regarded as a potential candidate for the lightweighting of automobiles. The high strain-hardening rate and superior mechanical properties of TWIP steels stem from deformation twinning and dynamic strain aging (DSA). To date, the influences of alloying elements, strain rates and temperatures on the twining behavior have received much attention. However, very limited research is focused on the DSA of TWIP steel. In this work, the main features of DSA, such as the critical strain and amplitude of serration, local strain concentration and dislocation density of Fe-22Mn-0.6C steel were investigated using OM, XRD and monotonic tensile tests during tensile deformation process. The results show that the critical strain of serration is about 0.03, and with increasing strains, the amplitude of serration and local strain concentration increase. The dynamic strain aging behavior becomes stronger in high strain range. XRD analyses show that the dislocation density and number of C-Mn clusters increases with increasing strains, which may be the main reason for the stronger dynamic strain aging phenomenon in the high strain range.

Key words: high manganese steel; twinning induced plasticity steel; dynamic strain aging; PLC band; dislocation density

7.2 管线钢

第二相粒子对压力容器钢和管线钢氢致开裂的影响研究进展

贾春堂，胡昕明，欧阳鑫，王 储，邢梦楠

（鞍钢集团钢铁研究院，辽宁鞍山 114000）

摘 要：对近年来关于夹杂物对压力容器钢和管线钢氢致开裂的影响研究进行了总结，主要从夹杂物种类、形状和分布三方面对压力容器和管线钢氢致开裂的影响进行了探究，对指导未来压力容器钢和管线钢的微观结构设计和合金应用提供参考。

关键词：氢致开裂；压力容器；微观组织；夹杂物；析出相

Research Progress of Effect of Second Phase Particles on Hydrogen Induced Cracking of Pressure Vessel Steel and Pipeline Steel

Jia Chuntang, Hu Xinming, Ouyang Xin, Wang Chu, Xing Mengnan

(Anshan Iron and Steel Group Steel Research Institute, Anshan 114000, China)

Abstract: the research on the influence of inclusions on hydrogen induced cracking of pressure vessel steel and pipeline steel in recent years is summarized, and the influence of inclusions on hydrogen induced cracking of pressure vessel steel and pipeline steel is mainly explored from three aspects of the type, shape and distribution of inclusions, so as to provide reference for guiding the microstructure design and alloy application of pressure vessel steel and pipeline steel in the future.

Key words: hydrogen induced cracking; pressure vessel; microstructure; inclusion; precipitated phase

南钢 X80MWϕ1422mm×35.2mm 感应加热弯管开发应用

翟冬雨，杜海军，吴俊平，姜金星，刘 帅

（南京钢铁股份有限公司技术研发处，江苏南京 210035）

摘 要：为了满足中俄东线 450 亿立方米/年超大输量或-40℃极寒地区服役的国家重大管道工程建设用钢要求，南钢结合自身设备特点，试验开发了具有高强韧性的 IB555/X80ϕ1422mm×35.2mm、R=5D 感应加热弯管用管线钢板，以中俄东线天然气管道工程管道技术规范要求设计了产品成分，确定了合适的组织类型；洁净钢冶炼技术结合高效浇铸工艺，获得了依据 YB 4003 进行低倍评级 C0.5 级高纯净度铸坯；试验确定了 TMCP 轧制工艺技术参数，获得了少量先共析铁素体+粒状贝氏体的软/硬相复合组织，多相组织的获得满足了具有高强韧性的 IB555/X80 ϕ1422 mm×35.2mm、R=5D 感应加热弯管用管线钢板的强韧性匹配，热机械轧制钢板理化性能检测，强韧性性能符合管道技术规范要求，实现了中俄东线低温超大输量管道用钢的批量工业生产。

关键词：中俄东线；管道技术；直缝埋弧焊管；多相复合组织；强韧性

Development and Application of Nangang X80MW ϕ1422mm×35.2mm Induction Heating Pipe

Zhai Dongyu, Du Haijun, Wu Junping, Jiang Jinxing, Liu Shuai

(Technology Research and Development Department, Nanjing Iron and Steel Co., Ltd., Nanjing 210035, China)

Abstract: In order to meet the requirements of the national major pipeline construction steel used for the large-scale transmission of 45 billion cubic meters per year in the East China-Russia line or in the extremely cold region of -40 °C, Nangang has developed the IB555/X80 with high strength and toughness in combination with its own equipment characteristics. Φ1422×35.2mmR=5D Induction heating bending pipeline steel plate, designed the product composition according to the technical specifications of the Sino-Russian East Line natural gas pipeline engineering pipeline, and determined the appropriate organization type; the clean steel smelting technology combined with the high efficiency casting process, obtained the basis YB 4003 carries low-grade C0.5 grade high purity casting billet; the technical parameters of TMCP rolling process are determined experimentally, and a small amount of soft/hard phase composite structure of pro-eutectoid ferrite + granular bainite is obtained. The organization has obtained the strong toughness matching of the IB555/X80 Φ1422mm×35.2mm R=5D induction heating elbow pipeline steel plate with high strength and toughness, the physical and chemical properties of the thermomechanical rolled steel plate, and the toughness and performance meet the requirements of the pipeline technical specifications. Mass industrial production of steel for low temperature and large volume pipelines on the Sino-Russian East Line.

Key words: Sino-Russian east line; pipeline technology; straight seam submerged arc welded pipe; multiphase composite structure; strong toughness

深海用 40.5mm 厚 L485FD 管线钢的开发

郭　斌，徐进桥

（宝钢股份中央研究院，湖北武汉　430080）

摘　要：本文介绍了 40.5mm 厚 L485 FD 管线钢的开发历程，包括成分设计、生产工艺、组织特征、力学性能等，阐述了 40.5mm 厚 L485 FD 管线钢开发过程中的一些关键控制技术，如洁净度、TMCP 和制管工艺等。结果表明，采用低 C-Mn-Nb 系的成分，以及高洁净度和高均质化冶炼连铸技术，奥氏体超细化控制和超快加速冷却技术，高精度制管工艺等技术，300mm 厚的连铸坯可以成功开发出了 40.5mm 厚 DNV SAWL485 FDU 管线钢板及钢管；40.5mm 厚 L485 FD 钢具有铁素体加贝氏体双相组织，其各项性能指标全面满足技术条件要求，强度控制适中，低温韧性优良，各向异性小，特别是落锤撕裂 DWTT 性能的韧脆转变温度低于–10℃。

关键词：L485；海底管线钢；TMCP；40.5mm 厚；DWTT

Development of 40.5mm Thick L485FD Submarine Pipeline Steel

Guo Bin, Xu Jinqiao

(Baosteel Central Research Institute of Baoshan Iron & Steel Co., Ltd., Wuhan 430080, China)

Abstract: This paper provides a detailed description of 40.5mm thick L485FD submarine pipeline steel, including its metallurgical design, manufacturing process, structural characteristics and mechanical properties. Some key issues such as the cleanliness, TMCP and manufacturing pipe parameters are addressed for the development. Results show that the steel can been produced by 300mm thick continuous cast slab, which it is used that the low C-Mn-Nb component series, the smelting and casting process of high cleanliness and high homogenization, the ultra fine austenite controlling and ultra fast accelerated cooling process, the high precision pipe making process. The steel has ferrite and bainite structure, suitable strength, excellent low toughness, small anisotropy, especially its ductile brittle transition temperature of DWTT is lower –10℃. Its properties meet the technology qualification.

Key words: L485; Submarine pipeline steel; TMCP; 40.5mm thick; DWTT

回火温度对高钒 X80 管线钢氢致塑性损失性能影响研究

李龙飞[1]，宋 波[2]，林腾昌[1]

（1. 钢铁研究总院冶金工艺研究所，北京 100081；
2. 北京科技大学冶金与生态工程学院，北京 100083）

摘 要：本文借助高分辨透射电镜、场发射扫描电镜及 Devanathan-Stachurski 双电解槽等设备和手段研究了热处理回火温度对高钒 X80 级管线钢显微组织、纳米尺度析出相等特征及氢扩散行为和氢致塑性损失性能的影响。结果表明：随着回火温度由 450℃升高至 650℃，细小纳米级碳化钒颗粒逐渐增多，有效氢扩散系数降低，管线钢抵抗氢致塑性损失能力提高。然而，回火温度为 700℃的实验钢中纳米级碳化钒颗粒发生粗化且数量降低，显微组织由粒状贝氏体+板条状铁素体+少量块状铁素体转变为粗大的块状铁素体，氢脆敏感性指数大幅上升。

关键词：管线钢；回火温度；纳米级析出相；有效氢扩散系数；氢致塑性损失

Study on Effect of Tempering Temperature on Hydrogen Induced Ductility Loss of High Vanadium X80 Pipeline Steel

Li Longfei[1], Song Bo[2], Lin Tengchang[1]

(1. Institute for Metallurgical Process, Central Iron and Steel Research Institute, Beijing 100081, China;
2. University of Science and Technology Beijing, School of Metallurgical and Ecological Engineering, Beijing 100083, China)

Abstract: In this paper, the effects of tempering temperature on microstructure, nano-scale precipitates, hydrogen diffusion behavior and hydrogen-induced ductility loss of high vanadium X80 pipeline steel were investigated by high resolution transmission electron microscopy, field emission scanning electron microscopy and Devanathan–Stachurski electrolytic cells. The results show that with increasing tempering temperature from 450 to 650℃, the number of fine nano-scale vanadium carbides increased, the effective hydrogen diffusion coefficient decreases and the resistance to hydrogen induced plastic loss is evidently improved. However, as the tempering temperature increases to 700℃, the vanadium carbide particles coarsen and the amount decreases, the microstructure changes from granular bainite + lath ferrite + a small amount of massive ferrite to coarse massive ferrite, and the hydrogen embrittlement susceptibility index increases significantly.

Key words: pipeline steel; tempering temperature; nano-scale precipitate; effective hydrogen diffusion coefficient; hydrogen-induced ductility loss

ASP 流程大壁厚 X80M 热轧卷板的研制

黄国建[1]，黄明浩[1]，王 杨[1]，董 洋[2]

（1. 鞍钢集团钢铁研究院，辽宁鞍山 114009；2. 鞍钢股份有限公司，辽宁鞍山 114021）

摘 要：针对中俄东线工程高钢级、大壁厚、大口径、高输送压力的要求，鞍钢开发出 22mm×1550mm×Cmm 规格 X80M 管线钢热轧卷板，并在辽阳石油钢管制造有限公司制成 22mm×1219mm 螺旋埋弧焊钢管，经第三方检验完全符合标准要求。产品采用 C-Mn-Mo-Cr-Nb 合金体系设计，适当添加 Ni、Cu，结合鞍钢 ASP 流程中薄板坯特色，采用纯净钢冶炼连铸技术、热机械轧制（TMCP）技术，综合利用固溶、细晶、位错、析出和相变等强化机制，最终组织为细小均匀的针状铁素体组织，产品具有高强度和良好的低温韧性。

关键词：管线钢；大壁厚；X80M

Research of ASP X80M Hot Rolled Strip with Maximum Thickness

Huang Guojian[1], Huang Minghao[1], Wang Yang[1], Dong Yang[2]

(1. Iron and Steel Research Institute of Ansteel Group, Anshan 114009, China;
2. Angang Steel Company Limited, Anshan 114021, China)

Abstract: With the requirements of high grade, large thickness, large caliber and high delivery pressure, Angang Steel Company Limited successfully developed 22mm×1550mm×Cmm X80M pipeline steel hot rolled plates and the plates were manufactured into 22mm×1219mm SSAW pipes in Liaoyang Petroleum Steel Pipe Manufacturing Co. Ltd, which fully meet the standard after third-party inspection. The product with C-Mn-Mo-Cr-Nb alloy system adding moderate Ni and Cu, featured with ASP process was developed by using melting process suitable to pure steel and thermo-mechanical rolling process. It has a fine and uniform microstructure featured as acicular ferrite (AF) with high strength and high ductility in low temperature, whose mechanism is gained through solution strengthening, grain-fining strengthening, dislocation strengthening, precipitation strengthening and phase transformation strengthening.

Key words: pipeline steel; maximum thickness; X80M

鞍钢冷轧搪瓷钢板的开发与应用

王永明[1]，吕家舜[1]，徐承明[1]，杨洪刚[1]，丁燕勇[2]，李　锋[2]

（1. 鞍钢集团钢铁研究院，辽宁鞍山　114009；2. 鞍钢股份公司制造管理部，辽宁鞍山　114009）

摘　要：本文简要介绍了鞍钢开发的冷轧搪瓷钢品种及生产工艺，并综合分析了不同产品的化学成分、力学性能及应用领域等，结果表明，鞍钢已完成搪瓷钢系列产品开发，覆盖市场需求所有牌号及规格，能够满足用户不同使用环境及涂搪和搪烧工艺的要求，产品具有优异的抗鳞爆性能、密着性能及成形性能，目前已成功应用于工业脱硫脱硝用波纹板、建筑装饰板和烧烤炉、热水器、电饭锅内胆制造等行业。

关键词：冷轧搪瓷钢；抗鳞爆性；密着性；开发应用

Development and Application of Cold Rolled Enamel Steel Sheet of Ansteel

Wang Yongming[1], Lü Jiashun[1], Xu Chengming[1], Yang Honggang[1], Ding Yanyong[2], Li Feng[2]

(1. AnSteel Group Iron and Steel Research Institute, Anshan 114009, China;
2. Department of Manufacture Management of AnSteel Co., Ltd., Anshan 114009, China)

Abstract: The categories and processing procedure of cold rolled enamel steel sheet of Ansteel were mainly introduced. The chemical compositions, properties and applications were analyzed. The result showed that the different chemical compositions of cold rolled enamel steel of Ansteel had been developed. The product has excellent properties of fish scale resistance, adhesion performance and formability, that can cover the demand of different enameling technology. The cold rolled enamel steel sheet of Ansteel have been used widely in air-gas heat exchangers, architectural ornament and household electrical appliances etc.

Key words: cold rolled enamel steel sheet; fish scale resistance; adhesion performance; development and application

低成本 X70M 管线钢的研制开发

徐海健[1,2]，沙孝春[1]，韩楚菲[1]，刘　留[1]，渠秀娟[1]，任　毅[1,2]

（1. 鞍钢股份有限公司，辽宁鞍山　114009；
2. 海洋装备用金属材料及其应用国家重点实验室，辽宁鞍山　114021）

摘　要：研究了不同加热温度、轧制工艺和冷却速率在内的低成本 X70M 管线钢组织及性能变化。结果表明：随着加热温度升高，奥氏体晶粒尺寸逐渐长大，当加热温度《1200℃时，奥氏体平均晶粒尺寸可控制在 50μm 以内；通过增加中间坯厚度，使其精轧阶段累计压下量增加，可显著细化钢板心部组织，同时增加析出相密度；通过提高轧后钢板的冷却速率，既抑制了先共析铁素体的转变，也促进了针状铁素体和粒状贝氏体形成。当冷却速率为25℃/s时，可得到有利于试验钢性能的由针状铁素体和粒状贝氏体为主复相组织，钢板的强韧性得到显著改善；生产的X70M级管线钢满足技术条件要求，可实现X70M级别管线钢降低成本生产。目前，鞍钢已可以实现无Mo低成本薄规格X70M管线钢稳定化批量生产。

关键词：加热温度；X70M管线钢；冷却速率；针状铁素体

Research and Development of Low-cost X70M Pipeline Steel

Xu Haijian[1,2], Sha Xiaochun[1], Han Chufei[1], Liu Liu[1], Qu Xiujuan[1], Ren Yi[1,2]

(1. Anshan Iron & Steel Co., Ltd., Anshan 114009, China; 2. State Key Laboratory of Metal Material for Marine Equipment and Application, Anshan 114021, China)

Abstract: The variations of microstructure and mechanical properties for low-cost X70M were studied, including heating temperature, rolling process and cooling rates. The results showed that the austenite grains growed continuously with enhancing temperatures. The average austenite grains can be controlled within 50μm by maintaining the heating temperature below 1200℃. The grains can be refined by increasing the thickness of intermediate slab for enhancing the cumulative reduction rates, and meanwhile increase the number density of precipitates. Eventually, the cooling rate is improved, which not only inhibits proeutectiod ferrite transformation, but also promotes the ultrafine granular bainite and acicular ferrite formation. The results show that when the cooling rate was 25℃/s, the microstructure consists the acicular ferrite and granular bainite, which was beneficial to improve the strength and toughness of steels. The X70M grade pipeline steel satisfies the requirements of the technicial agreements, which can provide a route to produce the low-cost X70M plates. Up to now, a batch of X70M plates without Mo addition was steady produced in Ansteel Company.

Key words: heating temperature; X70M pipeline steel; cooling rate; acicular ferrite

节约型X70M管线钢工业化试验研究

王　爽[1,2]，任　毅[1,2]，张　帅[1,2]，高　红[1,2]，徐海健[1,2]

（1. 海洋装备用金属材料及其应用国家重点实验室，辽宁鞍山　114009；
2. 鞍钢集团钢铁研究院，辽宁鞍山　114009）

摘　要：通过工业化试验研究了含Mo和无Mo两种成分厚壁X70M管线钢的组织性能差异，分析了终冷温度对厚壁无Mo X70M管线钢组织性能的影响。结果表明：采用无Mo成分设计，通过终冷温度的优化调整，可以生产出满足性能要求的厚度32mm X70M管线钢板。在相同工艺条件下，与含Mo成分相比，无Mo成分试验钢先析铁素体大量增加，强度明显下降，但是韧性变化不大。随着终冷温度降低，无Mo试验钢中的针状铁素体和贝氏体组织明显增加，强度上升。无Mo试验钢终冷温度在370~400℃获得的组织和力学性能，与含Mo成分的试验钢基本处于同一水平。

关键词：节约型管线钢；工业化试验；显微组织；性能

Industrial Experimental Investigation on the Conservation Oriented X70M Pipeline Steel

Wang Shuang[1,2], Ren Yi[1,2], Zhang Shuai[1,2], Gao Hong[1,2], Xu Haijian[1,2]

(1. State Key Laboratory of Metal Materials for Marine Equipment and Applications, Anshan 114009, China; 2. Iron and Steel Research Institute of Angang Group, Anshan 114009, China)

Abstract: By means of industrial experiment, the difference of microstructure and property between including Mo and excluding Mo heavy section X70M pipeline steels was investigated, effect of final cooling temperature on microstructure and property of heavy section X70M pipeline steels was analyzed. The results indicated that adopting excluding Mo chemical design, with measures of optimization and adjustment on final cooling temperature, 32mm thickness X70M pipeline steel plate which met mechanical property requirements could be produced.In the same technology condition, compared with the test plate including Mo, the pro-eutectoid ferrite was large increased in the test plate excluding Mo, the strength of the test plate declined obviously, but the toughness of the test plate was little changed. With the decrease of final cooling temperature, acicular ferrite and bainite in the test plates excluding Mo were markedly increased. when the final cooling temperature of the test plate excluding Mo produced at 370-400℃, the microstructure and properties were at the same level with the test plate including Mo.

Key words: conservation oriented pipeline steel; industrial experiment; microstructure; property

大口径强韧性 X70M 管线钢的研发

李中平，熊祥江，易春洪，陈　炼

（湖南湘潭钢铁有限公司，湖南湘潭　411101）

摘　要： 通过采用低碳成分设计、Nb-Ti-Mo-Ni 微合金化成分设计配合合理的控轧控冷工艺，开发了大口径 ϕ1422mm×31.75mmX70 管线钢，获得了以针状铁素体为主的显微组织结构的 X70M，所开发的管线钢具有良好的力学性能，强度均匀稳定，同板差小。$R_{t0.5}$：500~570MPa；R_m：600~690MPa；A_{50}≥35%；$R_{t0.5}/R_m$≤0.90；低温冲击值≥400J；–25℃落锤剪切面积比≥85%。

关键词： 管线钢；微观组织；控轧控冷；X70M

Development of Large Diameter Strength and Toughness X70M Pipeline Steel

Li Zhongping, Xiong Xiangjiang, Yi Chunhong, Chen Lian

(Xiangtan Iron & Steel Group Co., Ltd., Xiangtan 411101, China)

Abstract: By adopting low carbon composition design, Nb-Ti-Mo-Ni microalloying composition design and reasonable controlled rolling and cooling process, a large diameter 1422×31.75 mm X70M pipeline steel was developed. The X70M with acicular ferrite as the main microstructure was obtained. The developed pipeline steel has good mechanical properties and the properties of the steel plate fully meet the API 5L pipeline. Line standard requirements: yield strength 490-570 MPa, tensile strength 595-680 MPa, elongation (>35%) and yield-strength ratio (< 0.90); Charpy impact energy (>400J) at –25 (>85%) and drop hammer shear area ratio (>85%).

Key words: pipelines; microstructure; TMCP; X70M

回火温度对高钒 X80 管线钢氢致塑性损失性能影响研究

李龙飞[1]，宋　波[2]，林腾昌[1]

（1. 钢铁研究总院冶金工艺研究所，北京　100081；
2. 北京科技大学冶金与生态工程学院，北京　100083）

摘　要：本文借助高分辨透射电镜、场发射扫描电镜及 Devanathan-Stachurski 双电解槽等设备和手段研究了热处理回火温度对高钒 X80 级管线钢显微组织、纳米尺度析出相等特征及氢扩散行为和氢致塑性损失性能的影响。结果表明：随着回火温度由 450℃升高至 650℃，细小纳米级碳化钒颗粒逐渐增多，有效氢扩散系数降低，管线钢抵抗氢致塑性损失能力提高。然而，回火温度为 700℃的实验钢中纳米级碳化钒颗粒发生粗化且数量降低，显微组织由粒状贝氏体+板条状铁素体+少量块状铁素体转变为粗大的块状铁素体，氢脆敏感性指数大幅上升。

关键词：管线钢；回火温度；纳米级析出相；有效氢扩散系数；氢致塑性损失

Study on Effect of Tempering Temperature on Hydrogen Induced Ductility Loss of High Vanadium X80 Pipeline Steel

Li Longfei, Song Bo, Lin Tengchang

(1. Institute for Metallurgical Process, Central Iron and steel Research Institute, Beijing 100081, China;
2. University of Science and Technology Beijing, School of Metallurgical and Ecological Engineering, Beijing 100083, China)

Abstract: In this paper, the effects of tempering temperature on microstructure, nano-scale precipitates, hydrogen diffusion behavior and hydrogen-induced ductility loss of high vanadium X80 pipeline steel were investigated by high resolution transmission electron microscopy, field emission scanning electron microscopy and Devanathan–Stachurski electrolytic cells. The results show that with increasing tempering temperature from 450 to 650℃, the number of fine nano-scale vanadium carbides increased, the effective hydrogen diffusion coefficient decreases and the resistance to hydrogen induced plastic loss is evidently improved. However, as the tempering temperature increases to 700℃, the vanadium carbide particles coarsen and the amount decreases, the microstructure changes from granular bainite + lath ferrite + a small amount of massive ferrite to coarse massive ferrite, and the hydrogen embrittlement susceptibility index increases significantly.

Key words: pipeline steel; tempering temperature; nano-scale precipitate; effective hydrogen diffusion coefficient; hydrogen-induced ductility loss

森吉米尔轧机一中间辊失效分析

袁海永[1]，李岳锋[1]，张大伟[2]

（1. 宁波宝新不锈钢有限公司，浙江宁波 315807；
2. 宝钢轧辊科技有限责任公司，江苏常州 213019）

摘 要：某不锈钢冷轧厂森吉米尔一中间辊在线发生局部剥落。本文对一中间辊剥落区域及整个轧辊表面进行剥落形貌观察，结合磁粉检测、硬度检测、宏观酸蚀试验、化学成分分析、金相检验等有效的理化检测手段，通过对各项检测结果综合分析，阐述了其产生失效的机理。

关键词：不锈钢冷轧；森吉米尔辊；剥落；理化检测

Failureanalysis of the First Intermediate Roll of Sendzimir Mill

Yuan Haiyong[1], Li Yuefeng[1], Zhang Dawei[2]

(1. Ningbo Baoxin Stainless Steel Co., Ltd., Ningbo 315807, China;
2. Baosteel Roll Science & Technology Co., Ltd., Changzhou 213019, China)

Abstract: Local spalling of Sendzimir intermediate roll occurred on line in a stainless steel cold rolling plant. In this paper, the spalling morphology of an intermediate roll and the whole roll surface is observed. Combined with effective physical and chemical testing methods such as magnetic particle testing, hardness testing, macro acid etching test, chemical composition analysis and metallographic examination, the failure mechanism is elaborated through comprehensive analysis of various testing results.

Key words: cold rolling of stainless steel; sendzimir roll; spalling; physical and chemical detection

高强镀锌板脉冲 MAG 焊接工艺研究

刘效云[1,2]，房 超[1,2]，张 鹏[3]，肖 鹏[1,2]，周 博[1,2]，邹家奇[1,2]

（1. 河钢股份有限公司承德分公司，河北承德 067002；2. 河北省钒钛工程技术研究中心，
河北承德 067002；3. 河钢集团钢研总院，河北石家庄 050023）

摘 要：对 1.6mm 镀锌板（SGH340）采用不同的焊接参数和焊接工艺进行焊接试验，分析在不同焊接工艺下其力学性能及微观组织，通过试验可知：采用脉冲 MAG 焊接，焊接过程电弧稳定，无跳弧现象产生，焊缝与母材圆滑过渡，焊缝表面表面无断弧、咬边、气孔、未熔合等缺陷；焊缝熔合区存在部分降低性能的魏氏组织，熔合线区也出现不同程度轻微的魏氏组织；焊接接头热影响区的腐蚀耐蚀性能最差，烧损处的耐蚀性能由热影响区向两边逐渐降低；随着焊接电流、焊接速度的增加，焊接接头性能呈现先增大后减小的趋势。最佳的焊接参数为：焊接电流为 127A，焊接速度为 90cm/min。

关键词：镀锌板；脉冲 MAG；微观组织；力学性能

Study on Pulse MAG Welding Technology of High Strength Galvanized Sheet

Liu Xiaoyun[1,2], Fang Chao[1,2], Zhang Peng[3], Xiao Peng[1,2], Zhou Bo[1,2], Zou Jiaqi[1,2]

(1. Chengde Branch of Hebei Iron and Steel Group Co., Ltd, Chengde, 067002,China;
2. Hebei Vanadium Titanium Engineering Technology Research Center, Chengde, 067002, China;
3. Technology Research Institute, HBIS GROUP, Shijiazhuang, 050023, China)

Abstract: The welding test of 1.6mm galvanized steel plate (Gr50-DX) was carried out by using different welding parameters and welding processes. The mechanical properties and microstructure of the plate were analyzed under different welding processes. The results showed that: pulse MAG welding was adopted, the arc was stable during the welding process, there was no arc jump, the weld joint and the base metal were in smooth transition, and the surface of the weld joint was free of defects such as arc break, undercut, porosity and lack of fusion There are some widmanstatten structures in the fusion zone of the weld, and there are also slightly widmanstatten structures in the fusion line zone; the corrosion resistance of the heat affected zone of the weld joint is the worst, and the corrosion resistance of the burned area is gradually reduced from the heat affected zone to both sides; with the increase of the welding current and welding speed, the performance of the weld joint first increases and then decreases, and the best welding is achieved parameters: welding current is 127A, welding speed is 90cm/min.

Key words: galvanized sheet; pulse MAG; microstructure; mechanical properties

7.3 低合金钢

热处理对 40CrA、60Si2Mn 和 20CrMnTi-1 力学性能试验结果的影响

张成亮

（宝武集团鄂钢公司质检中心，湖北鄂州 436000）

摘 要：本文通过总结宝武集团鄂钢公司质检中心 40CrA、60Si2Mn 和 20CrMnTi-1 物理检验工作实践，分析了其力学性能试验结果与热处理工艺之间的关系，提出了提高其力学性能试验结果准确性的措施，取得了一定的成效。

关键词：热处理；工艺；40CrA 60Si2Mn 20CrMnTi-1；力学性能；试验结果

Influence on Heat Treatment to Test Result of 40CrA, 60Si2Mn and 20CrMnTi-1 Mechanical Property

Zhang Chengliang

(Quality Inspection Center, Echeng Steel Co., Baowu Group, Ezhou 436000, China)

Abstract: In this paper the work experience of physical examination to 40CrA, 60Si2Mn and 20CrMnTi-1 in Quality inspection center, Echeng steel Co., Baowu group was summarized. The relevance between test result of mechanical property and heat treatment process was analysed. The measures were proposed to improve accuracy on test result of mechanical property. The effect is remarkable.

Key words: Heat treatment; Process 40CrA 60Si2Mn 20CrMnTi-1; Mechanical property; Test result

正火和回火工艺对 3.5%Ni 钢微观组织和力学性能的影响

杜　林，张宏亮，朱莹光，侯家平

（鞍钢集团钢铁研究院，辽宁鞍山　114009）

摘　要：利用万能试验机、扫描电镜和透射电镜，研究正火＋回火工艺对 3.5%Ni 钢力学性能及微观组织的影响。结果表明：正火+回火处理后，3.5%Ni 钢微观组织为铁素体基体加粒状珠光体。粒状珠光体的形成、净化铁素体两者共同作用是低温韧性提高的主要原因；合金渗碳体的在回火过程析出起到了析出强化作用，保证了 3.5%Ni 钢回火后的强度。

关键词：3.5%Ni 钢；热处理；低温韧性

Effect of Normalizing and Tempering Process on Microstructure and Mechanical Properties of 3.5%Ni Steel

Du Lin, Zhang Hongliang, Zhu Yingguang, Hou Jiaping

(Ansteel Group Iron and Steel Research Institute, Anshan 114009, China)

Abstract: The effect of normalizing and tempering process on the mechanical properties and microstructure of 3.5%Ni steel was studied by using universal testing machine, scanning electron microscope and transmission electron microscope.The results show that the microstructure of 3.5%Ni steel after normalizing and tempering is ferrite matrix and granular pearlite. The formation of granular pearlite and the combined action of purifying ferrite are the main reasons for the improvement of toughness at low temperature. The precipitation of cementite during tempering plays a role of precipitation strengthening, which ensures the strength of 3.5%Ni steel after tempering.

Key words: 3.5%Nisteel; heat treatment; cryogenic toughness

氮含量对 V-N 微合金钢相变行为和组织的影响

赵宝纯[1]，赵　坦[1]，李桂艳[1]，鲁　强[2]

（1. 鞍钢股份有限公司技术中心，辽宁鞍山　114009；

2. 鞍钢股份有限公司鲅鱼圈钢铁分公司，辽宁营口　115007）

摘 要：采用Gleeble-3800热力模拟试验机对不同含氮量的钒氮微合金钢进行多道次轧制工艺模拟实验，研究了钒氮微合金钢热变形奥氏体连续冷却过程的转变行为及组织变化规律，应用热膨胀法结合金相法，建立了相应的连续冷却转变（CCT）图。分析了氮含量对实验钢转变行为和组织演变规律的影响。结果表明：三种含氮量的实验钢在连续冷却过程中均发生铁素体，珠光体和贝氏体转变，含氮量高的实验钢具有更高的铁素体开始相变温度和更高的临界冷却速度，且铁素体的晶粒更细小。在低冷却速度条件下，实验钢显微硬度值随氮含量的增加而增加。然而，在高冷却速度条件下，高含量氮实验钢的显微硬度值反而低于低含量氮实验钢的显微硬度值。

关键词：V-N微合金钢；相变行为；组织演变；硬度

Effect of Nitrogen on Transformation Behaviors and Microstructure of V-N Microalloyed Steel

Zhao Baochun[1], Zhao Tan[1], Li Guiyan[1], Lu Qiang[2]

(1. Technology Center of Angang Steel Co., Ltd., Anshan 114009, China;
2. Angang Steel Co., Ltd., Bayuquan Subsidiary Co., Yingkou 115007, China)

Abstract: Multi-pass deformation simulation tests were performed on V-N microalloyed steels with different nitrogen contents by using Gleeble-3800 thermo-mechanical simulator and the corresponding continuous cooling transformation (CCT) diagrams were determined by thermal dilation method and metallographic method. The deformed austenite transformation behaviors and resultant microstructures of the tested steels were studied. Furthermore, the effects of nitrogen content on the transformation behaviors and microstructure evolution were analyzed. The results show that in consideration of the three tested steels, there are three kinds of phase transformations, ferrite, pearlite and bainite transformation during the continuous cooling process. For the two tested steel with higher nitrogen content, higher ferrite start temperature and critical cooling rate is observed. Furthermore, higher nitrogen content contributes to smaller ferrite grain formation. Under low cooling rate, the hardness number of the tested steel increases with increasing nitrogen content. However, the hardness number of the steel with higher nitrogen content is smaller than that of the steel with lower nitrogen content under high cooling rate.

Key words: V-N microalloyed steel; transformation behavior; microstructure evolution; hardness number

高耐蚀型耐候钢在煤浸出液中的腐蚀行为研究

温东辉[1]，葛红花[2]，宋凤明[1]

（1. 宝山钢铁股份有限公司 研究院，上海 201900；
2. 上海电力大学 上海市电力材料防护与新材料重点实验室，上海 200090）

摘 要：对比了两种耐候钢和普碳钢在煤浸出液中的周浸腐蚀行为。分析了三种试验钢在模拟阴天及晴天条件下的电化学阻抗与试验时间的变化和极化曲线，并在扫描电镜下分析了表面锈层的形貌及组分。结果显示，在晴天条件下试验钢更易于形成保护性锈层，高耐蚀型耐候钢S450EW形成保护性锈层的时间更短，电化学阻抗值最大。两种耐候钢在试验过程中均形成了致密的保护性锈层，其中S450EW中含有的较多的α-FeOOH，BC550表面锈层中主要含有β-FeOOH和少量的α-FeOOH，而Q345B表面锈层中主要为β-FeOOH和γ-FeOOH，没有稳定的α-FeOOH形成。整体上S450EW在煤浸出液中表现出更好的耐腐蚀性能。

关键词：耐候钢；煤浸出液；运煤敞车；周浸腐蚀

Study on Corrosion Behavior of Weathering Steels in Coal Leaching Solution

Wen Donghui[1], Ge Honghua[2], Song Fengming[1]

(1. Research institute, Baoshan Iron and Steel Co., Ltd., Shanghai 201900, China;
2. Shanghai Key Laboratory of Materials Protection and Advanced Materials in Electric Power, Shanghai University of Electric Power, Shanghai 200090, China)

Abstract: The periodic immersion corrosion behavior of two kinds of weathering steels and carbon steel in leaching solution was compared and analyzed, and the electrochemical impedance spectroscopy and polarization curves of three kinds of tested steels under simulated cloudy and sunny conditions were analyzed. The morphology and composition of surface rust layer were analyzed by scanning electron microscope. The results show that the protective rust layer is easier to form in sunny day, the time of forming protective rust layer is shorter for S450EW, and the electrochemical impedance value is the highest. The protective rust layer of two kinds of weathering steels is more dense than that of carbon steel during the test, in which S450EW contains more α-FeOOH, and BC550 mainly contains β-FeOOH and a small amount of α-FeOOH, while Q345B mainly contains β-FeOOH and γ-FeOOH, without stable α-FeOOH was observed. So S450EW showed better corrosion resistance in coal leaching solution.

Key words: weathering steel; coal leaching solution; coal open wagon; periodic immersion corrosion

690MPa 级海洋平台用特厚齿条钢回火稳定性

刘　庚，李慧杰，王庆海，叶其斌

（轧制技术及连轧自动化国家重点实验室，辽宁沈阳　110819）

摘　要：研究了回火时间（2、4 和 8h）对海洋平台用齿条钢 A514 GrQ 显微组织及力学性能的影响。结果表明，回火过程中，淬火态马氏体板条组织发生回复，C、Cr、Mo 和 V 等元素扩散，析出 MC、M2C 和 M23C6 型碳化物，回火 2h 后碳化物类型趋于稳定，并发生 Ostwald 熟化现象；随着回火时间的增加，实验钢的抗拉强度由淬火态的 1142MPa 逐渐降低到回火 8h 时的 813MPa，屈服强度由淬火态的 931MPa 逐渐降低至回火 8h 时的 725MPa；淬火态实验钢回火 2h 后，−60℃冲击功由淬火态的 187J 增加至 238J，继续增加回火时间，冲击功趋于稳定，约为 245J；延伸率由淬火态的 15.2%增加至回火 2h 时的 16.7%，并且随着回火时间的增加，升高至回火 8h 时的 19.5%。本研究在实现了钢板良好的强韧性匹配的同时，又尽可能的缩短了回火时间。

关键词：特厚齿条钢；回火时间；碳化物；强韧性

Tempering Stability of a 690MPa Grade Ultra-heavy Rack Steel for Offshore Platform

Liu Geng, Li Huijie, Wang Qinghai, Ye Qibin

(State Key Laboratory of Rolling and Automation, Northeastern University, Shenyang 110819, China)

Abstract: The effects of tempering timeon microstructure and mechanical properties of as-quenched A514 GrQ rack steel for offshore platform were studied.The results show that the martensite lath structure recovers,and the elements of C、Cr、Mo and V diffused during tempering at 600℃. MC, M2C, and M23C6 type carbides precipitated in the martensite matrix, the carbides tend to be stable, Ostwald ripening occurs after tempering for 2h. The tensile strength and yield strength of the tested steel decreased with the increase of tempering time,and the tensile strength decreased from 1142MPa in as-quenched state to 813MPa at 8h tempering, and the yield strength decreased from 931 MPa in quenching state to 725MPa at 8h tempering. After tempering for 2h, the impact energy at −60℃ increased from 187J to 238J. The impact energy tends to be stable after tempering for 2 hours, about 245J. The elongation increased from 15.2% of quenched state to 16.7% at 2h tempering, And with the increase of tempering time, it increases to 19.5% of 8h tempering. The excellent matching of strength and toughness of the steel plate is realized, and the tempering time is shortened as much as possible.

Key words: ultra-heavy rack steels; tempering time; carbides; strength and toughness

在线直接淬火对 EH690 海工钢组织与性能的影响

宫晓兰[1]，朱 拓[1]，阚立烨[1]，叶其斌[1]，肖大恒[2]

（1. 东北大学 轧制技术及连轧自动化国家重点实验室，辽宁沈阳 110819；
2. 湖南华菱湘潭钢铁有限公司，湖南湘潭 411100）

摘 要：与离线淬火(RQ)工艺对比，研究了在线直接淬火(DQ)对 EH690 级海工钢组织与性能的影响。采用光学(OM)、扫描电子显微镜(SEM)分析了两种淬火及回火态(T)微观组织形貌，用电子背散射衍射(EBSD)表征了微观织构，并对比了室温拉伸和−40℃夏比冲击性能。结果表明：DQT 钢屈服强度、抗拉强度和冲击功分别为 1010MPa、1073MPa 和 52J，而 RQT 钢对应性能分别为 880MPa、921MPa 和 108J。DQT 钢马氏体板条细小，位错密度更高，因此强度显著提高。但 DQT 钢有利于韧性的织构强度和大角度晶界比例均较低，因此−40℃低温平均冲击功比 RQT 钢低了 56J。

关键词：直接淬火；强度；冲击韧性；织构；EH690 海工钢

Online Direct Quenching on Microstructure and Properties of Tempered EH690 Grade Steel for Offshore

Gong Xiaolan[1], Zhu Tuo[1], Kan Liye[1], Ye Qibin[1], Xiao Daheng[2]

(1. State Key Laboratory of Rolling and Automation, Northeastern University, Shenyang 110819, China;
2. Hunan Valin Xiangtan Iron and Steel Co., Ltd., Xiangtan 411100, China)

Abstract: The effect of online direct quenching (DQ) on the organization and properties of EH690 grade offshore steel was investigated in comparison with the offline quenching (RQ) process. The microstructure morphology of the two quenched and tempered states (T) was analyzed by optical (OM) and scanning electron microscopy (SEM), the microstructure weave was characterized by electron backscatter diffraction (EBSD), and the room temperature tensile and −40℃ Charpy impact properties were compared. The results showed that the yield strength, tensile strength and impact work of DQT steel were 1010Mpa, 1073MPa and 52J, respectively, while the corresponding properties of RQT steel were 880Mpa, 921MPa and 108J. The martensitic slats of DQT steel were fine and the dislocation density was higher, so the strength was significantly higher. However, the weave strength and the proportion of large-angle grain boundaries, both of which are beneficial to

toughness, are lower in DQT steel, so the average low-temperature impact work at −40℃ is 56J lower than that of RQT steel.
Key words: direct quenching; strength; texture; impact toughness; EH690 grade steel for offshore

回温变形对低碳低合金钢显微组织及变形过程的影响

王益民，李慧杰，苏元飞，徐晓宁，叶其斌

（东北大学轧制技术及连轧自动化国家重点实验室，辽宁沈阳 110819）

摘 要：针对低碳低合金钢，利用变形热膨胀相变仪及扫描电子显微镜，采用回温变形工艺主要研究了中间冷却温度和回温温度对实验钢微观组织演变及变形过程的影响。结果表明：回温变形工艺对实验钢的组织成分、晶粒尺寸、回温后的变形应力及终冷过程有较大影响。中间冷却温度为 500℃，回温温度为 800℃时，实验钢的变形应力低且 Ar_3 温度提高，组织中形成了大量超细晶铁素体，平均晶粒尺寸小，是最佳的回温变形工艺。

关键词：低碳低合金钢；回温变形；晶粒细化；组织演变；Ar_3

Effect of Temperature-reversion Deforming on the Microstructure and Deformation Process of Low-carbon Low-alloy Steel

Wang Yimin, Li Huijie, Su Yuanfei, Xu Xiaoning, Ye Qibin

(The State Key Laboratory of Rolling Automation, Northeastern University, Shenyang 110819, China)

Abstract: For low-carbon low-alloy steel, the effects of intermediate cooling temperature and re-temperature on the microstructure evolution and deformation process of the experimental steel were mainly investigated by using the deformation thermal expansion phase transducer and scanning electron microscope with the temperature-reversion deforming process. The results show that the temperature-reversion deforming process has a large effect on the microstructure composition, grain size, deformation stress and final cooling process of the experimental steels. When the intermediate cooling temperature is 500℃ and the retemperature is 800℃, the deformation stress of the experimental steel is low and the Ar_3 temperature is increased, a large amount of ultrafine grain ferrite is formed in the organization and the average grain size is small, which is the best temperature reverting rolling process.

Key words: low-carbon low-alloy steel; temperature-reversion deforming; grain refinement; microstructure evolution; Ar_3

正火轧制和正火工艺对 S355NL 钢板低温韧性的影响

潘中德，胡其龙，顾小阳，张 淼

（南京钢铁股份有限公司板材事业部，江苏南京 210035）

摘 要：以欧洲标准 EN10025-3 焊接细晶粒钢为基本依据，采用包晶钢、Nb+Ti 等微合金化成分设计，进行了正火轧制和正火热处理工艺生产 S355NL 低温结构钢板的试制开发。试验结果表明，对于厚度 60mm 及以下 S355NL 钢板，正火轧制、正火热处理工艺均可满足–40℃低温冲击性能要求，然而对于–50℃及以下低温冲击性能，则需要采用正火热处理工艺路线才能满足要求。金相组织分析，试制钢板均为铁素体+珠光体组织，但正火热处理工艺生产的钢板组织更为均匀和细小。

关键词： 正火轧制；正火；S355NL；低温韧性；金相组织

Effect of Normalizing Rolling and Normalizing Process on Low-temperature toughness of S355NL Steel Plate

Pan Zhongde, Hu Qilong, Gu Xiaoyang, Zhang Miao

(Plate Business Unit of Nanjing Iron & Steel Co., Ltd., Nanjing 210035, China)

Abstract: Based on the European standard EN10025-3 welding fine grain steel, S355NL low-temperature structural steel plate was produced by normalizing rolling and normalizing heat treatment process using peritectic steel, Nb + Ti and other microalloying components. The test results show that for S355NL steel plate with thickness of 60mm and below, both normalizing rolling and normalizing heat treatment processes can meet the requirements of low-temperature impact properties at –40℃. However, for low-temperature impact properties at –50℃ and below, normalizing heat treatment process is needed to meet the requirements. Metallographic analysis shows that the microstructure of S355NL steel plate is ferrite & pearlite, but the microstructure of steel plate produced by normalizing heat treatment process is more uniform and fine.

Key words: normalizing rolling; normalizing; S355NL; low-temperature toughness; metallographic structure

中/低碳富 Si 合金钢制备铁素体/贝氏体双相钢的组织及力学性能

贾 鑫[1]，王天生[1,2]

（1. 燕山大学亚稳材料制备技术与科学国家重点实验室，河北秦皇岛 066004；
2. 燕山大学国家冷轧板带装备及工艺工程技术研究中心，河北秦皇岛 066004）

摘 要： 本文将中碳和低碳富硅合金钢经温轧变形后，通过两相区退火+等温淬火处理得到铁素体/贝氏体双相组织，通过扫描电镜和透射电镜对热处理后得到的组织进行观察，并对其进行力学性能测试。实验结果表明，0.33C 钢经不同工艺处理后得到细小的等轴铁素体晶粒和板条贝氏体铁素体的双相组织。0.21C 钢中的铁素体则以细长条状为主，并且在 M_s 点以下等温转变后得到板条贝氏体铁素体，在 M_s 点以上等温转变后同时得到板条贝氏体铁素体和粒状贝氏体组织。0.33C 钢的抗拉强度和屈服强度均随着等温温度的升高而下降，冲击功随着等温温度的升高而升高，0.21C 钢拉伸性能的变化规律与 0.33C 钢一致，但冲击功呈现相反的趋势，随着等温温度的升高而下降。

关键词： 中/低碳富 Si 合金钢；两相区退火；等温淬火；铁素体/贝氏体双相组织；力学性能

Microstructure and Mechanical Properties of Ferrite/Bainite Dual Phase Steel Prepared from Medium/Low Carbon Si-rich Alloy Steel

Jia Xin[1], Wang Tiansheng[1,2]

(1. State Key Laboratory of Metastable Materials Science and Technology, Yanshan University, Qinhuangdao 066004, China; 2. National Engineering Research Center for Equipment and Technology of Cold Strip Rolling, Yanshan University, Qinhuangdao 066004, China)

Abstract: In this paper, the ferrite/bainite dual phase microstructure was obtained by intercritical annealing and austempering after warm rollingin medium carbon and low carbon silicon-riched alloy steels.The microstructure after heat treatment was observed by scanning electron microscopy and transmission electron microscopy, and the mechanical properties were investigated.The results show that dual-phase microstructures of fine equiaxial ferrite grains and lath bainite ferrite are obtained after different processing of 0.33C steel. The ferrite in 0.21C steel is mainly thin-striped shaped, and the lath bainite ferrite is obtained after austempering below-M_s, while the lath bainite ferrite and granular bainite structure are obtained after austempering above-M_s.The tensile strength and yield strength of 0.33C steel decrease with the increase of austempering temperature, and the impact energy increases with the increase of austempering temperature. The variation of tensile properties of 0.21C steel is consistent with that of 0.33C steel, but the impact energy presents the opposite trend, and decreases with the increase of austempering temperature.

Key words: medium/low carbon Si-rich alloy steel; intercritical annealing; austempering; ferrite/bainite dual phase microstructure; mechanical properties

100mm 厚度 EH47 高止裂韧性钢板组织与性能研究

王红涛[1]，田　勇[1,2]，叶其斌[1]，陈林恒[2,3]，赵晋斌[2,3]，邱保文[2,3]

（1. 东北大学轧制技术及连轧自动化国家重点实验室，辽宁沈阳　110819；

2. 南京钢铁股份有限公司，南钢研究院，江苏南京　210035；

3. 南京钢铁股份有限公司，江苏省高端钢铁材料重点实验室，江苏南京　210035）

摘　要： 利用光学显微镜、XRD 等显微组织分析手段和拉伸、冲击、落锤等力学性能测试方法，研究了超大集装箱船用 100mm 厚 EH47 高止裂韧性钢板的组织与性能。钢板采用新型控制轧制工艺生产，梯度温度型双重拉伸试验结果表明，钢板-10℃止裂韧性 *Kca* 高达 9041.5N/mm$^{3/2}$，远高于船级社规范要求。厚度规格和止裂韧性均达到国内最高水平，打破了国外企业的技术壁垒。全厚度细化多边铁素体和少量贝氏体混合组织、心部较强γ线织构和较弱{001}<110>织构是保证钢板优异止裂韧性的关键组织因素。特厚钢板高止裂韧性显微组织调控的关键在于低压缩比条件下的控制轧制工艺创新。钢板全厚度断裂时裂纹尖端呈“多峰”分布的“劈钉”特征有利于提高钢板的止裂韧性。

关键词： 100mm 厚止裂钢；高止裂韧性；NEU-Rolling 新型控轧工艺；显微组织

Microstructures and Properties of EH47 High Crack Arrest Toughness Steel with the Thickness of 100mm

Wang Hongtao[1], Tian Yong[1, 2], Ye Qibin[1], Chen Linheng[2, 3], Zhao Jinbin[2, 3], Qiu Baowen[2, 3]

(1. State Key Laboratory of Rolling and Automation, Northeastern University, Shenyang 110819, China;
2. Nanjing Iron and Steel Co., Ltd., Nanjing Iron and Steel Research Institute, Nanjing 210035, China;
3. Nanjing Iron and Steel Co., Ltd., Jiangsu Key Laboratory for Premium Steel Material, Nanjing 210035, China)

Abstract: The microstructures and properties of EH47 high crack arrest toughness steelwith a thickness of 100mm for mega container carriers were studied by optical microscope, XRD,tensile, impact, and drop-weight tests. The result of thetemperature-gradient double tension test shows that the crack arrest toughness (*K*ca) of the developed steel produced by a novel controlled rolling process at –10℃ is as high as 9041.5N/mm$^{3/2}$ which is far higher than the requirements of classification society. Besides, the thickness specification and crack arrest toughness have reached the highest level in China, breaking the technical barriers of foreign enterprises. The key factors to ensure the excellent crack arrest toughness of the ultra-heavy steel are achieving the fine ferrite grains accompanied a small amount of bainite, strong texture along γ-orientation, and weak {001}<110> texture at the 1/2 thickness through the innovation of controlled rolling process under the condition of low compression ratio. Besides, the "split nail" feature of multi-peaks distribution at crack tip is helpful to improve the crack arrest toughness when the steel plate is full thickness fractured.

Key words: steel with the thickness of 100mm; high crack arrest toughness; NEU-Rolling process; microstructures

热轧、冷轧搪瓷钢的强度机理及抗鳞爆性能研究

张志敏[1]，熊爱明[1]，黄学启[2]，刘再旺[1]，郭　敏[2]，梁立川[3]

（1. 首钢集团有限公司技术研究院，北京　100043；
2. 北京首钢股份有限公司，北京　100043；
3. 首钢京唐钢铁联合有限责任公司，河北唐山　063200）

摘　要：热轧搪瓷钢搪烧后屈服强度大幅下降，冷轧搪瓷钢搪烧后力学性能变化不大，主要原因为热轧搪瓷钢搪烧后钢板的晶粒尺寸变大和析出物粒子粗化，冷轧搪瓷钢搪烧后晶粒尺寸变化很小。热轧搪瓷钢搪烧后 TH 值降低，主要原因为热轧搪瓷钢中析出物粒子尺寸变大和晶界面积减小，冷轧搪瓷钢搪烧后 TH 值降低，主要原因为渗碳体的粗化导致渗碳体与铁素体界面减少。

关键词：搪瓷钢；强度；抗鳞爆；TH 值

Strength Mechanism and Fish-scaling Resistance of Hot and Cold Rolled Enamel Steel

Zhang Zhimin[1], Xiong Aiming[1], Huang Xueqi[2], Liu Zaiwang[1], Guo Min[2], Liang Lichuan[3]

(1. Shougang Group Co.,Ltd. Research Institute of Technology, Beijing 100043, China; 2. Beijing Shougang Co., Ltd, Beijing 100043, China; 3. Shougang Jingtang United Iron & Steel Co., Ltd, Tangshan 063200, China)

Abstract: Yield strength of hot rolled enamel steel decreased obviously during enameling process. The main reason is that grains of steel and precipitates in steel grow up when steel sheets were baked. Mechanical properties of cold rolled enamel steel after baking were nearly unchanged. The reason is that grains of the steel hardly grow up during enameling process. TH value of hot rolled enamel steel decreased during enameling process because the size of precipitated particles in steel increased and area of grain boundaries decreased when steel sheets were baked. TH value of cold rolled enamel steel after baking decreased due to a decrease in cementite-ferrite interfaces when cementites grew up.

Key words: enamel steel; strength; fish-scaling resistance; TH value

热处理对1000MPa超高强钢微观组织及力学性能影响

张学伟[1,2]，黄进峰[1]，杨才福[2]

（1. 北京科技大学新金属材料国家重点实验室，北京 100083；
2. 钢铁研究总院 工程用钢研究所，北京 100081）

摘　要：研究了淬火匹配不同回火温度对1000MPa超高强度钢微观组织及力学性能的影响，采用SEM扫描电镜、TEM透射电镜分析了马氏体组织结构以及基体中碳化物、富铜相等析出相特征，马氏体板条内高密度位错以及冲击断口形貌特征。研究结果表明，随着回火温度的升高，碳化物从马氏体板条内析出演变至板条间析出和晶界析出，同时发生粗化，而富铜相从长条状转变为椭球状，尺寸显著增大。随着回火温度的升高，马氏体内位错密度逐渐降低，试验钢的屈服强度逐渐降低，抗拉强度单调降低，伸长率与冲击吸收功具有相同的逐渐增加的趋势。在–84℃冲击试样的断裂区呈现脆断，呈现准解理断裂以及混合断裂形貌。

关键词：回火温度；力学性能；微观组织；碳化物

Effect of Heat Treatment on Microstructure and Mechanical Properties of the 1000MPa Ultra-high Strength Steel

Zhang Xuewei[1,2], Huang Jinfeng[1], Yang Caifu[2]

(1. State Key Laboratory for Advanced Metals and Materials, University of Science and Technology Beijing, Beijing 100083, China; 2. Institute for Engineering Steel, Central Iron and Steel Research Institute, Beijing 100081, China)

Abstract: Effects of quenching matching different tempering temperatures on microstructure and mechanical properties of 1000MPa ultra-high strength steel were studied. The characteristics of martensite structure, carbide and copper-rich precipitates in the martensite matrix, high density dislocation in the martensite lath and fracture characteristics of impact specimens were analyzed by scanning electron microscopy (SEM) and transmission electron microscopy (TEM). The results show that, with the increase of tempering temperature, the carbide precipitates from intra-lath to inter-lath and grain boundary, and gradually coarsens, the copper-rich phase from long strips to nearly spheroidizes and its size increases significantly. With the increasing of tempering temperature, the dislocation density in martensite decreases gradually, the yield strength and the tensile strength decrease gradually, at the same time, the elongation had the same increasing trend as the impact absorbed energy. The impact fracture of the specimens presents brittle fracture at –84℃ and takes on the appearance of quasi-cleavage fracture and mixed fracture morphology.

Key words: tempering temperature; mechanical property; microstructure; carbide

大型集装箱船用 EH47 钢止裂行为研究

刘朝霞，白　云，刘　俊，宁康康，刘　洋

（江阴兴澄特种钢铁有限公司研究院，江苏江阴　214400）

摘　要：研究 EH47 试验钢组织特征及其对止裂性能的影响。力学性能测试结果表明，3 种试验钢力学性能满足 EH47 钢交货标准，–40℃、–60℃冲击值≥200J。显微组织分析表明，终冷温度较低的 1#、2#试验钢主要为细小板条贝氏体及少量的针状铁素体组成，终冷温度稍高的 3#试验钢组织为针状铁素体、超细铁素体、MA 组元及少量的低碳板条贝氏体组织组成。平均晶粒尺寸大于 1#、2#试验钢。示波冲击试验表明，与止裂相关冲击值占总冲击值比例，受贝氏体片层或晶粒大小影响不明显，平均晶粒尺寸较大的 3#试验钢具有更高的止裂性。原位拉伸分析表明，裂纹启裂的位置微裂纹倾向于在<111>、<112>取向的晶粒位置处发生，避开了<110>取向的晶粒。在裂纹扩展过程中，遇到<110>的晶粒会发生 90°转角，遇到小角晶界穿晶沿直线通过，遇到大角晶界裂纹传播路径发生改变。

关键词：EH47 止裂；示波冲击；晶粒取向

Investigation on Crack-arrest Behavior of EH47 Steel for Large Container Ship

Liu Zhaoxia, Bai Yun, Liu Jun, Ning Kangkang, Liu Yang

(Research Institute, Jiangyin Xingcheng Special Steel Works Co., Ltd., Jiangyin 214400, China)

Abstract: The microstructure of EH47 experimental steels and its effect on crack-arrest properties were studied. The mechanical properties test results show that the mechanical properties of the three experimental steels meet the delivery requirements of EH47 steel, and the impact value at the temperature of –40℃ and –60℃ is not less than 200J. The microstructure of 1# and 2# experimental steels with a lower final cooling temperature is mainly composed of fine lath bainite and a small amount of acicular ferrite. The microstructure of 3# experimental steel with a higher final cooling temperature is composed of acicular ferrite, ultra-fine ferrite, MA component and a small amount of low carbon lath bainite. The average grain size 3# experimental steel is larger than other two. Instrumental impact test at –60℃ shows that the ratio of the impact value related to crack-arrest vs total impact value is hardly affected by the size of bainite lamellae or grain size. Actually, 3# experimental steel with a larger grain size has higher crack-arrest property. In situ tensile analysis shows that the microcracks tend to occur at the grain positions with <111>, <112> orientation, avoiding the grain with <110> orientation. During the crack propagating, it will turn with a 90° angle when the grain with <110> orientation is encountered; transgranular straightly will take place, when low small-angle grain boundaries are encountered; propagation changes the route when large-angle grain boundaries are encountered.

Key words: EH47 crack-arrest; instrumental impact test; grain orientation

7.4　特殊钢

航天用 17-4PH 不锈钢锻件热处理工艺与组织性能研究

李荣之，曹征宽，朱　斌，张全新

（重庆钢铁研究所有限公司，重庆　400084）

摘　要：本文通过航天用 17-4PH 不锈钢锻件的热处理实验，研究了不同热处理工艺对其微观组织和性能的影响规律。研究结果表明，17-4PH 不锈钢的固溶组织以马氏体为基体，锻件在不同方向和部位的组织和硬度均匀；一次时效处理后 17-4PH 钢的马氏体含量增加，随着时效温度升高，析出相数量和尺寸增加，锻件硬度先升后降，480℃时效的硬度最高；中间调整热处理有助于合金元素的析出，在相同的时效温度下，相对于一次时效，其组织更加细小，马氏体含量更高，沉淀相更多，但时效析出的合金元素不能形成高硬度的碳化物，时效析出强化效果有限，因此，相对于一次时效，其硬度更低。

关键词：17-4PH 不锈钢；热处理；金相组织；硬度

Study on the Relationship between Heat Treatment Process and Microstructure and Properties of 17-4PH Stainless Steel Forgings for Aerospace

Li Rongzhi, Cao Zhengkuan, Zhu Bin, Zhang Quanxin

(Chongqing Iron and Steel Research Institute Co., Ltd., Chongqing 400084, China)

Abstract: The effects of different heat treatment processes on microstructure and mechanical properties of 17-4PH stainless steel forgings for aerospace were studied by heat treatment experiments. The results show that martensite is the main solid solution microstructure of 17-4PH stainless steel, and the microstructure and hardness of forgings are uniform in all directions and parts, as the amount and size of precipitates increase, the hardness of the forgings increases first and then decreases, and the hardness of the forgings aged at 480℃ is the highest, martensite content is higher, precipitation phase more, but the precipitation of alloy elements can not form hard carbide, strengthening effect is limited, therefore, compared with a single aging, its hardness is lower.

Key words: 17-4PH stainless steel; heat treatment; structure; hardness

Fe-36Ni 因瓦合金带材与中厚板的开发

李大航[1,2]，赵 刚[1,2]，郑 欣[3]，刘 璇[1,2]，朱义轩[1,2]，张友鹏[1,2]，吴宇新[1,2]

（1. 鞍钢集团钢铁研究院，辽宁鞍山 114009；2. 鞍钢海洋装备用金属材料及其应用国家重点实验室，辽宁鞍山 114009；3. 鞍钢股份有限公司，辽宁鞍山 114021）

摘 要： 根据因瓦合金的特点，结合自身装备情况，利用冶炼+模铸+热连轧+冷轧+连续退火工艺和电渣重熔+热轧+热处理工艺，鞍钢成功开发了因瓦合金带材及中厚板产品。检验结果表明：合金带材和中厚板的组织均匀，为单一的奥氏体组织；合金带材的力学性能和膨胀性能均满足 LNG 船货舱围护系统的要求。中厚板产品的力学性能和膨胀性能指标优良，探伤满足 EN 10160 标准 S3/E4 级要求。

关键词： 因瓦合金；膨胀性能

Development and of Invar Alloy Strip and Plate

Li Dahang, Zhao Gang, Zheng Xin, Liu Xuan, Zhu Yixuan, Zhang Youpeng, Wu Yuxin

Abstract: according to the characteristics of invar alloy, combined with its own equipment, Angang has successfully developed invar alloy strip and plate by using smelting + die casting + hot continuous rolling + cold rolling + continuous annealing process and ESR + hot rolling + heat treatment process. The test results show that the microstructure of the alloy strip and plate is uniform and single austenite structure; the mechanical properties and expansion properties of the alloy strip meet the requirements of LNG ship cargo hold enclosure system. The mechanical properties and expansion properties of the plate are excellent, and the flaw detection meets the requirements of S3/E4 of EN 10160 standard.

Key words: Fe-36Ni invar alloy; expansion property

改善 1.2344 系列扁钢退火组织的工艺研究

孙盛宇，付 博，叶世圣

（东北特殊钢集团股份有限公司，辽宁大连 116105）

摘 要： 高端 1.2344 系列扁钢的退火组织应满足北美压铸协会标准《NADCA#207》或热作工具钢显微检验标准《SEP1614》的要求，某公司 1.2344 系列扁钢退火组织一次合格率仅为 79%，通过调整钢锭的高温扩散工艺、增大扁钢轧后在线冷却强度及延长等温球化退火保温时间等工艺措施成功改善了带状偏析、液析碳化物、退火组织均匀性等问题，一次合格率提升至 98%以上。

关键词： 1.2344；热作模具扁钢；高温扩散；在线冷却；等温球化退火；带状偏析；液析碳化物

Investigation on the Process of Improving the Annealing Structure of 1.2344 Die Flat Steel

Sun Shengyu, Fu Bo, Ye Shisheng

(Dongbei Special Steel Group Co., Ltd., Dalian 116105, China)

Abstract: The annealing structure of high quality 1.2344 flat steel should match the Standard 《NADCA#207》 of North American Die-casting Association or 《SEP1614》 of Microscopic Inspection Standard for Hot Tool Steel.The first time yield of 1.2344 flat steel annealing structure only 79%.The problems of Band Segregation,Primary Carbide and uniformity of annealing structure were improved successfully by improve the high temperature diffusion process of ingot, increasing the on-line cooling strength of flat steel after rolling and prolonging the time of isothermal spheroidizing annealing,The First Time Yield of annealing structure got up to 98% or more.

Key words: 1.2344; hot die flat steel; high temperature diffusion; on-line cooling; isothermal spheroidizing; band segregation; primary carbide

超高强含铜钢热变形 Arrhenius 型本构关系

阚立烨[1]，宫晓兰[1]，朱　拓[1]，叶其斌[1]，赵　坦[2]

（1. 东北大学 轧制技术及连轧自动化国家重点实验室，辽宁沈阳　110819；

2. 鞍钢集团 海洋装备用金属材料及其应用国家重点实验室，辽宁鞍山　114001）

摘　要：采用应变速率、温度修正方法构建了超高强含铜钢的 Arrhenius 本构关系模型。在温度范围为 850~1150℃、应变速率范围为 0.01~10s^{-1} 条件下，在 MMS-200 热力模拟实验机上进行热压缩试验，获得了实验钢真应力-真应变数据。利用应变速率及温度修正了 Arrhenius 型本构方程，并通过相关系数和绝对平均误差等统计参数对模型的准确性和可靠性进行了量化。结果表明，所提出的模型对给定变形条件下钢的流变应力具有良好的预测能力。

关键词：Arrhenius 型；本构关系；应变及温度补偿；超高强含铜钢

A Study on Constitutive Relationship of Cu-bearing Ultra-high-strength Steel during Hot Deformation Based on Arrhenius-type Model

Kan Liye[1], Gong Xiaolan[1], Zhu Tuo[1], Ye Qibin[1], Zhao Tan[2]

(1. The State Key Laboratory of Rolling and Automation, Northeastern University, Shenyang 110819, China; 2. The State Key Laboratory of Metal Material for Marine Equipment and Application, Ansteel Group, Anshan 114011, China)

Abstract: Constitutive relationship of Cu-bearing ultra-high-strength steel is investigated by the Arrhenius-type constitutive model incorporating the strain rate and temperature compensation. The experimental true stress–true strain data were obtained from hot compression tests on the MMS-200 thermo-mechanical simulator in the temperature range of 850–1150℃ and strain rate range of 0.01–10 s^{-1}. The Arrhenius type model is modified by means of strain rate compensation and temperature compensation. The accuracy and reliability of the model were quantified by employing statistical parameters such as the correlation coefficient and absolute average error. The results show that the proposed models have excellent predictabilities of flow stresses for the present steel in the specified deformation conditions.

Key words: Arrhenius-type; constitutive relationship; strain and temperature compensation; Cu-bearing ultra-high-strength steel

减免涂装钢结构用 1000MPa 级耐火耐候螺栓钢高温行为研究

罗志俊[1,2]，徐士新[1]，孙齐松[1,2]，王晓晨[1]，刘　锟[1]，田志红[1]

（1. 首钢集团有限公司技术研究院，北京　100043；
2. 绿色可循环钢铁流程北京市重点实验室，北京　100043）

摘　要：利用高温拉伸试验、扫描电镜和透射电镜对比分析了 1000MPa 级耐火耐候螺栓钢和传统普碳螺栓钢的高温耐火性能及其微观机理。结果表明：耐火耐候螺栓钢与传统普碳螺栓钢在具有相同室温力学性能的条件下，经过 600℃保温 3h，高温抗拉强度依然能保持 606MPa，较传统普碳螺栓钢折减率提高了 32.1%，具有优异的耐火性能；耐火耐候螺栓钢回火时马氏体板条间及板条上析出的纳米级 Nb、V 析出物以及含 Mo 合金渗碳体的二次硬化是其具有优良高温性能的主要原因，同时析出物的数量越多、尺寸越小，高温回火稳定性越强，材料的耐火性能越强。

关键词：绿色钢结构；耐火耐候螺栓钢；高温性能；纳米级析出物

Study on High Temperature Behavior of 1000MPa Grade Fire Resistant Weathering Bolt Steel for Reduce and No Paint Steel Construction

Luo Zhijun[1,2], Xu Shixin[1], Sun Qisong[1,2], Wang Xiaochen[1], Liu Kun[1], Tian Zhihong[1]

(1. Research Institute of Technology, Shougang Group Co., Ltd., Beijing 100043, China; 2. Beijing Key Laboratory of Green Recyclable Process for Iron and Steel Production, Beijing 100043, China)

Abstract: The high temperature refractory performance and microscopic mechanism of 1000MPa fire resistant weathering bolt steel and traditional carbon bolt steel were analyzed by using high temperature tensile test, scanning electron microscopy and transmission electron microscopy. The results show that, under the condition of the same mechanical properties at room temperature, after 600℃×3h, the tensile strength of the fire resistant weathering bolt steel can still be maintained at 606MPa, the reduction rate of fire resistant weathering bolt steel is 32.1% higher than that of traditional carbon bolt steel, which has excellent fire-resistant performance. The main reason for the excellent high temperature properties of fire-resistant bolt steel is the nanoscale Nb and V precipitates between and on the martensitic lath and the secondary hardening of Mo alloy cementite, at the same time, the more the amount and the smaller the size of the precipitate, the stronger the high temperature tempering stability and the stronger the fire resistance of the material.

Key words: green steel construction; fire resistant weathering bolt steel; high-temperature behavior; nanoscale precipitates

富 Si-H13 钢等温淬火及其回火组织和力学性能

孙晓文[1]，王岳峰[1]，王天生[1,2]

（1. 燕山大学亚稳材料制备技术与科学国家重点实验室，河北秦皇岛 066004；
2. 燕山大学国家冷轧板带装备及工艺工程技术研究中心，河北秦皇岛 066004）

摘　要：富 Si-H13 钢经两种调质预处理+350℃等温淬火后获得纳米贝氏体组织，然后分别进行 560℃×1h 一次回火、560℃×1h +580℃×1h 二次回火和 560℃×1h +580℃×1h +600℃×1h 三次回火处理。采用扫描电镜、透射电镜和硬度、拉伸及冲击试验方法，研究了回火过程显微组织和力学性能的演变。结果表明，一次回火后，薄膜状残余奥氏体开始分解，富 V 碳化物在贝氏体铁素体板条及其边界上析出，产生二次硬化和析出强化，而塑性和冲击韧性明显降低。二次回火后，抗拉强度和硬度变化不大，冲击韧性降低。三次回火后，等温淬火试样的贝氏体铁素体板条厚度仍保持 110nm 左右，硬度和抗拉强度略微降低，其具有良好的回火稳定性。

关键词：富 Si-H13 钢；等温淬火；纳米贝氏体；回火；组织；力学性能

Austempered and Tempered Microstructure and Mechanical Properties of Si-rich H13 Steel

Sun Xiaowen[1], WangYuefeng[1], WangTiansheng[1,2]

(1. State Key Laboratory of Metastable Materials Science and Technology, Yanshan University, Qinhuangdao 066004, China; 2. National Engineering Research Center for Equipment and Technology of Cold Strip Rolling, Yanshan University, Qinhuangdao 066004, China)

Abstract: The Si-rich H13 steel with nano-bainite structure was obtained after two quenching and tempering pretreatments +350℃ austempering. The microstructure and mechanical properties of the nano-bainite steel after different tempering processeswere investigated by scanning electron microscopy, transmission electron microscopy, and hardness, tensile and impact tests. The tempering processes included the first tempering of 560℃×1h, the twice tempering of 560℃×1h +580℃× 1h and the third tempering of 560℃×1h+580℃×1h +600℃×1h. The results show that after the first tempering, the film retained austenite decomposed, and V-rich carbides were precipitated on the bainitic ferrite lath and its boundary, resulting in secondary hardening and precipitation strengthening, and the plasticity and impact toughness decreasedsignificantly. After the second tempering, the tensile strength and hardness changed little, and the impact toughness decreased. After the third tempering, the thickness of the bainitic ferrite lath remained about 110nm, the hardness and tensile strength reduced slightly, therefore the austempered samples have good tempering stability.

Key words: Si-rich H13 steel; austempering; nanobainite; tempering; microstructure; mechanical properties

汽车减震器套管用钢 LAX340Y410T 质量提升工艺实践

张正波

（宝武杰富意特殊钢有限公司，广东韶关 512000）

摘 要：首次开发试制的汽车减震器套管用钢 LAX340Y410T，圆钢表面存在通条裂纹，烂钢等缺陷。表面漏磁探伤合格率极低，探伤初检合格率仅 3.13%，失效分析结果表明，缺陷产生于轧制之前。铸坯剥皮发现铸坯表面存在凹坑缺陷，铸坯凹坑缺陷剥皮去除后经加热轧制同样存在通条裂纹，烂钢缺陷，圆钢表面探伤合格率也仅 30%左右。铸坯剥皮与不剥皮的生产试验比对结果表明，LAX340Y410T 应力裂纹敏感性较强，未经缓冷的铸坯在加热过程中表面容易产生应力裂纹缺陷。铸坯表面的凹坑缺陷会在加热过程中因应力集中加剧裂纹的扩展。分析认为铸坯表面凹坑缺陷的产生与亚包晶钢中加入的 N 有关。LAX340Y410T 裂纹敏感性较强，铸坯进行缓冷可有效防止应力裂纹的产生。通过工艺优化调整，冶炼过程取消氮气环流，增加铸坯缓冷工序，可有效防止应力裂纹。试验结果表明圆钢表面探伤合格率由初次试制的 3.13%提升至 90.2%，产品质量大幅度提高。

关键词：LAX340Y410T；氮；缓冷；探伤合格率；质量；提高

Quality Improvement of Steel LAX340Y410T for Automobile Shock Absorber Casing raft Practice

Zhang Zhengbo

(Baowu JFE Special Steel Co., Ltd., Shaoguan 512000, China)

Abstract: The steel LAX340Y410T for automobile shock absorber casing was developed and trial-produced for the first time. The round steel has defects such as cracks and rotten steel on the surface of the round steel. The pass rate of surface magnetic flux leakage detection is extremely low, and the pass rate of initial inspection is only 3.13%. The failure analysis results show that the defects occurred before rolling. Slab peeling found that there were pit defects on the surface of the cast slab. After the pit defects were stripped and removed, there were also stub cracks and rotten steel defects. The pass rate of round steel surface flaw detection was only about 30%. The results of the production test comparison between the slab peeling and non-skinning show that LAX340Y410T is more sensitive to stress cracks, and the surface of the slab without slow cooling is prone to stress crack defects during the heating process. The pit defects on the surface of the cast slab will aggravate the propagation of cracks due to stress concentration during the heating process. It is analyzed that the occurrence of pit defects on the surface of the cast slab is related to the N added in the sub-peritectic steel. LAX340Y410T has strong crack sensitivity, and slow cooling of the billet can effectively prevent stress cracks from occurring. Through the optimization and adjustment of the process, the nitrogen circulation is eliminated during the smelting process, and the slab slow cooling process is added, which can effectively prevent stress cracks. The test results show that the pass rate of round steel surface flaw detection has increased from 3.13% in the initial trial production to 90.2%, and the product quality has been greatly improved.

Key words: LAX340Y410T; nitrogen; slow cooling; flaw detection qualification rate; quality; improvement

9Ni 钢断口纤维率与其组织和成分关系的研究

张宏亮[1]，杜　林[1]，朱莹光[1]，侯家平[1]，李文竹[1]，赵启斌[2]

（1. 鞍钢集团钢铁研究院，辽宁鞍山　114009；
2. 鞍钢股份有限公司鲅鱼圈钢铁分公司，辽宁营口　115007）

摘　要：应用扫描电镜观察了 9Ni 钢冲击断口形貌和显微组织，测定了 9Ni 钢冲击断口纤维率，应用电子探针分析了不同断口纤维率冲击试样的元素偏析，研究了 9Ni 钢断口纤维率与其组织和成分关系，结果表明，断口纤维率为 60%和 75%的试样放射区为韧脆混合的准解理形貌，成分偏析是导致 9Ni 钢冲击断口纤维率低的主要原因，最主要的偏析元素是 Ni 元素。

关键词：9Ni 钢；断口纤维率；偏析；准解理

Study on the Relationship between Fracture Fiber Ratio and Microstructure and Composition of 9Ni Steel

Zhang Hongliang[1], Du Lin[1], Zhu Yingguang[1],
Hou Jiaping[1], Li Wenhu[1], Zhao Qibin[2]

(1. Ansteel Group Iron and Steel Research Institute, Anshan 114009, China;
2. Angang Steel Company Limited, Bayuquan Iron & Steel Subsidiary, Yingkou 115007, China)

Abstract: The impact fracture morphology and microstructure of 9Ni steel were observed by scanning electron microscopy, and the fiber ratio of 9Ni steel was measured. The elemental segregation of impact samples with different fiber ratio at different fracture sections was analyzed by using electron probe. The relationship between the fiber ratio of 9Ni steel fracture and its microstructure and composition was studied. The sample with fracture fiber ratio of 60% and 75% has a quasi cleavage morphology of ductile-brittle mixture. The main reason for the low impact fracture fiber ratio of 9Ni steel is composition segregation, and the most important segregation element is Ni element.

Key words: 9Ni steel; fracture fiber rate; segregation; quasi-cleavage

基于先进核能技术需求的特种合金材料研制

徐长征，马天军，欧新哲，徐文亮，敖　影

（宝武特种冶金有限公司 技术中心，上海　200940）

摘　要：核电是先进的清洁能源，是实现国家节能减排目标的最重要举措之一，“十四五”期间核能将搭上碳中和的快车，进一步明确了“在确保安全的前提下积极有序发展核电”的思路，目前我国已能独立生产制造百万千瓦级核电核岛的绝大部分主设备，但某些部件的关键材料仍依赖进口，成为核电自主化建设的瓶颈之一。本文重点介绍宝武特种冶金有限公司（及其前身宝钢特钢有限公司）“十三五规划”以来在压水堆核电技术涉及的蒸汽发生器下封头水室隔板用镍基合金厚板、爆破阀剪切盖用镍基合金大截面棒材、屏蔽主泵屏蔽套用镍基合金冷轧薄板、镍基合金焊接材料和高温气冷堆、钍基熔盐堆涉及的镍基合金板、管、棒等关键材料研制进展。

关键词：核电站；镍基合金；水室隔板；主泵屏蔽套；爆破阀剪切盖；高温气冷堆

Development of Special Alloys based on Advanced Nuclear Energy Technology Requirements

Xu Changzheng, Ma Tianjun, Ou Xinzhe, Xu Wenliang, Ao Ying

(Baowu Special Metallurgy Co., Ltd., Shanghai 200940, China)

Abstract: As an advanced clear energy, the nuclear energy is one of the most important measures to achieve the target of energy conservation and emission reduction. The nuclear energy will ride the express train of carbon neutrality in the 14th Five-year Plan, and the the idea of “actively and orderly developing nuclear power under the premise of ensuring safety” is also further clarified. Chinese equipment manufacturing enterprises are able to independently produce most of the important equipment used in the million kilowatt nuclear power station. However, some of the key materials are still depends on the imports, which becomes one of the most important bottlenecks in the independent development of nuclear power. The present passage mainly introduced the important development progresses in the pressurized water reactor nuclear power technology such as the nickel-based alloy plate used as the steam generator divider plate, large size nickel-base alloy rod used as squib valve shear cap, cold-rolling nickel-based alloy thin trip used as coolant pump can and nickel-based welding materials produced in BAOWU since the 13th Five-year Plan period. Moreover, the development progress of nickel-based plates, pipes and bars used in the high temperature gas-cooled reactor and thorium based molten salt reactor are also illustrated in the present work.

Key words: nuclear power station; nickel-based alloy; divider plate; coolant pump can; squib valve shear cap; high temperature gas-cooled reactor

Nb、V 对 440C 不锈轴承钢析出相影响的热力学分析

鸡永帅，李　阳，姜周华，马　帅，孙　萌，李立业

（东北大学冶金学院，辽宁沈阳　110819）

摘　要：基于 Thermo-Calc 热力学计算软件，研究不同 Nb、V 含量下 440C 不锈轴承钢析出相的析出温度及析出质量，并考虑了 N 含量的影响。计算结果表明：440C 钢中主要析出相为 BCC_A2 (α-Fe)、$M_{23}C_6$（富 Cr 碳化物）、金属间化合物(G_PHASE 和 BCC_A2#2)和 AlN，添加 Nb 或 V 后，分别析出新相 FCC_A1#2(富 Nb 碳化物)和 Z_PHASE（富 V 氮化物），并且随着加入量的提升，呈现上升趋势；不同温度下合金元素 Nb 含量的变化对 $M_{23}C_6$ 碳化物的析出量影响显著，而 V 和 N 影响较小，但 V 的加入明显增加 M_7C_3 稳定区域；不同温度下随着 Nb 含量的增加，在钢中通常以大尺寸网状结构存在的 $M_{23}C_6$ 富 Cr 碳化物逐渐减少，Z_PHASE（富 V 氮化物）随 N 含量的增加而增加。

关键词：Thermo-Calc 热力学软件；440C 轴承钢；铌钒合金化；析出相

Thermodynamic Analysis of the Influence of Nb and V on the Precipitation Phase in 440C Stainless Bearing Steel

Ji Yongshuai, Li Yang, Jiang Zhouhua, Ma Shuai, Sun Meng, Li Liye

(School of Metallurgy, Northeastern University, Shenyang 110819, China)

Abstract: Based on Thermo-Calc thermodynamic calculation software, the precipitation temperature and precipitation quality of the precipitation phase of 440C martensitic stainless bearing steel under different Nb, V are studied, and the effect of N element contents was considered. The calculation results show that the main precipitated phases in 440C steel are BCC_A2 (α-Fe), $M_{23}C_6$ (Cr-rich carbides), intermetallic compounds (G_PHASE and BCC_A2#2) and AlN. After adding Nb or V, new phases are precipitated respectively FCC_A1 #2 (Nb-rich carbide) and Z_PHASE (V-rich nitride), and present an upward trend with the increase of the added amount; the change of alloy element Nb content at different temperatures has a significant effect on the precipitation of $M_{23}C_6$ carbide, V and N has little effect, but the addition of V significantly increases the M_7C_3 stable area; with the increase of Nb content at different temperatures, $M_{23}C_6$ Cr-rich carbides gradually decreases, Z_PHASE (V-rich nitride) increases with the N content increase.

Key words: Thermo-Calc software; 440C bearing steel; niobium-vanadium alloying; precipitation phase

轴承钢中大尺寸夹杂物的特征、来源及改进工艺

龙　鹄[1]，成国光[2]，丘文生[1]，曾令宇[1]，余大华[1]，赵　科[1]，刘　栋[1]

（1. 宝武集团广东韶关钢铁有限公司技术研究中心，广东韶关　512123；
2. 北京科技大学，北京　10083）

摘　要：本文以韶钢 BOF- ArS (氩站)-LF-RH-CC 轴承钢生产工艺为研究背景，采用水浸超声探伤与扫描电镜相结合的方法研究了大尺寸夹杂物的特征和来源，并提出改进工艺。研究结果表明：大尺寸的夹杂物主要有两类，一种是含 6%~7%SiO_2 的低熔点的 $CaO-Al_2O_3-SiO_2$(C)，主要源于出钢过程低碱度渣的卷入；另一种是不含 SiO_2 的 $CaO-MgO-Al_2O_3$(C)l 类夹杂，主要来源为精炼过程高铝渣的卷入。基于此改进造渣工艺，在出钢过程提前加入高铝渣和石灰，将炉渣碱度控制在 5~9 范围，Al_2O_3 质量分数控制在 23%～28%。改进后炉渣流动性好，有效减少因炉渣卷入形成的低熔点大尺寸夹杂，水口结瘤现象得到改善，轧材中主要为细小的 $MgO-Al_2O_3$ 尖晶石及复合硫化物类夹杂，成品探伤合格率得到有效提升。

关键词：轴承钢；超声探伤；大尺寸夹杂物；精炼渣；流动性

Characteristics, Sources and Optimization of Large Size Inclusions in Bearing Steel

Long Hu[1], Cheng Guoguang[2], Qiu Wensheng[1], Zeng Lingyu[1], Yu Dahua[1], Zhao Ke[1], Liu Dong[1]

(1. Baowu Grounp Guangdong Shaoguan Iron and Steel Co., Ltd., Shaoguan, 512123, China;
2. University of Science and Technology Beijing, Beijing 10083, China)

Abstract: Large size inclusions have an important influence on the fatigue life of bearing steel. Based on the BOF-Ar Stirlling-LF-RH-CC process of bearing steel production in Shaogang, the characteristics and sources of inclusions were explored through the method of ultrasonic test combined with scanning electron microscope. The refining slag was optimized with the application of the slag design model, a new slag forming practice was proposed, and its improvement effect was studied. Results showed that there were mainly two kinds of large inclusions, with the composition of $CaO-Al_2O_3-SiO_2-C$ and $CaO-MgO-Al_2O_3$, respectively. The former was attributed to the entrapment of low basicity slag during steel tapping, which was difficult to float up with the property of low melting point. The latter was formed by the charge of large bulk of calcium-aluminate slag during refining process, which was difficult to be melted rapidly. In order to control the large size inclusion, new slag formation technology was applicated by feeding the large bulk of calcium-aluminate slag during tapping instead of refining. The slag with fine fluidity was formed rapidly, which obviously decreased the quantity of large size inclusions, and the residues were mainly composed of micro $MgO-Al_2O_3$ and composite sulfides. The cleanliness of bearing steel products evaluated by high frequency water immersion UST was significantly improved.

Key words: bearing steel; UST; large size inclusions; refining slag; liquidity

等轴晶和柱状晶对铁素体不锈钢织构和性能的影响

杜 伟，陈 旭

（宝山钢铁股份有限公司中央研究院，上海 201900）

摘 要： 为了研究凝固组织对铁素体不锈钢织构和冲压性能的影响，作者采用 XRD 和 EBSD 等分析手段详细研究了宏观织构演化和成品板的微观取向。结果发现：相对于柱状晶的试样，等轴晶试样展示了更高强度的{111}织构，且不同取向的晶粒分布更加弥散。等轴晶要比柱状晶的起皱更低，这主要同柱状晶试样中存在明显的晶粒簇有关。另外，经最终的再结晶退火后，等轴晶试样也展示了更好的 r 值，这同等轴晶试样形成了强烈的 γ 纤维织构紧密相关。

关键词： 凝固组织；铁素体不锈钢；深冲性；起皱；织构

Effect of Equiaxed and Columnar Grains on Texture and Drawability of Ferritic Stainless Steel

Du Wei, Chen Xu

(Central Research Institute, Baoshan Iron & Steel Co., Ltd., Shanghai 201900, China)

Abstract: In order to understand the effect of solidified structure on texture and deep-drawing property, texture evolution were studied in detail using X-ray diffraction and Electron Back Scatter diffraction technique. The result showed that equiaxedgrain specimen had the more high intensity of {111}-textureand uniformly distributed grains with different orientationscompared to the columnar grain specimen. In addition, the sheet of equiaxed grains specimen showed excellent ridging resistance, which mainly attributed to the formation of grain colonies in columnar grain specimen. After final recrystallization annealing, more higher intensity γ-fiber texture and excellent r-value was obtained for equiaxed grain specimen.

Key words: solidification structure; ferritic stainless steel; drawability; ridging; texture

组织配分对双相不锈钢点蚀扩展的影响

汪毅聪，胡 骞，黄 峰，刘 静

（武汉科技大学 省部共建耐火材料与冶金国家重点实验室，
武汉科技大学 湖北省海洋工程材料及服役安全工程技术研究中心 湖北武汉 430081）

摘 要： 通过选取铁素体和奥氏体占比不同的 2205 双相不锈钢试样，采用不同外加电位下进行的恒电位极化测试方法，结合超景深三维显微镜对蚀坑形貌进行表征，研究了双相不锈钢的点蚀扩展方式及影响因素。结果表明，双相不锈钢点蚀在表面方向上沿轧向的扩展速度更快，呈现浅宽形形貌特点。随着外加电位升高，点蚀沿各方向的扩展速率均增大。但沿表面方向扩展速率的提升较大，深度方向扩展速率的提升较小。在相同的外加电位下，基体组织耐蚀性更好的试样，其点蚀花边盖破裂程度更高，点蚀更倾向沿表面方向扩展。

关键词： 双相不锈钢；组织配分；点蚀扩展

Effect of Microstructure Partition on Pitting Propagation of Duplex Stainless Steel

Wang Yicong, Hu Qian, Huang Feng, Liu Jing

(The State Key Laboratory of Refractories and Metallurgy, Huibei Engineering Technology Research Centre of Marine Materials and Service Safety, Wuhan University of Science and Technology, Wuhan, 430081, China)

Abstract: The pitting propagation behavior and influencing factors of 2205 duplex stainless steels with different proportions of ferrite and austenite were investigated by potentiostatic polarization test under different applied potentials and three-dimensional microscope. Results show that the pitting corrosion of duplex stainless steel propagates faster along the rolling direction on surface and presents the characteristics of shallow and wide shape. With the increase of applied potential, the pitting propagation rate increases along all directions. However, the increase of the propagation rate along the surface direction is larger than that in the depth direction. Under the same applied potential, the fracture degree of the pitting lace cover of the sample with better corrosion resistance is higher, and the pitting corrosion tends to propagate along the surface direction.

Key words: duplex stainless steel; microstructure partition; pitting propagation

大厚度超低温压力容器用 3.5Ni 钢板的开发

张 朋，龙 杰，庞辉勇，罗应明，张晓华

（河钢集团舞钢公司，河南舞钢 462500）

摘 要： 本文针对国内合成氨设备及石油加工制造压力容器设备的需要，对大厚度超低温压力容器用 3.5Ni 钢进行了开发，并进行了系列分析。结果表明，所开发的超低温压力容器用 3.5Ni 钢最大厚度达到 90mm，钢板的强度和 −100～−120℃低温韧性优良，同时低温落锤等试验结果低于−105℃，达到设备技术要求。工业大批量生产后性能稳定，满足了国内合成氨及石化加工低温容器的用钢需求。

关键词：超低温压力容器；3.5Ni；NDTT；低温韧性

The Development of Large Thickness Steel 3.5Ni for Ultra-low Temperature Pressure Vessels

Zhang Peng, Long Jie, Pang Huiyong, Luo Yingming, Zhang Xiaohua

(HBIS Wuyang Iron and Steel Co., Ltd., Wugang 462500, China)

Abstract: In this paper, large thickness steel 3.5Ni for ultra-low temperature pressure vessels is developed and analyzed, according to the needs of domestic ammonia equipment and pressure vessel equipment for petroleum processing. The results show that the maximum thickness of the developed steel used in the ultra-low temperature pressure vessel is 90mm. The strength and the low temperature toughness of −100 ~ −120 ℃ of the steel plate are excellent. The test results of NDTT is less than -105℃. The performance of the steel after large-scale industrial production is stable, which meets the technical requirements of domestic synthetic ammonia and petrochemical processing low temperature vessels.

Key words: ultra-low temperature pressure vessels; 3.5Ni; NDTT; low-temperature toughness

稀土钇对 654SMO 超级奥氏体不锈钢凝固组织影响

肖　俊，张　月，崔译夫，郄镕鉴，赵爱民

（北京科技大学钢铁共性技术协同创新中心，北京　100083）

摘　要：采用热力学计算与实验相结合，借助高温共聚焦显微镜、EBSD、光学显微镜和场发射扫描电镜重点探讨钢中添加稀土钇对超级奥氏体不锈钢凝固组织以及固态相变的影响，并对第二相 σ 析出进行了表征。结果表明，钢中主要析出相为 σ 相，在凝固过程中，Cr 和 Mo 元素偏析越来越严重，形成了 σ 相周围存在 Cr 和 Mo 元素贫化区。添加稀土 Y 能提高超级奥氏体不锈钢的等轴晶率，降低铸件边缘和 *R*/2 处共析组织体积分数。通过对 654SMO 凝固过程的原位观察，表明稀土 Y 可以提高形核温度，减小凝固温度范围，均匀并细化凝固组织，并发现了共析组织的原位析出过程。

关键词：超级奥氏体不锈钢；稀土钇；凝固；析出相

The Effect of Rare Earth Yttrium on the Solidification Structure of 654SMO Super Austenitic Stainless Steel

Xiao Jun, Zhang Yue, Cui Yifu, Qie Rongjian, Zhao Aimin

(University of Science and Technology Beijing, Beijing 100083, China)

Abstract: Combining thermodynamic calculations with experiments, with the help of high-temperature confocal

microscope, EBSD, optical microscope and field emission scanning electron microscope, the effect of adding rare earth yttrium in steel on the solidification structure and solid phase transformation of super austenitic stainless steel is focused on, and the effect of the second phase The σ precipitation was characterized. The results show that the main precipitated phase in the steel is the σ phase. During the solidification process, the segregation of Cr and Mo elements becomes more and more serious, revealing the existence of Cr and Mo element depleted areas around the σ phase. The addition of rare earth Y can increase the equiaxed crystal ratio of super austenitic stainless steel and reduce the volume fraction of eutectoid structure at the edge of the casting and at *R*/2. Through in-situ observation of the solidification process of 654SMO, it is shown that rare earth Y can increase the nucleation temperature, reduce the solidification temperature range, uniform and refine the solidification structure, and found the in-situ precipitation process of the eutectoid structure.
Key words: super austenitic stainless steel; rare earth yttrium; solidification; precipitation phase

热轧板退火温度对含 Sn 铁素体不锈钢力学及耐腐蚀性能的影响

白　杨[1]，贺　彤[2]，杨卫波[3]，吴　纯[1]，高　宇[1]

(1. 辽宁工程技术大学材料科学与工程学院，辽宁阜新　123000；2. 东北大学研究院，辽宁沈阳　110004；3. 中国建材检验认证集团股份有限公司，北京　100024)

摘　要：本文以含 Sn 铁素体不锈钢为研究对象，探究了热轧板退火温度对实验钢耐腐蚀性能和力学性能的影响规律。采用电化学工作站测定极化曲线和电化学阻抗谱，采用扫描电镜观察实验钢在 35℃、6wt.% $FeCl_3$ 溶液的腐蚀形貌；采用电子拉伸试验机测定抗拉强度、屈服强度和延伸率。结果表明：调整和控制热轧板退火温度，使成品板获得细小均匀的再结晶组织，是提高含 Sn 铁素体不锈钢耐腐蚀性能和力学性能的有效途径。热轧板退火温度在 900~1050℃范围内，当热轧板退火温度为 950℃时，冷轧退火板形成了均匀细小的等轴铁素体晶粒，有利于增强钝化膜的稳定性并改善力学性能，此时点蚀电位达到最大值 0.28V，抗拉强度达到最大值 531.8MPa，断后延伸率为 42.5%，实验钢兼具优异的耐腐蚀性能和力学性能。
关键词：铁素体不锈钢；耐蚀性；力学性能；热轧板退火

Effect of Hot Band Annealing Temperature on Mechanical Properties and Corrosion Resistance of Sn-containing Ferritic Stainless Steel

Bai Yang[1], He Tong[2], Yang Weibo[3], Wu Chun[1], Gao Yu[1]

(1. College of Materials Science and Engineering, Liaoning Technical University, Fuxin 123000, China;
2. Research Academy, Northeastern University, Shenyang 110004, China;
3. China building materials inspection and Certification Group Co., Ltd., Beijing 100024, China)

Abstract: In this paper, Sn-containing ferritic stainless steel was studied as a research object, and the effect of hot band annealing temperature on corrosion resistance and mechanical properties of Sn-containing ferritic stainless steel was explored. The polarization curve and electrochemical impedance spectrum of the experimental steel were measured by electrochemical workstation, and the corrosion morphology was observed by scanning electron microscope in 35℃、6wt.%

$FeCl_3$ solution; Thethe tensile strength, yield strength and elongation were measuredby using the electronic tensile testing machine. The results showed that it is an effective way to improve the corrosion resistance and mechanical properties of Sn-containing ferritic stainless steel to adjust and control the HBA temperature, and make the final sheet obtain fine and uniform recrystallization structure. When the HBA temperature is within the range of 900~1050℃, the uniform and fine ferrite grains were formed in cold rolled and annealed sheets with fewer surface defects when the HBA temperature is 950℃, which is conducive to enhancing the stability of passivation film and improving the mechanical properties. The pitting potential reached a maximum value of 0.28V, and the tensile strength reached a maximum value of 531.8MPa，the elongation reached 42.5%. The experimental steel had both excellent corrosion resistance and mechanical properties.

Key words: ferritic stainless steel; corrosion resistance; mechanical properties; hot band annealing

增材制造马氏体不锈钢性能预测

吴灵芝[1]，张　聪，尹海清[1]，苏　杰[2]，张瑞杰，王永伟，姜　雪

（1. 北京科技大学钢铁共性技术协同创新中心，北京　100083；
2. 钢铁研究总院特殊钢研究所，北京　100081）

摘　要：马氏体不锈钢作为超高强度钢在航空航天、核能等行业发挥着关键性作用。本文基于增材制造马氏体不锈钢的本征特性，研究组织特征及力学性能的预测模型，包括裂纹敏感性的预测、马氏体转变温度、硬度及屈服强度的定量预测，综合考虑本征应力及固溶强化、位错强化、细晶强化以及沉淀强化等多种强化机制的作用。开展增材制造粉床熔化打印实验，对模型进行验证，模型与实验值基本吻合，硬度模型相对误差≤4%之内，强度模型相对误差≤4%。

关键词：增材制造；马氏体不锈钢；组织特征；性能预测

Property Prediction of Martensitic Stainless Steel in Additive Manufacturing

Wu Lingzhi[1], Zhang Cong, Yin Haiqing[1], Su Jie[2], Zhang Ruijie, Wang Yongwei, Jiang Xue

(1. Institute of Collaborative Innovation Center of Steel Technology, University of Science and Technology Beijing, Beijing 100083, China; 2. Center Iron & Research Institute, Beijing 100081, China)

Abstract: As an ultra-high strength steel, martensitic stainless steel plays an increasingly critical role in aerospace, nuclear energy industries.In this paper, based on the intrinsic characteristics of additive manufacturing of martensitic stainless steela variety of performance prediction models have been studied:predictthe sensitivity of hot cracking (or hot tearing), the quantitative prediction of martensite transformation temperature, hardness and yield strength.The strength model comprehensively considers the combined effects of intrinsic stress, solid solution strengthening, dislocation strengthening, fine grain strengthening, and precipitation strengthening.Conduct material manufacturing powder bed melting experiment, compared with the model, the model is basically consistent with the experimental value. The relative error of the hardness model is within 4%, and the relative error of the strength model is less than 4%.

Key words: additive manufacturing; martensitic stainless steel; microstructure characteristics; property prediction

良好折弯成型性能专用车厢体轻量化耐磨钢 NM500 关键技术研发及应用

刘红艳[1,2]，陈子刚[1]，邓想涛[2]，管连生[1]，徐桂喜[1]

（1. 河钢集团邯钢公司，河北邯郸　056015；
2. 东北大学轧制技术及连轧自动化国家重点试验室，辽宁沈阳　110819）

摘　要：专用车车厢采用低合金耐磨钢 NM500 材料替代 NM400 车厢材料可减重 10%~30%。采用高压段+低压段两段式控温淬火冷却工艺，控制马氏体尺寸均匀性，获得良好成型性能。高压段大于临界冷速快速冷却至 500℃以下，发生马氏体相变，低压段中低冷速间隔式冷却至室温，在冷却集管开启段实现马氏体相变，在冷却集管关闭段实现内应力释放，使马氏体相变和内应力释放交替进行，实现"低压段"相变均匀、内应力释放充分。该工艺生产的专用车厢体轻量化 NM500 近表位置马氏体板条宽度范围 0.1~0.5μm，心部位置马氏体板条宽度范围 0.3~0.8μm，抗拉强度差值范围是 2~15MPa，延伸率差值范围 0.1%~1.5%，表面布氏硬度差值范围 2~8HBW，马氏体板条尺寸、力学性能通板分布均匀，钢板在拉伸、切割、U 型折弯时获得各向均匀的延伸，保证加工成型的一致性。切割窄条后钢板平直度范围 1~4mm/2m，U 型折弯后折弯端部直线度不大于 2mm，具有良好的成型性能。

关键词：折弯成型；专用车厢体；耐磨钢 NM500；轻量化

Research and Application of Key Technology of NM500 Special Car Body Lightweight Wear Resistant Steel with Good Bending Performance

Liu Hongyan[1,2], Chen Zigang[1], Deng Xiangtao[2], Guan Liansheng[1], Xu Guixi[1]

(1.HBIS Group Hansteel Company, Handan 056015, China; 2.State Key Laboratory of Rolling Technology and Continuous Rolling Automation, Northeast University, Shenyang 110819, China)

Abstract: Using low alloy wear-resistant steel NM500 instead of NM400 can reduce weight by 10%~30%. The high pressure section and low pressure section two-stage temperature control quenching process is adopted to control the size uniformity of martensite and obtain good formability. The martensitic transformation occurs when the high pressure section is cooled to below 500℃ faster than the critical cooling rate, and the low pressure section is cooled to room temperature with low cooling rate interval. The martensitic transformation is realized in the opening section of the cooling header, and the internal stress release is realized in the closing section of the cooling header, so that the martensitic transformation and internal stress release are carried out alternately, and the "low pressure section" has uniform transformation and sufficient internal stress release. The width range of martensite lath near the surface is 0.1~0.5μm, the width range of martensite lath at the center is 0.3~0.8μm, the difference range of tensile strength is 2~15MPa, the difference range of elongation is 0.1%~1.5%, the difference range of surface Brinell hardness is 2~8HBW, the size and mechanical properties of martensite lath are evenly distributed through the plate, and the steel plate is in the process of drawing, cutting and bending The uniform extension in all directions is obtained during U-bending to ensure the consistency of processing. After cutting narrow strip, the flatness range of steel plate is 1~4mm/2m, and the straightness of bending end is not more than 2mm after U-bending, which has good forming performance.

Key words: bending forming; special carriage body; wear resistant steel NM500; lightweight

大型低温储罐用 13MnNi6-3 钢的开发

牟　毓，谢章龙，潘中德，姜金星，吴俊平
（南京钢铁股份有限公司，江苏南京　210035）

摘　要：本文介绍了大型低温储罐用 13MnNi6-3 钢的研制和开发，南京钢铁股份有限公司通过优化成分设计和冶炼、轧制、热处理工艺，成功开发了 12~38mm 不同厚度规格的 13MnNi6-3 钢板，试验结果表明，试制钢板内部质量好，夹杂物含量低，金相组织为均匀细小的铁素体+少量珠光体组织，钢板及焊接接头均可以满足–60℃低温冲击韧性，产品具有优异的可焊接性能。

关键词：13MnNi6-3 钢；化学成分；生产工艺；低温冲击韧性；可焊性

Research and Development of 13MnNi6-3 Steel for Large Low Temperature Storage Tank

Mu Yu, Xie Zhanglong, Pan Zhongde, Jiang Jinxing, Wu Junping
(Nanjing Iron & Steel Co., Ltd, Nanjing 210035, China)

Abstract: This article introduces the research and development of 13MnNi6-3 steel for large-scale cryogenic storage tanks. Nanjing Iron and Steel Co., Ltd. has successfully developed 13MnNi6-3 steel plates with different thickness specifications from 12 to 38 mm through optimized composition design and smelting, rolling, and heat treatment processes. The test results show that the internal quality of the trial steel plate is good, the content of inclusions is low, and the metallographic structure is uniform and fine ferrite + a small amount of pearlite. Both steel plates and welded joints can meet the low temperature impact toughness of –60℃, and the products have excellent weldability.

Key words: 13MnNi6-3 steel; chemical composition; production process; low temperature impact toughness; weldability

7.5　高温合金

熔融制样-X 射线荧光光谱同时测定硅锰合金中锰、硅、磷和钛

张延新，李　京，刘　斌
（青岛特殊钢铁有限公司试验检测所，山东青岛　266043）

摘　要：熔融制样-X 射线荧光光谱法测定硅锰合金的关键点在于玻璃熔片的制备，需保证样品熔解完全和避免浸

蚀铂黄坩埚。本法有效避免熔片制备过程中对铂黄坩埚的浸蚀，简化操作步骤、提升熔片过程的自动化程度，具有安全环保、成本低、速度快和准确度高的优点。选取 12 种国家标准样品建立校准曲线，线性相关系数 0.998~0.999。对硅锰合金标准物质进行精密度试验，各组分测定结果相对标准偏差（RSD，n=5）为 0.31%~2.43%；应用该实验方法与化学法[1]同时对硅锰合金试样进行测定，结果再现性满足相应国标规定。

关键词：X 射线荧光光谱法（XRF）；硅锰合金；氧化剂；熔融制样

Determination of Manganese, Silicon, Phosphorus and Titanium in Silicomanganese by X-ray Fluorescence Spectrometry with Fusion Samplepreparation

Zhang Yanxin, Li Jing, Liu Bin

(Testing and Detection Institute of Qingdao Special Iron and Steel Co., Ltd., Qingdao 266043, China)

Abstract: The key to the determination of Silicomanganese by X-ray fluorescence spectrometry lies in the preparation of glass melting plate. It is necessary to ensure the melting of the sample completed and avoid the etching of platinum-gold crucible.That this method can effectively avoid the corrosion of platinum-gold crucible in the process of melting sheet preparation, simplify the operation steps, improve the automation degree of melting sheet process, and has the advantages of safety and environmental protection, low cost, fast speed and high accuracy. 12 kinds of national standard samples were selected to establish the calibration curve, and the linear correlation coefficient was 0.998~0.999. The relative standard deviation (RSD, n = 5) of the determination results of each component is 0.29%~5.62% .The experimental method and chemical method [1] were used to determine Silicomanganese samples at the same time, and the reproducibility of the results met the corresponding national standard.

Key words: X-ray fluorescence spectrometry(XRF); silicomanganese; oxidant; fusion

稀土渣料对电渣重熔高温合金夹杂物的影响研究

高小勇[1]，张立峰[1,2]，曲选辉[3]，章　林[3]

（1. 燕山大学机械工程学院，河北秦皇岛　066004；
2. 燕山大学亚稳材料制备技术与科学国家重点实验室，河北秦皇岛　066004；
3. 北京科技大学新材料技术研究院，北京　100083）

摘　要：采用含有不同含量稀土氧化物的渣料对镍基高温合金进行电渣重熔。采用带有能谱仪的扫描电子显微镜和夹杂物自动扫描仪对自耗电极和电渣锭中的夹杂物进行表征。研究结果表明：采用不含稀土氧化物的常规渣料时，夹杂物的数量增加，夹杂物的成分由电极中的 $MgO-Al_2O_3$ 转变为电渣锭中的 $MgO-Al_2O_3-CaO$，夹杂物的尺寸减小；采用稀土渣料时，夹杂物的数量减少，夹杂物的成分转变为 $MgO-Al_2O_3-CaO-Ce_2O_3$，夹杂物的尺寸进一步减小。稀土渣料可以有效减少高温合金中的夹杂物，提高纯净度。

关键词：高温合金；电渣重熔；夹杂物；稀土渣；纯净度

Effect of Rare Earth Bearing Slag on Inclusion of Electroslag Remelted Superalloy

Gao Xiaoyong[1], Zhang Lifeng[1,2], Qu Xuanhui[3], Zhang Lin[3]

(1. School of Mechanical Engineering, Yanshan University, Qinhuangdao 066004, China; 2. State Key Laboratory of Metastable Materials Science and Technology, Yanshan University, Qinhuangdao 066004, China; 3. Institute for Advanced Materials and Technology, University of Science and Technology Beijing, Beijing 100083, China)

Abstract: Electroslag remelting of nickel-based superalloy was carried out with slags containing different rare earth oxides. Inclusions in the consumable electrode and electroslag ingots were characterized by manual and automatic scanning electron microscope. The results show that the number of inclusions increased, the composition of inclusions changed from $MgO-Al_2O_3$ of the electrode to $MgO-Al_2A_3-CaO$, and the size of inclusions decreased when using conventional slags without rare earth oxide. However, the number of inclusions decreased, the composition of inclusions changed to $MgO-Al_2O_3-CaO-Ce_2O_3$, and the size of inclusions further decreased when using rare earth bearing oxide. Rare earth bearing slag can effectively remove the inclusions in superalloys and therefore improve the purity.

Key words: superalloy; electroslag remelting; inclusion; rare earth bearing slag; purity

$Ni_{1.5}CrMnFe_{0.5}Al_x$高熵合金的时效强化现象与机理

倪　倩，刘　意，甘章华，吴传栋，刘　静

（武汉科技大学材料与冶金学院，湖北武汉　430081）

摘　要：本文通过真空电弧熔炼制备 $Ni_{1.5}CrMnFe_{0.5}Al_x$（x=0,0.1,0.2,0.3）高熵合金，对合金样品进行物相分析、微观组织观察以及硬度测试，探究了不同的时效温度对不同铝含量的高熵合金微观组织结构及合金硬度的影响。研究结果表明，铸态 $Ni_{1.5}CrMnFe_{0.5}Al_x$ 高熵合金样中随着 Al 的添加产生 BCC 相。时效处理后的 $Ni_{1.5}CrMnFe_{0.5}Al_x$ 高熵合金产生 Ni-Al 析出相，合金中枝晶组织且随 Al 含量的增加而细化；随着时效温度的提高，合金的硬度相较于铸态有较大幅度增加，其中 $Ni_{1.5}CrMnFe_{0.5}Al_{0.3}$ 合金样品在 625℃、4h 时效处理后硬度最高达到 540.83HV，相比铸态提升 168%。

关键词：$Ni_{1.5}CrMnFe_{0.5}Al_x$ 高熵合金；显微结构；时效强化；析出相；显微硬度

Direct Aging Strengthening Phenomenon and Mechanism of $Ni_{1.5}CrMnFe_{0.5}Al_x$ High-entropy Alloys

Ni Qian, Liu Yi, Gan Zhanghua[1], Wu Chuandong, Liu Jing

(College of Materials and Metallurgy, Wuhan University of Science and Technology, Wuhan 430081, China)

Abstract: In this paper, $Ni_{1.5}CrMnFe_{0.5}Al_x$ (x=0,0.1,0.2,0.3) high entropy alloy was prepared by vacuum arc melting. The phase analysis, microstructure observation and hardness test of the alloy samples were carried out. The effects of aging temperature on the microstructure and hardness of the high entropy alloy with different aluminum content were investigated. The results show that the BCC phase is formed in the as-cast $Ni_{1.5}CrMnFe_{0.5}Al_x$ alloy with the addition of Al. After aging treatment, $Ni_{1.5}CrMnFe_{0.5}Al_x$ high entropy alloy produces Ni-Al precipitates, and the dendritic structure in the alloy is refined with the increase of Al content. With the increase of aging temperature, the hardness of the alloy increases greatly compared with that of the as-cast alloy, in which the hardness of the $Ni_{1.5}CrMnFe_{0.5}Al_{0.3}$ alloy sample reaches 540.83HV after aging at 625℃ for 4h, which is 168% higher than that of the as-cast alloy.

Key words: $Ni_{1.5}CrMnFe_{0.5}Al_x$ high-entropy alloy; microstructure; aging strengthening; precipitated phase; microhardness

7.6 电工钢

减轻高牌号无取向电工钢冷轧边裂的工艺措施

刘文鹏，高振宇，李亚东，陈春梅，孙　超，张智义，姜福健

（鞍钢集团钢铁研究院，辽宁鞍山　114009）

摘　要：本文结合生产实践，对高牌号无取向电工钢边裂产生原因进行分析，分析结果表明，边部不均匀性是边裂产生的主要诱因，针对分析结果，在各生产工序引入相应的工艺改进措施，有效解决了相关高牌号无取向电工钢的边裂问题，提升了冷轧工序成材率。

关键词：高牌号无取向电工钢；冷轧；边裂；成材率

The Measures to Reduce Edge Cracks of High Grade Non-oriented Electrical Steel in Cold Rolling

Liu Wenpeng, Gao Zhenyu, Li Yadong, Chen Chunmei, Sun Chao, Zhang Zhiyi, Jiang Fujian

(Iron and Steel Research Institutes of AnSteel Group Corporation, Anshan 114009, China)

Abstract: The paper has made analysis on the causes of edge cracks of high grade non-oriented electrical steel according to the production practice of Ansteel. Results shows that the inhomogeneities of edge part are the main cause of edge cracks. The edge crack problem is solved and the cold rolling yield is lifted after corresponding improvement measures are introduced in each production process.

Key words: high grade non-oriented electrical steel; cold-rolling; edge crack; yield

低温高磁感取向硅钢脱碳退火工艺与底层的研究

陈文聪[1]，喻　越[1]，桂　虎[1]，刘　敏[2]，李　伟[1]，骆新根[2]

（1. 宝钢股份武汉钢铁有限公司，湖北武汉　430080；
2. 国家硅钢工程技术研究中心，湖北武汉　430080）

摘　要：采用常规低温高磁感取向硅钢钢种，开展脱碳退火速度工艺试验，研究了不同脱碳退火速度及炉内气氛对脱碳板参数和成品底层质量的影响。结果表明，随着工艺速度的提升会导致退火钢板的氧化层减薄。通过对炉内气氛的适当调节，得到一种既能提高退火工艺速度，又能够优化最终底层质量的脱碳渗氮工艺，最终达到节能降耗的目的。

关键词：低温高磁感取向硅钢；脱碳退火；炉内气氛；退火工艺速度

Study on the Decarburization Annealing Process and Bottom Layer of Low Temperature High Magnetic Induction Oriented Silicon Steel

Chen Wencong[1], Yu Yue[1], Gui Hu[1], Liu Ming[2], Li Wei[1], Luo Xingen[2]

(1. Baosteel Wuhan Iron and Steel Co., Ltd., Wuhan 430080, China;
2. National Engineering Research Center for Silicon Steel, Wuhan 430080, China)

Abstract: The conventional low temperature and high magnetic induction oriented silicon steel grades were used to carry out the decarburization annealing speed process test. The effects of different decarburization annealing speeds and furnace atmosphere on the decarburization plate parameters and the quality of the finished bottom layer were studied. The results show that as the process speed increases, the oxide layer of the annealed steel sheet is thinned. Through proper adjustment of the atmosphere in the furnace, a decarburization and nitriding process which can improve the annealing process speed and optimize the quality of the final bottom layer is obtained, and finally achieves the purpose of energy saving and consumption reduction.

Key words: low temperature, high magnetic induction oriented silicon steel; decarburization annealing; furnace atmosphere; annealing process speed

大功率电磁炉趋肤及邻近效应分析

吴　胜，熊金乐

（武汉钢铁有限公司硅钢部，湖北武汉　430080）

摘　要：大功率电磁炉在硅钢轧制、热处理等生产工艺中逐渐被广泛采用，电磁炉带钢加热相对于传统加热方式无明火、无废气排放、无噪音、无污染，输出功率精确可控、升温快、热效高，绿色环保及节能特点十分突出。随着电磁炉功率增加电缆载流量随之增加，大电流密度情况下的趋肤效应及临近效应也随之凸显。本文结合具体案例采用有限元法对趋肤效应及临近效应进行仿真分析，总结出电缆布置需避免的情况以杜绝类似故障。

关键词：电磁感应加热；趋肤效应；临近效应；有限元法

Analysis of Skin Effect and Proximity Effect of High-power High Frequency Induction Heating Furnace

Wu Sheng, Xiong Jinle

(Wuhan Iron & Steel Co., Ltd. Silicon Steel Department, Wuhan 430080, China)

Abstract: High-power induction cooker is gradually widely used in the silicon steel rolling, heat treatment and other production processes. Compared with the traditional heating method, the induction cooker strip steel heating has no open fire, no exhaust gas emissions, no noise, no pollution, accurate and controllable output power, fast heating, high thermal efficiency, green environmental protection and energy saving features are very prominent. With the power of induction cooker increasing, the cable carrying capacity will increase, and the skin effect and proximity effect will also become prominent in the case of high current density. In this paper, the skin effect and proximity effect are simulated and analyzed by using finite element method combined with specific cases, and the cable layout is summarized to avoid similar faults.

Key words: induction heating; skin effect; proximity effect; finite element method

3.1%Si 取向硅钢显微组织和宏观织构演变研究

董丽丽[1,2]，黄　利[2]，刘宝志[3]，卢晓禹[2]，张　浩[3]，麻永林[1]

（1. 内蒙古科技大学材料与冶金学院，内蒙古包头　014010；
2. 内蒙古包钢钢联股份有限公司技术中心，内蒙古包头　014010；
3.包头市威丰稀土电磁材料股份有限公司，内蒙古包头　014010）

摘　要：本文以 3.1%Si 取向硅钢为研究对象，利用蔡司显微镜、X 射线衍射仪等检测设备分析检测取向硅钢一次冷轧、脱碳退火、二次冷轧、高温退火、拉伸平整退火阶段显微组织和宏观 XRD 织构。研究表明，取向硅钢显微组织类型为铁素体，脱碳退火完成初次再结晶后平均晶粒尺寸为 10.23μm，经高温退火后完成二次再结晶，晶粒尺寸开始达到厘米级，经拉伸平整退火后晶粒更加均匀，平均晶粒尺寸为 2.3cm。取向硅钢经过脱碳退火后的主要织构类型为γ纤维织构，有少量的高斯织构{110}<001>，高温退火后宏观织构为高斯{110}<001>织构，经拉伸平整退火后，高斯织构达到最强最锋锐的程度。

关键词：取向硅钢；显微组织；织构；脱碳退火；高温退火

Evolution of Microstructure and Macrotexture of 3.1%Si Oriented Silicon Steel

Dong Lili[1,2], Huang Li[2], Liu Baozhi[3], Lu Xiaoyu[2], Zhang Hao[3], Ma Yonglin[1]

(1. School of Material and Metallurgy, Inner Mongolia University of Science and Technology, Baotou 014010, China; 2. Technology Center of Inner Mongolia Baogang Steel Union Co., Ltd., Baotou 014010, China; 3. Baotou Winfiner Tombarthite Magnetism Material Co., Ltd., Baotou 014010, China)

Abstract: In this paper, the microstructure and macroscopic XRD texture of 3.1%Si oriented silicon steel were analyzed and detected by using Zeis microscope, X-ray diffraction and other testing equipment during the first cold rolling, decarbonization annealing, second cold rolling, high temperature annealing and tensile leveling annealing. The results show that the microstructure of oriented silicon steel is ferrite. The average grain size is 10.23μm after the first recrystallization after decarbonization annealing, and the grain size begins to reach centimeter level after the second recrystallization after high temperature annealing. The average grain size is 2.3cm after the tensile leveling annealing. The main texture type of oriented silicon steel after decarburization annealing is γ fiber texture, a small amount of Goss texture {110}<001>, high temperature annealing macroscopic texture is Goss {110}<001> texture, after tensile flat annealing, Goss texture reaches the strongest and sharpest degree.

Key words: oriented silicon steel; microstructure; texture; decarburization annealing; high temperature annealing

常化温度对无取向电工钢组织和性能的影响

姚海东，张保磊，胡志远，程　林，马　琳，李泽琳

（首钢智新迁安电磁材料有限公司，河北迁安　064404）

摘　要： 无取向电工钢广泛应用于电力、航空航天、新能源汽车等领域，为满足各行业对电工钢电磁性能的需求，常化技术已经应用到无取向电工钢的生产当中，本文通过实验设计，研究常化温度对无取向电工钢组织及性能的影响。研究结果表明：常化板织构特征呈现织构梯度分布，中心层分布的主要是{100}<021>、{114}<481>、{112}<421>等α*取向晶粒，Goss和黄铜等取向晶粒则主要分布于表层及次表层区域。经3min常化处理，常化温度由900℃提升到1000℃，常化板晶粒尺寸增加，成品板晶粒尺寸增加，铁损值P1.5/50降低约0.106W/kg，成品板中心层{100}取向晶粒体积分数增强，{111}和{112}取向晶粒体积分数降低，织构得以改善，磁感B5000升高约0.0167T。

关键词： 无取向电工钢；常化温度；组织；织构

Influence of Normalizing Temperature on Microstructure and Properties of Non-oriented Electrical Steel Sheets

Yao Haidong, Zhang Baolei, Hu Zhiyuan, Cheng Lin, Ma Lin, Li Zelin

(Shougang Zhixin Qian'an Electromagnetic Materials Co., Ltd., Qian'an 064404, China)

Abstract: Non-oriented electrical steels are widely used in electric power, aerospace, new energy vehicles and other fields. In order to meet the needs of various industries for the electromagnetic properties of electrical steels, normalization technology has been applied to the production of non-oriented electrical steels. Through experimental design, this paper studies the influence of normalization temperature on the microstructure and properties of non-oriented electrical steels.The results show that the texture characteristics of the normalized plate show gradient distribution , the center layer is mainly distributed in {100} < 021 >, {114} < 481 >, {112} < 421 > grain such as α orientation, Goss and brass orientation, such as grain size is mainly distributed in surface and subsurface areas. After 3min normalization treatment, the normalization temperature is raised from 900℃ to 1000℃, the grain size of normalized plate increased, the grain size of finished plate increased, the iron loss P1.5/50 decreased by about 0.106W/kg, the volume fraction of {100} oriented grains in the center layer of finished plate increased, the volume fraction of {111} and {112} oriented grains decreased, the texture was improved, and the magnetic induction B5000 increased by about 0.0167T.

Key words: unoriented electrical steel; normalizing temperature; organization; texture

磷酸盐类抗氧化剂对炭套耐氨气侵蚀性能的影响

何明生[1]，王雄奎[2]，谢文亮[2]，张　敬[2]，李胜金[2]，雷　勇[2]

（1. 武钢有限技术中心，湖北武汉　430080；2. 武钢有限硅钢部，湖北武汉　430083）

摘　要：在低温 HiB 钢连续退火过程中，炭套是支撑和输送钢带最好的炉底辊。但渗氮区炭套使用一段时间后表面侵蚀严重，表面粗糙度增大，出现边部磨损甚至结瘤，严重影响硅钢产品表面质量。根据炭套的运行条件，探讨了氨气对炭套侵蚀的原因和机理，以及抗氧剂对低温 HiB 钢表面质量的影响。磷酸盐对低温 HiB 钢连续退火炉炭套的耐氨气侵蚀性能较差，并不是最好的抗氧化剂选择。

关键词：炭套；耐氨气侵蚀；低温 HiB 钢；连续退火；渗氮

Effect of Phosphate Antioxidant on Resistance to Ammonia Erosion of Carbon Sleeve

He Mingsheng[1], Wang Xiongkui[2], Xie Wenliang[2], Zhang Jing[2], Li Shengjin[2], Lei Yong[2]

(1. R&D Center of Wuhan Iron & Steel Co., Ltd., Wuhan 430080, China; 2. Silicon Steel Division of Wuhan Iron & Steel Co., Ltd., Wuhan 430080, China)

Abstract: In continuous annealing furnace for low temperature grain-oriented silicon steel production, carbon sleeve is used as one kind of the best hearth rolls to support and convey steel strip. However, the surface of carbon sleeve in nitriding zone is seriously corroded after a period of time, the surface roughness of carbon sleeve increases, the edge wears and even buildups appear, which seriously affects the surface quality of products. Based on the working conditions of carbon sleeve, the causes and mechanism of corrosion on carbon sleeve by ammonia, and effects of antioxidants on surface quality of low temperature grain-oriented silicon steel are discussed. In terms of the resistance to ammonia, phosphate is not a good antioxidant for carbon sleeve in continuous annealing furnace for low temperature grain-oriented silicon steel production.

Key words: carbon sleeve; resistance to ammonia; low temperature HiB steel; continuous annealing; nitriding

取向硅钢新产品 B23R070 的开发和应用

马长松，章华兵，胡卓超

（宝山钢铁股份有限公司硅钢事业部，上海　201900）

摘　要：随着国家绿色发展、节能减排政策的落实，碳达峰、碳中和目标的推进和国家能效新标准的实施，这就要求变压器制造企业提高变压器能效等级，也促使取向硅钢生产企业提高产品等级。经过近几年快速发展，宝钢已经成为全球领先的取向硅钢制造商，形成六大系列、50 余个牌号取向硅钢产品系列，可以满足不同客户的使用需求，近年来宝钢开发出 B23R070 取向硅钢新产品，磁性能优异，目前已经实现批量化生产。本文对 B23R070 新产品性能和用户使用情况进行介绍，以期对变压器制造厂家选材起到参考作用。

关键词：宝钢；取向硅钢；B23R070

The Development and Application of New Grain Oriented Electrical Steel Product B23R070

Ma Changsong, Zhang Huabing, Hu Zhuochao

(Baoshan Iron & Steel Co., Ltd., Shanghai 201900, China)

Abstract: With the national green development, the promotion of energy conservation, action plan for peaking carbon dioxide emissions and carbon neutral and the implementation of new energy efficiency standards for transformers, the transformer manufacturers are required to improve the energy efficiency level of transformers and the grain oriented electrical steel manufacturers are also urged to improve the product level. After rapid development, Baosteel has become the world's leading manufacturer of grain oriented electrical steel, forming six series - more than 50 brands of products, which can meet the needs of different customers. In recent years, Baosteel has developed B23R070 new products with excellent magnetic properties. The new products have realized mass production. In this paper, the properties and transformer performance of B23R070 are introduced, expecting favorable for material selection of the transformer manufacturers.

Key words: baosteel; electrical steel; B23R070

全机架高速钢轧辊在无取向硅钢的应用技术研究

姜　南，何国赛，赵　敏，陈　猛

（武汉钢铁有限公司条材厂，湖北武汉　430083）

摘　要：经过多年研究和攻关，CSP 形成了系统性的硅钢板形控制技术，使硅钢原料板形得到了很好的保证。当出现板廓不良时，通过组织换辊改善断面质量，但是换辊频繁除增加轧辊消耗外，换辊开轧钢卷的稳定性也很难保证，增加异常钢卷比率。本文研究分析高速钢轧辊的特性，高速钢轧辊的磨削工艺调整，研究其对硅钢板廓的影响，大胆尝试在无取向硅钢上使用全机架高速钢轧辊，推进高速钢材质轧辊的应用对于提升硅钢质量具有重要意义，也是高精度硅钢轧制的方向。

关键词：硅钢；高速钢；硅钢板廓

Study on Application Technology of Full Frame HSS Roll in Non-oriented Silicon Steel

Jiang Nan, He Guosai, Zhao Min, Chen Meng

(Plant of Long Product, Wuhan Iron & steel Co., Ltd., Wuhan 430083, China)

Abstract: After many years of research, CSP has formed a systematic shape control technology of silicon steel, which ensures the shape of silicon steel raw material. When the plate profile is bad, the section quality can be improved through the organization of roll changing. However, in addition to increasing the consumption of rolls, frequent roll changing also makes it difficult to guarantee the stability of rolling coils, and increases the ratio of abnormal coils. In this paper, analyzing the characteristics of high speed steel roll, high speed steel roll grinding process adjustment, to study the influence on the silicon steel plate profile, bold attempt to use the whole frame of high speed steel roll in the absence of oriented silicon steel, to promote the application of high speed steel roll material is of great significance for improving the quality of silicon steel, is also the direction of high precision silicon steel rolling.

Key words: silicon stee; HSS; silicon steel plate profile

X 射线无损衍射断层三维晶体成像在电工钢织构表征中的应用

张　宁[1]，孙　骏[2]，孟　利[1]

（1. 钢铁研究总院冶金工艺研究所，北京　100081；
2. Xnovo Technology ApS，克厄，4600，丹麦）

摘　要： 作为首次利用实验室 X 射线无损衍射断层三维晶体成像（Lab-based DCT）技术表征取向硅钢超薄带，本工作旨在探讨取向硅钢超薄带的微观组织在不同热处理条件下的演变，借以论证 Lab-based DCT 技术在表征大范围样品体积的优势。利用 Lab-based DCT 技术对五个在不同退火条件下处理的硅钢超薄带样品的微观织构进行了对比分析，结果显示，不同退火条件下样品再结晶后的组织有明显差异。退火时间较短的样品晶粒尺寸分布均匀，Goss 和整体 η 线取向晶粒所占比例高。退火时间较长的样品中部分晶粒存在长大优势，导致微观组织不均匀，而且具有显著长大优势的晶粒多为不利于取向硅钢磁性能的杂取向。因此，在高性能取向硅钢超薄带的制备过程中，精确控制退火的参数条件是获得有利织构、提高磁性能的关键。

关键词： 电工钢；取向硅钢超薄带；X 射线无损衍射断层三维晶体成像；退火热处理；Goss 织构；η 线织构

Texture Characterization of Electrical Steels Using Laboratory-based Diffraction Contrast Tomography

Zhang Ning[1], Sun Jun[2], Meng Li[1]

(1. Metallurgical Technology Institute, Central Iron and Steel Research Institute, Beijing 100081, China;
2. Xnovo Technology ApS, Koege, 4600, Denmark)

Abstract: Being the very first study using Lab-based diffraction contract tomography (Lab-based DCT) for quantitative characterization of oriented electrical steels, the present work has investigated the microstructure and texture evolution of ultra-thin grain-oriented electrical steel samples with various annealing treatments. With the 3D grain maps obtained from Lab-based DCT, the macro-texture and grain structure of the five samples have been analyzed. It is revealed that different annealing treatments lead to significant difference in the microstructure. The samples with shorter annealing time exhibit homogeneous microstructure with strong Goss and η-fibre texture. With increasing annealing time, some grains showed growing advantage, resulting in non-homogeneous grain size across the sample. The grown grains also exhibit other orientations than η-fibre that can degrade the magnetic properties of ultra-thin oriented electrical steels.

Key words: electrical steel; ultra-thin oriented electrical steel; lab-based diffraction contrast tomography; annealing; Goss texture; η-fiber texture

含铜无取向电工钢退火过程中的显微组织和织构演变

石文敏[1]，李 准[1]，马金龙[2]，杨 光[1]，吕 黎[1]，黄景文[1]

（1. 宝钢股份中央研究院，湖北武汉 430080；2. 宝钢股份武钢有限硅钢部，湖北武汉 430080）

摘 要：对含≥1.0%铜无取向电工钢冷轧试样在700~980℃进行成品退火，采用背散射电子衍射对退火过程中的显微组织和织构演变进行了研究。试验结果表明含≥1.0%Cu 的高牌号无取向电工钢中，再结晶织构的形成主要取决于取向成核和孪晶成核机制。{110}<001>晶粒在{111}<112>和{112}<110>形变带内再结晶形核，在晶粒长大初期凭借其高角度晶界优势获得长大。随着近{111}取向晶粒在退火长大过程中逐渐形成晶团并最终贯穿整个厚度方向，其晶粒尺寸优势决定了其成为成品主要织构类型，而{110}<001>晶粒最终消失。

关键词：含铜无取向硅钢；退火；织构；EBSD

Microstructure and Texture Evolution during Annealing in Non-oriented Electrical Steel Containing Cu

Shi Wenmin[1], Li Zhun[1], Ma Jinlong[2], Yang Guang[1], Lü Li[1], Huang Jingwen[1]

(1. Baosteel Center Research Institute, Wuhan 430083, China;
2. Baosteel Silicon Business Unit, Wuhan 430080, China)

Abstract: The cold rolled non-oriented electrical steels containing ≥1.0%Cu were recrystallization annealing at temperatures varying from 700 to 980℃. The evolution of texture during annealing was investigated using electron backscatter diffraction (EBSD) techniques. The results show that the formation of recrystallization texture depends on orientation nucleation and twin nucleation mechanism in the high grade non-oriented electrical steels containing Cu. {110}<001>grains recrystallize and nucleate in {111}<112> and {112}<110> deformed bands, and grow by virtue of high angle boundaries advantage. The near {111} orientation grains grow gradually into grain colony and penetrate the thickness direction, and the grain size advantage makes them the main texture type in the finished samples. However, the grains with Goss orientation disappear finally.

Key words: non-oriented electrical steel containing Cu; annealing; texture; EBSD

高强度无取向电工钢疲劳性能及断裂机制

宋新莉，彭宇凡，李兆振，黄昌虎，潘子钦，刘 静

（武汉科技大学省部共建耐火材料与冶金国家重点实验室，湖北武汉 430081）

摘 要：本文测试了高强无取向电工钢的 *S-N* 曲线，并借助光学显微镜、扫描电子显微镜、透射电子显微镜分析了

试验钢组织，疲劳断口形貌和位错结构。结果表明：室温条件下，频率为 20Hz，应力比 *R* 为 0.1，循环 107 周次时，试验钢的疲劳强度为 360MPa，疲劳裂纹萌生于试验钢的次表面，裂纹萌生点附近有沿晶开裂现象，疲劳裂纹扩展区域有解理台阶与疲劳条纹，瞬间断裂区是韧性断裂，有大量韧窝。试验钢在循环应力作用下基体中产生了大量位错，并有驻留滑移带终止在晶界位置。

关键词：高强无取向电工钢；疲劳强度；断口形貌；位错

Fatigue Properties and Fracture Mechanism of High Strength Non-oriented Electrical Steel

Song Xinli, Peng Yufan, Li Zhaozhen, Huang Changhu, Pan Ziqin, Liu Jing

(The State Key Laboratory of Refractories and Metallurgy, Wuhan University of Science and Technology, Wuhan 430081, China)

Abstract: The *S-N* curve of a high strength non-oriented electrical steel was tested. The microstructure and fatigue fracture morphology and dislocation were analyzed by optical microscope, scanning electron microscope, transmission electron microscope. The results show that the fatigue limit of experimental steel is 360MPa at the condition of a frequency of 20Hz and stress ratio of 0.1 and room temperature. The fatigue crack initiation at the surface of the steel and the intergranular fracture phenomenon occurs. There are cleavage steps and fatigue stripes in the fatigue crack propagation region, and the instantaneous fracture region is ductile fracture with a large number of dimples. A large number of dislocations appear in the matrix under cyclic stress, and there resident slip band terminating at the grain boundary position.

Key words: high strength non-oriented electrical steel; fatigue strength; fracture morphology; dislocation

3.0%Si 无取向硅钢表面发暗缺陷研究及改进措施

杜　军，刘青松，裴英豪，施立发，胡　柯

（马鞍山钢铁股份有限公司技术中心，马鞍山　241000）

摘　要：研究了 3.0%Si 高牌号无取向硅钢表面发暗缺陷的产生原因，结果表明：发暗缺陷是由于带钢表面被氧化造成的，通过提升连续退火炉内氢气含量至 25%以上、缩短连续退火均热时间至 35s 以下、并配合温度的调整，显著改进了表面发暗缺陷的产生，同时可以使产品铁损 $P_{1.5/50}$ 达到 2.40W/kg 磁感 B_{50} 达到 1.675T。

关键词：高牌号无取向硅钢；表面发暗；连退；改进措施

Analysis and Countermeasures for Surface Darkening Defect of 3.0%Si Non-oriented Silicon Steel

Du Jun, Liu Qingsong, Pei Yinghao, Shi Lifa, Hu Ke

(Technical Center of Maanshan Iron and Steel Co., Ltd, Ma'anshan 241000, China)

Abstract: The cause of surface darkening defect of 3.0%Si non-oriented silicon steel of Masteel is studied. The results show that the darkening defect is caused by the surface oxidation of strip steel. By increasing the content of hydrogen in the continuous annealing furnace to more than 25%, shortening the soaking time of continuous annealing to less than 35s, and adjusting the temperature, the surface darkening defects can be significantly improved, and the magnetic properties of the products can be ensured which the iron loss $P_{1.5/50}$ is 2.40W/kg the magnetic induction intensity B_{50} is 1.675T.

Key words: high grade non-oriented silicon steel; surface darkening defect; continuous anneal; improvement measures

退火温度对 Fe-3.0%Si 薄规格无取向硅钢组织及性能的影响

刘青松，裴英豪，施立发，占云高，程国庆，祁　旋

（马鞍山钢铁股份有限公司技术中心，安徽马鞍山　243000）

摘　要：为探索退火温度对 Fe-3.0%Si 薄规格无取向硅钢电磁性能及组织的影响，借助多气氛连续式退火炉模拟不同退火温度对 0.25mm 厚度 Fe-3.0%Si 无取向硅钢电磁性能、组织以及织构的影响。结果表明：退火温度从 850℃提高到 975℃，铁损 $P_{1.5/50}$ 从 3.22W/kg 降低至 1.47W/kg，铁损 $P_{1.0/400}$ 从 17.63W/kg 降低至 12.55W/kg；退火温度超过 950℃时铁损 $P_{1.5/50}$ 基本稳定在 1.47W/kg 左右；退火温度从 975℃提高至 1000℃，铁损 $P_{1.0/400}$ 从 12.55W/kg 提高到 13.14W/kg；退火温度为 975℃时可以获得较低的铁损 $P_{1.5/50}$ 和 $P_{1.0/400}$，临界晶粒尺寸为 125μm，工业化生产中最佳退火温度在 950~975℃之间。

关键词：退火温度、无取向硅钢、铁损、磁感、组织

Effect of Annealing Temperature on Microstructure and Properties of Fe-3.0%Si Thin Gauge Non-oriented Silicon Steel

Liu Qingsong, Pei Yinghao, Shi Lifa, Zhan Yungao, Cheng Guoqing, Qi Xuan

(Technology Center, Ma′anshan Iron and Steel Co.,Ltd., Ma'anshan 243000, China)

Abstract: In order to explore the effect of annealing temperature on electromagnetic properties and microstructure of Fe-3.0%Si non-oriented silicon steel, the influence of annealing temperature on electromagnetic properties, microstructure and texture of Fe-3.0% Si non oriented silicon steel with 0.25mm thickness were simulated by multi atmosphere continuous annealing furnace. The results show that the annealing temperature increases from 850℃ to 975℃, the iron loss $P_{1.5/50}$ decreases from 3.22W/kg to 1.47W/kg, and the iron loss $P_{1.0/400}$ decreases from 17.63W/kg to 12.55W/kg; When the annealing temperature exceeds 950℃, the iron loss $P_{1.5/50}$ is about 1.47W/kg; The annealing temperature increases from 975℃ to 1000℃, and the iron loss $P_{1.0/400}$ increases from 12.55W/kg to 13.14W/kg; When the annealing temperature is 975℃, the core loss $P_{1.5/50}$ and $P_{1.0/400}$ are lower, and the critical grain size is 125 μm. The optimum annealing temperature in industrial production is 950 ~ 975℃.

Key words: annealing temperature; non-oriented silicon steel; iron loss; magnetic induction; microstructure

稀土 Y 对无取向 6.5wt.% Si 钢组织演变及拉伸性能的影响

李　民[1,2]，李昊泽[1]，马颖澈[1]

（1. 中国科学院金属研究所师昌绪创新中心，辽宁沈阳　110016；
2. 沈阳航空航天大学 材料科学与工程学院，辽宁沈阳　110136）

摘　要： 本文通过微观组织表征、高温拉伸试验和断口形貌分析，研究了稀土 Y 元素对无取向 6.5wt.% Si 钢组织演变和拉伸性能的影响。研究结果表明，添加 Y 元素可以在钢液中形成高熔点 Y 化物，促进异质形核，细化凝固组织。热轧组织不均匀，由表层至芯部分别形成等轴晶、等轴晶/拉长晶和拉长晶的混合组织。退火后，热轧变形组织完全由等轴晶组织取代，含 Y 实验钢的退火组织得到明显细化。500℃时效处理后，含 Y 实验钢具备较低的有序度，300℃的拉伸断口呈现韧性断裂特征，断后延伸率达到 20.2%。相反，无 Y 实验钢发生脆性断裂，断后延伸率仅为 2.1%。研究结果证实，添加 Y 元素可以通过细化组织和降低有序度改善无取向 6.5wt.% Si 钢的中温塑性。

关键词： 无取向 6.5wt.% Si 钢；稀土；微观组织；有序度；拉伸

Effect of Rare-earth Element Y on the Microstructure Evolution and Tensile Properties of Non-oriented 6.5wt.% Si Electrical Steel

Li Min[1,2], Li Haoze[1], Ma Yingche[1]

(1. Shi-changxu Innovation Center for Advanced Materials, Institute of Metal Research, Chinese Academy of Sciences, Shenyang 110016, China; 2. College of Material Science and Engineering, Shenyang Areospace University, Shenyang 110136, China)

Abstract: The effect of Y on the microstructure evolution and tensile properties of non-oriented 6.5wt.% Si electrical steel was investigated by means of the microstructure characterization, high-temperature tensile test and fracture morphology analysis. The results showed that the addition of Y resulted in the formation of high-melting Y-rich precipitates in the melt which dramatically enhanced heterogeneous nucleation, thus leading to a much refined solidification microstructure. The hot rolling microstructure was inhomogeneous through the thickness and demonstrated equiaxed grains, mixed equiaxed/elongated grains and elongated grains at the surface, subsurface and center layers, respectively. After annealing, the hot deformed microstructures were totally replaced by equiaxed structures. The annealing microstructure of the experimental steel doped with Y was significantly refined as compared to that undoped with Y. After aging treatment at 500℃, the experimental steel doped with Y possessed a lower ordered degree and underwent totally ductile fracture when tensile tested at 300℃. The corresponding elongation approached as high as 20.2%. On the contrast, the experimental steel undoped with Y showed completely brittle fracture morphology and the elongation was only 2.1%. The results verified that the addition of Y could improve the intermediate-temperature ductility of non-oriented 6.5wt.% Si electrical steel through microstructure refinement and lowering ordered degree.

Key words: non-oriented 6.5wt.% Si electrical steel; rare earth; microstructure; ordered degree; tensile test

7.7 不锈钢

厚板轧制不锈钢端部折叠缺陷发生机理初探

张敏文

（宝钢股份厚板部，上海 201900）

摘 要：本文对厚板轧制过程中不锈钢端部折叠缺陷进行诊断，解析了该缺陷分布规律特征，并结合过程数据与对比试验寻找到产生该缺陷的根源，同时指明了重点改善方向。

关键词：厚板轧制；不锈钢；端部折叠

Research on Head Fold Defect during Heavy Plate Rolling of Stainless Steel

Zhang Minwen

(Heavy Plate Mill Baosteel, Shanghai 201900, China)

Abstract: Head fold defect during heavy plate rolling of stainless steel has been researched on, and the distributing disciplinarian and controlling factors have been discussed in this paper, the formation mechanism of the defect has also been found through data analysis and experimental comparison study.

Key words: heavy plate rolling; stainless steel; head fold defect

430 不锈钢砂金（或金尘）缺陷产生机理与对策

贾 涛

（东北大学轧制技术及连轧自动化国家重点实验室，辽宁沈阳 110819）

摘 要：430 不锈钢经光亮退火后，通常进行覆膜处理以保护钢板表面。但当用户将表面覆膜揭下后，部分产品的表面存在有屑状物质脱落。如图 1 所示。通过 SEM 观察，可以看出试样表面呈现屑状或片状剥离，此类现象被视作砂金（或金尘）缺陷。砂金严重降低 430 不锈钢光亮退火板的表面光洁度，是其表面质量控制中的一个痛点。在工艺链条的追溯中发现，砂金缺陷与界面处链状碳化物的形成相关，即晶界或相界处的碳化物降低界面结合力，如图 2 所示。当钢板表面存在拉力时，表层晶粒或组成相剥离，最终导致砂金缺陷的产生。

针对热轧时近似连续冷却降温过程，开展 430 和 410s 对比实验与相变动力学计算研究。如图 3 所示，在 1200℃降温至 900℃过程中发生铁素体→奥氏体相变，相界面向左移；在 900℃以下继续降温时发生奥氏体→铁素体相变，相界面向右移。由于 Cr%与相变动力学行为直接相关，导致在 850℃时 430、410s 在奥氏体/铁素体相界面分别形成约 1.8%和 0.25%的偏聚峰（相对于铁素体基体），实测值与计算值吻合良好。为了研究相界面处的溶质偏聚对碳化物析出行为的影响，在连续冷却降温过程中不同温度淬火。如图 4 所示，在 850℃继续冷却至 200℃时，430 相界面处偏聚的 Cr%逐渐析出形成界面链状碳化物，而 410s 中未观察到这一现象。据此可以判断，与相变动力学相关的界面溶质偏聚是链状碳化物产生的主要原因，而如何控制或消除溶质偏聚是解决砂金缺陷的关键。

据此，设计了针对性的工艺改进方案，开展了实验室热模拟和热轧研究。如图 5，设计的三种工艺改进方案均可显著消除晶界链状碳化物。本研究为砂金缺陷的解决提供了理论与工艺指导。

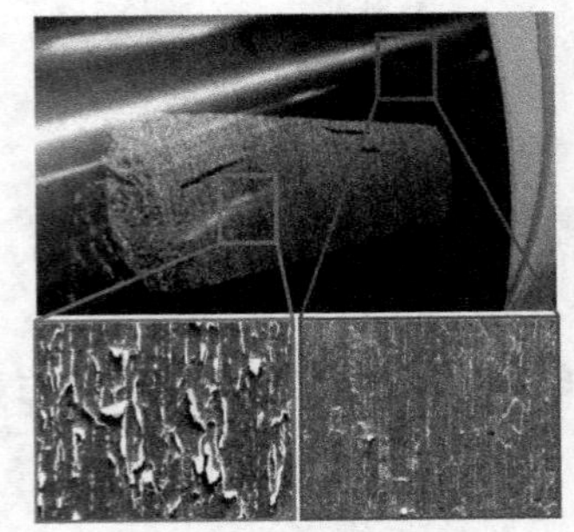

图 1　存在砂金缺陷的 430BA 钢卷

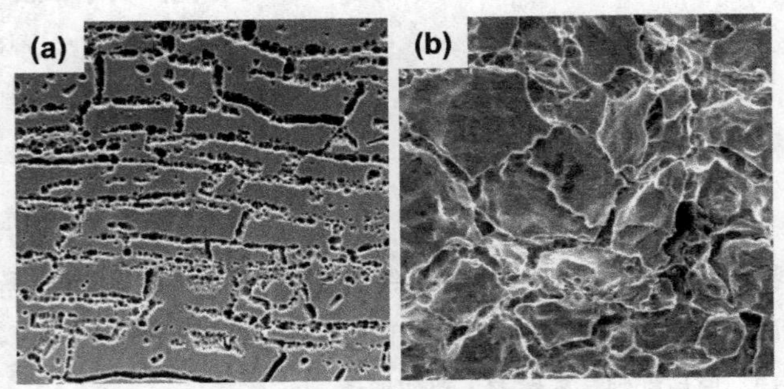

图 2　砂金缺陷追溯

（a）罩式退火后组织；（b）酸洗后组织

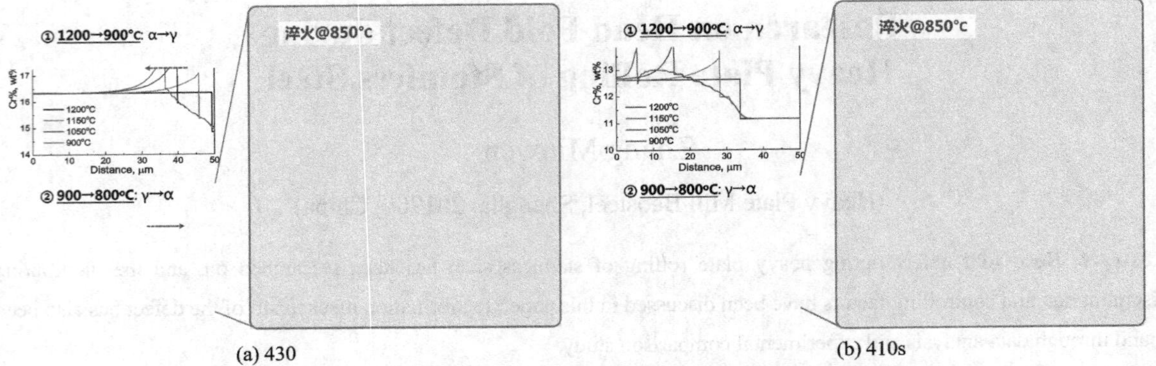

(a) 430　　　　(b) 410s

图 3　连续冷却过程的模拟实验与 Dictra 计算结果

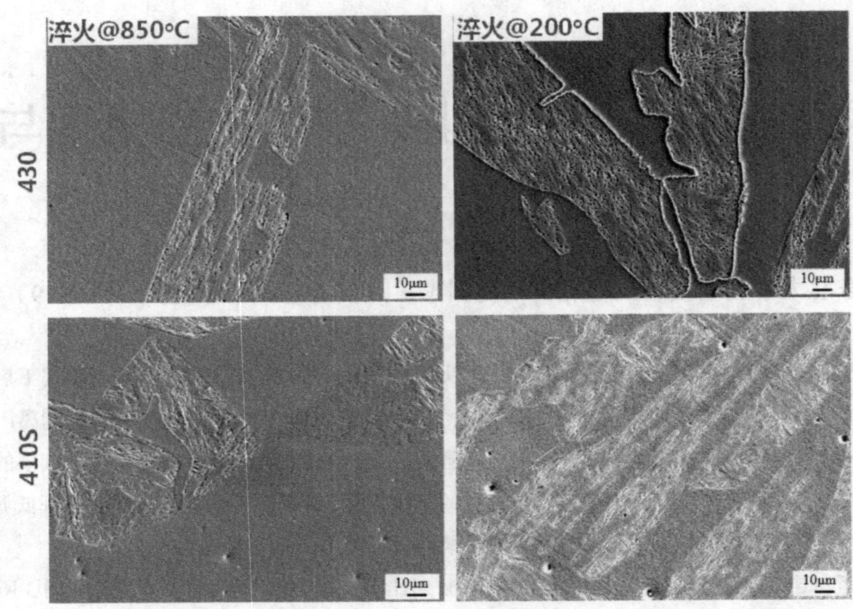

图 4　430 和 410s 连续冷却时在 850℃和 200℃的淬火组织

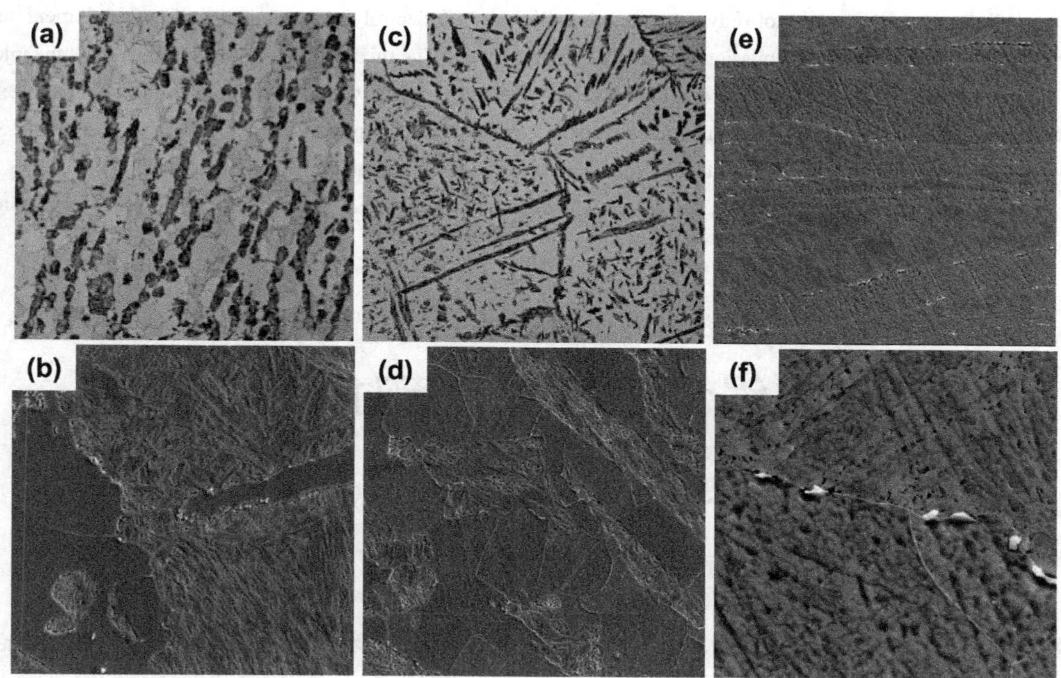

图5　工艺改进方案实施效果
（a, b）工艺一；（c, d）工艺二；（e, f）工艺三

高效低耗环保不锈钢板带酸洗新技术

贾鸿雷，廖砚林，李春明
（中冶南方工程技术有限公司，湖北武汉　430223）

摘　要：不锈钢板带酸洗存在处理流程复杂，涉及工艺介质品种多，废水有毒、腐蚀性、强氧化性且固含量高的特点。本文介绍了超声波和电解结合的新工艺处理带钢，并利用超声波方式输泥和能源介质循环利用措施，不仅提高了电解效率，延长了电极板等设备寿命，而且实现了工艺有效介质的零排放。混酸酸洗系统采取了酸洗槽底部喷射工艺和能源介质循环利用措施，提高了酸洗效率，延长了换热器等设备寿命，工艺有效介质回收利用，节省能源介质耗量，直接对酸泥泥饼进行了化学处理并输出环境友好的中性污泥。

关键词：不锈钢；板带；酸洗；中性盐电解；混酸；高效；低耗；环保

New Technology of Stainless Steel Strip Pickling for High-efficiency, Low-consumption and Environmental Protection

Jia Honglei, Liao Yanlin, Li Chunming
(WISDRI Engineering & Research Incorporation Limited, Wuhan 430223, China)

Abstract: The pickling of stainless steel strips has a complex treatment process, involves a variety of process media, and the waste water is toxic, corrosive, strongly oxidizing, and has the characteristics of high solid content. This article introduces a new process combining ultrasonic and electrolysis to process strip steel, and uses ultrasonic mud transport and energy media recycling measures, which not only improves the efficiency of electrolysis, extends the life of equipment such as electrode plates, and achieves zero emission of effective process media. The mixed acid pickling system adopts the injection process at the bottom of the pickling tank and the recycling of energy media, which improves pickling efficiency, extends the life of heat exchangers and other equipment, recovers and utilizes effective process media, and saves energy and energy media consumption. The mud cake is chemically treated and output environmentally friendly neutral sludge.

Key words: stainless steel; strip; pickling; neutral salt electrolysis; mixed acid; efficient; low consumption; environmental protection

铁铬铝合金焊接工艺对比研究

孙智聪，张文娟，胡　静

（北京首钢吉泰安新材料有限公司，北京　102206）

摘　要：通过制定合理的焊接工艺参数和焊接技术措施，采用氩弧焊以及冷焊填丝两种焊接方法，对铁铬铝合金0Cr21Al6Nb 进行焊接对比试验。研究结果表明：冷焊填丝焊接能量集中，焊接熔层与基体结合力好，可以实现小热输入量较大规格材料的焊接，焊缝和热影响区宽度也明显小于氩弧焊，可有效防止材料焊接后裂纹和断裂现象发生；冷焊填丝焊接部位焊口规则、致密、电阻较小，通电测试焊缝部位负载功率明显小于材料本身，发热量较小。

关键词：焊接；不锈钢；铁铬铝合金；冷焊；氩弧焊

Comparative Study of Fe-Cr-Al Alloy Welding Process

Sun Zhicong, Zhang Wenjuan, Hu Jing

(Beijing Shougang Gitane New Materials CO., LTD., Beijing 102206, China)

Abstract: By formulating welding parameters and welding techniques to develop a reasonable measure, Adopt two welding methods: argon arc welding and cold welding filler to product wireWelding comparison test on Fe-Cr-Al alloy 0Cr21Al6Nb. The results show that the welding energy of cold welding filler wire is concentrated, and the bonding force between the welding layer and the substrate is well, which can realize the welding of materials with small heat input and larger specifications, and the width of the weld and heat-affected zone is also significantly smaller than that of argon arc welding, which can effectively prevent the occurrence of cracks and fractures after the material is welded. The welding part of cold welding filler wire is regular, dense, and has low resistance. The load power of the weld part in the electrification test is significantly less than the material itself, and the heat generation is smaller.

Key words: welding; stainless steel; Fe-Cr-Al alloy; cold welding; argon arc welding

节镍奥氏体不锈钢其他鳞折问题研究

任建斌

（宝钢德盛不锈钢有限公司，福建福州　350600）

摘 要： 节镍奥氏体不锈钢具有优异的力学性能、一定的耐蚀性能以及较低的成本，在面板、制品和构件等领域应用广泛。由于其奥氏体相为亚稳定，在热轧时容易出现其它鳞折问题。本文研究了其他鳞折的产生机理，并分析了影响其他鳞折产生的主要影响因素。结果表明，热塑性不良是产生其他鳞折的主要原因，通过热轧工艺的优化可以控制其他鳞折的产生，同时与降氮等改善材料热塑性的措施可以进一步降级其他鳞折产生的风险。

关键词： 节镍奥氏体不锈钢；其他鳞折；热塑性

Research on the others Squamous Folding of Low Nickel Austenitic Stainless Steel

Ren Jianbin

(Baosteel Desheng Stainless Steel Ltc., Fuzhou 350600, China)

Abstract: Low-nickel austenitic stainless steel was widely used in board, kitchen ware and structure due to its excellent mechanism properties, certain corrosion resistance and low cost. The the others squamous folding was easily induced in hot-rolling process, due to its unstable phase. The paper investigated generation mechanism of the others squamous folding, and analyzed the main influence factors. The results indicated that the bad thermoplastic was the main reason, and the the others squamous folding could be controlled by the hot rolling technology. Moreover, the nitrogen content decreased were also beneficial for delayed cracking.

Key words: low nickel austenitic stainless steel; the others squamous folding; thermoplastic

CAS 精炼过程钢包流场的水模型研究

韩立浩[1]，李跃华[1]，韩立宁[2]，黄伟青[1]，刘燕霞[1]，齐素慈[1]

（1. 河北工业职业技术学院，河北石家庄 050091；
2. 安阳钢铁股份有限公司，河南安阳 455004）

摘 要： 针对 CAS 精炼在熔池混匀和夹杂物上浮以及合金收得率存在问题，本文在相似原理的基础上利用水模型对 CAS 法精炼过程钢包内流场进行了实验室研究，考察了底吹位置，底吹气量，渣层厚度和黏度，浸渍罩下插深度、位置及尺寸等因素对钢液混匀时间、排渣能力及夹杂物上浮的影响规律，并对合金的运动轨迹进行了研究。结果表明：最优底吹位置应设在 0.5r 处，底吹气量设置为 0.18~0.22m^3/h，浸渍罩插入深度控制为 0.060m 左右；随着底吹气量的增加，排渣直径逐渐增加。要使尽可能少的熔渣进入浸渍罩内，需使渣面直径与浸渍罩外径相当；为保证 CAS 精炼时合金收得率，在添加合金时，建议添加在浸渍罩中心和透气砖中心同轴附近的位置处。

关键词： CAS 精炼；水模型研究；示踪颗粒；运动轨迹；合金元素

Water Model Study on Flow Field in CAS Process

Han Lihao[1], Li Yuehua[1], Han Lining[2], Huang Weiqing[1], Liu Yanxia[1], Qi Suci[1]

(1. Hebei College of Industry and Technology, Shijiazhuang 050091, China;
2. Anyang Iron & Steel Co., Ltd., Anyang 455000, China)

Abstract: In view of the problems existing in bath mixing, inclusion floating and the alloy yield, a water model has been established to study the flow field of CAS process on the basis of similarity principle. In this paper, the influences of the bottom blowing location, bottom blowing rate, slag layer thickness, the inserting depth, the location and the size of the impregnation cover on molten steel mixing time, slag removal capacity and inclusion floating have been investigated in detail. The motion trajectory of alloy was also studied. The results show that with the increase of bottom blowing rate, the slag removal diameter increases. To ensure the amount of slag in the impregnation cover as little as possible, it would be better that the slag surface is equal to the diameter of impregnation cover. The optimum bottom blowing location is located at 0.5r, the bottom blowing rate is between $0.18m^3 \cdot h^{-1}$ and $0.22m^3 \cdot h^{-1}$, and the inserting depth of impregnation cover is about 0.060m. To ensure the yield of alloy elements, it is suggested that the alloy elements should be added near the coaxial position between the center of impregnation cover and porous plug.

Key words: CAS process; water model study; tracer particles; movement trajectory; alloy elements

超级奥氏体不锈钢 UNS N08367 耐腐蚀性能研究

余式昌，欧新哲

（宝武特种冶金有限公司，上海　200940）

摘　要：研究了超级奥氏体不锈钢 UNS N08367 耐晶间腐蚀、点腐蚀、缝隙腐蚀和电厂烟气脱硫模拟溶液下的腐蚀性能。结果表明：UNS N08367 具有优异的耐晶间腐蚀性能，其耐点腐蚀温度高于 50℃以上，高铬成分时耐缝隙腐蚀温度能达到 39℃以上，该合金在电厂烟气脱硫模拟溶液腐蚀速率较低；铬含量提高有助于提高 UNS N08367 的耐腐蚀性能。

关键词：超级奥氏体不锈钢；　晶间腐蚀；点腐蚀；缝隙腐蚀；模拟溶液腐蚀

Study on Corrosion Resistance of Super Austenitic Stainless Steel UNS N08367

Yu Shichang, Ou Xinzhe

(BAOWU Special Metallurgy Co., Ltd., Shanghai 200940, China)

Abstract: The corrosion resistance of super austenitic stainless steel UNS N08367 in intergranular corrosion, pitting corrosion, crevice corrosion and simulated FGD environment was studied. The results show that UNS N08367 has excellent resistance to intergranular corrosion, its resistance to pitting corrosion temperature is higher than 50℃, and its resistance to crevice corrosion temperature is higher than 39℃ with high chromium content; the corrosion rate of the alloy is low in the simulated solution of FGD in power plant; the increase of chromium content helps to improve the corrosion resistance of UNS N08367.

Key words: super austenitic stainless steel; intergranular corrosion; pitting corrosion; crevice corrosion; simulated solution corrosion

不锈钢复合输水管道腐蚀原因与对策研究

王小勇，刘立伟，王凤会，罗家明

（首钢集团有限公司技术研究院，北京　100043）

摘　要： 随着民众环保及公众安全意识的提高，城市供水安全及水质问题也日益受到大家的关注。近年来，常规配水、输水管道由于内壁腐蚀对水体二次污染以及产生渗漏、爆管问题频发，也增强了民众对市政管网安全的关注。不锈钢复合管作为新一代环保型管材，因为在卫生性和性价比方面的优势，使其成为改善居民生活用水的优选产品[1-2]。不锈钢复合管主要有机械复合管和冶金焊接复合管，其中冶金焊接复合管常用有大口径管道（ϕ400mm 以上），通过复合板直缝焊的方式生产，已经在地下管廊水管及工业输送管道中得以应用。但市政管网中供水管道，所需要直径多在 400mm 以下，很难通过直缝焊接方式实现，需要探索新的制管方法。

为了实现小口径复合管的生产及应用，某公司尝试采用螺旋焊制管方式将 304+Q235B 不锈钢复合板焊接成复合管（ϕ219mm），并将其用于生活用自来水输送。复合水管内侧为 304 不锈钢，管内输送介质为生活用水；外侧为 Q235B，与土壤接触，但在使用半年左右后，焊缝处开始出现多处腐蚀穿孔现象。对此，本文从复合管焊接接头的组织、化学成分、宏观形貌、微观形貌等方面对失效管段进行了失效分析，发现试验管腐蚀失效原因在于不锈钢层焊缝成分稀释。由于复层较薄、焊接热输入大，造成焊接过程中不锈钢侧熔池中存在成分稀释，焊缝中 Cr、Ni 元素远远低于 304 标准数值，与母材不锈钢相比耐蚀性能差，在水中氯离子作用下不锈钢焊缝处优先形成点蚀坑。点蚀坑形成后，与不锈钢母材形成腐蚀原电池，焊缝点蚀坑作为阳极，加速腐蚀，沿熔合线附近垂直向管外侧扩展；当扩展到母材基层碳钢时，由于碳钢母材不含耐蚀元素，电化学电位更负，碳钢母材与不锈钢母材及焊缝形成电偶腐蚀，碳钢母材将作为阳极，发生阳极溶解，不锈钢焊缝作为阴极，受到保护，因此腐蚀向碳钢母材加速扩展，腐蚀量最大，最终形成腐蚀穿孔[3-5]。

在腐蚀原因分析的基础上，通过对复合管不锈钢层焊道成分稀释程度的量化，获得当前复合层占比条件下 Ni、Cr 等元素的稀释率，就此对焊丝成分及工艺进行了优化；模拟螺旋制管焊接后结果显示，通过优化焊丝成分，不锈钢焊道 Ni、Cr 含量明显提高（与 304 母材相当），焊道耐蚀性良好，为这类薄规格不锈钢复合管道耐蚀性提高及后续应用提供了对策及思路。

430、444 不锈钢连铸坯的高温力学性能研究

周士凯[1]，黄　军[2]，曾　晶[1]，田　川[1]，王　蓉[1]

（1. 中国重型机械研究院股份公司，陕西西安　710032；
2. 内蒙古科技大学，内蒙古包头　014010）

摘　要： 本文通过 Gleeble-1500D 热/力模拟试验机对某钢厂生产的 430、444 两种不锈钢连铸坯进行高温力学性能测试，分析了其高温强度性能和高温热塑性。结果表明，两种不锈钢连铸坯的抗拉强度和屈服强度均随温度的升高而降低，在 1200℃以后抗拉强度<20MPa，屈服比高达 85%~90%；430 不锈钢连铸坯的第Ⅰ脆性温度区为 T_m~1338℃，第Ⅲ脆性温度区为 800~900℃，在 1050~1300℃的断面收缩率>60%，塑性较好；444 不锈钢的第Ⅰ脆性温度区为 T_m~1327℃，在 600~1050℃断面收缩率均>60%，塑性较好；另外分别用不同的应变速率进行拉伸比较实验，得

出两不锈钢的强度随应变速率的变化关系。

关键词：430、444 不锈钢；力学性能；应变速率

Study on High Temperature Mechanical Properties of 430 and 444 Stainless Steel Continuous Casting Billets

Abstract: In the paper, the high temperature mechanical properties of 430 and 444 stainless steel continuous casting billets produced by a steel mill were tested by Gleeble-1500D heat/force simulation test machine, and their high temperature strength properties and high temperature thermoplasticity were analyzed. The result shows, The tensile strength and yield strength of two stainless steel continuous casting billets decrease with increasing temperature, After 1200℃, the tensile strength is Less than 20MPa, the yield ratio is as high as 85%~90%, and the utilization rate is low; The first brittle temperature zone of 430 stainless steel continuous casting billet is T_m~1338℃, the third brittle temperature zone is 800~900℃, the section shrinkage of 1050~1300℃ is more than the 60%, and the plasticity is better; The first brittle temperature zone of 444 stainless steel is T_m~1327℃, and the shrinkage rate is more than the 60% at 600~1050℃, and the plasticity is better; In addition, tensile experiments were carried out with different strain rates respectively, the strength of the two stainless steels increased with the increase of strain rate.

Key words: 430 and 444 stainless steel; mechanical properties; strain rate

2507 超级双相不锈钢凝固过程 sigma 相析出研究

陈兴润[1]，钱张信[1]，潘吉祥[1]，王长波[1]，马敏敏[2]

（1. 酒钢集团宏兴钢铁股份有限公司，甘肃嘉峪关　735100；
2. 酒钢集团技术中心，甘肃嘉峪关　735100）

摘　要：2507 超级双相不锈钢中连铸坯、热轧黑卷和时效处理的样品中，会存在一些析出相。这些析出相会严重影响 2507 的各项性能，其中 sigma 相（σ）的危害最大。针对双相不锈钢 sigma 相的研究，多集中于双相不锈钢在热处理过程中的 σ 相的析出行为、影响因素及这些析出产物对产品各项性能的影响。而这些研究甚少关注双相不锈钢在连铸过程中的析出行为。而连铸过程中铸坯直接由液态钢水凝固而成。在凝固过程中，铸坯不同部位经历不同冷却速度，一直由熔化温度连续冷却到室温。在该过程中形成不同铸态组织结构（包括宏观和微观偏析），此外这个过程中还存在铁素体由高温到低温过程向奥氏体的转变、转变过程中的溶质元素再分配等等。所有这些因素都可能对连铸过程中的 σ 相等的析出造成影响，致使这些未知 σ 相等的析出行为有可能大大不同于前述析出相在纯热处理过程中的行为。控制 σ 相的形成对超级双相不锈钢的研究不但具有重要科学意义，而且会对酒钢等双相不锈钢的生产企业具有重要现实技术指导意义。因此，本文针对 2507 超级双相不锈钢板坯修磨过程中发生的脆断问题，通过显微组织观察、热力学计算等方法对脆断原因进行剖析。深入分析钢水化学成分、及其冷却速率对 sigma 相析出的影响。研究结果表明：

（1）2507 板坯脆断样品中存在大量微裂纹，sigma 相的面积分数平均值达到 24.5%。sigma 相的大量析出，导致 2507 内部体积增大，内应力集中。此外，sigma 相的大量析出导致微区各相硬度差别增大，材料塑性降低。因此，2507 板坯脆断的主要原因为：板坯下线到室温冷却过程中冷却速度偏慢造成 σ 相析出过多，板坯修磨过程在压应力作用下导致板坯脆断。

（2）在 2507 超级双相不锈钢板坯中 σ 相总是沿着 α/γ 的相界析出并向 α 相内部不断推进长大。2507 板坯中 σ 相的析出也是按照 α→σ +γ2 共析反应方式进行。

（3）在硅含量相同的情况下，钼含量越高板坯中 sigma 相的面积百分比也越高，sigma 相中钼元素的质量分数也越高。在钼含量相同的情况下，硅含量越低板坯中 sigma 相的面积百分比越低，sigma 相中钼元素的质量分数也越低。在钼含量和硅含量分别为 3.4%和 0.30%的情况下，板坯心部试样中 sigma 相的面积分数小于 3.0%。

（4）2507 板坯凝固冷却速率越大，凝固组织中 sigma 相越少；反之，凝固冷却速率越小，sigma 相析出量越大，板坯中心部 sigma 相的面积百分比最高。在 2507 凝固过程中，由于不同部位冷却速率不同，导致元素扩散不同，最终导致板坯中部分区域存在成分宏观偏析。成分宏观偏析也是影响 sigma 相析出的一个因素。

关键词：冶金；2507 超级双相不锈钢；sigma 相；凝固

红土镍矿混合冶炼生产镍铁的研究

马东来[1]，赵剑波[1]，王伦伟[1]，吕学明[1]，
余文轴[1,2]，游志雄[1,2]，白晨光[1,2]

（1. 重庆大学 材料科学与工程学院，重庆　400044；
2. 重庆大学 钒钛冶金及新材料重庆市重点实验室，重庆　400044）

摘　要：红土矿火法冶炼过程通常需配入一定的助熔剂进行造渣，使渣相具有优良的性能以促进镍铁颗粒的聚集和分离。本研究提出褐铁矿型和硅镁质型红土矿混合冶炼的新思路，通过配加褐铁矿型红土矿取代助熔剂的加入，控制铁氧化物的还原程度，部分转变为 FeO 调节渣相的性能，以促进金属聚集和渣金分离。理论计算表明，在配碳量为 6.8%，焙烧温度为 1380℃时，褐铁矿型红土镍矿的配比从 0 增加到 20%，可使还原后球团中渣相熔点降低、液相量增多；还原产物的微观结构也表明，镍铁颗粒的平均直径可从 7μm 增加至 45μm，明显促进了镍铁颗粒的聚集；磁选分离后可使镍回收率从 55.17%显著增加至 92.86%，本研究有利于实现两种红土镍矿的综合利用。

关键词：红土镍矿；混合冶炼；还原焙烧；磁选；镍铁

Study on the Preparation of Ferronickel by Smelting a Mixture of Two Types of Nickel Laterite Ore

Ma Donglai[1], Zhao Jianbo[1], Wang Lunwei[1], Lü Xueming[1], Yu Wenzhou[1,2],
You Zhixiong[1,2], Bai Chenguang[1,2]

(1.College of Materials Science and Engineering, Chongqing University, Chongqing 400044, China; 2. Laboratory of Vanadium–Titanium Metallurgy and New Materials, Chongqing University, Chongqing 400044, China)

Abstract: The traditional pyrometallurgical process of saprolitic laterite ore requires a certain amount of flux to form slag bearing excellent properties, which is beneficial to the aggregation and separation of metallic phase from slag. In this study, a novel method of smelting a mixture of two types of laterite ore to prepare ferronickel was put forward. By adding limonitic laterite ore to replace the addition of flux, the properties of the slag phase can be regulated via controlling the reduction degree of iron oxides. Theoretical calculation indicated that when the carbon content is 6.8% and the roasting temperature is 1380℃, increasing the ratio of limonitic laterite nickel ore from 0 to 20% would decrease the melting point of the slag phase and increase the amount of liquid phase in slag. The increase in liquid phase content is conducive to the aggregation and growth of ferronickel particles, whose average diameter increased from 7μm to 45μm. The nickel recovery was also increased significantly from 55.17% to 92.86%. The novel method of mixed smelting was favorable to comprehensively utilizing two types of laterite ore.

Key words: nickel laterite ore; mixed smelting; reduction roasting; magnetic separation; ferronickel

热处理条件对 SUS430 不锈钢热轧板带 MAG 焊接头组织和力学性能的影响

王长波[1]，胡昱轩[2]，王军伟[1]，李玉峰[1]

（1. 酒钢集团宏兴股份钢铁研究院，甘肃嘉峪关 735100；2. 兰州理工大学，甘肃兰州 730000）

摘　要： 采用 MAG 焊方法对 5mm 厚的 SUS430 不锈钢热轧板带进行焊接，并通过体视显微镜（SM）、光学显微镜（OM）、场发射扫描电镜（SEM）二次电子成像、X 射线衍射（XRD）、背散射电子衍射（EBSD）及室温拉伸试验等技术手段，研究了不同热处理条件对热影响区的组织转变以及力学性能变化规律。试验结果表明：在 SUS430 相变点温度以下进行一定时间的热处理，随热处理温度的升高可以明显改善热影响区的性能。其中 850℃保温 10min 能有效改善热影响区组织，同时降低其硬度最低至 214.8HV，硬度值略高于母材，从而提高该区域的力学性能。

关键词： SUS430 不锈钢；MAG 焊；焊后热处理；力学性能

Effect of Heat Treatment Conditions on Microstructure and Mechanical Properties of MAG Welded Joints of SUS430 Hot-rolled Strip

Wang Changbo[1], Hu Yuxuan[2], Wang Junwei[1], Li Yufeng[1]

(1. Iron and Steel Research Institute of Hongxing Iron and Steel Co., JISCO Group, Jiayuguan 735100, China; 2. Lanzhou University of Technology, Lanzhou 730000, China)

Abstract: The hot rolled SUS430 ferritic stainless steel strip with a thickness of 5mm was welded by MAG welding technology. The heat affected zone (HAZ) of the welding sample after different heat treatment conditions were studied by means of Zoom-stereo microscope (SM), optical microscope (OM), field emission scanning electron microscopy (SEM), X-ray diffraction (XRD), backscattered electron diffraction (EBSD) and room temperature tensile test. The changes of microstructure and mechanical properties were analyzed. The experimental results show that the performance of heat-affected zone can be improved obviously with the increase of heat treatment temperature below SUS430 phase transition point for a certain time. In the experimental range, the condition at Holding 850℃ for 10 min can effectively improve the microstructure of HAZ, and meanwhile reduce its hardnessto the lowest 214.8HV, slightly higher than the base material, thus improving the mechanical properties of the area.

Key words: SUS430 stainless steel; MAG welding; post-weld heat treatment; mechanical properties

S32760 超级双相不锈钢 χ 相时效析出研究

梁祥祥

（山西太钢不锈钢股份有限公司，山西太原 030003）

摘　要：为研究 S32760 超级双相不锈钢 χ 相，利用 Thermo-Calc 热力学软件计算相图，根据计算结果在 650℃进行 2h 时效，利用 SEM 和 TEM 分析 χ 相的成分和结构。结果表明，χ 相在 S32760 超级双相不锈钢中析出温度区间为 500~700℃，峰值温度为 650℃。χ 相在铁素体晶内弥散分布，主要成分为 Fe-Cr-Mo，电子衍射花样属于立方晶系（a=0.903），χ 相和铁素体基体位向关系为[023] χphase∥[$\bar{1}$11] Ferrite。实验样品 χ 相析出导致屈服强度、抗拉强度分别升高 35MPa、40MPa，硬度升高 30HBW，延伸率下降 4%，冲击功降低 40J。

关键词：S32760；相图；χ 相；力学性能

S32760 Super Duplex Stainless Steel χ Phase Aging Precipitation Research

Liang Xiangxiang

(Shanxi Taigang Stainless Steel Co., Ltd., Taiyuan 030003)

Abstract: To investing χ phase of S32760 super duplex stainless steel, using Thermo-Calc thermodynamic software calculate phase diagram. After aging at 650℃ for 120min as the calculation results, the structure and chemical component of χ phase were investigated by SEM and TEM. Results show that, the temperature of χ phase is 500~700℃, top temperature is 650℃. χ phase were dispersed in ferrite matrix, main chemical component is Fe-Cr-Mo, SAED of χ phase is tetragonal structure(a=0.903),the OR of χ phase、ferrite matrix could be determined as[023] χ phase∥[$\bar{1}$11] Ferrite. The sample yield strength and tensile strength increase 35MPa、40MPa, the hardness increase 30HBW, the elongation decrease 4%, the impact energy decrease 40J.

Key words: S32760; phase diagram; χ phase; mechanical property

HNS900 高氮不锈钢耐点蚀性能试验研究

吴月龙，王俊海

（山东泰山钢铁集团有限公司，山东莱芜　271100）

摘　要：为研究 HNS900 高氮奥氏体不锈钢耐点蚀性能，取样进行不同钢种对比试验并进行临界点蚀温度测定，发现 HNS900 不锈钢耐点蚀性能优于 06Cr19Ni10 不锈钢、022Cr17Ni12Mo2 不锈钢但较双相不锈钢 022Cr23Ni5Mo3N 差，临界点蚀温度为 32℃。

关键词：高氮不锈钢；点蚀；失重率

Experimental Study on Pitting Corrosion Resistance of HNS900 High Nitrogen Stainless Steel

Wu Yuelong, Wang Junhai

(Shangdong Taishan Iron and Steel Co., Ltd., Laiwu 271100, China)

Abstract: In order to study the pitting corrosion resistance of HNS900 high nitrogen austenitic stainless steel, the contrast tests of different steel types were carried out and the critical pitting corrosion temperature was measured. It was found that the pitting corrosion resistance of HNS900 stainless steel was better than that of 06Cr19Ni10 stainless steel and 022Cr17Ni12Mo2 stainless steel, but worse than that of duplex stainless steel 022Cr23Ni5Mo3N, and the critical pitting corrosion temperature was 32℃.

Key words: high nitrogen stainless steel; pitting; weightlessness rate

刃具用控氮马氏体不锈钢的开发

王宏霞，张　爽，吴月龙

（山东泰山钢铁集团有限公司技术研发中心，山东济南　271100）

摘　要：20Cr13N 是一种氮合金化型马氏体不锈钢，主要应用在刃具行业，对淬硬性和耐蚀性有一定的要求，本文通过合理的化学成分和生产工艺设计，采用 TSR 炉利用喷吹氮气进行氮合金化，该生产工艺简单，成本低，产品的纯净度高。通过氮合金化使产品淬硬性和耐腐蚀性能得到明显改善，实现该产品的成功开发。

关键词：马氏体不锈钢；氮合金化；力学性能；耐腐蚀性能

Development of Nitrogen-Controlled Martensitic Stainless Steel for Cutting Tools

Wang Hongxia, Zhang Shuang, Wu Yuelong

(Shandong Taishan Iron and Steel Group Co., Ltd., Technology R&D Center, Jinan 271100, China)

Abstract: 20Cr13N is a nitrogen-alloyed martensite stainless steel, which is mainly used in the cutting tool industry. It has certain requirements for harden ability and corrosion resistance. This article adopts reasonable chemical composition and production process design, and uses TSR furnace to use nitrogen injection. Nitrogen alloying, the production process is simple, the cost is low, and the purity of the product is high. Through nitrogen alloying, the harden ability and corrosion resistance of the product are significantly improved, and the successful development of the product is realized.

Key words: martensite stainless steel; nitrogen alloying; mechanical character; corrosion resistance

奥氏体不锈钢的新成分探索

张　爽[1]，王晓东[1]，张　琳[1]，邹存磊[1]，董　闯[1,2]

（1. 大连交通大学材料科学与工程学院，辽宁大连　116028；
2. 大连理工大学三束材料改性教育部重点实验室，辽宁大连　116024）

摘　要：奥氏体不锈钢以钝化元素 Cr 为基础成分特征，通过面心立方奥氏体组织将一定量的耐蚀元素均匀带入，从而实现高耐蚀性。在成分特征上，奥氏体不锈钢是基于 Fe-Cr-Ni 三元基础体系的多元合金；在组织特征上，奥氏体不锈钢属于面心立方固溶体合金。多元合金化为奥氏体不锈钢的成分解析带来了困难，而近程有序的固溶体结构为其成分解析提供了可能性。对于固溶体中的化学近程序结构，可引入“团簇加连接原子”模型进行描述，该模型将近程序简化为类分子结构单元，由第一近邻配位多面体团簇（简称为团簇）加上位于团簇间隙的若干连接原子所组成，表述成团簇式形式为：[团簇](连接原子)。在团簇式框架下，结合传统的当量方法，我们对现有的奥氏体不锈钢牌号进行了全面的梳理，结果发现成熟的不锈钢牌号均对应着特定的团簇式。尤其在高合金化区域，优质奥氏体不锈钢的成分式表现出了高熵合金化的特征，即团簇第一近邻位置不仅含有 Fe 元素，还必须含有其它合金化元素，也就是说溶剂不能视为单一的 Fe。因此，有望在目前奥氏体不锈钢的成分空白区，结合“团簇加连接原子”模型与当量控制方法，研发出具有高熵合金化特征的奥氏体不锈钢新成分。

关键词：“团簇加连接原子”模型；奥氏体不锈钢；高熵合金化；成分设计；组织调控

不锈钢的成分式

董　闯[1,2]，张　爽[1]，王　清[2]

（1. 大连交通大学材料科学与工程学院，辽宁大连　116028；
2. 大连理工大学三束材料改性教育部重点实验室，辽宁大连　116024）

摘　要：本报告通过引入“团簇加连接原子”模型，分析不锈钢的化学近程序结构，试图澄清不锈钢成分配方的根源，进而给出典型不锈钢的类分子化学成分式。首先介绍固溶体中的化学近程序结构模型，即“团簇加连接原子”模型，把近程序描述成第一近邻配位多面体团簇（简称为团簇）以及位于团簇之间的若干连接原子，看似复杂的合金成分可表述为简单的成分式：[团簇](连接原子)。我们已从理论上证实，这种描述方式可给出类似于分子的结构单元，满足特定结构单元的成分式实现了溶质的均匀分布。各种不锈钢虽然成分各异，均满足特定原子个数的团簇加连接原子成分式。我们把典型工业不锈钢的成分均做了解析，指出它们只是数种团簇加连接原子成分式的体现，在团簇成分式的框架下，通过控制当量，能实现钢的组织设计。尤其重要的是，我们发现，若干高 Cr 含量的不锈钢实质上就是一类高熵合金，按照我们发展的“团簇加连接原子”模型，这类不锈钢的特点在于第一近邻由多种元素构成，而普通不锈钢的第一近邻仅由 Fe 构成。

关键词：“团簇加连接原子”模型；不锈钢；成分式；结构单元

04Cr13Ni5Mo 不锈钢中厚板加工开裂原因分析

李　俊[1]，杜晓建[2,3]，成生伟[2,3]，刘承志[3]，张　利[2,3]

（1. 太原钢铁（集团）有限公司先进不锈钢材料国家重点实验室，山西太原　030003；
2. 中北大学机械工程学院，山西太原　030051；
3. 中北大学特种金属材料与装备研究院，山西太原　030051）

摘　要：04Cr13Ni5Mo 不锈钢中厚板越来越受到水电行业的需要，但是其加工仍然存在问题。本文对 04Cr13Ni5Mo 不锈钢中厚板加工开裂的问题进行分析，采用金相法和扫描电镜对比观察加工前后的组织结构。结果显示，加工后的 04Cr13Ni5Mo 不锈钢中厚板组织中有碳化物析出。因而，在 04Cr13Ni5Mo 不锈钢中厚板加工后需要及时进行回火处理，避免产生裂纹。

关键词：04Cr13Ni5Mo 不锈钢；中厚板；加工开裂

Analysis on the Crack during the Maching Process of 04Cr13Ni5Mo Stainless Steel Plate

Li Jun[1], Du Xiaojian[2,3], Cheng Shengwei[2,3], Liu Chengzhi[3], Zhang Li[2,3]

(1. State Key Laboratory of Advanced Stainless Steel Materials,Taiyuan Iron and Steel(Group)Co.,Ltd.,Taiyuan 030003, China; 2. School of Mechanical Engineering, North University of China, Taiyuan 030051, China; 3. Institute of Special Metal Materials and Equipment Research, North University of China, Taiyuan 030051, China)

Abstract: Thicknes of 04Cr13Ni5Mo stainless steel plate had be required by the hydropower industry more than before. However, there were still problems during the maching process. In this study, it was analyzed that the crack of 04Cr13Ni5Mo stainless steel plate. The microstructure before and after the pcocessing were observed comparatively through OM and SEM. The results shows that, more precipitated carbides were observed in the processed sample. Therefore, in order to avoid the formation of crack, it must be tempered after the processing of 04Cr13Ni5Mo stainless steel plate.

Key words: 04Cr13Ni5Mo stainless steel; plate; crack

430 不锈钢微观组织特征与性能对应关系的研究

王明涛

（东北大学材料科学与工程学院，辽宁沈阳　114001）

摘　要：本研究以工业 430 冷轧板为初始材料，基于不同热处理制度调控材料组织，形成不完全再结晶组织、完全再结晶组织、马氏体-铁素体双相组织及铁素体-马氏体-残余奥氏体三相组织，完成不同类型组织材料机械性能测试，同时利用课题组开发的工业应用范围内、真实时空条件下多晶相场模型模拟材料组织演化过程，探究不同组织特点及形成机理，最终利用基于粒子受力连续体理论构建的 Eshelby 模型计算不同微观组织材料的力学性能，探究强化机制。

研究发现，随着再结晶程度的增加，在畸变能的驱动下，低位错密度的再结晶晶粒形核长大、材料硬度、强度下降，延伸率提高，基于硬度计算再结晶完成程度，明确了残余应变（位错）是材料强化的主要原因；当发生完全再结晶时，随着加热时间延长在界面能的驱动下再结晶晶粒粗化，材料的屈服强度降低、延伸率增加，屈服强度与平均晶粒尺寸符合 Hall-Patch 关系，相场模拟研究发现碳化物通过阻碍晶粒长大细化晶粒，进而影响力学性能；热处理温度 900℃-1000℃范围内，在奥氏体/铁素体化学自由能驱动下组织表现为两相共存区，室温下为马氏体/铁素体双相组织，随着马氏体的含量增加、材料的屈服强度、抗拉强度升高、延伸率下降；通过 QP 热处理，由于配 C 配分的作用，材料室温组织由铁素体-马氏体-残余奥氏体三相构成，材料的屈服强度和抗拉强度较常规铁素体组织可提高 200MPa，延伸率略有降低达 20%，材料强度与塑性变化除了马氏体相的影响外，残余奥氏体的 TRIP 效应同样作用明显，基于 Eshelby 模型计算的应力应变曲线与实际材料拉伸结果基本一致，在此基础上进一步探究了该类三相组织中相比例、形貌等特征对材料性能的影响。

综上，本研究以 430 冷轧板为初始材料，基于合金组织热力学特征，通过不同热处理工艺调控组织，进行力学性能测试，并通过微观组织特征分析与相场、Eshelby 模型等模拟方法结合的方式探究了组织特征与材料性能间的构效关系，进一步提出了基于组织设计优化材料性能的可能性及方向，通过各种不同的组织调控，实现该材料不同的性能特征，最终满足各种需求，为材料的进一步应用提供基础。

高稳定高效率 18 辊轧机的开发与应用

李守卫，杨　威，张令琴
（中冶南方工程技术有限公司，湖北武汉　430223）

摘　要：18 辊型轧机从 21 世纪初引入国内，多年来在实际生产中普遍存在侧支系统稳定性差、背衬轴承消耗大、工作辊断裂、止推轴承损坏频繁等诸多问题，给生产企业造成重大损失。本文针对 18 辊轧机存在的多项问题进行了研究分析，提出相应解决措施，开发出多项专有技术，最终获得了一种高稳定高效率的 18 辊轧机，并得到良好推广应用。

关键词：18 辊轧机；高稳定；高效率

Development and Application of High Stability and High Efficiency 18 High Rolling Mill

Li Shouwei, Yang Wei, Zhang Lingqin
(WISDRI Engineering and Research Incorporation Limited, Wuhan 430223, China)

Abstract: The 18 high rolling mills have been introduced into China since the beginning of this century. Many problems, such as poor stability of side support system, large consumption of backup bearing, work roll fracture and frequent damage of thrust bearing, have been widely encountered in actual production for many years, which have caused heavy losses to manufacturers. In this paper, various problems existing in the 18 high rolling mills were studied and analyzed, and corresponding solutions were put forward. A number of proprietary technologies were developed, and finally a high stability and high efficiency 18 high rolling mill was obtained, which was well applied.

Key words: 18 high rolling mill; high stability; high efficiency

不锈钢复合管皮尔热轧工艺及熔合机理研究

双远华[1]，何宗霖[1,2]
（1. 太原科技大学 山西省冶金设备设计理论与技术重点实验室，山西太原　030024；
2. 山西工程职业学院 机械制造工程系，山西太原　030009）

摘　要：基于不锈钢复合管生产工艺复杂、成本高等实际生产情况，本文首次提出了皮尔格热轧无缝复合管工艺。建立了皮尔格热轧不锈钢复合管数值模型，分析了其冶金复合成形机理。结果表明：该工艺可行，能够生产出较好冶金复合状态下的无缝管。

关键词：不锈钢复合管；冶金复合；皮尔格热轧工艺

Study on Pilger Hot Rolling Process and Fusion Mechanism of Stainless Steel Clad Tube

Shuang Yuanhua[1], He Zonglin[1,2]

(1. Shanxi Provincial Key Laboratory of Metallurgical Equipment Design and Technology, Taiyuan University of Science and Technology, Taiyuan 030024, China; 2. Mechanical Engineering Department, Shanxi Engineering Vocational College, Taiyuan 030009, China)

Abstract: Based on the complex production process and high cost of stainless steel clad pipe, the pilger hot rolling seamless clad pipe process is proposed in this paper. The numerical model of Pilger hot rolled stainless steel clad pipe was established, and the metallurgical forming mechanism was analyzed. The results show that the process is feasible and can produce seamless pipe in better metallurgical composite state.

Key words: stainless steel clad pipe; metallurgy clad; pilger hot rolling process

GB/T 4334—2020《金属和合金的腐蚀奥氏体及铁素体-奥氏体（双相）不锈钢晶间腐蚀试验方法》国家标准的修订及说明

朱玉亮，丰　涵，宋志刚，郑文杰，何建国

（钢铁研究总院特殊钢研究所，北京　100081）

摘　要：由钢铁研究总院主持修订的新版国家标准 GB/T 4334—2020《金属和合金的腐蚀奥氏体及铁素体-奥氏体（双相）不锈钢晶间腐蚀试验方法》已于 2020 年 4 月 30 日发布。为便于使用者更好地理解该标准，本文从标准修订的过程、修订的主要内容两个方面对新版标准进行说明。

关键词：不锈钢；晶间腐蚀；国家标准修订

Revision and Introduction for the National Standard GB/T 4334—2020 *Corrosion of metals and alloys-Test methods for intergranular corrosion of austenitic and ferritic-austenitic(duplex) stainless steels*

Zhu Yuliang, Feng Han, Song Zhigang, Zheng Wenjie, HeJianguo

(Institute for Special Steels, Central Iron and Steel Research Institute, Beijing 100081, China)

Abstract: The new edition of national standard GB/T 4334—2020 *Corrosion of metals and alloys-Test methods for intergranular corrosion of austenitic and ferritic-austenitic(duplex) stainless steels* had been revised by CISRI and issued on April 30th, 2020. To enable users to understand the standard more profoundly,the revision procedure and the main difference betweenthe two editions are introduced by the drafter of the standard in this paper.

Key words: stainless steel; intergranular corrosion; national standard revision

SUS301 不锈钢热轧钢带氧化铁皮难酸洗问题研究

魏海霞，陈兴润，纪显彬，刘天增

（酒钢集团宏兴钢铁股份有限公司，甘肃嘉峪关 735100）

摘 要：针对 SUS301 奥氏体不锈钢热轧钢带难酸洗问题，使用 SEM、XRD 对不同酸洗程度热轧钢带氧化皮厚度、结构进行了实验室分析，并进行了机理分析。研究结果表明：SUS301 不锈钢经炉卷轧机轧制的高温钢卷下线后，因为冷却方式及冷却时间不同，造成难酸洗程度的不同。冷却速度慢，氧化皮以 Fe_2O_3 为主，易酸洗且酸洗后表面较好；冷却速度快，氧化皮以 Fe_3O_4、FeO 为主，较难酸洗，表面氧残、边部黑带等酸洗不良缺陷严重。控制黑卷下线冷却速率和冷却方式，可有效避免表面酸洗不良缺陷发生，提高酸洗效率。

关键词：SUS301 奥氏体不锈钢；氧化铁皮；酸洗；冷却

Research on the Problem of Difficult Pickling of Oxide Scale in SUS301 Stainless Steel Hot-rolled Steel Strip

Wei Haixia, Chen Xingrun, Ji Xianbin, Liu Tianzeng

(Hongxing Iron & Steel Co., Ltd., Jiuquan Iron and Steel Group Corporation, Jianyuguan 735100, China)

Abstract: Aiming at the problem of difficult pickling of SUS301 austenitic stainless steel hot-rolled strip,the thickness and structure of oxide scale in SUS301 stainless steel hot rolled strip were analyzed by SEM and XRD in laboratory, and the mechanism of different pickling degrees was analyzed.The results showed thatthe pickling degree of SUS 301 coil rolled by steckel mill is different because of the different cooling methods and cooling time.When the cooling rate is slow, the oxide scale is Fe_2O_3, easy to pickle and the surface is better after pickling;while the cooling rate is fast, the oxide scale is mainly Fe_3O_4 and FeO, which is difficult to pickle, the surface oxygen residue, the edge black belt, and other bad pickling defects are serious.The control of cooling rate and cooling mode of the black coil can effectively avoid the defects of the surface pickling and improve the pickling efficiency.

Key words: SUS301 austenitic stainless steel; oxide scale; pickling; cooling

SUS445J2 高铬铁素体不锈钢耐蚀性评价

任娟红[1]，高仁强[1]，陈安忠[1]，潘吉祥[2]

（1. 酒钢集团钢铁研究院，甘肃嘉峪关 735100；
2. 酒钢集团不锈钢分公司，甘肃嘉峪关 735100）

摘 要：通过盐雾试验、电化学试验、点腐蚀试验和应力腐蚀试验，对比研究了 445J2 超纯铁素体不锈钢和 316L 奥氏体不锈钢的耐腐蚀性能。结果表明：445J2 超纯铁素体不锈钢点腐蚀速率小于 316L 奥氏体不锈钢，具有比 316L 更优异的耐氯离子腐蚀性能；445J2 不锈钢和 316L 不锈钢均不具有晶间腐蚀敏感性；316L 不锈钢有可能发生应力腐蚀断裂，445J2 不锈钢没有这种危险。可见，445J2 超纯铁素体不锈钢是 316L 奥氏体不锈钢理想的替代材料，可广泛用于沿海地区耐蚀性要求较高的建筑屋面、幕墙等众多领域。

关键词： SUS445J2 超纯铁素体不锈钢；盐雾试验；电化学试验；点腐蚀；应力腐蚀试验

Corrosion Resistance Evaluation of SUS445J2 Ultra-pure Ferritic Stainless Steel

Ren Juanhong[1], Gao Renqiang[1], Chen Anzhong[1], Pan Jixiang[2]

(1. Research Institute, JISCO, Jiayuguan 735100, China;
2. Stainless Steel Branch, JISCO, Jiayuguan 735100, China)

Abstract: The corrosion behavior of 445J2 ultra-pure ferritic stainless steel and 316Laustenitic stainless steel has been compared bysalt spray test, electrochemical test, $FeCl_3$ point corrosion test and stress corrosion test. The result shows the Pitting corrosion rate of 445J2 ultra-pure ferritic stainless steel is lower than 316Laustenitic stainless steel, and it has better corrosion resistance than 316L. 445J2 and 316L stainless steel are sensitive to intergranular corrosion. It can be seen that 445J2 ultra-pure ferritic stainless steel is an ideal substitute for 316Laustenitic stainless steel,which can be widely used in many fields such as building roofs and curtain walls with high corrosion resistance requirements in coastal areas.

Key words: SUS445J2 ultra-pure ferriticstainless steel; salt spray test; electrochemical test; pitting corrosiontest;stress corrosion test

J441 超纯铁素体不锈钢表层破皮缺陷的分析

徐　斌，潘吉祥，李具仓

（酒泉钢铁集团宏兴股份公司，甘肃嘉峪关　735100）

摘　要： 通过金相显微镜、扫描电镜、能谱等设备检测了 J441 超纯铁素体不锈钢卷板的表层破皮缺陷及其内部的各类夹杂物类型。研究表明，产生表层破皮缺陷的夹杂物类型为包含 K、Na 的 Ca-Si-AL 的复合氧化物，与连铸机保护渣组分基本一致；卷板中主要包括 TiN 颗粒、NbC 包裹 TiN 复合夹杂物、NbC 以 TiN-($MgO\cdot Al_2O_3$)为析出核心的复合夹杂物，与造成表层破皮缺陷的夹杂物类型存在明显差异。

关键词： 超纯铁素体不锈钢；表层破皮；夹杂物；保护渣

Analysis of Surface Skin-breaking Defect of J441 Ultra-Pure Ferritic Stainless Steel

Xu Bin, Pan Jixiang, Li Jucang

(Hongxing Co., Ltd., Jiuquan Iron and Steel Group Corporation, Jiayuguan 735100, China)

Abstract: The surface cracking defects of J441 ultra-pure ferritic stainless steel coil plate and its inclusion types were examined by means of Metallographic Microscope, Scanning Electron Microscope and Energy Spectrum Scanning.The results showed that the inclusion type of surface cracking defect is Ca-Si-Al oxide containing K and Na, which is consistent with the composition of mold powder.NbC coated TiN compound inclusions and NbC compound inclusions with TiN-(MgO-AL_2O_3) as precipitating core are mainly included in the Coiling plate.

Key words: ultra-pure ferritic stainless steel;surface skin-breaking;inclusions; mold powder

建筑装饰用 J443 超纯铁素体不锈钢夹杂物控制研究

李鸿亮

（酒钢集团宏兴股份钢铁研究院，甘肃嘉峪关 735100）

摘 要： J443 超纯铁素体不锈钢容易出现可浇性恶化、铸坯夹杂物多等问题，造成产品表面质量缺陷。为了提高冶金质量，作者通过缺陷取样，利用光学显微镜、扫描电镜对其进行观察和检测。同时，还进行了大量生产数据统计分析。研究发现：造成产品线鳞缺陷的主要原因是连铸中包水口结瘤引起结晶器液面波动，进而发生结晶器液面卷渣；而铸坯内部夹杂主要是聚集分布的 TiN、Mg-Al-O 的复合夹杂物。为此，从工艺因素和化学反应机理两方面进行分析，进而制定了相应控制措施，经过生产实践验证，产品冶金质量获得了明显提升。

关键词： J443 不锈钢；夹杂物；水口结瘤；质量缺陷

Study on Inclusions of J443 Ultra-pure Ferrite Stainless Steel for Architectural Decoration

Li Hongliang

(Hongxing Iron & Steel Jiuquan Iron and Steel Co., Ltd., Jiayuguan 735100, China)

Abstract: J443 ultra-pure ferrite stainless steel is prone to product surface quality defects, Because of the deterioration of castability and internal inclusions. In order to improve the metallurgical quality, the author observed and detected it by defect sampling, by use of optical microscope and scanning electron microscope. At the same time, a large number of production data had been statistically analyzed. It is found that the main cause of the product line scale defects are the fluctuation in the mould caused by the submerge nozzle clogging in the continuous casting, and the slag involved in slab. While the inclusions in the billet are mainly the composite inclusions of TiN、Mg-Al-O. To this end, the process factors and chemical reaction mechanism have been analyzed, and then, the corresponding control measures have been formulated. After the production practice verification, the product metallurgical quality has been significantly improved.

Key words: J443 stainless steel; inclusions; nozzle clogging; quality defects

不锈钢板材在盐酸溶液中的腐蚀行为研究

靳塞特

（酒钢集团宏兴钢铁股份有限公司，甘肃嘉峪关 735100）

摘　要：利用化学浸泡法研究分析 430、409L、410S 铁素体不锈钢热轧板材在不同温度、不同浓度盐酸溶液中的腐蚀行为，并与 304 奥氏体不锈钢热轧板材做对比。结果表明：耐盐酸腐蚀性能 304>430>409L>410S，奥氏体不锈钢的耐盐酸腐蚀性能优于铁素体不锈钢，较低的碳含量和 Ti 元素的加入提高了 409L 的耐蚀性，410S 中的马氏体降低了它的耐蚀性；由于氯离子的存在，410S 和 430 发生了明显的点腐蚀形貌，304 和 409L 以全面腐蚀为主。盐酸溶液腐蚀性 10%HCl>3%HCl>1%HCl，盐酸的浓度越高其腐蚀能力越强。各钢种在 10% HCl 溶液中的微观腐蚀形貌与其金相组织具有对应性，这说明材料的微观组织对其耐腐蚀性能有直接影响。

关键词：不锈钢；化学浸泡腐蚀；盐酸溶液；微观腐蚀形貌

Study on Corrosion Behaviors of Stainless Steel Plates in Hydrochloric Acid Solution

Jin Saite

(Hongxing Iron & Steel Jiuquan Iron and Steel Co., Ltd., Jiayuguan 735100, China)

Abstract: Corrosion behaviors of hot-rolled ferrite stainless steel plates (430, 409L and 410S) were studied by chemical immersion methods in hydrochloric acid solution at different concentrations and temperatures, and compared with 304 austenite stainless steel hot-rolled plates in same conditions. The study results signify that: descending order for corrosion resistance in hydrochloric acid solution are 304,430,409L and 410S. Austenite stainless steel is superior to ferrite stainless steel on sulfuric acid proofness. Corrosion resistance of 409L is increased by low carbon content and titanium added. Because of martensite phase, corrosion resistance of 410S is lower. Causticity of hydrochloric acid solution with different concentrations can be ranked as 10%HCl, 3%HCl and 1%HCl in descending order. Due to the presence of chloride ion, 410S and 430 have obvious pitting corrosion morphology, and 304 and 409L are mainly based on comprehensive corrosion. According to correspondence between microcosmic pitting morphology and metallurgical structure for these steels at 10% HCl solution, the microstructures of steel plates directly influence its corrosion resistance.

Key words: stainless steel; chemical immersion; hydrochloric acid solution; microcosmic pitting morphology

淬火冷却温度对低铬铁素体不锈钢组织和性能的影响

罗　刚，李　俊，李国平

（太原钢铁（集团）有限公司 先进不锈钢材料国家重点实验室，山西太原　030003）

摘　要：对一种低铬铁素体不锈钢进行了淬火-配分（Quenching and Partitioning，Q&P）处理，研究了淬火冷却温度（20~300℃）对试验材料组织和性能的影响。结果表明，经 Q&P 处理后，试验材料具有铁素体、马氏体和少量残余奥氏体的复合组织，其中马氏体位于铁素体边界，尺寸大小为数微米至数十微米之间。淬火冷却温度对材料的组织形貌影响不大，但对残余奥氏体含量有一定的影响。当淬火冷却温度为 200~250℃时，残余奥氏体含量达到最大（约 5%）。随着淬火冷却温度升高，材料的抗拉强度由 824MPa 降低至 780MPa 后又升高至 812MPa，而断后伸长率由 16.5%升高至 20.5%后又下降至 17.6%。从改善高强钢塑性的角度考虑，低铬铁素体不锈钢在进行淬火-配分（Q&P）处理时，存在一个最佳淬火冷却温度区间。

关键词：淬火-配分；淬火冷却温度；残余奥氏体；马氏体相变；不锈钢

Effect of Quenching Cooling Temperature on Microstructure and Properties of a Low-Chromium Ferritic Stainless Steel

Luo Gang, Li Jun, Li Guoping

(State Key Laboratory of Advanced Stainless Steel, Taiyuan Iron and Steel Group Company Limited, Taiyuan 030003, China)

Abstract: A low-chromium ferritic stainless steel was treated by the Quenching and Partitioning (Q&P) process, and the effect of quenchingcooling temperature(from 20 to 300℃) on the microstructure and properties of the steelwas investigated.The results showed that the steel has a complex structure consisting of ferrite, martensite and a small amount of retained austeniteafter the Q&P treatment. The martensiteis located at the boundary of ferrite, and its size is amongthe range of several microns to tens of microns. The temperature of quenching coolinghas little effect on the microstructure, but has an effect on the content of retained austenite. And when the temperature is 200~250℃, the content of retained austenite reaches the maximum (about 5%). With the increase of the temperature of interrupting quenching, the tensile strength decreases from 824MPa to 780 MPa and then increases to 812MPa, while the elongation after fracture increases from 16.5% to 20.5% and then decreases to 17.6%. From the perspective of improving the plasticity of a low-chromium ferritic stainlesssteel with high strength, there was an optimal temperature rangeof interrupting quenching during the Quenching and Partitioning (Q&P) treatment.

Key words: quenching and partitioning; quenchingcoolingtemperature; retained austenite; martensite transformation; stainless steel

2101 双相不锈钢表面缺陷机理分析

纪显彬，陈兴润，钱张信，任培东

（酒泉钢铁（集团）有限责任公司，甘肃嘉峪关 735100）

摘 要：研究加热温度和时间对双相不锈钢表面缺陷的影响，结果表明，在 1245℃加热的条件下，随着加热时间的延长内氧化程度加剧，氧化向基体内延伸，基体内氧含量较低，优先形成硅锰的氧化物，沿着氧化层向基体内部延伸，在经过热轧后形成双相不锈钢的固有的表面缺陷。

关键词：双相不锈钢；加热温度；在炉时间；起皮；裂纹

Mechanism Analysis of Surface Defects of Duplex Stainless Steel

Ji Xianbin, Chen Xingrun, Qian Zhangxin, Ren Peidong

(Jiuquan Iron & Steel (Group) Co., Ltd., Jiayuguan 735100, China)

Abstract: The effect of heating temperature and time on the surface defects of duplex stainless steel is studied. The results show that the degree of internal oxidation intensifies with the extension of heating time at 1245℃, and the oxidation extends to the matrix. The oxygen content in the matrix is low. The silicon manganese oxide is preferentially formed and extends along the oxide layer to the interior of the matrix. After hot rolling, the inherent surface defects of duplex stainless steel are formed defects.

Key words: keywords duplex stainless steel; heating temperature; in furnace time; peeling; crack

冷轧压下量和时效处理对沉淀硬化不锈钢 630 组织性能的影响

黄俊霞，毕洪运，常　锷

（宝武中央研究院不锈钢技术中心，上海　201901）

摘　要： 本文研究了固溶温度、冷轧压下量和时效处理对沉淀硬化不锈钢 630 的组织性能影响。结果表明：沉淀硬化不锈钢 630 在固溶状态下的组织为板条马氏体和未转变的残余奥氏体。随着固溶温度的升高，残余奥氏体的含量增加。沉淀硬化不锈钢 630 在冷轧条件下具有急剧的加工硬化效应，压下量为 10%时，残余奥氏体几乎全部转变为形变马氏体，强度达到较高的水平，继续增加压下量强度缓慢增加。沉淀硬化不锈钢 630 在 480℃时效后，强度和塑性提高，时效 1h 具有最高的强度和最好的延伸率。

关键词： 固溶温度；冷轧压下量；时效处理；强度

The Influence of Cold Rolling Reduction and Aging Treatment on the Microstructure and Properties of Precipitation Hardening Stainless Steel 630

Huang Junxia, Bi Hongyun, Chang E

Abstract: The influence of solution temperature, cold rolling reduction and aging time on microstructure and properties of precipitation hardening stainless steel 630 was studied.The results is shown that the microstructure of PH stainless steel 630 at room temperature is composed oflath martensite and retained austenite. The percentage of retained austenite is increased with increasing solution temperature. PH stainless steel 630 has anobvious work hardening effect after cold rolling.The retained austenite transformed into deformed martenite under 10% cold reduction. As a result, the strength increased greatly.When the reduction is greater than 20%, the strength increased slowly. The stength and ductility are improved after aging treatment. The best combination of strength and ductility occured after aging at 480℃ for 1 hour.

Key words: solution temperature; cold rolling reduction; aging treatment; strength

高氮奥氏体不锈钢强韧化研究进展

彭翔飞[1]，吕梦楠[1]，李　俊[2]，杨　阳[3]，李国平[2]，刘燕林[3]，王　宇[1]

（1. 中北大学材料科学与工程学院，山西太原　030051；2. 先进不锈钢材料国家重点实验室，山西太原　030003；3. 兵器科学研究院宁波分院，浙江宁波　315103）

摘 要：高氮奥氏体不锈钢（High-Nitrogen Austenitic Stainless Steel，简称 HNASS）是一种目前正在蓬勃发展的非常新颖的材料。节镍高氮的奥氏体不锈钢相比于传统奥氏体不锈钢，其具有优良的综合力学性能，如：高强度、高韧性、大的蠕变抗力、良好的耐腐蚀性能。氮在 HNASS 中各种强化作用对于钢强度的提升具有十分显著的效果，同时它也具有十分优异的吸能和冲击硬化性，尤其是动态冲击硬化性能，故其抗弹性能比较突出。但目前 HNASS 的氮元素强化机理以及动态防护性能研究比较零散，缺少系统性的综述，故本文综述了 HNASS 在氮强化机理和防护领域的研究现状以及进展，主要包括高氮奥氏体不锈钢的强化机理、动态力学性能以及其抗弹性能等几个方面。最后结合国内外研究现状，提出了高氮奥氏体不锈钢在防护领域的未来发展思路。

关键词：高氮奥氏体不锈钢（HNASS）；强化机理；动态力学性能；抗弹性能

Research Progress on Strengthening Mechanism of High Nitrogen Austenitic Stainless Steels

Peng Xiangfei[1], Lü Mengnan[1], Li Jun[2], Yang Yang[3], Li Guoping[2], Liu Yanlin[3], Wang Yu[1]

(1. School of Material Science and Engineering, North University of China, Taiyuan 030051, China;
2. Technology Center,Taiyuan Iron and Steel (Group) Co., Ltd., Taiyuan 030003, China;
3. Ningbo Branch of China Academy of Ordnance Sciences, Ningbo 315103, China)

Abstract: High-Nitrogen Austenitic Stainless Steel (HNASS) is a novel material which is developing rapidly at present. Compared with traditional austenitic stainless steels, nickel saving and high nitrogen austenitic stainless steels have excellent comprehensive mechanical properties, such as high strength, high toughness, large creep resistance and good corrosion resistance. The various strengthening effects of nitrogen in HNASS have a very significant effect on the improvement of steel strength. At the same time, it also has a very excellent energy absorption and impact hardening property, especially the dynamic impact hardening property, so its elastic resistance is more outstanding. But now HNASS nitrogen enhancement mechanism and dynamic protective performance research is scattered and lack of systematic review, therefore, the paper summarizes the HNASS nitrogen in strengthening mechanism and the research status and progress in the field of protection, including the reinforcement mechanism of high nitrogen austenitic stainless steel, dynamic mechanical properties and its resistance to elastic energy. Finally, the future development of high nitrogen austenitic stainless steel in the field of protection is put forward based on the research status at home and abroad.

Key words: high nitrogen austenitic stainless steel (HNASS); strengthening mechanism; dynamic mechanical properties; ballistic performance

Fe-C 相图中奥氏体最大固溶度、共晶点、共析点的成分解析

张 琳[1]，王晓东[1]，张 爽[1]，董 闯[1,2]

（1. 大连交通大学材料科学与工程学院，辽宁大连 116028；
2. 大连理工大学三束材料改性教育部重点实验室，辽宁大连 116024）

摘 要：本报告重点关注 Fe-C 相图中的奥氏体最大固溶度、共晶点及共析点的成分特征，并引入“团簇加连接原

子”模型，对其进行成分解析。该模型将近程有序结构简化为类分子结构单元，可表述成团簇式的形式，即[团簇](连接原子)。由于 C 间隙固溶于奥氏体组织中，因此可确定 C 在奥氏体中以八面体间隙构型的团簇[C-Fe_6]存在，由此可写出团簇式[C-Fe_6]Fe_4≈$Fe_{90.91}C_{9.09}$，解释了奥氏体的最大固溶度 9.06at.%C。根据铁素体与渗碳体的晶体结构，可将其团簇分别确定为[Fe-Fe_{14}]和[C-Fe_9]。因此，Fe-C 相图中的共晶点可采用来自于奥氏体和渗碳体的双团簇式进行解析，即[C-Fe_6+C-Fe_9]C_2Fe_4 =$Fe_{19}C_4$≈ $Fe_{82.61}C_{17.39}$，该结果与德国的 R.Ruer 在 1920 年利用实验所确定的 17.3at.%C 非常接近；共析点可采用来自于铁素体和渗碳体的双团簇式进行解析，即[Fe-Fe_{14}+C-Fe_9]Fe_4=$Fe_{28}C$≈ $Fe_{96.55}C_{3.45}$，该结果与 1961 年德国 E.Scheil 通过热力学计算所得到的 3.44at.%C 几乎一致。

关键词： Fe-C 相图；“团簇加连接原子”模型；奥氏体固溶度；共晶点；共析点；成分解析

基于“团簇加连接原子”模型的新型奥氏体不锈钢成分设计

王晓东[1]，张　琳[1]，张　爽[1]，邹存磊[1]，董　闯[1,2]

（1. 大连交通大学材料科学与工程学院，辽宁大连　116028；
2. 大连理工大学三束材料改性教育部重点实验室，辽宁大连　116024）

摘　要： 奥氏体不锈钢在冶炼过程中需要严格控制其元素含量，以求达到产品所需的性能。然而，要想对奥氏体不锈钢成分进行精确设计，需理解其成分特征，并揭示其成分根源。为此，我们引入“团簇加连接原子”模型，对奥氏体不锈钢进行了成分解析及设计。首先，根据类分子结构单元的思想，可计算出奥氏体不锈钢的 16 原子成分通式，即[(Si, Mn, Ni)-(Fe, Ni)$_{12}$]Cr_3，其中，[(Si, Mn, Ni)-(Fe, Ni)$_{12}$]表示面心立方奥氏体组织的团簇部分，具有立方八面体构型，而连接原子部分是 3 个 Cr，以满足耐蚀性要求。然后，基于该成分通式，我们解析了现有的全部奥氏体不锈钢牌号成分，并结合 Schaeffler 组织图，根据 Cr 当量和 Ni 当量对奥氏体不锈钢成分特征进行了深入探索，验证了成分通式的合理性。最后，在团簇式指导下，同时考虑敏化因素，我们进一步调整了 C 含量和 Cr 含量，设计出一系列不同于现有奥氏体不锈钢牌号的新成分，并实施结构表征和性能测试实验，将所获得的组织特征与性能参数与常用的 304 和 316L 不锈钢进行对比，确定出综合性能更优异的新成分，从而实现了新型奥氏体不锈钢的研发。

关键词： 奥氏体不锈钢；“团簇加连接原子”模型；成分式；当量

铬系耐蚀塑料模具钢预硬化工艺及组织研究

王勇胜，孙盛宇，付　博

（东北特殊钢集团股份有限公司技术中心，辽宁大连　116105）

摘　要： 分析了塑料模具钢的行业现状及 4Cr13 型耐蚀塑料模具钢的特性，对不同热处理工艺的 4Cr13 模具钢的显微组织、硬度及成分进行了对比分析。结果显示，模铸的 4Cr13 钢经在线预硬、一次回火后，显微组织呈条带状分布，其原因为钢中成分偏析导致，经二次回火后可减轻。另外，对不同冶炼状态下的 4Cr13 模具钢组织进行了对比分析，结果显示，电渣钢经二次回火后，组织细致均匀，可用于具有高要求的塑料模具生产制造。

关键词： 预硬化；显微组织；耐蚀塑料模具钢；4Cr13

Study on the Pre-hardening Process and Microstructures of Cr-corrosion Resistant Plastics Die Steel

Wang Yongsheng, Sun Shengyu, Fu Bo

(RD Center of Dongbei Special Steel Group, Dalian　116105, China)

Abstract: The present situation of die and mould industry and the corrosion resistant plastic die steel 4Cr13 are analyzed, the microstructures, hardnesses and chemical compositions of different heat treatment methods are compared. Result shows that, after online pre-hardening process and once tempered, the microstructures of ingot cast 4Cr13 steel are distribute on band-pattern, caused by chemical composition segregation, which could be reduced by twice temper treatment. What's more, the microstructures of die steel 4Cr13 of different smelt method are compared too. Result shows that, after ESR treatment, the microstructures are fine and uniform, the steel can be used on the manufacture of high quality plastic die.

Key words: pre-hardening process; microstructures; corrosion resistant plastic use mould; 4Cr13

海洋工程用超级奥氏体不锈钢的发展

李乒乒，田志凌，郎宇平，陈海涛，屈华鹏，冯翰秋

（钢铁研究总院，北京　100081）

摘　要：发展海洋油气产业是我国未来建设海洋强国的重要战略，也是建设能源强国的战略需求，但是与陆地油气开采相比，海水流动剧烈、海温和压力随深度变化大、海底岩层结构与陆地井迥异，且海洋油气中 H_2S、CO_2 和 Cl^-等含量普遍较高，存在固液气三相腐蚀，海底微生物种类复杂，易发生腐蚀。超级奥氏体不锈钢有着无磁性、具有优异的耐腐蚀性、有良好的冷热成型性和焊接性能，被广泛应用于此环境中，但目前研发超级奥氏体不锈钢的厂商均为国外，我国超级奥氏体不锈钢任重而道远。

关键词：海洋工程；不锈钢；发展

Development of Super Austenitic Stainless Steels for Marine Engineering

Abstract: Development of ocean oil and gas industry is the important strategy of the future construction of Marine power in our country, and the strategic requirements of building energy powers, but compared with land oil and gas, the sea water flow sharp, SST and pressure changes with the depth of the great rock strata, the bottom of the sea and land Wells, and Marine oil and gas such as H_2S, CO_2 and Cl^- in the content is higher, the solid liquid gas three phase corrosion, The species of microorganisms on the seafloor are complex and prone to corrosion. Super austenitic stainless steel has no magnetism, has excellent corrosion resistance, has good cold and hot forming and welding performance, is widely used in this environment, but the current research and development of super austenitic stainless steel manufacturers are foreign, China's super austenitic stainless steel has a long way to go.

Key words: marine engineering; stainless steel; development

444 铁素体不锈钢高温氧化时微量元素的表现行为

段秀峰[1,2]，张晶晶[1]，王　斌[1]，李国平[1]

（1. 太原钢铁（集团）有限公司先进不锈钢材料国家重点实验室，山西太原　030003；
2. 太原科技大学材料与科学工程学院，山西太原　030024）

摘　要：本工作研究了 3 种成分体系 444 不锈钢在 1100℃下短时（4h）高温氧化行为，利用 X 射线衍射、扫描电子显微镜和能谱仪分析氧化膜物相、微观形貌及化学成分分布。结果表明，1100℃高温下 444 样品短时氧化增重明显，氧化膜结构为$(Mn,Cr)_3O_4/Cr_2O_3/SiO_2$。Mn 元素的快速扩散促进了 Nb 元素向氧化膜与金属界面处扩散，形成内生的 Nb 氧化物增大了内应力和热应力，造成氧化膜剥落。界面处 SiO_2 势垒层，阻止了金属离子向外扩散和氧原子向内扩散，降低氧化速率，避免异常氧化发生。Al、Ti 元素发生了明显的内氧化，内氧化层深度与 SiO_2 厚度成反向关系。

关键词：铁素体不锈钢；高温氧化；微量元素；氧化膜剥落；扩散；内氧化

Behavior of Minor Elements in 444 Ferritic Stainless Steel during High Temperature Oxidation

Duan Xiufeng[1,2], Zhang Jingjing[1], Wang Bin[1], Li Guoping[1]

(1. Taiyuan Iron and Steel (Group) Co., Ltd State Key Laboratory of Advanced Stainless Steel, Taiyuan 030003, China; 2. College of Materials Science and Engineering, Taiyuan University of Science and Technology, Taiyuan 030024, China)

Abstract: We investigated the short-term (4h) high-temperature oxidation behavior of 444 stainless steel with three component systems at 1100℃. X-ray Diffraction, Scan Electron Microscopy and Energy Dispersive Spectrometer were used to analyze the phase, microscopic morphology and chemical composition distribution of the oxide film. The results show that the short-term oxidation weight gain of the 444 sample at a high temperature of 1100℃ is obvious, and the oxide film structure is $(Mn,Cr)_3O_4/Cr_2O_3/SiO_2$. The rapid diffusion of Mn element promotes the diffusion of Nb element towards the interface between the oxide film and the metal, and the formation of interstitial Nb oxide increases the internal stress and thermal stress, causing the oxide film to peel off. The SiO_2 barrier layer at the interface prevents the outward diffusion of metal ions and the inward diffusion of oxygen atoms, reducing the oxidation rate and avoiding abnormal oxidation. The Al and Ti elements have undergone obvious internal oxidation, and the depth of the internal oxide layer has an inverse relationship with the thickness of SiO_2.

Key words: ferritic stainless steel; high temperature oxidation; minor elements; spallation of oxide film; diffusion; internal oxidation

Micro/Nano-structure Leads to Super Strength and Excellent Plasticity in Nanostructured 304 Stainless Steel

Sheng Jie[1,2], Wei Jiafu[1], Li Yufeng[2], Ma Guocai[1,2], Li Zhengning[3], Meng Qian[2,3], Zheng Yuehong[1], La Peiqing[1]

(1. State Key Laboratory of Advanced Processing and Recycling of Nonferrous Metal, Lanzhou University of Technology, Lanzhou 730050, China; 2. Jiuquan Iron and Steel Group, Jiayuguan 735000, China; 3. School of Materials Science and Engineering, Lanzhou Jiaotong University, Lanzhou 730070, China)

Abstract: Micro/nano-structure is one of the important types of multilayer construction, which could make nanocrystalline metals and alloys obtained excellent comprehensive mechanical properties. By means of Aluminothermic Reaction (AR) followed by cold rolling at room temperature and annealing treatment, a heterogeneous composite structure of micro/nano-structured 304 stainless steel is obtained. Such a stainless steel specimen exhibits a tensile strength as high as 1023MPa and an elongation-to-failure of about 27.3%. In the process, the much elevated strength originates from the martensite strengthening, austenite grain refinement and dislocation density increasing, while ductility from the recrystallization of deformed austenite and recovery of strain-induced martensite. Superior strength-ductility combination achieved in micro/nano-structured 304 stainless steel demonstrates a novel approach for optimizing the mechanical properties in engineering materials.

Key words: 304 stainless steel; micro/nano-structure; strength and plasticity; heterogeneous composite structure

硼对超级双相不锈钢棒材热轧裂纹的影响

郑立春，黄斯琦，胡涵哲，李滢玉

（东北大学冶金学院，辽宁沈阳　110819）

摘　要：为研究硼微合金化对超级双相不锈钢棒材表面热轧裂纹的影响，对比了含硼与不含硼的超级双相不锈钢热轧棒材表面质量，发现不含硼的热轧棒材表面局部存在三角裂纹，而含硼的棒材表面未见明显裂纹。裂纹微观组织分析表明该三角裂纹属于应力开裂，与冷却过程 σ 相的析出有关。不含硼的热轧棒材中 σ 相面积占比为 12.64%，而含硼热轧棒材中 σ 相面积占比为 10.66%。进一步的等温析出表明，硼抑制初期 σ 相的析出速度，延长时间和提高温度均会减弱该抑制效果。

关键词：超级双相不锈钢；热轧裂纹；硼微合金化；σ 相

Effect of Boron on Hot Rolling Cracks of Super Duplex Stainless Steel Bars

Zheng Lichun, Huang Siqi, Hu Hanzhe, Li Yingyu

(School of Metallurgy, Northeastern University, Shenyang 110819, China)

Abstract: To study the effect of boron microalloying on surface cracks of super duplex stainless steel, the surface quality of hot rolled super duplex stainless steel bars was compared. It was found that triangular cracks existed locally on the surface of hot rolled bars without boron. However, no obvious cracks were found on the surface of boron-containing bars. The microstructure analysis of the crack shows that the triangular crack belongs to stress cracking, which is related to the precipitation of σ phase during cooling. The proportion of σ phase area in boron-free hot rolled bars is 12.64%, while that in boron-containing hot rolled bars is 10.66%. Further isothermal precipitation shows that boron inhibits the precipitation rate of σ phase in initial stages. Prolonging time and increasing temperature will weaken the inhibition effect.

Key words: super duplex stainless steel; hot rolling crack; boron microalloying; σ phase

AISI 301 精密带钢厚度及退火处理对耐蚀性能和力学性能的影响

李亚季[1]，施爱娟[2]，李俊[1,3]，王剑[1]，韩培德[1]

（1. 太原理工大学材料科学与工程学院，山西太原　030024；
2. 烟台汽车工程学院车辆运用工程系，山东烟台　265500；
3. 太原钢铁集团太钢技术中心，山西太原　030000）

摘　要：采用 SEM、EBSD 和 TEM 对不同厚度退火前后 AISI 301 精密带钢的微观组织进行表征，并讨论了对耐蚀性能及力学性能的影响。结果表明：厚度 0.02mm 的试样变形带明显，晶粒尺寸处在 0.1~0.49μm 亚微米区域，变形带中含有大量马氏体及高密度位错；而 0.15mm 和 0.3mm 的试样晶粒尺寸处在 10~28μm 区域。厚度越小耐蚀性越差，这要是由于冷变形组织中马氏体含量不同引发的。由于超薄带钢延伸率极低，对其进行了退火处理，退火处理有利于使马氏体转变为奥氏体，并消除轧制变形带来的高密度位错等。0.02mm 试样退火后延伸率得到大幅提升，尤以退火温度达 625℃时具有较好的耐蚀性和强韧性。

关键词：超薄精密带钢；AISI 301 不锈钢；耐蚀性能；力学性能；退火

Effect of Thickness and Annealing Treatment on Corrosion Resistance and Mechanical Properties of AISI 301 Precision Strip Steel

Li Yaji[1], Shi Aijuan[2], Li Jun[1,3], Wang Jian[1], Han Peide[1]

(1. Department of Materials Science and Engineering Taiyuan University of Technology, Taiyuan 030024, China; 2. Yantai Automobile Engineering Professional College, Yantai 265500, China; 3. Taiyuan Iron and Steel Group Taiyuan Iron and Steel Technology Center, Taiyuan 030000, China)

Abstract: The microstructure of AISI 301 precision strip before and after annealing with different thickness was characterized by SEM, EBSD and TEM, and the effects on corrosion resistance and mechanical properties were discussed. The results show that the deformation band of the sample with thickness of 0.02 mm is obvious, whose grain size is between 0.1-0.49μm sub micron region, the deformation zone contains a large number of martensite and high-density dislocations. Besides, the grain size of 0.15mm and 0.3mm samples is 10-28μm area. The smaller the thickness, the worse the corrosion resistance, which is caused by the different content of martensite in the cold deformed structure. The elongation of ultra-thin strip is very low, so it was annealed. Annealing treatment is conducive to transforming martensite into austenite and eliminating high-density dislocations caused by rolling deformation. After annealing, the elongation of 0.02mm sample is greatly improved, especially when the annealing temperature is 625℃, it has better corrosion resistance and strength and toughness.

Key words: ultra thin precision strip steel; AISI 301 stainless steel; corrosion resistance; mechanical properties; annealing

退火温度对超纯铁素体不锈钢 441 组织和力学性能的影响

武 敏，李国平，邹 勇

（太原钢铁（集团）有限公司先进不锈钢材料国家重点实验室，
山西太钢不锈钢股份有限公司技术中心，山西太原 030003）

摘 要：为探索超纯铁素体不锈钢热轧板材在退火过程中组织和力学性能的演变，对 441 进行了 900-1050℃的退火实验，利用 OM、SEM 和 TEM 表征了 441 在退火过程中显微组织的变化规律，并通过拉伸实验和冲击实验研究了退火温度对力学性能的影响。结果表明，随着退火温度升高，轧制组织发生再结晶，且晶粒逐渐长大。退火后 441 热轧板材中存在三种析出相，初生(Ti, Nb) (C, N)、二次 Nb(C, N)和 Laves 相。Laves 相仅在 900～950℃退火样品中大量析出，尺寸约为几百纳米。441 的屈服强度随着退火温度的升高先减小再增大，抗拉强度逐渐降低，而延伸率逐渐升高。形变强化对材料的屈服强度具有最大贡献，固溶强化次之，析出强化最小。冲击实验结果显示，1000℃退火后 441 具有最低的韧脆转变温度，第二相与晶粒长大对材料韧性均有显著的不良影响。

关键词：超纯铁素体不锈钢；退火温度；Laves 相；晶粒尺寸；韧脆转变温度

Effect of Annealing Temperature on Microstructure and Mechanical Properties of Ultra-purified Ferritic Stainless Steel 441

Wu Min, Li Guoping, Zou Yong

(Taiyuan Iron and Steel (Group) Co., Ltd. State Key Laboratory of Advanced Stainless Steel,
ShanxiTaigang Stainless SteelCo., Ltd., Technical Center, Taiyuan 030003, China)

Abstract: The present study is to explore the microstructure evolution and mechanical properties of 441 ultra-purified ferritic stainless steel during annealing. 441 hot-rolled plates were annealled at 900-1050℃, and the microstructure and precipitates during annealing were characterized by OM, SEM and TEM. The effects of annealing temperature on mechanical properties were studied by tensile test and impact test. The results show that the rolling structure recrystallizes and the grains grow gradually with increasing annealing temperature. There are three precipitates in 441 hot-rolled-annealed plate, primary (Ti,Nb) (C,N), secondary Nb(C,N) and Laves phase. A large number of Laves phase, with the size of hundreds nanometers, precipitates in samples annealed at 900-950℃. As the annealing temperature increases, the yield strength of 441 firstly decreases and then increases, the tensile strength decreases gradually, while the elongation increases continually. Deformation strengthening has the greatest contribution to the yield strength, followed by solution strengthening and precipitation strengthening. The impact test indicates that 441 plate has the lowest ductile to brittle transition temperature after annealing at 1000°C, and Laves phase and grain growth are detrimental to the impact toughness of this material.

Key words: ultra-purified ferritic stainless steel; annealing temperature; Laves phase; grain size; ductile to brittle transition temperature (DBTT)

超级双相不锈钢 S32750 热压缩组织演变的 EBSD 研究

张寿禄，武　敏

（山西太钢不锈钢股份有限公司技术中心，山西太原　030003）

摘　要：通过不同条件的高温热压缩试验，研究了 S32750 钢的高温压缩变形流变力学行为，采用扫描电镜（SEM）和电子背散射衍射（EBSD）技术分析了热压缩变形过程中的组织演变规律。结果显示：流变应力随着变形程度的增加而先增加后降低，流变应力随温度的升高而降低，随着变形温度的升高，峰值应力向应变减小的方向移动；两相的动态再结晶比例均随着变形温度的提高而提高，而奥氏体相的变化幅度要明显的大；应变速率的提高有利于奥氏体相的动态再结晶；变形温度对动态再结晶的影响要比应变速率的影响显著；减小应变速率，有利于铁素体向奥氏体相的转变；在热压缩试样中观察到典型的通过大角度晶界迁移实现的“不连续动态再结晶”现象。

关键词：超级双相不锈钢 S32750；热压缩；EBSD；显微组织；动态再结晶

EBSD Investigation on Microstructure Evolution of Super Duplex Stainless Steel S32750 during Hot Deformation

Zhang Shoulu, Wu Ming

(Technology Center, Shanxi Taigang Stainless Steel Co., Ltd., Taiyuan 030003, China)

Abstract: In the present study, the hot deformation behavior of super duplex stainless steel S32750 was investigated by hot compression tests at different conditions, and the microstructure evolution was analyzed by SEM and EBSD. The results showed that the flow stress first increased and then decreased with increase of stain. With increasing deformation temperature, the flow stress decreased and the peak stress moved to smaller strain. The dynamic recrystallization fraction of both phases increased at higher temperature, while the change in austenite phase is significantly larger. The increase of strain rate promoted the dynamic recrystallization of austenite. For the dynamic recrystallization of S32750 steel during hot working, the effect of deformation temperature was more significant than that of strain rate. Decreasing strain rate was beneficial to the transformation of ferrite to austenite. The typical phenomenon of "discontinuous dynamic recrystallization" was observed in the hot compression specimens.

Key words: super duplex stainless steel S32750; hot compression; EBSD; microstructure; dynamic recrystallization

浅析不锈钢绿色生产中感应炉的作用

钱红兵

（应达工业（上海）有限公司，上海　201203）

摘 要：钢铁行业可持续发展必须依靠绿色生产，适应日益严格的环保要求，可以提高资源利用效率和资源投入产出比，有效降低钢铁企业生产成本。短流程钢铁生产可以打幅度降低能演消耗，减少温室气体的排放，应该是不锈钢优先采用的工艺。炼钢的所用的电炉包括电弧炉与感应炉，感应炉具有能耗低和温室气体排放量少的特点，在不锈钢的绿色生产中起特殊的作用。应达公司的 VIP Power-Trak 电源，具有输出频率自动跟踪和自动调节电源的输出频率的功能，大直径的“矮胖”炉型能够减少水冷热损失。

关键词：绿色生产；不锈钢；电弧炉；感应炉；短流程

Analysis of the Role of Induction Furnace in Green Production of Stainless Steel

Qian Hongbing

(Inductotherm Industrial (Shanghai) Co., Ltd., Shanghai 201203, China)

Abstract: The sustainable development of iron and steel industry must rely on green production to meet the increasingly stringent requirements of environmental protection. It can improve resource utilization efficiency and resource input-output ratio, and effectively reduce the production cost of iron and steel enterprises. Short process steel production can greatly reduce energy consumption and greenhouse gas emissions, which should be the preferred process for stainless steel. The electric furnace used in steelmaking includes electric arc furnace and induction furnace. The induction furnace has the characteristics of low energy consumption and low greenhouse gas emissions, which plays a special role in the green production of stainless steel. VIP Power-Trak supplied by Inductotherm has the functions of automatic tracking of output frequency and automatic adjustment of output frequency of power supply. Large diameter "dwarf" furnace type can reduce water cooling heat loss.

Key words: green production; stainless steel; electric arc furnace; induction furnace; short process

低铬铁素体不锈钢 Cr12 焊接工艺研究

赵振铎[1,2]，李　莎[1,2]，范光伟[1,2]，张心保[1,2]

（1. 太原钢铁（集团）有限公司先进不锈钢材料国家重点实验室，山西太原　030003；
2. 山西太钢不锈钢股份有限公司技术中心，山西太原　030003）

摘 要：本文分别采用常规熔化和高频焊对低铬铁素体不锈钢焊接接头组织及韧性进行了研究，结果表明：低铬铁素体不锈钢 Cr12 经焊接热循环后热影响区脆化倾向比较严重，这与热影响区铁素体晶粒严重长大有关；低铬铁素体不锈钢 Cr12 HAZ 粗晶区保温时间达到 20s 以后晶粒长大仍然继续，在焊接过程中温度是影响低铬铁素体不锈钢 Cr12 晶粒长大的最主要因素，减小焊接热输入是抑制低铬铁素体不锈钢 Cr12 晶粒长大的有效手段之一；采用高频焊焊接低铬铁素体不锈钢 Cr12，焊接接头的焊缝及热影响区组织细小，这种焊接接头组织性能有利焊接接头性能的改善；采用高频焊工艺制备的低铬铁素体不锈钢 Cr12 焊管焊接接头的焊缝及热影响区显微组织分布均匀且晶粒细小，接头的综合力学力性良好，焊管经压扁后，焊缝无开裂；对 Φ219mm×4mm 低铬铁素体不锈钢 Cr12 焊管进行抗瓦斯煤层爆破冲击性能试验评价，经过两次瓦斯煤尘爆炸冲击测试，不锈钢瓦斯管抽采管均未发生位置移动，法兰连接完好，螺栓未松动，管材表面完整、无裂隙、未发生变形，该项铁素体不锈钢抗瓦斯煤层爆破冲击性能评价工作的开展，为今后铁素体不锈钢焊管在瓦斯抽采领域的推广应用提供了重要基础试验数据支撑。

关键词：低铬铁素体不锈钢；Cr12；MAG；高频焊；显微组织；韧性

Study on Welding Process of Low Chromium Ferritic Stainless Steel Cr12

Zhao Zhenduo[1,2], Li Sha[1,2], Fan Guangwei[1,2], Zhang Xinbao[1,2]

(1. State Key Laboratory for Advanced Stainless Steel Materials, Taiyuan Iron & Steel (Group) Co., Ltd., Taiyuan 030003, China; 2. Technology Center of Shanxi Taigang Stainless Steel Co., Ltd, Taiyuan 030003, China)

Abstract: In this paper, the microstructure and toughness of welded joints of low chromium ferritic stainless steel were studied by the conventional melting and high frequency welding, respectively. The results show that the embrittlement tendency of heat affected zone of low chromium ferritic stainless steel Cr12 is serious after the welding thermal cycle, which is related to the serious growth of ferrite grains in HAZ; The grain growth of Cr12 HAZ coarse grain zone of low chromium ferritic stainless steel continues after 20s holding time. Therefore, the temperature is the most important factor affecting the grain growth of Cr12 HAZ coarse grain zone of low chromium ferritic stainless steel during the welding. Therefore, reducing the welding heat input is one of the effective means to inhibit the growth of Cr12 grain in the low chromium ferritic stainless steel; High frequency welding is used to weld Cr12 low chromium ferritic stainless steel. The microstructure of the weld and heat affected zone of the welded joint is fine, which is beneficial to the improvement of the properties of the welded joint; The results show that the microstructure of welded joint and heat affected zone of low chromium ferritic stainless steel Cr12 welded pipe prepared by high frequency welding process is uniform and the grain size is fine. The joint has good comprehensive mechanical properties. After flattening, the weld has no cracking; Φ219mm × 4mm low chromium ferritic stainless steel Cr12 welded pipe was tested and evaluated for its anti gas coal seam explosion impact performance. After two times of gas coal dust explosion impact test, the stainless steel gas drainage pipe did not move, the flange connection was intact, the bolts were not loose, and the pipe surface was complete, without cracks and deformation, The evaluation results of the blast impact performance of the ferritic stainless steel in gas resistant coal seam provide important basic test data support for the popularization and application of the ferritic stainless steel welded pipe in the field of gas drainage in the future.

Key words: low chromium ferritic stainless steel; Cr12; MAG; high frequency welding; microstructure; toughness

8Cr13MoV 钢电渣锭共晶碳化物的分布与控制

张　杰[1]，李　晶[1]，史成斌[1]，李首慧[1]，李积回[2]

（1. 北京科技大学钢铁冶金新技术国家重点实验室，北京　100083；
2. 阳江十八子集团有限公司，广东阳江　529500）

摘　要：解剖并分析了 300kg 工业级 8Cr13MoV 钢电渣锭，明确了电渣锭中共晶碳化物的类型、形貌及分布，为共晶碳化物的控制奠定基础。电渣锭顶部和底部冷却条件的差异导致沿电渣锭轴向共晶碳化物的尺寸和分布不均匀。为提高电渣锭共晶碳化物轴向尺寸和分布的均匀性，提出电渣重熔过程逐渐减小电流的技术，与恒电流电渣重熔工艺相比，电渣锭顶部中心处共晶碳化物的体积分数由 1.8%降低到 0.9%，顶部边缘处体积分数由 1.5%降低到 0.2%。

关键词：共晶碳化物；电渣重熔；马氏体不锈钢；分布

Distribution and Control of Eutectic Carbide of 8Cr13MoV Electroslag Ingot

Zhang Jie[1], Li Jing[1], Shi Chengbin[1], Li Shouhui[1], Li Jihui[2]

(1. State Key Laboratory of Advanced Metallurgy, University of Science and Technology, Beijing 100083, China; 2. Yangjiang Shibazi Group Co., Ltd., Yangjiang 529500, China)

Abstract: Industrial scaled 8Cr13MoV electroslag remelting ingot with 300kg in weight was dissected and analyzed. The type, morphology and distribution of eutectic carbide in electroslag ingot were made clear, which laid the foundation of controlling eutectic carbide. The difference of cooling conditions at the top and bottom of electroslag ingot resulted in the inhomogeneity of size and distribution of eutectic carbide along the axis of electroslag ingot. The axial uniformity of the size and distribution of eutectic carbide was improved by gradually reducing the electroslag remelting current. Compared with constant current electroslag remelting, the eutectic carbide in the electroslag remelting ingot with decreasing current obviously gained better control. The volume fractions of eutectic carbide were 0.9% at the top center and 0.2% at the top edge of the modified electroslag remelting ingot while they were 1.8% and 1.5% in the traditional electroslag remelting ingot, respectively.

Key words: eutectic carbide; electroslag remelting; martensitic stainless steel; distribution

低铬铁素体不锈钢的局部腐蚀特性研究

惠　恺，李具仓

（酒钢集团宏兴钢铁股份有限公司，甘肃嘉峪关　735100）

摘　要：低铬铁素体不锈钢以其低廉的价格，优良的冷加工性和适中的耐腐蚀能力，广泛应用在日常制品加工，集装箱外壳制造及汽车排气系统冷端的制造中。此类钢种由于耐蚀性一般，一旦在服役过程中发生局部腐蚀，则会迅速导致制品失效。本文对四种常见的低铬铁素体钢种进行了 $FeCl_3$ 溶液浸泡腐蚀测试，极化曲线测试，及动电位再活化（EPR）测试，最终发现高碳氮元素含量及不含稳定化元素的低铬铁素体不锈钢不存在明显的局部腐蚀倾向，而低碳氮元素含量及含 Nb，Ti 等稳定化元素的低铬铁素体不锈钢具有明显的局部腐蚀倾向，且 Nb，Ti 双稳定的钢种耐局部腐蚀能力优于单 Ti 稳定的钢种。

关键词：低铬铁素体不锈钢；局部腐蚀；点腐蚀；极化曲线；动电位再活化

不同脱氧方式对 440C 不锈轴承钢夹杂物的影响

马　帅[1]，李　阳[1]，姜周华[1]，孙　萌[1]，鸡永帅[1]，刘　航[2]

（1. 东北大学 冶金学院，辽宁沈阳　110819；2. 中国科学院金属研究所，辽宁沈阳　110016）

摘　要：本文通过 $MoSi_2$ 炉以不同脱氧方式制备了 3 炉 440C 不锈轴承钢，采用扫描电镜、全氧分析等手段研究了不同脱氧方式对 440C 不锈轴承钢中 T.O 含量及夹杂物的尺寸、分布、形貌和成分影响。得出如下结论：3 炉实验钢中 T.O 含量由低到高依次为：先 Al 后 Ce 脱氧<Al 脱氧<先 Si 后 Ce 脱氧，其中采用先 Al 后 Ce 脱氧可将 T.O 含量控制在 13 ppm。先 Al 后 Ce 脱氧的钢中夹杂物平均直径最低，为 1.96μm，夹杂物密度为 3 炉最高的 164 个/mm^2。采用 FactSage 热力学计算软件对夹杂物形成现象进行了验证，并分析了三种脱氧方式对 440C 不锈轴承钢夹杂物的影响。在本实验条件下，先 Al 后 Ce 脱氧对 440C 中 T.O 含量及夹杂物优化较为明显，为本实验条件下最佳脱氧剂。

关键词：440C；脱氧方式；夹杂物；FactSage 热力学计算

Effect of Different Deoxidation Methods on Inclusions in 440C Stainless Bearing Steel

Ma Shuai[1], Li Yang[1], Jiang Zhouhua[1], Sun Meng[1], Ji Yongshuai[1], Liu Hang[2]

(1. School of Metallurgy, Northeastern University, Shenyang 110819, China;
2. Institute of Metal Research, Chinese Academy of Sciences, Shenyang 110016, China)

Abstract: In this paper, 3 furnaces of 440C stainless bearing steel were prepared by $MoSi_2$ furnace with different deoxidation methods. The effects of different deoxidation methods on T.O content and the size, distribution and shape of inclusions in 440C stainless bearing steel were studied by means of scanning electron microscopy and total oxygen analysis. Appearance and composition influence. The following conclusions are drawn: the T.O content in the three furnaces of test steels from low to high is: Al first and then Ce deoxidation<Al deoxidation<First Si and then Ce deoxidation, and the T.O content can be controlled at 13 ppm by adopting Al first and then Ce deoxidation. In the steel deoxidized by Al and Ce, the average diameter of inclusions is the lowest, 1.96μm, and the density of inclusions is 164/mm^2, the highest among three furnaces. FactSage thermodynamic calculation software was used to verify the formation of inclusions, and the influence of three deoxidation methods on inclusions in 440C stainless bearing steel was analyzed. Under the experimental conditions, the optimization of T.O content and inclusions in 440C is more obvious for Al first and then Ce deoxidation, which is the best deoxidizer under the experimental conditions.

Key words: 440C; deoxygenation method; inclusions; FactSage thermodynamic calculation

高氮奥氏体不锈钢液中氮行为的热力学模拟

游志敏，姜周华，李花兵

（东北大学冶金学院，辽宁沈阳　110819）

摘　要：为了考察高氮奥氏体不锈钢液中氮的行为，本研究对 Fe-Mn-Cr-Si-N 体系的液相进行了热力学模拟。基于对所有可用实验数据的批判性评价，本模拟利用修正的准化学模型（MQM）对 Fe-Mn-Cr-Si-N 体系范围内所有含氮液相的吉布斯自由能进行了热力学优化。与假设熵随机混合和所有体系统一应用 Muggianu 内插技术的传统 Bragg-Williams 随机混合模型（BWRMM）相比，基于配对近似的 MQM 考虑了原子对自由分布的构型混合熵，因此它对液相的描述更加符合实际情况。并且，为了描述液相的短程有序现象，MQM 中引入了可随成分变化的配位数。此外，MQM 还可根据所涉及二元体系液相的性质选用合适的内插技术，旨在同时提高对三元和多元体系液相的预测准确度和减少模型参数的数量。本研究按照一元、二元、三元、多元体系的顺序对 Fe-Mn-Cr-Si-N 体系液相进行了系统的热力学模拟，并根据吉布斯自由能最小化原理，利用 FactSage 软件确定了整个五元体系液相的一组吉布斯自由能，其中该吉布斯自由能为温度、压力和成分的函数。另外，通过研究氮在纯铁液的热力学性质和锰、铬、硅对纯铁液中氮溶解行为的影响发现，锰、铬元素可以大大促进氮在铁液中的溶解，而硅元素对铁液的氮的溶解则呈相反的作用，根据该结果还可以反推出不同锰、铬、硅含量下铁液中氮的活度系数。通过对比实验数据可知，只需少量的模型参数可以准确地再现不同温度和压力下氮在Fe-Mn-Cr-Si-N体系各液态合金内的溶解度。总的来说，本模拟对理解高氮奥氏体不锈钢冶炼过程中氮的行为具有重要意义。特别地，本研究所建立的热力学数据库可用于计算特定条件下氮在高氮奥氏体不锈钢液中的溶解度，而且还可以用于预测任意化学成分、温度和压力条件下Fe-Mn-Cr-Si-N体系范围内未被研究的液相相平衡关系和热力学性质。

关键词： 热力学模拟；Fe-Mn-Cr-Si-N液相；高氮奥氏体不锈钢；修正的准化学模型（MQM）；氮溶解度

Thermodynamic Modeling of the Behavior of Nitrogen in Liquid High-nitrogen Austenitic Stainless Steels

You Zhimin, Jiang Zhouhua, Li Huabing

(School of Metallurgy, Northeastern University, Shenyang 110819, China)

Abstract: Thermodynamic modeling of the liquid Fe-Mn-Cr-Si-N solution was performed in this study to investigate the behavior of N in molten high-N austenitic stainless steels. In the modeling, Gibbs energies of all N-containing liquid solutions within the Fe-Mn-Cr-Si-N system were thermodynamically optimized using the Modified Quasichemical Model (MQM) based on critical evaluation of all available experimental data. Compared to conventional Bragg-Williams Random Mixing Model (BWRMM) with assumption of random mixing of entropy and an intrinsic Muggianu interpolation technique for all the systems, the MQM in the pair approximation gives more realistic description of the liquid solution by considering the configurational entropy of mixing given by the random distribution of atom pairs, and composition-dependent coordination numbers are introduced in MQM for describing the short-range ordering exhibited in the liquid solution. Besides, proper interpolation techniques depending on the nature of involved binary liquids are also employed in MQM for improving the predictive accuracy on descriptions of ternary and higher-order liquids and reducing the number of model parameters simultaneously. In the present study, a systematic thermodynamic modeling of the liquid Fe-Mn-Cr-Si-N solution was carried out in the order of unary, binary, ternary, multi-component systems, and one set of Gibbs energy as a function of temperature, pressure and composition for the entire five-component liquid solution was determined in the Gibbs energy minimization routine with assist of FacSage software. Besides, investigation was made on thermodynamic properties of N in pure liquid Fe and effects of Mn, Cr, Si on the dissolution of N in liquid Fe, and it was found that Mn and Cr enhance the dissolution of nitrogen in liquid iron significantly while Si behaves a contrary effect, from which the activity coefficient of N in liquid Fe affected by the concentration of Mn, Cr and Si can be back-calculated as well. Comparing to experimental data, the solubility of N in various molten Fe-Mn-Cr-Si-N alloys of different temperatures and pressures were accurately reproduced with a small number of model parameters. Overall, the present modeling is valuable for comprehending the behavior of N during the smelting process of high-nitrogen austenitic stainless steels. In particular, the thermodynamic database developed in this work can be applied to calculate the solubility of N in molten high-N austenitic stainless steels of designated conditions. Furthermore, it can also be used to predict unexplored phase equilibria and thermodynamic properties of liquid alloys within the Fe-Mn-Cr-Si-N system at any chemical composition, temperatures and pressures.

Key words: thermodynamic modeling; liquid Fe-Mn-Cr-Si-N solution; high-N austenitic stainless steels; modified quasichemical model (MQM); solubility of nitrogen

特殊用途 316 不锈钢中厚板轧制晶粒度控制研究

杨相歧

（山西太钢不锈钢股份有限公司临汾分公司，山西临汾　041000）

摘　要：本文针对 316 不锈钢中厚板晶粒度控制问题在实验室进行了一系列的轧钢试验，分别从钢坯原始组织状态、总轧制压缩比、单道次变形率三个因素进行分析。试验结果表明：当轧制压缩比超过 6 时，钢坯原始组织状态对中厚板全厚度晶粒均匀性无明显影响；钢坯加热温度、道次压下量相同时，总压缩比为 6 生产工艺能够轧制出全厚度晶粒均匀的钢板；当轧制总压缩比为 4 时，单道次压下率超过 30%时，钢板表面晶粒度为 2 级 7 级混晶组织；单道次轧制变形量均小于 10%时，即使轧制总压缩比足够大，钢板热轧态晶粒度依然不均匀。

关键词：特殊用途；316 不锈钢中厚板；轧钢；晶粒度；变形率

Esearch on Grain Size Control of Special Purpose 316 Austenitic Stainless Steel Medium Plate

Yang Xiangqi

(LISCO Branch Company, Shanxi Taigang Stainless Steel Co., Ltd., Linfen 041000, China)

Abstract: In this paper, the influence of the status of billet steel, rolling compression ratio, deformation rate per pass on grain of 316 austenitic stainless steel medium plate was done in laboratory. The results explained that the total compression ratio is more than 6, the original structure of steel billet has no influence on grain size of hot rolling plate, when the total compression ratio is more than 6, the homogeneous grain in the whole thickness was obtained in condition of same billet steel heating-up temperature and deformation rate per pass. But when the total compression ratio is more than 6, the 2 grade and 7 grade duplex grain was obtained even though the deformation rate per pass was more than 30%, and when the total compression ratio is less than 10%, the grain was heterogeneous even though the total compression ratio is big enough.

Key words: special purpose; 316; stainless steel medium plate; hot rolling; gain; deformation rate per

2507 超级双相不锈钢热轧黑卷脆断原因分析

钱张信，陈兴润，纪显彬

（酒泉钢铁（集团）有限责任公司，甘肃嘉峪关　735100）

摘　要：利用 EBSD 技术分析了 2507 双相不锈钢黑卷脆断处的组织，采用拉伸试验、硬度测试、EBSD 等手段研究了精轧终轧温度对 2507 双相不锈钢力学性能以及 σ 相析出的影响。结果表明，脆断区域存在 σ 相，σ 相比例偏高是其脆断的主要原因。随精轧终轧温度的降低，σ 相析出量增加，材料塑性变差，终轧温度低于 950℃伸长率仅 6.5%，终轧温度高于 970℃其伸长率大于 20%。

关键词：脆断；超级双相不锈钢；终轧温度；力学性能；σ 相

Reason of Brittle Fracture in Hot Coil of 2507 Super Duplex Stainless Steel

Qian Zhangxin, Chen Xingrun, Ji Xianbin

(Jiuquan Iron and Steel Co., Ltd., Jiayuguan 735100, China)

Abstract: The microstructure of brittle fracture in hot coil of 2507 steel were investigated by EBSD, the effect of finishing temperature on mechanical properties and σ-phase precipitation were studied by tensile, hardness tests and EBSD. The results show that there exists σ phase in brittle fracture zone, the higher proportion of the sigma phase is the main reason for the brittle fracture of 2507 steel. With the decrease of finishing temperature, the precipitation of the σ-phase increases and the plasticity of the material becomes worse; the elongation was only 6.5% under the condition of temperature below 950°C; while the finishing temperature over 970°C, the elongation was higher than 20%.

Key words: brittle fracture; super duplex stainless steel; finishing temperature; mechanical properties; σ-phase

宽幅不锈钢极薄带头部厚度超差原因分析与改进

李　实[1]，刘亚军[1]，包玉龙[1]，冷仙勇[1]，张勃洋[2]

（1. 宁波宝新不锈钢有限公司设备部，浙江宁波　315000；
2. 北京科技大学机械学院，北京　100083）

摘　要：针对某钢厂森吉米尔轧机生产不锈钢过程中出现的带头部分厚度超差的问题进行分析，确定导致同等轧制力条件下厚度超差的主要原因是低速轧制过程中润滑不足。采用摩擦润滑理论分析和轧件辊系有限元计算相结合的方法，分别建立了轧制区油膜厚度和摩擦系数的计算模型，并分析了高速轧制和低速轧制条件下对摩擦分布的影响规律，将上述计算得到的摩擦系数代入到森吉米尔轧机轧件辊系一体化计算模型，分析不同轧制速度下出口带钢厚度，证明了同等轧制力条件下低速轧制润滑不足会导致出口厚度超差。

关键词：森吉米尔轧机；不锈钢；摩擦；乳化液润滑

Analysis and Improvement of Thickness out of Tolerance for Wide Stainless Steel Head

Li Shi[1], Liu Yajun[1], Bao Yulong[1], Leng Xianyong[1], Zhang Boyang[2]

(1. Ningbo Baoxin Stainless Steel Co.,Ltd., Equipment Department, Ningbo 315000, China;
2. School of Mechanical Engineering University of Science and Technology Beijing, Beijing 100083, China)

Abstract: This paper analyzes the thicknesst error of the part in the production of ultra-thin stainless steel strip by Sendzimir mill, It is determined that the main reason of thickness overrun under the same rolling force conditon is the lack of lubrication in low rolling. The method of combining friction lubrication theory with finite eiement calculation of rolling is adopted, the calculation model of oil film thickness and friction coefficient in rolling zone is established, the above results were substituted into the rolling model to simulate the outlet thickness at different rolling speeds. The above results prove our theoretical analysis and are consistent with the practice.

Key words: sendzimir mill; stainless steel; friction; lubrication

双相不锈钢高温 N 迁移作用下微观组织及热塑性演化

黎　旺，李静媛

（北京科技大学材料科学与工程学院，北京　100083）

摘　要：热加工中双相不锈钢（Duplex Stainless Steel，DSS）处在退火与形变的相互耦合中，加剧了合金元素在两相间的迁移及再分配，尤其间隙型 N 原子。研究基于 N 可兼顾对 DSS 力学性能及耐蚀性能的多重改善作用，探究了高温下两相间 N 扩散与回迁行为及多组元协同作用下析出行为，揭示了升温、冷却及形变中组织演变规律对热变形性能的影响机制，进而阐明了通过调控热变形参数诱导元素再分配来改善热轧开裂缺陷的可行性。

关键词：双相不锈钢；N 迁移；组织演变；热塑性；析出物

Microstructure and Thermoplastic Evolution of Duplex Stainless Steel under High Temperature N Migration

Li wang, Li Jingyuan

(School of Materials Science and Engineering,University of Science and Technology Beijing, Beijing 100083, China)

Abstract: Duplex Stainless Steel (DSS) undergoes in the coupling of annealing anddeformation, which aggravates the migration and redistribution of alloyingelements between the two phases, especially the interstitial-type N atoms. Basedon the multiple improvement effects of N on the mechanical properties andcorrosion resistance of DSS, the diffusion and migration behavior of N in twophases at high temperature and precipitation behavior under the synergisticaction of multiple components were explored, revealing the influence mechanismof microstructure evolution law in heating, cooling and deformation on thermaldeformation properties. Furthermore, the feasibility of adjusting the thermaldeformation parameters to induce element redistribution to improve the crackingdefect of hot rolling is illustrated.

Key words: duplex stainless steel; N migration; microstructure evolution;thermoplasticity; precipitate

稀土 Y 对 PH13-8Mo 粉末特性及 SLM 打印件性能的影响

王长军，梁剑雄，刘振宝，杨志勇，张梦醒，马聪慧

（钢铁研究总院，特殊钢研究所，北京　100081）

摘　要：通过 SEM、EMPA、粉体特性测试仪等实验手段，研究了稀土 Y 对 PH13-8Mo 粉末特性及 SLM 打印件性能的影响规律。结果表明，稀土 Y 可以显著改善 PH13-8Mo 高强不锈钢的粉末特性与微观组织结构，同时改善了 SLM 打印件的脆性夹杂物结构，并由此导致了 PH13-8Mo 钢的综合力学性能，特别是塑韧性显著升高，可借鉴于未来增材制造用高强不锈钢的合金设计依据。

关键词：稀土 Y；SLM；PH13-8Mo；粉末特性；塑韧性

Influence of Yttrium on the Characteristics of Powders and Mechanical Properties of the PH13-8Mo Specimens Printed by Selective Laser Melting

Wang Changjun, Liang Jianxiong, Liu Zhenbao,Yang Zhiyong,
Zhang Mengxing, Ma Conghui

(Institute of Special Steels, Central Iron and Steel Research Institute, Beijing 100081, China)

Abstract: The effect of rare earth Yttrium on the characteristics of PH13-8Mo powder and the specimens of Selective Laser Melting (SLM)print was studied by SEM, EMPA, powder property tester and other experimental means.The results show that Y can significantly improve the powder properties and microstructure of PH13-8Mo high-strength stainless steel, and at the same time improve the brittle inclusion structure of SLM printed parts, and thus lead to the comprehensive mechanical properties of PH13-8Mo steel, especially the significant increase in plastic toughness can be used for reference in the alloy design basis of high-strength stainless steel for future additive manufacturing.

Key words: rare earth Y; SLM; PH13-8Mo; powder characteristics; plastic toughness

S31254 超级奥氏体不锈钢第二相表征及演变规律

丰　涵，任建斌，宋志刚，朱玉亮，何建国

（钢铁研究总院特殊钢研究所，北京　100081）

摘　要：利用热力学计算了 S31254 超级奥氏体不锈钢在 500～1200℃温度范围内的平衡态析出相，并结合热模拟试验、扫描电镜、透射电镜等方法，对不同析出物的析出行为进行了表征和分析。结果表明，S31254 不锈钢奥氏体基体中可存在的第二相包括 σ、χ、Laves 等金属间相，Cr_2N、π 型氮化物相以及 $M_{23}C_6$ 型碳化物相，高 Mo、高 N、高 Cr 含量是该钢析出相种类复杂的主要原因；实验钢具有高的第二相析出倾向，σ 相开始析出温度约为 1150℃，而在 900~800℃区间可发现 χ 相和 σ 相的转变，χ 相更易作为一种稳定相存在；析出相的析出位置和形貌呈现不同特点，晶界析出主要为 σ 相、χ 相和 Laves 相，而晶内主要有呈针状和块状分布的 χ 相和呈棒状析出的 Cr_2N 相。

关键词：S31254；超级奥氏体不锈钢；析出相；等温处理

Evolution and Microstructure Characterization of Precipitate Phases in S31254 Super Austenitic Stainless Steel

Feng Han, Ren Jianbin, Song Zhigang, Zhu Yuliang, He Jianguo

(Institute for Special Steels, Central Iron and Steel Research Institute, Beijing 100081, China)

Abstract: The equilibrium precipitates of S31254 super austenitic stainless steel in the temperature range of 500-1200℃ were calculated by thermodynamics. The precipitates of different precipitates were characterized and analyzed by thermal simulation test, scanning electron microscopy (SEM), transmission electron microscopy (TEM) and other methods.The results show that the second phases in the austenitic matrix of S31254 stainless steel include intermetallic phases such as σ, χand Laves, Cr_2N, π-type nitride phases and $M_{23}C_6$ type carbide phases. The high Mo, high N and high Cr contents are the main reasons for the complexity of precipitates.The experimental steel has a high tendency to precipitate the second phase. The initial precipitation temperature of σ phase is about 1150℃, and the conversion of χ phase and σ phase can be found in the range of 900~800℃. The χ phase is more likely to exist as a stable phase. The location and morphology of the precipitates are different. The grain boundary precipitates are mainly σ phase, χ phase and Laves phase, while the intergrain precipitates are mainly acicular and massiveχphase and rod-shaped Cr_2N phase.

Key words: S31254; super austenitic stainless steel; precipitated phase; isothermal treatment

典型不锈钢应力腐蚀失效行为与机理研究

刘智勇[1,2]，李晓刚[1,2]，杜翠薇[1,2]，董超芳[1,2]

（1. 北京科技大学国家材料腐蚀与防护科学数据中心，北京 100083；
2. 教育部腐蚀与防护重点实验室，北京 100083）

摘　要：应力腐蚀开裂（SCC）是不锈钢结构或承压设备运行安全的主要威胁之一。由于机理复杂、影响因素众多，迄今为止仍难以得到有效控制，在诸多行业和领域时有发生，因此，其长期以来一直受到工业界和学术界的广泛关注。不锈钢应力腐蚀是断裂行为、钝化膜失效行为和电化学行为的交互过程，其裂纹的二维特征和动态扩展特点决定了其电化学过程的非稳态特性，该过程受到敏感环境、成分体系、微观缺陷和组织状态等因素及其交互作用的强烈影响。因此，高性能不锈钢的开发设计须以其失效微观机制与影响因素的深入解读为基础。本工作课题组长期以来对不锈钢应力腐蚀失效机理及其耐蚀调控设计开展了系统的研究。在国际上首次提出了应力腐蚀非稳态电化学理论和耐应力腐蚀不锈钢的二元法调控方法。本报告分别汇报了课题组在典型不锈钢应力腐蚀失效分析、典型体系的应力腐蚀行为机理以及不锈钢应力腐蚀新型评价技术等方面的研究进展。这些研究不仅有助于增进对不锈钢应力腐蚀机理的认识，还能为防控应力腐蚀和提升或优化不锈钢耐蚀性提供参考，具有重要科学意义和实用价值。

关键词：不锈钢；应力腐蚀开裂；耐蚀设计；评价技术

Study on Stress Corrosion Cracking Failure Behavior and Mechanism of Typical Stainless Steels

Liu Zhiyong[1,2], Li Xiaogang[1,2], Du Cuiwei[1,2], Dong Chaofang[1,2]

(1. National Materials Corrosion and Protection Scientific Data Center, Beijing 100083, China;
2. Key Laboratory of Corrosion and Protection of Ministry of Education, Beijing 100083, China)

Abstract: Stress corrosion cracking (SCC) is one of the main threats to the operation safety of stainless steel structures or pressure equipments. Due to the complex mechanism and multifarious influencing factors of SCC, it is still difficult to get effective control so far, which occurs from time to time in many industries and fields. Therefore, it has been widely concerned by the industrial and academic circles since the very beginning. SCC of stainless steel involves interactive processes of fracture behavior, passive film rupture and electrochemical behavior. The two-dimensional characteristics and dynamic propagation characteristics of cracks determine the non-steady characteristics of electrochemical process. The process is strongly affected by sensitive environment, composition system, micro defects and microstructure state and their interactions. Therefore, the development and design of high performance stainless steel must be based on the in-depth understanding of its failure micro mechanism and influencing factors. For a long time, our research group has carried out systematic research on stress corrosion failure mechanism and corrosion control design of stainless steel. We put forward the theory of non-steady state electrochemistry of stress corrosion and the design methodology of binary composition method for stress corrosion resistant stainless steel. This report summarized our research progresses in the aspects of stress corrosion failure analysis of typical stainless steel, stress corrosion behavior mechanism in typical environments and new evaluation technology of stress corrosion of stainless steel. These studies not only help to improve the understanding of the stress corrosion mechanism of stainless steel, but also provide reference for the prevention & control of stress corrosion and the improvement or optimization of corrosion resistance of stainless steel, which has important scientific significance and practical value.

Key words: stainless steel; stress corrosion cracking; anti-corrosion design; evaluation technology

Fractographic Studies on the Ductile-to-Brittle Transition of Super Duplex Stainless Steels UNS S32750

Feng Han[1], Wang Baoshun[2,4], Wu Minghua[3],
Wu Xiaohan[1], He Jianguo[1], Song Zhigang[1]

(1. Special Steel Institute, Central Iron & Steel Research Institute, Beijing 100081, China; 2. Zhejiang Jiuli Hi-Tech Metals Co., Ltd., Huzhou 313028, China; 3. Yongxing Special Stainless-Steel Co., Ltd., Huzhou 313005, China; 4. Engineering Research Center of High Performance Nuclear Power Pipe Forming of Zhejiang Province, Huzhou 313028, China)

Abstract: The impact fracture behavior of super duplex stainless steels S32750 at different test temperatures is studied, and the composition of the steel specimens' absorbed impact energy is analyzed in this paper. By employing the optical and scanning electron microscopy, the fracture of the specimens during the ductile to brittle transition process is observed to study the crack propagation behavior, and the relationship between the fracture surface and microstructure is established. The results show that the steel specimen's absorbed impact energy in the temperature range of 20 ~ -100 ℃ decreases in an S shape with the decrease of the test temperature, and the two basically follow the Boltzmann function relationship: KV=211.1+(-186.8)/(1+e^((T-(-39.9))/11.4)). With the decrease of the test temperature, the ratio of fiber percentage to the

crack initiation area in the impact fracture of the specimens gradually declines, and the steel specimens show different fracture behaviors in the US range, DBT range, and LS range. The steel specimens tested in the DBT range show both α-phase cleavage and γ-phase quasi-cleavage, which is the main reason for the low-temperature brittleness of the steels.
Key words: super duplex stainless steels; fractography; microstructure; impact toughness; ductile-to-brittle transition

FeCrAl 耐热钢的组织稳定性及力学性能

李　朋[1]，刘　洁[2]，王　剑[1]，李玉平[1]，李国平[3]，韩培德[1]

（1. 太原理工大学材料科学与工程系，山西太原　030024；2. 太原科技大学材料科学与工程学院，山西太原　030024；3. 太原钢铁（集团）有限公司技术中心，山西太原　030003）

摘　要： 利用光学显微镜、扫描电子显微镜、透射电镜研究了 Fe-13Cr-5Al-0.3Ti 耐热钢在 700~1250℃分别保温 1h 后析出相、晶粒尺寸的变化以及对耐热钢力学性能的影响。结果表明：试样在 700~1250℃保温 1h 后均表现为韧性断裂，试样的强度以及塑性随温度的升高先增加后减小，晶界处的 Cr23C6 随温度的升高逐渐减少，但并未完全消失，而 TiC 颗粒则极为稳定，晶粒尺寸没有明显变化。在 1000~1200℃分别保温 1h 后，Cr23C6 完全回溶，由于晶界处 TiC 数量逐渐减少，晶粒开始缓慢长大；在 1250℃，晶界处 TiC 回溶，晶粒长大明显。Fe-13Cr-5Al-0.3Ti 在 700~1200℃具有良好的组织稳定性。

关键词： FeCrAl 耐热钢；Cr23C6；TiC；拉伸性能；组织稳定性

Microstructure Stability and Mechanical Properties of FeCrAl Heat Resistant Steels

Li Peng[1], Liu Jie[2], Wang Jian[1], Li Yuping[1], Li Guoping[3], Han Peide[1]

(1. School of Materials Science and Engineering, Taiyuan University of Technology, Taiyuan 030024, China; 2. School of Materials Science and Engineering, Taiyuan University of Science and Technology, Taiyuan, 030024, China; 3. Technology Center, Shanxi Taigang Stainless Steel Co., Ltd., Taiyuan, 030003, China)

Abstract: The changes of precipitates and grain size and their effects on the mechanical properties of Fe-13Cr-5Al-0.3Ti heat-resistant steel after being held between 700-1250℃ for 1h were studied by Optical Microscopy, Scanning Electron Microscopy and Transmission Electron Microscopy. The results show that the specimens were ductile fracture and the strength and plasticity first increase and then decrease with the increase of temperature. $Cr_{23}C_6$ at grain boundary gradually decreases but does not completely disappear with the increase of holding temperature between 700 -900℃,TiC particles are extremely stable and the grain size does not change significantly. Cr23C6 was completely dissolved at 1000-1200℃ held for 1h and the grains began to grow slowly because the amount of TiC at the grain boundary gradually decreased. TiC particles dissolved back at grain boundaries at 1250℃, the grains grew rapidly. Fe-13Cr-5Al-0.3Ti has good Microstructure stability between 700-1200℃.

Key words: FeCrAl heat-resistant steel; $Cr_{23}C_6$; TiC; tensile property; microstructure stability

7.8 非晶合金

双辊薄带铸轧制备 Zr-Cu 基非晶合金板带

张晨阳，袁 国，张元祥，王国栋

（东北大学 轧制技术及连轧自动化国家重点实验室，辽宁沈阳 110819）

摘 要：非晶合金具有高强度、高硬度等优异的力学性能，作为结构材料具有巨大的应用前景。然而，目前非晶合金产品的尺寸和形状都很受限，且结构材料中常用的薄板材仍然没有有效的生产方式。本研究基于双辊薄带铸轧技术，对 Zr-Cu 基非晶合金薄板进行了铸轧制备。通过调整布流方式与铸轧力大小以增大合金冷却速度与出辊温度，成功制备出了 $Zr_{55}Cu_{30}Al_{10}Ni_5$、$(Zr_{55}Cu_{30}Al_{10}Ni_5)_{99.8}Y_{0.2}$、$Cu_{47}Zr_{45}Al_8$ 和 $Cu_{64}Zr_{36}$ (at.%)非晶合金带，充分证明了双辊薄带铸轧制备非晶合金板带的可行性。该研究为制备大面积非晶合金板材提供了一种有效的方法。

关键词：非晶合金；双辊薄带铸轧；力学性能；组织

Zr-Cu-based Amorphous Alloy Sheets Fabricated by Twin-roll Strip Casting

Zhang Chenyang, Yuan Guo, Zhang Yuanxiang, Wang Guodong

(The state key laboratory of rolling automation, Northeastern University, Shenyang 110819, China)

Abstract: Amorphous alloys exhibit diverse mechanical properties, including high strength and high hardness, with important application prospects as structural materials. However, amorphous alloy products have size and shape limitations. And there is no mature process for continuously producing sheet products of amorphous alloys, despite the importance of these products as structural materials. In this work, $Zr_{55}Cu_{30}Al_{10}Ni_5$, $(Zr_{55}Cu_{30}Al_{10}Ni_5)_{99.8}Y_{0.2}$, $Cu_{47}Zr_{45}Al_8$, and $Cu_{64}Zr_{36}$ (at.%) amorphous alloy sheets were successfully fabricated by twin-roll strip casting. This work proves the feasibility of continuously preparing amorphous alloys via twin-roll strip casting, and it offers a significant guide for the industrial production of large amorphous alloy sheets.

Key words: amorphous alloy; twin-roll strip casting; mechanical properties; microstructure

FeSiBPNbCuC 纳米晶软磁复合材料的组织结构及性能调控

董亚强，吴 悦，贾行杰，王新敏

（中国科学院宁波材料技术与工程研究所，浙江宁波 315000）

摘　要：本工作采用DSC、XRD、SEM、VSM、TEM等测试分析和表征手段，系统研究了C元素含量对$Fe_{74.75}(Si, B, P)_{23-x}Nb_{1.5}Cu_{0.75}C_x$ (x = 0-3 at. %)纳米晶复合材料的晶化行为、显微形貌、组织结构以及软磁性能的影响规律。结果表明，组分为$Fe_{74.75}(Si, B, P)_{21}Nb_{1.5}Cu_{0.75}C_2$的粉末同时具有高非晶度、宽热处理区间以及高饱和磁感应强度等特点，以2%环氧树脂为绝缘粘结剂，在1800 MPa下压制成型，随后在510 ℃下热处理1小时，在非晶基体中均匀析出了平均粒径为20 nm左右的α-Fe(Si)纳米晶粒。在50 mT，100 kHz的条件下，纳米晶软磁复合材料的损耗最低，为121.45 mW/cm^3；在100 kHz的条件下，其磁导率达到51；在施加100 Oe的外加直流偏置场时，其磁导率仍能保持初始值的52.1 %，表现出较高的稳定性。总之，制备所得的纳米晶软磁粉芯具有优异的综合磁性能，在高频电力电子领域具有良好的应用前景。

高 B_s Co 基块体非晶合金的制备及其性能研究

张　伟

（大连理工大学材料科学与工程学院，大连　116024）

摘　要：软磁性Co基非晶合金具有低矫顽力（H_c）、高磁导率、低铁损、接近于零的磁致伸缩系数等特性，尤其是其高频软磁性能极佳，在电力、电子产业及5G通信技术领域具有越来越重要的应用价值。但Co基合金的玻璃形成能力（GFA）相对较低，且饱和磁感应强度（B_s）不高，限制了它们的应用范围。通过添加前过渡金属元素（ETM）及稀土元素（RE）合金化，可有效提高Co基合金的GFA，研发出了一系列Co基块体非晶合金（BMGs）体系。但由于这些BMGs含有较多量ETM和RE而使它们B_s明显降低，甚至失去磁性。最近，我们研制出了不含ETM和RE的高B_s Co基Co-Fe-B-Si-P BMGs，其B_s、H_c、有效磁导率、压缩屈服强度和塑性应变分别在1.02-1.24 T、0.8-4.6 A/m、12700-18500、3343-3590 MPa和0.2-2.7 %之间。此外，还将介绍我们最近研发出的不含ETM的高B_s Co基(Co, Fe)-RE-B BMG体系，并探讨Co基BMGs成分和GFA及磁性能的相关性。

关键词：钴基非晶合金；合金元素置换；玻璃形成能力；软磁性能；力学性能

高频低损耗 Fe 基纳米晶软磁粉芯的制备与性能研究

吴　悦，董亚强，贾行杰，王新敏

（中国科学院宁波材料技术与工程研究所）

摘　要：传统的金属软磁粉芯凭借其优良的软磁特性被广泛应用于电力电子领域。但随着社会需求的不断发展，传统的金属软磁粉芯难以满足电子元器件高频化与小型化的发展需求，本文通过感应熔炼制备得到FeSiBPNbCuC合金，并利用真空气雾化装置制备得到球形非晶粉末，在粉末中添加2wt.%的树脂作为包覆剂并在室温下以1800MPa的压力压制成环，退火得到纳米晶软磁粉芯，通过调控其热处理工艺，探究不同退火温度及升温速率对磁性能的影响规律。研究表明，在510℃下热处理1小时后，样品中析出单一的α-Fe(Si)相，证明510℃为其最佳的热处理温度；进一步调控其升温速率，当升温速率为100K/min时，样品的综合磁性能进一步优化：在50 mT，500kHz的高频环境下，损耗低至 760.5mW/cm^3；在 100kHz 的条件下，其磁导率为 61.7，表现出优异的频率稳定性；在施加 100 Oe 的外加直流偏置场时，其磁导率仍能保持初始值的 55%。本文所制备的纳米晶软磁粉芯在高频下具有远低于传统粉芯的损耗，能够应用于高频电子元器件的开发与制备，在未来具有较高的应用价值。

关键词：Fe 基；磁粉芯；热处理；损耗

机器学习方法在非晶合金中的应用

刘晓悌，沈　军

（深圳大学）

摘　要：随着大数据与人工智能技术在材料科学领域的发展，诸多材料科学问题可以很好地利用机器学习方法研究和解决。例如，非晶合金新材料的设计、非晶态物质的结构认识、非晶合金的物理、化学、力学性质研究等。对于非晶合金的形成能力，现有研究从多个角度提出了一些经验准则和参数来评估合金的玻璃形成能力。由于大多数参数是由特征温度组合运算得到，因此只能在实际制备出玻璃态样品之后进行测量，进而评估其玻璃形成能力。利用这些参数难以在制备样品之前对玻璃形成能力进行预测，缺乏对非晶合金成分设计的指导作用。本研究基于人工神经网络运算模型建立了预测玻璃形成能力的方法。利用该方法可以预测未知成分的玻璃形成能力，指导非晶合金的成分设计。

非晶晶化/半固态烧结制备双尺度结构钛合金

陈　涛，杨　超

（华南理工大学国家金属材料近净成形工程技术研究中心，广东广州　510640）

摘　要：提出了一种基于单相熔化形成液相的新型半固态烧结技术，利用非晶晶化/半固态烧结非晶粉末成功制备高强韧新型双尺度核壳结构 $Ti_{68.8}Nb_{13.6}Co_6Cu_{5.1}Al_{6.5}$ 合金。机械合金化制备的非晶粉末由非晶基体和纳米 β-Ti 相组成（图 1），在约 500℃时非晶晶化成 $CoTi_2$ 和 β-Ti 相，在 1138℃时 $CoTi_2$ 相熔化。其双尺度结构由微米晶 β-Ti 和晶内纳米针状 α'马氏体组成，超细晶 $CoTi_2$ 孪晶沿 β-Ti 晶界分布构成核壳结构（图 2）。此结构源于半固态烧结独特的组织演变：（1）颗粒重排；（2）非晶晶化；（3）$CoTi_2$ 相熔化；（4）双尺度结构形成。合金的压缩屈服强度、抗拉强度和塑性分别达 1611MPa，3139MPa 和 38.7%，优于已报道双尺度结构钛合金，这归因于 $CoTi_2$ 核壳结构的强化及纳米 α'马氏体的加工硬化作用。此研究为制备新型高强韧钛合金提供思路。

关键词：钛合金；非晶粉末；半固态烧结；双尺度核壳结构；力学性能

Ti-based Composites with Bimodal Architecture Processed via Novel Semi-Solid Sintering from Amorphous Powder

Chen Tao, Yang Chao

(National Engineering Research center of Near-Net-Shape Forming for Metallic Materials, South China University of Technology, Guangzhou 510640, China)

Abstract: We report a novel approach to fabricate $Ti_{68.8}Nb_{13.6}Co_6Cu_{5.1}Al_{6.5}$ composites with a bimodal core-shell microstructure and reveal the underlying mechanism of their microstructure evolution and mechanical properties. The amorphous powder fabricated by mechanical alloying was composed of amorphous matrix and nanostructured β-Ti phase. The amorphous powder was crystallized at approximately 500℃ to form $CoTi_2$ and β-Ti phase while $CoTi_2$ phase started to melt at 1138℃. Herein, the bimodal core-shell microstructure was attained via semi-solid sintering caused by the melting of $CoTi_2$ phase. Specifically, the bimodal matrix consisted of nanostructured α' martensite and micro-size β-Ti grains, while the micron-sized $CoTi_2$ twins distributed along β-Ti matrix constituted the core-shell architecture. Fundamentally, the unique structure stemmed from the specific mechanism of microstructural evolution which was divided into four stages: (1) spatial rearrangement of nanocrystalline/amorphous composite powders; (2) nucleation and growth of β-Ti and $CoTi_2$ phases; (3) formation of liquid phase; (4) formation of bimodal core-shell structure. In particular, such bimodal core-shell composite exhibited high compressive yield strength of 1611MPa, ultimate strength of 3139MPa with large compressive plasticity of 38.7%, which represent the highest values reported thus far in the literature. The high yield strength was attributed to the blocking effect on dislocation from $CoTi_2$ twins and nanostructured α' martensite. In addition, the micro-sized β-Ti matrix and the strain-hardening effect from α' martensite contributed the ductility of composite. This work provides fundamental insight into the development and fabrication of novel structural materials with a high melting point for demanding structural applications.

Key words: titanium alloys; amorphous powder; semi-solid sintering; bimodal core-shell structure; mechanical properties

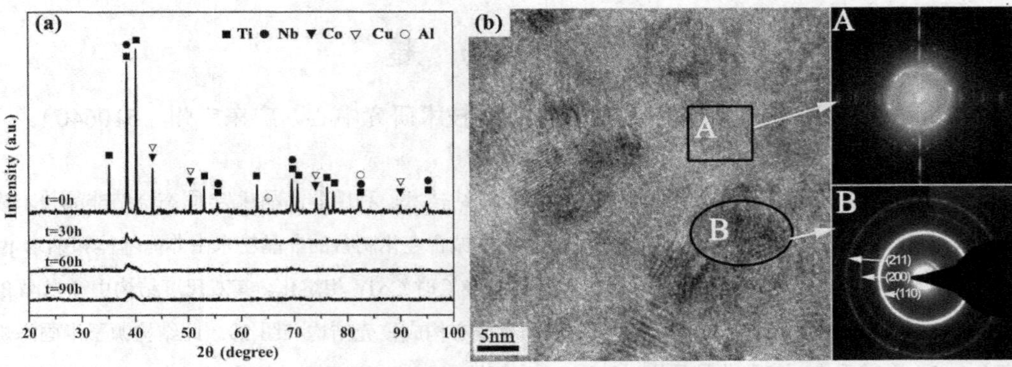

图 1　非晶粉末的 XRD 谱和 TEM 结果

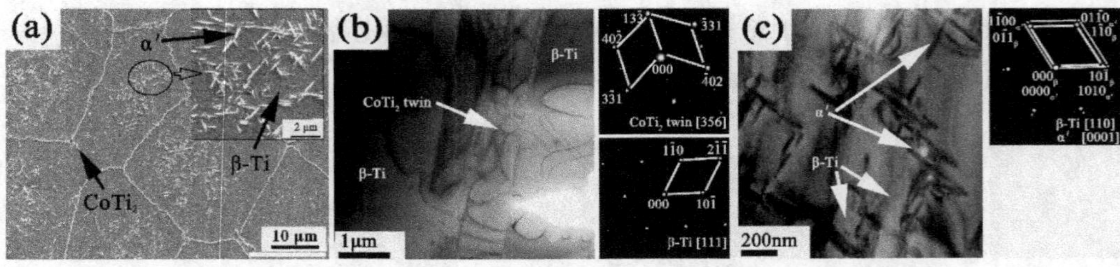

图 2　半固态烧结 $Ti_{68.8}Nb_{13.6}Co_6Cu_{5.1}Al_{6.5}$ 合金 SEM 像（a）和对应的 $CoTi_2$（b）及 α′马氏体（c）TEM 像

Cu-Zr-Al 三元非晶合金原子级结构的第一性原理研究

卢文飞[1]，王之略[1]，项红萍[1]，沈 军[2]

（1. 同济大学材料科学与工程学院；上海 201804；
2. 深圳大学机电与控制工程学院；广东深圳 518060）

摘 要：用第一性原理研究了 Cu-Zr-Al 三元非晶合金的原子级结构。发现了结构稳定性规律:具有较好的化学有序或五重对称的结构具有更好的稳定性。Al 原子与 Cu、Zr 原子之间的连接性较强，键缩短程度明显。当 Al 原子作为二十面体中心原子时，对结构稳定性贡献更大。这意味着只有在 Cu-Zr 二元体系中加入少量的 Al 原子，才能显著提高二十面体结构的稳定性，最终提高合金的非晶形成能力。Al 原子的持续加入对结构稳定性的改善不大。

TiCuMo 非晶合金去合金化制备高性能 HER 电催化剂

Aneeshkumar K.S.，田锦森，沈 军

（深圳大学机电与控制工程学院，深圳 518060）

摘 要：制备非贵金属高性能催化剂是解决能源和环境危机的重大挑战。本文中我们报道一种去合金化制备的自支撑纳米孔结构的双相晶体/非晶合金催化剂 Cu60Ti37Mo3。该催化剂在较高的电流密度下具有和商业 Pt 催化剂相当的性能，并且即使在更高的电流密度下也能保持稳定。去合金化形成了大量的催化活性位点，而 Mo 的加入提高了 H2 吸附/去吸附动力学，从而获得了较低的过电位和塔菲尔斜率。

放电等离子烧结核壳结构非晶/晶体复合材料的组织与性能研究

朱家华，沈 军

（深圳大学，广东深圳 518060）

摘 要：粉末冶金法制备复合材料是改善材料性能重要手段，然而增强相的弥散程度（或团聚）和异质界面的扩散严重影响复合材料的性能。本文首先利用气雾化方法制备了一种具有核壳结构的富 Fe 非晶/富 Cu 晶体的复合粉体，随后利用放电等离子烧结技术在 900℃下烧结，成功得到了一类致密的颗粒增强 Cu 基复合材料。其中，富 Fe 第二相颗粒弥散分布在基体中，没有偏聚现象；此外，在整个烧结过程中只有基体-基体界面，避免了异质界面的出现；同时，该类复合材料具有优良的力学性能，其压缩强度可达 1000MPa，压缩塑性约为 30%。该方法可为制备高强

度粉末冶金复合材料提供参考。

关键词：粉末冶金；非晶合金；复合材料；力学性能

高通量计算助力开发置换型固溶体成分与晶格常数关系普适模型

王明旭，王　丽

（山东大学（威海）机电与信息工程学院，山东 威海　264209）

摘　要：长久以来，固溶强化是工程材料最常用强化方式之一，而强化程度往往与晶格常数密不可分。因此，准确描述固溶体晶格常数与成分的依赖关系具有重要意义。然而，现有的相关模型不能很好的描述固溶体成分与晶格常数的非线性依赖关系。在此，我们基于虚晶近似并综合考虑尺寸效应和电子效应提出了一个新模型。对于一个 N 元系置换型固溶体，以相应单质和 N 个固溶体参考成分点的晶格常数/体弹模量为输入，该模型即可对固溶体全成分区间的晶格常数进行预测。系统采用第一性原理高通量计算获得的大量数据以及报道的实验数据证实了该模型的准确性和普适性。该模型有望加深对材料成分和性能关系的认识。

关键词：虚晶近似；体弹模量；内应力平衡；电子耦合效应；第一性原理高通量计算

A Generally Reliable Model for Composition-dependent Lattice Constants of Substitutional Solid Solutions

Wang Mingxu, Wang Li

(School of Mechanical and Electrical Engineering, Shandong University (Weihai), Weihai 264209, China)

Abstract: Solid solutioning has long been employed to improve the performance of enegineering materials, the degree of improvement generally correlates closely with the resultant lattice parameters. It is therefore of great importance for materials design to describe accurately the composition-dependent lattice constants of the solid solutions (SSs). However, existing models could hardly reproduce the usually non-linear relationship between the compositions and the lattice constants. Herein, we present a new model within the framework of virtual crystal approximation by taking into account both the size factor and the electronic effect. The model takes inputs as simple as the fundamental property parameters of the elementary substances and N referential SSs for an N-component system, and can then predict the lattice constant of SS with any composition within the system. Systematical validation using datasets obtained from high-throughput first-principles calculations and available experiments confirmed the high reliability and general applicability of our model for various substitutional SSs. It is expected that this model will deepen the understanding of the relationship between the composition and the properties of materials.

Key words: virtual crystal approximation; bulk modulus; internal stress equilibration; electronic coupling effect ; high-throughput first-principles calculations

8 粉末冶金

大会特邀报告

分会场特邀报告

矿业工程

炼铁与原燃料

炼钢与连铸

轧制与热处理

表面与涂镀

金属材料深加工

先进钢铁材料及其应用

★ 粉末冶金

能源、环保与资源利用

冶金设备与工程技术

冶金自动化与智能化

冶金物流运输

其他

碳酸钠强化还原钛精矿的等温动力学研究

陈 丹，侯有玲，吕晓东，辛云涛，吕学伟

（材料科学与工程学院，重庆大学，重庆 400044）

摘 要：本文采用等温动力学研究方法对碳酸钠强化还原低品位钛精矿获得优质氯化 UGS 渣原料的新工艺中的强化还原动力学进行研究，结果如下：整个还原过程分为缩核反应（0-10min）、扩散反应（10-30min）以及成核长大（30-90min）三个阶段，第一阶段主要受缩核模型控制，第二阶段主要受三维扩散控制；添加碳酸钠可以降低控制还原反应过程中的表观活化能，第一阶段由 13.20kJ/mol 降低至 7.89kJ/mol，第二阶段由 20.18kJ/mol 降低至 10.90kJ/mol；反应过程中产生的金属铁会通过熔融相迁移长大，避免形成包裹着反应核的致密金属壳，阻碍反应的进行，熔融相中也能增加反应核与还原剂之间反应次数和几率，从而促进反应的进行，强化还原效果。

关键词：钛精矿；碳酸钠；强化还原；等温动力学

Isothermal Kinetics Study on Enhancement of Reduction Ilmenite Concentrate by Na_2CO_3

Chen Dan, Hou Youling, Lv Xiaodong, Xin Yuntao, Lv Xuewei

(College of Materials Science and Engineering, Chongqing University, Chongqing 400044, China)

Abstract: Isothermal kinetics method was used to study enhanced reduction kinetics, which is a new process for the enhanced reduction of low-grade ilmenite concentrate by Na_2CO_3 to obtain high quality chlorinated UGS slag. The results are as follows: the whole reduction process is divided into three stages: nuclear shrinkage reaction (0-10min), diffusion reaction (10-30min) and nucleation growth (30-90min). The first stage is mainly controlled by the shrinking core model, and the second stage is mainly controlled by the three-dimensional diffusion. Adding Na_2CO_3 can reduce the apparent activation energy in the reduction reaction. It decreased from 13.20kJ/mol to 7.89kJ/mol in the first stage and from 20.18kJ/mol to 10.90kJ/mol in the second stage. During the reaction, the metal iron produced will migrate and grow through the molten phase to avoid the formation of a dense metal shell wrapped around the reaction core, which hinders the progress of the reaction. The molten phase can also increase the reaction times and probability between the reaction core and the reducing agent, thus promoting the progress of the reaction and strengthening the reduction effect.

Key words: ilmenite concentrate; sodium carbonate; enhanced reduction; isothermal dynamics

溶液燃烧法用于 AlON 透明陶瓷的制备

张一铭，秦明礼，吴昊阳，贾宝瑞，曲选辉

（北京科技大学新材料技术研究院，北京 100083）

摘 要：作为 Al_2O_3-AlN 二元体系中重要的单相化合物，γ-AlON 陶瓷在紫外至中红外范围内具有优异的透光性，良好的机械强度和耐腐蚀性，因此在军事和商业上具有广阔的应用范围。通过溶液燃烧法获得的超细前驱体，可在低温下快速氮化，并可用于 AlON 透明陶瓷的制备。

以硝酸铝、尿素和葡萄糖配制的混合溶液在加热过程中释放出大量的气体，获得具有极高的反应活性前驱体，该前驱体在 1700℃煅烧 10min 即可合成单相 AlON 粉体。本文详细讨论了尿素/硝酸铝的比例，煅烧温度以及保温时间对粉体性能的影响。烧制的粉体以 Y_2O_3 和 La_2O_3 为烧结助剂，在 1930℃常压烧结 20h 得到光学透过率为 85%的透明 AlON 陶瓷。

注射成型生物医用钛铌基合金的烧结行为

赵大鹏[1]，Ebel Thomas[2]，Willumeit- Römer Regine[2]，Pyczak Florian[2]

（1. 湖南大学生物学院，湖南长沙 410082；2. Helmholtz-Zentrum Hereon，Max-Planck Str.1，Geesthacht，Germany，21502）

摘　要：采用元素粉通过注射成型制备了生物医用 Ti-Nb 基合金，并采用 XRD、热膨胀仪、SEM 和 EDS 等研究了合金的烧结行为与显微组织演化。在 700℃到 1500℃之间，合金的均匀化与致密化过程可以分为三段，即 Ti 扩散控制段，Ti-Nb 扩散段与基体扩散段。在低温下，主要通过 Ti 颗粒而不是 Nb 颗粒的扩散消除孔隙。随着温度升高，Ti-Nb 之间的扩散逐渐加剧，出现明显均匀化和进一步致密化。当成分均匀化完成后，致密化则完全由基体扩散来实现。

Sintering Behavior of Metal-injection Molded Ti-Nb-based Biomedical Materials

Zhao Dapeng[1], Ebel Thomas[2], Willumeit- Römer Regine[2], Pyczak Florian[2]

(1. College of Biology, Hunan University, Changsha 410082, China;
2. Helmholtz-Zentrum Hereon, Max-Planck Str.1, Geesthacht, Germany, 21502)

Abstract: Sintering behavior and microstructure evolution of Ti-Nb-based alloys processed by metal injection molding (MIM) technology using elemental powders were investigated in this work by optical microscopy, X-ray diffraction (XRD), dilatometer, scanning electron microscopy (SEM) and energy-dispersive spectroscopy (EDS). It was found that from 700℃ to 1500℃ the homogenization and densification process of Ti-Nb alloy consisted of three steps, i.e., Ti-diffusion-controlled step, Ti-Nb-diffusion step and Matrix-diffusion step. At low temperatures, mainly the diffusion between Ti particles contributed to the pore elimination, while, the Nb particles acted as diffusion barriers. With increasing temperature, the diffusion between Ti and Nb powders became notable, leading to apparent homogenization and further densification. At the Matrix-diffusion step, the homogenization had finished. The final densification was only dependent on the diffusion in the Ti-Nb matrix.

激光选区熔化 WMoTaTi 高熵合金的组织与性能研究

刘　畅，朱科研，陈佳男，丁旺旺，陈　刚，曲选辉

（北京科技大学新材料技术研究院，北京　100083）

摘 要：本研究采用机械合金化结合流化技术制备出近球形 WMoTaTi 高熵合金粉末，并进行激光选区熔化（SLM）成形研究，制备 WMoTaTi 高熵合金，对成形件的组织性能进行表征与分析。结果表明，通过流化技术制备的 WMoTaTi 高熵合金粉末具有良好的流动性，能够很好地适应 SLM 成形工艺。此外，SLM 制备的 WMoTaTi 高熵合金为单一 BCC 相，晶粒形貌为树枝状，成形件致密度达到 95.4%，显微硬度为 617.2HV。本研究提供了一种低成本制备增材制造用难熔高熵合金粉末的新技术，并初步研究了激光选区熔化 WMoTaTi 高熵合金的成形工艺与组织性能。

关键词：高熵合金；粉末制备与改性；激光选区熔化；显微组织

Microstructure and Properties of WMoTaTi High Entropy Alloy by Selective Laser Melting

Liu Chang, Zhu Keyan, Chen Jianan, Ding Wangwang, Chen Gang, Qu Xuanhui

(Institute for Advanced Materials and Technology, University of Science and Technology Beijing, Beijing 100083, China)

Abstract: In this study, mechanical alloying combined with the gas-solid fluidization technology was used to prepare quasi-spherical WMoTaTi high entropy alloy powder. The WMoTaTi alloy was fabricated via selective laser melting (SLM) using the as-prepared quasi-spherical powders. Microstructure and properties of WMoTaTi high entropy alloy by SLM were analyzed. The results show that the as-prepared WMoTaTi high entropy alloy powder exhibits good fluidability that is suitable for the SLM processing. Single BCC phase and dendrite crystal morphology were identified for the WMoTaTi alloy made by SLM, while its relative density and microhardness reach 95.4% and 617.2HV, respectively. In a word, this work provides a novel technology for low-cost preparation of refractory high entropy alloy powders for additive manufacturing. Preliminary research was also demonstrated in terms of processing and microstructure properties of WMoTaTi alloy by SLM.

YbB_6 对 Ti-6Al-4V 钛合金组织和性能的影响

罗铁钢[1]，高春萍[2]，元想胜[2]

（1. 广东省科学院材料与加工研究所，广东广州 510650；2. 长安大学，陕西西安 710061）

摘 要：运用放电等离子烧结（SPS）技术系统研究了 YbB_6 对 Ti-6Al-4V 合金显微组织和性能的影响，采用 OM、SEM、EDS 和 TEM 等方法进行了表征与分析。结果表明，随着 YbB_6 的加入，钛合金显微组织发生转变，晶粒明显细化，原位反应生成的 TiB 晶须和 Yb_2O_3 颗粒有利于力学性能的提高。添加 0.6wt%YbB_6 后，烧结件的相对密度、显微硬度、拉伸屈服强度、极限强度和延伸率分别达到 99.43%、403HV、903MPa、1148MPa 和 3.3%，与 Ti-6Al-4V 试样相比，分别提高了 0.37%、13.8%、38.07%、17.14%和 32%。强化机制主要是晶粒细化和弥散强化。

关键词：钛合金；放电等离子烧结；微观结构；力学性能；硼化镱

Effect of YbB_6 on Microstructure and Mechanical Properties of Ti-6Al-4V Titanium Alloy

Luo Tiegang[1], Gao Chunping[2], Yuan Xiangsheng[2]

(1. Institute of Materials and Processing, Guangdong Academy of Sciences,Guangzhou 510650,China; 2. Chang'an University, Xi'an 710061, China)

Abstract: In this paper, the effects of YbB_6 additions on microstructure and mechanical properties of Ti-6Al-4V alloy were systematically investigated by spark plasma sintering (SPS).The microstructure and properties of Ti-6Al-4V alloy were characterized by OM, SEM, EDS and TEM. The results show that with the content of YbB_6 increases, the microstructure of the titanium alloy has transformed, the grains were obviously refined, and TiB whiskers and Yb_2O_3 compound particles are formed through in-situ reaction, which are beneficial to the improvement of mechanical properties of the composites. The results show that when added 0.6wt% YbB_6, the relative density, microhardness, tensile yield strength, ultimate strength and elongation of sintered parts reach 99.43%, 403HV, 903Mpa, 1148Mpa and 3.3% respectively, which are 0.37%, 13.8%, 38.07%, 17.14% and 32% higher than those of Ti-6Al-4V samples. The strengthening mechanism is mainly grain refinement and dispersion strengthening.

Key words: titanium alloy; spark plasma sintering; microstructure; mechanical properties; YbB_6

碳热还原法制备高质量纳米 WC 粉末及其机理研究

王倩玉，吴昊阳，秦明礼，贾宝瑞，曲选辉

（北京科技大学新材料技术研究院，北京　100083）

摘　要：具有优异性能的纳米晶 WC 硬质合金在电子信息、计算机、汽车工业、航空航天、国防军工等众多领域有着巨大的应用潜力。纳米 WC 粉末的制备是生产纳米晶 WC 硬质合金的基础和关键。本研究以低价、易得的 WO_3 和碳黑为原料，采用简便、短流程的碳热还原工艺合成了粒径小(≤100nm)，分散性好、粒度分布窄、纯度高(＞99.9wt.%)、化合碳含量(6.10wt.%)接近理论值且游离碳含量低（0.09wt.%）的纳米 WC 粉末，并详细研究了纳米 WC 粉末的合成过程、成核和生长机制。本方法不仅显著降低了加工温度(1100℃)，缩短了保温时间(3h)，还厘清了各要素的耦合作用机制，实现了对 WC 粉末特性的精确控制，并提供工艺技术和理论依据。

关键词：纳米 WC 粉末；碳热还原；制备；影响因素；机理

Study on Preparation and Mechanism of High-Quality WC Nanopowders via Carbothermal Reduction

Wang Qianyu, Wu Haoyang, Qin Mingli, Jia Baorui, Qu Xuanhui

(China Institute for Advanced Materials and Technology, University of Science and Technology Beijing, Beijing 100083, China)

Abstract: Nanocrystalline WC cemented carbide with excellent properties has great potential in many fields, such as electronics, computer, automobile, aerospace, national defense and military industry. The preparation of WC nanopowders is the basis and key to produce nanocrystalline WC cemented carbide. In this study, low-cost, easily available WO_3 and carbon black were used as raw materials, and a simple, short-flow carbothermal reduction process was used to synthesize WC nanopowders with small particle size (≤100nm), good dispersibility, narrow particle size distribution, high purity (>99.9wt.%), combined carbon content (6.10wt.%) close to the theoretical value and low free carbon content (0.09wt.%). The synthesis process, nucleation and growth mechanism of WC nanopowders were also investigated in detail. The research not only results in lower processing temperature (1100℃) and shorter holding time (3h), but also clarifies the coupling mechanism of each element, realizes precise control of the characteristics of WC powders, and provides process technology and theoretical basis.

Key words: WC nanopowders; carbothermal reduction; preparation; influencing factors; mechanism

氧含量及其分布对氮化铝陶瓷热导率的影响

张智睿[1]，刘　昶[1]，韩丽辉[2]，贾宝瑞[1]，吴昊阳[1]，秦明礼[1]，曲选辉[1]

（1. 北京科技大学新材料技术研究院，北京　100083；2. 北京科技大学冶金学院，北京　100083）

摘　要：氮化铝的高导热性长期受限于杂质氧元素，并且氧的存在形式在对导热机制有着不同程度的影响。本文介绍了一种氮化铝陶瓷中氧元素含量及其分布的测试方法，制备了不同晶格氧，晶界氧的氮化铝陶瓷，讨论了两者对热导率的影响。结果表明 Y_2O_3 作为烧结助剂时有固氧作用，添加量达到 3wt.%时达到固氧极限。高温退火能够净化晶格，使氧含量进一步降低到 0.036wt.%，热导率从 165W·m^{-1}·K^{-1} 升高到 218W·m^{-1}·K^{-1}。

关键词：氮化铝；热导率；氧含量；粉末冶金

Effect of Oxygen Content and Distribution on Thermal Conductivity of Aluminum Nitride Ceramics

Zhang Zhirui[1], Liu Chang[1], Han Lihui[2], Jia Baorui[1], Wu Haoyang[1], Qin Mingli[1], Qu Xuanhui[1]

(1. Institute for Advanced Materials and Technology, University of Science and Technology Beijing, Beijing 100083, China; 2. School of Metallurgical and Ecological Engineering, University of Science and Technology Beijing, Beijing 100083, China)

Abstract: The high thermal conductivity of aluminum nitride has been limited to the impurity oxygen for a long time. And the existing form of oxygen has different degrees of influence on the thermal conductivity mechanism. In this article, a method for measuring the content and distribution of oxygen in aluminum nitride ceramics is introduced. Aluminum nitride ceramics with various oxygen content, which dissolve in the lattice and remain at grain boundary oxygen, are prepared. The influence of both upon thermal conductivity is discussed. The results show that Y_2O_3, as a sintering agent, has the effect of oxygen binding, and the oxygen binding limit is reached when the added quantity reaches 3wt.%. High temperature annealing can purify the lattice, further reduce the oxygen content to 0.036wt.%, and increase the thermal conductivity from 165W·m^{-1}·K^{-1} to 218W·m^{-1}·K^{-1}.

增材制造 WC-Co 硬质合金研究进展及挑战

李晓峰[1,2]，郭子傲[1]，王　行[1]，赵宇霞[1]，刘　斌[1]，白培康[1]，刘　咏[2]，马　前[3]

（1. 中北大学材料科学与工程学院，山西太原　030051；2. 中南大学粉末冶金国家重点实验室，湖南长沙　410083；3. 墨尔本皇家理工大学，澳大利亚墨尔本　3000）

摘　要：硬质合金因其具有较高的硬度、强度、耐磨及耐腐蚀等性能，被广泛地应用在切削刀具、钻探设备、模具和量具等领域。本文分析了不同成形方式成形 WC-Co 硬质合金的现状以及未来的研究方向，采用选区激光熔化（SLM）与激光融化沉积（LMD）技术、利用不同基板制备了硬质合金复合材料，探索了不同工艺参数条件下合金微观组织以及力学性能的演变规律，讨论了不同基板对成形硬质合金试样摩擦摩损和耐腐蚀性能的影响。

关键词：硬质合金；粉末；选区激光熔化；激光融化沉积；基板

Additive Manufacturing of WC-Co Cemented Carbide: Research Progress and Challenges

Li Xiaofeng[1,2], Guo Ziao[1], Wang Hang[1], Zhao Yuxia[1],
Liu Bin[1], Bai Peikang[1], Liu Yong[2], Ma Qian[3]

(1. School of Materials Science and Engineering, North University of China, Taiyuan 030051, China;
2. The State Key Laboratory of Powder Metallurgy, Central South University, Changsha 410083, China;
3. Centre for Additive Manufacturing, School of Engineering, RMIT University, Melbourne, VIC, 3000, Australia)

Abstract: Cemented carbide is widely used in cutting tools, drilling equipment, molds and measuring tools because of its high hardness, strength, wear resistance and corrosion resistance. In this paper, the current status and the future research direction of WC-Co cemented carbide fabricated by different forming methods were analyzed, selective laser melting (SLM) and laser melting deposition (LMD) technologiesand different substrates were used to preparecemented carbide composite materials. The evolution ofthe alloy microstructure and mechanical propertiesunder different process parameters was explored, and the influence ofdifferent substrateson the friction, wear and corrosion resistance of cemented carbide was discussed.

Key words: cemented carbide; powder; selective laser melting; laser melting deposition; substrate

高性能软磁复合材料及其关键技术研究

刘　辛[1,2]，徐　佳[1,2]，王　健[1,2]，卢克超[1,2]

（1. 广东省科学院新材料研究所，广东广州　510650；2. 国家钛及稀有金属粉末冶金工程技术研究中心，广东广州　510650）

摘　要：金属基软磁复合材料广泛应用于新能源汽车、电力电子、5G 通讯等战略新兴产业。针对目前常用金属软磁材料高频性能差的问题，提出高耐热绝缘材料的设计理念，通过高耐热绝缘材料对磁粉体表面改性，构建新型绝缘复合结构改善材料的高频性能。发展新型化学合成法，系统研究纳米金属氧化物对金属磁粉芯磁导率、电阻率和致密度的影响，探索高温下界面结构稳定性、析出相形成及软磁性能的演化过程，探讨界面结构和界面相析出对涡流损耗的影响；研究热处理工艺参数对材料微观组织结构、静态磁性以及动态磁性的影响规律，建立微观组织结构、工艺参数和软磁性能之间的关联机制，获得具有优良高频软磁性能金属基复合材料的微观结构和性能的调控方法，制备出兼有高磁导率、高频超低损耗的软磁复合材料。

关键词：金属基磁粉芯；耐高温绝缘材料；微观结构；软磁性能；界面结构

钼铼合金箔材制备技术研究

王广达[1]，熊 宁[1]，阚金锋[2]

（1. 安泰科技股份有限公司，北京 100081；2. 安泰天龙钨钼科技有限公司，天津 301899）

摘 要：钼铼合金具有优良的室温塑性，是应用于电子、半导体等行业的重要元器件材料。通过粉末冶金工艺方法制备坯料，经过轧制变形和热处理，得到具有各向同性的弯曲或深冲性能的钼铼合金箔材。本文研究了钼铼（Mo-35wt%Re）合金坯料制备中粉末粒度的配比对于烧结性能的影响，进行了不同轧制变形工艺试验，并对箔材进行去应力退火、部分再结晶退火和完全再结晶退火等不同退火制度的研究。通过金相观察、力学性能测试和杯突试验，研究发现，交叉轧制可以提高钼铼箔材的性能，并且改变交叉轧制的方式可以得到更好的效果；部分再结晶退火处理后，材料具有最优的塑性和杯突值，可以获得较好的力学性能和各向同性。

关键词：钼铼合金；箔材；粉末；轧制；退火；显微组织

Study on the Preparation Technology of Molybdenum-Rhenium Alloy Foil

Wang Guangda[1], Xiong Ning[1], Kan Jinfeng[2]

(1. Advanced Technology & Materials Co., Ltd., Beijing 100081, China;
2. ATTL Advanced Materials Co., Ltd., Tianjin 301899, China)

Abstract: Molybdenum-Rhenium Alloy has excellent room-temperature plasticity, and is an important component material used in electronics, semiconductor and other industries. Molybdenum-rhenium alloy foils with isotropic bending or deep drawing properties were prepared by rolling deformation and heat treatment using the powder metallurgy process. In this paper, the effect of powder size ratio on sintering properties of Mo-35wt% Re alloy billets was studied, and different rolling deformation processes were carried out, the different annealing systems, such as stress relief annealing, partial recrystallization annealing and complete recrystallization annealing, were studied. Through metallographic observation, mechanical property test and cupping test, it is found that the properties of rhenium foil can be improved by cross rolling, and the better effect can be obtained by changing the way of cross rolling, the material has the best plasticity and cupping value, and can obtain better mechanical properties and isotropy.

Key words: molybdenum-rhenium alloy; foil; powder; rolling; annealing; microstructure

应用于极端动高压的 W-Cu 体系梯度复合材料设计与制备

张 建，罗国强，沈 强，张联盟

（武汉理工大学材料复合新技术国家重点实验室，湖北武汉 430070）

摘　要：材料在极端动高压加载条件下的响应特性与物性研究一直是高压物理和材料科学与工程领域关注的焦点。本文针对战略武器内爆复杂动力学加卸载过程实验模拟的重大需求，重点研究 W-Cu 体系梯度材料的设计与制备，发展熔点差异悬殊梯度材料的粉体包覆、共烧技术，研制多物系、宽组分梯度材料，建立梯度结构的形成动力学过程、梯度结构精细控制方法，实现梯度材料按设计要求的精密梯度化，为梯度材料设计、构筑过程中梯度结构的形成与控制提供技术支撑。

关键词：梯度材料；动高压加载；W 基材料；材料设计与制备

Design, Preparation and Application in Dynamic High-pressure of W-based Graded Composites

Zhang Jian, Luo Guoqiang, Shen Qiang, Zhang Lianmeng

(State Key Lab of Advanced Technology for Materials Synthesis and Processing, Wuhan University of Technology, Wuhan 430070, China)

Abstract: The response characteristics and physical properties of materials under extreme dynamic high-pressure loading have always been the focus of attention in the field of high-pressure physics, material science and engineering. Aiming at the great demand of experimental simulation of complex dynamic loading and unloading process of strategic weapon implosion, this paper focuses on the design and preparation of two kinds of W-based graded materials, develops powder coating and co-fired technology of graded materials with wide melting point difference, develops multi-system and wide component graded materials, establishes the formation dynamic process of graded structure andthe fine control method of graded structure, realizes the precise gradient of graded material according to the design requirements, and provides technical support for the formation and control of graded structure in the design and construction of graded materials.

Key words: graded materials; dynamic high-pressure loading; W-based materials; material design and preparation

W-5Re 合金废料中 Re 的回收及其机理研究

陈鹏起[1,2]，许　荡[1,2]，程继贵[1,2]

（1. 合肥工业大学材料科学与工程学院，安徽合肥　230009；

2. 安徽省粉末冶金工程技术研究中心，安徽合肥　230009）

摘　要：铼（Re）资源稀缺，价格昂贵，因具有高硬度、高强度以及高耐热等优良性能，在航空航天、石油化工等领域受到高度重视。随着 Re 需求量的不断增长，从有限的矿物中提取 Re 往往难以满足需要，而从含铼废料中回收 Re 被认为是重要的提取 Re 的途径。另一方面，W-Re 合金性能优异，随着其使用量的增加，产生的废料也较多。因此，开展 W-Re 合金废料的高效分离，特别是从其中回收 Re，具有重要的经济价值。由于 Re_2O_7 气体的挥发率和收集率对 Re 的回收率有明显的影响，目前关于火法回收 Re 的研究存在回收率低、对易挥发的 Re_2O_7 气体收集率不高以及还原过程多反应难以精确控制等问题。因此，本文首先设计制备了一种适用于从 W-Re 合金废料中回收 Re 的火法回收装置，并采用该装置开展从 W-5wt.%Re(W-5Re)合金废料中火法回收 Re 的实验研究，探索高回收率、高纯度 Re 粉的火法回收途径。

实验中，从 W-5Re 合金废料中回收 Re 的过程包括四个步骤：（1）W-5Re 合金废料的氧化；（2）Re_2O_7 的氨水浸出；（3）NH_4ReO_4 粉末的制备；（4）Re 粉的制备。考察了火法工艺参数对 Re_2O_7 挥发率、Re 的回收率以及纯度的影响，对回收制备的 NH_4ReO_4 粉末和 Re 粉的组织结构和性能进行表征，同时对回收过程过中氧化还原反应的热力学及动力学机理进行了研究分析。结果表明，通过此方法成功地从 W-5Re 合金废料中回收 Re。在 675℃的空气气氛中保温 3h 可使 Re_2O_7 完全挥发并得到高效收集。Re_2O_7 经氨水浸出、蒸发结晶后制得的 NH_4ReO_4 粉末于 550℃的 H_2 气氛中保温 90min 可制备出平均粒径为 14.17μm、纯度为 99.9634%的 Re 粉，Re 的总回收率达 96.24%。此外，研究了回收过程中氧化还原反应的热力学和动力学机理，发现 NH_4ReO_4 粉末的氢还原是多反应同时发生的过程，且以 NH_4ReO_4 粉末与 H_2 的直接反应为主。

关键词：W-Re 合金废料；铼；火法冶金；回收；高铼酸铵

硬质合金辊环材料研究进展及应用

龙坚战[1,2]，孟湘君[2]，阳建宏[2]，杨开明[2]，
徐 涛[1,2]，崔焱茗[1,2]，魏修宇[1,2]

（1. 硬质合金国家重点实验室，湖南株洲 412000；
2. 株洲硬质合金集团有限公司，湖南株洲 412000）

摘 要：硬质合金辊环是钢铁工业线材轧制所需的重要耗材，其质量水平直接影响轧制线材的品级。本文从硬质合金辊环的应用背景出发，简述了硬质合金辊环材料的最新研究成果，分析了汽车用帘线钢、弹簧钢等优特线材轧制、高速棒材控温轧制以及“新国标”高强度螺纹钢多切分轧制领域对硬质合金辊环材料的性能要求，讨论了硬质合金 WC 晶粒形貌调控技术和粘结相强化技术对硬质合金辊环材料性能的影响规律，提出未来应从全架次解决方案出发，优化辊环材料性能，为高线全架次配辊方案提供支持和依据。

关键词：辊环；硬质合金；组织；性能；轧制

Research Progress and Application of Carbide Roll Materials

Long Jianzhan[1,2], Meng Xiangjun[2], Yang Jianhong[2], Yang Kaiming[2],
Xu Tao[1,2], Cui Yanming[1,2], Wei Xiuyu[1,2]

(1. Cemented Carbide of State Key Laboratory, Zhuzhou 412000, China;
2. Zhu Zhou Cemented Carbide Group Co., Ltd., R&D Center for Hard Materials, Zhuzhou 412000, China)

Abstract: Carbide rolls are important consumable materials required for wire or rod rolling in industry, and their quality level directly affects the grade of the rolled wire or rod. Starting from the application background of carbide rolls, the latest research results of carbide roll materials were briefly described in this paper. It was also analyzed that performance requirements of carbide roll materials in three areas, namely, the rolling of high-quality special wire or rods such as automotive cord steel and spring steel, high-speed bar temperature-controlled rolling and the "new national standard" high-strength rebar multi-slit rolling field. Meanwhile, the influence of WC grain morphology control technology and the binder phase strengthening technology on the properties of carbide roll materials were discussed. It is proposed that in the future, we should start from the solution of the whole stands, optimize the performance of the carbide roll materials, and provide support and basis for the whole stands distribution program of the high-speed wire or bar.

Key words: roll ring; cemented carbide; structure; performance; rolling

基于粉末烧结的高锰铝钢多孔化研究

徐志刚，王传彬，沈　强

（武汉理工大学，湖北武汉　430070）

摘　要：多孔钢作为一种典型的结构功能一体化材料，具有轻质高强、耐腐蚀、耐高温、比表面积大、吸能减震、隔音等诸多优良特性，是应用于吸能缓冲、过滤分离、生物医疗以及催化等领域的理想材料。钢基体高熔点的特性使得大量用于多孔铝的造孔方法难以复制于多孔钢的制备。粉末固相烧结法制备温度低，能够克服钢基体熔点高的缺点，是多孔钢制备的一种重要方法。为此，本文以高锰铝高强钢为研究对象，提出了一种综合利用柯肯达尔效应、反应烧结和锰升华效应协同高效造孔的一种新策略。在高锰铝高强钢多孔化的过程中，本文系统分析和讨论了高锰铝多孔钢的孔隙形成机制、显微结构演变以及力学性能的变化规律，为高锰铝多孔钢显微结构和力学性能优化提供坚实的理论依据。

关键词：高强钢；多孔材料；粉末烧结

Mo-$ZrTiO_4$钼合金第二相塑韧化作用机制

胡卜亮，王快社，胡　平

摘　要：钼合金脆性大，难加工一直是制约其发展和应用的主要原因，Mo-La 合金的发展虽然提高了合金的塑性，但也带来一定的环境问题。本研究通过传统第二相错配度理论与固体电子理论的对比，摒弃了错配度理论在钼合金第二相界面设计的应用。基于固体电子理论，创新性设计了钼合金塑韧化 $ZrTiO_4$ 第二相，其界面电子密度差与钼基体在三个低指数晶面上平均二维错配度达到 31.46%，而界面电子密度达到 6.5×10^5e/nm^3，远高于 TiC、ZrC、TiO_2、ZrO_2（约 1.5×10^5e/nm^3），表明 $ZrTiO_4$ 与钼基体具有很强的结合力。经过溶胶凝胶法合成 20~40nm 的 $ZrTiO_4$ 粉末后，优化成分，当 Mo-2.5wt.% $ZrTiO_4$ 通过传统简单的固-固混料工艺绿色化制备后，材料加工性能明显提高，轧制态抗拉强度~1100MPa，在其强度不降低的情况下，轧制退火后伸长率达到 23.1±2.3%，高于文献报道的 TZM 合金的综合力学性能。

Cu-16wt.%Sn 合金粉体的组织形貌及烧结研究

石　林，邹军涛，孙利星，王宇轩，黄安卓，张志伟，宋旭航

（西安理工大学材料科学与工程学院，陕西西安　710048）

摘　要：采用电极感应熔炼气雾化法制备了 Cu-16wt.%Sn 合金粉体，研究了合金粉体的微观组织形貌以及热压烧结后合金中析出相的分布情况。结果表明，合金粉体的球形度较好，随着冷却速率的增加，异形粉数量减少，合金粉体逐渐由树枝晶凝固向胞状晶凝固转变，粉体表面缩孔尺寸逐渐减小；合金粉体中除 a-Cu 相外，还存在均匀分布

的金属间化合物δ相；相较于传统铸造法，热压烧结后的合金中δ相尺寸相对细小、含量占比相对较低，且在合金组织中均匀分布。

关键词： Cu-16wt.%Sn 合金粉体；电极感应熔炼气雾化法；金属间化合物δ相；热压烧结

Study on Microstructure and Sintering of Cu-16wt.%Sn Alloy Powder

Shi Lin, Zou Juntao, Sun Lixing, Wang Yuxuan,
Huang Anzhuo, Zhang Zhiwei, Song Xuhang
(School of Materials Science and Engineering, Xi'an University of Technology, Xi'an 710048, Chnia)

Abstract: The Cu-16wt.%Sn alloy powder was prepared by electrode induction melting-gas atomization method. The microstructure of the alloy powder and the distribution of precipitated phases in the alloy after hot pressing and sintering were studied. The results show that the sphericity of the alloy powder is good. With the increase of the cooling rate, the number of special-shaped powder decreases, and the alloy powder gradually transforms from dendritic solidification to cellular solidification, and the size of the shrinkage cavity on the surface of the powder decreases gradually. In addition to the a-Cu phase, there is also a uniform distribution of intermetallic compound δ phase in the alloy powder. Compared with the traditional casting method, the size of the δ phase in the alloy after hot-press sintering is relatively fine, the content of the δ phase is relatively low, and it is evenly distributed in the microstructure of the alloy.

Key words: Cu-16wt.% Sn alloy powder; electrode induction melting-gas atomization method; intermetallic compound δ phase; hot-press sintering

选区激光熔化制备 Fe、Ni 基连续梯度合金研究进展

温耀杰[1,2]，张百成[1,2]，曲选辉[1,2]
（1. 北京材料基因工程高精尖创新中心，北京 100083；
2. 新材料技术研究院，北京科技大学，北京 100083）

摘 要： 多材料一体化制备是近年来新兴的研究领域，要求通过控制成分分布来实现零件不同部位具有不同的性能。在以往研究中，不同材料之间的连接区域往往在服役过程中最先失效。梯度材料(FGM, functional graded material)作为一种新型的复合材料，在两种材料之间采用成分逐渐变化的梯度区间来代替直接界面结合，进而避免不同材料由于热物性等的突变而发生的失效。铁基合金与镍基合金是目前应用最广的两类的合金，几乎所有的机械工程结构都含有铁镍基合金。不同铁镍元素含量配比的合金会展现出不同的特性并以此衍生出一系列常用合金，如马氏体时效钢、模具钢、因瓦合金、坡莫合金、高温合金等等。因此将不同成分的铁镍合金进行组合，将大幅扩展零件的功能性。上述提到的合金大多是以铁镍元素为主，再加入其他合金元素，通过组织调控来进一步提高其性能。在铁镍基梯度合金中，由于铁镍元素与其他合金元素的相互作用，特别是需要热处理强化的合金，其不同合金元素配比在热处理作用后合金内部相的析出情况以及组织演化情况非常复杂，也缺乏相关的热力学数据与热力学模型。本研究以典型的铁镍合金牌号 316L 不锈钢和 Inconel718 高温合金为研究对象，使用自研的连续梯度合金选区激光熔化设备制备 316L-In718 连续梯度合金(CGA, compositionally graded material)，试样致密度超过 98.5%，表面质量好。沿成分梯度分析表征了不同成分的组织结构、相组成及力学性能，并探究了热处理对其影响。进一步我们使用 CALPHAD (CALculation of PHase Diagrams)热力学计算对梯度合金内部的相组成进行了预测并比对了实验表征结果。本研究所使用的连续梯度合金制备方法将是一种高效的高通量制备方法，通过实验表征结果对热力学数据库进

行补充并对热力学模型进行改进，也将有助于我们理解多材料制备过程中成分-加工方式-组织结构-性能之间的关系。

关键词：连续梯度合金；铁镍合金；选区激光熔化；力学性能；相图计算

Research Progress of Compositionally Graded Alloy Prepared by Laser Powder Bed Fusion

Wen Yaojie[1,2], Zhang Baicheng[1,2], Qu Xuanhui[1,2]

(1. Beijing Advanced Innovation Center Materials Genome Engineering, Beijing 100083, China; 2. Institute for Advanced Materials and Technology, University of Science and Technology Beijing, Beijing 100083, China)

Abstract: Multi-material integrated preparation is an emerging research field in recent years, by controlling the distribution of components to achieve different performance in different positions of the part. In previous studies, the connection area between different materials often failed first in the service process. FGM as a new type of composite material that the gradient interval with gradually changing composition is used to replace the direct interface bonding between the two materials. to avoid the failure of different materials due to the sudden change of thermophysical properties. Iron base alloy and nickel base alloy are the two most widely used alloys at present, almost all mechanical engineering structures contain iron nickel base alloys. Alloys with different iron nickel content ratio will show different characteristics, and a series of common alloys will be derived, like maraging steel, die steel, invar alloy, permalloy, superalloy and so on. Therefore, the combination of Fe-Ni Alloys with different components will greatly expand the functionality of parts. Most of the alloys mentioned above are mainly iron and nickel, and then other alloy elements are added to further improve their properties through microstructure regulation. In Fe-Ni based gradient alloys, due to the interaction between Fe, Ni and other alloy elements, especially the alloys that need heat treatment strengthening, the precipitation and microstructure evolution of internal phases of the alloys with different alloy element ratios after heat treatment are very complex, and there is also a lack of relevant thermodynamic data and thermodynamic models. In this work, the stainless steel 316L and Inconel 718 are selected to representative the Fe-based alloy and Ni-based alloy respectively. And the 316L-In 718 CGA (compositionally graded alloy) are fabricated by the self-developed multi-material LPBF manufacturing equipment. The relative density of the 316L-In718 CGAs are higher than 98.5% with good surface quality. The microstructure, phase evolution and mechanical properties along the composition gradient are investigate, and the effect of heat treatment on it was also explored. The phase evolution are also predicted by CALPHAD (CALculation of PHase Diagrams). The compositionally graded alloy preparation method used in this study will be an efficient and high-throughput preparation method. The experimental characterization results will supplement the thermodynamic database and improve the thermodynamic model, which will also help us to understand the relationship between composition-processing mode- microstructure-properties in the process of multi material preparation.

Key words: compositionally graded alloy; Fe-Ni based alloy; laser powder bed fusion; mechanical properties; CALPHAD

高铁含量铜铁合金的制备及性能研究

张陈增，陈存广，李沛，郭志猛

（北京科技大学新材料技术研究院，北京　100083）

摘　要：采用水雾化法制备了超细的高铁含量（Fe>5wt.%）铜铁合金粉末，以此粉末为原料通过冷等静压和真空烧结工艺制备了大尺寸铜铁合金坯锭。并利用扫描电子显微镜(SEM)、X 射线衍射(XRD)、拉伸测试、四探针测电阻法等测试手段研究了铜铁合金粉末及坯锭的形貌组织与性能。结果表明，粉末形貌呈近球形，颗粒细小，D50<10μm。烧结坯中铁相分布均匀无偏析，铁颗粒呈近球形，尺寸细小，平均粒径为 1μm，烧结坯相对密度均大于 98%。该制备方法解决了传统熔铸法难以制备铁颗粒尺寸细小、成分均匀、无偏析的高铁含量铜铁合金的难题。力学及物理性能测试结果表明以此方法制备的铜铁合金较传统熔铸法相比具有更加优异的综合力学物理性能。最后，讨论了铜铁合金的微观组织及力学物理性能之间的联系。

关键词：铜铁合金；水雾化；粉末冶金；合金粉末；冷等静压

Study on Preparation and Performance of Cu-Fe Alloy with High Ironcontent

Zhang Chenzeng, Chen Cunguang, Li Pei, Guo Zhimeng

(Institute for Advanced Materials and Technology, University of Science and Technology Beijng, Beijng 100083, China)

Abstract: Ultra-fine Cu-Fe alloy powders with high iron content (Fe>5wt.%) were prepared by water atomization method, and large-size Cu-Fe alloy ingots were prepared by cold isostatic pressing and vacuum sintering using the powders as raw material. The morphology and performance of Cu-Fe alloy powders and billet were investigated scanning electron microscopy (SEM), X-ray diffraction (XRD), tensile test, four-probe resistance measurement method and other testing methods. The results show that the powder has a nearly spherical morphology, fine particles, and D50<10μm. The Fe phase in the sintered compact is uniformly distributed without segregation, the Fe particles are nearly spherical, small in size, with an average particle size of 1μm, and the relative density of the sintered compact is higher than 98%. The preparation method solved the problem that the traditional casting method was difficult to prepare a Cu-Fe alloy with high iron content with small iron particle size, uniform composition and no segregation. The test results of mechanical and physical properties show that the Cu-Fe alloy prepared by this method has better comprehensive mechanical and physical properties than the traditional casting method. Finally, the relationship between the microstructure and mechanical and physical properties of Cu-Fe alloys was discussed.

Key words: Cu-Fe alloy; water atomization; powder metallurgy; alloy powders; cold isostatic pressing

基于高通量增材制造的 AlxCoCrFeNi 高熵合金性能筛选

郭朝阳[1]，张百成[1,2]

（1. 北京材料基因工程高精尖创新中心，新材料技术研究院，北京科技大学，北京　100083；
2. 现代交通金属材料与加工技术北京实验室，北京　100083）

摘　要：高熵合金(HEAs)因其独特的设计和原子结构而显示出许多优异的性能。然而，如何探索其广阔的成分空间和复杂的微观结构是一个巨大的挑战。本项研究提出了一种新的高通量激光粉末床熔融(LPBF)系统，用于制备大尺存的连续成分梯度 HEAs。本研究通过制备 AlxCoCrFeNi(0<x<20)成分梯度合金，系统地研究了该合金的相选择、

织构和力学性能的演变。随着 Al 含量的增加，打印态 AlxCoCrFeNi HEAs 的晶体结构逐渐由 fcc→fcc+bcc→bcc 转变，显微硬度和抗拉强度也随之提高。有趣的是，与传统的 HEAs 制备技术相比，激光处理的 AlxCoCrFeNi 的相变成分区间明显提前和变窄，归因于 LPBF 过程中的超快冷却速度和热循环过程。在 $Al_{0.52}$CoCrFeNi 处发现双相区，沿构建方向呈 bcc/fcc 层状类阶梯式分布，晶体取向与全 fcc 或全 bcc 的<001>方向显著不同。双相区组织呈细小的类等轴晶，没有明显的织构。拉伸结果表明，打印态的双相合金保持了 748MPa 的抗拉强度，延伸率约为 22.8%。强度和塑性的权衡归因于韧性面心立方相和脆性体心立方相的独特耦合共晶分布。最后，本文分析了两相 HEA 相变终点提前的基本机理和高铝含量下非平衡条件下的开裂行为。

关键词：高熵合金；高通量激光粉末床熔融；双相；强塑性权衡

Dual Phase AlxCoCrFeNi with Trade off Strength/Ductile Developed by High-throughput Laser Powder-bed Fusion Technology

Guo Chaoyang[1], Zhang Baicheng[1,2]

(1. Beijing Advanced Innovation Center Materials Genome Engineering, Advanced Material & Technology Institute, University of Science and Technology Beijing, Beijing 100083, China; 2. Beijing Laboratory of Metallic Materials and Processing for Modern Transportation, Beijing 100083, China)

Abstract: High entropy alloys (HEAs) exhibit multiple excellent properties because its unique design and atom structure. However, the exploring of its vast compositional space composition and complex microstructure is a grand challenge. A new self-developed high-throughput laser powder bed fusion (LPBF) system is proposed for large-scale microstructurally and compositionally graded HEAs preparation. In this study, an AlxCoCrFeNi ($0<x<20$) compositionally graded alloy component is prepared. The phase selection, texture and mechanical properties evolution are systematically studied. The crystal-structures of printed AlxCoCrFeNi HEAs were gradually transformed from FCC→FCC+BCC→BCC, with increasing Al content, accompanied with an increase in microhardness and tensile strength. Interestingly, compared to traditional HEAs preparation techniques, the phase transition interval from FCC to BCC of laser processed AlxCoCrFeNi is significantly advanced and narrowed, due to the ultrafast cooling rate and thermal cycles during LPBF. Moreover, the dual-phase region is found at $Al_{0.52}$CoCrFeNi with BCC/FCC layer-wise distribution. Distinguish with the significant <001> orientation from FCC or BCC regions from printed AlxCoCrFeNi component, the dual-phase structure shows a fine equiaxed grain without obvious texture. Tensile results demonstrate that the as-printed dual-phase alloy maintains 748 MPa tensile strength with about 22.8% elongation. The trade-off in strength and ductility is attributed to the unique coupled eutectic-like distribution of the ductile FCC and brittle BCC phases with fine densely arranged grains. At last, the fundamental mechanisms for the advancement of phase transformation endpoints in dual-phase HEAs and cracking behavior with high aluminum content under nonequilibrium conditions were analyzed.

Key words: high entropy alloys; high-throughput laser powder-bed fusion; dual-phase; strength and ductility trade-offs

元素 Al/Ti 对 TaNbVTi 难熔高熵合金组织和性能的影响

廖　涛，刘　彬，付　遨

（中南大学粉末冶金研究院，湖南长沙　410000）

摘　要：近年来，难熔高熵合金以其优异的高温力学性能引起了广泛的关注。本文使用粉末冶金方法制备了

TaNbVTiAl$_x$ 和 Ta$_x$Nb$_x$V$_x$Ti 两个体系难熔高熵合金。本文详细研究了两种合金随元素含量变化的组织与性能变化情况，并对其强化的机制进行了探索。实验结果表明，两个系列难熔高熵合金具有均匀的单相 BCC 微观组织。TaNbVTiAlx 合金随着 Al 含量增加，合金强度先增加后降低，其中 Al$_{0.2}$ 合金综合性能最好。Ta$_x$Nb$_x$V$_x$Ti 合金中随 Ti 含量的增加，合金强度降低，塑性提高，其中 x=0.4 时合金的室温塑性超过 50%；x=0.8 时合金具有最优综合性能，其室温压缩屈服强度和塑性分别达到 1396MPa 和 28%，1200℃时还能保持 144MPa 的屈服强度，远高于现有镍基高温合金。

铜铁复合粉末绿色制备技术工艺研究

班丽卿[1,3]，张　将[3]，汪礼敏[2,3]，李楠楠[1,3]，王　蕊[1,3]，
王林山[1,3]，贺会军[2,3]

（1. 北京有研粉末新材料研究院有限公司，北京　101407；2. 有研粉末新材料股份有限公司，北京　101407；3. 金属粉体材料产业技术研究院，北京　101407）

摘　要： 铜包铁作为一种 Fe/Cu 核壳型复合粉末，兼具 Fe 的高强度与 Cu 的耐蚀性等优点，同时与青铜相较有较高的强度和明显的成本优势，故在含油轴承、金刚石工具领域应用广泛。目前市场主要采用化学置换方法制备，存在环保、资源浪费等问题，而物理混合方法又存在成分偏析、力学性能不稳定等缺点。通过本技术可制备出包覆层均匀完整、成分含量易控、成形性强度好（生坯密度 5.85g/cm^3，强度 40~45MPa）、烧结尺寸稳定（收缩率≤0.2%），含油率高（18.7%）的高性能产品。

关键词： 机械化学法；铜铁复合粉末；压溃强度；成形性；含油率

Study on Green Preparation Technology of Copper Iron Composite Powder

Ban Liqing[1,3], Zhang Jiang[3], Wang Limin[2,3],Li Nannan[1,3], Wang Rui[1,3],
Wang Linshan[1,3], He Huijun[2,3]

(1. GRIPM Research Institute Co., Ltd., Beijing 101407, China;
2. GRIPM Advanced Materials Co., Ltd., Beijing 101407, China;
3. Industrial Research Institute for Metal Powder Material, Beijing 101407, China)

Abstract: As a Fe/Cu core-shell composite powder, copper coated iron powder makes the composite material combines the advantages of both iron (Fe) and copper (Cu), such as the high strength of Fe and the corrosionof Cu. At the same time, it has higher strength and obvious cost advantage compared with bronze,therefore, it is widely used in the fields of oil-bearing bearings and diamond tools. At present, the chemical replacement method is mainly used to prepare copper coated iron powder in the market, but it has some problems, such as environmental protection and the waste of resources, meanwhile, the physical mixing method has disadvantages such as component segregation and unstable mechanical properties. With this technology, for the product, the coating layer can be made uniform and complete, and the component content is easy to control, which has excellent formability (green density 5.85g/cm^3, green strength 40-45MPa), stable sintered dimension (shrinkage rate≤0.2%), and high oil content (18.7%).

Key words: mechanochemical method; copper-iron composite powder; crushing strength; suppressing performance; oil content

气雾化工艺参数对 TC11 合金 3D 打印专用粉末粒度的影响

周　舸，田　辰，张志鹏，张浩宇，朱晓飞，张晓洁，陈立佳

（沈阳工业大学材料科学与工程学院，辽宁沈阳　110870）

摘　要：采用 JIGA-2 型真空气雾化制粉设备，对 TC11 钛合金开展了不同工艺制度下的 3D 打印球形粉末制备试验，并对 EIGA 法制备过程中加热功率、进给速度和气雾化压力对粉末粒径的影响进行了研究。结果表明：加热功率对该合金粉末粒径的影响不大，而提高气雾化压力能够显著提高细粉收得率，母合金棒材进给速度与加热功率的匹配性是影响该合金粉末球形度的重要因素。TC11 合金 3D 打印专用粉末（粒径分布区间为 45-55μm）制备的最佳工艺参数为：加热功率为 29kW，气液流量比 0.5，气雾化压力为 6MPa，此时细粉收得约为约为 35%，粉末形貌为球形且质量较好，未出现表面为星球及凹坑现象。

关键词：EIGA 气雾化；TC11 合金；粉末粒径；3D 打印金属粉末

Effects of Atomization Parameters on Particle Size of 3D Printing Exclusive Powder from TC11 Alloy

Zhou Ge, Tian Chen, Zhang Zhipeng, Zhang Haoyu,
Zhu Xiaofei, Zhang Xiaojie, Chen Lijia

(School of Materials Science and Engineering, Shenyang University of Technology, Shenyang 110870, China)

Abstract: Metal powder of TC11 alloy in different particle sizes was prepared on a JIGA-2 vacuum atomization powder manufacturing apparatus. The effects of power, feeding speed, and atomization pressure on the powder particle size were studied. It was found the heating power did not largely affect the alloy powder particle size, but the atomization pressure considerably enhanced the fine powder yield. The key influence factor on the alloy powder sphericity was the matching rate between the feeding speed and heating power. The optimal technological parameters on preparation of 3D printing exclusive powder (particle size within 45-55μm) from TC11 alloy were the heating power at 29kW, gas-liquid flow ratio at 0.5, and atomization pressure at 6Mpa. The fine powder yield under the optimal condition was 35%, and the powder morphology was sphere, but no star or pit.

Key words: EIGA atomization; TC11 Alloy; powder particle size; 3D printing metal powde

钼合金单晶材料研制进展

高选乔

（西北有色金属研究院，陕西西安　710016）

摘　要：钼合金单晶材料具有一系列优异且独特的物理化学性能、力学性能而被广泛应用于航空、航天、核能等高技术领域和固体物理等基础理论研究领域。本报告主要介绍钼合金单晶的制备技术，对电子束悬浮区域熔炼技术和等离子弧熔炼技术进行了比较，讨论了单晶材料发展现状，从成分体系、微观结构、物理力学性能及形变机理等方面介绍钼合金单晶的发展历史和研究进展。此外，还对单晶材料制备技术的发展前景提出一些建议。

渗铜烧结钢的研究现状

林鹏程[1,2,3]，王林山[1,2,3,4]，梁雪冰[1,2,3,4]，胡　强[1,2,3,4]，汪礼敏[1,2,3,4]

（1. 北京有研粉末新材料研究院有限公司，北京　101407；2. 有研粉末新材料股份有限公司，北京　101407；3. 北京有色金属研究总院，北京　100088；4. 有研科技集团有限公司金属粉体材料产业技术研究院，北京　101407）

摘　要：铁基粉末冶金制品（烧结钢）是粉末冶金行业生产量最大的一类，采用传统压制成形的工艺制备的产品存在8%~15%的孔隙，影响了产品性能和应用。渗铜工艺是最常用的消除残留孔隙的工业方法之一，可大幅提高材料的密度和力学性能（特别是动态力学性能）。本文介绍了铁基粉末冶金制品渗铜的原理、性能和特点，综述了渗铜在3D打印工艺的应用，并对渗铜烧结钢的发展提出了建议。

关键词：烧结钢；孔隙；渗铜；动态力学性能

Research of Copper-infiltrated Sintered Steel

Lin Pengcheng[1,2,3], Wang Linshan[1,2,3,4], Liang Xuebing[1,2,3,4], Hu Qiang[1,2,3,4], Wang Limin[1,2,3,4]

(1. Beijing GRIPM Advanced Materials Research Institute Co., Ltd., Beijing 101407, China; 2. GRIPM Advanced Materials Co., Ltd., Beijing 101407, China; 3. General Research Institute for Nonferrous Metals, Beijing 100088, China; 4. Metal Powder Materials Industrial Technology Research Institute of GRINM, Beijing 101407, China)

Abstract: Iron-based powder metallurgy products (sintered steel) are one of the largest production in the powder metallurgy industry. The products prepared by the traditional press forming process have 8%~15% pores, which affects the performance and application of the products. Copper infiltration process is one of the most commonly used industrial methods to eliminate residual pores, which can greatly improve the density and mechanical properties (especially dynamic mechanical properties) of materials. This paper introduces the principle, properties and characteristics of copper infiltration of iron-based powder metallurgy products, summarizes the application of copper infiltration in 3D printing process, and puts forward some suggestions on the development of copper infiltration sintered steel.

Key words: sintered steel; pore; copper infiltration; dynamic mechanical properties

FeAlCrBY 多孔材料的制备及其抗高温碳化性能研究

张惠斌，王龙飞，余　航

（浙江工业大学 材料科学与工程学院，浙江杭州　310014）

摘　要： FeAlCr 合金多孔材料广泛应用于高温煤气和冶炼荒煤气净化，其中由 CH_4、CO 等构成的富碳气氛会引起材料的碳化，从而严重恶化材料的力学性能。由于碳难以在氧化物中扩散，因此如何在低氧势气氛下构建稳定的氧化膜对提高 FeAlCr 合金多孔材料的抗碳化性能具有重要意义。本文在 Fe-Al-Cr 中引入 1.0wt.%的 YB_4，通过反应合成方法制备了 FeAlCrBY 合金多孔材料，并研究了 B、Y 合金化对材料抗高温碳化腐蚀的影响。结果表明，FeAlCr 多孔材料在经 900℃高温碳化后，表面产生了以$(Fe,Cr)_7C_3$为主的渗碳层，而 B、Y 合金化促进了 FeAlCrBY 多孔材料在低氧势下生成稳定的 Al_2O_3 层，抑制渗碳层厚度的增加，显著减轻力学性能的下降趋势。

关键词： 多孔材料；高温过滤；碳化；反应合成；力学性能

Preparation of FeAlCrBY Porous Material and its Resistance to High-temperature Carbonization

Zhang Huibin, Wang Longfei, Yu Hang

(College of Materials Science and Engineering, Zhejiang University of Technology, Hangzhou 310014, China)

Abstract: FeAlCr alloy porous materials are widely used in the purification of high-temperature coal gas and raw coke oven gas. The carbon-rich atmosphere composed of CH_4 and CO will cause the carbonization of the filter materials, which can seriously deteriorate their mechanical properties. Building a stable oxide film under the low oxygen potential atmosphere is of great significance to improve the carbonation resistance of FeAlCr porous materials due to the sufficiently low diffusivity of carbon in oxides. In this paper, FeAlCrBY alloy porous material was prepared by reactive synthesis method by introducing 1.0wt.% YB_4 into Fe-Al-Cr mixture, and the effect of B-Y alloying on the high-temperature carbonization corrosion resistance of the material was studied. The results showed that the carburized layer mainly made of $(Fe, Cr)_7C_3$ was produced on the surface of FeAlCr porous material after high temperature carbonization at 900℃, while B-Y alloying promotes the formation of stable Al_2O_3 layer under low oxygen potential, thus inhibiting the thickening of the carburized layer and retarding the deterioration of mechanical properties.

Key words: porous materials; high-temperature filtration; carbonization; reactive synthesis; mechanical properties

粉末热机械固结法制备的 Fe-22Mn 棒材的显微组织研究及铁锰氧化物对力学性能的影响

张有鋆，赵晓丽，Valladares L.De Los Santos，张德良

（东北大学材料学院先进粉末冶金与技术实验室，辽宁沈阳　110819）

摘　要： 通过粉末热机械固结法直接制备钢材有低能耗，短流程，成分和微观组织均匀细化的优势。本团队制备出相对密度为 99.9%的 Fe-22Mn(wt.%)棒材，其组织包括 γ-奥氏体、ε-马氏体，α′-马氏体和硬而脆的铁锰氧化物，材料的力学性能优异。当应变量超过 0.09 时，加工硬化的主要贡献由 γ→ε相变转化为ε→α′相变，经过均匀化处理后，铁锰氧化物颗粒中的铁含量大幅降低、硬度下降。一些铁锰氧化物颗粒会在拉伸过程中与基体发生脱离，促进了裂纹的萌生与扩展，从而严重损害材料的塑性。

关键词： 粉末冶金；高锰钢；应变诱导马氏体相变；铁锰氧化物

Microstructure of Fe-22Mn Extruded Rods Prepared by Thermomechanical Powder Consolidation and Effects of Ferrous Manganese Oxides on Mechanical Properties

Zhang Youyun, Zhao Xiaoli, Valladares L. De Los Santos, Zhang Deliang

(APM-Lab, School of Materials Science and Engineering, Northeastern University, Shenyang 110819, China)

Abstract: Steel prepared by the thermomechanical powder consolidation (TMC) method has the advantages of low energy consumption, short process, uniform composition and refined organization. Our team has prepared Fe-22Mn(wt.%) rods with a relative density of 99.9%. Their microstructures includes γ-austenite, ε-martensite, α'-martensite, hard and brittle ferrous manganese oxides (FMOs). The mechanical properties of the material are excellent. When the amount of strain exceeds 0.09, the main contribution of work hardening arises to ε→α′ phase transition instead of γ→εphase transition. Through homogenization, the Fe content in FMOs is greatly reduced and the hardness decreases. Decohesion between some FMOs particles and the matrix occurs during the tensile test, and promotes the initiation and propagation of cracks, causing the tensile ductility of the material to deteriorate.

Key words: powder metallurgy; high manganese steel; strain-induced martensitic transformation; ferrous manganese oxide

气相沉积技术在钨基材料表面改性的研究介绍

陈树群，谭翠翠，周文元，王金淑

（北京工业大学，材料与制造学部，稀有金属研究所，北京　100124）

摘　要：难熔金属钨由于具有高熔点、高硬度、高密度、良好的导热性等诸多优点，在电子信息、医疗、机械制造、核工业等领域得到广泛应用，但也存在中高温有氧环境下易氧化的突出问题。针对这一问题，结合钨基材料在铝合金压铸模具上的应用需求，本研究采用化学气相沉积技术在在纯钨表面制备了 $TiN/TiCN/Al_2O_3$ 陶瓷梯度涂层，测试样品在 700-1000℃条件下的氧化特性，分析相应的氧化机制。研究发现梯度陶瓷涂层在纯钨表面具有高结合力和高硬度，在 700℃经 100h 氧化后的样品单位面积增重仅 0.56mg/cm^2（纯钨为 124.12mg/cm^2），1000℃经 5h 氧化后的样品单位面积增重为 7.5mg/cm^2（纯钨为 218.4mg/cm^2）。对 1000℃条件下的涂层结构演变分析发现，钨的外扩散是影响涂层使用效能的关键因素（而非氧的内扩散），这一发现对难熔金属表面抗氧化涂层的设计具有积极的指导意义。

关键词：难熔金属；化学气相沉积；陶瓷涂层；抗氧化性能

铜包铁复合粉的制备及其包覆层结构表征

庞建明[1]，李石稳[1,2]，韩　伟[1]，罗林根[1]，赵志民[1]

（1. 中国钢研科技集团有限公司，北京　100081；2. 钢铁研究总院，北京　100081）

摘　要：利用机械力化学作用将铁粉和超细氧化铜粉进行复合，得到前驱体 Fe@CuO 复合粉，经过氢气还原与烧结获得 Fe@Cu 复合粉。研究了氧化铜粉在球磨细化过程中的机械力化学效应，探讨了 Fe 与 CuO 颗粒在机械力作用下的复合过程和后续氢气还原与烧结过程中铜包覆层的形成及其致密化过程。研究表明：球磨细化后的氧化铜粉具有更高的表面缺陷密度与表面能，这为 Fe 与 CuO 颗粒之间的复合提供了良好的动力学条件；粒径比（D_{Fe}/D_{CuO}）越大，Fe 与 CuO 颗粒之间越容易复合；制备出的 Fe@Cu 复合粉的铜包覆率大于 95%，包覆层厚度 2~5μm，含铜量大于 15%。本文的研究结果有助于机械化学法制备包覆型粒子的研究，也为探索环保、经济的铜包铁复合粉制备工艺做出了新的尝试。

关键词：铜包铁粉；机械力化学；粒径比；包覆层

Preparation and Coating Structural Characterization of Copper Cladded Iron Composite Powder

Pang Jianming[1], Li Shiwen[1,2], Han Wei[2], Luo Lingen[1], Zhao Zhimin[1]

(1. China Iron & Steel Research Institute Group, Beijing 100081, China;
2. Central Iron & Steel Research Institute, Beijing 100081, China)

Abstract: Iron powder and ultrafine copper oxide powder were compounded by mechanochemical action to obtain Fe@CuO of precursors and then Fe@Cu Composite powder was obtained by hydrogen reduction and sintering. The mechanochemical effect of copper oxide powder during ball-milling was studied. The composite process of Fe and CuO particles under mechanical force and the formation and densification of copper coating during hydrogen reduction and sintering were discussed. The results show that copper oxide powder by ball-milling has higher surface defect density and surface energy， which provides better kinetic conditions for the recombination between Fe and CuO particles; In addition, the larger the particle size ratio (D_{Fe}/D_{CuO}), the easier it is to compound between Fe and CuO particles. The copper coating rate is greater than 95%，the coating thickness 2-5 μm, and copper content greater than 15%. This paper contributes to the research on the preparation of coated particles by mechanochemical method and also makes a new attempt to explore the environmentally friendly and economical preparation process of Fe@Cu composite powder.

Key words: copper cladded iron powder; mechanochemistry; particle size ratio; coating

粉末冶金钛基层状结构材料

张卫东[1]，曹远奎[2]，吴正刚[1]

（1. 湖南大学材料科学与工程学院，湖南长沙　410082；
2. 粉末冶金国家重点实验室，湖南长沙　410083）

摘　要：日益严苛复杂的服役环境对金属结构材料的性能提出了更高的要求。层状结构复合设计能够充分发挥金属/金属复合材料的性能潜力，实现性能指标的最优化配置。本研究提出运用粉末冶金技术结合热、冷加工获取高性能金属层状结构材料的学术思路。并以钛基层状结构材料为例，验证了混粉与铺粉两类粉末冶金方法制备高强韧金属层状结构材料的可行性。混粉末式 Ti/Mo 材料，经热旋锻后 Ti/Mo 复合材料抗拉强度高达 1600MPa，这种超高强度的获得主要依赖于各组元细晶强化与层状结构的背应力强化效果；铺粉式 Ti/β-Ti 层状结构材料，经冷轧与不完全再结晶退火后该材料兼具高强度与高塑性，其抗拉强度明显高于纯钛，约 832MPa，延伸率与纯钛相当，约 20%。这种强塑性的良好匹配主要与层状结构设计及界面层对裂纹的阻碍作用有关。上述实验结果充分证实了粉末冶金技术是高性能钛基层状结构材料的有效制备方法，其具有过程简单、原料利用率高、易于实现组元调控与材料性能优化等优点。

关键词： 粉末冶金；层状结构；钛；高强度；高塑性

Laminated Titanium Matrix Materials Processed by Powder Metallurgy

Zhang Weidong[1], Cao Yuankui[2], Wu Zhenggang[1]

(1. College of Materials Science and Engineering, Hunan University, Changsha 410082, China;
2. State Key Laboratory of Powder Metallurgy, Central South University, Changsha 410083, China)

Abstract: The increasingly harsh and complex service environment puts forward higher requirements on the performance of metal structural materials. The composite with lamellar structure could give full play to the potential of metal/metal composite materials and realize the optimal configuration of properties. This work combines powder metallurgy technology with hot and cold processing to fabricate high-performance laminated metal matrix materials with high performance. Taking laminated titanium-matrix materials as examples, the feasibility of fabricating high-strength and tough metal laminated structure materials by powder metallurgy (mixing powder and spreading powder) is verified. For mixed-powder Ti/Mo material after hot rotary forging, its tensile strength of Ti/Mo composite material is as high as 1600MPa. The achievement of the ultra-high strength mainly depends on the fine-grain strengthening of each component and the back stress strengthening effect of the lamellar structure; For spread-powder Ti/β-Ti laminated structure material after cold rolling and incomplete recrystallization annealing, the material has both high strength and high plasticity. Its tensile strength is significantly higher than that of pure titanium, about 832MPa. And it has equivalent elongation to pure titanium, about 20%. The excellent balance between strong and plasticity is mainly related to the lamellar structure design and the hindering effect of the interface layer on crack propagation. The above experimental results fully confirm that powder metallurgy technology is an effective method to prepare laminated titanium-matrix structural materials with high-performance. Besides, this method has the advantages of simple process, high material utilization rate, easy component control and material properties optimization.

Key words: powder metallurgy; laminated structure; titanium; high strength; high plasticity

9　能源、环保与资源利用

大会特邀报告
分会场特邀报告
矿业工程
炼铁与原燃料
炼钢与连铸
轧制与热处理
表面与涂镀
金属材料深加工
先进钢铁材料及其应用
粉末冶金
★ 能源、环保与资源利用
冶金设备与工程技术
冶金自动化与智能化
冶金物流运输
其他

9.1 冶金能源

某钢铁厂一种新型干燥器均压控制系统

王延明
（首钢京唐钢铁联合有限责任公司能源与环境部，河北唐山 063200）

摘 要：压缩空气为钢铁厂生产所必需的介质之一，空压站的运行效率是否合理，对钢铁企业的稳定生产有一定的影响。以首钢京唐公司压缩空气系统中的干燥器为主要研究对象，建立了一种新型的干燥器均压控制系统。应用实践表明，此控制系统提高了干燥器的运行效率，降低压缩空气露点，有助于提高钢铁厂的管理水平，增加公司的经济效益。

关键词：干燥器；均压；控制系统；节能

A New-type Iron & Steel Plant Drayer Pressure-sharing Control System

Wang Yanming
(Department of Energy, Shougang Jingtang Iron & Steel Co., Ltd., Tangshan, 063200, China)

Abstract: Compressed air is one of the necessary medium for steel plant production. Whether the operation efficiency of air compressor station is reasonable has a certain influence on the stable production of iron and steel enterprises. Taking the dryer in the compressed air system of Shougang Jingtang Company as the main research object, a new pressure equalizing control system for the dryer was established. The application practice shows that the control system improves the operation efficiency of the dryer, reduces the dew point of the compressed air, helps to improve the management level of the steel plant, and increases the economic benefit of the company

Key words: dryer; pressure-sharing; control system; energy saving

电站锅炉省煤器腐蚀失效特征研究及预防措施

李丛康[1]，贾丽娣[2]，王东山[2]，张天赋[2]，何 嵩[1]，刘柏寒[1]，孙 亮[1]
（1. 鞍钢股份有限公司能源管控中心，辽宁鞍山 114002；
2. 鞍钢股份有限公司技术中心，辽宁鞍山 114021）

摘 要：针对发生腐蚀失效的 220t 燃煤锅炉省煤器管，利用 XRD、SEM、EDS 等手段对试样中不同区域的相组成及元素分布进行表征，并对腐蚀失效特征进行了分析。结果表明：采用SNCR燃煤锅炉省煤器外壁腐蚀产物主要元素为Fe、S、O。腐蚀产物的主要成分为Fe_2O_3、$Fe(NH_4)_2(SO_4)_2 \cdot 6H_2O$；结垢层主要成分为$CaSO_4$、$SiO_2$、$Fe(NH_4)_2(SO_4)_2 \cdot 6H_2O$、$(NH_4)_2SO_4$。省煤器管在氧、硫酸、硫酸氢铵耦合作用下发生了腐蚀失效。采用光管鳍片式换热管、镀搪瓷膜的ND钢管、增加冲洗除灰次数、减少尿素喷射量，降低氨逃逸等方法能有效减轻腐蚀，减少积灰，降低锅炉尾部受热面爆管次数。

关键词：省煤器；腐蚀；SNCR

The Research on the Corrosion Failure Characters of Economizer in Power Plant and Preventive Measures

Li Congkang[1], Jia Lidi[2], Wang Dongshan[2], Zhang Tianfu[2], He Song[1], Liu Baihan[1], Sun Liang[1]

(1. Energy Management Centre of Angang Company Limited, Anshan 114002, China;
2. The General Iron Plant of Angang Steel Company Limited, Anshan 114021, China)

Abstract: Tube samples have been taken from dismantled economizer for a 220t coal fired boiler with the corrosion failure characters. XRD and SEM（EDS）tests were carried out to featuring the content of elements and the phases in different area and the corrosion failure characters also analyzed. The results show that the main elements of the corrosion product in the out layer of the aforementioned tube are Fe, S and O. The main gradients in the corrosion product are Fe_2O_3 and $Fe(NH_4)_2(SO_4)_2 \cdot 6H_2O$, and main gradients for dust layer are $CaSO_4$, SiO_2 , $Fe(NH_4)_2(SO_4)_2 \cdot 6H_2O$ and $(NH_4)_2SO_4$. The corrosion failure occurred due to combination effect of oxygen corrosion, sulf-corrosion and ABS (ammonium bisulfate) corrosion. Methods, such as finned tube, enamel coating ND steel tube, increasing dust flush frequency, decreasing the amount of urea jet and reducing volatilizing ammonia, could alleviate corrosion, reduce dust accumulation and decrease the frequency of tube explosion on the heating side in tailing part of boiler.

Key words: economizer; corrosion; SNCR

钢铁厂引进新型燃料-混空轻烃的可行性探讨

陶有志[1,2]，韩渝京[1,2]，李　鹏[1,2]，曹勇杰[1,2]

（1. 北京首钢国际工程技术有限公司，北京　100043；
2. 北京市冶金三维仿真设计工程技术研究中心，北京　100043）

摘　要：由于钢铁厂副产煤气产销出现不平衡，其中富含CO的副产煤气作为化工原料可用于产品深加工，创造新的附加值，其缺口日益扩大，引进的天然气受到国家政策调控，故需要全新审视引进新的能源介质，保证钢铁厂正常生产。新型燃料-混空轻烃属于石油化工行业副产品，通过物理加工后达到相当于天然气品质，因此引进混空轻烃作为钢铁厂的补充，纳入到钢铁厂煤气系统有两种方式进行可行性探讨。

关键词：钢铁厂副产煤气不平衡；引进；混空轻烃；供气方式

Probe into Introducing of New Type Fuel-mixed Light Hydrocarbon in Iron and Steel Plant

Tao Youzhi[1,2], Han Yujing[1,2], Li Peng[1,2], Cao Yongjie[1,2]

(1. Beijing Shougang International Engineering Technology Co., Ltd., Beijing 100043, China;
2. Beijing Metallurgy 3D Simulation Design Engineering Technology Research Center, Beijing 100043, China)

Abstract: The production and marketing of by-product gas in iron and steel plants are unbalanced. The by-product gas rich in CO is used as chemical raw materials for deep processing of products, creating new added value. The gap is widening day by day. The imported natural gas is regulated by the state policy. Therefore, it is necessary to examine the introduction of new energy media to ensure its normal production. The new type of fuel-air-mixed light hydrocarbon is a by-product of petrochemical industry. It is equivalent to the quality of natural gas after physical processing. Therefore, the introduction of air-mixed light hydrocarbon as a supplement to iron and steel plants and its incorporation into the gas system can be considered in two ways.

Key words: iron and steel plant by-product gas imbalance; introduction; mixed light hydrocarbon; gas supply mode

内蒙古某地区光伏发电站设计及效果评估分析

彭扬东，桂树强，李海峰，黄晓明

（中冶集团武汉勘察研究院有限公司，湖北武汉　430080）

摘　要：2021 年是“十四五”规划开局之年，全面落实碳达峰、碳中和工作，实现绿色低碳发展将是“十四五”期间重要的发展基调，“十四五”期间我国国内年均新增光伏装机规模可达 70GW，乐观预计可达 90GW，因此光伏电站的设计、施工、运维将成为未来五年的重点工作之一。笔者结合内蒙古某地区光伏扶贫电站项目进行了设计分析及发电效果评价，为其它光伏电站设计提供了一定参考意义。

关键词：能源；光伏；电站设计；发电效果

Photovoltaic Power Station Design and Effect Evaluation Analysis in a Certain Area of Inner Mongolia

Peng Yangdong, Gui Shuqiang, Li Haifeng, Huang Xiaoming

(Wuhan Surveying-Geotechnical Research Institute Co., Ltd. of MCC, Wuhan 430080, China)

Abstract: 2021 is the first year of the "14th Five-Year Plan". Full implementation of carbon peaking and carbon neutrality, and achieving green and low-carbon development will be an important development keynote during the "14th Five-Year Plan" period. The average annual domestic installed photovoltaic capacity in my country can reach 70GW, and it is optimistically expected to reach 90GW. Therefore, the design, construction, and operation and maintenance of photovoltaic power stations will become one of the key tasks in the next five years. The author carried out design analysis and power generation effect evaluation in conjunction with a photovoltaic poverty alleviation power station project in a certain area of Inner Mongolia, which provided a certain reference for the design of other photovoltaic power stations.

Key words: energy; photovoltaic; power station design; power generation effect

热轧加热炉炉压控制与燃耗改善技术研究

谢家振，俞立勋，杨永红

（宝钢湛江钢铁有限公司热轧厂，广东湛江　524072）

摘　要：国内某 2250 热轧产线加热炉在投用前期由于锚固件材质等问题，导致炉壳出现较为严重的发红现象[1]，为了避免此类情况进一步恶化，实际生产过程均采用较低的炉压设定值进行控制。2019 年 5 月对炉墙锚固件改造后，炉压的控制具备进一步提升的条件。本文就该产线炉压控制优化过程对加热炉单耗、材料烧损等方面进行了研究和分析，并对相关的问题制定了改善措施。

关键词：炉压控制系统；加热炉燃耗；成材率；烟道闸板

Study on Furnace Pressure Control and Burnup Improvement Technology

Xie Jiazhen, Yu Lixun, Yang Yonghong

(Baosteel Zhanjiang Iron & Steel, Zhanjiang 524072, China)

Abstract: In the early stage of putting into operation, the furnace shell of a 2250 Hot Rolling Line in China was seriously red due to the problems of anchor material. In order to avoid further deterioration of this situation, the actual production process was controlled by a lower furnace pressure setting value. After the modification of the furnace wall anchor in May 2019, the furnace pressure control has the conditions for further improvement. In this paper, the furnace pressure control optimization process of the production line, the unit consumption of heating furnace, material burning loss and other aspects are studied and analyzed, and the improvement measures for related problems are formulated.

Key words: furnace pressure control system; burnup of heating furnace; material yield; flue shutter

鞍钢股份 3 号高炉降低燃料比生产实践

张大伟，张　磊，赵迪平，刘宝奎，杨长亮

（鞍钢股份有限公司炼铁总厂，辽宁鞍山　114021）

摘　要：对鞍钢 3 号高炉降低燃料比生产实践进行了总结。通过提高焦炭质量，完善上下部调剂，提高焦丁比，采取高风温、高顶压操作，以及加强炉前管理等措施，使 3 号高炉燃料比长期稳定在 490kg/t 以下，焦比低于 300kg/t。

关键词：高炉；燃料比；焦比

Low Fuel Ratio Operation of Ansteel Blast Furnace No.3

Zhang Dawei, Zhang Lei, Zhao Diping, Liu Baokui, Yang Changliang

(General Ironmaking Plant of Angang Steel Co., Ltd., Anshan 114021, China)

Abstract: The low fuel ratio operation of Ansteel blast fumace (BF) No.3 Was carried out. By taking some measures, such as improving coke quality, consummating upper and lower transfers, raising the nut coke ratio, high blast temperature and high top gas pressure, strengthening cast house operation management, Ansteel BF No.3 has kept the fuel ratio below 490 kg/t and the coke ratio below300 kg/t for along period.

Key words: blast furnace; fuel ratio; coke ratio

液压系统油温控制设定值与节能效果的研究

王广吉

（鞍钢股份有限公司设备工程部，辽宁鞍山 114021）

摘　要：液压系统广泛应用于制造业的各个环节，尤其在冶金等重工业中发挥着极其重要的作用。众所周知，由于其工作原理及元件设计等限制，液压系统普遍存在总效率低下的问题，节能因此成为液压系统研究的重要方向。本文主要针对液压系统油温控制目标值及控制方法进行研究比较，以冷轧厂2130产线激光焊机液压系统为研究对象，试找出液压系统在不同的设定温度（区间）下工作时的加热、冷却损失功率，研究其对于能量消耗及液压系统工作能力的影响，提出新的较为合理的液压系统油温控制理念，提高节能水平。

关键词：液压系统；温度控制；损失功率；节能

The Research of the Hydraulic Oil Temperature Setting and Energy Saving Strategy

Wang Guangji

(Department of Equipment and Engineering, Ansteel Co., Ltd., Anshan 114021, China)

Abstract: The hydraulic system plays a very important role in many fields, especially in the steel melting and products. We all know the hydraulic system is lower than other mechenical transfer system because of its principle and component.It is important to research for the energy saving methods.The thesis is mainly about the hyadraulic system of the welding mechine in cold rolling mill company, to find a way to set a proper oil temperature and lower the power lossing.

Key words: hydraulic system; temperature control; power loss; energy saving

高炉-转炉生产流程能源结构及节能潜力分析

张天赋，马光宇，孙守斌，徐 伟，李卫东

（鞍钢股份技术中心，辽宁鞍山 114009）

摘 要：粗钢生产以高炉-转炉流程为主，具有能耗高、能源系统复杂的显著特征。高炉-转炉产过程能源消耗实质是煤为主的碳物质和电能的转化和应用，优化生产过程、能源结构和能源转化过程是降低能源消耗的核心。某生产过程能耗最好水平和平均水平差表明：正常工况与最好水平相比较，整个生产流程节能潜力约为12%。

关键词：高炉-转炉；能源结构；节能潜力；碳物质

Analysis on Energy Structure and Energy Saving Potential for BF-BOF Production Process

Zhang Tianfu, Ma Guangyu, Sun Shoubin, Xu Wei, Li Weidong

(Technology Center of Ansteel Co., Ltd., Anshan 114009, China)

Abstract: The crude steel production is dominated by BF-BOF route which known as high energy intensity and complicated in energy system. The nature of energy consumption is the transfer and utilization of carbon mass mainly refer to coal and electricity. The core to energy saving for BF-BOF iron and steel production process is to optimum the production process ,energy structure and energy transfer process. The difference between the best and average energy consumption level for a BF-BOF production process shows that the energy saving potential is near to 10%.

Key words: BF-BOF; energy structure; energy saving potential; carbon mass

5000 吨/天小粒径石灰石筒仓预热工艺技术研究

董武斌，赵国成，闫伟超

（山西太钢鑫磊资源有限公司生产技术部，山西阳泉 045000）

摘 要：通过理化试验室焙烧和小规模工业试验，确立了减弱或消除易粉难焙烧石灰石爆裂率工艺技术路径；因地制宜实施技术改造，利用窑尾废气余热对圆筒料仓直径约 10~40mm 小粒径石灰石进行预热处理，实现冶金回转窑（1000 吨/窑·日）煅烧奥陶纪马家沟熔剂灰岩达产、达效。

关键词：小粒径；石灰石；预热工艺；技术研究

Study on Preheating Technology of 5000t/d Small Size Limestone Silo

Dong Wubin, Zhao Guocheng, Yan Weichao

(Shanxi TISCO Xinlei Resources Co., Ltd., Production Technology Department, Yangquan 045000, China)

Abstract: Through physical and chemical laboratory roasting and small-scale industrial test, the technological path to reduce or eliminate the burst rate of limestone which is easy to powder and difficult to bake is established; The technical transformation was carried out according to local conditions, and the small-size limestone with a diameter of about 10-40mm in the cylindrical silo was preheated by using the waste heat of kiln tail gas, so as to achieve the production and efficiency of calcining Ordovician Majiagou flux limestone in the metallurgical rotary kiln (1000tons/kiln·day).

Key words: small particle size; limestone; preheating process; technical research

实施能源“零距离”管理，加快实现“双碳”目标

宋万强

（本钢集团矿业辽阳贾家堡铁矿有限责任公司，辽宁辽阳 111219）

摘 要：能源管理是高耗能企业永恒的主题，通过创新管理模式，降低碳排放，是加快实现“双碳”目标的有效途径。本文以选矿企业能源管理为例，建立了以一线生产职工队伍为核心的“零距离”管理模式，改善管理结构，减少管理层级，职能部门服务生产工人，提高生产控制水平，减轻生产工人额外负担，激发生产工人积极性和创新性，实现节能减排。

关键词：“双碳”目标；“零距离”管理；生产工人；服务

Implementing the “Zero-distance” Energy Management Model and Moving Faster to Achieve the “Double Carbon” Targets

Song Wanqiang

(Jia jiapu Iron Mine, Liaoyang, Mining Company, Benxi Steel, Liaoyang 111219, China)

Abstract: The energy management is the eternal theme of the business administration of the high energy consumption enterprises. Innovated the mode of energy management to reducing carbon emissions is an effective way to accelerate the realization of the “double carbon” target. This paper takes the energy management of mineral processing enterprises as an example, the “zero distance” management model was set up which the front-line production staff is as the core. With this model, the management structure is improved, the management layers are reduced. The functional departments of the enterprise serve production workers to improve the level of production control, alleviate the extra burden on production workers and stimulate enthusiasm and creativity of production workers. The targets of energy conservation and emission reduction of enterprise are achieved.

Key words: “double carbon” targets; “zero distance” management model; production workers; service

新一代轧钢加热炉燃烧监控系统研发及应用

周劲军，翟　炜，黄　敏，张　停，胡玉畅，曹曲泉，刘自民

（马鞍山钢铁股份有限公司，安徽马鞍山　243000）

摘　要：为有效解决轧钢加热炉内燃烧气氛精确控制问题，马钢自主研发了新一代轧钢加热炉燃烧监控系统，主要包括在线煤气成分法燃烧监控、炉顶取气式分段燃烧监控、加热炉 NOx 分段监测预警等三项核心技术。通过在 2250 热轧产线加热炉上的成功应用，实现了烧损下降 15%以上、燃耗下降 4%以上、烟气中 NOx 有效控制等效果，达到了预期目标。

关键词：轧钢加热炉；燃烧监控系统；研发；应用

Development and Application of a New Generation Combustion Monitoring System for Steel Rolling Heating Furnace

Zhou Jinjun, Zhai Wei, Huang Min, Zhang Ting, Hu Yuchang, Cao Ququan, Liu Zimin

(Ma'anshan Iron & Steel Co., Ltd., Ma'anshan 243000, China)

Abstract: In order to effectively solve the problem of accurate control of combustion atmosphere in steel rolling heating furnace, Masteel independently developed a new generation of combustion monitoring system for steel rolling heating furnace, which mainly includes three core technologies: on-line combustion monitoring by gas composition method, gas-taking staged combustion monitoring at the top of furnace, and NOx staged monitoring and early warning of heating furnace. Through the successful application in the heating furnace of 2250 hot rolling line, the burning loss is reduced by more than 15%, the fuel consumption is reduced by more than 4%, and the NOx in flue gas is effectively controlled, which achieves the expected goal.

Key words: rolling reheating furnace; combustion monitoring system; research and development; application

两起能源介质污染事件的处理及反思

蔡树梅，宁健民

（新疆八一钢铁股份有限公司能源中心，新疆乌鲁木齐　830022）

摘　要：通过对两起能源介质产品质量不达标造成用户停产事件的分析、处理过程的回顾。阐述了能源介质管网系统相互联通，在方便使用的同时也带来许多风险。能介系统介质污染危害性大，影响面广。结合现状对能源介质管网系统及联通系统提出了技术和管理措施。

关键词：能介；污染；措施

The Treatment and Reflection of Two Pollution Incidents of Energy Medium

Cai Shumei, Ning Jiamin

(Xinjiang Bayi Iron & Steel Co., Ltd. Energy Center, Urumqi 830022, China)

Abstract: Through the analysis of two cases of energy medium product quality is not up to standard caused by the user to stop production of the event, the review of processing process. The interconnection of energy medium pipe network system is not only convenient to use, but also brings many risks. Energy medium system pollution hazards, influence a wide range. Combined with the present situation, the technology and management measures are put forward for the energy medium pipe network system and the interconnection system.

Key words: can lie; pollution; measures

钢铁厂低品位余热蒸汽的利用

王　成

（新疆八一钢铁股份有限公司能源中心，新疆乌鲁木齐　830022）

摘　要：钢铁厂各工序回收余热蒸汽，在回收方式的工艺技术方面都比较成熟，富裕余热蒸汽主要用来发电，但存在蒸汽发电效率低的问题，热效率不足20%，造成较大的能源利用损失，能否更多元的利用低品位余热蒸汽，提高能源利用效率。

关键词：低品位余热蒸汽；发电利用效率低；多元利用；提高效率

Utilization of Low Grade Waste Heat Steam in Steel Works

Wang Cheng

(Xinjiang Baiyi Iron & Steel Co., Ltd. Energy Center, Urumqi 830022, China)

Abstract: Steel recycling waste heat steam, each process in the way of recycling process technology is mature, the rich are mainly used for power generation, waste heat steam but exists the problem of low efficiency of the steam power generation, thermal efficiency is less than 20%, larger energy loss caused by, more diversity, whether the use of low grade waste heat steam that improve energy efficiency.

Key words: low grade waste heat steam; low efficiency of power generation; multivariate utilization; improve efficiency

低品位热能利用助力钢铁企业碳达峰

墙新奇

（新疆八一钢铁股份有限公司能源中心，新疆乌鲁木齐 830022）

摘 要：钢铁企业属于高能耗行业，随着碳达峰限期临近，二氧化碳减排压力日益增加。钢铁企业大量的低品位热能以废气、热水等形式被排放，回收利用潜力巨大。本文通过全新的制冷循环的建立，人为制造冷源，摆脱低品位热能利用环境温度限制。新的制冷循环由气动压缩循环子系统、冷量增益循环子系统、低压补冷循环子系统三个相对独立的子系统组成，以低品位热能作为主要能源，电能作为辅助能源。制冷循环产生的冷量阶梯利用，用于品质稍高的低品位热能扩大可利用温度区间。通过低品位热能的阶梯利用，提高低品位热能利用率，从而助推碳达峰的进程。

关键词：碳达峰；低品位热能；冷源；制冷循环

The Utilize of Low-grade Thermal Energy Help Iron and Steel Enterprise Reach the Peek of Carbon

Qiang Xinqi

(Xinjiang Baiyi Iron & Steel Co., Ltd. Energy Center, Urumqi 830022, China)

Abstract: Iron and steel enterprise blong to high energy consumption industry,as the carbon peak deadline is approaching, the pressure of carbon dioxide emission reduction is increasing. A large number of low-grade thermal energy in iron and steel enterprises is discharged in the form of waste gas, hot water and so on, which has great potential for recycling and utilization. In this paper, through the establishment of a new refrigeration cycle, artificial cold source is made to get rid of the environmental temperature limit of low grade thermal energy utilization. The new refrigeration cycle is composed of three relatively independent subsystems, namely, pneumatic compression cycle subsystem, cooling gain cycle subsystem and low pressure cooling supplement cycle subsystem. Low grade thermal energy is used as the main energy and electric energy as auxiliary energy. The step-by-step utilization of cold capacity generated by refrigeration cycle is used to expand the available temperature range of slightly higher quality low grade heat energy. Through the step utilization of low grade thermal energy, improve the utilization rate of low grade heat energy, so as to boost the process of carbon peak.

Key words: carbon peek; low grade thermal energy; cooling source; refrigeration cycle

利用工业废热进行岩土跨季节储能供热技术研究

桂树强，臧中海，胡纯清，彭扬东，董利军

（中冶集团武汉勘察研究院有限公司，湖北武汉 430080）

摘 要：利用工业废热进行岩土跨季节储能供热技术，满足城市增长供暖需求已渐趋成熟。笔者通过地埋管系统和热库的设计结合岩土储能与城市热网耦合技术并建立物理模型和数学模型，通过不同工况的全年模拟，验证不同热库温度下所需取热量和蓄热量及对应的时间。该技术对促进该市高新区建设清洁低碳、安全高效的能源体系有着积极的推动作用，也有利于节能减排和城市雾霾的治理。

关键词：能源；工业废热；储能；蓄热；跨季

Research on the Technology of Cross-Season Energy Storage and Heating in Rock and Soil Using Industrial Waste Heat

Gui Shuqiang, Zang Zhonghai, Hu Chunqing, Peng Yangdong, Dong Lijun

(Wuhan Surveying-Geotechnical Research Institute Co., Ltd. of MCC, Wuhan, 430080, China)

Abstract: The use of industrial waste heat for geotechnical cross-season energy storage heating technology has gradually matured to meet the increasing heating demand of cities. The author combines the geotechnical energy storage and urban heating network coupling technology through the design of the buried pipe system and the thermal storage, and establishes the physical model and mathematical model. Through the year-round simulation of different working conditions, it verifies the heat extraction and the required heat extraction at different thermal storage temperatures. Heat storage and corresponding time. This technology has a positive role in promoting the construction of a clean, low-carbon, safe and efficient energy system in the city's high-tech zone, and is also conducive to energy conservation, emission reduction and urban smog management.

Key words: energy; industrial waste heat; energy storage; thermal storage; cross-season

20000m³/h 制氧机氩气产量不达标的原因与措施

龚喜凤

（新疆八一钢铁股份有限公司能源中心，新疆乌鲁木齐 830022）

摘　要：分析空分设备制氩系统优化操作的理论依据，介绍制氩过程遇到的问题及处理方法；简述了提高氩提取率的正确操作方法。

关键词：空分设备；制氩系统；粗氩塔；氩提取率；优化操作；回流比；组分平衡

Reasons and Measures for the Unqualified Argon Output of 20000 m³/h Oxygen Generator

Gong Xifeng

(Xinjiang Baiyi Iron & Steel Co., Ltd. Energy Center, Urumqi 830022, China)

Abstract: The theoretical basis of optimal operation of argon production system of air separation equipment is analyzed, and the problems encountered in argon production process and their treatment methods are introduced. The correct operation method to increase the yield of argon extraction is briefly described.

Key words: air separation equipment; argon system; crude argon column; argon extraction rate; optimized operation; reflux ratio; composition balance

空分装置氧气量变负荷的探讨

李　祥

（新疆八一钢铁股份有限公司能源中心，新疆乌鲁木齐　830022）

摘　要：从理论上定量分析了空分装置变氧气量生产的可行性以及可实现的负荷调节范围。指出：空分装置变氧气量生产是可行的；对于膨胀空气进上塔、氧气外压缩流程的空分装置，其变负荷范围 70%~110%，可减少氧气放散率。

关键词：大型空分设备；氧气产品；变量；生产；可行性；调节范围

Making an Approach to Variable Oxygen Production Capactiy of an Air Separation Plant

Li Xiang

(Xinjiang Baiyi Iron & Steel Co., Ltd. Energy Center, Urumqi 830022, China)

Abstract: Quantitatively analyzes the feasibility of the variable oxygen production of the air separation plant and the achievable load adjustment range.Piont out that the variable oxygen production of the air separation unit is feasible.For the air separation unit where the expanded air enters the upper tower oxygen external compress, the variable load range of 70%-110% canreduce the emission rate.

Key words: large air separation equipment; oxygen product; variable; production; feasibility; adjustment range

煤气峰谷发电一体化联动系统的开发与应用

赵素仿，马　志，徐甲庆

（河钢集团邯钢公司，河北邯郸　056015）

摘　要：基于钢铁企业煤气柜及锅炉汽机发电一体化联动系统，根据峰谷平时段及煤气压力自动升降煤气柜位、自动调整自备电厂锅炉汽机负荷，达到煤气柜、发电及管网大系统的动态平衡，提高发电效率，降低用电成本，实现发电效益最大化。

关键词：峰谷发电；自动控制；节能降耗；能源利用

Development and Application of Coal Gas Peak-Valley Integrated Power Generation Linkage System

Zhao Sufang, Ma Zhi, Xu Jiaqing

(Hebei Iron and Steel Group Hansteel Company, Handan 056015, China)

Abstract: Based on the integrated linkage system of gas tank and boiler turbine power generation in iron and steel enterprises, according to the peak, trough and peacetime section and the gas pressure automatically rise and fall the gas tank, automatically adjust the boiler and turbine load of the power plant to achieve the dynamic balance of gas tank, power generation and pipe network system, improve the power generation efficiency, reduce the cost of electricity, and realize the maximum benefit of power generation.

Key words: peak power; automatic control; energy saving; energy utilization

提高河钢邯钢发电机组发电效率技术创新集成

范　杰

（河钢集团邯钢公司设备动力部，河北邯郸　056002）

摘　要： 河钢邯钢现有发电机组 23 台套，总装机容量为 764.7MW，因此如何提高现有发电机组发电效率是企业节能减排创效长久课题，采用一种钢铁企业低压生产蒸汽公共管网压力稳定的调整方法，增加邯钢西区低压蒸汽回收，减少汽轮机抽汽；对邯钢西区能源中心干熄焦发电机组循环水处理技术升级改造，提高机组真空度；对能源中心 1 号 60MW 发电机组采用冷端优化技术提高发电机组发电效率。通过以上技术创新实施，2020 年共计多发电 2095 万千瓦时。

关键词： 发电机；汽轮机；真空度；蒸汽

Improve the Efficiency of HBIS Group Hansteel Company's Generator Power Generation Technology Innovation Integration

Fan Jie

(Equipment Power Department oof HBIS Group Hansteel Company, Handan 056002, China)

Abstract: HBIS Group Han Steel Company has 23 generator sets with a total installed capacity of 764.7MW.Therefore, how to improve the efficiency of the existing generating units is a long-term issue of energy conservation and emission reduction.In order to increase the recovery of low-pressure steam from Handbo Corporation and reduce the extraction of steam from steam turbine, a method to adjust the pressure stability of public pipe network for low-pressure steam production in iron and steel enterprises was adopted.In order to upgrade the circulating water treatment technology of the dry quenching coking generator set of Handbo Energy Center to improve the vacuum degree of the unit; The cold end optimization technology is adopted to improve the generating efficiency of 1#60MW generator set in the energy center. Through the implementation of the above technological innovations, a total of 20.95 million kWh of electricity will be generated in 2020.

Key words: generator; steam turbine; vacuum degree; steam

Feasibility Evaluation of Terminated Waste Energy In-situ Conversion Strategy Toward Carbon Neutralization in Metallurgical Process

Dong Jianping[1], Zhang Huining[1], Wei Chao[1], Yang Liang[1], Cao Caifang[1], Yang Shaohua[1], Zhang Zuotai[2]

(1. Faculty of Materials, Metallurgy and Chemistry, Jiangxi University of Science and Technology, Ganzhou 341000, China; 2. School of Environmental Science and Engineering, Southern University of Science and Technology, Shenzhen 518055, China)

Abstract: Integrated consideration of carbon dioxide reduction, terminated energy conversion and wastes like plastics resourceful disposal broadens a new horizon for green metallurgical process. Here, chlorine-free plastics gasification process in-situ utilizing converter slag and flue gas heat was proposed as an example, and this complicated system superiority was comprehensively evaluated in terms of syngas preparation efficiency and carbon dioxide reduction depth through thermodynamics. Specifically, the effects of process parameters such as temperature, pressure,H_2O/P, CO_2/P and SS/P ratio on both were discussed in detail, moreover, two sets of theoretical operation conditions were explored for maximum syngas preparation efficiency or carbon dioxide reduction, respectively. Furthermore, the effect of waste plastics and gasification agent amount on reaction system heat balance were discussed when H_2/CO ratio in syngas gas equaled 3:1or 2:1 for CH_4 or CH_3OH preparation, in purpose of terminated product quality regulation. The results showed thatH_2, CO yield and CO_2 conversion rate increased as temperature increased but decreased as pressure increased, H_2 yield was promoted by H_2O/P ratio instead of CO yield and CO_2 conversion rate, and CO_2/P ratio had a positive effect on CO yield except for H_2 yield and CO_2 conversion rate, CH_4 and $CaCO_3$ yield increased as SS/P ratio increased. Meanwhile, chlorine-free plastics consumption increased as temperature increased, and less gasification agent was need. Otherwise, this system had the maximum syngas yield consisting of 44.82%-64.13%H_2 and 88.88%-135.69%CO when temperature, pressure,H_2O/P ratio,CO_2/P ratio and SS/P ratio were 800-1000℃, 1.0atm, 1-2.0, 1.2-3.0 and 1.0, respectively, and as the aspect of CO_2 reduction efficiency, the capacity of 75.17%-90% can be obtained when 800-1000℃, 1.0atm, 0-0.8, 0.4-1.2 and 1.0were adopted, respectively. Definitely, the strategy of in-situ converter slag and flue gas heat conversion utilizing chlorine-free plastics was indeed feasible for industrial application.

Key words: converter slag heat conversion; chlorine-free plastics gasification; flue gas;syngas preparation; carbon dioxide reduction

全流程钢铁企业碳排放核查分析及减碳路径研究

易 祥

（武汉钢铁有限公司能源环保部，湖北武汉 430000）

摘 要：本文以武汉钢铁有限公司为例分析了其碳核查数据源及核查边界，及其碳排放总量与特点，提出了结合数据分析结果的减碳路径，为今后的碳减排管理和参与碳市场交易提供了思路和参考。

关键词：钢铁行业；碳减排；核查；数据分析；路径

The Verification and Analysis for the Whole Process of Steel Enterprises Carbon Emission, and the Research of Carbon Reduction Path

Yi Xiang

(Energy and Environmental protection Department of Wuhan Iron and Steel Co., Ltd., Wuhan 430000, China)

Abstract: Taking Wuhan Iron and Steel Co. Ltd. as an example, first, this paper analyses the carbon verification data source and verification boundary. Then the carbon reduction path is presented combined with the data analysis results. Finally, the idea and reference for the future carbon emission reduction management and participation in carbon market trading are provided.

Key words: iron and steel industry; carbon emission reduction; verification; data analysis; path

高炉炉顶放散碳回收工艺及应用

翟玉龙，张　昌，王学利

（宝钢股份上海梅山钢铁股份有限公司炼铁厂，江苏南京　210039）

摘　要：高炉炉顶均压放散回收系统对碳中和具有积极的贡献，减排降耗同时也会产生经济效益。该系统国外应用较早，国内借鉴改进后逐步推广，近十年来应用广泛，并且对各组成工序进行了适应性改进。对该系统工序分解后，详细分析各组成工序的功能，介绍适用的场合。提出该回收设施选用时可能存在的问题，并介绍不同的解决方案。

关键词：碳回收；炉顶放散；均压煤气；高炉

The Recovering Process and Application of Carbon from Top of Blast Furnace

Zhai Yulong, Zhang Chang, Wang Xueli

(Baosteel Shanghai Meishan Iron & Steel Co., Ltd., Nanjing 210039, China)

Abstract: Blast furnace top pressure distribution recovery system has positive contribution to carbon neutrality, emission reduction and consumption reduction at the same time will produce economic benefits. The system was used earlier in foreign countries, and gradually promoted in China after reference and improvement. It has been widely used in recent ten years, and has carried on adaptive improvement to each component process. After decomposing the working procedure of the system, the functions of each working procedure are analyzed in detail, and the applicable occasions are introduced. The possible problems in the selection of the recycling facility are proposed and the different solutions are introduced.

Key words: carbon neutrality; blast furnace top gas emission; equalizing gas; blast furnace

9.2 环保与资源利用

通风机在烧结烟气超净排放系统中存在的若干问题

朱红兵

（上海宝钢节能环保技术有限公司，上海 201900）

摘 要：通风机应用在特定的工艺系统中，由于工艺运行条件的特殊性，通风机必须进行专门设计和开发。本文详细论述了应用在钢铁烧结烟气净化工艺流程中几种工艺风机的流程布置、烟气特性和设计要求，指出了原设计存在的问题。结合实例，从结构设计、材料选择、表面处理等方面给出了解决问题的方法，并指出了工艺风机在工作管网中气动性能与设计性能产生偏差的原因，提出了科学、实用的工程解决方法。

关键词：烧结烟气；脱硝净化；脱硫净化；热引风机；配风风机；主引风机

Problems of Ventilator in Ultra-clean Sintering Flue Gas System

Zhu Hongbing

(Shanghai Baosteel Energy Saving Technology Co., Ltd., Shanghai 201900, China)

Abstract: The fan is applied in a specific process system. Because of the particularity of process operating conditions, the fan must be specially designed and developed. In this paper, the flow arrangement, flue gas characteristics and design requirements of several kinds of process fans applied in the process of purifying flue gas of steel sintering are discussed in detail, and the problems existing in the original design are pointed out. Combining with the examples, this paper gives the methods to solve the problems from the aspects of structure design, material selection and surface treatment, and points out the reasons for the deviation between the aerodynamic performance and the design performance of the process fan in the working pipe network, and puts forward the scientific and practical engineering solutions.

Key words: sintering flue gas; denitration purification; desulfurization purification; heat induced draft fan; mix the wind fan; main induced draft fan

硫酸钠回收技术在不锈钢酸洗废液处理过程中的应用实践

吴 磊

（天津太钢天管不锈钢有限公司，天津 300301）

摘 要：本文详细介绍了硫酸钠回收技术的工艺流程，及其在宝钢集团太钢天管不锈钢酸洗废液处理过程中的应用情况，阐明该技术生产应用的成熟性，其通过对重金属六价铬的处理有效保护水源，实现硫酸钠的再利用，对企业可持续发展的重要性，并对于硫酸钠回收工艺在实际应用过程中出现的问题，提出了具体的工艺改进方案。

关键词：硫酸钠回收；六价铬处理；不锈钢酸洗；不锈钢

Application of Sodium Sulfate Recovery Technolgy in Treatment of Spent Picking Solution of Stainless Steel

Wu Lei

(Tianjin TTSS Stainless Steel Co., Ltd., Tianjin 300301, China)

Abstract: The technical process of the sodium sulfate recovery technology is stainless steel of TTSS corporation is introduced. The technology has been successfully used in production and it plays an important role in sustainable development of the enterprise. The problems in the practical application of sodium sulfate recovery process a specific process improvement scheme is proposed.

Key words: sodium sulfate recovery; Cr^{6+} treatment; pickling of stainless steel; stainless steel

鞍钢西区非传统水资源利用的探索研究

王　飞[1]，龙海萍[2]，胡绍伟[1]，陈　鹏[1]，刘　芳[1]，王　永[1]，马光宇[1]

（1. 鞍钢股份有限公司技术中心，辽宁鞍山　114009；
2. 鞍钢股份能源管控中心，辽宁鞍山　114009）

摘　要：雨水是一种优质的非传统资源，雨水资源化利用是缓解水资源危机的重要途径之一。通过对鞍钢西区雨水资源现状的分析，针对性地提出实施雨水资源化利用的措施，为鞍钢西区雨水的收集利用以及在全厂的推广提供了参考价值。

关键词：钢铁企业；雨水利用；地渗水；径流

The Exploration of Nontraditional Water Source Recycling and Application in West Section of Ansteel

Wang Fei[1], Long Haiping[2], Hu Shaowei[1], Chen Peng[1], Liu Fang[1], Wang Yong[1], Ma Guangyu[1]

(1. Ansteel Technology Center, Anshan 114009, China;
2. Energy Management and Control Center, Anshan 114009, China)

Abstract: Rainwater is a kind of high quality nontraditional water source and could be one of important way for reliving the crisis of water resource. Through the analysis of current rainwater status in ANSTEEL, corresponding measures has been suggested and could be valuable reference for rainwater collecting and use and promoting application.

Key words: iron and steel industry; rainwater utilization; groundwater seepage; runoff

热风炉废气循环利用系统腐蚀成因及解决方案研究

钱　峰[1]，侯洪宇[1]，王　永[1]，张荣军[2]，蔡光富[2]，孟凡双[2]

（1. 鞍钢股份有限公司技术中心，辽宁鞍山　114009；
2. 鞍钢股份有限公司炼铁总厂，辽宁鞍山　114009）

摘　要：热风炉废气主要成分、性质与氮气相近，近年来被用作氮气的替代品广泛应用到高炉中，如供高炉炉顶阀箱罩封、探尺密封、布袋除尘器反吹使用，或引入高炉风口喷吹使用。热风炉废气再利用技术不但节约生产成本，更重要的是减少了废气排放量，很大程度上缓解了环保超低排放要求给企业带来的生存压力，具有较好的市场应用前景。然而，该工艺在实际应用过程中出现了一系列问题，鞍钢 4#高炉热风炉废气循环利用系统经过几年的使用后，出现了严重腐蚀和泄漏，不得不被迫停止使用。本文以鞍钢高炉热风炉废气代替氮气系统为例，对鞍钢厂区内煤气含尘量、煤气中粉尘成分、煤气水封水成分、热风炉废气成分、废气管路冷凝水成分等开展了全面的分析，对系统腐蚀成因和解决方案进行了研究和探讨，为热风炉废气循环利用工艺优化提供依据和参考。

关键词：高炉热风炉；系统腐蚀；废气代替氮气；资源利用

Study on Corrosion Causes and Solutions of Waste Gas Recycling System of Hot Blast Furnace

Qian Feng[1], Hou Hongyu[1], Wang Yong[1], Zhang Rongjun[2], Cai Guangfu[2], Meng Fanshuang[2]

(1. Technical Center, Angang Steel Co., Ltd., Anshan 114009, China;
2. General Steelmaking Plant of Angang Steel Co., Ltd., Anshan 114009, China)

Abstract: The main composition and properties of hot blast stove exhaust gas are similar to nitrogen. In recent years, it has been used as a substitute for nitrogen gas and widely used in blast furnace, such as for blast furnace top valve box cover sealing, sounding rod sealing, back blowing of cloth dust collector, or injection into blast furnace tuyervent. Hot blast stove waste gas reuse technology not only saves production cost, but more importantly reduces waste gas emissions, which to a large extent eases the survival pressure of enterprises brought about by the requirements of ultra-low emission of environmental protection, and has a good market application prospect. However, a series of problems appeared in the process of practical application. After several years of use, serious corrosion and leakage appeared in the waste gas recycling system of No.4 BF hot blast stove in Anshan Iron and Steel Co., Ltd. and it had to be stopped. Taking the system of replacing nitrogen with exhaust gas of hot blast stove in Anshan Iron and Steel Co., Ltd as the research object, this paper has carried out a comprehensive analysis on the dust content of gas, dust composition of gas, water seal composition of gas, exhaust gas of hot blast stove, condensate composition of exhaust gas pipeline in Anshan Iron and Steel Co., Ltd., and studied and discussed the causes and solutions of corrosion of the system. It provides basis and reference for optimization of exhaust gas recycling process of hot blast stove.

Key words: hot blast furnace; system corrosion; exhaust gas instead of nitrogen; resource utilization

生物强化技术在鞍钢焦化废水的应用

王 永，胡绍伟，陈 鹏，王 飞，刘 芳
（鞍钢股份有限公司技术中心，辽宁鞍山 114009）

摘 要：目前，生物处理法是焦化废水处理的主要方法，生物强化技术能提高好氧池活性污泥降解污染物的效率，降低后续的处理焦化废水的费用。本文对焦化厂好氧池出水的苯酚指标较高的特点，向好氧池生化池中加入源位提取筛选出的菌剂，强化苯酚的去除效果，在不改变整个工艺的情况下提高 COD 降解能力。结果表明，添加菌剂以后，好氧池中的苯酚浓度去除率提高了 63.4%，COD 去除率提高了 52.8%。

关键词：焦化废水；生物强化技术；菌种

The Application of Bio-augmentation Technique on Coking Waste-water Treatment in Ansteel

Wang Yong, Hu Shaowei, Chen Peng, Wang Fei, Liu Fang
(Technology Center of Angang Steel Co., Ltd., Anshan 114009, China)

Abstract: Currently, the biological treatment method is primarily used to treat coking waste-water. Bio-augmentation technique can improve the efficiency of pollutants degradation in the aerobic tank which use activated sludge to gradate pollutants and lowing the cost in the following coking waste-water treatment process. Taking the aerobic tank influent in a coking plant as research target, based on the characteristic of the high content of phenol in effluent, adding the selected bacteria agent, keep the whole process unchanged, the effect of removal phenol is intensified and biodegradation capacity of COD is improved. The results show that, with the adding of bacteria agent, in the compared tank, the removal rate of phenol and COD is improved by 63.4% and 52.8% respectively.

Key words: coking wastewater; bioaugmentation technique; bacterial

高级氧化法处理焦化废水中重金属的实验研究

刘 芳，胡绍伟，陈 鹏，王 永，王 飞，徐 伟
（鞍钢集团钢铁研究院，辽宁鞍山 114009）

摘 要：通过外加弱磁场条件下的零价铁高级氧化工艺处理焦化废水中的重金属污染物，对工艺相关参数进行了优化，并进行了反应的动力学分析。实验结果表明：在废铁屑投加量 0.8g、初始溶液 pH 为 3.0、氧化剂过氧化氢质量浓度为 1500ppm、外加弱磁场强度 10mT、室温条件下反应 30min，铜、砷、镉、铬的去除率分别达到了 95.9%、94.2%、90.5%和 89.9%，去除效果显著优于无磁条件下的 61.3%、67.1%、65.6%和 72.3%；零价铁/过氧化氢高级氧化法去除重金属的反应过程遵循一级动力学规律。

关键词：弱磁场；Fenton；焦化废水；重金属

Experimental Study on the Treatment of Heavy Metals in Coking Wastewater by Advanced Oxidation Process

Liu Fang, Hu Shaowei, Chen Peng, Wang Yong, Wang Fei, Xu Wei

(Iron & Steel Research Institute of Angang Group, Anshan 114009, China)

Abstract: Under the condition of external weak magnetic, the oxidation process of zero-valent iron was used to remove heavy metal pollutants in coking wastewater. The process parameters were optimized and the reaction kinetics was analyzed. The results showed that: under the condition of waste iron filings was 0.8g, initial solution pH was 3.0, mass concentration of hydrogen peroxide was 1500ppm, weak magnetic field strength was 10mT and reaction 30min at room temperature, the removal rate of copper, arsenic, cadmium and chromium reached 95.9%, 94.2%, 90.5% and 89.9%. The removal effect was significantly better than that of non-magnetic conditions (61.3%, 67.1%, 65.6% and 72.3%). The removal of heavy metals by high oxidation of zero-valent iron/hydrogen peroxide follows the first-order kinetics.

Key words: weak magnetic; Fenton; coking wastewater; heavy metal

循环水系统微油污染处理技术研究

刘　芳[1]，胡绍伟[1]，陈　鹏[1]，王　永[1]，王　飞[1]，周　航[2]

（1. 鞍钢集团钢铁研究院，辽宁鞍山　114009；2. 鞍钢工程技术有限公司，辽宁鞍山　114009）

摘　要：采用活性炭吸附处理泄漏在密闭式循环水系统中的微量油污染。通过影响因素考察实验、吸附动力学实验和吸附平衡实验对吸附过程进行了深入的探讨，确定了吸附效果最优活性炭。对初始浓度范围为 5～25mg/L 的含油微污染废水，得出果壳炭、煤质炭和椰壳炭的饱和吸附容量分别为：17.094mg/g、15.625mg/g 和 16.807mg/g。

关键词：活性炭；油；微污染；循环水

Study on Treatment Technology of Micro-oil Pollution in Circulating Water System

Liu Fang[1], Hu Shaowei[1], Chen Peng[1], Wang Yong[1], Wang Fei[1], Zhou Hang[2]

(1. Iron & Steel Research Institute of Angang Group, Anshan 114009, China;
2. Anshan Iron& Steel Engineering Technology Co., Ltd., Anshan 114009, China)

Abstract: The experiment adopted active carbon absorption method to deal with closed-circle system oil micro-pollution wastewater. Through the experiment of influence factors, adsorption dynamics and adsorption equilibrium, adsorption process was studied systematically. Then determined the adsorption optimal activated carbon. The largest adsorption capacity of activated carbon (nut shell carbon, coal carbon and coconut shell carbon) are 17.094mg/g,15.625mg/g and 16.807mg/g in the oil initial concentration ranges of 5~25mg/L.

Key words: active carbon; oil; micro-polluted; recycled water

鞍钢高炉热风炉排放水平及减排方向

孟凡双，张 磊，李建军，李林春，张恒良

（鞍钢股份有限公司炼铁总厂，辽宁鞍山 114021）

摘 要：阐述鞍钢本部八座高炉热风炉排放量情况，分析高炉热风炉排放现状，针对不同结构热风炉特点，分析热风炉废气排放的特点，同时根据当前环保的要求，提出降低热风炉排放的一些措施和下一步高炉热风炉减排工作方向。

关键词：高炉；热风炉；减排

Emission Level and Emission Reduction Direction of Hot Blast Stove in Angang

Meng Fanshuang, Zhang Lei, Li Jianjun, Li Linchun, Zhang Hengliang

(Ironmaking General Factory of Angang Co., Ltd., Anshan 114021, China)

Abstract: Expounds the discharge of hot-blast stoves in eight blast furnaces in Angang headquarters, analyzes the emission status of blast furnaces, analyzes the characteristics of exhaust emission of hot stoves according to the characteristics of hot stoves with different structures, and puts forward some measures to reduce the emission of hot stoves according to the current environmental protection requirements, as well as the next work direction.

Key words: blast furnace; hot blast stove; emission reduction

硫酸钡重量法测定烧结烟气脱硫灰中硫酸根、亚硫酸根含量

邓军华[1,2]，亢德华[1,2]

（1. 海洋装备金属材料及应用国家重点实验室，辽宁鞍山 114009；
2. 鞍钢集团钢铁研究院，辽宁鞍山 114009）

摘 要：建立了硫酸钡重量法测定烧结烟气脱硫灰中硫酸根和亚硫酸根含量的方法。试样采用盐酸进行分解（亚硫酸根需要预氧化），试样中硫酸根转变成可溶性的硫酸盐。试液氨水分离、加入掩蔽剂消除干扰离子，在弱酸性条件下利用硫酸根与氯化钡作用生成溶解度很小的硫酸钡白色沉淀。沉淀经灼烧后称重，再换算得出硫酸根和亚硫酸根的含量。硫酸根测定结果相对标准偏差（n=6）为 0.42%~1.65%，亚硫酸根测定结果相对标准偏差（n=6）为 0.46%~0.73%，加标回收验证方法的准确性，回收率 98.4%~103.5%。

关键词：脱硫灰；硫酸根；亚硫酸根；硫酸钡；重量法

Determination of Sulfate and Sulfite Content in Sintering Flue Gas Desulfurization Ash by Barium Sulfate Gravimetric Method

Deng Junhua[1,2], Kang Dehua[1,2]

(1. State Key Laboratory of Marine Equipment Made of Metal Material and Application, Anshan 114009, China; 2. Iron & Steel Research Institute of Ansteel Group, Anshan 114009, China)

Abstract: A barium sulfate gravimetric method was established to determine the content of sulfate and sulfite in sintering flue gas desulfurization ash.The sample is decomposed by hydrochloric acid (sulfite needs to be pre-oxidized), the sulfate in the sample is converted to soluble sulfate.Test solution was separated by ammonia add masking agent to eliminate interfering ions, under weakly acidic conditions, a white precipitate of barium sulfate with little solubility is generated by the interaction of sulfate and barium chloride. The precipitate is weighed after being burned, and then converted to obtain the content of sulfate and sulfite. The relative standard deviation of the determination result of sulfate (n=6) is 0.42%~1.65%, the relative standard deviation of the determination result of sulfite (n=6) is 0.46%~0.73%. The accuracy of the standard recovery verification method is 98.4%~103.5%.

Key words: desulfurization ash; sulfate; sulfite; barium sulfate; gravimetric method

转炉副产炉气的回收与综合利用现状

宋翰林，程功金，刘建兴，张金鹏，薛向欣

（东北大学冶金学院，辽宁沈阳 110819；辽宁省冶金资源循环科学重点实验室，辽宁沈阳 110819；东北大学钒钛产业技术创新研究院，辽宁沈阳 110819；辽西地区钒钛磁铁矿资源综合利用产业创新研究院，辽宁朝阳 122000）

摘　要： 转炉炼钢在我国占比 90%左右，是国内炼钢生产的主要形式。而炼钢过程中产生的转炉炉气是典型的钢铁企业副产炉气之一，具有 1400~1600℃的高温和相对纯净的化学组成，其中 CO 和 CO_2 含量占 50%~90%，是优质的碳源和热源，其热值一般是高炉炉气的两倍，具有非常高的回收和综合利用价值。本文整理了国内部分企业的转炉炉气回收条件和综合利用的现状，根据宝钢的转炉炉气回收及放散数据进行了估算，发现转炉炉气的年放散量及损失价值巨大，有必要在目前转炉炉气的综合利用的基础上，对转炉放散炉气进一步开展“近零排放”和低热值煤气的综合利用的研究。

关键词： 转炉；副产炉气；回收；综合利用；近零排放

Current Status of Recovery and Comprehensive Utilization of By-Product Gas from Basic Oxygen Furnace

Song Hanlin, Cheng Gongjin, Liu Jianxing, Zhang Jinpeng, Xue Xiangxin

(School of Metallurgy, Northeastern University, Shenyang 110819, China; Liaoning Key Laboratory of Recycling Science for Metallurgical Resources, Shenyang 110819, China; Innovation Research Institute of Vanadium and Titanium Resource Industry Technology, Northeastern University, Shenyang 110819, China; Innovation Research Institute of Comprehensive Utilization Technology for Vanadium-Titanium Magnetite Resources in Liaoxi District, Chaoyang 122000, China)

Abstract: Basic oxygen furnace (BOF) steelmaking accounts for about 90% in China. It is the main form of domestic steelmaking production. The BOF gas produced in the steelmaking process is one of the typical by-product gases of iron and steel enterprises. It has a high temperature of 1400~1600℃ and a relatively pure chemical composition. The content of CO and CO_2 accounts for 50%~90%, which is high-quality calorific value of carbon source and heat source. Its calorific value is generally twice that of blast furnace gas, which has a very high recovery and comprehensive utilization value. This paper sorts out the BOF gas recovery conditions and the status quo of comprehensive utilization of some domestic enterprises. According to Baosteel's BOF gas recovery and emission data, it is estimated that the annual emission and loss value of BOF gas are huge. On the basis of comprehensive utilization of furnace gas, further research on "net-zero emission" and comprehensive utilization of low-calorific value coal gas will be carried out on abandoned BOF gas.

Key words: basic oxygen furnace; by-product gas; recovery; comprehensive utilization; net-zero emissions

工业废水 COD、氨氮去除技术探讨

孔丽丹

（内蒙古包钢钢联股份有限公司，内蒙古包头　014010）

摘　要：钢铁企业污水处理系统担负整个企业污水集中处理的重任，其外排水源主要有反渗透浓水和厂区回用水，水中 COD、氨氮含量较高。而 COD、氨氮含量较高则会给水体带来富营养化，水中耗氧污染物、有机物增多，毒害鱼类及一些水生生物。为达到当前日益严格的环保外排指标要求，针对外排水中 COD、氨氮值较高的现状，积极开展 COD、氨氮降解剂投加效果试验，摸索 COD、氨氮降解剂的投加效果与适宜的药剂投加方案。

焦炉煤气废脱硫剂高炉处置试验研究

饶　磊，刘自民，桂满城，马孟臣，张耀辉，刘英才

（马鞍山钢铁股份有限公司技术中心，安徽马鞍山　243000）

摘　要：本文对废脱硫剂特性进行了分析，开展了废脱硫剂实验室实验及高炉处置工业试验，并对废脱硫剂返回高

炉处置对相关指标的影响进行了分析，提出了一种高炉法处置废脱硫剂的参考方案。

关键词：废脱硫剂；高炉；处置；试验

Experimental Study on Blast Furnace Disposal of Coke Oven Gas Waste Desulfurizer

Rao Lei, Liu Zimin, Gui Mancheng, Ma Mengchen, Zhang Yaohui, Liu Yingcai

(Technology Center of Ma'anshan Iron and Steel Co., Ltd., Ma'anshan 243000, China)

Abstract: This paper analyzes the characteristics of waste desulfurizer, carries out theoretical analysis and laboratory test of waste desulfurizer disposal in blast furnace, application test of blast furnace disposal, and analyzes the influence of waste desulfurizer returned to blast furnace disposal, and puts forward a reference scheme of waste desulfurizer disposal by blast furnace method.

Key words: waste desulfurizer; blast furnace; disposal; experiment

碳热还原法协同处置含锌粉尘和含铬尘泥的热力学及动力学研究

郑睿琦，陈　卓，堵伟桐，居殿春，邱家用

（江苏科技大学张家港校区冶金与材料工程学院，江苏张家港　215600）

摘　要：含锌粉尘和含铬尘泥为钢铁厂主要固体废弃物，其资源的高效利用有利于推动钢铁行业的绿色循环生产。本论文通过热力学分析了高温焙烧温度区间混合球团中各组分的反应吉布斯自由能，并结合高温焙烧实验研究还原温度和还原时间对碳热还原的影响规律，同时采用非等温热分析法对混合球团的还原动力学进行了研究，通过TG-DSC 热重-差热法测试升温速率对还原过程的影响，分析了混合球团的还原机理。结果表明，随着还原温度升高，铁氧化物的还原过程遵循逐级还原规律，最终被还原为金属铁；相比于铁氧化物，铬氧化物需在较高的还原温度下才能实现还原。随着升温速率的增大，反应速率也增大，反应速率曲线的峰值也呈增大趋势。

关键词：含锌粉尘；含铬尘泥；动力学；热力学

Thermodynamics and Kinetics Study on Co-processing of Chromium-containing Sludge with Zinc-bearing Dusts by Carbothermic Reduction Method

Zheng Ruiqi, Chen Zhuo, Du Weitong, Ju Dianchun, Qiu Jiayong

(School of Metallurgy and Material Engineering, Jiangsu University of Science and Technology, Zhangjiagang 215600, China)

Abstract: Zinc-bearing dust and chromium-containing dust are the main solid wastes of steel mills, and the efficient utilization of them is conducive to promoting green and circular production for the ironmaking and steelmaking. This paper analyzes the reaction gibbs free energy of each component in the mixture at theroasting temperature range, and analyze the variation in the gas-liquid-solid phase and reduction lawof the oxidesduring the carbothermal reduction process. Meanwhile, high temperature roasting experiment to study the influence law of reduction temperature and reduction time on carbothermal reduction.The reduction kinetics of the mixed pellet is studied by the non-isothermal thermal analysis method. The effect of the heating rate on the reduction is tested by the thermogravimetric-differential thermal method (TG-DSC), and the reduction mechanism of the mixed pellets is analyzed. The results show that as the reduction temperature increases, the reduction process of iron oxides follows a gradual reduction law, and is finally reduced to metallic iron.Chromium oxides need to be reduced at a higher reduction temperature compared with iron oxides.As the heating rate increases, the reaction rate also increases, and the peak value of the reaction rate curve also shows an increasing trend.

Key words: zinc-bearing dust; chromium-containing sludge; kinetics study; thermodynamics

海绵城市建设背景下钢铁园区雨水资源利用的研究

秦洋洋

（重庆科技学院冶金与材料工程学院，重庆 401331）

摘 要：随着我国的城市建设对资源高效利用的重视，海绵城市理论逐步走向成熟，在部分城市展开建设，取得较为理想的成效。海绵城市理论是雨水资源化利用的一种方式，主要是将雨水收集储蓄起来，并利用到景观植物上，以提升城市绿化环境。我国有些居住区、校园都已经引入海绵城市理论并开始建设，形成了完整的雨水收集利用系统，但大多钢铁企业目前还未将此理论引入到钢铁园区之中。我国水资源匮乏，是海绵城市建设的重点区域，钢铁园区布局较多，占地面积较大。因此，钢铁园区引入海绵城市理论并展开建设显得尤其重要。本文在海绵城市建设的大的背景，针对华北地区钢铁园区雨水资源化利用到景观的注意事项以及建设措施进行分析。

关键词：海绵城市；钢铁园区；雨水；资源化利用；景观

Study on the Utilization of Rain Water Resources in Iron and Steel Park under the Background of Sponge City Construction

Qin Yangyang

(Chongqing Institute of Science and Technology, Chongqing 401331, China)

Abstract: With the emphasis on the efficient use of resources in the urban construction of our country, the sponge city theory has gradually matured, carried out construction in some cities and achieved satisfactory results. Sponge city theory is a way of resource utilization of Rain Water, which mainly collects and saves Rain Water and applies them to landscape plants in order to improve the urban green environment. Some residential areas and campuses in China have introduced the sponge city theory and begun to build, forming a complete Rain Water collection and utilization system, but most iron and steel enterprises have not introduced this theory into the iron and steel park. The shortage of water resources in China is the key area of sponge city construction, with more iron and steel parks and large area. Therefore, it is particularly important for the Iron and Steel Park to introduce the sponge city theory and carry out the construction. Under the background of sponge city construction, this paper analyzes the matters needing attention and construction measures for the landscape utilization of Rain Water in North China Iron and Steel Park.

Key words: sponge city; iron and steel park; rain water; resource utilization; landscape

浅析中水回用在钢厂的应用

吕洋洋

（青岛特殊钢铁有限公司，山东青岛　266409）

摘　要：青岛是一个水资源匮乏之城市，随着水处理技术的发展，青岛钢铁有限公司环保搬迁设计时充分考虑了水资源综合利用，本文描述了中水回用在钢厂成功应用的案例。

关键词：中水回用；全膜法；反渗透；节水

Application of Reclaimed Water Reuse in Steel Plant

Lv Yangyang

(Qingdao Special Steel Co., Ltd., Qingdao 266409, China)

Abstract: Qingdao is a city with a shortage of water resources. With the development of water treatment technology, comprehensive utilization of water resources is fully considered in the design of environmental protection relocation of Qingdao Iron and Steel Co., Ltd. This paper describes the successful application of reclaimed water reuse in steel works.

Key words: reclaimed water reuse; full membrane method; reverse osmosis; water conservation

焦炉荒煤气强化蒸汽重整制氢的研究

张津宁，谢华清，家丽非，于震宇，王潘磊，王征宇

（东北大学冶金学院，辽宁沈阳　110819）

摘　要：本文提出一种吸附强化焦炉荒煤气蒸汽重整制氢的工艺概念，利用焦炉荒煤气自身尚未回收利用的高温显热在 CO_2 吸附剂的参与下强化其焦油等大分子有机组分蒸汽重整制氢进程，在实现焦炉煤气余热利用、在线焦油脱除的同时，达到焦炉煤气增质增量化制氢的目的。论文通过热力学分析和实验研究对该工艺重整制氢过程进行探究，结果表明：焦炉荒煤气经过吸附强化蒸汽重整反应后，其焦油组分能够完全转化为 H_2 等小分子气体，重整产气中氢气产量和氢气浓度均得到显著提升，荒煤气中的绝大部分碳以高纯 CO_2 的形式被捕集，体现出显著的 CO_2 减排效益，有效验证了该工艺的可行性和先进性，将有力推动我国钢铁行业的低碳绿色转型升级。

关键词：焦炉荒煤气；蒸汽重整；吸附强化；氢气

Study on Hydrogen Production by Steam Reforming with Raw Coke Oven Gas

Zhang Jinning, Xie Huaqing, Jia Lifei, Yu Zhenyu, Wang Panlei, Wang Zhengyu

(Northeastern University School of Metallurgy, Shenyang 110819, China)

Abstract: This paper presents an improved coke oven waste gas adsorption of hydrogen production from steam reforming process concept, itself has not been recycled with coke oven waste gas of high temperature sensible heat under CO_2 adsorbent in strengthening its tar process, hydrogen production from steam reforming of large molecules such as organic components in the implementation of coke oven gas waste heat utilization, tar removal online at the same time, achieve the goal of quality quantitative increasing hydrogen production from coke oven gas. Through thermodynamic analysis and experimental research, this paper explores the hydrogen production process of the process reforming, and the results show that: Shortage of coke oven gas after adsorption improved steam reforming reaction, the tar component can completely into small molecule gas such as H_2, hydrogen production and hydrogen concentration in the reforming gas were significantly increased, for the most part of the waste gas carbon were arrested in the form of high purity CO_2 sets, reflects the significant CO_2 emission reduction benefits, effective verify the feasibility of the technology and advanced nature, It will strongly promote the low-carbon green transformation and upgrading of China's steel industry.

Key words: raw coke oven gas; steam reforming; absorbent; hydrogen

还原时间对不锈钢粉尘碳热还原产物形貌和尺寸的影响

郑立春，陈孝琦，任宏雨，娄　健，姜周华

（东北大学冶金学院，辽宁沈阳　110819）

摘　要：为探索碳热还原不锈钢粉尘制备高附加值球形超细金属粉末的可能性，本文系统研究了 1350℃下不锈钢粉尘碳热还原产物金属颗粒的微观结构、形貌和尺寸，以及还原时间对其的影响。还原时间越长，反应越彻底，金属颗粒球形度越好，但是金属颗粒尺寸越大。金属颗粒存在两种明显不同的形貌，即球形金属颗粒和不规则金属颗粒。不规则金属颗粒中包含一次和二次 M_7C_3 碳化物相。一次 M_7C_3 碳化物以及周围高熔点残渣的共同作用导致金属颗粒形貌不规则。

关键词：不锈钢粉尘；碳热还原；金属颗粒；微观结构；形貌；尺寸

Effect of Reduction Time on Morphology and Size of Carbothermal Reduction Products of Stainless Steel Dust

Zheng Lichun, Chen Xiaoqi, Ren Hongyu, Lou Jian, Jiang Zhouhua

(School of Metallurgy, Northeastern University, Shenyang 110819, China)

Abstract: To explore the possibility of preparing spherical ultrafine metal powder with high added value by carbothermal reduction of stainless steel dust, the microstructure, morphology and size of metal particles produced by carbothermal reduction of stainless steel dust at 1350℃ and the effect of reduction time on it were systematically studied in this paper. The longer the reduction time, the more thorough the reaction, the better the sphericity of metal particles, but the larger the size of metal particles. There are two obviously different morphologies of metal particles, namely spherical metal particles and irregular metal particles. Irregular metal particles contain primary and secondary M_7C_3 carbide phases. The combined action of primary M_7C_3 carbide and surrounding high melting point residual slag leads to irregular morphology of metal particles.

Key words: stainless steel dust; carbothermic reduction; metallic particles; microstructure; morphology; size

宣钢焦炉烟气脱硫脱硝技术实践

庞 江，佟 超

（河钢集团宣钢公司 焦化厂，河北宣化 075105）

摘 要：探讨对比脱硫脱硝技术特点，简要介绍采用 SDS 干法脱硫、SCR 中低温脱硝技术优势，根据在宣钢焦炉烟气超低排放改造项目的实际应用情况，对脱硫脱硝工艺操作和设备进行改进优化，一、分析复热式焦炉采用不同煤气种类加热时，烟气超标排放的原因，进一步优化煤气置换及加热调节的管理；二、从设备设施方面入手，优化改进系统脱硫剂投加路径，确保磨机系统检修时烟气出口 SO_2 排放可控，提高烟气排放合格率。同时，跟踪学习焦化行业内烟气超低排放的先进设计理念，不断完善提高焦化有组织、无组织烟尘处理效率，确保宣钢在绿色转型发展中逐步走向成熟。

关键词：超低排放；脱硫脱硝；改进

Practice of Coke Oven Flue Gas Desulfurization and Denitrification Technology in Xuanhua Iron and Steel Co., Ltd.

Pang Jiang, Tong Chao

(Coking Plant of HBIS Group Xuansteel Company, Xuanhua 075105, China)

Abstract: This paper discusses and compares the characteristics of desulfurization and denitrification technology, and briefly introduces the advantages of adopting SDS dry desulfurization and SCR medium and low temperature denitration technology. According to the actual application situation in Xuanhua Steel's coke oven ultra-low emission transformation project, the desulfurization and denitrification process operation and equipment are improved and optimized. First, the reasons for excessive flue gas emission when different types of gas are used for reheating coke oven are analyzed and further optimized Second, from the aspect of equipment and facilities, optimize and improve the system desulfurization agent dosing path, ensure that the SO_2 emission at the flue gas outlet can be controlled during the maintenance of the mill system, and improve the qualified rate of flue gas emission. At the same time, tracking and learning the advanced design concept of ultra-low emission of flue gas in coking industry, continuously improve the efficiency of organized and unorganized flue gas treatment in coking industry, and ensure the gradual maturity of Xuanhua Steel's green transformation and development.

Key words: ultra low emission; desulfurization and denitrification; improvement

基于碳热还原法回收不锈钢粉尘制备铁铬镍碳合金研究

刘培军[1]，柳政根[1]，储满生[2]，闫瑞军[1]，唐　珏[1]，李　峰[1]

（1. 东北大学冶金学院，辽宁沈阳　110819；2. 东北大学轧制技术及连轧自动化国家重点实验室，辽宁沈阳　110819）

摘　要：不锈钢粉尘作为现代不锈钢冶炼产生的主要固体废弃物，其内部含有大量 Fe、Cr、Ni 的金属氧化物，开发绿色高效回收不锈钢粉尘的新方法可实现二次资源的高效利用，促进钢铁行业的可持续性发展。本文对不锈钢粉尘进行碳热还原冶炼铁铬镍碳合金的工艺进行试验研究。主要研究还原温度、还原时间、配碳比对不锈钢粉尘中 Fe、Cr、Ni 回收率和铁铬镍碳合金中金属品位的影响。试验结果表明：还原温度 1450℃，还原时间 20min，配碳比 0.8，优化工艺参数条件下，金属还原回收率较高，Fe、Cr、Ni 的回收率分别为：91.5%、90.19%、92.4%。合金中金属品位较高，Fe、Cr、Ni 的金属品位分别为：65.5%、18.96%、4.03%。合金颗粒可直接作为冶炼不锈钢的原料，实现了不锈钢粉尘的高效回收利用。

关键词：不锈钢粉尘；固废回收利用；碳热还原；铁铬镍碳合金

Study on Synthesis Fe-Cr-Ni-C Alloy from Stainless Steel Dust based on Carbon-Thermal Reduction

Liu Peijun[1], Liu Zhenggen[1], Chu Mansheng[2], Yan Runjun[1], Tang Jue[1], Li Feng[1]

(1. School of Metallurgy, Shenyang 110819, China; 2. State Key Laboratory of Rolling and Automation, Northeastern University, Shenyang 110819, China)

Abstract: Stainless steel dust, as the main solid waste produced by stainless steel smelting, contains a large number of metal oxides of Fe, Cr and Ni. The development of a new green and efficient method for recovering stainless steel dust can realize the efficient utilization of secondary resources and promote the sustainable development of iron and steel industry. In this paper, the smelting process of Fe-Cr-Ni-C alloy from stainless steel dust by carbon-thermal reduction was studied. The effects of reduction temperature, reduction time and carbon ratio on the recovery of Fe, Cr and Ni and the grade of metal in Fe-Cr-Ni-C alloy were studied. The results showed that the recovery of Fe, Cr and Ni was 91.5%, 90.19% and 92.4% respectively under the conditions of reduction temperature 1450℃, reduction time 20min, carbon ratio 0.8 and optimized process parameters. The metal grades of Fe, Cr and Ni are 65.5%, 18.96% and 4.03% respectively. Alloy particles can be directly used as raw materials for smelting stainless steel, which can realize the efficient recycling of stainless steel dust.

Key words: stainless steel dust; solid waste recycling; carbon-thermal reduction; Fe-Cr-Ni-C alloy

轧钢加热炉高温烟气余热回收利用浅析

陈有为

（新疆八一钢铁股份有限公司能源中心，新疆乌鲁木齐　830022）

摘　要：钢铁企业在生产过程中需要投入大量的外购能源，这些能源通过企业内部的循环利用，最终除附加于产品（包括副产品、固废等）本身及外卖等形式外，其余绝大部分以热量的形式散失到环境中，包括物料冷却散热、炉窑排烟散热、蒸汽冷凝水散热、冲渣水散热、循环冷却水散热等，在外部节能减排和企业内部降本双重压力下，企业越来越重视低品味余热余压回收利用，通过对轧钢加热炉外排废气余热回收，新建余热锅炉，产生蒸汽并网运行。采暖季节所产蒸汽全部用于生活区供暖，非采暖季节汽轮发电机组所发电回用于热轧区域生产。

关键词：高温废气余热；回收利用

The Utilize of Low-grade Thermal Energy Help Iron and Steel Enterprise Reach the Peek of Carbon

Chen Youwei

(Xinjiang Baiyi Iron & Steel Co., Ltd. Energy Center, Urumqi 830022, China)

Abstract: Iron and steel enterprises need to invest a large amount of energy in the production process. These energy resources are recycled within the enterprises, and are eventually added to the products (including by-products, solid waste, etc.) themselves and to take-out food, most of the rest is lost to the environment in the form of heat, including material cooling heat, furnace flue gas heat, steam condensation water heat, slag water heat, circulating cooling water heat, etc. , under the pressure of external energy-saving and emission-reducing and internal cost-reducing, enterprises pay more and more attention to the recovery and utilization of low-taste waste heat and pressure. The steam produced in heating season is used for heating of living area, while the electricity generated by steam turbine in non-heating season is used for electric generator production.

Key words: waste heat of high temperature exhaust gas; recycling

钢铁混合煤气锅炉低氮燃烧技术应用

王　丹，陈有为，王立元，关　馨，赵　增，吴新平

（新疆八一钢铁股份有限公司能源中心热力分厂，新疆乌鲁木齐　830022）

摘　要：NOx 是大气中的重要污染源之一，燃气锅炉污染物排放主要是 NOx，目前对于锅炉 NOx 排放治理工艺主要为低氮燃烧器、SCR、SNCR、氧化脱硝等，但后者均有新增能耗问题，本文通过在钢铁行业中混合煤气锅炉低氮燃烧方式优化燃气与空气比例，达到降低锅炉热力性 NOx，降低 NOx 的排放量，达到排放标准。

关键词：混合煤气锅炉；低氮燃烧器；NOx

Application of Low Nitrogen Combustion Technology in Steel-iron Mixed Gas Boiler

Wang Dan, Chen Youwei, Wang Liyuan, Guan Xin, Zhao Zeng, Wu Xinping

(Thermal Branch of Energy Center of Xinjiang Bayi Steel Co., Ltd., Urumqi 830022, China)

Abstract: NOx is one of the important pollution sources in the atmosphere, the main pollutant emission from gas-fired boiler is NOx. At present, the control technologies for NOx emission from boiler are mainly low-nitrogen burner, SCR, SNCR, oxidation and denitrification, but the latter all have the problem of increasing energy consumption, this paper optimizes the ratio of gas to air through the low nitrogen combustion mode of the mixed gas boiler in the Iron and steel industry to reduce the boiler thermodynamic Nox, reduce the NOx emission and meet the emission standard.

Key words: mixed gas boiler; low nitrogen burner; NOx

垃圾干燥热风炉换热系统

赵 增

（新疆八一钢铁股份有限公司能源中心，新疆乌鲁木齐 830022）

摘 要：自改革开放以来中国垃圾产量每年都在变化，当然现在处于一种越来越多的形式。垃圾焚烧可以极大地降低垃圾的体积，让垃圾变成灰烬，最后掩埋在地下。国外的发达国家采用的也是垃圾焚烧技术，他们在垃圾处理上面要领先中国很多，而且由于国外的垃圾种类不一样，天气气候也不一样，所以他们采用的垃圾焚烧设备也是不同的。众所周知，在垃圾焚烧的过程中有的垃圾会产生浓烟，浓烟的味道闻者并不是很好，而且还会破坏环境和对人体造成伤害。想要通过焚烧处理垃圾那就需要先要烘干垃圾，不然焚烧过程将会进行得极为缓慢。我们要做的就是通过热风炉产生热交换，让通过热风炉的冷空气变成热空气，再将热空气通入需要干燥的垃圾当中。

关键词：垃圾处理；热风炉；结构设计；数值模拟

Waste Drying Hot Blast Stove Heat Exchange System

Zhao Zeng

(Xinjiang Bayi Iron & Steel Co., Ltd., Energy Center, Urumqi 830022, China)

Abstract: Since the reform and opening up, China's garbage output has been changing every year, of course, it is now in a more and more form. Garbage incineration can greatly reduce the volume of garbage, so that the garbage into ashes, and finally buried in the ground. Foreign developed countries also use waste incineration technology, they are much ahead of China in garbage disposal, and because the types of garbage abroad are different, the weather and climate are also different, so the waste incineration equipment they use is also different. As we all know, in the process of garbage incineration, some garbage will produce thick smoke, the smell of the smoke is not very good, but also can damage the environment and cause harm to the human body. If you want to dispose of the garbage through incineration, you need to dry the garbage first, otherwise the incineration process will be extremely slow. What we need to do is to generate heat exchange through the hot air stove, turn the cold air through the hot air stove into hot air, and then pass the hot air into the garbage that needs to be dried. The hot air stove designed in this paper is a hot air stove that heats cold air and then dries garbage. In the design process, the hot air stove is divided into three parts: the internal heat exchanger, the intermediate heat exchanger and the shell of the hot air stove.

Key words: waste treatment; hot blast stove; structure design; numerical simulation

四氯化钛除钒尾渣钠化焙烧动力学研究

姜丛翔，堵伟桐，陈 卓，居殿春

（江苏科技大学张家港校区冶金与材料工程学院，江苏张家港 215600）

摘 要：针对氯化法生产钛白粉工艺中除钒固体废弃物难以有效再利用的问题，本文采用非等温热重技术研究钠盐添加量、升温速率等因素对含钒尾渣氧化的影响规律，采用 Kissinger-Akahira-Sunose(KAS)法求解含钒尾渣氧化过程活化能，并通过焙烧实验进一步研究含钒尾渣氧化过程的规律。结果表明：含钒尾渣完全氧化的温度为 700℃左右，随着 Na_2CO_3 的增加，表观活化能逐渐降低，氧化速率提高，$NaVO_3$ 生成量增多；当钠盐添加量超过 20%后，钒氧化过程中各个反应的竞争作用随 Na_2CO_3 的继续增加而逐渐减弱，表观活化能开始逐渐增大，Na_2SiO_3 对 $NaVO_3$ 的包覆加重，进而影响钒的氧化速率。

关键词：含钒尾渣；非等温热重；氧化过程；活化能

Study on Sodium Roasting Kinetics of Titanium Tetrachloride Vanadium Removal Slag

Jiang Congxiang, Du Weitong, Chen Zhuo, Ju Dianchun

(School of Metallurgy and Material Engineering, Jiangsu University of Science and Technology, Zhangjiagang 215600, China)

Abstract: In the process of producing titanium dioxide by the chlorination method, it is difficult to effectively reuse the vanadium-containing solid waste. This paper uses non-isothermal thermogravimetric technology to study the influence of sodium salt addition, heating rate, and other factors on the oxidation of vanadium-containing tailings. The Kissinger-Akahira-Sunose (KAS) method was used to calculate the activation energy of the vanadium-containing tailings oxidation process, and the laws of the oxidation process of the tailings were further studied through roasting experiments. The results show that the complete oxidation temperature of vanadium-containing tailings is about 700℃. With the increase of Na_2CO_3, the apparent activation energy gradually decreases, the oxidation rate increases, and the amount of $NaVO_3$ generated increases. When the addition of sodium salt exceeds 20%, the competitive effect of each reaction in the vanadium oxidation process gradually weakens with the continuous increase of Na_2CO_3, and the apparent activation energy begins to gradually increase. The coating of Na_2SiO_3 on $NaVO_3$ is aggravated and affects the oxidation rate of vanadium.

Key words: vanadium-containing tailings; non-isothermal thermogravimetry; oxidation process; activation energy

废油脂用于工业炉窑喷吹的实验室研究

竺维春[1]，王金花[1]，范正赟[1]，张 彦[1]，
刘文运[1]，王 喆[2]，黄坤鹏[2]，庄 辉[2]
（1. 首钢技术研究院，北京 100041；2. 首钢京唐钢铁联合有限责任公司，河北 唐山 063200）

摘 要：A 公司废乳化液泥中水分 12.74%，干基发热值与神华煤接近。废磨削液中 98.18%为水。乳化液泥与废磨削液混合液(3:2)、加水 30%的乳化液泥其黏度分别为 34.3 泊、29.1 泊。废乳化液泥燃点 366℃，废磨削液燃点 466℃。高炉单个风口日喷 6t，对高炉论燃烧温度及炉腹煤气量影响较小。在不改变回转窑内氧气含量的前提下，喷吹 0.5t/h 的废油脂混合溶液需要补充 1700.4m^3/h 空气；喷吹废油脂降低综合燃料消耗约 0.29kg/t；废油脂喷吹范围不超过窑头处 2m，在此状况下，喷管直径应不低于 5mm；喷吹废油脂后烟气中水分相对增加了 0.14%，对电除尘和管道腐蚀、回转窑结圈、烟气中 SO_2 含量、球团有害元素含量影响小。

关键词：废油脂；高炉；回转窑；喷吹；黏度

Laboratory Study on Waste Oil Injection into Industrial Furnace

Zhu Weichun[1], Wang Jinhua[1], Fan Zhengyun[1], Zhang Yan[1],
Liu Wenyun[1], Wang Zhe[2], Huang Kunpeng[2], Zhuang Hui[2]
(1. Shougang Research Institute of Technology, Beijing 100041, China;
2. Shougang Jingtang Steel Co., Ltd., Tangshan 063200, China)

Abstract: The moisture content in waste emulsion mud of A company is 12.74%, and the dry basis calorific value is close to Shenhua coal. 98.18% of the waste grinding fluid is water. The viscosity of emulsion mud mixed with waste grinding fluid (3:2) and emulsion mud with 30% water is 34.3 poise and 29.1 poise respectively. The ignition point of waste emulsion mud is 366℃, and that of waste grinding fluid is 466℃. The daily injection of 6 tons per tuyere has little effect on the combustion temperature and bosh gas volume. Under the premise of not changing the oxygen content in the rotary kiln, 1700.4m^3/h air is needed to inject 0.5t/h waste grease mixed solution; the comprehensive fuel consumption is reduced by about 0.29kg/t by injecting waste grease; the injection range of waste grease is not more than 2m at the kiln head, under this condition, the nozzle diameter should not be less than 5mm; after injecting waste grease, the moisture in the flue gas increases by 0.14%, which is harmful to electrostatic precipitator and pipeline corrosion ,the influence of ring formation, SO_2 content in flue gas and harmful elements content in pelletizing is small.

Key words: waste grease; blast furnace; rotary kiln; injection; viscosity

工业污水处理技术在长钢公司的应用实践

闫伟巍，李小婷，申强强
（首钢长治钢铁有限公司，山西长治 046031）

摘　要： 介绍了工业污水处理技术在长钢公司的应用实践，对建设背景、工艺流程、原水水质进行了阐述，重点讲述了运行过程中出现的问题及采取的措施，并根据实践制定了关键控制参数和应急管理措施。

关键词： 电导率；膜污染；纳滤；运行管理

The Application Practice of Industrial Sewage Treatment Technology in Changgang Company

Yan Weiwei, Li Xiaoting, Shen Qiangqiang

(Shougang Changzhi Iron & Steel Co., Ltd., Changzhi 046031, China)

Abstract: This paper introduces the application practice of industrial sewagetreatment technology in Changgang Company, expounds the constructionbackground, process process and raw water quality, focuses on the problems inthe operation process and measures, and formulates key control parameters andemergency management measures according to the practice.

Key words: electrical conductivity; membrane pollution; nano - filter; operation andmanagement

复合改性沥青混合料制备与性能研究

徐　兵[1]，李　帅[2]

（1. 宝武集团环境资源科技有限公司，上海　201900；

2. 同济大学道路与交通工程教育部重点实验室，上海　201804）

摘　要： 将烧结脱硫灰与橡胶粉这两种固体废弃物应用于沥青混合料改性，以提高沥青混合料耐久、抗疲劳等路用性能与降低噪声、改善行车舒适性，同时解决烧结脱硫灰和废弃轮胎堆放带来的环境与经济问题。研究结果表明，烧结脱硫灰与橡胶粉对沥青混凝土进行复合改性，有效提高沥青混凝土的高温、水稳定性能，且湿法制备工艺下，烧结脱硫灰与橡胶粉复合改性的沥青混凝土的高温性能要优于干法工艺下制备的沥青混合料高温性能。

关键词： 烧结脱硫灰；胶粉；复合改性；沥青混合料

Preparation and Properties of High Durable and Low Noise Composite Modified Asphalt Mixture

Xu Bing[1], Li Shuai[2]

(1. Baowu Group Environmental Resources Technology Co., Ltd, Shanghai 201900, China;

2. The Key Laboratory of Road and Traffic Engineering, Ministry of Education, Shanghai 201804, China)

Abstract: Two kinds of solid wastes, sintered desulfurization ash and rubber powder, are used in asphalt mixture modification to improve the durability, fatigue resistance, noise reduction and driving comfort of asphalt mixture, and to solve the environmental and economic problems caused by sintered desulphurization ash and waste tire stacking. The results show that the performance of sintered desulphurizer and rubber powder is better than that of asphalt mixture prepared by dry process.

Key words: sintering desulfurization ash; rubber powder; compound modification;asphalt mixture

冶金固废混合技术研究

钱 峰[1]，李志斌[2]，侯洪宇[1]，刘沛江[2]，徐鹏飞[1]

（1. 鞍钢股份有限公司技术中心，辽宁鞍山 114009；
2. 鞍钢股份有限公司炼铁总厂，辽宁鞍山 114021）

摘 要：钢铁企业每年产生大量的固体废弃物，种类杂多且数量巨大。这些固体废弃物中含有一定量的 TFe、C、CaO、MgO 等有价值元素，可返回烧结工序进行资源化回收利用。由于这些固体废弃物来自于不同工序，各种固体废弃物物化性能差异较大，必须进行充分混和均匀后再参与烧结配料，以保证原料稳定性，进而保证烧结矿质量及生产顺行。一些企业对这些固体废弃物的预处理只是简单粗放地采用铲车堆混，不但影响混匀效果，又污染环境。本文以鞍钢厂区产生的固体废弃物为研究对象开展了大量试验，采用卧式、立式强力混合机分别开展了一系列不同参数的混匀工业试验，对比了不同填充率、混合时间、转数等参数下的混匀效果，得到最优化的混匀参数条件，为钢铁企业多种固体废弃物混匀工艺提供数据支持和参考。

关键词：固体废弃物；混匀；资源利用

Research Progress of Metallurgical Solid Waste Mixing Technology

Qian Feng[1], Li Zhibin[2], Hou Hongyu[1], Liu Peijiang[2], Xu Pengfei[1]

(1. Technical Center, Angang Steel Co., Ltd., Anshan 114009, China;
2. General Steelmaking Plant of Angang Steel Co., Ltd., Anshan 114021)

Abstract: Iron and steel enterprises produce a large number of solid waste every year. These solid wastes contain a certain amount of TFe, C, CaO, MgO and other valuable elements, which can be returned to the sintering process for recycling. Since these solid wastes come from different processes, and their physical and chemical properties vary greatly, they must be fully mixed and even before participating in sintering battering, so as to ensure the stability of raw materials and the quality of sinter and the smooth production. Some enterprises for the pretreatment of these solid waste is simply extensive use of forklift heap mixing, not only affect the mixing effect, but also pollution of the environment. In this paper, a large number of experiments have been carried out on the solid waste produced in Anshan Iron and Steel Plant. A series of industrial mixing tests with different parameters were carried out by using horizontal and vertical power mixers. By comparing the mixing effects under different filling rates, mixing time, revolutions and other parameters, the optimal mixing parameters were obtained, which provided data support and reference for the mixing process of various solid wastes in iron and steel enterprises.

Key words: solid waste; mixing; resource utilization

赤泥冶金固废资源提铁研究

经文波，彭洁丽，卢朝贵

（百色学院材料科学与工程学院，广西百色 533000）

摘　要：本研究目的是提高赤泥冶金固废资源利用率，提高产品铁品位和金属回收率。实验中发现焙烧时间及焙烧温度都会影响产品铁品位以及回收率，基本呈现正相关关系，但并不是焙烧时间越久、焙烧温度越高，产品铁品位以及回收率就越高。实验发现：1300℃时铁品位最高，为95.78%；1280℃铁回收率最高，为86.41%。合理恒温焙烧时间为30min左右，合理恒温焙烧温度为1280～1300℃。

关键词：赤泥；提铁；含铁品位；回收率；研究

Iron Extraction from Solid Waste Resources of Red Mud Metallurgy

Jing Wenbo, Peng Jieli, Lu Chaogui

Abstract: The purpose of this study is to improve the utilization rate of solid waste resources in red mud metallurgy and improve the iron grade and metal recovery of the products. In the experiment, it was found that both roasting time and roasting temperature would affect the iron grade and recovery of the product, and basically there was a positive correlation. However, the longer the roasting time and the higher the roasting temperature, the higher the iron grade and recovery of the product. It was found that the highest iron grade was 95.78% at 1300℃. The highest iron recovery was 86.41% at 1280℃. Reasonable constant temperature roasting time is about 30min, and reasonable constant temperature roasting temperature is about 1280-1300℃.

Key words: red mud; iron; iron grade; recovery; research

冶金过程协同处置城市固废的思考—废塑料的粒化与深度粒化试验探索

王　广，张宏强，聂志睿，王静松，薛庆国

（北京科技大学 钢铁冶金新技术国家重点实验室，北京　100083）

摘　要：为了解决城市固废的污染问题和实现资源的循环利用，开创冶金服务美好生活的新时代愿景，助力钢铁工业碳达峰碳中和，本研究针对炼铁过程协同处置城市固废的一个关键点——“粒化”开展研究，提出了“基于热处理的城市固废废塑料深度粒化”新工艺。以聚乙烯和聚氯乙烯薄膜塑料为研究对象，通过开展实验室基础研究，证明了该工艺的可行性。我国应加强相关技术的基础研究，促进钢铁企业与城市的可持续低碳共融发展。

关键词：城市固废；钢铁；协同处置；资源化；粒化

Thoughts on Metallurgical Co-disposal of Municipal Solid Waste-Pulverization of Waste Plastics and Deep Pulverization Experiment

Wang Guang, Zhang Hongqiang, Nie Zhirui, Wang Jingsong, Xue Qingguo

(State Key Laboratory of Advanced Metallurgy, University of Science and Technology Beijing, Beijing 100083, China)

Abstract: In order to solve the pollution problem of municipal solid waste and realize the recycling of resources, create a new era of metallurgy to serve a better life, and help the iron and steel industry to achieve carbon emission reduction, a key point of co-disposal of domestic waste in the iron-making process —— "pulverization" has been studied and a new process of deep pulverization of municipal solid waste plastics based on heat treatment" has been proposed in the present paper. The feasibility of the process was demonstrated by conducting basic laboratory studies using polyethylene and polyvinyl chloride film plastics as the raw materials. China should strengthen the basic research of this technology and promote the sustainable low-carbon co-integration development of the steel industry and cities.

Key words: municipal solid waste; iron and steel making; co-disposal; comprehensive utilization; pulverization

利用尖峰平谷调整水源系统

刘　剑

（武钢有限能源环保部，湖北武汉　430081）

摘　要：介绍了水源系统的组成，江边水站、12 号水站工艺，明渠、沉淀池的主要作用，水源系统的调度，对尖峰平谷进行了简介，就能源调度如何利用尖峰平谷调整水源系统进行了探讨和分析，并进行了成本核算。

关键词：水源系统；尖峰平谷；降本增效；水源系统调度

Abstract: Water supply system composition is introduced. The technology of river side water station and number 12 water station are introduced. The main function of open channel and settling tank is introduced. The dispatching of water supply system is introduced. Peak and valley is briefly introduced. How energy dispatch using peak and valley to adjust water supply system is discussed and analysed. Costing is accounted.

Key words: water supply system; peak and valley; cost reduction; dispatching of water supply system

降低皮带机系统无组织排放实践与探讨

柴学良，曾　辉，殴书海，宋福亮，吴建海

（北京首钢股份有限公司炼铁作业部，河北唐山　064400）

摘　要：首钢股份坚定不移推动绿色、循环、低碳发展，经过艰苦努力，在钢铁行业内率先实现了全流程超低排放，是全国首家环保 A 类企业，用实际行动守护绿水青山，本文讲述首钢股份在皮带机系统中降低无组织排放的实践与探索。

关键词：无组织排放；粉尘；皮带机；实践；探讨

基于响应面分析的SCR催化剂单孔脱硝反应研究

徐继法，朱　繁，邱明英，崔　岩，王建华

（中冶京诚工程技术有限公司，北京　102600）

摘　要：以催化剂单孔孔道内的脱硝反应作为研究对象，利用响应面分析法设计试验方案，分析空速、温度以及氨氮比对脱硝效率的影响。结论如下：对三个因素利用响应面分析得到了脱硝效率和三因素之间的多项式，同时发现温度对脱硝效率的影响最大，空速次之，氨氮比影响最弱；另外对因素之间的两两相互作用对脱硝效率影响从高到低依次为：空速和温度、温度和氨氮比、空速和氨氮比。故在运行过程中，应当优先调节温度和空速，此时能够在最短路程内达到最大的脱硝效率。

关键词：催化剂；反应；RSM；脱硝；SCR

Research on Single Hole Denitration Reaction of SCR Catalyst based on Response Surface Analysis

Xu Jifa, Zhu Fan, Qiu Mingying, Cui Yan, Wang Jianhua

(Capital Engineering & Research Incorporation Limited, Beijing 102600, China)

Abstract: Taking the denitrification reaction in the single-pore channel of the catalyst as the research object, the response surface analysis method is used to design the test plan to analyze the influence of space velocity, temperature and ammonia-nitrogen ratio on the denitrification efficiency. The conclusion is as follows: Using response surface analysis of the three factors, the denitrification efficiency and the polynomial between the three factors are obtained. At the same time, it is found that the temperature has the greatest influence on the denitrification efficiency, followed by the space velocity, and the ammonia-nitrogen ratio has the weakest influence; The effect of pairwise interaction on the denitrification efficiency from high to low is: space velocity and temperature, temperature and ammonia nitrogen ratio, space velocity and ammonia nitrogen ratio. Therefore, during operation, the temperature and airspeed should be adjusted first, and the maximum denitrification efficiency can be achieved in the shortest distance.

Key words: catalyst; reaction; RSM; denitration; SCR

超低标准下的除尘系统低成本合理优化

江建平，靳　娟

（太原钢铁（集团）有限公司能源环保部，山西太原　030003）

摘　要：超低排放政策出台后，现有钢铁行业首当其冲，在刚刚完成上一轮实现特别排放限值改造的基础上，除尘设施如何再提效这个问题又摆在了企业面前，在没有更好技术支持的前提下，只能通过提升滤料材质、提高电源性能、增加过滤精度等方面思考和实施，用最少的投入、最小的影响，保证最佳的效果，履行企业的社会责任，保证企业绿色和谐发展。

关键词：除尘系统；低成本；优化

Low-cost and Reasonable Optimization of Dust Removal System under Ultra-low Standards

Jiang Jianping, Jin Juan

(Taiyuan Iron and Steel Group Co., Ltd., Taiyuan 030003, China)

Abstract: After the introduction of the ultra-low emission policy, the existing steel industry was the first to bear the brunt. On the basis of just completing the last round of transformation to achieve special emission limits, the issue of how to improve the efficiency of dust removal facilities is again in front of the company. In the absence of better technical support Under the premise of this, we can only think and implement by improving the material of the filter material, improving the performance of the power supply, and increasing the filtering accuracy, with the least investment and the least impact to ensure the best results, fulfill the corporate social responsibility, and ensure the green and harmonious corporate development of.

Key words: dust removal system; low-cost; optimization

钒渣钙镁复合提钒工艺研究

王 鑫[1]，向俊一[1,2]，裴贵尚[1]，吕学伟[1,2]

（1. 重庆大学材料科学与工程学院，重庆 400044；
2. 钒钛冶金及新材料重庆市重点实验室，重庆 400044）

摘 要：针对传统钙化焙烧—酸浸提钒工艺中钒回收率低的问题，提出了CaO/MgO复合焙烧—酸浸提钒工艺。本研究以转炉钒渣为原料，研究了MgO/(CaO+MgO)摩尔比和焙烧浸出参数对钒回收的影响。结果表明：CaO完全被MgO替代后，钒的浸出率由88%下降到81%；而CaO/MgO复合焙烧可以改善提钒效果。MgO/(CaO+MgO)摩尔比为0.5:1时，钒的浸出率可达94%。X射线衍射仪(XRD)和扫描电镜(SEM)表明CaO/MgO复合焙烧提高了酸溶性钒酸盐的生成速率，抑制硫酸钙沉淀生成，加快浸出动力学，并提高钒的浸出率。

关键词：钒；钒渣；焙烧；酸浸；回收

Study on the Technology of Vanadium Extraction from Vanadium Slag with Calcium and Magnesium

Wang Xin[1], Xiang Junyi[1,2], Pei Guishang[1], Lv Xuewei[1,2]

(1. College of Materials Science and Engineering, Chongqing University, Chongqing 400044, China;
2. Chongqing Key Laboratory of Vanadium-titanium Metallurgy and Advanced Materials, Chongqing 400044, China)

Abstract: Comparedwiththetraditionaltechnology ofvanadiumextractionfromvanadiumslag, a novel process of composite roasting with CaO/MgO and subsequent acid leaching was proposed. In this study, Linz-Donawiz (LD) converter vanadium slag was used as raw material, and the effects of the MgO/(CaO+MgO) molar ratio and the roasting and leaching parameters on the recovery of vanadium were studied. The results showed that the leaching efficiency of vanadium decreased from 88% to 81% when CaO was replaced completely by MgO; however, it could be improved by roasting with the composite of CaO/MgO. The leaching rate of vanadium can reach 94%under the optimal conditions of MgO/(CaO+MgO) molar ratio of 0.5:1.The results from X-ray diffractometry (XRD) and scanning electron microscopy (SEM) confirm that the formation rate of acid-soluble vanadates can be enhanced during roasting with the composite of CaO/MgO and that the leaching kinetics can be accelerated owing to the suppression of calcium sulfate precipitation.

Key words: vanadium; vanadium slag; roasting; acid leaching; recovery

碳酸钠对钛精矿碳热还原行为的影响研究

侯有玲，陈丹，吕晓东，辛云涛，吕学伟

（重庆大学材料科学与工程学院，重庆 400044）

摘　要：随着钛工业的快速发展，世界各地开始重视低品位钛精矿资源，低品位钛精矿直接利用会导致产能低污染重的问题，通过电炉熔炼富集获得钛渣的过程则耗能严重效率低，通过低温半熔融还原冶炼生产钛渣则能够节能高效，本文中，研究了添加剂 Na_2CO_3 对钛精矿半熔融还原行为的影响。研究结果表明：钛渣产物中主要物相为 Fe、Ti_3O_5、$Na_{0.23}TiO_2$、$Na_2Fe_2Ti_6O_{16}$ 和 M_3O_5（M：Fe、Mg、Ti 等）。添加 Na_2CO_3 能参与钛精矿还原过程，从而以反应物的形式提高反应物的活性，降低反应的起始温度和活化能，促进还原的进行和金属铁相的生长，有利于后续渣铁的分离，导致金属铁的特征峰随着 Na_2CO_3 的增加而逐渐减弱。

关键词：钛精矿；碳热还原；碳酸钠；还原机理

Study on the Effect of Sodium Carbonate on Carbothermal Reduction Behavior of Titanium Concentrate

Hou Youling, Chen Dan, Lv Xiaodong, Xin Yuntao, Lv Xuewei

(College of Materials Science and Engineering, Chongqing University, Chongqing 400044, China)

Abstract: With the rapid development of titanium industry, the world begins to attach importance to the low-grade titanium iron ore resources. The direct utilization of low-grade ilmenite concentrateleads to the problem of low productivity and heavy pollution. The electric furnace smelting process of obtaining titanium slag is low in efficiency and consumes plenty of energy. However, the process of low temperature semi-melting reduction is low energy-consumption and high efficiency. The influence of additive Na_2CO_3 on the semi-melting reduction behavior of ilmenite concentratewas studied in this article. The results show that main phases in the titanium slag products are Fe, Ti_3O_5, $Na_{0.23}TiO_2$, $Na_2Fe_2Ti_6O_{16}$ and M_3O_5 (M: Fe, Mg, Ti, etc.). Na_2CO_3 participated in the reduction process of ilmenite concentrate in the form of reactant, improving the activity of reactants and reducing the initial reaction temperature and activation energy of the reaction. The addition of Na_2CO_3 promotes the growth of metal iron phase, which is beneficial to the subsequent separation of iron and slag. The characteristic peak of metal iron is gradually weakened with the increase of Na_2CO_3.

Key words: ilmenite; carbothermal reduction; sodium carbonate; reduction mechanism

热轧层流冷却系统水质净化及冷却能力提升

吕中付

（武汉钢铁有限公司热轧厂，湖北武汉 430083）

摘 要：通过在层流冷却系统全流程各环节采取水质净化措施，解决积渣堵塞问题，提高水质水平，提高冷却均匀性，钢带横向温差获得显著改善。通过工艺参数和测量标准化，设备和工艺同步采用措施，以多种新方法提高系统冷却能力，突破限制，基本解决了高温季节高强结构钢生产过程中发生的停顿间歇问题，并提出彻底性解决方案。

关键词：层流冷却；水质净化；均匀性；冷却能力

Water Purification and Improvement of Hot Rolling Laminar Cooling System

Lv Zhongfu

(Wuhan Iron & Steel Co., Ltd., Hot Rolling Mill, Wuhan 430083, China)

Abstract: By adopting water purification measures in the whole process of laminar cooling system, the problem of slagging blockage can be solved, the water quality and cooling uniformity can be improved, and the transverse temperature difference of steel strip can be significantly improved. Through the standardization of process parameters and measurement, and the simultaneous measures of equipment and process, the cooling capacity of the system was improved by a variety of new methods, and the limitation was broken through. The pause and intermittent problem occurred in the production of high strength structural steel in high temperature season was basically solved, and a thorough solution was put forward.

Key words: laminar cooling;water quality purification; uniformity; cooling capacity

利用氧化铁皮制备多孔金属铁的方法及性能研究

明守禄[1,2]，张 芳[1,2]，李斐斐[1,2]

（1. 内蒙古科技大学材料与冶金学院，内蒙古包头 014010；2. 内蒙古自治区先进陶瓷材料与器件重点实验室，内蒙古包头 014010）

摘 要：钢铁工业是我国国民经济的基础产业和支柱产业，其生产过程伴有大量氧化铁皮，氧化铁皮具有全铁含量高，杂质含量低，便于提纯等优点。本文以废弃的氧化铁皮为原料制备多孔金属铁，并对氧化铁皮的还原机理开展多角度研究。利用 Factsage 热力学分析软件对还原反应进行了模拟计算，并采用正交试验分析方法确定最佳配碳还原工艺为配碳比(90 %)→气体流量(Ar,150 mL/min)→保温时间(2h)→温度(1150 ℃)→升温速度(20 ℃/min)。实验表明，通 H_2 进行二次还原后的产物中的[C]皆低于 0.03 %，且[O]含量也达到了极低的目的；在相同配碳比下，通 H_2 焙烧样品的孔隙率＞通 H_2 前焙烧样品的孔隙率。

关键词：多孔金属铁；热力学分析；配碳还原；正交试验

Study on the Method and Performance of Preparing Porous Metallic Iron from Iron Oxide Scale

Ming Shoulu[1,2], Zhang Fang[1,2], Li Feifei[1,2]

(1. School of Material and Metallurgy, Inner Mongolia University of Science and Technology, Baotou 014010, China; 2. Inner Mongolia Key Laboratory of Advanced Ceramic Materials and Devices, Baotou 014010, China)

Abstract: The iron and steel industry is the basic industry and pillar industry of my country's national economy, and its production process is accompanied by a large amount of iron oxide scale. Iron oxide scale has the advantages of high total iron content, low impurity content, and ease of purification. This paper uses waste iron oxide scale as raw material to prepare porous metallic iron, and conducts a multi-angle study on the reduction mechanism of iron oxide scale. Factsage thermodynamic analysis software was used to simulate the reduction reaction, and the orthogonal test analysis method was used to determine the optimal carbon distribution reduction process as carbon distribution ratio (90%) → gas flow (Ar, 150 ml/min) → holding time (2h) → temperature (1150℃) → heating rate (20℃/min). Experiments show that the [C] in the product after the second reduction with H_2 is less than 0.03%, and the [O] content is also extremely low; under the same carbon ratio, the porosity of the sample calcined with H_2＞The porosity of the calcined sample before passing H_2.

Key words: porous metal iron; thermodynamics research; carbon reduction; orthogonal test

真空还原轧钢铁皮制备多孔不锈钢热力学研究

赵立杰，张　芳，彭　军，黄　兰，明守禄

（内蒙古科技大学 材料与冶金学院，内蒙古包头　014010）

摘　要： 采用热力学分析软件 FactSage 与实验验证相结合方法，对在真空条件下、利用石墨粉还原氧化铁皮制备多孔 316 不锈钢的还原热力学进行了研究。探究在真空对 Cr 的存在形式、含量以及转变温度的影响。结果发现 Cr 元素存在 $FeCr_2O_4+Cr_3O_4 \to Cr_2O_3 \to Cr_{23}C+Cr_7C_3 \to Cr_{(FCC)} \to Cr_{(L)}$转变。大气压从 1atm 降低至 10^{-3}atm 时，Cr（FCC）由 2.96 g 增加至 16.812g，含 Cr 化合物的生成和消失温度降低，使的在低温条件下可以反应。液相生成温度由 1153.5℃增加至 1473.5℃，更加有利于以固相反应的方式制备多孔不锈钢。杂质氧化物之间只存在化合反应、不参与氧化还原反应，对还原过程没有影响。相同温度条件下，样品尺寸的增加，比表面积大的试样还原的更快。制样压力对样品的还原几乎没有影响。根据热力学计算结合实验得到利用氧化铁皮直接制备多孔不锈钢材料工艺的最佳制度为：1200℃、高真空、保温 3h。热力学模拟结果可用于开发利用冶金生产废料的新技术，为后续更多新技术的开发提供理论依据。

关键词： 氧化铁皮；还原热力学；多孔不锈钢

Study on Thermodynamics of Porous Stainless Steel Prepared by Vacuum Reduction Rolling Steel Skin

Zhao Lijie, Zhang Fang, Peng Jun, Huang Lan, Ming Shoulu

(School of Materials and Metallurgy, Inner Mongolia University of Science and Technology, Baotou 014010, China)

Abstract: The thermodynamic analysis software FactSage is combined with experimental verification to study the reduction thermodynamics of porous 316 stainless steel prepared by graphite powder reduction of iron oxide scale under vacuum conditions. Explore the effect of vacuum on the existence form, content and transformation temperature of Cr. The results found that the Cr element has a $FeCr_2O_4+Cr_3O_4 \rightarrow Cr_2O_3 \rightarrow Cr_{23}C+Cr_7C_3 \rightarrow Cr_{(FCC)} \rightarrow Cr_{(L)}$ transition. When the atmospheric pressure is reduced from 1atm to 10^{-3}atm, $Cr_{(FCC)}$ will increase from 2.96 to 16.812g, and the formation and disappearance temperature of Cr-containing compounds will decrease, making it possible to react under low temperature conditions. The liquid phase formation temperature is increased from 1153.5℃ to 1473.5℃, which is more conducive to the preparation of porous stainless steel by solid-phase reaction. The impurity oxides only have a compound reaction, do not participate in the oxidation-reduction reaction, and have no effect on the reduction process. Under the same temperature conditions, as the sample size increases, the sample with a larger specific surface area will be reduced faster. The sample preparation pressure has almost no effect on the reduction of the sample. According to thermodynamic calculations and experiments, the best system for the process of directly preparing porous stainless steel using oxide scale is: 1200℃, high vacuum, and heat preservation for 3h. The results of thermodynamic simulation can be used to develop new technologies for the utilization of metallurgical production waste, and provide a theoretical basis for the development of more new technologies in the future.

Key words: scale; vacuum reduction thermodynamics; porous stainless steel

水热反应条件对冷轧油泥中油相分离行为的影响规律研究

阙志刚[1]，付尹宣[1]，刘自民[2]，石金明[1]，艾仙斌[1]

（1. 江西省科学院能源研究所，江西南昌 330096；
2. 马鞍山钢铁股份有限公司技术中心，安徽马鞍山 243003）

摘 要：冷轧油泥是一种油、水、铁屑混合组成的危险固体废弃物，其直接排放不仅会严重危害生态环境，而且会造成油和铁资源的浪费，如何无害资源化处置冷轧油泥已成为钢铁企业迫切需要解决的技术难题。鉴于临界状态下水的独特理化特性，本研究在高压反应釜中以水做溶剂，分别研究了水热反应温度和冷轧油泥质量占比对其水热反应过程油、铁屑资源化回收效率的影响规律。研究结果表明，水热反应温度由325℃增加到375℃时，油相的分离效率呈现先增大后减小的趋势，且在350℃下达到最高，约为78%。随着冷轧油泥的质量占比从5%升高到15%时，油相分离效率逐渐降低。冷轧油泥在水热反应温度为350℃、反应时间为30 min、占比为5%的条件下，油相分离效率达89.28%，有效实现冷轧油泥中油和铁屑的资源化回收。

关键词：冷轧油泥；水热；反应温度；油泥占比；分离效率

Effects of the Hydrothermal Reaction Conditions on the Separation Behavior of the Oil Phase in the Oily Cold Rolling Sludge

Que Zhigang[1], Fu Yinxuan[1], Liu Zimin[2], Shi Jinming[1], Ai Xianbin[1]

(1. Institute of Energy Research, Jiangxi Academy of Science, Nanchang 330096, China;
2. Technology Center, Ma'anshan Iron and Steel Co., Ltd., Ma'anshan 243003, China)

Abstract: Oily cold rolling sludge is composed of oil, water and fine metal powders, which is a dangerous solid residue. Not only it seriously harms to the ecological environment, but it also causes a waste of oil and iron resources. How to harmlessly and resourcefully treat the oily cold rolling sludge is a technical problem for iron and steel industry at present. Due to the unique physical and chemical properties of water in the critical state, the influences of the hydrothermal reaction temperature and the mass ratios of oily cold rolling sludge on the recovery efficiency of oil and iron metals in the hydrothermal reaction process was studied with the solvent of water in the high-pressure reactor, respectively. The results showed that when the hydrothermal reaction temperature increased from 325°C to 375°C, the separation efficiency of the oil phase increased firstly and then decreased. It reached the peak value about 78% at 350°C. Meanwhile, as the mass ratios of oily cold rolling sludge increased from 5% to 15%, the separation efficiency of the oil phase gradually decreases. The separation efficiency of oil phase was 89.28% when the hydrothermal reaction temperature was 350℃, the reaction time was 30 min, and the mass ratios of oily cold rolling sludge was 5%. The oil and iron metals were effectively and resourcefully recycled in the oily cold rolling sludge.

Key words: oily cold rolling sludge; hydrothermal; reaction temperature; proportion of oily sludge; separation efficiency

高炉煤气布袋除尘输灰系统优化设计实例

杨 贝，王亚楠，张红磊，徐 茂

（中冶京诚工程技术公司环保与暖通工程技术所，北京 100176）

摘 要：设计了一种高炉煤气布袋除尘输灰系统，包括四条刮板输送机及智能加湿搅拌机。刮板运输机上部端口与布袋除尘器卸灰系统相连，收集到各个除尘器排放的高炉灰，最终通过带角度刮板输送机，输送到智能加湿机，省去了斗式提升机的设置。通过智能加湿机对物料的称重控制水量的补充，能够最大程度的避免高炉灰到拉灰车的二次扬尘，同时也能起到节水节电的作用。本系统优化了高炉布袋除尘装置机械输灰过程物料的提升存贮，省去了灰仓的设置，为实际的工程建设节省了资金投入。

关键词：高炉煤气；输灰系统；节能；卸灰输灰方式

Optimization Design Example of Bag-type Dust Removal and Transportation System for Blast Furnace Gas

Yang Bei, Wang Yanan, Zhang Honglei, Xu Mao

(Capital Engineering & Research Incorporation Limited, Beijing 100176, China)

Abstract: A bag dust removal and conveying system for blast furnace gas is designed, including four scraper conveyors and intelligent humidification mixer. The upper port of the scraper conveyor is connected with the dust discharging system of the bag filter, and the blast furnace dust discharged by each dust collector is collected. Finally, it is transported to the intelligent humidifier through the scraper conveyor with an angle, eliminating the setting of the bucket elevator. Through the intelligent humidifier to the material weighing control water supplement, it can avoid the secondary dust from the blast furnace ash to the ash truck to the greatest extent, and also can play the role of water and electricity saving. The system optimizes the lifting and storage of materials in the process of mechanical ash conveying of bag dust removal device of blast furnace, saves the setting of ash bin, and saves capital investment for the actual engineering construction.

Key words: blast furnace gas; ash conveying system; energy saving; ash unloading and conveying mode

污泥的资源化利用及工业炉窑协同处置技术

平晓东[1,2]，王　锋[1,2]，王海风[1,2]

（1. 钢铁研究总院先进钢铁流程及材料国家重点实验室，北京　100081；
2. 钢研晟华科技股份有限公司，北京　100081）

摘　要：污泥作为污水处理的副产物，在污水处理率不断提高地同时，污泥产生量也与日俱增，所以如何减量化、稳定化、无害化以及资源化处理污泥成为了企业关注的焦点。本文针对污泥的组成、资源化利用途径以及处理处置技术进行了综述，重点分析了利用冶金炉窑协同处理污泥的处置技术，明确了冶金炉窑协同处理污泥具有很大的发展前景。

关键词：污泥；资源化利用；工业炉窑；冶金炉窑

Resource Utilization of Sludge and the Co-treatment Technology in Industrial Furnace

Ping Xiaodong[1,2], Wang Feng[1,2], Wang Haifeng[1,2]

(1. State Key Laboratory of Advanced Steel Processes and Products, Central Iron and Steel Research Institute, Beijing 100081, China; 2. CISRI Sunward Technology Co., Ltd., Beijing 100081, China)

Abstract: Sluge is a by-product of sewage treatment, the amount of sludge is increasing with the continuous improvement of sewage treatment rate. How to reduce, stabilize, harmless and resource treatment of sludge has become the focus of enterprises. This paper reviews the sludge composition, resource utilization and treatment technologies. The co-treatment of sludge in metallurgical furnaces is mainly analized, which has a great development prospect in future.

Key words: sludge; resource utilization; industrial furnace; metallurgical furnace

喷雾焙烧盐酸再生尾气超低排放工艺的探索与应用

吴宗应，赵金标，郭金仓，王　军

（中冶南方工程技术有限公司，湖北武汉　430081）

摘　要：钢铁、机械加工、化工企业产生的酸洗废液通常采用喷雾焙烧工艺进行再生，但是废液再生过程中产生的尾气中含有大量的粉尘，需要净化达标后排放。本文根据喷雾焙烧盐酸废液再生工艺尾气中粉尘的物理、化学特性，提出了一种超低排放工艺。本文提出的工艺具有能耗低、效率高的特点，并已成功应用于多套喷雾焙烧盐酸再生装置，效果显著。本工艺可以为钢铁行业盐酸再生装置实施超低排放改造提供可靠的技术解决方案。

关键词：喷雾焙烧；盐酸再生；超低排放

Exploration and Application of Ultra-low Emission Process for Spraying & Roasting Hydrochloric Acid Regeneration Process

Wu Zongying, Zhao Jinbiao, Guo Jincang, Wang Jun

(WISDRI Engineering & Research Incorporation Ltd., Wuhan 430081, China)

Abstract: The pickling waste acid produced by ironand steel, mechanical processing and chemical enterprises is usually regeneratedby spray roasting process, however，theexhaust gas generated in spray roasting process contains a lot of dust, whichis very difficult to remove. In this paper, ,a better of ultra-low emissionprocess ,been developed based on the physical and chemical characteristics ofthe dust , of the exhaust gas which is produced by the acid regeneration process is been proposed. The processproposed in this paper possess the characteristics of low energy consumptionand high efficiency, and has been successfully applied to several sets of acid regeneration system with remarkable results. Thisprocess can provide a reliable technical solution for the implementation ofultra-low emission transformation of acid regeneration system in iron and steelenterprises.

Key words: spray roasting process; acid regeneration; ultra-low emissions

10　冶金设备与工程技术

高炉炉顶密封阀稳定长寿研究及应用

邓振月，王仲民，张　华，陈文彬，薛理政，张志宽
（北京首钢股份有限公司，北京　100043）

摘　要：本课题重点阐述了高炉炉顶核心设备密封阀常出故障，由于密封阀作业环境在料罐内部，同时下密封阀又与高炉炉内直接相通，粉尘大、温度高，压力环境作业等原因长期使用后造成阀板胶圈与阀座阀口压不实出现跑风现象，曾多次被迫停风处理该类设备故障，对高炉的顺稳生产带来了严重的影响。通过发明高炉炉顶核心设备密封阀调整方法，攻克了国内首发炉顶煤气全回收系统长期运行条件下对核心设备带来的严重影响，延长了使用周期15天以上，保证了高炉炼铁炉顶设备的稳定长寿运行。

关键词：高炉；料罐；密封阀；压痕

Study and Application of Stability and Longevity of Sealing Valve of Blast furnace Roof

Deng Zhenyue, Wang Zhongmin, Zhang Hua, Chen Wenbin, Xue Lizheng, Zhang Zhikuan
(Beijing Shougang Co., Ltd., Beijing 100043, China)

Abstract: This topic focuses that the sealing valve of the blast furnace roof often fails. Since the operating environment of the sealing valve and the lower sealing valve are directly connected with the blast furnace, with large dust, high temperature, pressure environment and other reasons, it has a serious impact on the smooth production of blast furnace.Through the adjustment method of sealing valve of blast furnace roof, the serious impact of the long-term operation of the gas recovery system on the core equipment, extended the service period for more than 15 days and ensured the stable and long operation of the blast furnace roof equipment.

Key words: blast furnace; material tank; sealing valve; indentation

转炉煤气加压机轮毂裂纹故障分析与研究

朱红兵
（上海宝钢节能环保技术有限公司，上海　201900）

摘　要：本文针对三台转炉煤气加压风机叶轮轮毂出现同样的裂纹故障，结合断裂力学原理，采取严密的计算和分析方法，从叶轮应力计算、焊接热影响区、疲劳断裂特点、焊接工艺、锻造工艺、锻造毛坯设计、锻造轮毂优化设计、转炉煤气腐蚀等八个方面进行了全面的阐述分析。在金相分析的基础上，分析了断口的微观特征和宏观形貌，

确定了断裂源在于轮毂结构设计不当导致加工过程中存在残余应力、使用过程中存在应力腐蚀的共同作用所致，为断裂原因的确定与机理的研究提供了有力的证据，并提出了后续改进方案。

关键词： 转炉煤气；加压机；轮毂；裂纹；应力腐蚀；金相分析

Fault Analysis and Research on Wheel Hub Crack of Converter Gas Pressurized Fan

Zhu Hongbing

(Shanghai Baosteel Energy Saving Technology Co., Ltd., Shanghai 201900, China)

Abstract: In this paper three sets of converter gas pressurized air blower impeller wheel cracks appear the same fault, combined with the fracture mechanics principle, adopt the method of layer upon layer stripping cocoon, from impeller stress calculation, the welding heat affected zone, fatigue fracture characteristics, welding process, the forging process and forging blank design, forging hub optimization design, converter gas corrosion, and eight aspects has carried on the comprehensive analysis of the paper. On the basis of the metallographic analysis, this paper analyzes the macro and micro characteristics of the fracture morphology, fracture source lies in the design of the wheel hub structure determines the improper result in residual stresses are present in the machining process, and exists in use process caused by a combination of stress corrosion, to determine the cause of fracture and mechanism of research provides a favorable evidence, and subsequent improvement scheme was put forward.

Key words: converter gas; pressure machine; wheel hub; crack; stress corrosion; metallographic analysis

热连轧机精轧轧辊冷却水系统优化改造与实施

吴长杰[1]，张会明[1]，东占萃[1]，王艺霖[1]，张志桥[1]，王香梅[2]

（1.北京首钢股份有限公司，河北迁安　064406；
2.北京首钢机电有限公司迁安电气分公司，河北迁安　064406）

摘　要： 热连轧机轧辊冷却不足，不仅导致其表面氧化膜过早脱落，造成带钢表面点状或者条状氧化铁皮压入缺陷，还会造成轧辊热凸度过大，从而影响凸度计算的准确性，及产生中间浪，甚至是有断辊的风险。通过对比分析国内先进企业精轧机轧辊冷却水设计，针对首钢热轧产品需求，对精轧机轧辊温度高的问题进行了对比分析，确定了产品温度高、原设计冷却不足是工作辊温度高的主要原因，并提出了轧辊工作辊冷却水优化改造方案，对流量需求重新计算、管路校核、阀门能力校核，流量计选型等工作，实现了轧辊冷却水流量的提升，同时对工作辊冷却水喷嘴进行了优化，使得轧辊冷却更加均匀，冷却效率得到明显提升。

关键词： 热连轧机；冷却水；辊温；喷嘴

Optimal Modification and Implementation of Cooling Water System for Finishing Rolls of Hot Strip Mill

Wu Changjie[1], Zhang Huiming[1], Dong Zhancui[1], Wang Yilin[1], Zhang Zhiqiao[1], Wang Xiangmei[2]

(1. Beijing Shougang Co., Ltd. Qian'an 064406, China;
2. Beijing Shougang Electromechanical Co., Ltd., Qian'an 064406, China)

Abstract: Insufficient cooling of the rolls of hot strip mills not only causes the surface oxide film to fall off prematurely, but also causes the spot or strip-shaped oxide scale on the surface of the strip steel to be pressed into defects, but also causes the rolls to have excessive thermal convexity, which affects the accuracy of crown calculation , And produce intermediate waves, or even risk of broken rolls Through comparative analysis of the cooling water design of the finishing mill rolls of domestic advanced enterprises, the problem of the high temperature of the finishing mill rolls was compared and analyzed according to the demand for hot-rolled products of Shougang Co., Ltd., the main reason for the high temperature of the work rolls was determined, and the roll work was proposed. Roll cooling water optimization and transformation plan, through flow demand calculation, pipeline check, valve capacity check, flowmeter selection, etc., the roll cooling water flow is increased, and the work roll cooling water nozzle is optimized to make The roll cooling is more uniform, and the cooling efficiency is significantly improved.

Key words: hot strip mill; cooling water; roll temperature; nozzle

米巴赫激光焊机喂丝系统稳定性分析与机构改造

赵新宝[1]，玄利剑[1]，李冠良[1]，王 明[2]，兰晓栋[1]，唐晓宇[1]，王承刚[1]

（1.首钢智新迁安电磁材料有限公司，河北迁安 064400；
2.首钢迁安新能源汽车电工钢有限公司，河北迁安 064400）

摘 要：本文以首钢智新迁安电磁材料有限公司常化酸洗机组配备的激光焊机为研究对象，针对生产运行过程中发生的典型喂丝系统故障，对影响喂丝稳定性的因素进行分析，制定一系列控制改善措施，经现场试验证明，这些措施为焊机喂丝系统稳定运行提供重要保障。

关键词：焊机；喂丝系统；措施；改造

Study on Improving the Stability of Wire Feeding System of Miebach Welding Machine

Zhao Xinbao[1], Xuan Lijian[1], Li Guanliang[1], Wang Ming[2], Lan Xiaodong[1], Tang Xiaoyu[1], Wang Chenggang[1]

(1. Shougang Zhixin Qian'an Electromagnetic Material Co., Ltd., Qian'an 064400, China;
2. Shougang Qian'an New Energy Automobile Electrical steel Co., Ltd., Qian'an 064400, China)

Abstract: In ShougangZhixinQian'an Electromagnetic Material Co., Ltd., a normative pickling line equipped with Laser Welder as the research object, in view of the production run during a typical thread feeding system fault, analyze the factors affecting the stability of wire feeding. Formulated a series of improvement measures, control the field test proved that these measures to provide important guarantee for the welding wire feed system and stable operation.

Key words: laser welder; wire feed system; measure; modification

一种滑车在炼钢厂行车滑线检修的应用

张银洲，刘 峰

（鄂钢公司转炉炼钢厂，湖北鄂州 436000）

摘 要：行车作为物料搬运机械，应用十分广泛、钢厂物料搬运尤其依赖行车。对钢厂中行车的应用及控制系统进行研究分析后，对行车安全滑触线进行研究分析，解决实际检修时对安全滑触线打磨问题。结合实际对炼钢厂过渡跨和加料跨在检修时对安全滑触线打磨圆角时，发现打磨出来的质量不好，并且员工的自身安全得不到良好保护，为了解决这一问题，我们从根源问题认真思考，制造一种能保护员工并且舒适打磨的安全装置。

关键词：滑触线打磨；安全屏障；改进措施

The Application of a Carriage in the Maintenance of a Steel Mill's Driving Cable

Zhang Yinzhou, Liu Feng

(Egang Converter Steelmaking Plant, Ezhou 436000, China)

Abstract: driving as a material evacuation machinery, the application is very extensive, steel mill material handling is particularly dependent on driving. After studying and analyzing the application and control system of driving in the steel mill, the paper studies and analyzes the driving safety slip contact line to solve the problem of grinding the safety slip contact line during the actual maintenance. Combined with the actual transition span of the steel mill and the filling cross in the maintenance of the safety slip contact line grinding fillet, found that the quality of grinding out is not good, and the employee's own safety is not well protected, in order to solve this problem, we seriously consider from the root cause of the problem, to create a protection of employees and comfortable grinding safety device.

Key words: touches the line polish slippery; safety barrier; improvement measure

鞍钢中厚板厂 2575kW 直流电机升级改造

苏胜勇

（鞍钢电气有限责任公司，辽宁鞍山 114005）

摘 要：特大型直流电机在测绘、设计、工艺、安装上都和常规直流电机不同，鞍钢电气有限责任公司在原鞍钢股份中厚板厂直流电机的基础上对 2575kW 直流电机电磁设计、制造工艺进行优化改造，通过对电机励磁绕组、合口连线、机座结构三大部分的技术改造，同时改进线圈制造工艺，使新制电机满足生产要求的同时，性能指标得到了大大的改善。

关键词：特大型电机；技术改造；工艺优化

Upgrade of 2575kW DC Motor in Medium and Heavy Plate Plant of Ansteel

Su Shengyong

(Anshan Electric Co., Ltd., Anshan 114005, China)

Abstract: The extra large DC motor is different from the conventional DC motor in surveying, design, technology and installation, angang Electric Co., Ltd. optimized the electromagnetic design and manufacturing process of 2575kW DC motor on the basis of the DC motor of Angang Plate plant, through the technical transformation of three major parts of the motor's excitation winding, closing connection and frame structure, and the improvement of the coil manufacturing process, the performance index of the new motor is greatly improved while meeting the production requirements.

Key words: oversize motor; technical transformation; process optimization

汽轮发电机组推力轴承故障分析与检修工艺改进

徐从庆，赵　宇，金　涛，王　仲，金宇翔，李雪松

（鞍钢股份有限公司能源管控中心，辽宁鞍山　114021）

摘　要：鞍钢能源管控中心 4#汽轮发电机发生推力轴承瓦块温度超限报警、不能满负荷运行的故障。通过设备拆解，发现了汽轮机设备原因。针对推力轴承需要大量、经常性维修的实际情况，采取提高维修效率、安装精度的一系列措施，取得了很好的效果。

关键词：汽轮发电机；推力轴承；推力轴瓦；轴承维修

Fault Analysis and Maintenance Process Improvement of Thrust Bearing of Steam Turbine Generator Unit

Xu Congqing, Zhao Yu, Jin Tao, Wang Zhong, Jin Yuxiang, Li Xuesong

(Energy Control Center of Ansteel, Anshan 114021, China)

Abstract: the 4# steam turbine generator of Angang energy management and control center has the fault that the temperature of thrust bearing pad exceeds the limit alarm and can not operate at full load. Through equipment disassembly, the causes of steam turbine equipment were found. In view of the actual situation that thrust bearings need a lot of frequent maintenance, a series of measures to improve maintenance efficiency and installation accuracy have been taken, and good results have been achieved.

Key words: turbogenerator; thrust bearing; thrust bearing bush; bearing maintenance

轧机液压辊缝振动的分析解决办法

张凤亮，赵连江，迟　异，廉法勇，商　融
（鞍钢股份有限公司冷轧厂，辽宁鞍山　114000）

摘　要：本文以鞍钢冷轧厂 1450 酸轧联合机组轧机液压辊缝控制系统在带钢轧制过程中遇见的辊缝异常振动作为研究对象，以解决辊缝振动现象为出发点，综合考虑轧钢工艺、电气控制、液压、机械设备等多重因素，结合多种专业知识以严谨的方法探寻振动产生原因，并制定出消除振动的方法，旨在避免液压辊缝控制系统振动造成的轧制精度下降，轧机断带，液压设备管线焊口开焊漏油等生产设备事故，为冶金设备维护提供参考信息，为智能维护模型的建立提供案例依据。

关键词：轧机；液压；辊缝；振动

Analysis and Solution of Hydraulic Gap Vibration of Rolling Mill

Zhang Fengliang, Zhao Lianjiang, Chi Yi, Lian Fayong, Shang Rong
(Cold Strip Works of Ansteel, Anshan 114000, China)

Abstract: In this paper, the abnormal vibration of the hydraulic roll gap control system of 1450 acid rolling combined unit in Angang Cold Rolling Mill is taken as the research object, the problem of roll gap vibration is solved as the starting point, the multiple factors such as rolling process, electrical control, hydraulic and mechanical equipment are considered comprehensively, and the causes of vibration are explored by rigorous methods combined with a variety of professional knowledge The method of eliminating vibration is put forward to avoid the production equipment accidents such as rolling accuracy decrease, rolling strip breaking and oil leakage caused by vibration of hydraulic roll gap control system. It provides reference information for intelligent maintenance of equipment and data basis for intelligent maintenance model.

Key words: rolling mill; hydraulic pressure; roll gap; vibration

入口横剪剪刃间隙机构调整原理分析

李建文
（武钢有限硅钢部，湖北武汉　430083）

摘　要：本文对入口横剪剪刃间隙调整机构组成分析，描述各机构部件相互作用关系，结合剪刃间隙调节装置安装尺寸链原理，制定剪刃间隙机构调节程序方法，指导生产操作正确使用设备，避免入口横剪机构部件产生松动，减少故障的发生。

关键词：活动剪身；凹槽滑道固定；剪刃间隙；楔形机构调整

Analysis on the Adjustment Principle of the Clearance Mechanism of the Entrance Cross Shear Blade

Li Jianwen

(Silicon Steel Business Division of WISCO, Hu Bei Wuhan 430083, China)

Abstract: This paper analyzes the composition of the blade clearance adjustment mechanism of the entrance cross shear, describes the interaction between the components of each mechanism, and combines with the principle of the installation dimension chain of the blade clearance adjustment device, formulates the adjustment program method of the blade clearance mechanism, guides the production operation, correctly uses the equipment, avoids the looseness of the components of the entrance cross shear mechanism, and reduces the occurrence of faults.

Key words: the activity cuts the body; the groove slideway is fixed; blade clearance; wedge adjustment

大型低速重载回转设备的故障诊断

胡申辉，陆炜煜

（宝武装备智能科技有限公司，上海 201900）

摘 要： 运输部马迹山港回转大轴承是典型的大型低速重载回转设备，属堆、取原料关键设备，为了确保该轴承的可靠运行是非常重要的。由于常规的振动传感器很难拾取如此低频的信号致使振动分析无法进行，因此我们利用润滑脂的光谱和铁谱分析方法对其进行状态监测。在此研究过程中，开发出新的润滑脂专用溶剂。研究结果表明，利用润滑脂的光谱和铁谱分析技术来评价该类型低速重载轴承磨损状态是可行的。

关键词： 滚动轴承；状态监测；润滑脂；铁谱技术

Fault Diagnosis of Low-speed and Heavy Load Slewing Equipment

Hu Shenhui, Lu Weiyu

(Baowu Equipment Intelligment Technology Co., Ltd., Shanghai 201900, China)

Abstract: The slewing big bearing of Majishan port is a typical large low-speed and heavy load slewing equipment,which belongs to the key equipment of piling and taking raw materials,so it's very important to ensure the reliable operation of the bearing.Because it is difficult for conventional vibration sensor to pick up such low frequency signal,so we use the spectral and ferrographic methods to monitor the grease.In this study,a new special solvent for grease was developed.The results show that it is feasible to evaluate the wear state of this type of bearing with low speed and heavy load by using spectrum and ferrography of grease.

Key words: rolling bearing; condition monitoring; grease; ferrography technique

冶金企业原燃料智能取制样系统的设计开发与应用

张　阳，彭国仲，王　莉，肖阳华，何小琴

（首钢京唐钢铁联合有限责任公司，河北唐山　063000）

摘　要：介绍了冶金企业原燃料智能取制样系统工艺设计及应用情况。系统以工业机器人为中心，设计开发了平面化的智能取制样系统、智能制样系统、样品智能转运系统、样品智能存储系统，实现原燃料取样、制样、运样、存样全过程无人值守。精密度和偏倚检验符合要求，煤制样效率提升 25%，矿制样效率提升 33%，熔剂制样效率提升 33%。

关键词：原燃料；取制样；无人值守；设计应用

Design, Development and Application of Intelligent Sampling System for Raw and Fuel in Metallurgical Enterprises

Zhang Yang, Peng Guozhong, Wang Li, Xiao Yanghua, He Xiaoqin

(Shougang Jingtang United Iron & Steel Co., Ltd., Tangshan 063000, China)

Abstract: The process design and application of intelligent sampling and preparation system for raw and fuel in metallurgical enterprises are introduced Taking industrial robot as the center, the system designed and developed a planar intelligent sampling and preparation system, intelligent sample preparation system, intelligent sample transfer system and intelligent sample storage system, which realized the whole process of raw and fuel sampling, preparation, transportation and storage unattended.The precision and bias test met the requirements, the efficiency of coal sampling is increased by 25%, the efficiency of ore sampling is increased by 33%, and the efficiency of flux sampling is increased by 33%.

Key words: raw and fuel; sample preparation; unattended; design application

固体激光在冷轧高强钢焊接中的应用研究

贺海清，苗　锋，邱　天，李　斯，赵永平，李　浩

（河钢唐钢冷轧事业部，河北唐山　063000）

摘　要：冷轧生产厂目前主要采用气体激光焊机焊接高强钢，焊接成本高。本文以固体激光应用到冷轧高强钢焊接为研究对象，利用焊接实验，分析焊接参数对焊缝质量的影响及焊接过程中飞溅原因，通过优化焊接工艺参数，焊接出质量好且质量稳定的焊缝，实现低成本焊接。

关键词：固体激光器；飞溅；参数；质量

Research on Application of Solid State Laser in Welding of Cold-rolled High-strength Steel

He Haiqing, Miao Feng, Qiu Tian, Li Si, Zhao Yongping, Li Hao

(HBIS Tangshan Iron and Steel Co., Ltd., Cold Rolling Division, Tangshan 063000, China)

Abstract: Cold rolling plant currently mainly use gas laser welder to weld high-strength steels, which have high welding costs. In this paper, the application of solid laser to cold-rolled high-strength steel welding is used as the research object. Welding experiments are used to research welding parameters on the quality of the weld during the welding process and the cause of spatters. By optimizing the welding process parameters, good quality and stable welding seams are produced. Achieve low cost welding.

Key words: solid state laser; spatter; parameters; quality

电缆在线监测及故障预警测距定位系统应用解析

李德生，阚 颂，徐 强

（鞍钢股份鲅鱼圈钢铁分公司能源动力部，辽宁营口 115007）

摘 要：本文简要介绍电缆在线监测及故障预警测距定位系统的功能及工作原理，对电缆如何做到在线监测、选线及故障预警，以及出现事故如何做到精准测距定位。在鞍钢股份鲅鱼圈分公司的 10kV 电缆中应用过程中遇到的电缆事故实际应用案例，做好事故前的预判及事故后及时抢修工作事先准备。

关键词：在线监测；选线；故障预警；测距定位

Application Analysis of Cable Online Monitoring and Fault Warning Location System

Li Desheng, Kan Song, Xu Qiang

(Energy and Power Department of Bayuquan Iron and Steel Branch of Angang, Yingkou 115007, China)

Abstract: This paper briefly introduces the function and working principle of the cable online monitoring and fault warning ranging and positioning system, how to achieve on-line monitoring, line selection and fault warning of cables, and how to achieve accurate distance measurement and positioning in case of accidents. In the application of 10kV cable in Bayuquan branch of Anshan Iron and Steel Co., Ltd., the actual application cases of cable accidents encountered in the application process should be well prepared before the accident and timely repair after the accident.

Key words: online monitoring; line selection; fault warning; location and location

TS_{120B}型铁水车研制

项克舜

（武汉钢铁有限公司 运输部，湖北武汉　430083）

摘　要： 介绍了TS_{120B}型铁水车的用途、技术参数、结构特点、试验及运用情况，对车体结构的强度、刚度进行了有限元分析。

关键词： 铁水罐；铁水车；结构；有限元分析

Development of TS_{120B} Molten Iron Car

Xiang Keshun

(Transportation Department of Wuhan Iron and Steel Co., Ltd., Wuhan 430083, China)

Abstract: The usage, technical parameters, structure and performance features of TS_{120B} molten iron car are expounded, and the tesr and operation, etc, are described, The finite element analysis on the strenth and stiffness of the carbody structure is made.

Key words: hot-metal ladle; molten iron car; structure; finite element analysis

320t 筒型混铁车倾翻电机故障原因分析

项克舜

（武汉钢铁有限公司运输部，湖北武汉　430083）

摘　要： 通过对一起 320t 筒型混铁车在倒灌站不能正常倒铁故障的处理，分析了产生倾翻故障的主要原因，提出了预防倾翻故障的措施。

关键词： 筒型混铁车；倾翻；电机；故障；分析

Analysis of Fault Causes of Tilting Motors of the 320t Cylindrical Torpedo Ladle Car

Xiang Keshun

(Transportation Department of Wuhan Iron and Steel Co., Ltd., Wuhan 430083, China)

Abstract: Through the treatment of the failure of the 320t cylindrical torpedo ladle car that fails to pour iron normally at an

inversion station, the main reasons for the tipping fault are analyzed, and measures to prevent tipping faults are put forward.
Key words: cylindrical torpedo ladle car; tipping; motors; failure; analysis

ICP-AES 法测定金属钙线中镁、铝、硅、锰、铁

郭　魏，沈　凯

（重庆钢铁制造管理部（技术中心）检测中心，重庆　401258）

摘　要：金属钙线采用化学法分析钙准确性不高，每个元素单独分析花费时间长。使用 ICP-AES 法测定金属钙线中镁、铝、硅、锰、铁杂质元素的含量，采用差量法测量钙含量，此方法较化学法操作更简便。本方法采用多组分拟合技术消除干扰，各元素曲线相关系数大于 0.999，回收率为 93%~105%，相对标准偏差低于 10%，分析结果准确可靠，能满足日常检测要求。
关键词：ICP-AES 法；金属钙线；基体；杂质元素

Determination of Magnesium, Aluminum, Silicon, Manganese and Iron in Calcium Wire by ICP-AES

Guo Wei, Shen Kai

(Testing Center of Chongqing Iron and Steel Manufacturing Management Department (Technical Center), Chongqing 401258, China)

Abstract: The accuracy of calcium analysis by chemical method is not high, and it takes a long time to analyze each element separately. The contents of magnesium, aluminum, silicon, manganese and iron impurities in metal calcium wire are determined by ICP-AES, and the content of calcium is measured by differential method. This method is simpler than chemical method. The method adopts multi-component fitting technology to eliminate interference. The correlation coefficient of each element curve is greater than 0.999, the recovery is 93%-105%, and the relative standard deviation is less than 10%. The analysis results are accurate and reliable, and can meet the requirements of daily detection.
Key words: ICP-AES method; calcium wire; matrix; impurity element

几种典型扇形段辊子结构的比较及应用

谭志强，向忠辉，熊　钢

（宝钢股份武汉有限公司炼钢厂，湖北武汉　430080）

摘　要：介绍连铸机扇形段几种典型辊子的结构及特点，对几种辊子的各自的特点进行比较和分析。结合某钢厂多年现场的实际应用情况，分析几种辊子各自的优点和缺点。
关键词：连铸机；扇形段；辊子

Comparison and Application of Several Typical Sector Roller Structures

Tan Zhiqiang, Xiang Zhonghui, Xiong Gang

(Equipment Maintenance Plant of WISCO, Wuhan, 430080, China)

Abstract: The structure and characteristics of several typical rollers in the sector of continuous caster are introduced, and their respective characteristics are compared and analyzed. Combined with the practical application of a steel plant for many years, the advantages and disadvantages of several rollers are analyzed.

Key words: caster; segment; roller

棒线材无头焊接轧制系统及经济效益

洪荣勇

（福建三钢闽光股份有限公司，福建三明　365000）

摘　要：当今棒线材的生产机仍主要属于单坯轧制，间断的坯料进给不仅给设备额外增加了许多载荷，还会造成堆钢现象，进而影响产能和管理。无头焊接轧制系统将两根坯料进行头尾焊接，可实现产品的无间断轧制生产，大幅度降低了成本。因此，本文就无头焊接轧制系统的设备组成及其创造的经济效益进行了阐述，因经济效益明显，具有推广价值。

关键词：无头焊接轧制系统；移动焊机；经济效益

Structure and Economic Benefit of Endless Rodand Wire Welding Rolling System

Abstract: Nowadays, the production mechanism of rod and wire still mainly belong to single billet rolling. Theintermittent billet feeding not only adds a lot of additional load to the equipment, but also causes the phenomenon ofsteel piling, which in turn affects production capacity and management. The endless welding rolling system can realize the uninterrupted rolling production of products by welding the head and tail of the two blanks, which greatly reduces the cost. Therefore, this paper expounds the equipment composition of endless welding rolling system and its economic benefits. Because of the obvious economic benefits, it has popularization value.

Key words: endless welding rolling; moving welder; economic benefit

冷轧拉矫机直传式张力辊齿轮箱改进设计

周为民

（宝山钢铁股份有限公司冷轧厂，上海　201900）

摘　要：冷轧的张力拉矫机传统上采用行星差动齿轮箱来控制延伸率，最近十年内随着电气控制技术的发展，开始用马达直接传动的方式来控制延伸率，直传式具有机构简单、维护方便等优势。但是在实际使用中发现反拖齿轮箱的轴承寿命较短。本文通过对轴承失效模式的分析，找到了影响轴承寿命的关键因素，并在齿轮箱的设计上进行了针对性的改进，对将来直传式拉矫机齿轮箱的设计选型具有指导意义。

关键词：拉矫机；直接传动；齿轮箱

Improved Design for Gearbox of Cold Rolling Tension Leveller

Zhou Weimin

(Baoshan Iron and Steel company, Shanghai 201900, China)

Abstract: In the past decade, with the development of electrical control technology, direct drive motor has been used to control elongation for cold rolling tension leveller. Direct drive type has the advantages of simple mechanism and convenient maintenance. However, in practical use, it is found that the bearing life of reverse drag gearbox is short. Through the analysis of bearing failure mode, this paper finds out the key factors that affect the bearing life, and makes targeted improvement on the design of gear box, which has guiding significance for the design and selection of gear box of direct drive tension leveller in the future.

Key words: tension leveller; direct drive; gear box

飞剪剪切控制系统及应用

庞慧玲，高　鑫，沈益钊，曾龙华

（湛江钢铁有限公司热轧厂，广东湛江　524000）

摘　要：本文介绍了飞剪剪切设备在轧钢生产过程中的应用，主要是针对 2250 热轧 TMEIC 模式飞剪剪切控制系统的组成及剪切原理进行研究。主要阐述了飞剪剪切控制的过程时序及头部剪切和尾部剪切的步骤，分析影响飞剪剪切及系统运行的关键因素，结合精轧入口区域辊道速度对前期出现过的飞剪异常进行总结归纳，提出进一步优化飞剪剪切控制系统的改进措施。

关键词：双曲柄式飞剪；辊道速度；2250 热轧；TMEIC 模式飞剪

Crop Shear Control System And Application

Pang Huiling, Gao Xin, Shen Yizhao, Zeng Longhua

(Hot Rolling Mill of Zhanjiang Iron & Steel Co., Ltd., Zhanjiang 524000, China)

Abstract: This paper introduces the application of crop shear cutting equipment in the process of steel rolling, which mainly studies the composition and shear principle of the crop shear cutting control system in the TMEIC mode of 2250 hot strip mill. It mainly describes the process time sequence of crop shear control and the steps of head cut and tail cut, analyzes the key factors affecting crop shear and system operation, summarizes the crop shear anomalies in the early stage based on the speed of roller table in the inlet area of finishing rolling, and puts forward the improvement measures to further optimize the crop shear control system.

Key words: double crank crop shear; roller speed; 2250 hot strip mill; crop shear mode of TMEIC

120t 转炉环缝装置结垢分析及改进

李鹏程，张志刚，石运亚，马　磊

（八钢公司第二炼钢厂，新疆乌鲁木齐　830022）

摘　要： 八钢公司第二炼钢厂三座 120t 转炉烟气净化及煤气回收系统采用的是传统的 OG 法。这是目前国内转炉烟气湿法除尘常用的方法，但该种除尘方式会容易造成除尘器环缝装置结垢。随着环保要求日益提高及原材料粉末量多、铁水成分不稳定等原因，近几年除尘器环缝清理频次由以前的三个月一次增加到每个月一次，每次清理时间 12~16h，严重制约了转炉产能发挥，对比国内其他钢厂（建龙、燕钢、日钢、八钢、新兴铸管等），该种装置均存在此问题，只是清理周期有所差别。因此需对 OG 系统环缝结垢原因进行分析及改进。通过对喷淋系统及环缝装置优化，有效解决了环缝结垢频繁问题。

关键词： 环缝；结垢；除尘

Analysis and Improvement of Scale Formation in Circular Joint Unit of 120t Converter

Li Pengcheng, Zhang Zhigang, Shi Yunya, Ma Lei

(No. 2 Steelmaking Plant of Eight Iron & Steel Corporation, Urumqi 830022, China)

Abstract: The flue gas purification and gas recovery system of the three 120-ton converters in the second steelmaking plant of BISC adopts the traditional OG method. This method is commonly used for wet dust removal of converter flue gas in China at present, but it is easy to cause scaling in the ring seam device of dust collector. With environmental protection requirements is increasing day by day and raw material powder quantity more, the reason such as the composition of the molten iron is not stable, dust catcher girth cleaning frequency in recent years by the previous three months an increase to once every month, every time cleaning time about 12 to 16 hours, severely restricted the converter capacity play, therefore need to OG system girth scaling reason analysis and improvement. By optimizing the spray system and ring joint device, the frequent scaling problem of ring joint is effectively solved.

Key words: annular seam; scale; dust removal

ICP-AES 测定铝铁中铝、磷、硅、锰元素

孙　倩

（首钢长治钢铁有限公司质量监督站，山西长治　046000）

摘　要：铝铁合金在炼钢生产中是一种新型脱氧剂，主要成分有铝、铁、磷、锰、硅等元素，本文讨论了用电感耦合等离子体发射光谱仪（ICP-AES）测定铝铁合金中铝、硅、磷、锰的分析条件并建立了测定方法。铝铁合金经稀王水分解，以 ICP-AES 法测定溶液中的铝、硅、磷、锰四个元素。通过选择没有干扰的谱线作为被测元素的分析线，背景干扰采用仪器自动背景校正方法扣除。我化验室通过此方法对铝铁试样进行分析检测，分析结果准确，同时此方法可以多元素同时检测，大大缩短了分析周期，提高了工作效率。

关键词：电感耦合等离子体发射光谱仪；铝铁合金；铝；硅；磷；锰

Abstract: Al-Fe alloy is a new type of deoxidizer. Its main components are aluminum, iron, phosphorus, manganese, silicon and other elements. This paper discusses the analytical conditions for the determination of aluminum, silicon, phosphorus and manganese in Al-Fe alloy by ICP-AES and establishes the determination method. The aluminum-iron alloy was decomposed by hydrochloric acid and nitric acid, and the four elements of aluminum, silicon, phosphorus and manganese in the solution were determined by ICP-AES. By selecting the spectral line without interference as the analysis line of the measured element, the background interference is deducted by the automatic background correction method of the instrument. This method is used in our laboratory to analyze and detect aluminum and iron samples, and the analysis results are accurate. At the same time, this method can detect multiple elements simultaneously, greatly shortening the analysis period and improving the work efficiency.

Key words: inductively coupled plasma emission spectrometer; Al-Fe alloy; Al; Si; P; Mn

利用声发射研究 Q235 钢氧化铁皮弯曲失效过程

段晶晶，周存龙，周晓泉，周　瑾，樊铭洋，郭　瑞

（太原科技大学机械工程学院，山西省冶金设备设计理论与技术重点实验室，山西太原　030024）

摘　要：对于无酸除鳞工艺参数难以确定的问题，掌握氧化铁皮破裂机理尤为重要。本文利用声发射、扫描电镜等技术研究了 Q235 钢氧化铁皮在弯曲过程中的失效行为。通过绘制声发射特征参数经历图、关联图及载荷经历图，并结合氧化铁皮截面形貌进行观察，发现下列结果：（1）受压时，氧化铁皮首先发生界间裂纹的萌生，然后进行扩展，最终向表面偏转形成贯穿性裂纹。界间裂纹萌生的幅值集中在 58~73dB；（2）受拉时，裂纹首先在氧化铁皮表面萌生，随后形成贯穿裂纹且不断增多，最后界间裂纹产生并发生剥落。表面裂纹产生的幅值集中在 62~74dB；（3）通过计算裂纹萌生的临界应变可知受压侧氧化铁皮更易剥落，并计算得到氧化铁皮的断裂韧性。该试验结果有助于掌握氧化铁皮破裂机理。

关键词：声发射；氧化铁皮；弯曲；裂纹

The Bending Failure Process of Oxide Scale of Q235 Steel was Studied by Acoustic Emission Properties

Duan Jingjing, Zhou Cunlong, Zhou Xiaoquan, Zhou Jin, Fan Mingyang, Guo Rui

(School of Mechanical Engineering, Taiyuan University of Science and Technology, Shanxi Provincial Key Laboratory of Metallurgical Equipment Design Theory and Technology, Taiyuan 030024, China)

Abstract: For the problem that the process parameters of acid-free descaling are difficult to determine, it is particularly important to grasp the cracking mechanism of the scale. This paper uses acoustic emission, scanning electron microscopy and other technologies to study the failure of Q235 steel oxide scale during the bending process. Acoustic emission characteristic parameter history graph, correlation graph and load history graph were drawn , these figures are combined with the cross-sectional morphology of the oxide scale to observe.The following results are found: (1) When under compressive stress, the oxide scale first initiates the initiation of inter-boundary cracks, then expands, and finally deflects to the surface to form penetrating cracks.The amplitude of the acoustic emission signal generated by the crack initiation between the boundaries is concentrated between 58~73dB; (2) When subjected to tensile stress， the cracks first initiated on the surface of the oxide scale, and then formed through cracks and increased, and finally the inter-boundary cracks occurred and peeled. The amplitude of the acoustic emission signal with cracks on the surface of the oxide scale is mainly concentrated in 62~74dB; (3) The calculated critical strain for crack initiation shows that the oxide scale on the compression side is more likely to peel off, and the fracture toughness of the oxide scale is calculated. The test results are helpful to grasp the cracking mechanism of the oxide scale.

Key words: acoustic emission; oxide scale; bending; crack

提高棒材锯切断面质量的研究与应用

孙国栋，魏恩选，李春祥，窦耀广，郭　强，汪　洋

（凌源钢铁股份有限公司，辽宁凌源　122500）

摘　要： 热锯机是热轧车间里用的最多的设备，切割钢材的工作零件是锯片，对于某些生产线热锯机的作业率非常高，通常在高温状态下锯切钢材，用于单根或编组多根的钢材定尺分段、切头、切尾和取样，切钢过程中主要通过水对锯片进行冷却，但是在切钢时锯片冷却水控制不好会造成被切钢材表面产生马氏体或贝氏体，降低钢材表面质量，一旦产生马氏体或贝氏体钢材表面硬度有所提高，造成锯切阻力增加、锯切电流提高从而加剧了锯片的磨损，另外受冷却水的影响造成切割过程中形成的毛刺内部组织的综合性能也有所提高，难以清理，因此受到冷却水的影响造成了钢材切割后的端面质量降低，可以通过对设备的改进降低冷却水对切割端面的影响。

关键词： 锯切；冷却水；断面质量；毛刺；锯罩；整体式辊道盖板

Research and Application of Improving the Quality of Bar Sawing Section

Sun Guodong, Wei Enxuan, Li Chunxiang, Dou Yaoguang, Guo Qiang, Wang Yang

(Lingyuan Iron and Steel Co., Ltd., Lingyuan 122500, China)

Abstract: The hot sawing machine is the most used equipment in the hot rolling workshop. The working part for cutting steel is the saw blade. For some production lines, the operation rate of the hot sawing machine is very high. It is usually used for sawing steel at high temperature. It is used for sizing, sectioning, head cutting, tail cutting and sampling of single or grouped steel. During steel cutting, the saw blade is mainly cooled by water, However, poor control of the cooling water of the saw blade during steel cutting will cause martensite or bainite on the surface of the steel to be cut and reduce the surface quality of the steel. Once martensite or bainite is generated, the surface hardness of the steel will increase, resulting in increased sawing resistance and sawing current, which will aggravate the wear of the saw blade, In addition, due to the influence of cooling water, the comprehensive performance of burr internal structure formed in the cutting process is also improved and difficult to clean. Therefore, due to the influence of cooling water, the quality of steel end face after cutting is reduced. The influence of cooling water on cutting end face can be reduced through the improvement of equipment.

Key words: sawing; cooling water; section quality; skin needling; saw cover; integral roller deck

中厚板轧机油膜轴承密封技术的研究

徐建翔

（宝钢集团新疆八一钢铁股份有限公司，新疆乌鲁木齐 830022）

摘 要：中厚板轧机油膜轴承进水问题突出，造成油膜系统故障频发，针对问题具体开展现场调查，分析了密封失效的原因，进行技术改进，先后从支承辊端面的修复方法、新增挡水防护装置的设计及使用，最终解决了油膜轴承进水问题，保障生产的顺行。

关键词：油膜轴承；进水量控制；技术改进

The Inflow of the Oil Film Bearing in Plate Mill Control Research

Xu Jianxiang

(Baosteel Group Xinjiang Bayi Iron and Steel Co., Ltd., Urumqi 830022, China)

Abstract: Pick to: in the heavy plate mill of oil film bearing water problems, causing frequent oil film system failures, in view of the problems to carry out the analysis of technical improvements, finally solve the water problem of the oil film bearing, ensure production along the line.

Key words: the oil film bearing; water inflow control; technical improvement

酸洗线平整机辊印攻关实践

梁文朋，党文文，王广英，宋明明，王宏民

（承德钢铁集团有限公司，河北承德 067000）

摘 要：平整机在酸洗产线应用越来越广泛，但是在平整机投用时,经常会出现辊印缺陷，给企业造成了较大经济损失。本文以实际生产为基础，通过对出现的缺陷进行测量、统计以及对轧辊异物进行取样分析等手段，排查出造成缺陷的各种原因。并从工艺、设备、原料等方面着手解决，使辊印缺陷得到有效控制。

关键词：酸洗；平整机；异物；辊印

Practice on rolling marks of pickling Line SPM

Liang Wenpeng, Dang Wenwen, Wang Guangying,
Song Mingming, Wang Hongmin

(Chengde Iron and Steel Group Co., Ltd., Chengde 067000, China)

Abstract: The skin pass mill is widely used in the pickling line, but when the skin pass mill is put into use, the roll mark defects often appeared, which cause great economic losses to the enterprises.Based on actual production, this article found the causes of the roll mark defects by measuring, statistics, and sampling of roller extraneous matter. And we from the process, equipment, raw materials and other aspects to make the roll mark defects are effectively controlled.

Key words: pickling; skin pass mill; extraneous matter; roll mark

电路板芯片级修复在钢厂实验室仪器维修中的应用

王明利，岳海丰，宣星虎，高　贺，周明雷

（北京首钢股份有限公司质量检验部，河北迁安　064404）

摘　要： 随着大型精密仪器在钢厂实验室的广泛使用，仪器电路板损坏的故障也越来越多，如何实现损坏短路板的修复再利用也成实验室设备维护管理的重要环节。本文以在国内钢厂实验室中广泛应用的 ARL4460 直读光谱仪、ARL9900 荧光光谱仪为例，介绍其电路板上芯片和元件损坏的故障故障现象、分析和处理方法。为实验室维护人员提供一种实验仪器电路板故障的修复思路，帮助维护人员快速恢复电路板故障，节约备件成本。

关键词： ARL4460；ARL9900；电路板故障；电路分析；电路板修复

Application of Circuit Board Chip Level Repair in Laboratory Instrument Maintenance of Steel Plant

Wang Mingli, Yue Haifeng, Xuan Xinghu, Gao He, Zhou Minglei

(Beijing Shougang Co., Ltd., Qian'an 064404, China)

Abstract: With the wide use of large-scale precision instruments in the laboratory of steel plants, t more and more Circuit board failure. How to repair and reuse the damaged circuit board has become an important part of laboratory equipment maintenance and management.In this paper, taken the ARL4460 direct reading spectrometer and ARL9900 fluorescence spectrometer as examples, which are widely used in domestic steel plant laboratories. Introduces the failure phenomenon, analysis and treatment method of chip and component damage on the circuit board.Provide a way to repair the circuit board fault of the instrument for the laboratory maintenance personnel, and helps them to quickly recover the circuit board fault and save the cost of spare parts.

Key words: ARL4460; ARL9900; circuit board fault; circuit analysis; circuit board repair

工业电子称重系统相关问题论述

张书文

（新疆八一钢铁股份有限公司能源中心，新疆乌鲁木齐　830022）

摘　要： 本文分析论述了通常情况下各种电子称重计量设备在使用过程遇到和关注的相关问题，从理论上进行分析论述，对计量校验结算误差的来源从力/电转换原理进行测试剖析，用于指导现场电子称重计量设备功能/精度调试，确保计量准确性。

关键词：电子称重系统；相关问题分析

Discussion on Related Problems of Industrial Electronic Weighing System

Zhang Shuwen

(Energy Center of Xinjiang Bayi Iron and Steel Co., Ltd., Urumqi 830022, China)

Abstract: This paper analyzes and discusses the sources of various electronic calibration and settlement errors from the principle of force/electricity conversion, which is used to guide the on-site electronic weighing and measuring equipment and ensure the accuracy of measurement.

Key words: electronic weighing system; analysis of related problems

流体计量节流装置通用性使用研究分析

张书文

（新疆八一钢铁有限责任公司能源中心，新疆乌鲁木齐 830022）

摘　要：根据流体力学的伯努力方程,主要分析研究流体密度变化时，对使用特定设计制造的节流装置测量流体流量产生的“误差”或“差异”影响，揭示流体在通过节流件时的动、静压能量转换关系和流体流量采用节流原理计量的关系。

关键词：能源流体介质；计量节流装置；通用性应用研究

Research and Analysis of Universal Use of Fluid Metering Throttle Device

Zhang Shuwen

(Energy Center, Xinjiang Bayi Iron and Steel Co., Ltd., Urumqi 830022, China)

Abstract: According to Bernhard's equation of fluid mechanics, the influence of “error” or “difference” caused by the change of fluid density on the measurement of fluid flow with a specially designed throttling device is mainly analyzed, and the relationship between the dynamic and static pressure energy conversion when fluid passes through the throttling element and the measurement of fluid flow by throttling principle is revealed.

Key words: energy fluid medium; metering throttle device; research on universal application

液化装置高、低温膨胀机异常原因分析及处理

顾　巍，刘　杰

（新疆八一钢铁股份有限公司，新疆乌鲁木齐　830022）

摘　要：针对液化装置高、低温膨胀机安装调试、及运行过程出现的多个问题，提出了一系列基于液化装置高低温增压透平膨胀机异常改进、解决方法。使得液化装置达标运转，优化设计得以提使用效率，实现节能减排。

关键词：液化装置；增压透平膨胀机；液氧；液氮

Analysis and Treatment of Abnormal Causes of High and Low Temperature Expanders in Liquefaction Unit

Gu Wei, Liu Jie

(Xinjiang Bayi Iron and Steel Co., Ltd., Urumqi, 830022, China)

Abstract: Aiming at many problems in the installation, commissioning and operation of high and low temperature expanders in liquefaction unit, a series of abnormal improvements and solutions of high and low temperature turboexpanders based on liquefaction unit are proposed. Make the liquefaction unit operate up to standard, optimize the design, improve the use efficiency and realize energy conservation and emission reduction.

Key words: liquefaction unit; booster turbine expander; liquid oxygen; liquid nitrogen

110kV GIS 开关试验过程合闸不到位的故障分析

张建冬

（新疆八一钢铁股份有限公司能源中心，新疆乌鲁木齐　830022）

摘　要：某厂在发电机并网前试验时发现，开关不能合闸到位处理过程，以及危险点分析，找出原因，提出预防措施。

关键词：开关；弹簧机构；储能

Fault Analysis of Inadequate Closing during 110kV GIS Switch Test

Zhang Jiandong

(Thermal Branch of Energy Center of Xinjiang Bayi Steel Co., Ltd., Urumqi 830022, China)

Abstract: During the pre grid test of generator in a power plant, it was found that the switch could not be closed in place, and the dangerous points were analyzed to find out the causes and put forward preventive measures.

Key words: electrical switches; spring mechanism; energy storge

高压开关柜断路器触头发热原因分析及预防措施

何 刚，何春利，唐 勇

（新疆八一钢铁股份有限公司能源中心，新疆乌鲁木齐 830022）

摘 要：本文以八钢热力分厂鼓风作业区10kV背压发电机出口断路器出现的触头发热故障，对高压开关柜断路器发热的产生的各种原因进行了分析，针对断路器的发热，采取了有效的预防措施;在高压开关柜动静触头加装无线测温模块；通过模块采集的温度趋势有效做出预防。

关键词：高压开关柜；断路器；触头；无线测温

Cause Analysis and Preventive Measures of Hot Contact Hair of Circuit Breaker in High Voltage Switch

He Gang, He Chunli, Tang Yong

(Xinjiang Baiyi Iron & Steel Co., Ltd. Energy Center, Urumqi 830022, China)

Abstract: In this paper, the heat failure of the 10kV back-pressure generator outlet circuit breaker in the blast operation area of Bayi Iron and Steel Co., Ltd., combined with the operating conditions of the field equipment, analyzes the various reasons for the heat generation of the circuit breaker of the high-voltage switch gear. Several measures have been taken for the heating of the circuit breaker. The most effective preventive measure is to install a wireless temperature measuring device for the high-voltage switch gear; the heating can be judged according to the temperature trend.

Key words: high voltage switch cabinet; breaker; contact; wireless temperature measurement

红外热成像技术在变电站发热点检测中的应用

张 健

（新疆八一钢铁股份有限公司能源中心热力分厂，新疆乌鲁木齐 830022）

摘 要：发热常常是设备损坏或功能故障的早期征兆,应用红外热成像技术定期对关键设备的温度进行检查，跟踪设备的运行状况，分析、识别、诊断电气设备存在的故障及隐患，提高对设备缺陷判断的预知性、准确性，进而指导设备检修，降低事故风险，避免电气设备因过热故障引起突发性事故。

关键词：变配电；电气连接点；测温；红外成像仪

Application of Infrared Thermai Imaging Technology in Substation Hot Spot Detection

Zhang Jian
(Baosteel Group Xinjiang Bayi Iron and Steel Co., Ltd., Energy Center, Urumqi 830022, China)

Abstract: Heating is often the early sign of equipment demage or functional failure.Infrared thermal imaging technology is used to regularly check the temperature of key equipment ,tarck the operation status of equipment,analyze,indentify and diagnose the faults and hidden dangers of electrical equipment,improve the predictability and accuracy of equipment defect judgment,and then guide equipment maintenance and reduce accident risk,Avoid sudden accidents caused by overheating of electrical equipment.

Key words: transformation and distribution; electrical connection point; temperature measurement; infrared imager

欧冶炉 TRT 转子、静叶磨损原因分析及应对措施

姜新河
（新疆八一钢铁股份有限公司能源中心，新疆乌鲁木齐 830022）

摘　要： 欧冶炉 TRT 解体后发现转子、静叶冲刷磨损，对冲刷磨损原因进行分析及制定应对措施。

关键词： TRT；原因；应对措施

消弧线圈在变电站 10kV 系统的应用

吕　争，李建军
（新疆八一钢铁股份有限公司能源中心，新疆乌鲁木齐 830022）

摘　要： 本文分析了中性点不接地系统的特点和运行中存在的一系列问题，阐述了消弧线圈在低压配电网中应用的必要性，从消弧线圈的工作原理、容量选择、接地变压器的选择等方面进行了简要说明。并在此基础上介绍了微机控制消弧线圈自动跟踪补偿装置的工作原理及其在配电系统实际应用中的使用效果。

关键词： 中性点接地；消弧线圈；电容电流

Application of Arc Suppression Coil in 10kV Substation System

Lv Zheng, Li Jianjun
(Xinjiang Baiyi Iron & Steel Co., Ltd., Energy Center, Urumqi 830022, China)

Abstract: This paper analyzes the characteristics of neutral ungrounded system and a series of problems existing in operation, expounds the necessity of arc suppression coil application in low voltage distribution network, and briefly explains the working principle of arc suppression coil, capacity selection, grounding transformer selection and other aspects. On this basis, the working principle of the automatic tracking and compensating device for arc suppression coil controlled by microcomputer and its application effect in distribution system are introduced.

Key words: neutral grounding; arc suppression coil; capacitive current

SWRCH35KE 冷镦开裂原因分析

夏　冰，赵志海，后宗保，王艳阳，王怀伟，付声丽

（马鞍山钢铁股份有限公司检测中心，安徽马鞍山　243000）

摘　要：利用宏观检验、化学成分分析及金相检测等方法对 SWRCH35KE 冷镦开裂产品进行了分析。结果表明，冷镦开裂原因是由于坯料表面氧化过度，在后续的生产中没有彻底去除氧化区域，在线材轧制过程中产生表面缺陷。提出轧后对盘条的酸洗检验，对轧线的调整进行指导是至关重要的。

关键词：SWRCH35KE；冷镦开裂；表面缺陷；质量控制

Reason Analysis on Cold Heading Crack of SWRCH35KE

Xia Bing, Zhao Zhihai, Hou Zongbao, Wang Yanyang, Wang Huaiwei, Fu Shengli

(Testing Center of Maanshan Iron & Steel Co., Ltd., Ma'anshan 243000, China)

Abstract: In order to analyze the cause of cold heading cracking of SWRCH35KE, the cracking products were analyzed by means of macroscopic examination, chemical composition analysis and metallographic examination.The results show that the reason of cold heading cracking is that the surface of billet is oxidized excessively, and the oxidized area is not completely removed in the subsequent production, and the surface defects are produced in the process of wire rolling.It is very important to guide the adjustment of the rolling line by pickling inspection of the rod after rolling.

Key words: SWRCH35KE; cold heading cracking; surface defects; quality control

低碳齿轮钢带状组织评定的热处理工艺影响分析

王怀伟，赵志海，后宗保，夏　冰，王艳阳，汪良俊

（马鞍山钢铁股份有限公司检测中心，安徽马鞍山　243000）

摘　要：试验分析了 18CrNiMo、19CN5 和 20CrMnTiH3 三种低碳齿轮钢在热轧、完全退火、等温正火工艺处理后的显微组织以及使用 GB/T 34474.1—2017 和 GB/T 13299—1991 的带状检验评级情况。结果显示低碳齿轮钢在不同

热处理制度下的带状组织存在明显差异。热轧态下，18CrNiMo、19CN5 和 20CrMnTiH3 含较多贝氏体，使用 GB/T 34474.1—2017 和 GB/T 13299—1991 均难以评定带状组织；经等温正火处理后 20CrMnTiH3 属于平衡态组织，用 GB/T 34474.1—2017 和 GB/T 13299—1991 都可完成带状组织评定，18CrNiMo、19CN5 存在少量马氏体，与 GB/T 13299—1991 要求的平衡态组织不相符，但是马氏体较少，可在 GB/T 34474.1—2017 标准下评定；经完全退火处理后三者都属于平衡态组织，使用 GB/T 34474.1—2017 和 GB/T 13299—1991 都可完成带状组织评定；完全退火与等温正火处理后钢材的带状级别存在明显差别。

关键词： 低碳齿轮钢；带状组织；完全退火；等温正火

Effect of Heat Treatment Process on Evaluation of Banded Structure of Low Carbon Gear Steel

Wang Huaiwei, Zhao Zhihai, Hou Zongbao, Xia Bing,
Wang Yanyang, Wang Liangjun

(Testing Center of Maanshan Iron & Steel Co., Ltd., Ma'anshan 243000, China)

Abstract: The microstructure of 18CrNiMo, 19CN5 and 20CrMnTiH3 low carbon gear steels treated by hot rolling, complete annealing and isothermal normal process and the band test rating of GB/T 34474.1—2017 and GB/T 13299—1991 were analyzed. The results show that the banded structure of low carbon gear steel is different under different heat treatment regimes. In the hot-rolled state, 18CrNiMo, 19CN5 and 20CrMnTiH3 contain more bainite, and it is difficult to evaluate the banded microstructure using GB/T 34474.1—2017 and GB/T 13299—1991. After isothermal normalization treatment, 20CrMnTiH3 is an equilibrium tissue, and the banded tissue evaluation can be completed with GB/T 34474.1—2017 and GB/T 13299—1991. A small amount of martensite exists in 18CrNiMo and 19CN5. It is not consistent with the equilibrium microstructure required by GB/T 13299—1991, but there is less martensite, which can be evaluated under GB/T 34474.1—2017 standard. After complete annealing treatment, the three tissues belong to the equilibrium state, and GB/T 34474.1—2017 and GB/T 13299—1991 can be used to complete the ribbon tissue evaluation. There is obvious difference in strip grade between fully annealed steel and isothermal normalizing steel.

Key words: low-carbon gear steel; banded structure; complete annealing; isothermal normalizing

三种彩涂板涂层厚度的测量方法适用性分析

汪良俊，陈　斌，后宗保，王晓敏，王怀伟，夏　冰

（马鞍山钢铁股份有限公司检测中心，安徽 马鞍山　243000）

摘　要： 钻孔破坏式显微镜法、磁性测厚仪法、磁性涡流测厚仪法均属于涂层厚度的测量方法。本文针对三种方法在彩涂板涂层厚度测量的适用性开展试验分析。结果表明，三种方法有不同的特点，根据本次试验分析内容，可为方法的选取提供参考依据。

关键词： 彩涂板；涂层厚度；测量方法

Applicability Analysis of Three Measurement Methods for Coating Thickness of Color Coated Sheet

Wang Liangjun, Chen Bin, Hou Zongbao, Wang Xiaomin, Wang Huaiwei, Xia Bing

(Physics Station of Testing Center of Ma'anshan Iron and Steel Co., Ltd., Ma'anshan 243000, China)

Abstract: Borehole destructive microscope method, magnetic thickness gauge method and magnetic eddy current thickness gauge method all belong to the measurement methods of coating thickness. In this paper, the applicability of three methods in color coated plate coating thickness measurement is tested and analyzed. The results show that the three methods have different characteristics. According to the analysis content of this test, it can provide a reference basis for the selection of methods.

Key words: color coated plate; coating thickness; measuring method

条钢初轧孔型轧辊槽底裂纹快速定量评价

张国星[1]，夏杨青[2]，陈 雄[3]

（1. 宝武集团中央研究院智能制造所 上海 201900；2. 宝武集团宝山钢管条钢事业部条钢部，上海 201900；3. 复旦大学，上海 200434）

摘 要： 本文介绍了一种针对孔型轧辊槽底裂纹深度、能够实现快速定量探测的技术原理，及其设备，该技术在条钢厂初轧轧辊车间的使用，可开展收集全部轧辊及其全生命周期内开裂深度信息的工作，形成数据库和工艺分析基础，助力磨辊智慧制造。该技术采用对称映射声场测深，精确控制超声波束指向性及其在远场开裂面上的分布宽度、轴心能量分布，根据接收声能量值和槽底固定高度探测面的开裂深度定量模型，实时获取开裂深度值，可在 30s 内完成轧辊全周身开裂深度测定，目前已用于消除在线断辊事故和轧辊优化车削。

关键词： 孔型轧辊；裂纹深度；定量评价；超声波；设备

Rapid Quantitative Evaluation of Bottom Cracking Depth of Shape-steel Roll in Blooming Factory

Zhang Guoxing[1], Xia Yangqing[2], Chen Xiong[3]

(1. Intelligent Manufacturing Institute, R & D Center of Baowu Steel Group, Shanghai 201900, China; 2. Blooming Factory of Baoshan Tube&Bar Division, Baowu Steel Group, Shanghai 201900, China; 3. Fudan University, Shanghai 200434, China)

Abstract: This paper introduces a technical principle and equipment which can detect the cracking depth at the bottom of shape-steel roll quickly and quantitatively. The application of this technology in the roll-maintaining workshop of bar type steel mill can collect the cracking depth information of all rolls, and their whole life cycle, thus to form a database, to create an analytic basis, and to help the intelligent maintaining of grinding rolls. This technology utilizes symmetric mapping sound field to detect crack depth, which can control the directivity of ultrasonic beam, its distribution width and axial energy distribution on the far-field of cracking surface accurately; obtain the value of cracking depth in real time according to the value of received acoustic energy and the cracking depth quantitative model of the fixed height detection surface at the bottom of the groove, and can complete the determination of the cracking depth of the whole roll within 30 seconds. At present, this technique is used to eliminate the online broken roll accident and optimize the maintaining of the roll.

Key words: shape-steel roll; cracking depth; quantitative evaluation; ultrasound; equipment

锤头形状对板坯定宽过程的影响研究

陈 兵，李士庆，韩烬阳

（北京科技大学机械工程学院，北京 100083）

摘 要：本文针对某钢企板坯经定宽压力机不同锤头定宽后出现边部质量问题，为实现锤头的选型及降低边部缺陷可能性之目的，利用有限元软件 Deform-3D 建立了板坯定宽的有限元模拟分析模型，并将仿真结果与现场侧压试验的板坯“狗骨”变化结果进行对比验证了模型的准确性。进而研究了倒角锤头不同侧压量、板坯初始宽度和锤头形状对板坯定宽过程侧压力、位移场及应力应变场的影响。结果表明：侧压力与板坯“狗骨”最大高度随侧压量增大而增大，随板坯初始宽度增大而减小，孔型锤头最大侧压力、板坯“狗骨”最大高度及应力峰值均大于倒角锤头。因此，倒角锤头相对孔型锤头效果更好，更有利于板坯变形均匀性，为定宽压力机锤头形状的选择提供了理论指导。

关键词：定宽压力机；Deform-3D；有限元模拟；锤头形状；板坯变形

Study on the Influence of Hammer Shape on Width Fixing Process of Slab

Chen Bing, Li Shiqing, Han Jinyang

(School of Mechanical Engineering, University of Science and Technology Beijing, Beijing 100083, China)

Abstract: In this paper, the quality of the side of a slab in a steel enterprise appears after the width of different hammers is fixed by a constant width press, in order to realize the selection of hammers and reduce the possibility of edge defects, the finite element simulation analysis model of slab width determination was established by using the finite element software Deform-3D, and the simulation results were compared with the “dog bone” changes of slab in the field lateral side pressing test, which verified the accuracy of the model. Furthermore, the effects of different side pressure of chamfering hammer, initial slab width and hammer shape on the side pressure, displacement field and stress-strain field during slab sizing process were studied. The results show that the side pressure and the maximum height of “dog bone” of slab increase with the increase of side pressure, and decrease with the increase of initial width of slab. The maximum side pressure, the maximum height of “dog bone” of slab and the peak stress of the pass hammer are greater than those of chamfering hammer. Therefore, the effect of chamfering hammer is better than that of pass hammer, which provides theoretical guidance for the selection of hammer head shape of slab sizing press.

Key words: slab sizing press; Deform-3D; finite element simulation; slab deformation; hammer head shape

X 荧光光谱仪故障分析与处理

修成博

（首钢长治有限公司质量监督站，山西长治　046031）

摘　要：X 荧光光谱仪已被广泛应用于冶金分析和即时测定领域，对缩短分析周期，提升对生产的指导性作出了巨大的贡献。我公司使用日本理学的 Simultix14 型 X 荧光光谱仪分析生铁、矿渣类样品，在操作使用中，往往会发生仪器故障，对分析工作造成妨碍。下面介绍一些常见的故障及分析处理方法，以供参考。

关键词：X 荧光光谱仪；Simultix14；故障；分析处理

Fault Analysis and Treatment of X Fluorescence Spectrometer

Xiu Chengbo

(The Quality Supervision Station of Shougang Changzhi Co., Ltd., Changzhi 046031, China)

Abstract: X fluorescence spectrometer has been widely used in the field of metallurgical analysis and timely determination, It has made great contributions to shortening the analysis cycle and promoting the guidance of production. Our company is using the Japanese RIGAKU X fluorescence spectrometer to analyze cast iron and slag samples, instrument failures often occur in operation, and hinder analysis. Here are some common faults and analysis processing for feference.

Key words: X fluorescence spectrometer; Simultix14; breakdown; analysis processing

水平连续加料电弧炉废钢预热效率的研究

何志禹，潘　涛，王　宇，姜周华，李花兵，倪卓文，朱红春

（东北大学冶金学院，辽宁沈阳　110819）

摘　要：水平连续加料电弧炉具有出钢时间短、生产能耗低、余热回收等优点，近年来得到了迅速发展。然而水平连续加料电炉中，废钢预热温度低一直是技术难题。本研究建立了水平连续加电弧炉废钢预热模型，研究了烟气与废钢之间的传热行为。由于废钢的阻力，气体向上流动，降低了烟气与废钢间的传热效率。烟气运动也导致了废钢上部的预热温度高于底部的预热温度。在水平连续加料系统中，烟气初始速度和废钢孔隙率均能提高废钢预热效果。为提高废钢预热效果，设计了 4 种加料方式，结果表明，方式 4 的废钢整体预热效果最好。最佳的加料方法是选择 3 种不同孔隙率的废钢，孔隙率最大的排在废钢层的顶部，孔隙率最小的排在中间，孔隙率介于中间的布置在底部区域。

关键词：水平连续加料电弧炉；废钢预热；加料方式；数值模拟

Research on the Scrap Preheating Efficiency in Horizontal Continuous Feeding Electric Arc Furnace

He Zhiyu, Pan Tao, Wang Yu, Jiang Zhouhua, Li Huabing, Ni Zhuowen, Zhu Hongchun

(School of Metallurgy, Northeastern University, Shenyang, 110819, China)

Abstract: Horizontal continuous feeding electric arc furnace (EAF) has the advantages of short tap-to-tap time, low production energy consumption, and waste heat recovery, which makes it develop rapidly in the field of EAF steelmaking in recent years. However, low scrap preheating temperature has always been a technical problem in horizontal continuous feeding EAF. This study established the scrap preheating model for the horizontal continuous feeding EAF to study the heat transfer behavior between gas and scrap. The calculation results showed that the gas in the continuous feeding system flowed upward due to the resistance of the scrap, which resulted in poor heat transfer between gas and scrap. The movement of gas also caused the preheating temperature above the scrap layer to be higher than that at the bottom. The initial gas velocity and scrap porosity could both improve the preheating effect of scrap in horizontal continuous feeding system. Four feeding methods were designed for improving scrap preheating effect, and the research results showed that the scrap overall preheating effect of Method 4 was the best. The best scrap feeding method was to select three kinds of scrap with different porosity, with the largest porosity arranged at the top of the scrap layer, the smallest porosity arranged at the middle position, and another kind arranged at the bottom area.

Key words: horizontal continuous feeding electric arc furnace; scrap preheating; feeding method; simulation

ICP-AES 测定铝铁中铝、磷、硅、锰元素

孙　倩

（首钢长治钢铁有限公司质量监督站，山西长治　046000）

摘　要：铝铁合金在炼钢生产中是一种新型脱氧剂，主要成分有铝、铁、磷、锰、硅等元素，本文讨论了用电感耦合等离子体发射光谱仪（ICP-AES）测定铝铁合金中铝、硅、磷、锰的分析条件并建立了测定方法。铝铁合金经稀王水分解，以 ICP-AES 法测定溶液中的铝、硅、磷、锰四个元素。通过选择没有干扰的谱线作为被测元素的分析线，背景干扰采用仪器自动背景校正方法扣除。我化验室通过此方法对铝铁试样进行分析检测，分析结果准确，同时此方法可以多元素同时检测，大大缩短了分析周期，提高了工作效率。

关键词：电感耦合等离子体发射光谱仪；铝铁合金；铝；硅；磷；锰

Abstract: Al-Fe alloy is a new type of deoxidizer. Its main components are aluminum, iron, phosphorus, manganese, silicon and other elements. This paper discusses the analytical conditions for the determination of aluminum, silicon, phosphorus and manganese in Al-Fe alloy by ICP- AES and establishes the determination method. The aluminum-iron alloy was decomposed by hydrochloric acid and nitric acid, and the four elements of aluminum, silicon, phosphorus and manganese in the solution were determined by ICP-AES. By selecting the spectral line without interference as the analysis line of the measured element, the background interference is deducted by the automatic background correction method of the instrument. This method is used in our laboratory to analyze and detect aluminum and iron samples, and the analysis results are accurate. At the same time, this method can detect multiple elements simultaneously, greatly shortening the analysis period and improving the work efficiency.

Key words: inductively coupled plasma emission spectrometer; Al-Fe Alloy; Al; Si; P; Mn

大型钢铁企业 U 型辐射管烧嘴的自主研究及应用

何 可

（宝钢股份武汉钢铁有限公司设备管理部，湖北武汉 430081）

摘 要：U 型辐射管式燃烧系统可低热值燃烧、利用废气对空气进行预热，节约能源，相对减少 NO_x 排放量；本文介绍了我国自主设计制造的超低 NO_x F-L 形辐射管鼓入式烧嘴的基本结构和工作原理；对该型烧嘴在加热炉使用中存在加热能力不足的问题进行了详细分析，结合实际提出了多种有效措施以缩短火焰长度，实现对钢坯进行间接加热进行热处理，提高了产品质量,并取得了令人满意的效果。为加快推进钢铁绿色低碳高质量发展做出了贡献。

关键词：辐射管烧嘴；加热能力；火焰长度；碳达峰

Independent Research and Application of U-shaped Radiant Tube Burner in Large Iron and Steel Enterprises

He Ke

(Equipment Mange Department of Wuhan Iron and Steel Co., Ltd., Baosteel Group, Wuhan 430081, China)

Abstract: The basic structure and working principle of the ultra-low NO_x type F-L shape radiant burner push type tube made of Steel enterprises are introduced. The problems of insufficient heating ability in heating furnace is analyzed. Some effective measures are put forward to shorten the flame length with the actual, and get a good performance.

Key words: radiant tube burner; heating ability; flame length; carbon peak

粗苯中三苯含量气相色谱法测量试验条件的研究

易小琴

（重庆钢铁制造管理部（技术中心）检测中心，重庆 401220）

摘 要：使用气相色谱法检测粗苯中三苯含量时，通过采用分流比不同来探讨分析条件对面积归一法测定粗苯中三苯含量结果的影响，确定出最佳分流比，来快速准确的分析出粗苯中的三苯含量，保证产品的质量，提高粗苯的回收利用率。

关键词：粗苯；三苯；分流比；面积归一法

Study on Experimental Conditions for Measuring Toluene Content in Crude Benzene by Gas Chromatography

Yi Xiaoqin
(Testing center of Chongqing Iron and steel manufacturing management department
(Technical Center), Chongqing 401220, China)

Abstract: when using gas chromatography to detect the content of triphenyl in crude benzene, through the use of different split ratio to explore the influence of analysis conditions on the results of determining the content of triphenyl in crude benzene by area normalization method, determine the best split ratio, to quickly and accurately analyze the content of triphenyl in crude benzene, ensure the quality of products, and improve the recovery and utilization rate of crude benzene.

Key words: crude benzene; triphenyl; split ratio; area normalization method

"操检合一"在首钢股份炼钢精整区域的探索和实践

成建峰，韩士洋，张小辉，刘　勇，安兰松，双占博
（北京首钢股份有限公司炼钢作业部，河北迁安　064404）

摘　要：本文介绍了"操检合一"在首钢股份有限公司炼钢作业部精整区域推行的背景和条件。重点阐述了精整区域基于"全员生产保全 TPM"管理理念和"点检定修"设备管理体制的基础上所探索和实践的"操检合一"创新管理模式的定位和定义，推行的具体做法，建立的管理体系及取得的效果（设备功能优恢复和优化，产品质量提升，管理效率提效）。

关键词：操检合一；管理创新；探索实践；扁平化；管理体系；推行方法；一岗多能

The Exploration and Practice of "Operation and Maintenance Combination" in the Slab Finishing Area of Shougang Co., Ltd.

Cheng Jianfeng, Han Shiyang, Zhang Xiaohui,
Liu Yong, An Lansong, Shuang Zhanbo
(Steelmaking Department of Beijing Shougang Co., Ltd., Qian'an 064404, China)

Abstract: This paper introduces the background and conditions of "operation and maintenance combination" management mode that implemented in the Slab Finishing area of the Steelmaking Department of Shougang Co., Ltd. It focuses on the positioning and definition of the innovative management mode of "operation and maintenance combination", which is explored and practiced in the Slab Finishing area that based on the management concept of "total productive maintenance TPM" and the equipment management system of "scheduled maintenance system". It also states the "operation and maintenance combination" about the specific methods of how to implement, and the management system of how to build, and the good results of how to achieve (equipment function recovery and optimization, accuracy increase、product quality improvement, management efficiency raise).

Key words: operation and maintenance combination; management of Innovation; exploration and practice; flat management; management system; implementation method; one major with multi-abilities

轧钢厂计划检修模式的优化研究与探索

任乐乐，冯　飞，于　杨，邵忠文，刘井泉，李罗扣

（首钢长治钢铁有限公司轧钢厂，山西长治　046000）

摘　要：为保证公司 2021 年下达的各项生产经营任务顺利完成，轧钢厂各产线日产量提升。产量任务的增加导致设备定修周期被延长、检修时间被压缩，工序成本指标的提高对设备运行的稳定性提出了更高的要求。因此，为保证设备稳定运行，强化短周期的设备点检、保证检修计划的准确性和检修质量，优化检修模式，势在必行。

关键词：日修；定修；点检；全员；培训

Research and Exploration on Optimization of Maintenance Mode in Rolling Mill

Ren Lele, Feng Fei, Yu Yang, Shao Zhongwen, Liu Jingquan, Li Luokou

(Steel Rolling Plant of Shougang Changzhi Iron and Steel Co., Ltd., Changzhi 046000, China)

Abstract: In order to ensure the successful completion of the production and operation tasks assigned by the company in 2021, the daily output of each production line in the rolling mill will be increased. The increase of production tasks leads to the extension of equipment regular maintenance cycle and the compression of maintenance time. The improvement of process cost index puts forward higher requirements for the stability of equipment operation. Therefore, in order to ensure the stable operation of the equipment, strengthen the short period of equipment point inspection, ensure the accuracy and quality of the maintenance plan, and optimize the maintenance mode, it is imperative.

Key words: daily maintenance; regular maintenance; spot check; all staff; train

高压水除鳞系统多线除鳞改造实践

于　杨，詹卫金，胡　洪

（首钢长治钢铁公司轧钢厂，山西长治　046031）

摘　要：轧钢厂产线布局为棒线等 3 条生产线并排设计，原有高线生产线及高棒生产线共用 1 套高压水除鳞系统，在实际生产中，只能实现单线除鳞，且压力不稳定。随着市场对成品表面质量的关注，需对棒材生产线、高线生产线、高棒生产线均完善除鳞设施，改善除鳞效果，通过改造，实现了多线除鳞的生产需要，同时改善了除鳞效果。

关键词：棒线；线材；高压水除鳞

Exploration and Practice of High-pressure Water Descaling System

Yu Yang, Zhan Weijin, Hu Hong
(Shougang Changzhi Iron & Steel Company, Changzhi 046031, China)

Abstract: The layout is a side-by-side design for three wire and bar production lines in the roll mill plant. It originally shares a high pressure water descaling system for the high speed wire and the high speed bar production line. In actual production, only single-line descaling can be achieved, and the pressure is unstable. As the market pays more attention to the surface quality of finished products, it is necessary to improve the descaling facilities for the descaling effect in the bar, high speed wire and high speed bar production line. Through the reform, it realizes multiline descaling for production requirement, and the descaling effect is improved at the same time.

Key words: wire and rod; wire rod; high-pressure descaling

现代带钢厂内部钢卷运输的现状及发展趋势

韦富强[1,2,3]，张　建[1,2,3]，杨建立[1,2]，郑江涛[1,2]，
王　超[1,2]，孟祥军[1,2]

（1. 北京首钢国际工程技术有限公司云翔装备分公司，北京　100043；
2. 北京市冶金三维仿真设计工程技术研究中心，北京　100043；
3. 北京首钢云翔工业科技有限责任公司，北京　100043）

摘　要：本文概述了近 20 年来带钢厂钢卷运输技术的演变，分析了常用的钢卷运输方式及其技术特点，介绍了超级电容供电的钢卷运输车，对比了该技术方案和传统的托盘式运输技术方案，简述了新型技术的工程案例及实际应用效果，得出了轨道式运输是带钢厂内部钢卷运输发展趋势等结论。

关键词：智能化；超级电容；重载；钢卷运输；轨道式；界面技术

Current Situation and Development Trend of Coil Transportation in Modern Strip Mill Plant

Wei Fuqiang[1,2,3], Zhang Jian[1,2,3], Yang Jianli[1,2], Zheng Jiangtao[1,2],
Wang Chao[1,2], Meng Xiangjun[1,2]

(1. Beijing Shougang International Engineering Technology Co., Ltd., Beijing 100043, China;
2. Beijing Metallurgical 3-D Simulation Design Engineering Technology Research Center, Beijing 100043, China; 3. Beijing Shougang Yunxiang Industrial Technology Co., Ltd., Beijing 100043, China)

Abstract: This paper summarizes the evolution of steel coil transportation technology in strip mills in recent 20 years, analyzes the common steel coil transportation modes and their technical characteristics, introduces the steel coil transportation vehicle powered by super capacitor, compares this technical scheme with the traditional pallet type technology, and briefly describes the engineering case and practical application effect of the new technology, It is concluded that rail transportation is the development trend of steel coil transportation in steel plant.

Key words: intelligent; super capacitor; heavy load; steel coil transportation; rail type; interface technology

影响 TAYLOR 窄搭接电阻焊机焊接稳定性的设备因素

李建文，尹晓菲，石若玉

（武汉钢铁有限公司硅钢部，湖北武汉 430083）

摘 要：硅钢部技术人员在多年应用美国 WINFELD TAYLOR 公司窄搭接电阻焊机过程中，逐步根据该焊机焊接工艺原理、机械设备结构、焊机自动化控制、焊机气动系统等设备运行特性，找出焊机设备中影响运行稳定性和焊接焊缝可靠性降低的设备因素，为生产操作人员正确操作和维护焊机，设备人员根据焊机发生故障问题快速排故指出明确的方向，以减少同类焊机焊接重焊率和断带事故发生。

关键词：影响；TAYLOR 电阻焊机；焊接稳定性；设备因素

Analysis of Equipment Factors Affecting Welding Stability of Taylor Narrow Lap Resistance Welding Machine

Li Jianwen, Yin Xiaofei, Shi Ruoyu

(Silicon Steel Business Division of WISCO, Wuhan 430083, China)

Abstract: Inthe process of using winfeld Taylor narrow lap resistance welding machine for many years, the technicians of silicon steel department gradually find out the equipment factors that affect the operation stability and reduce the reliability of welding seam according to the welding process principle, mechanical equipment structure, automatic control and pneumatic system of the welding machine, In order to correctly operate and maintain the welding machine, the equipment personnel can quickly solve the problem according to the welding machine failure, and point out a clear direction, so as to reduce the re welding rate and the broken belt accident of the same kind of welding machine.

Key words: influence; Taylor resistance welding machine; welding stability; equipment factors

检修过程管理信息化实践

陈建平[1]，乔 栋[1]，桂小龙[2]，李 红[3]

（1. 武汉钢铁有限公司设备管理部，湖北武汉 430083；2. 宝信软件武汉有限公司，湖北武汉 430083；3. 宝山钢铁股份有限公司中央研究院炼铁所，湖北武汉 430083）

摘 要：针对传统设备管理系统中存在纸质资料不易留存、数据统计分析困难、人为突击补填记录、检修人员细化不到位等问题，介绍了一种检修过程管理信息化的实现方法，通过开发移动检修工业 APP，利用手机相机扫码功能，实现检修人员与检修委托单快速绑定，最终实现项目检修过程管理，为检修作业标准化和流程固化提供一种智慧管理工具。

关键词：检修过程管理；信息化；智慧管理；工业 APP

Practice of Informatization in Maintenance Process Management

Chen Jianping[1], Qiao Dong[1], Gui Xiaolong[2], Li Hong[3]

(1. Equipment Management Department of Wuhan Iron & Steel Co., Ltd., Wuhan 430083, China; 2. Wuhan Baosight Software Co., Ltd., Wuhan 430083, China; 3. Institute of Ironmaking Technology, Central Research Institute, Baoshan Iron & Steel Co., Ltd., Wuhan 430083, China)

Abstract: In view of the problems existing in traditional equipment management system, such as paper data is not easy to be retained, data statistical analysis is difficult, artificial shock filling records, maintenance personnel refinement is not in place and so on, this paper introduces a realization method of maintenance process management informatization. Through the development of mobile maintenance industry app, using mobile phone camera code scanning function, maintenance personnel and maintenance order can be quickly bound, Finally, the project maintenance process management is realized, which provides an intelligent management tool for maintenance standardization and process solidification.

Key words: maintenance process management; promotion of information technology; smart management; industrial app

高效高炉铸铁冷却壁开发与应用

胡　伟[1,2]，王智政[1,2]，梅丛华[1,2]

（1. 北京首钢国际工程技术有限公司，北京　100043；
2. 北京市冶金三维仿真设计工程技术研究中心，北京　100043）

摘　要：随着高炉喷煤比、球团比例、富氧率提高，炉腹、炉腰和炉身下部冷却壁的寿命成为高炉长寿的限制性环节。本文基于某高炉大修改造工程，分析上代炉役较短的原因是原设计冷却壁交接处凸入炉内和冷却壁过长，并开发了高效高炉铸铁冷却壁。开发过程中应用多种设计软件，保证了设计过程中的精准性，为将来类似工程提供经验；高炉投产后冷却壁壁体温度、炉体热流强度和热负荷低于设计值，说明高效高炉铸铁冷却壁技术应用良好，值得进一步推广运用。

关键词：高炉；冷却壁；大修；热负荷

Development and Application of High Efficiency Blast Furnace Cast Iron Stave

Hu Wei[1,2], Wang Zhizheng[1,2], Mei Conghua[1,2]

(1. Beijing Shougang International Engineering Technology Co., Ltd., Beijing 100043, China; 2. Beijing Metallurgical Three-Dimensional Simulation Design Engineering Technology Research Center, Beijing 100043, China)

Abstract: With the increase of coal injection ratio, pellet ratio and oxygen enrichment rate of blast furnace, the service life of bosh, belly and lower cooling stave of blast furnace has become the limiting link of blast furnace longevity. Based on the overhaul and reconstruction project of a blast furnace, this paper analyzes that the reason for the short service of the previous generation is that the junction of the original design cooling stave protrudes into the furnace and the cooling stave is too long, and develops a high-efficiency blast furnace cast iron cooling stave. A variety of design software are applied in the development process to ensure the accuracy of the design process and provide experience for similar projects in the future; After the blast furnace is put into operation, the stave temperature, furnace heat flow intensity and heat load are lower than the design value, indicating that the high-efficiency blast furnace cast iron stave technology is well applied and worthy of further popularization and application.

Key words: blast furnace; cyclone dust collectors; blast furnace overhaul; thermal load

多通道凸度仪在2250mm热连轧产线的应用研究

张 强[1]，陈 刚[2]，李洪波[1]

（1. 北京科技大学机械工程学院，北京 100083；
2. 宝钢股份武汉钢铁有限公司，湖北武汉 430083）

摘 要：介绍了武钢2250mm热连轧生产线精轧机组F7机架出口设置的检测带钢凸度指标的多通道凸度仪IMS的测量原理，并根据现场实际生产带钢的板廓特点，分析了凸度仪工作过程中其测量间隔、多项式拟合等因素对不同带钢板廓特征的适应性问题，讨论了由于凸度仪在线凸度值与实际离线凸度值的偏差对生产造成的影响，有利于现场根据凸度仪影响因素的分析研究，提出适用于现场实际生产情况的带钢凸度调控策略。

关键词：热轧带钢；多通道凸度仪；板廓；凸度

Application Study of Multi-channel Convexity Meter in 2250mm Hot Strip Rolling Line

Zhang Qiang[1], Chen Gang[2], Li Hongbo[1]

(1. School of Mechanical Engineering, University of Science and Technology Beijing, Beijing 100083, China; 2. Baosteel Wuhan Iron & Steel Co., Ltd., Wuhan 430083, China)

Abstract: The measurement principle of the multi-channel convexity meter IMS installed at the exit of the F7 stand of the finishing mill in the 2250mm hot strip mill production line of WISCO is briefly introduced, and the adaptability of the measurement interval, polynomial fitting and other factors of the convexity meter to different strip profile characteristics during its operation is analyzed according to the characteristics of the actual strip profile produced on site. The impact on production due to the deviation of convexity values between the convexity meter online and the actual convexity values offline is discussed, which is beneficial to the analysis and research of the factors influencing the convexity meter in the field, and the strip convexity control strategy applicable to the actual production situation in the field is proposed.

Key words: hot rolled steel strip; multi-channel convexity meter; strip profile; crown

SCE500发电机组汽轮机叶片断裂失效机理分析

廖礼宝[1]，蔡 青[2]，龚 明[2]

（1. 宝钢股份设备部，上海 201900；2. 宝钢股份能环部，上海 201900）

摘 要：某SCE500发电机组汽轮机运行中末级13个叶片断裂。对叶片断口分别进行宏观和微观分析，发现所有叶片断口特征均相似。断口宏观分析均有金属光泽、较为新鲜，各区域均无氧化腐蚀特征。叶片均断裂于叶顶侧拉筋根部处，两侧表层位置可见台阶状冲击韧断特征。断口微观 SEM 分析均发现局部可见条带台阶特征及冲击韧断的典型形貌特征，进汽侧和出汽侧断口微观特征也均有典型的韧窝形貌。对 13 个断裂叶片试样的金相组织、非金属夹杂物、化学成分、布氏硬度、冲击韧性和室温拉伸性能等各项指标分别进行检测，结果所有数据均满足 0Cr17Ni4Cu4Nb（SUS630 钢）的要求。综上所述，可排除材质及加工质量、运行工况、冲蚀和腐蚀等因素导致的断裂。断裂性质为典型的大应力作用下的双向弯曲低周高应冲击断口。

关键词： SCE500 发电机组；汽轮机；叶片断裂；失效机理；分析

Failure Mechanism Analysis on Blade Fracture of SCE500 Generator-set Related Steam Turbine

Liao Libao[1], Cai Qing[2], Gong Ming[2]

(1. Equipment Department, Baoshan Iron & Steel Co., Ltd., Shanghai 201900, China;

2. Energy and Environment Department, Baoshan Iron & Steel Co., Ltd., Shanghai 201900, China)

Abstract: Once 13pieces of steam turbine blades were fractured in production in SCE500 generator-set.This article is aimed at systematical analysis about impact cracks and abnormal steam turbine blades fractures formed by all possible factors, and from the perspective of materials and equipment technology, failure research and analysis of this type of steam turbine blades are carried out .Based on fracture of steam turbine blades, the failure mechanism was analyzed, from the micro and macro morphology, chemical composition, hardness gradient distribution, mechanical properties, metallurgical structure, high temperature oxidation and scanning electron microscope (SEM) and so on. The fracture properties may be a impact fracture pattern under heavy stress with the typical characteristics of “low cycle high stress”.

Key words: SCE500 generator set; steam turbine; blade fracture; failure mechanism; analysis

薄板热连轧中板廓局部高点的形成演变和控制策略

郑旭涛，喻　尧，季伟斌，王　骏，陈传敬

（日照钢铁控股集团有限公司，山东日照　276800）

摘　要： 板带热连轧生产中，当厚度规格较薄时（≤2.0mm），板廓局部高点出现概率显著增加，从而在卷取过程中因局部增厚的叠加效果呈现出起鼓/起筋现象。针对该问题，建立了轧辊轧件局部变形模型和缺陷遗传模型。通过模型分析发现：针对某一机架分析，从前到后，单机架的轧件局部高点影响系数逐渐减小，轧辊局部磨损影响系数逐渐增加；且带钢变形抗力越大，上述趋势越明显。而针对连轧过程，后段机架的遗传效应要显著大于前段机架；但随着带钢变形抗力增加，遗传系数权重会向前移动。基于该遗传演变规律，给出了薄板热连轧过程中板廓局部高点的控制策略，并在日钢薄板坯无头轧制线上取得良好效果。

关键词： 板带热连轧，薄板，局部高点，轧辊磨损，遗传效应

Formation Evolution and Control Strategy of Local High Point in Thin Plate Hot Rolling

Zheng Xutao, Yu Yao, Ji Weibin, Wang Jun, Chen Chuanjing

(Rizhao Steel Holding Group Co., Ltd., Rizhao 276800, China)

Abstract: In hot strip rolling production, when the thickness specification is thin (≤2.0 mm), the probability of local high point of plate profile increases significantly, so the superposition effect of local thickening in coiling process presents the phenomenon of bulging / reinforcement, which seriously affects the quality of products. To solve this problem, the local deformation model and defect heredity model of roll rolling parts are established. Through the model analysis, it is found that the influence coefficient of the local high point of the single stand decreases gradually, the influence coefficient of the roll local wear increases gradually, and the greater the deformation resistance of the strip steel, the more obvious the above trend is. For the continuous rolling process, the hereditary effect of the downstream stand is significantly larger than that of the upstream stand, but with the increase of the deformation resistance of the strip, the weight of the hereditary coefficient will move forward. Based on the evolution law, the control strategy of local high point of plate profile during hot rolling is given.
Key words: hot strip rolling; thin plate; local high point; work roll wear; hereditary effect

薄板坯无头轧制生产线轧机工作辊热行为研究

喻　尧，郑旭涛，王　骏，杜建全

（日照钢铁控股集团有限公司，山东日照　276806）

摘　要： 薄板坯连铸连轧线，在无头生产模式中，轧辊处于持续与轧辊接触状态，不存在轧制间隙冷却过程，从而造成轧辊的温度场和热膨胀将与其他热连轧生产线块轧模式下存在显著差异。由此而带来的板形控制策略将需要特殊设定，才能保证批量薄规格生产时的板形质量稳定。本文通过实际跟踪测量轧辊下线温度场、热膨胀量，并通过MATLAB 程序构建了轧辊热行为模型，对无头轧制过程中轧辊的温度场变化、热膨胀变化进行了仿真研究。分析结果表明，轧辊温度场和热膨胀达到稳定的时间和程度与常规块轧模式存在显著不同，热凸度变化趋势的特殊性进而决定了轧辊原始辊形、负荷、弯辊等的设定变化趋势。并基于此对轧辊原始辊形、冷却策略、轧制计划和其他工艺参数设定进行了优化，为热轧薄板生产板形控制提供了有效的指导。

关键词： 薄板坯连铸连轧；无头模式；工作辊；热行为

Analysis of Work Roll Thermal Behavior in Endless Mode for Thin Slab Casting and Rolling Line

Yu Yao, Zheng Xutao, Wang Jun, Du Jianquan

Abstract: In endless mode of thin slab casting and rolling line, the work roll is keeping in contact with the rolling material, and there is no pure cooling process. For that reason, the roll temperature field and thermal expansion will exist significant difference with batch mode in other strip production line. Therefore, the strip profile and flatness control strategy will need special setting to ensure the stability of the flatness quality during the production of thin strips. Based on the actual roll temperature field and thermal expansion offline tracking measurement, the roll thermal behavior model is constructed and optimized by MATLAB. Using this model, roll thermal behavior has carried simulation research in the process of endless rolling. Analysis results shows that the interval of reaching stability state of roll temperature field and thermal expansion has significant different with the normal batch rolling model，and the characteristics of the changing trend of hot roll expansion have determined the changing trend of the roll original shape, load distribution, work roll bending force and so on. Based on the analysis results, the roll original shape, roll cooling strategy, rolling schedule and other process parameters were optimized , and the strip profile and flatness control was improved effectively.

Key words: thin slab casting and rolling; endless mode; work roll; thermal behavior

300t RH 钢液脱碳模型及工艺研究

付有彭，赵占山，郑旭涛，任　涛，朱韶哲，赵梓云

（日照钢铁控股集团有限公司，山东日照　276800）

摘　要： 为提高生产效率，降低生产成本。基于碳氧平衡，根据冶金动力学及热力学理论，建立了 RH 真空处理生产低碳钢时钢水脱碳数学模型。并通过现场生产数据验证模型的可靠性。利用模型分析 RH 脱碳速率的影响因素，认为将初始碳控制在 300~400ppm，初始[O]/[C]控制在 1.4~1.5，初始温度控制在 1575~1595℃，处理 2~3min 时提高提升气体的流量，5min 后降低提升气体流量，吹氧强制脱碳时，吹氧开始时机应在 2~3min 既可以提高生产效率又可以降低生产成本。

关键词： RH 精炼；脱碳模型；低碳钢；强制脱碳；生产效率；终点控制

Study on Decarburization Model and Process of 300t RH Molten Steel

Fu Youpeng, Zhao Zhanshan, Zheng Xutao, Ren Tao, Zhu Shaozhe, Zhao Ziyun

(Rizhao Steel Holding Group Co., Ltd., Rizhao 276800, China)

Abstract: In order to improve production efficiency and reduce production costs. Based on the balance of carbon and oxygen,according to the theory of metallurgical kinetics and thermodynamics,a mathematical model of decarburization of molten steel during RH vacuum treatment of low carbon steel was established.The reliability of the model is verified by field production data.The model was used to analyze the influencing factors of RH decarburization rate.It was considered that when the initial carbon was controlled at 300-400ppm,the initial [O]/[C] was controlled at 1.4-1.5,the initial temperature was controlled at 1575-1595℃,the flow rate of lifting gas was increased after 2-3min treatment,and the flow rate of lifting gas was decreased after 5min treatment.When oxygen blowing was used for forced decarburization, the start time of oxygen blowing should be 2min-3min,which can improve the production efficiency and reduce the production cost Production cost.

Key words: RH refining; decarburization model; low carbon steel; forced decarbonization; production efficiency; end point control

11 冶金自动化与智能化

大会特邀报告

分会场特邀报告

矿业工程

炼铁与原燃料

炼钢与连铸

轧制与热处理

表面与涂镀

金属材料深加工

先进钢铁材料及其应用

粉末冶金

能源、环保与资源利用

冶金设备与工程技术

★ 冶金自动化与智能化

冶金物流运输

其他

高效、稳定的鞍钢热轧板坯库管理系统

高　松，车志良，蔡　博
（鞍钢股份有限公司信息化管理中心，辽宁鞍山　114000）

摘　要：1780 板坯库管理系统是热轧 MES 系统的重要组成部分，主要负责 1780 线板坯库的板坯入库管理、板坯库内管理、板坯出库管理、板坯辊道跟踪、板坯实绩数据收集，同时与公司 ERP 系统、二级过程控制系统、一级 PLC 系统进行通讯，收集和传递生产数据，保证各级系统信息的数据一致性。该系统建立在 Stratus 容错服务器平台上，相比传统服务器集群，具有更高的安全性和可靠性。该系统投入运行以来，充分发挥出板坯库管理的指导性作用，将鞍钢 1780 生产线的板坯库管理和 L1 基础自动化、L2 的过程控制、ERP 管理等功能紧密地连接在一起，大大地提高了板坯库生产管理的自动化水平，优化了生产线人力资源配备，对促进企业的信息化进程具有重要的理论和实际意义。
关键词：鞍钢；板坯库管理系统；MES；stratus

The Management System of Hot-rolling Slab-yard in Ansteel Corp. with High-efficiency and Stability

Gao Song, Che Zhiliang, Cai Bo
(Information Management Center, Ansteel Corp., Ltd., Anshan 114000, China)

Abstract: The management system of 1780 slab-yard is the key composition part of MES of Hot Strip Rolling Mill. It mainly takes charge of the management of slab in-yard, management of slab-yard, management of slab out-yard, tracing of slab roller, slab act data collection of 1780mm product line. Meanwhile, it communicates with company's ERP system and L2 process control system and L1 PLC system, to transfers and gathers various production data, keep the information data consistency at all levels system. This system is built on the platform of stratus fault-tolerant server, compared with traditional cluster server, it will takes us higher safety and reliability.Since the system was brought into operation, it takes guidance for the management of slab-yard,and realizes the high-class technology by firmly connecting the production planning management in 1780mm product line of Ansteel, L1 basic automation and L2 process control and the production management of ERP.The system greatly promotes the automation level of production.It has an important theoretical as well as actual meaning by accelerating the informatization process of Ansteel group.
Key words: Ansteel; management system of slab-yard; MES; stratus

基于大数据的冷轧深加工产品研发与质量管控一体化平台开发

宋志超[1]，武智猛[2]，何　方[1]，张才华[1]
（1. 河钢股份有限公司邯郸分公司技术中心，河北邯郸　056015，
2. 河钢集团河钢工业技术服务有限公司，河北石家庄　050023）

摘　要：目前，河钢邯钢已经建立起工业大数据云平台，具备从炼钢到冷轧工序全流程的关键过程数据采集与存储能力，但现有质量管理平台对海量的数据资源的利用不足。为了充分利用大数据解决产品研发，过程控制、成本管理和质量管理等方面的问题，开发了适用于研发人员和质量管理人员的一体化平台，实现对大数据的充分利用。

关键词：产品研发；性能预测；成本测算；过程质量控制；自动判级；大数据；人工智能；Python；TensorFlow；Django

Development of an Integrated Platform for Cold Rolling Product Development and Quality Control based on Big Data

Song Zhichao[1], Wu Zhimeng[2], He Fang[1], Zhang Caihua[1]

(1. HBIS Group Handan Iron & Steel, Handan 056015, China;
2. HBIS Group HBIS Industrial Tech., Shijiazhuang 050023, China)

Abstract: At present, HBIS Group Hansteel has established an industrial big data cloud platform, which has the ability to collect and store key process data from steelmaking to cold rolling process, but the existing quality management platform has insufficient utilization of massive data resources. In order to make full use of big data to solve the problems of product development, process control, cost management and quality management, an integrated platform for R&D personnel and quality management personnel is developed to make full use of big data.

Key words: product development; mechanical performance prediction; cost estimation; process quality control; automatic grading; big data; artificial intelligence; Python; TensorFlow; Django

1450 酸连轧机组卡罗塞尔卷取机倒卷技术的研究与应用

王宇鹏，颜廷洲，周晓琦，刘　磊，兰晓栋，范正军，王承刚

（首钢智新迁安电磁材料有限公司，河北迁安　064400）

摘　要：本文结合卡罗塞尔卷取机的结构特点，充分利用卡罗塞尔卷取机双芯轴的优势，突破传统的卷取模式，开发了 1450 酸连轧机组卡罗塞尔卷取机倒卷技术。通过优化卷取侧芯轴的反转速度，调整助卷侧芯轴的正转速度，匹配两个芯轴之间的速度和张力关系，保证钢卷重新卷取过程中张力稳定，完成抽芯钢卷的重新卷取。最终实现钢卷在卡罗塞尔卷取机双芯轴之间的切换。

关键词：卡罗塞尔卷取机；倒卷技术；速度张力匹配

Research and Application of Rewinding Technology of Carrousel Coiler for 1450 Cold Tandem Mill

Wang Yupeng, Yan Tingzhou, Zhou Xiaoqi, Liu Lei, Lan Xiaodong, Fan Zhengjun, Wang Chenggang

(Shougang Zhixin Qian'an Electromagnetic Material Co., Ltd., Qianan 064400, China)

Abstract: In this paper, combining with the structural characteristics of Carrousel coiler, making full use of the advantages of double mandrel of Carrousel coiler and breaking through the traditional coiling mode, the rewinding technology of Carrousel coiler in 1450 cold tandem mill is developed. By optimizing the reverse speed of the coiling side mandrel, adjusting the forward rotation speed of the mandrel at the auxiliary coiling side, and matching the speed and tension relationship between the two spindles, the stable tension in the rewinding process of the steel coil is ensured, and the rewinding of the core pulling steel coil is completed. Finally, the coil can be switched between the two spindles of Carrousel coiler.
Key words: carrousel coiler; rewinding technology; speed tension matching

基于大数据分析技术的智能连续退火炉实践与探讨

王 鲁

（宝钢日铁汽车板公司生产部，上海 201900）

摘 要：随着中国汽车业迅速发展，汽车板市场前景非常广阔，国内各大钢厂都在相继建设汽车板生产线。连续退火炉作为汽车板生产线的主要核心设备，与退火炉相关的工艺参数、设备状态直接影响着产品质量。本文介绍了智能连续退火炉技术，采用大数据分析的方法建立工业炉大数据平台，将退火炉的工艺、操作、设备等过程主要参数和各种制造资源连接在一起形成统一的资源池，在人机物共同决策下作出智能的响应，并综合炉子的气氛参数、能耗指数，环保排放数据实现炉子健康度可视化，对趋势异常的数据实现智能预警和远程诊断，防止重大质量事故或者设备故障的发生。
关键词：连续退火炉；大数据分析；智能；模型；预测

Practice and Discussion of Intelligent Continuous Annealing Furnace based on Big Data Analysis Technology

Wang Lu

(Baosteel-Nippon Steel Automotive Steel Sheets Co., Ltd., Shanghai 201900, China)

Abstract: With the rapid development of China's automobile industry, the prospect of automobile plate market is very broad. Continuous annealing furnace is the main core equipment of automobile plate production line, and the process parameters and equipment status related to annealing furnace directly affect the product quality. This paper introduces the establishment of industrial furnace big data platform by using big data analysis method, and introduces the main process of annealing furnace process, operation, equipment and so on parameters and various manufacturing resources should be connected together to form a unified resource pool, and intelligent response should be made under the joint decision of human, machine and material, so as to realize remote diagnosis and intelligent early warning for abnormal trend data. Prevent the occurrence of major quality accidents or equipment failures.
Key words: continuous annealing furnace; big data analysis; intelligent; model; prediction

转炉炼钢动态控制技术的研究与应用

魏春新，苏建铭，孙 涛，曹 祥

（鞍钢集团鞍山钢铁鞍钢股份有限公司炼钢总厂三分厂，辽宁鞍山 114000）

摘 要：社会经济的飞速发展对钢铁企业的生产效率以及产品质量提出了更高的要求，而转炉炼钢动态控制技术的发展是限制转炉炼钢生产效率的主要因素。本文分析了限制转炉炼钢动态控制技术发展的主要矛盾，并对当前投入使用的动态控制技术和动态控制模型进行了综述和分析，指出了各技术和模型的优势以及存在问题，探索实现转炉炼钢动态精确控制的方法。

关键词：转炉炼钢；动态控制技术；动态控制数学模型；人工智能算法

Research and Application of Dynamic Control Technology for Steelmaking

Wei Chunxin, Su Jianming, Sun Tao, Cao Xiang

(The Third Branch of Steelmaking Plant, Anshan Iron and Steel Group Co., Ltd., Anshan 114000, China)

Abstract: The rapid development of social economy puts forward higher requirements for the production efficiency and product quality of iron and steel enterprises, and the dynamic control technology of converter steelmaking is the main factor limiting the production efficiency of converter steelmaking. This paper summarizes the current technology and mathematical model of dynamic control, analyzes the main contradictions restricting the development of converter steelmaking dynamic control technology. The advantages and problems of the technologies and the models have been pointed out. And the method of dynamic control on converter steelmaking has been explored.

Key words: basic oxygen furnace; dynamic control technology; mathematical model of dynamic control; artificial intelligence algorithm

智能制造在宝钢炼钢工序实施路径的思考

阎建兵，舒建春

（宝山钢铁股份有限公司设备部，上海 201900）

摘 要：本文分析了宝钢股份直属炼钢厂信息系统的现状、不足，并对宝钢炼钢工序智能制造实施路径进行了思考，并提出了实施建议。

关键词：炼钢；智能制造；数据字典；模型

Study on the Implementation Path of Intelligent Manufacturing in Baosteel's Steelmaking Process

Yan Jianbing, Shu Jianchun

(Department of Equipment ,Baosteel Co., Ltd. ,Shanghai 201900, China)

Abstract: This paper analyzes the present situation of the information system of the directly affiliated steelmaking plant of Baosteel, and study about the implementation path of the intelligent manufacturing process in the steelmaking process of Baosteel, and puts forward the implementation suggestions.

Key words: steelmaking; intelligent manufacturing; data dictionary; model

激光测速装置在大盘卷生产线的应用

胡占民

（广东韶钢松山股份有限公司特轧厂，广东韶关 512122）

摘 要：利用激光测速装置对轧件进行实时速度检测信息和轧件长度检测信息，可以解决大盘卷生产线卷取机上游飞剪由于轧件速度检测偏差大导致的控制不稳定，消除飞剪切头和切尾时产生的短钢，有效避免短钢进入轧线通道引起的生产中断事故。同时，利用激光测速装置的轧件长度信息，可以精确地实现轧件头部的跟踪，使轧件头部在卷取机的定位准确。

关键词：激光测速装置；轧件跟踪；飞剪控制；卷取机头部定位

Application of the Laser Velocity Measurement Device on Big Bar Coil Line

Hu Zhanmin

(Special Bar Rolling Plant of SGIS Songshan Co., Ltd., Shaoguan 512122, China)

Abstract: Taking advantage of the real-time rolling speed and the rolling material length from the laser velocity measurement device. We can solve the instability of the flying shear due to the big offset of the detected rolling speed, avoid the undesirable short rolling piece into the rolling through pass. And utilize the rolling piece length from the laser velocity measurement device, can accurately realize the material head tracking and head positioning.

Key words: the laser velocity measurement device; material tracking; flying shear control; head positioning control of the coiling machine

微机联锁远程控制系统的设计

李士斌[1]，曲 峰[1]，刘福东[1]，姜增涛[1]，李 红[1]，王 慧[2]

（1. 鞍山钢铁集团有限公司铁运分公司，辽宁鞍山 114021；
2. 鞍钢集团公司工程技术板块鞍钢工程技术有限公司，辽宁鞍山 114021）

摘 要：本文对鞍钢厂区的工艺线运输及车站如何实现远程操纵集中控制的过程进行论证。随着通讯技术、网络技术、控制技术、电子技术、计算机软硬件技术、人工智能技术的快速发展，计算机联锁系统在世界范围内得到广泛应用，特别是可靠性和容错技术的深入研究。计算机联锁系统的安全技术在不断更新、完善和发展，使得微机联锁远程集中控制框架构成、系统设置、光缆铺设及补救措施、西门子组网方式的设计成为可能。设计旨在作业过程简化、减员增效。

关键词：微机联锁；远程控制；二级网络；光缆环网

Design of Microcomputer Interlocking Remote Control System

Li Shibin[1], Qu Feng[1], Liu Fudong[1], Jiang Zengtao[1] Li Hong[1], Wang Hui[2]

(1. Railway Transportation Branch of Anshan Iron and Steel Group Co., Ltd., Anshan 114021, China;
2. Angang Engineering Technology of Angang Group Co., Ltd., Anshan 114021, China)

Abstract: Demonstrate process line tranasportation in Angang plant and the process of how to realize the centralized control of remote operation. As the rapid development of communication technology, network technology, control technology, electronic technology, computer software and hardware technology, artificial intelligence technology. Computer interlocking system is widely used in the world, especially in the deep research of reliability and fault-tolerance technology. Constantly updating, improving and developing in the safety technology of computer interlocking system make it possible for the structure of remote centralized control frame for microcomputer interlocking, system settings, optical cable laying and remedial measures, design of Siemens networking mode. The design aims to simplify the operation process, reduce staff and increase efficiency.

Key words: microcomputer interlocking; remote control; secondary network; optical fiber loop network

冷轧连续退火线焊缝跟踪系统的研究及应用

李全鑫，柳 军，徐 涛，张 哲，李轶轩

（鞍钢股份冷轧厂，辽宁鞍山 114000）

摘 要：本文研究的主要内容是冷轧厂连续退火线带钢跟踪系统的设计与实施应用，主要工作包括以下几点：

（1）针对现场的实际情况设计了带钢跟踪的功能规格，对焊缝跟踪等功能做了较为详细的研究。

（2）根据生产线工艺参数及响应速度的实际需求，完成了对控制器选型及网络组态工作。实现了与二级及 HMI 服务器的 TCP/IP 通讯，与其他一级控制器的 EGD 通讯。

（3）利用过程数据采集软件对跟踪系统的实际应用效果进行了测评，其中焊缝跟踪的偏差比率低于 1/1000，实现了精确的焊缝定位。

本文为依靠国内技术力量自主研发并应用大型连续生产线的跟踪控制系统积累了宝贵的经验，具有较高的借鉴和推广价值。

关键词：连续退火生产线；Alstom HPC；焊缝跟踪

Research and Application of Weld Tracking System in Cold Rolling Continuous Annealing Line

Li Quanxin, Liu Jun, Xu Tao, Zhang zhe, Li Yixuan

(The Cold Rolling Plant of Ansteel, Anshan 114000, China)

Abstract: The main contents of the thesis are the research and application of the continuous annealing line strip tracking system of cold rolling mill. The main contributions of this thesis are as follows:

(1) According to the actual situation of the site, the function specification of strip tracking is designed, and the function of weld tracking and material tracking are studied in detail.

(2) According to the actual requirements of the production process parameters and response speed, the controller selection and network configuration work are completed. To achieve TCP/IP communication with the Level 2 and the HMI server, and EGD communication with other Level 1 controllers.

(3) The actual application effect of the tracking system is evaluated by the process data acquisition software. The deviation radio of the weld tracking is less than 1/1000, and the precise weld positioning is realized. The strip data of the material tracking is accurately transmitted and sent, and the set value is precisely switched and executed.

The thesis has accumulated valuable experience for the independent research and development and application of tracking control system in large continuous production line by domestic technical force. The project has high reference and promotion value.

Key words: continuous annealing line; Alstom HPC; weld track

钢铁行业智能运维研究

王剑虎，黄冬明，龚敬群，戴宛辰

（宝武装备智能科技有限公司，上海　201900）

摘　要：随着第四次工业革命浪潮的到来及中国制造 2025 的提出，钢铁行业传统运维模式在效率、能耗、库存、质量、生产率和成本等方面不能满足数字化和智能化时代设备管理的新需求。本文以钢铁行业为背景，围绕钢铁行业设备管理新需求，基于云计算、物联网、大数据、移动互联网和人工智能等先进技术，提出了钢铁行业智能运维新定义，研究了智能运维的本质特征，分析了钢铁行业智能运维发展等级，介绍了智能运维产品图谱，阐述了基于智能运维模式的设备服务体系和服务模式，最后介绍了智能运维的两种服务模式在宝武某热轧生产线、高线生产线、某基地风机群和齿轮箱上的实施效果，验证了智能运维模式的可行性、可靠性和有效性，为流程行业设备管理变革和智能运维模式实施提供了参考。

关键词：智能运维；运维模式；故障预测与诊断；健康管理；大数据；人工智能；服务体系和模式

Study on Intelligent Maintenance in Steel Industry

Wang Jianhu, Huang Dongming, Gong Jingqun, Dai Wanchen

(Baowu Equipment Intelligent Technology Co., Ltd., Shanghai 201900, China)

Abstract: With the emerging of the fourth industrial revolution and proposing of China's manufacturing 2025, the traditional maintenance mode couldn't satisfy the new era's equipment management requirements, such as efficiency, energy consumption, inventory, quality, productivity and cost. This paper takes the steel industry as study background, centers on steel industry equipment's new requirements, based on cloud technology, IOT, big data, mobile net and AI and so on, puts forward the new definition of intelligent maintenance, studies the essence of intelligent maintenance, analyzes the intelligent maintenance development level, introduces the products mapping, elaborates the equipment service system and modes. Finally, this paper demonstrates the practical applications of service system, modes and products in a hot rolling line, a wire rod line, some blowers group and a gearbox bearing, which validates the feasibility, reliability and effectiveness of those intelligent maintenance products. Besides, this paper also provides the basis for equipment service revolution and implementation of intelligent maintenance modes in process industry.

Key words: intelligent maintenance; maintenance mode; prognostic and diagnostic; health management; big data; AI;

service system and modes

钢轧界面板坯出入库规划模型研究与应用

刘雪莹，郭　亮，宋海洋

（首钢京唐钢铁联合有限责任公司信息计量部，河北唐山　063200）

摘　要： 产销一体化系统上线后，炼钢板坯库实现了低库存运转。目前板坯出入库码垛调度由人工指定，缺乏合理规划，精细化管控力度不足，从而造成无效倒垛次数的增加。随着公司产能不断提高，板坯库上料能力需要在现有高水平上增加约 30 块/天，因此通过研发更稳定、高效的板坯库出入库规划模型，强化板坯库存调度精准度，提高库存周转效率，加快上料节奏，提升钢轧界面衔接效率。

关键词： 出入库规划；逻辑分区；板坯分类

Research and Application of the Plan Model for Slab Inventory between Steel and Rolling

Liu Xueying, Guo Liang, Song Haiyang

(Information Department of Shougang Jingtang Iron and Steel Co., Ltd., Tangshan 063200, China)

Abstract: After the integrated production and marketing system was launched, the slab warehouse realized a low inventory operation. At present, the palletizing scheduling of slabs in and out of the warehouse is manually designated, lacking reasonable planning, and insufficient fine-grained management and control, resulting in an increase in the number of invalid stacks. As the company's production continues to increase, the slab storage capacity needs to be increased by about 30 pcs/day from the current high level. Therefore, developing a more stable and efficient slab inventory model to strengthen the accuracy of slab scheduling, to improve the efficiency ofinventory, and improve the efficiency of steel-rolling connection.

Key words: inventory decision-making; logical partition; slab classification

转炉废钢料槽车号智能识别系统研究与应用

吴　政

（宝钢股份湛江钢铁炼钢厂，广东湛江　524000）

摘　要： 炼钢转炉是炼钢生产的重要设备，转炉在生产过程中需要加入铁水和废钢，废钢通过牵引车从废钢堆场拉到炼钢厂，每炉的废钢与钢种计划相匹配，废钢的重量及种类需与现冶炼品种相匹配，但由于炼钢生产复杂性，往往伴生各种生产异常引起钢种计划变更，计划变更后，导致实际入炉废钢信息与 L2 二级机信息不匹配，给操作工带来极大不便，有时直接导致磷高、硫高质量异常。本项目通过研究转炉废钢料槽车号识别系统，成功开发了转炉废钢料槽车号智能识别系统，解决了由于计划变更造成入炉废钢与 L2 二级机信号不匹配问题。

关键词： 转炉；废钢槽号；视频识别；图像识别

Research and Application of Intelligent Recognition System for Converter Scrap Tank Number

Wu Zheng

(Steelmaking Plant of Baosteel Zhanjiang Iron & Steel Co., Zhanjiang 524000, China)

Abstract: Converter is an important equipment in steelmaking production. Iron and scrap need to be added in converter production. Scrap steel is pulled from scrap yard to steelmaking plant by tractor. Scrap steel in each furnace is matched with steel grade plan. The weight and type of scrap steel need to be matched with current smelting varieties. However, due to the complexity of steelmaking production, all kinds of production anomalies are often accompanied, resulting in the change of steel grade plan. After the plan change, the actual scrap information does not match the information of L2 secondary machine. This brings great inconvenience to operators, sometimes leading directly to high phosphorus and sulfur quality anomalies.This paper successfully developed the intelligent recognition system for scrap tank number by studying the recognition system for scrap tank number, and solved the problem of mismatch between the signal of the scrap into the furnace and the L2 secondary machine due to the change of the plan.

Key words: converter; scrap tank number; video recognition; image recognition

基于深度学习的辊道电机故障诊断分析方法

葛 超，杨奇睿，刘佳伟，吴晓宁，臧理萌，陈 亮

（鞍钢集团自动化有限公司设备诊断业务部，辽宁鞍山 114000）

摘 要：针对辊道电机中单个电机故障，根据所有电机正常运行状况和故障状况下收集的电机电流数据，提出一种基于深度学习的智能化电机故障诊断分析策略。首次提出了考虑用协同工作的其他电机的电流来对被监测电机进行故障诊断。该方案选择采用 Lenet-5 模型进行分类预测训练,以诊断电机中的故障，在生产现场实际应用此方案并进行数据结果验证。结果表明，所提出的方法在电机故障诊断中可行且有效。

关键词：机电故障诊断；辊道电机；电机电流；深度学习；Lenet-5 模型

Fault Diagnosis and Analysis Method of Roller Motor based on Deep Learning

Ge Chao, Yang Qirui, Liu Jiawei, Wu Xiaoning, Zang Limeng, Chen Liang

(Device Diagnosis Business Department, Ansteel Information Industry Co., Ltd., Anshan 114000, China)

Abstract: In order to avoid frequent unwarned shutdowns of motors, diagnosis and analysis of electromechanical equipment is a very important task. Aiming at a single motor fault in the roller motor, an intelligent motor fault diagnosis and analysis strategy based on deep learning was proposed according to the motor current data collected under normal operating conditions and fault conditions of all motors. Considering the current of other motors working in coordination to diagnose the fault of the monitored motor was proposed for the first time. In this scheme, the Lenet-5 model was used for classification prediction training to diagnose faults of the motor. In addition, the actual deployment and implementation of this program and the collection of experimental data showed that the proposed method was feasible and effective in motor fault diagnosis.

Key words: electromechanical fault diagnosis; roller motor; motor current; deep learning; Lenet-5; model

自适应数字滤波的旋转机械动平衡的边缘计算实现

刘佳伟，陈　亮，葛　超，朱晋锐，杨奇睿，吴晓宁

（鞍钢集团自动化有限公司，辽宁鞍山　114009）

摘　要：针对目前对振动信号进行检测分析的平台都是在服务器云平台上，提出一种在边缘计算端实现的办法。通过自适应数字滤波分析出旋转机械不平衡故障，利用中心频率自适应带通数字滤波器算法，并且将算法固化到嵌入式单片机平台。将采集的信号通过中心频率带通滤波器，利用数字化滤波器传递函数，将数据进行计算分析。同时利用嵌入式单片机采集转速信号，让滤波器的中心频率随着转子基频变化，去除噪声，得到不平衡振动信号。通过计算所得的不平衡振动信号，利用 FFT 提取故障频率，得出旋转机械动平衡故障。

关键词：带通数字滤波；嵌入式单片机边缘计算；快速傅里叶变换；旋转机械动平衡；中心频率自适应

Edge Calculation Method for Dynamic Balance of Rotating Machinery based on Adaptive Digital Filter

Liu Jiawei, Chen Liang, Ge Chao, Zhu Jinrui, Yang Qirui, Wu Xiaoning

(Ansteel Information Industry Co., Ltd., Anshan 114009, China)

Abstract: In view of the fact that the current platforms for vibration signal detection and analysis are all on the server cloud platform, this paper proposes a method to realize it on the edge computing end. The unbalance fault of rotating machinery is analyzed by adaptive digital filter, and the algorithm of adaptive band-pass digital filter based on center frequency is applied to embedded MCU platform. The collected signal is passed through the center frequency band-pass filter, and the digital filter transfer function is used to calculate and analyze the data. At the same time, the embedded microcontroller is used to collect the speed signal, so that the center frequency of the filter changes with the fundamental frequency of the rotor, the noise is removed, and the unbalanced vibration signal is obtained. Through the unbalanced vibration signal obtained by calculation, the fault frequency is extracted by FFT, and the dynamic balance fault of rotating machinery is obtained.

Key words: band pass digital filter; embedded MCU edge calculation; FFT; dynamic balance of rotating machinery; self-adaption of center frequency

冶金轧辊磨床参数监测系统开发

杨　波[1,3]，顾晓辉[2]，容逸颉[2]，吴　琼[1]，李世辉[2]，李鸿光[3]

（1. 宝钢股份中央研究院，上海　201900；2. 宝武重工上海江南轧辊厂，上海　201999；
3. 上海交通大学机械与动力工程学院，上海　200240）

摘 要：基于工业 4.0 的基础架构，融合边缘计算采集、工业物联网技术、云平台智能存储和客户端展示各个模块，开发出了一套冶金轧辊磨床多参数监控系统。该系统能实时监测轧辊磨床头架、尾架，床身和砂轮驱动端的加速度，驱动电流、轴承温度，以及磨削液的 pH 值，为磨床的预防式维护、工厂智慧制造，工业大数据分析提供基础的数据驱动保障。该系统被安装在险峰 MKD84125 多功能型数控轧辊磨床上进行测试，测试结果显示系统设计合理，数据特征明显，并具有足够的冗余性和鲁棒性。

关键词：智能制造；工业物联网；边缘计算；轧辊磨床

Development of the Parameters Monitoring of the Roll Grinder

Yang Bo[1,3], Gu Xiaohui[2], Rong Yijie[2], Wu Qiong[1], Li Shihui[2], Li Hongguang[3]

(1. Baoshan Iron & Steel Co., Ltd., Research Institute, Shanghai 201900, China;
2. Baowu Steel Co., Ltd., Shanghai Jiangnan Roller Factory, Shanghai 201999, China;
3. School of Mechanical Engineering, Shanghai Jiao Tong University, Shanghai 200240, China)

Abstract: Based on the infrastructure of Industry 4.0, a multi-parameter monitoring system of roller grinder was developed by integrating edge computing collection, industrial internet, cloud intelligent platform and client display modules. The system can continuously monitor the acceleration of roller grinder include head frame, tail frame, main girder, and drive unit of grinding wheel. Moreover, the drive current, bearing temperature and pH value of grinding fluid can be collected by industrial WIFI network. These data provide will support to predictive maintenance, intelligent manufacturing, and industrial big data analysis. The system is installed on a roller mill-Type MKD84125. Test results show that the system is reasonably designed with good characteristic data separation and robustness.

Key words: Intelligent manufacturing; Industrial internet; Edge computing; roll grinder

钢企基于工业互联网在能源管理绩效提升的探索与研究

许 斌[1]，皮 坤[1]，贾鹏杰[1]，邝昌云[1]，范心怡[1]，湛自丽[2]

（1. 云南昆钢电子信息科技有限公司，云南安宁 650302；
2. 昆明钢铁控股有限公司能源环保部，云南安宁 650302）

摘 要：随着云计算、大数据等技术将实现大规模应用，钢铁企业正向智能制造迈进，但还处于起步阶段。为实现企业用能的精准管理，创新了工具与方法，开发能源大数据，助力公司有效落地能源的精准化管理，实现企业降耗减排，推进生产和消费双向互动，优化资源配置，提高能源利用效率，降低企业用能成本，提出工业互联网平台探索与实践在能源管控智能化方向发展，帮助企业建立“能源价值管理”（由能耗转变能源成本）。本文对能源管理未来基于工业互联网平台的建设实践，提供建设性参考意见。

关键词：能源管理；工业互联网；能源大数据；智慧能效

Steel Enterprises' Exploration and Research of Increasing Energy Management Performance based on Industrial Internet

Xu Bin[1], Pi Kun[1], Jia Pengjie[1], Kuang Changyun[1], Fan Xinyi[1], Zhan Zili[2]

(1. Yunnan KISC Electronic Information Technology Co., Ltd., Anning 650302, China;
2. Kunming Iron & Steel (Group) Co., Ltd. Energy-environment Department, Anning 650302, China)

Abstract: As cloud computing, big data and other technologies are to be deployed in a large scale, steel enterprises are moving towards intelligent manufacturing, but still in a starting period. In order to deliver the precise management of enterprises' energy usage, this study innovates tools and methods and develops energy big data, so as to help enterprises effectively implement the precise management of energy, reduce their consumption and emission, boost the two-way interaction of production and consumption, optimize resource allocation, increase energy utilization efficiency, and lower energy costs. This proposes the development of industrial Internet platforms' exploration and practice in the direction of intelligent energy management and control, so as to help enterprises establish "energy value management" (changing from energy consumption to energy costs). This study provides some constructive references for the construction and practice of energy management based on the industrial Internet platform in the future.

Key words: energy management; industrial internet; energy big data; intelligent energy efficiency

钢铁行业工业控制系统信息安全防护解决方案

金　鹏[1]，李文嵩[2]，苏凯旋[2]

（1. 鞍钢集团信息产业有限公司，辽宁鞍山　114009；
2. 鞍钢集团自动化有限公司，辽宁鞍山　114009）

摘　要：根据《工业控制系统信息安全防护指南》及《信息安全技术 网络安全等级保护基本要求》中的相关要求，对钢铁行业工业生产网络进行网络安全防护建设，抵御两网间相互的非法入侵与攻击，对网络安全攻击行为进行事前监测、发现与应急处置。

关键词：信息安全；工控安全；工业控制系统；安全域；网络隔离；高级威胁；等级保护

Information Security Protection Solution of Industrial Control System in Iron and Steel Industry

Jin Peng[1], Li Wensong[2], Su Kaixuan[2]

(1. Ansteel Information Industry Co., Ltd., Anshan 114009, China;
2. Ansteel Automation Co., Ltd., Anshan 114009, China)

Abstract: According to the requirements of《Industrial control system information security protection guide》and《Information security technology basic requirements for classified protection of network security》, we have implemented network security protection construction for industrial production network of iron and steel industry, and have also kept resisting the illegal intrusion and attack between two networks, monitoring, discovering and dealing with network security attacks in advance.

Key words: information security; industrial control security; industrial control system; security domain; network quarantine; high-ranking threat; classified protection

中厚板热喷印字符检测识别系统设计

罗 全，袁 野

（鞍钢集团信息产业有限公司，辽宁鞍山 114009）

摘 要：针对中厚板生产过程中对板材热喷码字符识别的需求，设计了一套实时光学字符检测识别系统。采用工业相机拍摄喷印区域图像，采用 MSER 方法进行文本区域检测，提取字符区域，然后再用 SVM 算法，根据字符的特征对样本进行分类，生成相应的字符库，并调用该字符库读取字符，进而识别出字符。

关键词：图像处理；字符检测；字符识别

Design of Detection and Recognition System for Plate Thermal Spray Code Characters

Luo Quan, Yuan Ye

(Ansteel Information Industry Co., Ltd., Anshan 114009, China)

Abstract: A real-time optical character detection and recognition system is designed to meet the requirements of plate thermal spray code character recognition in plate production. Industrial camera is used to capture the image of spray printing area. MSER method is used to detect the text area, extract the character area, and then SVM algorithm is used to classify the samples according to the characteristics of the characters, generate the corresponding character library, and call the character library to read the characters, and then recognize the characters.

Key words: image processing; character detection; character recognition

电气设备预警系统及预防性维护的实施

谭秋生，刘国松，刘同朋，林 健，宗 璐，杨维涛，杜晓东，解 兵

（青岛特殊钢铁有限公司线材事业部电气作业区，山东青岛 266400）

摘 要：随着钢铁企业盈利水平逐渐好转，市场需求的逐步提升，要求高线厂通过技术进步、精益化管理、精细化操作，从设备角度出发，提高设备性能，降低设备故障率，充分释放产能。在此背景条件下，电气设备的预警及预防性维护实施，降低设备事故率是目前尤为重要的工作。各大钢铁企业应对标挖潜，对设备改进优化，通过开展攻关工作，在稳定原有生产水平的基础上，充分释放产能，实现产线产量最大化的目标。

关键词：设备故障率；电气设备预警；预防性维护；设备改进优化

Implementation of Early Warning System and Preventive Maintenance for Electrical Equipment

Tan Qiusheng, Liu Guosong, Liu Tongpeng, Lin Jian, Zong Lu, Yang Weitao, Du Xiaodong, Xie Bing

(Qingdao Speacial Iron and Steel Co., Ltd., Qingdao, 266400 China)

Abstract: With the gradual improvement of the profitability of iron and steel enterprises and the gradual improvement of the market demand, it is required that the high-wire factory should improve the equipment performance, reduce the equipment failure rate and fully release the capacity through technological progress, lean management and fine operation from the perspective of equipment. In this context, the early warning and preventive maintenance of electrical equipment to reduce the equipment accident rate is particularly important work. Major iron and steel enterprises should explore the potential of the standard, improve and optimize the equipment, and fully release the capacity on the basis of stabilizing the original production level to achieve the goal of maximizing the output of the production line through tackling key problems.

Key words: equipment failure rate; electrical equipment early warning; preventive maintenance; equipment improvement and optimization

LZW 压缩方法在工业数据采集系统中的应用

刘　刚

（宝钢股份冷轧厂，上海　200941）

摘　要： 本文结合钢厂对工业生产数据的采集系统的需求和要求，采用了一类适合于封闭系统基于 FPGA 的数据采集系统基础框架，作为 PLC 的 CPU 和数据库服务器之间的分隔，隔离式模块采集数据可以实现从网络封闭、比较孤立的自动化系统中采集数据，实现了高速处理采集数据的技术。针对数据压缩问题，通过对多种压缩技术的比较，提出了基于 LZW 算法的数据压缩技术，给出了基于 LZW 算法实现数据压缩的功能结构图和硬件设计原理，包括并验证了算法的可行性与正确性。

关键词： 数据采集；现场可编辑逻辑门电路（FPGA）；总线隔离板；LZW 算法

Application on Industrial Data Acquisition based on LZW Compression

Liu Gang

(Baoshan Iron & Steel Co., Ltd., Cold Mill Plant, Shanghai 200941, China)

Abstract: Based on the requirements of data acquisition system in iron steel work, a data acquisition system framework based on FPGA was designed to collect data from network closed and isolated automation systems by the communication-isolating module which can monitor the data from the PLC as a separator between CPU and Database server. By comparing variable data compression algorithms, the LZW algorithm block diagram is given to achieve data compression function. The hardware implementation of FPGA data acquisition system is presented to verify the LZW algorithm.

Key words: data acquisition; field programmable gate array; bus isolated board; LZW algorithm

无人化行车在热轧钢卷库区的应用研究与实践

赵　毅，曾龙华，李传喜，王海楠，庞慧玲

（宝钢湛江钢铁有限公司热轧厂，广东湛江　524072）

摘　要：为推进宝钢湛江钢铁热轧厂钢卷库无人化、智能化、高效率作业，湛钢热轧厂在钢卷库首次进行无人化行车改造作业，这也是国内无人化行车在热轧厂钢卷库区域的首次应用作业，本文主要介绍无人化行车系统的车上局、地下局构建原理和实现方式，同时剖析无人化行车自动吊运时逻辑、作业效率问题及解决方案，使无人化行车在湛钢热轧钢卷库成功投用运行。

关键词：无人化行车；自动吊运；湛钢热轧钢卷库

Application Research and Practice of Unmanned Vehicle in Hot Rolling Coil Storage Area

Zhao Yi, Zeng Longhua, Li Chuanxi, Wang Hainan, Pang Huiling

(Hot Rolling Mill of Zhanjiang Iron & Steel Co., Ltd., Zhanjiang 524072, China)

Abstract: To promote 2250 hot-rolled coil library a kind, intelligent, efficient operation, 2250 hot-rolled steel coil library first unmanned driving renovation work, it is also the first time domestic unmanned driving in the region of the hot rolled steel coil finished-parts storage application, this paper mainly introduces unmanned driving system of the car bureau, bureau of underground construction principle and implementation approach, at the same time, this paper analyzes the logic, operation efficiency and solutions of the automatic hoisting of the unmanned vehicle, so that the unmanned vehicle can be successfully put into operation in the 2250 hot-rolled steel coil warehouse.

Key words: unmanned driving; automatic hoisting; 2250 hot rolled steel coil warehouse

第二炼钢连铸运钢坯平板车无人自动控制系统

王　强，刘　园

（新疆八一钢铁股份公司炼钢厂，新疆乌鲁木齐　830022）

摘　要：根据平板车质量获取行走条件，分手动、自动、半自动，能够实现手动和自动控制相结合，基于二炼钢厂现有设备，利用自动化技术削减人员的重复劳动，结合计算机 intouch 画面和 PLC 实现远程自动控制，减少了人员劳动成本和后续资金投入，加装安全护栏门连锁急停、现场急停和画面急停，可进行现场无人操作，大大减少了设备运行的安全风险。

关键词：自动行走；激光检测器；称重传感器；PLC

Abstract: According to platform car quality for walking condition, break up, automatic, semi-automatic, can realize the combination of manual and automatic control, based on existing equipment No.2 steelmaking plant, use of automation technology to cut staff, rework, combined with computer intouch screen and PLC to realize remote automatic control, reduces labor costs and follow-up funding, equipped with safety guardrail door chain emergency stop, on-site emergency stop and picture emergency stop, can carry out on-site unmanned operation, greatly reducing the safety risk of equipment operation.

Key words: automatic walking; laser detector; weighing sensor; PLC

三维数字化钢厂构建研究与实践

岑子政，臧中海，袁怀月，汪　畅，高启洋，王　妮

（中冶集团武汉勘察研究院有限公司，中冶智诚（武汉）工程技术有限公司，湖北武汉　430075）

摘　要：本文提出了一种聚焦钢厂的三维数字化构建方案，并分析方案在生产项目中实践应用成果，验证方案的必要性、可行性与成熟度。对于既有钢厂采用逆向建模方式建立三维空间地理信息模型，运用于钢厂内公辅和产线数字化建设；对于新建钢厂使用数字化设计成果通过数字化交付手段形成统一的数字化竣工成果资料，并运用于钢厂产线数字化建设。利用空间信息结合具有三维可视化表达能力、地理空间信息管理能力、数字化交付能力、传感器接入管理能力、云渲染能力、大数据处理能力、AI 视频智能分析能力的数字钢厂基础信息平台打造钢厂基础数字底座。为钢厂打造具有信息化、智慧化、数字化的应用提供了数据支撑和平台基础。数字钢厂基础信息平台是具有大数据处理、云计算、虚拟仿真、数字孪生能力的企业级钢铁信息基础应用平台，赋能钢厂实现全面智能制造。

关键词：大数据；地理信息系统；工业物联网；虚拟仿真；数字孪生

Research and Practice of 3D Digital Steel Plant Construction

Cen Zizheng, Zang Zhonghai, Yuan Huaiyue, Wang Chang, Gao Qiyang, Wang Ni

(MCC Smart City (Wuhan) Engineering Technology Co., Ltd., Wuhan Surveying-geotechnical Research Institute Co., Ltd. of MCC, Wuhan 430075, China)

Abstract: This paper proposed a 3D digital construction project focusing on steel plant, and analyzed the practical application results of the project in production projects, and verifies the necessity, feasibility and maturity of the scheme. As for existing steel plant, the reverse modeling method is adopted to establish a three-dimensional spatial geographic information model, which is used in the construction of public auxiliary and production line digitization in the steel plant; as for the newly-founded steel plant, the forward modeling method: the digital design results are used to form a unified digital data set through digital delivery means, which is commonly used in the digital construction of steel plant production line. The digital steel plant basic information platform, which includes 3D visualization expression ability, geospatial information management ability, digital delivery ability, sensor access management ability, cloud rendering ability, big data processing ability and AI video intelligent analysis ability, can provide data support and platform foundation for steel plants to build informatization, intelligence and digitalization applications. he digital steel plant basic information platform is an enterprise level steel information basic application platform with the ability of big data processing, cloud computing, virtual simulation and digital twin realization., which makes steel plants to build comprehensive intelligent manufacturing possible.

Key words: big data; geographic information system; industrial internet of things; virtual simulation; digital twin

钢铁企业安全生产智能监控平台

黄　爽，臧中海，袁怀月，汪　畅，高启洋，明平寿

（中冶集团武汉勘察研究院有限公司，中冶智诚（武汉）工程技术有限公司，湖北武汉　430075）

摘　要：随着工业信息化的发展，钢铁企业安全生产逐步迈入智能监控时代。本文基于三维地理空间信息技术、数字孪生技术、物联网技术、视频智能识别技术等，针对公辅安全、生产安全和产线安全提出了可视化的智能监控方法。通过对厂房设备的精确建模，结合传感器监测数据，实时展示煤气浓度、煤气压力和流量，消防设施状态，铁包、钢包位置和液态金属状态，高炉冶炼状况，焦化产线生产状况以及相关的人员车辆位置等信息。实现重点关注领域报警信息的可视化联动，提高钢铁企业安全监控智能化水平。

关键词：安全生产；三维可视化；煤气安全；液态金属；数字孪生；消防集控

Visualized Intelligent Management Platform for Safety Production in Iron and Steel Enterprises

Huang Shuang, Zang Zhonghai, Yuan Huaiyue, Wang Chang, Gao Qiyang, Ming Pingshou

(MCC Smart City (Wuhan) Engineering Technology Co., Ltd., Wuhan Surveying-geotechnical Research Institute Co., Ltd. of MCC, Wuhan 430075, China)

Abstract: With the development of industrial informatization, the safety production of iron and steel enterprises has gradually entered the times of intelligent monitoring and management. Based on 3D geographic information technology, digital twin technology, internet of things technology and video intelligent identification technology, this paper proposed a visual intelligent management method for the safety problems in the fields of public auxiliary, production process and production line in steel plants. Based on the precise modeling of plants and equipments and sensor monitoring data, the platform displays real-time information such as gas concentration, gas pressure and flow rate, fire facilities status, ladle location and liquid metal status, blast furnace smelting status, production status of coking production line and related personnel and vehicle location. This platform realized visual linkage of alarm information and relevant facilities in key areas and improved the intelligent level of safety management in iron and steel enterprises.

Key words: safety production; 3D visualization; gas safety; liquid metal; digital twin; centralized control system of fire safety

智能化技术在烧结系统的建设和应用

何　杰

（宁波钢铁有限公司，浙江宁波　315800）

摘　要：在智能化、信息化的条件下，为了提高烧结矿质量，保证生产的稳定，针对烧结生产现场的实际情况，开发应用了一套智能烧结控制系统。首先利用 Kepware 软件同关系型与非关系型数据库的组合，采集接收存放来自基

础自动化的数据；然后，将生产工艺的操作流程、计算方法、逻辑判断编写前、后反馈模型程序控制生产；最后，对生产过程中产生的数据进行分析学习，以提高控制模型的精度与效果。

关键词： 关系型数据库与非关系型数据库；读写分离；系统架构；数据采集接口

Construction and Application of Intelligent Technology in Sintering System

He Jie

(Ningbo Steel Iron and Steel Co., Ltd., Ningbo 315800, China)

Abstract: Under the conditions of intelligence and information, in order to improve the quality of sintered ore and ensure the stability of production, a set of intelligent sintering control system has been developed and applied according to the actual situation of the sintering production site. First, use the combination of Kepware software and relational and non-relational databases to collect, receive and store data from basic automation; then, compile pre- and post-feedback model programs for the production process, calculation methods, and logical judgments to control production; finally, Analyze and learn the data generated in the production process to improve the accuracy and effect of the control model.

Key words: relational and non-relational databases; read-write separation; system architecture; data acquisition interface

基于工业互联网的钢铁企业能源智能化管控系统开发与应用

白　雪[1]，刘　伟[2]，曲泰安[1]，刘常鹏[3]，王艳飞[1]，高大鹏[1]

（1. 鞍钢集团自动化有限公司，辽宁鞍山　114009；2. 鞍钢股份有限公司鲅鱼圈钢铁分公司，辽宁营口　115007；3. 鞍钢集团钢铁研究院，辽宁鞍山　114009）

摘　要： 利用大数据、云计算、人工智能、知识图谱、5G 等新一代信息技术建立具有自感知、自分析、自决策、自优化能力的钢铁企业能源智能化管控系统。系统充分运用冶金流程学、热力学等机理知识，从物质流和能源流协同优化的角度出发为钢铁企业能源管控提供全站所室集中操控、多能源流供需平衡优化、全流程能耗评价优化、全流程碳排放分析等“看得见、说得清、管得住”的全流程一站式智能服务。通过系统的有效应用，能够提高钢铁企业能源使用效率，降低能耗损失，改善钢铁企业周边环境，助力绿色钢铁生态化发展，为早日实现钢铁企业“碳中和”“碳达峰”提供有力支撑。

关键词： 大数据；能源管控；协同优化；碳达峰；碳中和

Intelligent Energy Management and Control System for Iron and Steel Enterprises based on Industrial Internet

Bai Xue[1], Liu Wei[2], Qu Taian[1], Liu Changpeng[3], Wang Yanfei[1], Gao Dapeng[1]

(1. Ansteel Automation Co., Ltd., Anshan 114009, China;
2. Bayuquan Iron and Steel Subsidiary Company of Ansteel Co., Ltd., Yingkou 115007, China;
3. Iron and Steel Institutes of Ansteel Group Corporation, Anshan 114009, China)

Abstract: A new generation of information technology, such as big data, cloud computing, artificial intelligence, knowledge mapping, 5G, is used to establish an energy intelligent management and control system for iron and steel enterprises with the ability of self perception, self analysis, self decision-making and self optimization. From the perspective of collaborative optimization of material flow and energy flow, the system makes full use of metallurgical process science, thermodynamics and other mechanism knowledge to provide "visible, clear and manageable" one-stop intelligent services for energy management and control of iron and steel enterprises, such as centralized control of the whole station, supply and demand balance optimization of multi energy flow, energy consumption evaluation optimization of the whole process, and carbon emission analysis of the whole process. Through the effective application of the system, it can improve the energy efficiency of iron and steel enterprises, reduce the loss of energy consumption, improve the surrounding environment of iron and steel enterprises, assist the ecological development of green iron and steel, and provide strong support for the early realization of "carbon neutralization" and "carbon peak" of iron and steel enterprises.

Key words: big data; energy management and control; collaborative optimization; carbon peak; carbon neutralization

关于煤焦自动检测系统在钢铁行业中的应用

黄 波，吴超超，朱学良

（宝武集团广东韶关钢铁有限公司检测中心，广东韶关 512123）

摘 要：韶钢检测中心引进一套煤焦自动检测系统，在钢铁行业中为首次应用，该系统可以自动检测煤中挥发分、灰分、内水分、全硫、发热值、碳、氢、氮指标。该系统的使用极大提高了现场检测效率，同时在使用过程中也发现该系统较多的问题，针对发现的问题对该系统的硬件配置及软件开发进行了提高和完善，同时还对用该系统检测焦炭进行了尝试。

关键词：煤焦；自动检测；钢铁；工业分析

Application of Coal and Coke Automatic Detection System in Iron and Steel Industry

Huang Bo, Wu Chaochao, Zhu Xueliang

(Testing Center of Baowu Group Guangdong Shaoguan Iron and Steel Co., Ltd., Shaoguan 512123, China)

Abstract: Shaoguan Iron and steel testing center has introduced a set of coal and coke automatic detection system, which is the first application in the iron and steel industry. The system can automatically detect the indexes of volatile matter, ash, internal moisture, total sulfur, calorific value, carbon, hydrogen and nitrogen in coal. The use of the system greatly improves the efficiency of on-site detection. At the same time, many problems are found in the process of using the system. Aiming at the problems found, the hardware configuration and software development of the system are improved and improved. At the same time, the coke detection with the system is also attempted.

Key words: coal char; automatic detection; steel; industrial analysis

冷轧热镀锌机组检测系统数据与数字钢卷匹配方法

俞鸿毅，王 劲，王学敏

（宝山钢铁股份有限公司，上海 201900）

摘　要： 热镀锌机组安装许多表面质量检测系统，如锌层厚度检测仪、表缺仪等，这些仪表往往自成体系，有自己的计长编码器，并将检测到的表面质量信息落位到对应的长度位置上，但对于同一钢卷，不同的编码器计长会有误差，即不同的检测系统输出的钢卷长度不同。钢卷的计长与数据的落位，一方面长度是指机组内的母卷长度，而非出口剪切后的子卷长度；另一方面，不同的编码器计长存在误差。如何将各独立的仪表检测数据统一起来，建立归一化模型，是本论文的核心内容。

关键词： 数字钢卷；数字工厂；智慧制造；大数据与人工智能

Matching Method Between Data of Detection System of Cold Rolling and Hot Galvanizing Line and Digital Coil

Yu Hongyi, Wang Jin, Wang Xuemin

(Baoshan Iron & Steel Co., Ltd., Shanghai 201900, China)

Abstract: Many surface quality detection systems are installed in the hot galvanizing unit, such as zinc layer thickness detector, meter missing instrument, etc. these instruments often have their own systems, have their own length encoder, and drop the detected surface quality information onto the corresponding length position, but for the same steel coil, different encoder length will have errors, that is, the length of steel coil output by different detection systems is different. On the one hand, the length refers to the length of the master coil in the unit, rather than the length of the sub coil after the exit shearing; on the other hand, different encoder length has errors. How to unify the independent instrument test data and establish the normalization model is the core content of this paper.

Key words: data coil; digital factory; intelligent manufacturing; big data and artificial intelligence

鞍钢汽车钢智联云平台推动智慧营销的实践

李红雨，曹　刚，倪克仁，张烜赫

（鞍钢集团汽车钢营销（服务）中心，辽宁鞍山　114041）

摘　要： 作为新中国第一个恢复建设的大型钢铁联合企业和最早建成的钢铁生产基地，鞍钢被誉为“新中国钢铁工业的摇篮”“共和国钢铁工业的长子”。鞍钢目前已形成跨区域、多基地、国际化的发展格局，成为国内布局完善的钢铁企业。鞍钢 800 余个品种、50000 多个规格钢铁产品，以及近 40 种焦化产品，被广泛应用于各行各业，并远销海外，拥有 500 多家国内外客户及合作伙伴。作为世界级精品钢铁钒钛和装备技术输出企业，鞍钢将始终秉承“制造更优材料，创造更美生活”的企业使命，为人类发展和社会进步提供科技含量更高、产品质量更优、环境更友好的钢铁、钒钛及其他新材料，不断满足和引领人类对美好生活的追求，做服务全球、最可信赖的合作伙伴。“鞍钢汽车钢智联云平台”以构建中心管控平台、钢加中心加工配送平台、智慧营销平台、客户自助服务平台为主体，在传统信息技术的基础上，运用混沌神经网络技术、数据挖掘技术、动态数据监控技术、移动互联技术、车辆识别技术、多智能体技术等创新技术，充分向客户展示从订单、生产、物流、质量、财务等综合信息的平台，消除各信息孤岛，构建向自动化智能化工业 4.0 靠近的管理模式。

关键词： 汽车钢；云平台；智慧营销；实践

Ansteel Automotive Steel Intelligent Joint Cloud Platform Romotes the Practice of Intelligent Marketing

Li Hongyu, Cao Gang, Ni Keren, Zhang Xuanhe

Abstract: "Angang automobile steel Zhilian cloud platform" takes the construction of central control platform, steel processing center processing and distribution platform, intelligent marketing platform and customer self-service platform as the main body. On the basis of traditional information technology, it uses chaotic neural network technology, data mining technology, dynamic data monitoring technology, mobile Internet technology, vehicle identification technology, multi-agent technology and other innovative technologies, fully display the platform of comprehensive information from order, production, logistics, quality and finance to customers, eliminate various information islands, and build a management mode close to automation and intelligent industry 4.0.

Key words: automotive steel; cloud platform; smart marketing; practice

全链接模式下网络与信息安全系统建设与应用

乔振华

（河钢宣钢计控中心，河北张家口 075100）

摘 要：为实现宣钢与集团总部统一信息平台、集中运营管控，满足宣钢数据中心新的超融合系统、新的虚拟化系统的网络链接及数据访问需求，满足安全等级保护要求，进行了安全设备配置、策略调整等网络安全等级保护测评系统建设、数据中心网络架构集群化改造和设备链路升级万兆网络建设、病毒监测平台和安全感知平台构建网络安全系统建设。

关键词：全链接；网络；信息安全；建设与应用

Construction and Application of Network and Information Security System under Full Link Mode

Qiao Zhenhua

(Information Center of Xuangang, Zhangjiakou 075100, China)

Abstract: In order to realize the unified information platform and centralized operation control between Xuangang and the group headquarters, meet the network link and data access requirements of the new hyper fusion system and the new virtualization system of Xuangang data center, and meet the requirements of security level protection, the network security level protection evaluation system including security equipment configuration and strategy adjustment was constructed data center network architecture cluster transformation and equipment link upgrade, 10 Gigabit network construction, virus monitoring platform and security awareness platform construction, network security system construction.

Key words: full link; network; information security; construction and application

BCS-燃烧优化热风炉控制技术在宣钢高炉的应用

李锦龙，闫新卓

（河钢宣钢机电公司，河北张家口　075100）

摘　要：采用先进的软测量技术、多变量解耦技术、过程优化控制技术、故障诊断与容错控制技术、先进的软件接口来实现高炉热风炉的自动优化控制，使高炉热风炉的运行更加安全、稳定和经济。PLC 系统与 BCS 通讯是通过 OPC 通讯完成的，并实现双机切换功能。

关键词：软测量技术；多变量解耦技术；过程优化控制技术；故障诊断与容错控制技术；先进的软件接口

Application of BCS Combustion Optimization Hot Blast Stove Control Technology in Blast Furnace of Xuanhua Steel

Li Jinlong, Yan Xinzhuo

(Hegang Xuangang Electromechanical Company, Zhangjiakou 075100, China)

Abstract: The advanced soft measurement technology, multivariable decoupling technology, process optimization control technology, fault diagnosis and fault tolerance control technology and advanced software interface are used to realize the automatic optimization control of blast furnace, which makes the operation of blast furnace safer, stable and economical. PLC system and BCS communication is completed through OPC communication, and realizes the function of dual machine switching.

Key words: soft sensing technology; multivariable decoupling technology; process optimization control technology; fault diagnosis and fault tolerant control technology; advanced software interface

河钢宣钢炼钢超融合虚拟化系统创新应用

刘怡生，张文耀，刘　涛

（河钢宣钢，河北张家口　075100）

摘　要：本文介绍了宣钢 150t 炉区炼钢二级数据采集服务器虚拟化平台的创新建设与应用，主要从应用背景、总体思路、超融合技术方案、超融合系统构架、超融合技术、超融合升级数据迁移、实施效果等方面阐述。项目成功地将超融合技术、虚拟化网络、虚拟化存储和虚拟化服务器技术、大数据技术、云平台技术引入了炼钢二级综合管控系统，促进了炼钢生产自动化、智能化、数据化、信息化支撑体系建设。

关键词：超融合；服务器；虚拟化；数据化和智能化

The Construction and Application of Virtualization based on Super-fusion

Liu Yisheng, Zhang Wenyao, Liu Tao

(HBIS Group Xuansteel Company, Zhangjiakou 075100, China)

Abstract: This paper introduces the Innovative Construction and application of steelmaking server virtualization platform, mainly from the scheme, framework, super-fusion technology, upgrade data migration, implementation effect. The project has successfully introduced super-fusion technology, virtual network, virtual storage and virtual server technology, big data technology and cloud platform technology into the secondary integrated management and control system of steel-making, with good results, it has promoted the construction of steel-making production automation, intelligence, data and information supporting system.

Key words: super fusion; server; virtualization; data and intelligence

5G+智慧炼钢系统应用实践

翟宝鹏[1]，于 洋[2]，金 鹏[3]，宋伟豪[1]，孟婷婷[1]

（1. 鞍钢集团自动化有限公司，辽宁鞍山 114009；2. 鞍钢集团有限公司，辽宁鞍山 114001；3. 鞍钢集团信息产业有限公司，辽宁鞍山 114009）

摘 要：基于5G通信技术的智慧炼钢系统的建设，其主要目的，即为提升炼钢环节生产效率与钢材质量，助力炼钢环节向智能化精细生产转型升级，降低生产成本，提升环保与经济效益，提升公司核心竞争力。基于5G通信技术的智慧炼钢系统，在自动控制系统，数据通信，数据采集计算，分析决策等不同维度上发力，多个系统融为一体。以信息化、智能化为抓手，为炼钢精细化高质量生产赋能。在打造智慧炼钢系统平台之上，助力公司的绿色发展远景规划，在中远期，为公司减少碳排放，实现碳达峰、碳中和奠定良好基础。

关键词：智慧炼钢；碳排放；自动化控制；大数据分析；人工智能

Application Practice of Intelligent Steel-Making System based on 5G Communication Technology

Zhai Baopeng[1], Yu Yang[2], Jin Peng[3], Song Weihao[1], Meng Tingting[1]

(1. Ansteel Information Industry Co., Ltd., Anshan 114009, China;
2. Ansteel Group Co., Ltd., Liaoning Anshan 114001, China;
3. Ansteel Information Industry Co., Ltd., Anshan 114009, China)

Abstract: The main purpose of the construction of intelligent steel-making system based on 5G communication technology in to improve the production efficiency and steel quality of steel-making link, help steel-making link to transform and upgrade to intelligent fine production, reduce production costs, improve environmental protection and economic benefits, and enhance the company's core competitiveness.Intelligent steel-making system based on 5G communication technology works in different dimensions such as automatic control system, data communication, data acquisition and calculation, analysis and decision-making, and integrates multiple systems. With information and intelligence as the starting point, it can enable the fine and high-quality production of steel-making. On the basis of building a smart steel-making system platform, it helps the company's long-term green development plan, and lays a good foundation for the company to reduce carbon emissions and achieve carbon peak and carbon neutralization in the medium and long term.

Key words: intelligent steel-making; carbon emissions; automated control; data analysis; artificial intelligence

轧钢过程质量溯源与过程诊断技术开发

郭　薇，张爱斌，谈　霖，王凤琴

（首钢集团有限公司技术研究院冶金过程研究所，北京　100041）

摘　要：目前钢铁工业各工序遗留的质量问题通常属于多变量耦合问题，现有系统缺乏高效的质量诊断、分析与优化技术。为快速、精准定位并发现问题根源，热轧质量数据分析系统利用机理模型、数据驱动算法和可视化方式，开发了面向质量稳定性和高精度提升的过程诊断优化集群系统工具和一键式解决方案，将大量、长期工艺模型积累的专家经验转化成智能化控制手段，完善对产线问题的预警、诊断、分析、优化的定制化功能，着力实现工序间或工序内部窗口的智能决策，避免质量缺陷频繁、重复发生，助力企业对产品质量进行控制与提升。

关键词：热膨胀；凸度控制；浪形准则；比例凸度分配；诊断技术

Process Quality Traceability and Diagnosis Technology Development of Steel Rolling Process

Guo Wei, Zhang Aibin, Tan Lin, Wang Fengqin

(Metallurgical Process Department, Shougang Research Institute of Technology, Beijing 100041, China)

Abstract: At present, the quality problems left by various processes in iron and steel industry usually belong to multivariable coupling problems, and the existing system lacks efficient quality diagnosis, analysis and optimization technology. For positioning and finding the root of the problems fastly and accurately, cluster system tools and one-click solutions for process diagnostics optimization for quality stability and high precision improvement are developed based on the hot rolling quality data analysis system by using the mechanism model, data-driven algorithm and visual way. This technology converts a lot of expert experience into intelligent control methods, improving the early warning, diagnosis, analysis and optimization of customized functions for industrial production line, and striving to achieve intelligent decision-making between process or process internal window, in order to avoid frequent and repeated quality defects, and help enterprises control and improve product quality.

Key words: thermal expansion; profile controlling; wave shape criterion; proportional profile distribution; diagnosis technology

新区 2 号空压站自动控制简述

邓洪辉

（新疆八一钢铁股份有限公司能源中心，新疆乌鲁木齐　830022）

摘　要：介绍新疆八一钢铁股份公司能源中心热力分厂新区 2 号空压站自动控制简析,同时简单分析了系统的一些缺陷及相应的处理对策。

关键词：离心式空压机；双电源单母线式主接线；露点温度

关于电弧炉自动控制系统的应用

李　兵

（新疆八一钢铁股份有限公司能源中心热力分厂，新疆乌鲁木齐　830022）

摘　要：电弧炉是利用电弧的能量来熔炼金属的一种电炉。本文以电弧炉计算机控制系统为研究对象，在查阅了大量国内外相关文献的基础上，综述了电弧炉控制技术的发展历程、研究现状及今后的发展趋势。根据电弧炉熔炼工艺对控制系统的控制要求，给出了控制系统总体设计方案，对电弧炉计算机控制系统的硬件系统配置作了详细的说明。在电极调节器控制方案设计的基础上给出了实用的控制算法的实现方法，电弧炉炼钢过程一般是以物料、热量和化学平衡为基础，采用理论与经验相结合的方法，建立的超高功率电弧炉冶炼工艺静态控制模型和动态控制模型，并对主要参数进行了计算。静态模型的建立能确定主要操作参数的计算，并对钢质量影响较大的终点温度及碳的含量进行神经网络预报。

关键词：电弧炉；电极调节；变压器保护；液压控制；水冷事故

Application of Automatic Control System for Electric Arc Furnace

Li Bing

(Xinjiang Bayi Iron and Steel Incorporated Company Energy Center Heating Power Branch, Urumqi 830022, China)

Abstract: Electric arc furnace is the use of electric arc energy to a metal melting furnace. Papers are EAF computer control system for the study, in view of the large number of relevant literature, based on an overview of the development process of electric arc furnace control technology, research status and future trends. EAF melting process under the control of the control system requirements, given the overall design scheme of the control system, computer control system for electric arc furnace hardware system configuration been described in detail. The electrode regulator control scheme is given based on the design of practical control algorithm method, including electrode regulator control unit, hydraulic control unit, vacuum switch, sub-gate control unit, the shift control unit regulating transformers and other logic control unit. EAF steelmaking process is generally based on the material, heat and chemical balance basis using a combination of theoretical and empirical ways to build ultra-high power electric arc furnace smelting process control model of static and dynamic control model, and the main parameters were calculated . Static model can determine the main operating parameters of the calculation, and the greater impact on the quality of steel endpoint temperature and carbon content of the neural network prediction.

Key words: electric arc furnace; electrode regulator; transformer protection; hydraulic control; water accident

钢铁行业燃气锅炉燃烧自动控制对策研究

王立元

（新疆八一钢铁股份有限公司能源中心，新疆乌鲁木齐　830022）

摘　要： 煤气燃烧锅炉越来越广泛地应用于钢铁行业能源综合利用项目上，高炉产线生产中产生的富余高炉煤气，通过锅炉燃烧被转化为电能使用。这种方式能够充分回收利用废弃资源，节能减排。因此，针对作为项目核心工艺部分的锅炉系统，设计先进可靠的自动化控制系统来监控炉膛安全，实现锅炉安全可靠、连续高效生产，是对该发电工程最终实现节能减排目标的重要保障。

关键词： 自动控制；热值；燃烧优化

Research on the Automatic Control of Gas Boiler Combustion Iron and Steel Industry

Wang liyuan

(Xinjiang Baiyi Iron & Steel Co., Ltd., Energy Center, Urumqi 830022, China)

Abstract: Gas fired boilers are more and more widely used in the comprehensive utilization of energy in the iron and steel industry. The surplus blast furnace gas produced in the production line of blast furnace is converted into electric energy through boiler combustion. This way can make full use of waste resources, energy conservation and emission reduction. Therefore, for the boiler system as the core process part of the project, the design of advanced and reliable automatic control system to monitor the furnace safety and realize the safe, reliable, continuous and efficient production of the boiler is an important guarantee for the final realization of the energy saving and emission reduction goal of the power generation project.

Key words: automatic control; calorific value; combustion

二十辊轧机板形控制精度问题分析及应对措施

李　胤

（宝钢股份武钢有限硅钢部，湖北武汉　430080）

摘　要： 板形控制即平直度控制，是带钢生产的核心控制功能。本文介绍了板形缺陷的表现形式及形成机理，板形控制系统中径向调整机构、轴向调整机构、板形测量仪等装置设备，进一步分析了基于神经网络、模糊控制的板形控制思想和控制策略，同时，根据现场生产存在的问题，从生产工艺、操作、零级、一级、二级等方面提出了切实可行的应对措施。生产实践证明，改进后效果良好，可以有效提高板形控制的精度和产品合格率，满足生产需求。

关键词： 板形控制；平直度；神经网络；模糊控制；应对措施；精度

Analysis and Countermeasures of Flatness Control Accuracy of 20 High Mill

Li Yin

(Baosteel Co., Ltd. WISCO Silicon Steel Department, Wuhan 430080, China)

Abstract: Flatness control is the core control function of strip production. This paper introduces the form and formation mechanism of flatness defects, the radial adjustment mechanism, axial adjustment mechanism, flatness measuring instrument and other devices in the flatness control system, and further analyzes the flatness control idea and control strategy based on neural network and fuzzy control. At the same time, according to the problems existing in the field production, from the aspects of production process, operation, zero level control, automatic control, etc. Practical measures are put forward in the first level and second level. The production practice shows that the effect is good after the improvement, which can effectively improve the accuracy of shape control and product qualification rate, and meet the production demand.

Key words: shape control; straightness; neural network; fuzzy control; countermeasures; accuracy

燃气锅炉智能燃烧系统的研发与应用

霍广平，杨　铮

（河钢集团邯钢公司自动化部，河北邯郸　056015）

摘　要：钢铁行业的燃气锅炉由于受燃料的压力、热值的影响燃烧变得较为复杂，而各参数之间又有较大的关联性，常规的控制系统无法实现全自动控制，目前国内各钢铁生产企业中有 90%以上的燃气锅炉的燃烧操作都还处在手动操作或单回路自动控制。基于上述原因我们采用智能燃烧控制系统实现了燃气锅炉的全自动烧炉。

关键词：燃气；锅炉；控制；智能燃烧；全自动

The Development and Application of Intelligent Gas Boiler Combustion System

Huo Guangping, Yang Zheng

(HBIS Group Hansteel Company Automation Department，Handan, 056015, China)

Abstract: Gas boiler in iron and steel industry due to the influence of pressure and calorific value of fuel combustion becomes more complex, and between various parameters and has greater relevance, the conventional control system can't achieve automatic control, the current domestic each iron and steel production enterprises, the operation of more than 90% of gas boiler combustion is still in manual or automatic single loop control. Based on the above reasons we adopt intelligent combustion control system to realize the automatic burning furnace gas boiler.

Key words: gas; boiler; control; intelligent combustion; fully automatic

浙大中控 DCS 在综合废水处理系统中的开发及应用

张　恒，申献民，李春华

（河钢集团邯钢公司自动化部，河北邯郸　056015）

摘　要：河钢邯钢综合废水处理系统是“十三五”国家科技水专项的示范工程，通过浙大中控 DCS 系统将综合废水经过催化氧化、超滤和生物脱氮处理，并用于炼钢回用或达标排放。综合废水处理系统主要包括催化氧化系统、超滤系统和高效生物脱氮系统组成；浙大中控DCS控制系统根据系统工艺由两套CPU控制系统组成。本文主要介绍了浙大中控DCS控制系统的结构、组成及控制程序的功能和应用。

关键词：DCS；冗余；超滤；生物脱氮

Development and Application of Supcon DCS in Comprehensive Wastewater Treatment System

Zhang Heng, Shen Xianmin, Li Chunhua

(Hebei Iron and Steel Group Handan Iron and Steel Co., Ltd., Handan 056015, China)

Abstract: Handan Iron and Steel's comprehensive wastewater treatment system is a demonstration project of the "13th Five-Year" national science and technology water project. The comprehensive wastewater treatment system mainly consists of a catalytic oxidation system, an ultrafiltration system and a high-efficiency biological denitrification system. The DCS control system of Supcon is composed of two sets of CPU control systems according to the system technology. This article mainly introduces the structure and composition of the DCS control system of Supcon and the function and application of the control program.

Key words: DCS; redundancy; ultrafiltration; biological denitrification

基于PCA模型的高炉风口燃烧带温度仿真研究及分析

张秀春

（飞马智科信息技术股份有限公司，安徽马鞍山 243000）

摘　要：本文从图像处理的角度出发，首先建立高炉风口燃烧带的图像数据库；其次，采用主成分分析（PCA）算法对高炉风口燃烧带不同温度的图像进行识别；最后，结合识别结果进行分析；实验结果验证本文模型算法具有一定的可行性和有效性。本文从图像处理的视角为实现高炉风口燃烧带温度的稳定性提供相应的信息。研究相关结果为某钢厂高炉的稳定性提供相关参考与指导。

关键词：高炉风口燃烧带；主成分分析；图像识别；温度

Simulation Research and Analysis on Temperature of Tuyere Combustion Zone of Blast Furnace based on Principal Component Analysis Model

Zhang Xiuchun

(PHIMA Intelligence Technology Co., Ltd., Ma'anshan 243000, China)

Abstract: This paper researches the blast furnace from the view of image processing, In the first chapter, we establish the database of the images of blast furnace tuyere combustion zone, and clear the data we collected. In the second chapter we establish the mathematical model of PCA, and then we analyze the images of different temperatures in tuyere combustion zone of blast furnace to identify. In the final chapter we use computer software to realize the above methods. The model algorithm proposed in this paper is verified tobe feasible and effective. This paper provides the corresponding information for realizing the temperature stability of tuyere combustion zone of blast furnace.The research results of this paper provide reference and guidance for the stability of Masteel blast furnace.

Key words: blast furnace tuyere combustion zone; principal component analysis; image recognition; temperature

面向钢铁企业的生产计划与排程系统设计

李志伟[1]，施灿涛[2]，张博睿[1]，栾治伟[2]

（1. 宁波钢铁有限公司制造部，浙江宁波　315807；
2. 冶金工业规划研究院工业智能研究中心，北京　100711）

摘　要：国内大多数钢铁企业信息化系统都具备五级架构体系，从设备控制、PLC、MES、ERP 到五级的商务智能分析。在这五级架构体系中企业资源规划系统(ERP)主要负责整体资源规划，用户订单接收等，MES 负责接收生产计划并执行，偏向于执行层面，而需要精细化管控的生产计划层面往往存在缺失。生产计划与排程系统就是弥补 ERP 和 MES 之间的计划空白，使得两者无缝集成。本文基于国内某钢铁企业现有生产组织模式，分析了生产计划组织过程中面临的问题以及生产计划与排程系统对钢铁企业智能化升级的作用。

关键词：钢铁行业；生产计划与排程；生产优化；智能化

Research on Production Planning and Scheduling System for Iron and Steel Enterprise

Li Zhiwei[1], Shi Cantao[2], Zhang Borui[1], Luan Zhiwei[2]

(1. Ningbo Iron and Steel Co., Ltd., Ningbo 315807, China;
2. China Metallurgical Industry Planning and Research Institute, Beijing 100711, China)

Abstract: This paper studies the role of production planning and scheduling system in the five level architecture system of iron and steel enterprises. Most of the information systems of iron and steel enterprises have five levels of architecture, from equipment control, PLC, MES, ERP to five levels of business intelligence analysis. In the five level architecture system, ERP is mainly responsible for the overall resource planning and receiving orders from users. MES is responsible for receiving and executing production plans, which is inclined to the implementation level, while the refined production planning level is often missing. Production planning and scheduling system is to make up for the planning gap between ERP and MES, making them seamless integration. Based on the existing production organization mode of a domestic iron and steel enterprise, this paper analyzes the role of production planning and scheduling system in iron and steel enterprises, and designs a complete production planning and scheduling system.

Key words: iron and steel industry; artificial intelligence; production optimization; intelligent manufacturing

列车自动驾驶系统在钢坯运输线的设计应用

王小萍[1]，段祥玉[2]，张柏文[1]

（1. 中国宝武集团广东韶钢松山有限公司物流部，广东韶关　512000；
2. 河南思维轨道交通技术研究院有限公司，河南郑州　450000）

摘　要：本文以建立智能化钢坯运输线路为目标，设计了一套完整的列车自动驾驶系统，并着重描述了列车自动驾驶系统的组成、功能，以及车载设备的详细设计构成、工作原理和自动驾驶场景。该系统具备了无人化、自动化、智能化运输能力，实现了降本、减员、减负、增效、高安全生产的目标。

关键词：钢坯运输线；智能化；列车自动驾驶；减负

Abstract: In this paper, a set of completed automatic train operation system is designed to build intelligent billet transportation lines. The composition and function of the automatic train operation system are described, and the detailed design, working principle and automatic operation scene of the vehicle equipment are described. The system has unmanned, automated, and intelligent transportation capabilities, and has achieved the goals of cost reduction, staff reduction, burden reduction, efficiency increase, and high safety production.

Key words: billet transportation lines; intellectualized; automatic train operation system; alleviate burdens on sb

ITSS 标准在钢铁企业信息化运维中的应用

张沛理，董瑞柯

（昆仑科技，广东韶关　512000）

摘　要：在这个各行各业信息化程度越来越高的时代，运行维护已经变成软件与系统全生命周期中持续时间最长、最重要的阶段，用户对信息系统的满意度很大程度上取决于信息系统上线后的运维服务。本文阐述了两化深度融合背景下，如何对标 ITSS 运维体系标准为钢铁企业庞杂信息系统提供高质量、高效率的 IT 运维服务，保证业务连续性、稳定性。

关键词：ITSS；IT 运维服务；钢铁

基于板号识别的物料跟踪系统的研发与应用

田　勇[1]，张庆超[1]，王丙兴[1]，张　田[2]，王国栋[1]，钱亚军[3]

（1. 东北大学轧制技术及连轧自动化国家重点实验室，辽宁沈阳　110819；2. 沈阳建筑大学机械工程学院，辽宁沈阳　110168；3. 湖南华菱湘潭钢铁有限公司，湖南湘潭　411104）

摘　要：国内钢铁企业生产厂的信息化物料跟踪都依赖于钢板号，由于生产流程的复杂度较高，急需高准确率的板号在线识别。自然场景下机器喷号的识别技术较成熟，但复杂场景下的手写板号难以实现自动识别。本文针对复杂工作场景下钢板表面手写板号特点提出一种以 BiLSTM-Attention 为主体结构的深度学习算法：首先结合复杂场景，对图像数据进行预处理，保证模型输入图片质量；然后利用残差神经网络(ResNet)提取图片特征；双向长短期记忆网络(Bi-directional Long Short-Term Memory, BiLSTM)提取基于图像的序列特征；最后基于注意力机制(Attention Mechanism)捕获序列内的信息流，对每个字符的特征进行整合形成文本特征向量以预测输出序列。经过现场测试，实现钢板表面手写板号识别任务准确率达到 90.15%，结果表明算法可行有效，满足实际生产需求。

关键词：双向长短期记忆网络；注意力机制；神经网络；手写钢板号；文本识别

Development and Application of Production Tracking System based on Plate ID Recognition

Tian Yong[1], Zhang Qingchao[1], Wang Bingxing[1], Zhang Tian[2], Wang Guodong[1], Qian Yajun[3]

(1. State Key Laboratory of Rolling and Automation, Northeastern University, Shenyang 110819, China;
2. School of Mechanical Engineering, Shenyang Jianzhu University, Shenyang 110168, China;
3. Xiangtan Iron & Steel Co., Ltd., of Hunan Valin, Xiangtan 411104, China)

Abstract: The production tracking of steel rolling depends on the plate ID number. Due to the high complexity of the production process, the on-line identification of plate ID number with high accuracy is urgently needed.The recognition technology of machine spray ID number in natural scene is mature, but it is difficult to realize automatic recognition of handwritten plate ID number in complex scene. In this paper, a deep learning algorithm based on BiLSTM-Attention is proposed for the characteristics of handwritten plate ID numbers located on the surface of steel plate in complex working scenes. Firstly, image data are preprocessed with complex scenes to ensure the quality of the input images in the model. Then, the image features are extracted by ResNet. And the image-based sequence features are extracted by Bi-directional Long Short-Term Memory. Finally, the information flow in the sequence is captured based on the attention mechanism, and the characteristics of each character are integrated to form a text feature vector to predict the output sequence. After actual testing, the accuracy of the recognition task of the handwritten plate ID number on the surface of the steel plate reached 90.15%. The results show that the algorithm is feasible and effective and meets the actual production requirements.

Key words: bi-directional long short-term memory; attention mechanism; neural networks; handwritten plate number; text recognition

检修高危智能管控系统的设计应用与实践

胡　明，张　达，李发展，张明剑

（广东韶钢松山股份有限公司，广东韶关 512123）

摘　要：智能管控系统是由现代通信与信息技术、计算机网络技术、智能控制技术汇集而成，是针对某一个方面的应用的智能集合。随着现代信息技术的不断发展，智能管控系统逐渐应用到各行各业中。本文阐述了韶钢自行设计检修高危智能管控系统，该系统包括计算机软硬件、网络设施、视频监控、报警系统及相关辅助设备等。同时也阐述了该系统的建设过程及实践应用情况。

关键词：布控球；通信与信息技术；智能管控系统；检修安全

Design Application and Practice of High-risk Maintenance Intelligent Control System

Hu Ming, Zhang Da, Li Fazhan, Zhang Mingjian

(Shao Iron & Steel Song Hill Co., Ltd., Shaoguan 512123, China)

Abstract: Intelligent control system is made form communication & information technology、computer network technology and intelligent control technology. Intelligent control system is an application for aspect. With the development of modern information technology, intelligent control system was applied to all walks of life.The article expounds a self-designed high-risk maintenance intelligent control system by Shao Iron & Steel Song hill Co., Ltd.It includes computer hardware and software, network facilities, video surveillance, alarm system and related auxiliary equipment, etc.The article expounds construction process and practical application of the system, yet.

Key words: ball monitor; communication & information technology; intelligent control system; maintenance safety

缩短处理高炉鼓风机急停回路故障时间的措施

梁静丽，刘　强，王春福，鲁绍军

（北京首钢股份有限公司能源部，河北唐山　063000）

摘　要： 通过在急停回路中增加安全继电器，用来准确判断是否为急停按钮动作或相应的线路故障，将接入急停回路中压力开关的台数增加为3个，通过DCS逻辑判断三选二后接入急停回路控制程序中，修改急停回路的DCS控制程序、增加上位监视等措施来解决急停回路故障不能快速准确地锁定故障点的问题及减少压力开关误动作造成异常停机的概率。

关键词： 急停回路；安全继电器；DCS控制；压力开关

Analysis and Improvement Measures of Blast Furnace Blower Emergency Stop Circuit

Liang Jingli, Liu Qiang, Wang Chunfu, Lu Shaojun

(Department of Energy, Beijing Shougang Co., Ltd., Tangshan 063000, China)

Abstract: By adding safety relay in emergency stop circuit to accurately judge whether. It is emergency stop button action or corresponding line fault. The number of pressure switches connected to emergency stop circuit is increased to 3, and then connected to the emergency stop loop control program after three choices are judged by DCS logic, and DCS control program of emergency stop circuit is modified. The measures of upper monitoring are added to solve the problem that the emergency stop circuit failure can not lock the fault point quickly and accurately and reduce the probability of abnormal shutdown caused by the misoperation of pressure switch.

Key words: emergency stop circuit; safety relay; DCS control;press switch

SMS3500 钢水临线智能化改造项目

修成博

（首钢长治有限公司质量监督站，山西长治　046031）

摘　要：本文针对我化验室凸显的质量问题，利用 5M1E 分析法查找当前化验室内质量管理的症结所在。因岗制宜，优化流程，整合资源，更新装备，实施实验室改造。通过对智能化改造的可行性分析，最终确定 SMS3500 钢水临线智能化改造项目，根据现场实际设计布局，逐步实施。通过智能化钢水分析工艺流程的正式运行，一举消除化验室内质量管理的诸多症结，提升化验室整体检验水平。

关键词：5M1E；质量管理；SMS3500 系统；智能化改造

SMS3500 Intelligent Renovation Project Nearing the Production Line of Molten Steel

Xiu Chengbo

(The Quality Supervision Station of Shougang Changzhi Co., Ltd., Changzhi 046031, China)

Abstract: In view of the quality problems in our laboratory, this paper uses 5M1E analytical method to find out the current crux of the quality management. According to the post, optimize the process, integrate resources, update equipment, to enforce laboratory renovation. Passing feasibility analysis of intelligentized transformation, SMS3500 Intelligent renovation project nearing the production line of molten steel is finally determined and implemented step by step according to the actual site layout. Through the official running of the intelligent renovation project, many sticking points of the quality management in the laboratory are eliminated in one fell swoop, and the overall assay level of the laboratory is improved.

Key words: 5M1E; quality management; SMS3500 system; intelligent renovation

韶钢 110kV 演山变电站的智能化建设

陶瑞基，孟　辉

（广东韶钢松山股份有限公司，广东韶关　512122）

摘　要：本文介绍韶钢 110kV 演山变电站的智能化建设的相关背景，以及智能化变电站的相关技术，项目价值，达到的效果等，为同行建设智能化变电站提供一些经验借鉴。

关键词：智能化变电站；智辅系统；智能组件

一起发电机联络线非事故跳闸的分析

陶瑞基，成　霞

（广东韶钢松山股份有限公司，广东韶关　512122）

摘　要：线路差动保护正常情况下在保护区域内无短路故障时不应该动作，但本文讨论的线路却在发生了线路区域外短路故障，而非区域内发生短路故障的情况下，导致了线路差动保护跳闸的案例。这个案例值得我们去做原因分析及采取合理的措施加以防范，防止类似的事故发生，做到保护该动作时准确动作，不该动作时不应动作，避免事故的扩大。

关键词：继电保护；差动保护；电流互感器；防范

一起开关动、静触头发热引起的事故分析

陶瑞基，孙　健

（广东韶钢松山股份有限公司，广东韶关　512122）

摘　要：金属铠装式高压开关柜广泛应用于供配电系统的受电、配电等，实现对供电线路、设备的控制、保护、检测，具有防止误操作断路器、防止带负荷拉手车、防止带电关合接地开关、防止接地开关在接地位置时送电和防止误入带电隔离等“五防”功能。金属铠装式高压开关柜有其优点，但也有其缺点，就是难于用常规的方法检查其内部的元器件运行情况，比如动、静触头是否存在发热现象等。本文对一起开关柜的动、静触头发热导致短路的事故进行了分析并制订了措施，为同行分析处理类似的事故提供一些借鉴。

关键词：断路器；触头发热；互感器套管；弧光短路

Analysis of an Accident Caused by the Heating of the Static and Dynamic Contact of the Switch

Tao Ruiji, Sun Jian

(Guangdong Shaogang Songshan Co., Ltd., Shaoguan 512122, China)

Abstract: Metal armored high-voltage switchgear is widely used in power supply and distribution systems to control, protect and test power supply lines and equipment, the utility model has the five functions of preventing misoperation of the circuit breaker, pulling the handcart with load, closing the earthing switch with electricity, sending electricity when the earthing switch is in the earthing position, and preventing the earthing isolation, etc. Metal armored high-voltage switchgear has its advantages, but it also has its disadvantages, that is, it is difficult to check the operation of its internal components by conventional methods, such as dynamic and static contacts, such as whether there is heating phenomenon. In this paper, an accident of short circuit caused by the heat of the dynamic and static contact of the switchgear is analyzed and some measures are worked out.

Key words: circuit breaker; contact heating; transformer bushing; arc short circuit

基于大数据的热轧精轧机组轴向力分析

余丹峰，布昭元

（宝武钢铁集团宝钢股份武钢有限热轧厂，湖北武汉 430081）

摘 要：轴向力是轧机状态管理中对轧制稳定性重要的衡量依据，但由于轴向力产生的机理较复杂，且影响因素较多，给轧机轴向力问题的定位和改善带来极大的困难。本文从轧制生产工艺过程数据、轧机牌坊精度、轧辊更换等几个维度入手，结合大数据方法通过历史数据分析和评估各项因子对轴向力的影响，同时对轴向力的劣化情况进行分类和归因，提出了基于大数据技术对轧机轴向力问题进行判断处理的新思路。

关键词：设备状态管理；工艺过程数据；轧机；轴向力；大数据

Axial Force Analysis of Hot Finishing Mill Group based on Big Data

Yu Danfeng, Bu Zhaoyuan

(Hot Rolling Mill of Wuhan Iron & Steel Co., Ltd., Baosteel Group, Wuhan 430081, China)

Abstract: Axial force is an important measurement basis for rolling stability in rolling mill condition management. However, because of the complicated mechanism of axial force and many influencing factors, it is very difficult to locate and improve the axial force problem of rolling mill. This paper from the rolling production process data, the accuracy of the memorial arch, roll changing dimensions, and combined with large data method through the historical data analysis and evaluate the effects of various factors on axial force, and classifying the deteriorating situation of the axial force and attribution, based on the technology of data to determine the mill axial force problem of the new ideas.

Key words: equipment state management; process data; rolling mill; axial force; big data

基于 PLC 的卸料小车无人化作业控制系统设计与实现

皮 坤，刘洪具，王合宽

（云南昆钢电子信息科技有限公司，云南昆明 650302）

摘 要：在钢铁冶金炼铁工艺中，卸料小车是原料从料场输送至烧结机进行烧结的必经环节，主要承担皮带输送过来的原料经过卸料小车分别卸料至各个存储料仓，当前操作方式均在小车两侧机盘操作，操作人员长期处于高粉尘工作环境中，本系统设计将实现卸料小车远程集控控制，全自动精准卸料，以改善工作环境及效率。但料仓处于密闭、高粉尘的环境之中，精确测量料仓料位，精准实时定位小车位置是核心所在，本文提出了详细设计与实现方法，提供建设性参考意见。

关键词：卸料小车；精准定位；智能化；PLC 控制系统

Design and Implementation of Unmanned Operation Control System for Discharging Car based on PLC

Pi Kun, Liu Hongju, Wang Hekuan

(Yunnan KISC Electronic Information Technology Co., Ltd., Kunming 650302, China)

Abstract: In ironmaking process of iron and steel metallurgy, discharging car is raw material from the yard to sintering the inevitable part of sintering machine, mainly for belt conveyor of the raw material after unloading the car unloading respectively to each storage bin, the current operating mode are operating in the car on either side of the machine set, an operator at a high dust work environment for a long time. The system design will realize the remote centralized control of the unloading trolley, automatic accurate unloading, in order to improve the working environment and efficiency. However, the silo is in a closed and high-dust environment, so it is the core to accurately measure the material level of the silo and accurately locate the car position in real time. This paper proposes a detailed design and implementation method to provide constructive reference.

Key words: unloading the car; accurate positioning; intelligent; PLC control system

钢铁企业智能电力调度技术探讨与展望

郝　飞[1]，燕　飞[2]，汪国川[2]，刘迎宇[3]，黄源烽[1]

（1. 南京南瑞继保电气有限公司系统软件研究所，江苏南京　211102；
2. 首钢京唐钢铁联合有限责任公司能环部，河北唐山　0632102；
3. 鞍钢股份鲅鱼圈钢铁分公司能源动力部，辽宁营口　115007）

摘　要： 低碳排放，绿色发展，已经成为钢铁企业发展的共识。本文首先介绍了智能电力调度的发展过程，并对智能调度的不同模式进行了对比，针对钢铁企业的电力系统发展的实际需求，通过态势感知、智能运维、新能源优化调度、虚拟电厂，为构建全新的智能电力调度系统，为构建以新能源为主体的新型电力系统，提供负荷侧响应和互动的强有力支撑。

关键词： 虚拟电厂；态势感知；智能运维；电网运行驾驶舱；知识决策分析

Discussion and Prospect of Intelligent Power Dispatching Technology in Iron and Steel Enterprises

Hao Fei[1], Yan Fei[2], Wang Guochuang[2], Liu Yingyu[3], Huang Yuanfeng[1]

(1. System Software Institute, NR Electric Co., Ltd., Nanjing 211102, China; 2. Energy and Environment Department of Shougang Jingtang United Iron & Steel Co., Ltd. Tangshan 063210, China; 3. Energy and Power Department, Bayuquan Steel Branch of Angang Steel, Yingkou 115007, China)

Abstract: Low carbon emission and green development have become the consensus of iron and steel enterprises. This paper first introduces the development process of intelligent power dispatching, and compares different modes of intelligent dispatching. According to the actual needs of power system development of iron and steel enterprises, through situation awareness, intelligent operation and maintenance, new energy optimal dispatching and virtual power plant, this paper aims to build a new intelligent power dispatching system and a new power system with new energy as the main body, provide strong support for load side response and interaction.

Key words: virtual power plant; situation awareness; intelligent operation and maintenance; power grid operation cockpit; knowledge decision analysis

一种连续退火（连续镀锌）钢卷生产时间新算法

李用存，沈奇敏

（马钢（合肥）板材有限公司，安徽合肥 230011）

摘　要： 运用数学建模，提出一种连续退火（CAL）和连续镀锌（CGL）钢卷生产时间新算法，更准确地反映每个钢卷在生产线的实际时间，对指导生产工艺和控制系统具有重要意义。

关键词： 数学模型；生产时间；连续退火；连续镀锌；智慧制造

An New Calculation Method for Production Time of CAL(CGL) Coil

Li Yongcun, Shen Qimin

(Masteel (Hefei) Strip Rolling & Processing Co., Ltd., Hefei 230011, China)

Abstract: To use mathematical modeling,to put forward an new calculation method for production time of continuous annealing line (CAL) and continuous galvanizing line (CGL) coil, which can reflect the real time needed during production of each coil, thus make it of great value to instructing productive technology and control system.

Key words: mathematical models; production time; continuous annealing line (CAL); continuous galvanizing line (CGL); intelligent manufacturing

宁钢铁前优化配矿模型研发及应用

夏志坚[1]，徐凌霄[1]，张军霞[1]，春铁军[2]，龙红明[2]，王　平[2]

（1. 宁波钢铁有限公司制造管理部，浙江宁波 315807；
2. 安徽工业大学冶金工程学院，安徽马鞍山 243032）

摘　要： 根据宁波钢铁铁前原料及工艺参数，开发了网页版铁前优化配矿模型，模型界面简洁、操作简单。模型包括基础数据模块、局部优化配料和一体化优化配料。模型兼具手动配矿和自动配矿两种模式，可以根据效益最大或者成本最低为目标进行优化配矿。模型使用后，局部优化配料成本较现有人工配矿方案低 1.0 元/t 以上，一体化配料优化的铁水理论成本较现有人工方案低 2.0 元/t 铁以上，经济效益显著。

关键词：烧结；高炉；一体化配矿；烧结矿成本

Development and Application of Optimized Ore Blending Model in Ningbo Steel Co., Ltd.

Xia Zhijian[1], Xu Lingxiao[1], Zhang Junxia[1], Chun Tiejun[2], Long Hongming[2], Wang Ping[2]

(1. Manufacturing Management Department of Ningbo Steel Co., Ltd., Ningbo 315807, China;
2. School of Metallurgical Engineering, Anhui University of Technology, Ma'anshan 243032, China)

Abstract: According to the raw materials and process parameters of Ningbo Steel Co., Ltd., a web version of the ore blending optimization model is developed, of which the model interface is simple, and the operation is convenient. The model includes basic data module, local optimization and integrated optimization. The model has two modes of manual ore blending and automatic ore blending, which can optimize ore blending for the goal according to the maximum benefit or the minimum cost. After using the model, the cost of partially optimized proportioning is 1.0 yuan/t lower than that of the existing manual proportioning scheme, the theoretical cost of hot metal optimized by integrated batching is 2.0 yuan/t lower than that of the existing manual scheme, and it makes remarkable economic benefits.

Key words: sinter; blast furnace; integrated ore blending; sinter cost

首钢股份装备智能化建设探索与实践

李文晖，辛鹏飞，周广成，王承刚，张余海

（北京首钢股份有限公司，河北迁安　064404）

摘　要：近年来，首钢股份公司设备管理紧密围绕《中国制造 2025》开展面向全流程的装备智能化建设集成创新，通过国家智能制造示范工厂智能装备单元应用实践，在传统装备的基础上有效应用集控平台、物联网、大数据、无线通讯、工业机器人、视觉识别等先进技术开展产线设备的智能化提升和改造，为公司智能制造提供持续推动力，更为全面建立“设备智能化、操作集控化、运维智慧化、物流信息化、工厂数字化”的首钢智能制造新格局打下基础。

关键词：智能工厂；物联网；机器人；集控

PCA 系统在热轧生产中的研究

庞慧玲，吴真权，陈宇翔，沈益钊，曾龙华

（宝钢湛江钢铁 2250 热轧厂，广东湛江　524000）

摘　要：为进一步推进设备一贯管理，提升设备管理能力，实现以设备管理为基础，以产品为主线的管理理念，提出了 PCA 系统的应用。本文主要通过介绍热轧厂 PCA 系统的组成、系统功能、技术方案和应用场景，并阐述实际应用效果。以生产过程数据挖掘为基础，结合设备控制原理和生产工艺分析，将设备人员管理经验转化为固化规则实现设备在线监测异常报警，打破传统的设备管理模式，提升劳动效率，促进点检智能化。

关键词：在线监控；预警报警；规则；热轧

Research of the Production Condition Analyzer System in the Production of HSM

Pang Huiling, Wu Zhenquan, Chen Yuxiang, Shen Yizhao, Zeng Longhua

(Baosteel Zhanjiang Iorn & Steel Co., Ltd., Hot Rolling Mill, Electrical Engineer of Finish Rolling and Coiling, Zhanjiang 524000, China)

Abstract: In order to further promote consistent equipment management, improve equipment management ability, and realize the management philosophy of taking equipment management as the foundation and product as the main line, the application of production condition analysis system (PCA system for short) was put forward. This paper mainly introduces the composition, development idea, technical scheme and application scene of the production condition analysis system of hot rolling mill, and expounds the practical application effect. Based on the production process data mining, combined with the equipment control principle and production process analysis, the management experience of equipment personnel is transformed into solidified rules to realize the equipment online monitoring abnormal alarm, break the traditional equipment management mode, improve labor efficiency, and promote intelligent spot inspection.

Key words: online monitoring; advance warning message; rule; hot still mill

智能制造在宁波钢铁的规划与实践

郑文艳，应东海

（宁波钢铁有限公司运营改善部，浙江宁波 315807）

摘 要：中国钢铁行业要打造面向制造业的“互联网+”产业生产体系，构建钢铁全流程智能制造系统，力促钢铁转型升级。面对国家、地方政府发展导向及行业发展趋势，在集团公司的指导要求下，宁波钢铁有限公司（以下简称宁钢）结合自身未来发展需要，为保持宁钢过去十年在同类企业信息化建设及应用方面的领先优势，必须推动宁钢智能制造迈上新台阶，通过全面推行两化融合，推进数字化建设，努力做好转型升级，全方位提升宁钢经营能力和综合竞争力，争取在制造业由大变强的过程中走在前列，为宁钢“高质量发展”赋能。

关键词：智能制造；数字化；转型升级；高质量发展

Planning and Practice of Intelligent Manufacturing in Ningbo Iron and Steel

Zheng Wenyan, Ying Donghai

(Operation Improvement Department, Ningbo Iron & Steel Co., Ltd., Ningbo 315807, China)

Abstract: China’s steel industry should build an “Internet plus” industrial production system oriented to manufacturing, build a whole-process intelligent manufacturing system for steel, and promote the transformation and upgrading of steel. In the face of the country, the development orientation of local government and industry development trend, combined with Group company requirements and Ningbo Steel future development plan, to maintain the Ningbo Steel over the past decade in the similar enterprise information construction and application of lead, focus on promoting Ningbo Steel intelligent manufacturing to a new level, fully implementing the two fusion, intelligent manufacture, promote digital construction to completes the transformation and upgrading of products, omni-directional Ninggang management ability and the comprehensive competitiveness, strive for in the process of manufacturing by the big teams at the forefront, for Ningbo Steel can "quality development".

Key words: intelligent manufacturing; digital; transformation and upgrading; high-quality development

浅谈宝钢全天候码头智能化改革与展望

郜 祺

（宝钢股份运输部成品出厂中心，上海 200000）

摘 要： 随着第四次工业革命以不可阻挡之势席卷全球，人工智能、5G、大数据等前瞻性技术已逐步改变了我们的工作、生活。如今的钢铁行业面临着产能过剩、利润低、行业竞争激烈等问题，原本劳动密集型结构、传统单一的生产理念以及无法实现交互运作的设备设施都与这个时代格格不入。宝钢早在 2014 年便提出了“智慧制造”的发展战略。要求各部门打破传统观念、转变思维方式、在原有设备的基础上引入新科技、新技术，从自动化生产领域跃入到智能化制作的范畴。而在众多改革项目中，宝钢全天候码头改造便是其中的一个缩影。

关键词： 改革；智慧制造；宝钢；全天候码头

Discussion on the Intelligent Reform and Prospect of Baosteel's All-weather Wharf

Gao Qi

(Finished Product Delivery Center of Baosteel Transportation Department, Shanghai 200000,China)

Abstract: With the fourth industrial revolution sweeping the world irresistibly, AI, 5G, big data and other forward-looking technologies have gradually changed our work and life.Today,the steel industry is faced with overcapacity, low profits, fierce competition and other issues,the original labor-intensive structure,the traditional single production concept, equipment and facilities that can not realize interactive operation are out of tune with this era. Baosteel put forward the development strategy of "intelligent production" as early as 2014. All departments are required to break the traditional concept, change the way of thinking, introduce new technology on the basis of the original equipment, and leap from the field of automatic production to the field of intelligent production. Among the many reform projects, Baosteel's all-weather wharf renovation is one of the epitomes.

Key words: reform; intelligent production; Baosteel; all-weather wharf

基于严密逻辑的 MES 数据归档系统的设计与实现

林一丁，谢 莹，刘 湃，于永涛

（本钢板材股份有限公司冷轧总厂，辽宁本溪 117000）

摘　要： 随着企业信息化建设的不断深化，企业的信息管理系统产生了大量的历史数据。这些数据具有访问频率极低、空间占有率高、存储周期长、冗余度极高的特点，对历史数据的存储和管理成为了企业数据管理维护的一个难题，并且该问题带来的矛盾也是日益凸显。本文根据本钢冷轧 MES 项目的实际需要，设计了数据归档系统。对归档表策略配置管理功能、归档链路策略配置管理功能、归档程序结构配置功能和归档表结构同步管理等功能的分析、设计和实现进行了描述。

关键词： 钢铁企业；数据归档；信息系统

Design and Implementation of MES Data Archive System based on Strict Logic

Lin Yiding, Xie Ying, Liu Pai, Yu Yongtao

(Bengang Steel Plates Co., Ltd., The Cold Rolling Mill, Benxi 117000, China)

Abstract: With the continuous deepening of enterprise information construction, the enterprise information management system has produced a large number of historical data. These data have the characteristics of very low access frequency, high occupancy of space, long storage cycle and high redundancy. The storage and management of historical data has become a difficult problem in enterprise data management and maintenance, and the contradiction brought by this problem is increasingly prominent. In this paper, according to the actual needs of MES project, a data archiving system is designed. This paper describes the analysis, design and implementation of archive table policy configuration management function, archive link policy configuration management function, archive program structure configuration function and archive table structure synchronization management function.

Key words: iron and steel enterprises ; data archiving ; information system

12　冶金物流运输

大会特邀报告

分会场特邀报告

矿业工程

炼铁与原燃料

炼钢与连铸

轧制与热处理

表面与涂镀

金属材料深加工

先进钢铁材料及其应用

粉末冶金

能源、环保与资源利用

冶金设备与工程技术

冶金自动化与智能化

★　冶金物流运输

其他

夹轨器夹臂有限元分析及其转动副的优化方案

周仲元，谭康超，毛家彦

（宝钢湛江钢铁有限公司物流部设备管理室，广东湛江 524094）

摘 要：本文通过利用 Creo Parametric 5.0 软件对夹臂三维建模、有限元分析等手段，探寻宝钢湛江钢铁物流部桥卸自动夹轨器夹臂转动副失效的根源。并分别对销孔加自润滑铜套及销孔转动副加润滑油道两种优化方案进行可行性分析。可为设计厂家、现场技术人员提供一种借助 CAD（计算机辅助设计，Computer Aided Design）构件失效根源探寻、设计优化提供案例经验借鉴。

关键词：夹轨器；有限元分析；转动副

Finite Element Analysis of Clamp Arm of Rail Clamp and Optimization Scheme of Its Rotating Pair

Zhou Zhongyuan, Tan Kangchao, Mao Jiayan

(Baosteel Zhanjiang Iron and Steel Co., Ltd., Department of Logistics, Equipment Management Room, Zhanjiang 524094, China)

Abstract: In this paper, by means of Creo Parametric 5.0 software, three-dimensional modeling and finite element analysis of the clamping arm, the root cause of failure of the rotating pair of the clamping arm of the bridge unloading automatic rail clamping device in Zhanjiang Iron and Steel Logistics Department of Baosteel is explored. The feasibility analysis of the two optimization schemes of pin hole plus self-lubricating copper sleeve and pin hole rotating pair plus lubricating oil passage is carried out respectively. It can provide a kind of case experience for design manufacturers and field technicians to use CAD (Computer Aided Design) forstructural failure root exploration anddesign optimization.

Key words: rail clamp; finite element analysis; rotating pair

构建循环送铁运输组织新模式，提高 100t 罐周转率

龚超胜

（武钢有限运输部铁运分厂，湖北武汉 430080）

摘 要：宝武融合以来，武钢有限青山基地全面对标宝山基地，而铁钢比、混铁车周转率的提高成为重中之重，本文在武钢有限钢铁工序对标找差提升极致效率的背景下，通过研究武钢有限 100t 罐运输组织模式的现状和特点，构建循环送铁的运输组织新模式，最大程度地提高 100t 罐周转率，满足炼钢、炼铁的需求。

关键词：100t 铁水罐；运输组织；周转率；铁钢比

自动化驼峰控制系统安全防护功能的完善

王占桥，徐　莉，赵允涛，韩庆宇

（鞍山钢铁铁路运输公司，辽宁鞍山　114009）

摘　要：针对自动化驼峰控制系统存在的安全隐患进行分析，采取有效措施，对系统故障报警、自动关闭信号、重复开放信号等功能进行完善，使系统的安全防护能力得到提升。

关键词：自动化；驼峰；控制系统；安全防护；功能；完善

Improvement of Control System Safety Protection Function of Automatic Hump

Wang Zhanqiao, Xu Li, Zhao Yuntao, Han Qingyu

(Railway Transportation of Anshan Iron and Steel Group Co., Ltd., Anshan 114009, China)

Abstract: Hidden trouble in security for hump automation control system were analyzed, and take effective measures, the system fault alarm and automatically shut down signals, repetitive open function to perfect, make the system safety protection ability get promoted.

Key words: automation; hump; control; system safety protection; function; perfect

鞍钢炼钢区域铁路运输作业研究

荀　涛，林传山，刘彦栋

（鞍山钢铁集团铁路运输分公司，辽宁鞍山　114021）

摘　要：分析炼钢区域目前产线分布数量、生产节奏变化、铁路运输机车配备及作业情况。重点对核心区域 3 号炼钢产线转炉生产节奏计算出渣罐的数量变化。再根据 2 号炼钢产线钢水机车、3 号炼钢产线渣罐机车和 5 号炼钢产线渣罐机车的作业效率，提出将 3 号炼钢产线渣罐机车减掉的运输组织方案，在保证钢厂生产的前提下，改变运输模式，提高机车作业效率。

关键词：生产节奏；铁路运输；运输组织方案；机车作业效率

Study on Railway Transportation in Steelmaking Area of Ansteel

Xun Tao, Lin Chuanshan, Liu Yandong

(Railway Transportation Company of Anshan Iron & Steel Group Co., Ltd., Anshan 114021, China)

Abstract: The distribution quantity of production lines, the change of production rhythm, the equipment and operation of railway transport locomotives in steelmaking area are analyzed. Focus on the core area of No.3 steelmaking production line converter production rhythm, calculate the number of slag pot change. Then according to the operation efficiency of molten steel locomotive of No.2 steelmaking production line, slag pot locomotive of No.3 steelmaking production line and slag pot locomotive of No.5 steelmaking production line, the transportation organization scheme of reducing slag pot locomotive of No.3 steelmaking production line is put forward. On the premise of ensuring steel plant production, the transportation mode is changed to improve locomotive operation efficiency.

Key words: production rhythm; railway transportation; transportation organization scheme; locomotive operation efficiency

浅析我国交通运输发展对钢铁企业物流管理的启示

杨 舒

（欧冶工业品股份有限公司华中大区，湖北武汉 430080）

摘 要：我国交通运输基础设施规模质量、技术装备和科技创新能力的快速发展，为钢铁企业物流发展提供了坚实的物质基础；与此同时，其发展思路和规划，给予钢铁企业物流管理以新的启示。钢铁企业物流依托我国交通运输发展，借鉴其发展思路，变革传统模式，将供应链管理、绿色物流、智慧物流理念融入物流管理，必将促进钢铁企业物流实现迭代升级。

关键词：交通运输；物流管理；供应链；一带一路；绿色物流；智慧物流

On the Enlightenment of China's Transportation Development to Logistics Management of Iron and Steel Enterprises

Yang Shu

(Obei Co., Ltd., Wuhan 430080, China)

Abstract: The rapid development of the scale and quality of China's transportation infrastructure, technical equipment and scientific and technological innovation capability has provided a solid material foundation for the logistics development of iron and steel enterprises; meanwhile, its development ideas and planning also give new enlightenment. Based on the development of China's transportation, logistics of iron and steel enterprise can effectively profit from its development ideas, change the traditional mode, and integrate the concepts of supply chain management, environmental logistics and intelligent logistics into logistics management, which will surely promote its iterative upgrading.

Key words: transportation; logistics management; supply chain; the Belt and Road; environmental logistics; intelligent logistics

浅析小运转机车综合成本极致降本优化措施

彭 翔

（武钢有限运输部，湖北武汉 430080）

摘　要：武钢运输部铁运分厂在线机车中有 4 台机车承担着铁运小运转运输作业，其中 2 台机车为在线运用机车，2 台为周转预备机车。机调作业区自 2018 年成立以来，一直承担着 4 台小运转机车的操作运用及设备维保工作。随着公司成本压力的增大，作业区不断调整小运转机车的运用管理，在成本使用及设备维检上追求极致降本目标。

关键词：小运转；油脂消耗；维保；运量

关于某钢铁基地冷轧厂钢卷内部倒运的分析

杨　迪，徐行青

（中国宝武集团宝山钢铁股份有限公司，湖北武汉　433000）

摘　要：某钢铁基地冷轧厂下设三个分厂，主要生产普冷、镀锌、电镀锌、酸洗钢及彩涂卷等冷材，用于汽车面板、家电板、产业设备及建筑材料等，年产量约 400 万吨。

受冷轧设备布局及酸轧、退火等生产工艺要求，该基地冷轧厂各分厂之间、分厂内均需要采用汽车倒运钢卷。通过采用钢卷汽运内倒的方式，在一定程度上保证了生产的顺利进行，但从公司减少无效运输、降低汽车费用及厂区内环保要求的角度来讲，提高汽运效率、降低汽车倒运量是大势所趋。本文通过对比近两年该基地冷轧厂钢卷汽车倒运量，结合汽车运力配置进行分析，并从费用管控的角度提出合理化建议。

关键词：钢卷；汽运；分析；效率

Analysis of Coil Internal Transportation in Cold Rolling Mill of a Steel Production Base

Yang Di, Xu Hangqing

(Bao Steel Stock Company, Baowu Group, Wuhan 433000, China)

Abstract: The cold rolling mill can provide a variety of production including cold rolled roil，galvanized coil, electric galvanized coil, acid pickling coil and color coated coil, mainly applied to automobile manufacturing and household appliance manufacturing .The annual output of this mill is about 4000,000tons.

Be affected by equipment layout and production process of cold rolling mill, transporting roils through trucks between plants in the mill is necessary. But it is the trend to reduce cost of internal truck transportation in future, so this article laid special stress on analyzing volume and efficiency of roils internal truck transportation in last two years, and this paper is to table a proposal on cost control of tuck transportation.

Key words: roil; tuck transportation; analysis; efficiency

钢铁企业物质流-能量流协同优化的生产运行管控技术

郑　忠[1]，连小圆[1]，张　开[1]，王永周[1]，沈薪月[1]，高小强[2]

（1. 重庆大学材料科学与工程学院，重庆　400044，

2. 重庆大学经济与工商管理学院，重庆　400044）

摘 要：从钢铁企业生产流程特点，以及生产管控中物质流-能量流的协同关系出发，进行生产运行的智能管控技术探讨，针对生产批量计划与能源日计划、钢厂调度排程及动态调度为核心的生产运行管控问题，阐述了物质流-能量流协同优化的方法，明确了生产批量计划与能源日计划、钢厂调度排程与能源管控的协同方法，可以实现企业生产效率和能源效率的多目标优化，为钢铁企业在生产运行管控和能源管理不同方面实现物质流-能量流的协同优化提供技术支撑。

关键词：钢铁企业；生产计划调度；物质流-能量流；协同优化

Production Operation Technology for Collaborative Optimization of Material Flow and Energy Flow in Iron and Steel Enterprises

Zheng Zhong[1], Lian Xiaoyuan[1], Zhang Kai[1], Wang Yongzhou[1], Shen Xinyue[1], Gao Xiaoqiang[2]

(1. School of Materials Science and Engineering, Chongqing University, Chongqing 400044, China; 2. School of Economics and Business Administration, Chongqing University, Chongqing 400044, China)

Abstract: Based on the characteristics of the production process in iron and steel enterprises and the collaborative relationship between material flow and energy flow in production, the intelligent control technology of production operation is discussed. In view of the production operation problems with the batch planning and energy daily planning, steelmaking plant scheduling and dynamic scheduling, the method for collaborative optimization of material flow and energy flow are described, and the coordination method between batch planning and energy daily planning, steelmaking plant scheduling and energy control is clarified. The method can be used for the multi-objective optimization of production efficiency and energy efficiency, and for a technological means for iron and steel enterprises to realize the collaborative optimization of material flow and energy flow in different aspects of production operation control and energy management.

Key words: iron and steel enterprise; production planning and scheduling; material flow-energy flow; collaborative optimization

铁路智能运输系统技术在冶金铁路中的应用

董 炜，高 彬

（马鞍山钢铁股份有限公司运输部，安徽马鞍山 243000）

摘 要：本文简要介绍了国铁铁路智能运输系统概念、实施目的、划分的三个层次以及现阶段发展状况等；通过与国铁运输的对比，概况总结了冶金铁路运输的宗旨与特点，指出了冶金铁路智能系统发展的必要性。参照国铁智能运输系统概念，对冶金铁路智能运输系统的涵义及层次划分提出了建议，并对冶金铁路智能系统的现状及存在的问题进行了总结，在此基础上，提出了冶金铁路智能运输系统的发展建议。

关键词：冶金铁路；智能运输；现状与问题；发展建议

Railway Intelligent Transportation System Technology Application in Metallurgical Railways

Dong Wei, Gao Bin

(Transportation Department, Ma'anshan Iron & Steel Co., Ltd., Ma'anshan 243000, China)

Abstract: This paper briefly introduces the concept, the purpose of implementation, the three levels of division and the current development status of the intelligent transportation system of national railway. By comparing with national railway transportation, this paper summarizes the purpose and characteristics of metallurgical railway transportation, and points out the necessity of developing metallurgical railway intelligent system. Referring to the concept of national railway intelligent transportation system, this paper puts forward suggestions on the connotation and hierarchy division of the intelligent transportation system of metallurgical railway, and summarizes the current situation and existing problems of the intelligent transportation system of metallurgical railway, on the basis of which, the development suggestions of the intelligent transportation system of metallurgical railway are put forward.

Key words: metallurgical railway; intelligent transportation; status quo and problems; development proposals

冶金企业机车乘务员“操检维”整合探索与实践

华　勇，成　鹏，曹　峰

（马鞍山钢铁股份有限公司运输部，安徽马鞍山　243000）

摘　要： 本文简要介绍了马钢公司实行检修业务专业化管理后，对铁路运输设备保供产生的不利影响。为积极应对变化，机务分厂立足自身在原有岗位上进行“操检维”的初步实践，在此基础上持续进行设备“操检维”整合的探索。

关键词： 乘务员；操检维；实践

Exploration and Practice on the Integration of “Operation、Spot inspection、Maintenance” of Locomotive Steward in Metallurgical Enterprises

Hua Yong, Cheng Peng, Cao Feng

(Transportation Department, Ma’anshan Iron & Steel Co., Ltd., Ma’anshan 243000, China)

Abstract: This paper briefly introduces the adverse effect on the guarantee of railway transportation equipment after the professional management of overhaul business in Ma’anshan Iron & Steel Co., Ltd.. In order to cope with the changes actively, the maintenance branch plant based on its own in the original position to carry out the preliminary practice of “operation、spot inspection、maintenance”, on this basis, continue to carry out the exploration of equipment “operation、spot inspection、maintenance” integration.

Key words: train attendant; operation、spot inspection、maintenance; practice

马钢铁路路基处理应用改良技术探讨

江宏法，邵义兵，田　伟

（马鞍山钢铁股份有限公司运输部，安徽马鞍山　243000）

摘　要： 本文介绍近年来马钢铁路路基加固处理应用改良创新技术成果，包括：水泥改良砟土混合料、石灰改良砟土混合料以及高炉水渣改良砟土混合料等项应用技术，从原材料收集、方案技术路线及实际应用效果等方面进行阐述，该项应用改良技术既解决了铁路路基填料来源问题，又有效实现砟土固废的合理经济利用，大量推广应用，具有良好的经济和社会效益。

关键词：路基填料；砟土；改良；换填

A Discussion on the Technological Improvements and Innovations of Masteel for Application Technologies in Railway Subgrage Reinforcement

Jiang Hongfa, Shao Yibing, Tian Wei

(Ma'anshan Iron & Steel Co., Ltd., Department of Transportation, Ma'anshan 243000, China)

Abstract: This paper introduces the technological improvements and innovations of Masteel for application technologies in railway subgrade reinforcement in recent years, including a cement-modified ballast stone and soil mixture, lime-modified ballast stone and soil mixture, granulated blast furnace slag-modified ballast stone and soil mixture, and the like, from the perspectives of raw material collection, technical roadmaps, and application effect. These technological improvements and innovations resolve the issue of sources for railway subgrade fillings, implement the rational and economic utilization of soil waste from ballast stone and soil mixtures, and will bring significant economic and social benefits once widely applied.

Key words: railway subgrage fillings; ballast stone and soil; improvements; replacement

浅谈铁路运输安全管理的探索与思考

王忠清

（宝武集团广东韶关钢铁有限公司，广东韶关　512123）

摘　要：2019 年 8 月韶钢物流部进行厂管作业区改革，将原普车作业区、铁水作业区、机务作业区三个作业区进行优化整合，成立了铁运作业区。铁路运输是韶钢生产运营中的一个重要环节，它是各生产单位的“桥梁和纽带”，共有铁路线路 70 余公里，纵横交错，分支遍布各个角落，外至马坝路局，内至各个货场、炼钢、炼铁、焦化等只要有生产的地方就有铁路线路，范围广，线路长，物的不安全因素多，加上普铁调车人员全部协力化，协力人员存在安全意识较差、技能水平较低、人员流动性较强等问题，给铁路运输安全增加不确定性。当前韶钢发展与改革并存的关键时期，如何强化铁路运输安全管理，固化和推动铁路运输安全显得尤为重要。

关键词：铁路运输；安全；管理

On the Exploration and Thinking of Railway Transportation Safety Management in Plant Management Area

Abstract: In August 2019, the logistics department of Shaoguan Iron and Steel Co., Ltd. carried out the reform of the plant management and operation area, optimized and integrated the area, the hot metal operation area and the locomotive operation area, and set up the iron transportation operation area. Railway transportation is an important link in the production and operation of shaogang. It is the "bridge and link" of various production units. It has more than 70 kilometers of railway lines, crisscrossed and branches all over the corner. At present, in the critical period of development and reform of Shaoguan Iron and Steel Co., it is very important to strengthen the safety management of railway transportation, solidify and promote the safety of railway transportation.

Key words: railway transportation; safety; management

韶钢六号高炉铁水运输组织的方案设计及优化

刘志雄，唐育刚，王忠清

（宝武集团广东韶钢松山股份有限公司物流部铁运作业区，广东韶关 512000）

摘 要：本文主要围绕保证韶钢六号高炉生产顺行，在铁路线路复杂、运输路程远、运输时间长、炉下配包时间紧等条件下，设计及优化六号高炉铁水运输组织方案，做好高炉保产、运输安全、铁钢生产平衡等工作，实现安全、高效、准点、有序的铁水运输服务，满足韶钢炼铁、炼钢的生产需求。

关键词：铁水；运输组织；高炉

Scheme Design and Optimization of Hot Metal Transportation Organization of No.6 BF

Liu Zhixiong, Tang Yugang, Wang Zhongqing

(Railway Operation Area of Logistics Department of Guangdong Shaoguan Iron and Steel Co., Ltd., of Baowu Group, Shaoguan 512000, China)

Abstract: This paper mainly focuses on ensuring the smooth production of No.6 blast furnace in Shaoguan Iron and Steel Group Co., Ltd. under the conditions of complex railway, long transportation route, long transportation time and tight ladle distribution under the furnace, designs and optimizes the organization scheme of hot metal transportation of No Sufficient Shaosteel iron, steel production demand.

Key words: molten iron; transportation organization; blast furnace

浅议止轮器在高炉下铁路应用

陈克白

（酒钢集团宏兴股份公司运输部，甘肃嘉峪关 735100）

摘 要：本文通过冶金企业铁水罐车在高炉下的制动模式试验，通过与传统制动方式对比，研究更安全、高效、科学的制动方式。结果表明：钢轨自动止轮器，其制动性最好、最安全、最有效。

关键词：制动轨；止轮器；高炉下应用

Study on the Application of "Rail Clamp" in the Lower Railway of Blast Furnace

Chen Kebai

(Jiuquan Iron and Steel Group Company Hongxing Iron and Steel Co., Ltd., Jiayuguan 735100, China)

Abstract: In this paper, through the experiment of the braking mode of the hot metal tank car in the blast furnace of the metallurgical enterprise, the more efficient and more scientific braking mode is studied by comparing with the traditional one. The results show that the scientific use of the "clamping rail device" has the best braking, the safest and the most effective.

Key words: brake rail; rail clamp; application

对提高冶金企业铁路运输效率的探讨

漆　敏

（甘肃酒钢集团宏兴钢铁股份有限公司运输部，甘肃嘉峪关　735100）

摘　要：在我国，铁路以其大运量、适合大宗物资运输、可实现原料及产品与国家铁路直通运输等优势，在冶金企业物流运输中占据着重要位置。随着我国经济发展进入新常态，钢铁行业也从以前的以规模和资源取胜，逐步转向依靠高质量、低成本取胜。在国家新型产业布局和打赢蓝天保卫战的要求下，扩增钢铁产能的可能性已不大，取而代之的是新旧产能的置换，酒钢对小高炉系统的升级、采用新炼钢工艺设施的建设的规划，都表明产能置换已进入正式启动阶段。

关键词：冶金企业；铁路运输；效率；提升

Discussion on Improving Railway Transport Efficiency of Metallurgical Enterprises

Qi Min

(Transportation Department of Gansu Jiugang Group Hongxing Iron & Steel Co., Ltd., Jiayuguan, 735100, China)

Abstract: In our country, railway plays an important role in the logistics transportation of metallurgical enterprises because of its large volume of transportation, suitable for bulk material transportation, raw materials and products can realize the direct transportation with the national railway and other advantages. As China's economic development has entered a new normal, the steel industry has gradually shifted from relying on scale and resources to relying on high quality and low cost.Under the requirements of the national new industrial layout and winning the battle to protect the blue sky, the possibility of expanding steel production capacity has been limited, replaced by the replacement of new and old production capacity. JISCO plans for the upgrading of low-capacity blast furnaces and the construction of new steelmaking facilities indicate that capacity replacement has officially started.

Key words: metallurgical enterprises; railway transport; efficiency; ascension

13　其他

双重预防机制建设的构建与实施

张立国，刘运桥，张淑艳，马 浩

（河钢股份有限公司承德分公司安全生产监督管理部，河北承德 067000）

摘 要：自双重预防机制推行以来，企业普遍存在风险辨识不全面，岗位不清楚本岗位存在的风险，隐患排查流于形式，现场隐患随处可见，风险管控与隐患排查职责不清晰等问题。本文以河钢股份有限公司承德分公司深入推进双重预防机制和信息化系统建设为例，阐述双重预防机制建设的方式方法，为同行提供有意义的参考和借鉴。

关键词：冶金企业；安全管理；双重预防机制

Construction and Implementation of Double Prevention Mechanism

Zhang Liguo, Liu Yunqiao, Zhang Shuyan, Ma Hao

(Hesteel Company Limited Chengde Branch, Chengde 067000, China)

Abstract: Since the implementation of the dual prevention mechanism, there are many problems in enterprises, such as incomplete risk identification, unclear post risks, mere formality of hidden danger investigation, on-site hidden danger everywhere, unclear responsibility of risk control and hidden danger investigation, etc. Taking Chengde Branch of Hesteel Co., Ltd. as an example, this paper expounds the ways and methods of double prevention mechanism construction, so as to provide meaningful reference for peers.

Key words: metallurgical enterprises; security management; dual prevention mechanism

基于企业安全文化建设经验的浅析

温和达，朱叶风

（宁波钢铁有限公司炼钢厂，浙江宁波 315807）

摘 要：本文主要结合本企业在开展班组安全文化建设方面的一些经验，从企业安全文化建设框架的搭设、安全承诺、安全行为规范与程序、自主学习与安全事务参与、安全行为激励与安全行为信息传播、评价与改进等 6 个方面，浅析了企业阶段性安全文化建设的实施方法，仅供同行借鉴参考。

关键词：安全文化；安全行为；安全承诺

Based on the Experience of Enterprise Safety Culture Construction

Wen Heda, Zhu Yefeng

(Ningbo Iron & Steel Co., Ltd., Ningbo 315807 China)

Abstract: In this paper, combined with the enterprise in the team, the experiences of safety culture construction in enterprise safety culture construction of the framework build-up, security commitments and safety norms and procedures, participate in autonomous learning and security affairs, safety behavior motivation and safety behavior information transmission from six aspects, such as, evaluation and improvement of enterprise periodic safety culture construction are analyzed implementation method, only for peer reference.

Key words: safety culture; safety behavior; safety commitment

迁钢公司备件库存管理对策的研究

李少龙

（北京首钢股份有限公司采购中心，北京　100043）

摘　要：迁钢公司拥有世界先进水平的钢铁产线，不断加快的生产节奏使备件库存管理成为不可忽视的重要部分。本文从库存管理现状入手，剖析问题原因，进而提出对策。首先根据备件特性进行科学分类；进而针对不同库存的特点及供给服务水平，施以定期订货与具有安全库存的随机型库存控制模型进行管理并验证效果。结论有二：第一，相应 ABC 分类法合理且可行性强；第二，库存控制模型的应用显著削减了库存资金与成本。

关键词：库存管理；分类方法；安全库存；库存控制模型

Research on the Countermeasures of Spare Parts Inventory in Qian'gang Company

Li Shaolong

(Purchasing Center of Beijing Shougang Co., Ltd., Beijing 100043, China)

Abstract: Qian'gang company of Shougang Co., Ltd., owns the world's advanced level of iron and steel production. Continuously improved accelerating rhythm of production are moving the warehousing and distribution for its spare parts storage growing higher.Firstly, we took the introduction of basic situation in Qian'gang company, and pointed out the reasons for the dilemma. Then, we put forward the solution. As the first step, this article presented ABC classification. Secondly, we set process on safety stock. As the third, we treated with regular inventory control model and stochastic inventory model with safety stock control management.The conclusion contains two points: As the first one, the ABC classification method is reasonable and feasible. Secondly, the inventory control model reduced the spare parts inventory capital occupation and costs.

Key words: stock manage; classification; safety stock; inventory control model

公司进口设备的采购管理问题探析

路　光

（欧冶工业品股份有限公司武汉分公司，湖北武汉　430080）

摘 要：由于全球经济一体化和中国市场经济的进一步发展，市场竞争更加激烈，公司提高进口设备采购管理水平既能保证产品质量又有效降低成本，对其创效越来越重要。本文以某公司进口设备采购管理为研究对象，分析其采购管理问题，阐述了公司进口设备采购管理定义、重要性、现状分析和对策。

关键词：进口设备；采购管理；短板；对策

The Problem Analysis of Procurement Management of Imported Equipment of the Company

Lu Guang

(Wuhan Branch of Obei Co., Ltd., Wuhan 430080, China,)

Abstract: Due to the global economic integration and the further development of China's market economy, the market competition is more and more intense. It is more and more important to improve the import equipment procurement management level, the company can not only ensure product quality and but also effectively reduce costs. This paper takes the purchase management of imported equipment of a company as the research object, analyzes its purchase management problems and expounds the definition, importance, current situation analysis and countermeasures of the purchase management of imported equipment of the company.

Key words: imported equipment; purchasing management; weakness; countermeasures

工程公司科技创新体系重构与科技创新能力提升

颉建新[1,2]

（1. 北京首钢国际工程技术有限公司战略技术部，北京 100043；
2. 北京市冶金三维仿真设计工程技术研究中心，北京 100043）

摘 要：分析国内外企业科技创新体系的发展现状，针对首钢工程科技创新体系存在的主要问题，提出首钢工程科技创新体系重构与实施内涵和主要做法，形成以新技术（新产品）研发为核心与研究工具、研究组织形式、研究方法有机融合的科技创新体系，具有突出的管理创新性，实施后取得显著的经济效益、管理效益和社会效益，具有广泛的推广与应用价值。

关键词：工程公司；科技创新；体系重构；能力提升

Scientific and Technological Innovation System Reconstruction and Ability Improvement of Engineering Company

Xie Jianxin[1,2]

(1. Strategy & Technology Department, BSIET, Beijing 100043, China; 2. Beijing Metallurgical 3-D Simulation Design Engineering Technology Research Center, Beijing 100043, China)

Abstract: The development situation of the scientific and technological innovation system of domestic and overseas enterprises is analyzed in this essay. Targeting at the problems existing in the scientific and technological innovation system of Shougang Engineering, the contents of implementation of restructuring the scientific and technological innovation system of Shougang Engineering and main measures are put forward, forming the scientific and technological innovation system with the research and development of new technologies or new products as its core and integrating research tools, forms of organization and methods. The management of the system is outstandingly innovative, remarkable economic, management and social benefits have been made after its implementation, meaning it has widespread value of promotion and application.

Key words: engineering company; scientific and technological innovation; system reconstruction; ability improvement

论企业采购绩效评价体系创新实证

梁　磊[1]，贾启超[1]，马克晶[2]

（1. 鞍钢股份有限公司原燃料采购中心，辽宁鞍山　114033；
2. 鞍钢股份有限公司企业管理部，辽宁鞍山　114033）

摘　要：绩效就是有价值的成效。传统的平均主义的绩效评价体系已经不能适应现代企业的发展，奖优罚劣已成为现代绩效评价体系的主流。采购过程是延伸企业能力的关键一环，采购绩效管理体系要同企业的总体战略和成熟度相适应。创建企业新采购绩效评价体系，特别要灵活使用目标和关键结果等集成工具来进行沟通、量化。利用项目管理来激发员工潜能，提高组织管理水平与员工素质，强化采购效率，实现组织经济效益目标，降低采购成本，并获得更多更优的产品和服务。

关键词：采购；绩效评价；创新；企业管理；人力资源

Demonstration of Improving the Enterprise Procurement Performance Evaluation System

Liang Lei[1], Jia Qichao[1], Ma Kejing[2]

(1. Angang Steel Company Limited Raw Material & Fuel Purchasing Center, Anshan 114033, China;
2. Angang Steel Company Limited Management Department, Anshan 114033, China)

Abstract: Performance is valuable results. The traditional egalitarian performance evaluation system has been unable to adapt to the development of modern enterprises, distinct rewards and punishment has become the mainstream of modern performance evaluation system. Procurement process is a key link to extend the ability of an enterprise. Procurement performance management system should adapt to the overall strategy and the development degree of an enterprise. Create enterprise new procurement performance evaluation system, in particular, flexible use of the target and key results integration tools to communicate, quantification.The use of project management to inspire potential employees, improve the level of organizational management and the quality of employees, to strengthen the procurement efficiency and achieve organization economic benefits, reduce procurement costs, and gain more better products and services.

Key words: procurement; performance evaluation; innovation; enterprise management;human resources

数字经济时代钢铁企业检测计量部门如何发挥作用的探索

张永丰，杨生田，刘仁丰，乔万有，岳 帅

（鞍钢股份鲅鱼圈钢铁分公司，辽宁营口 115007）

摘 要：进入数字经济时代，数据成为社会的生产要素。钢铁企业的检测计量部门是企业生产经营数据提供的核心部门，应充分发挥数据要素在企业发展中的作用，保证数据真实、准确，做到数据高效、全面、引领；挖掘数据价值，发挥企业发展要素资源潜力，为公司生产运营提供支撑，为企业发展注入新的动能。

关键词：钢铁企业；检测计量；数据

Exploration on How to Play the Role in Detection and Measurement Department of Iron and Steel Enterprises in the Digital Economy Era

Zhang Yongfeng, Yang Shengtian, Liu Renfeng, Qiao Wanyou, Yue Shuai

(Bayuquan Iron and Steel Branch of Ansteel Co., Ltd., Yingkou, 115007, China)

Abstract: In the era of digital economy, data has become a social factor of production. The detection and measurement department of iron and steel enterprises are the core department of providing production and operation data, and should give full play to the role of data elements in enterprise development. Ensure that the data is true, accurate. high efficient, comprehensive, and leading;mining the value of data, provide resources potential to the enterprise development factor, and provide support for the company's production and operation, and inject new momentum for the enterprises development.

Key words: iron and steel enterprise; measurement; data

设备管理新模式及标准体系的应用与实践

胡 明，张明剑，李发展

（广东韶钢松山股份有限公司，广东韶关 512122）

摘 要：为全面实现设备管理目标，持续提升基础管理能力，发挥团队力量，以设备管理体系标准实施为动力源，推动标准化作业，促体系能力提升，提高设备管理水平，保障生产稳定运行，全面推动钢铁工业高质量发展。

搭建设备管理三大体系建设，实现管理体系项目化覆盖，能精度适用韶钢目前的四大标准，动态的修订机制及无形的技术资产，为技术储备提供资源，经营管理层据此做出阶段性的调整决策，并起到防止再发，教育训练的作用。

关键词：设备管理体系；三大体系；四大标准

Application and Practice of New Model and Standard System of Equipment Management

Hu Ming, Zhang Mingjian, Li Fazhan

(SGIS Songshan Co., Ltd., Shaoguan 512122, China)

Abstract: In order to fully realize the equipment management goal, continue to improve the basic management ability, give play to the team strength, to equipment management system standard implementation as power source, promote standardized operations, promote the system ability, improve the level of equipment management, ensure the stable operation of production, and comprehensively promote the high quality development of the iron and steel industry.

The construction of three major equipment management systems, the realization of project-oriented coverage of the management system, can accurately apply the current four standards of SISG , dynamic revision mechanism and intangible technical assets, to provide resources for the technical reserve, management and management to make periodic adjustment decisions, and play a role in preventing recurrence, education and training.

Key words: equipment management system; three systems; the four criteria

宁钢协力单位安全“四同”管理实践

张　忠

（宁波钢铁有限公司安全保卫部，浙江宁波　315807）

摘　要：该文通过对钢铁企业协力单位的安全管理不规范、协力人员素质和技能低、作业过程中违章多的特点,分析协力人员在使用过程中存在安全管理难点,提出协力安全管理的“四同”措施,建立协力单位安全管理“计划、布置、检查、总结、评比”工作机制，协同推进协力员工的安全教育与培训，以提升协力员工的安全意识和技能，从而提高协力单位的自主安全管理水平，避免和减少安全事故的发生,进而达到安全生产的目的。

关键词：协力单位；安全管理；实践

Practice of Safety “Four Tongs” Management in Cooperative Units of Ningbo Iron & Steel Co., Ltd.

Zhang Zhong

(Ningbo Iron & Steel Co., Ltd., Ningbo 315807, China)

Abstract: This article through to the iron and steel enterprise cooperation unit safety management is not standard, low cooperation unit personnel quality and skills, work illegally operation more characteristics, in the process of analysis to personnel safety management difficulty in use process, put forward to safety management same with “four measures”, establish cooperation unit safety management “plan, layout, check, summary, comparison” working mechanism, jointly promote the safety education and training of cooperative employees, so as to enhance the safety awareness and skills of cooperative employees, so as to improve the independent safety management level of cooperative units, avoid and reduce

the occurrence of safety accidents, and then achieve the purpose of safe production.
Key words: cooperation unit; safety management; practice

熔融制样-X 射线荧光光谱法测定硅钙合金中硅钙磷

李艳霞[1,2]，王子超[1]，贾东涛[1,2]，王富扬[3]

（1. 邢台钢铁有限责任公司，河北邢台 054027；2. 河北省线材工程技术创新中心，河北邢台 054027；3. 洛阳鼎辉特钢制品股份有限公司，河南洛阳 471322）

摘 要：采用四硼酸锂挂壁制备熔剂坩埚，以四硼酸锂和碳酸锂作熔剂，用硝酸钾作氧化剂对样品进行处理，用 X 射线荧光光谱仪对硅钙合金中 Si、Ca、P 进行测定。本法避免了传统湿法过程繁琐、不易掌握、周期长的问题，同时，实验先在马弗炉中对样品进行预氧化，样品中的单质元素在预氧化处理中转化成氧化物，解决了单质元素与铂形成低温共熔体而腐蚀铂金坩埚的难点。采用本分析方法制得的玻璃片均匀、透亮、无气泡，通过对样品精确度和准确度分析，结果满足分析要求。
关键词：X 射线荧光光谱法；硅钙合金；熔融制样；熔剂坩埚

X-ray Fluorescence Spectrometry for the Determination of Silicon, Calcium and Phosphorus in Silicon-calcium Alloy with Fusion Sample Preparation

Li Yanxia[1,2], Wang Zichao[1], Jia Dongtao[1,2], Wang Fuyang[3]

(1. Xingtai Iron and Steel Co., Ltd., Xingtai 054027, China; 2. Wire Engineering Technology Innovation Center, Xingtai 054027, China; 3. Luoyang Dinghui Special Stee Co., Ltd., Luoyang 471322, China)

Abstract: The wall of a flux crucible was attached with lithium tetraborate. Consequently, the determination of silicon, calcium, and phosphorus was realized by X-ray fluorescence spectrometer with lithium tetraborate-lithium carbonate as flux and potassium nitrate as oxidizer. The method avoids the problems of tedious process, time-consuming and large reagent consumption of conventional wet method. The sample was pre-oxidized in muffle furnace. The elementary elements in silicon-calcium alloy were converted into oxides to avoid the corrosion of platinum alloy crucible due to the formation of low-temperature eucectic formed by elementary elements and platinum at high temperature. The results showed that the prepared glass piece was uniform and transpatent without bubbles. The precision and accuracy tests indicates that the proposed method could completely satisfy the requirements of routine analysis.
Key words: X-ray fluorescence spectrometry; silicon-calcium alloy; fusion sample preparation; flux crucible

无缝金属纯钙线中钙含量的不确定度评定

郝志奎，胡乐明，邢文青

（宝武集团广东韶关钢铁有限公司，广东韶关 512000）

摘　要：无缝金属纯钙线能使钢水中 Al_2O_3 夹杂完全变性，减少连铸钢水口堵塞概率。钙含量是无缝金属纯钙线品质的重要指标，因此准确测定其含量的意义重大。本文对 EDTA 容量法测定无缝金属纯钙线中钙含量的不确定度展开评定，分析整个实验过程中不确定度的来源，并对各个不确定度进行计算。结果表明：EDTA 容量法测定无缝金属纯钙线中钙含量的差异，主要由标定标准溶液导致的。当测定无缝金属纯钙线中钙含量质量分数为 98.76%时，其拓展不确定度为 0.52%。

关键词：EDTA 容量法；无缝金属纯钙线；钙含量；不确定度

Uncertainty Evaluation of Calcium Content in Seamless Metal Pure Calcium Wire

Hao Zhikui, Hu Leming, Xing Wenqing

(Guangdong Shaoguan Iron and Steel Co., Ltd., Baowu Group, Shaoguan 512000, China)

Abstract: The seamless pure calcium wire can completely denature the Al_2O_3 inclusion in molten steel and reduce the probability of nozzle clogging in CC steel. Calcium content is an important index for the quality of seamless pure calcium wire, therefore it has great significance to accurately determine its content. This paper evaluates the uncertainty of EDTA titration method for the determination of calcium content in seamless metal pure calcium wire, analyzes the source of uncertainty in the whole experimental process, and calculates various uncertainty components. The results show that the calibration of standard solution causes the difference of calcium content in seamless metal pure calcium wire determined by EDTA titration method. When the mass fraction of calcium in seamless pure calcium wire is 98.76%, the expanded uncertainty is 0.52%.

Key words: EDTA titration; seamless pure calcium wire; uncertainty; calcium content

电感耦合等离子体质谱法测定钼铁中铅锡砷锑铋

张　杰，田秀梅，刘冬杰，李　颖，王一凌，王　伟

（鞍钢集团钢铁研究院，海洋装备金属材料及应用国家重点实验室，辽宁鞍山　114009）

摘　要：研究了应用电感耦合等离子体质谱法（ICP-MS）测定钼铁中铅、锡、砷、锑和铋的分析方法。用硝酸溶解样品，并且加入 Sc、La 和 Re 混合内标溶液，使用基体匹配工作曲线来校正基体效应对结果的影响，使样品的基体效应和仪器漂移得到了很好的补偿。选取了最佳的仪器分析条件，在优化仪器工作参数的基础上，选择合适同位素避免质谱干扰。本方法经大量实践检验，分析周期短，结果可靠，加标回收率在 94.00%～106.67%之间，相对标准偏差 RSD 在 0.43%～6.45%之间。

关键词：钼铁；铅；锡；砷；锑；铋；电感耦合等离子体质谱法（ICP-MS）

Determination of Pb, Sn, As, Sb and Bi Content in Ferromoly Bdenum by ICP-MS

Zhang Jie, Tian Xiumei, Liu Dongjie, Li Ying, Wang Yiling, Wang Wei

(Iron & Steel Research Institute of Ansteel Group, State Key Laboratory of Marine Equipment Made of Metal Material and Application, Anshan 114009, China)

Abstract: In this paper, researched this method for determination of trace Pb,Sn,As,Sb and Bi by inductively coupled plasma mass spectrometry. The samples were dissolved with nitric acid,and used Sc, La and Re mixed internal solution to correct the effect of the matrix for the results, compensating the matrix effect and the drift of the instrument. The best instrument analysis conditions were selected and the appropriate isotopes were selected to avoid mass spectroscopic interference. The experimental results showed that the recovery rate was between 94.00%~106.67%, and the relative standard deviation was between 0.43%~6.45%. This method can meet the needs of scientific research and production.

Key words: bismuth; erromolybdenum; lead; stin; arsenic; antimony; inductively coupled plasma-mass spectrometry

大体积混凝土压缩机基础施工质量控制措施

朱登明

（新疆八一钢铁股份有限公司能源中心，新疆乌鲁木齐　830022）

摘　要： 结合工程实例，介绍了在气温较高，昼夜温差较大的环境下施工大体积混凝土的技术措施及施工质量保证手段，从混凝土原材料的选用、配合比的设计、混凝土的拌制、运输、浇捣、养护等方面进行管控，并采用预埋冷却水管的方式对大体积混凝土内外温差进行有效的控制，确保了大体积混凝土动力设备基础的施工质量。

关键词： 大体积混凝土；水化热；温差控制

Quality Control Measures for Mass Concrete Compressor Foundation Construction

Zhu Dengming

(Energy Center, Xinjiang Bayi Iron & Steel Stock Co., Ltd. , Urumqi 830022, China)

Abstract: Combined with engineering example, introduced in the temperature is higher, larger temperature difference between day and night environment construction of mass concrete technical measures and construction quality assurance measures, from raw material selection, mix proportion of concrete design, concrete mixing, transportation, vibration control, at the same time with the method of embedded cooling water pipe for effective control of mass concrete of inside and outside temperature difference, to ensure the construction quality of mass concrete power equipment.

Key words: mass concrete; hydration heat; temperature difference control

三钛酸钠 $Na_2Ti_3O_7$ 的高温热容

杨利连[1,2]，裴贵尚[1,2]，侯有玲[1,2]，吕学伟[1,2,3]

（1. 钒钛冶金及新材料重点实验室，重庆大学，重庆　400044；2. 材料科学与工程学院，重庆大学，重庆　400044；3. 机械传动国家重点实验室，重庆大学，重庆　400044）

摘　要： 钠离子电池具有比能量高，资源丰富，安全性好，成本低廉等优点，在储能电池领域有望成为锂离子电池的替代品。其中，三钛酸钠($Na_2Ti_3O_7$)电势低、理论电容高，是一种极具发展前景的钠离子电池阳极材料。为了更好地了解 $Na_2Ti_3O_7$ 的使用性能，对其热力学性质的研究是必不可少的。然而，文献表明，$Na_2Ti_3O_7$ 的高温与低温热容在室温附近存在较大的差异，且测量高温数据使用的样品纯度较低。因此，本文以 Na_2CO_3 和 TiO_2 为原料，通过固相反应合成高纯度的 $Na_2Ti_3O_7$ 样品，并采用高温综合量热仪在 573~1323K 温区内对其高温热容进行了测量与研究。其热容与温度的拟合关系式为：C_p=255.51073+0.06059T−3.86912×106T^{-2} (J·mol^{-1}·K^{-1}) (298.15~1403K)，然后计算得到该温度范围的摩尔焓变、熵变和吉布斯自由能变等热力学函数，并给出其在 0~1403K 范围内的热容方程。
关键词：三钛酸钠；固相反应；热容；热力学性质

Heat Capacity of Sodium Trititanate ($Na_2Ti_3O_7$) at High Temperatures

Yang Lilian[1,2], Pei Guishang[1,2], Hou Youling[1,2], Lu Xuewei[1,2,3]

(1. Chongqing Key Laboratory of Vanadium-Titanium Metallurgy and Advanced Materials, Chongqing University, Chongqing 400044, China; 2. College of Materials Science and Engineering, Chongqing University, Chongqing 400044, China; 3. State Key Laboratory of Mechanical Transmissions, Chongqing University, Chongqing 400044, China)

Abstract: Sodium-ion batteries (NaIBs) have attracted extraordinary attentions as a promising scalable energy storage alternative to current lithium-ion batteries (LIBs), owing to their natural abundance and low costs. Sodium trititanate ($Na_2Ti_3O_7$) is a promising material as the material of NaIBs with low potential and high theoretical capacitance. In order to better understand their service performance, studies on thermodynamic properties of $Na_2Ti_3O_7$ are indispensable. However, an extensive literature review revealed that the heat capacity of $Na_2Ti_3O_7$ are established with divergences, especially for high-temperature region. Therefore, the high purity of $Na_2Ti_3O_7$ powder was first synthesized through solid-state reaction with Na_2CO_3 and TiO_2 as raw materials. The as-prepared samples wereused to measure the heat capacityfrom 573 to 1323K which was carried outwith Multi-high temperature calorimeter (MHTC) 96 line. The temperature dependence of heat capacity was modeled as a function: C_p=255.51073+0.06059T−3.86912×10$^6T^{-2}$ (J·mol^{-1}·K^{-1}) (298.15-1403K), and then used for computing changes in enthalpy, entropy, and Gibbs free energy. Heat capacity of $Na_2Ti_3O_7$ from 0 to 1403K was given for future application of $Na_2Ti_3O_7$ in rechargeable batteries.
Key words: sodium trititanate; solid-state reaction; heat capacity; thermodynamic functions

钢铁企业噪声职业危害防控研究

刘冠中，张保忠

（杭钢集团宁波钢铁有限公司，浙江宁波 315807）

摘　要：热轧现场噪声职业危害的防控研究是热轧厂 2019 年开展的一项科技攻关项目，本项目通过研究分析热轧厂主要噪声危害的分布及控制现状，分析各类噪声产生的原因以及对接触人员的危害状况，研究和试验各类噪声源的控制措施。消除或有效降低噪声源的危害，减少员工接害时间，保护员工的职业健康。
关键词：噪声；危害；分析；防控

鞍钢鲅鱼圈分公司燃气系统智能化改造应用

金　鹏[1]，高大鹏[2]，张　鎏[2]

（1. 鞍钢集团信息产业有限公司，辽宁鞍山　114009；
2. 鞍钢集团自动化有限公司，辽宁鞍山　114009）

摘　要：燃气能源管理系统的建设，其主要目的，即为提升燃气能源的生产、使用效率，提升环保、经济效益，提升公司竞争力。鲅鱼圈分公司能源燃气系统，在基础设备层、控制工艺层、网络通信层、分析决策层不同维度上发力，进行系统化的协同提升。以智能化、信息化为抓手，为燃气能源系统赋能。在打造智慧能源系统平台之上，助力于公司的绿色能源远景规划。为公司减少碳排放，实现碳达峰、碳中和的中长期目标，奠定坚实有力的基础。

关键词：燃气能源系统；碳排放；自动化控制；数据分析

Application of Intelligent Transformation of Gas System in Angang Steel Company Limited

Jin Peng[1], Gao Dapeng[2], Zhang Liu[2]

(1. Ansteel Information Industry Co., Ltd., Anshan 114009, China;
2. Ansteel Automation Co., Ltd., Anshan 114009, China)

Abstract: The main purpose of the construction of the gas energy management system is to improve the efficiency of gas energy production and use, improve environmental protection and economic benefits, and enhance the company's competitiveness. The system of energy and gas enhances overall capabilities in different perspectives such as basic equipment, control technology, network communication, analysis and decision-making. It uses intelligence and information as the starting point to empower the gas energy system. On the platform of building an intelligent energy system, it will help the company's green energy vision plan. In the medium and long term, it will lay a good foundation for the company to reduce carbon emissions, achieve carbon peaks and carbon neutrality.

Key words: gas energy system; carbon emissions; automated control; data analysis

软磁材料用高纯磷铁制备研究

庞建明[1]，林保全[1,2]，宋耀欣[1]

（1. 中国钢研科技集团有限公司，北京　100081；2. 钢铁研究总院，北京　100081）

摘　要：本文根据金属热还原工艺原理，采用磷矿物、含铁料和工业硅（还原剂），分别以同样的加热功率在高温实验炉和微波加热炉内进行高纯磷铁合金的制备试验研究，研究结果表明两种加热炉所制得的磷铁合金 C、Ti、Al、V 等含量均低于 0.05%，满足软磁材料用高纯磷铁合金质量要求。通过正交实验等分析，高温炉内高纯磷铁合金制备的最佳工艺条件为冶炼温度 1390℃，保温时间 60min，配硅系数 1.2，碱度 1.1，该条件下 P 组分收得率为>96%。

微波炉内高纯磷铁合金制备的最佳工艺条件为冶炼温度 1360℃，保温时间 30min，配硅系数 1.2，碱度 1.1，该条件下 P 组分收得率>98%。对比可知，微波炉内最佳冶炼温度较高温实验炉最佳冶炼温度降低 30℃，保温时间缩短 30min。采用微波加热方式制备高纯磷铁有效降低了冶炼能耗，符合当前低碳冶金和节能减排的发展方向，为软磁材料用高纯磷铁合金的制备提供了新的思路。

关键词：高纯磷铁；金属热还原；微波；正交实验

Experimental Research on Preparation of High-purity Ferrophosphorus Alloy for Soft Magnetic Materials

Pang Jianming[1], Lin Baoquan[1,2], Song Yaoxin[1]

(1. China Iron & Steel Research Institute Group, Beijing 100081, China;
2. Central Iron & Steel Research Institute, Beijing 100081, China)

Abstract: According to the principle of metal thermal reduction process, the preparation of high-purity ferrophosphorus alloy was studied in high-temperature experimental furnace and microwave oven with the same heating power by using phosphorus mineral, iron containing material and industrial silicon (reducing agent). The results show that the contents of carbon, titanium, aluminum and vanadium of ferrophosphorus alloy prepared in the two heating furnaces are less than 0.05%, which can meet the quality requirements of high-purity ferrophosphorus alloy for soft magnetic materials. Through the analysis of orthogonal experiment, the optimum process conditions for the preparation of high-purity ferrophosphorus alloy in high-temperature furnace are smelting temperature of 1390°C, holding time of 60min, silicon ratio of 1.2 and alkalinity of 1.1. Under these conditions, the recovery rate of phosphorus is more than 96%. The optimum technological conditions for the preparation of high-purity ferrophosphorus alloy in microwave oven are smelting temperature of 1360°C, holding time of 30min, silicon ratio of 1.2 and alkalinity of 1.1. Under these conditions, the recovery rate of phosphorus is more than 98%. The comparison shows that the optimal smelting temperature in the microwave oven is 30°C lower than that in the high-temperature experimental furnace, and the holding time is shortened by 30min. The preparation of high-purity ferrophosphorus by microwave heating effectively reduces the smelting energy consumption, which is in line with the current development direction of low-carbon metallurgy, energy conservation and emission reduction, and provides a new idea for the preparation of high-purity ferrophosphorus alloy for soft magnetic materials.

Key words: high-purity ferrophosphorus; metal thermal reduction; microwave; orthogonal experiments